Lecture Notes in Mathematics

Edited by A. Dold and B. Eckmann

T0213400

843

Functional Analysis, Holomorphy, and Approximation Theory

Proceedings of the Seminário de Análise Funcional, Holomorfia e Teoria da Aproximação, Universidade Federal do Rio de Janeiro, Brazil, August 7 – 11, 1978

Edited by Silvio Machado

Springer-Verlag
Berlin Heidelberg New York 1981

Editor

Silvio Machado
Instituto de Matemática
Universidade Federal do Rio de Janeiro
Caixa Postal 1835
21910 Rio de Janeiro RJ
Brazil

AMS Subject Classifications (1980): 32-XX, 41-XX, 46-XX

ISBN 3-540-10560-3 Springer-Verlag Berlin Heidelberg New York
ISBN 0-387-10560-3 Springer-Verlag New York Heidelberg Berlin

Printing and binding: Beltz Offsetdruck, Hemsbach/Bergstr.
2141/3140-543210

FOREWORD

This volume contains the proceedings of the Seminário de Análise Funcional, Holomorfia e Teoria da Aproximação held at Instituto de Matemática, Universidade Federal do Rio de Janeiro (UFRJ) in August 7-11, 1978. It includes papers either of a research or of an advanced expository nature. Some of them could not be actually presented during the Seminar, and are being included here by invitation. The participant mathematicians came from Belgium, Brazil, Chile, France, Germany, Ireland, Spain, United States and Uruguay.

The members of the organizing committee were J.A. Barroso, S. Machado (coordinator), M.C. Matos, J. Mujica, L. Nachbin, D. Pisanelli, J.B. Prolla and G. Zapata. For direct financial support thanks are due to Conselho Nacional de Desenvolvimento Científico e Tecnológico (CNPq); to Conselho de Ensino para Graduados e Pesquisa (CEPG) of UFRJ, and here we also thank particularly Dr. Sergio Neves Monteiro for his understanding; to IBM of Brazil, with special thanks to Dr. José Paulo Schiffini. Travelling grants for some participants were individually provided by Fundação de Amparo à Pesquisa do Estado de São Paulo (FAPESP) and Universidade de Campinas (UNICAMP), São Paulo, Brazil.

Professor Radiwal Alves Pereira, then Director of the Instituto de Matemática, collaborated beyond the line of his duty in the organizational details of the meeting; the Coordinator emphasizes his heartfelt thanks to him. Professor Paulo Emidio Barbosa made available the facilities of the Centro de Ciências Matemáticas e da Natureza (CCMN) of UFRJ, of which he is Dean and to which belongs the Instituto de Matemática; it is a pleasure to offer our thanks to him. A special word of appreciation on the part of the Coordinator goes to

Professor Leopoldo Nachbin for making available his experience and unfailing moral support. We also thank Wilson Góes for a competent typing job.

Rio de Janeiro, August 1978

Silvio Machado

CONTENTS

AN EXAMPLE OF A QUASI-NORMABLE FRÉCHET FUNCTION SPACE WHICH IS NOT A SCHWARTZ SPACE

J.M. Ansemil and S. Ponte

Departamento de Teoría de Funciones
Facultad de Matemáticas
Universidad de Santiago de Compostela
Spain

1. INTRODUCTION AND PRELIMINARIES

Let E and F be complex Banach spaces, U an open subset of E and $(\mathcal{H}_b(U;F),\tau_b)$ the vector space of the mappings $f: U \to F$ which are holomorphic of bounded type on U endowed with its natural topology τ_b. It is clear that $(\mathcal{H}_b(U;F),\tau_b)$ is a Fréchet space.

In [6] it has been shown that when U is balanced then the topological dual of $(\mathcal{H}_b(U;F),\tau_b)$ is isomorphic to a certain space of sequences $S(U;F)$. Moreover, assuming that U is, either all of E, or else a bounded balanced convex open subset of E, then $S(U;F)$ has been endowed with a natural topology τ_ℓ which has been shown to be finer than the strong topology τ_β on the dual space. We shall now show that if U is the open ball $B(0,R)$, $0 < R \leq \infty$ (the case $R = \infty$ corresponds to $U = E$), then the above isomorphism is topological, and that $(\mathcal{H}_b(U;F),\tau_b)$ is a quasi-normable Fréchet space which is not a Schwartz space unless $\dim E < \infty$ and $\dim F < \infty$.

Finally, we want to acknowledge Profs. J.M. Isidro, J. Mujica and L. Nachbin for their help and suggestions while preparing this paper.

2. THE DISTINGUISHED CHARACTER OF $(\mathcal{H}_b(U;F),\tau_b)$.

<u>Definition</u>. For each r, $0 < r < R$, we define $S_r(U;F)$ to be the Banach space of sequences $\mu = (\mu_n) \subset \prod_{n=0}^{\infty} P({}^nE;F)'$ for which there is a constant $C \geq 0$ such that

$$\|\mu_n\| \leq C\, r^n$$

for all $n \in N$, endowed with the norm

$$\|\mu\|_r = \sup_{n\in N} \frac{\|\mu_n\|}{r^n}, \qquad \mu = (\mu_n) \in S_r(U;F).$$

We define $S(U;F)$ as the vector space $\bigcup_{0<r<R} S_r(U;F)$, and τ_ℓ is defined to be the corresponding inductive limit topology on $S(U;F)$.

The following theorem is Propositions 8 and 9 of [6].

<u>Theorem 1</u>. For each $\mu = (\mu_n) \in S(U;F)$ the mapping

$$(*) \qquad \mu: f \to \langle f,\mu\rangle = \sum_{n=0}^{\infty} \langle \frac{1}{n!}\hat{d}^n f(0),\mu_n\rangle, \qquad f \in \mathcal{H}_b(U;F)$$

defines an element $\mu \in (\mathcal{H}_b(U;F),\tau_b)'$. Conversely, if $\mu \in (\mathcal{H}_b(U;F),\tau_b)'$, then the sequence (μ_n), $\mu_n = \mu|_{P({}^nE;F)}$, of its restrictions to the subspaces $P({}^nE;F) \subset \mathcal{H}_b(U;F)$ defines an element of $S(U;F)$ whose associated functional is μ by $(*)$.

Moreover, if we identify the dual space $(\mathcal{H}_b(U;F),\tau_b)'$ with $S(U;F)$ by means of the (algebraic) isomorphism given above, we have ([6], Proposition 12) that τ_ℓ is finer than the strong topology τ_β.

We shall repeat here for further use the proof of Proposition 13, [6].

<u>Lemma 1</u>. For each τ_β-bounded subset χ of $S(U;F)$ there is an r, $0 < r < R$, such that χ is contained and bounded in the Banach space $S_r(U;F)$.

<u>Proof</u>. Since $(\mathcal{H}_b(U;F),\tau_b)$ is a barreled space, each τ_β-bounded

subset of the dual space $S(U;F)$ is equicontinuous. Hence, given χ, there is a neighbourhood W of the origin in $(\mathcal{H}_b(U;F),\tau_b)$ such that

$$\sup_{(\mu_n)\in\chi, f\in W} \left|\sum_{n=0}^{\infty} \langle \frac{1}{n!}\hat{d}^n f(\theta),\mu_n\rangle\right| \leq 1.$$

We may assume that

$$W = \{f \in \mathcal{H}_b(U;F) : \sup_{\|t\|\leq r} \|f(t)\| \leq \delta\}$$

for some r, $0 < r < R$, and some $\delta > 0$. Now, for each $n \in \mathbb{N}$ and each $P \in \mathcal{P}(^nE;F)$ with $\|P\| = 1$ we have

$$\frac{\delta}{r^n} P \in W$$

therefore

$$(*) \qquad \left|\frac{\delta}{r^n} \langle P,\mu_n\rangle\right| \leq 1$$

for all $n \in \mathbb{N}$, all $P \in \mathcal{P}(^nE;F)$ with $\|P\| = 1$ and all $(\mu_n) \in \chi$. From (*) it is easy to see that χ is contained and bounded in $S_r(U;F)$.

<u>Proposition 1</u>. τ_β and τ_ℓ induce the same topology on each τ_β-bounded subset of $S(U;F)$.

<u>Proof</u>. Let χ be a τ_β-bounded subset of $S(U;F)$. Without loss of generality we may assume that χ is balanced and convex. Because of Lemma 1, χ is contained and bounded in $S_r(U;F)$ for a suitable r, $0 < r < R$. Take ρ such that $r < \rho < R$ and let $(\mu^\lambda)_{\lambda\in\Lambda} \subset \chi$ be a net in χ which is τ_β-convergent to zero. Then $(\mu^\lambda)_{\lambda\in\Lambda}$ converges to zero in $S_\rho(U;F)$. Indeed, since $(\mu^\lambda)_{\lambda\in\Lambda}$ is bounded in $S_r(U;F)$, there is $C \geq 0$ such that

$$\|\mu_n^\lambda\| \leq C\, r^n$$

for all $n \in N$, and all $\lambda \in \Lambda$, hence

$$\frac{\|\mu_n^\lambda\|}{\rho^n} \leq C\left(\frac{r}{\rho}\right)^n$$

for all $n \in \mathbb{N}$ and all $\lambda \in \Lambda$. Therefore, given any $\varepsilon > 0$, there is an index $M \in \mathbb{N}$ such that

$$\sup_{n \geq M} \frac{\|\mu_n^\lambda\|}{\rho^n} \leq \varepsilon$$

for all $\lambda \in \Lambda$. Since the closed unit ball of $\mathcal{P}({}^nE;F)$, $n \in \mathbb{N}$, is a bounded subset of $(\mathcal{H}_b(U;F), \tau_b)$ and $(\mu^\lambda)_{\lambda \in \Lambda}$ is τ_β-convergent to zero, we have

$$\mu_n^\lambda \to \theta$$

for each $n \in \mathbb{N}$. Hence, there is a $\lambda_o \in \Lambda$ such that

$$\frac{\|\mu_n^\lambda\|}{\rho^n} \leq \epsilon$$

for all $\lambda \geq \lambda_o$ and $n = 0,1,\ldots,M$ so that we have

$$\sup_{n \in \mathbb{N}} \frac{\|\mu_n^\lambda\|}{\rho^n} \leq \varepsilon$$

for all $\lambda \geq \lambda_o$. This shows that $(\mu^\lambda)_{\lambda \in \Lambda}$ converges to zero in $S_\rho(U;F)$. Now, we can apply the result of Grothendieck ([5], p.105, Lemme 5), to conclude that the topology induced on χ by τ_β is finer than the one induced by $S_\rho(U;F)$. The converse is obvious and this completes the proof.

Lemma 1 and the proof of Proposition 1 proves the following.

Corollary 1. The inductive limit

$$(S(U;F), \tau_\ell) = \lim_{0<r<R} S_r(U;F)$$

is boundedly retractive, that is, every bounded set χ is contained and bounded in some $S_r(U;F)$, $0 < r < R$ and $S(U;F)$ and $S_r(U;F)$ induce on χ the same topology.

Theorem 2. τ_β and τ_ℓ coincide on $S(U;F)$. Therefore, $(S(U;F), \tau_\ell)$ and $((\mathcal{H}_b(U;F), \tau_\beta)', \tau_\beta)$ are isomorphic as topological vector spaces.

Proof. It suffices to prove that the identity mapping

$$(*) \qquad (S(U;F), \tau_\beta) \to (S(U;F), \tau_\ell)$$

is continuous. Since $(\mathcal{H}_b(U;F), \tau_b)$ is a metrizable space,

$(S(U;F),\tau_{8})$ is a $\mathcal{DF}$ space, hence, by ([5], p.69, Corollaire 1), in order to prove that (*) is continuous all we have to do is to show that the restriction of (*) to each bounded subset χ of $(S(U;F),\tau_{8})$ is continuous, and this is true by Proposition 1. This completes the proof.

Corollary 2. $(\mathcal{H}_b(U;F),\tau_b)$ is a distinguished space.

Proof. Its strong topological dual is an inductive limit of Banach spaces, hence barreled.

3. OTHER PROPERTIES OF $(\mathcal{H}_b(U;F),\tau_b)$.

It is known ([5], p.108, Proposition 14) that every quasi-normable metrizable locally convex space is distinguished, hence, the following proposition improves Corollary 2.

Proposition 2. The Fréchet space $(\mathcal{H}_b(U;F),\tau_b)$ is quasi-normable.

Proof. Because of Theorem 2, the strong dual of $(\mathcal{H}_b(U;F),\tau_b)$ is a boundedly retractive inductive limit of Banach spaces, hence, it satisfies the strict Mackey's convergence condition. But it is known ([5], p.106) that if the strong dual of an infrabarreled space satisfies such a condition, then it is quasi-normable, so, the result.

Proposition 3. $(\mathcal{H}_b(U;F),\tau_b)$ is a Montel, Schwartz, or a nuclear space if and only if E and F are finite dimensional.

Proof. It is well know that when $\dim E < \infty$ and $\dim F < \infty$, $(\mathcal{H}_b(U;F),\tau_b)$ is a Montel, Schwartz, nuclear space. Conversely, if $(\mathcal{H}_b(U;F),\tau_b)$ is a Montel, Schwartz or a nuclear space, then every topological subspace (in particular E' and F) have the corresponding property, and since both E' and F are normed spaces, they are finite dimensional.

Proposition 4. The bidual of $(\mathcal{H}_b(U;F),\tau_b)$ can be identified with the space of the sequences $\psi = (\psi_n) \in \prod_{n=0}^{\infty} \mathcal{P}(^nE;F)''$ such that

(*) $$\limsup_{n\in\mathbb{N}} \|\psi_n\|^{\frac{1}{n}} \le \frac{1}{R}$$

Proof. Assume $\psi = (\psi_n) \in \prod_{n=0}^{\infty} P({}^nE;F)''$ satisfies condition (*). Then, the mapping $\psi: S(U;F) \to C$ given by

$$\mu = (\mu_n) \mapsto \langle \mu, \psi \rangle = \Sigma \langle \mu_n, \psi_n \rangle, \quad \mu \in S(U;F)$$

is well defined, linear, and its restriction to each of the $S_r(U;F)$, $0 < r < R$, is continuous. Hence ψ is an element of the bidual space.

Let $\mu = (\mu_n) \in S(U;F)$ be given, and let us write $\hat{\mu}_n = (0,\ldots,0,\mu_n,0,\ldots)$ for $n \in \mathbb{N}$. If we take any r, $0 < r < R$, satisfying

$$\limsup_{n\in\mathbb{N}} \|\mu_n\|^{\frac{1}{n}} < r$$

then, it is easy to see that we have

$$\mu = \Sigma \hat{\mu}_n$$

in $S_r(U;F)$.

Now, assume $\psi \in (S(U;F), \tau_\ell)'$ and let (ψ_n), $\psi_n = \psi|_{P({}^nE;F)'}$, $n \in \mathbb{N}$, be the sequence of the restrictions of ψ to $P({}^nE;F)'$. Then, $\psi_n \in P({}^nE;F)''$ for all $n \in \mathbb{N}$ and

$$\limsup_{n\in\mathbb{N}} \|\psi_n\|^{\frac{1}{n}} \le \frac{1}{R}.$$

Indeed, if we had

$$\limsup_{n\in\mathbb{N}} \|\psi_n\|^{\frac{1}{n}} > \frac{1}{R}$$

then there would be an $\alpha > \frac{1}{R}$, a sequence $(n_j)_{j\in\mathbb{N}}$ and a $\mu_{n_j} \in P({}^{n_j}E;F)'$ with $\|\mu_{n_j}\| = 1$ such that

$$|\langle \mu_{n_j}, \psi_{n_j} \rangle| > \alpha^{n_j}$$

for all $j \in \mathbb{N}$. Let us define

$$\mu^*_{n_j} = \frac{|\langle \mu_{n_j}, \psi_{n_j} \rangle|}{\langle \mu_{n_j}, \psi_{n_j} \rangle} \mu_{n_j}$$

for $j \in \mathbb{N}$, and let $\nu = (\nu_n)$ be given by

$$\nu_{n_j} = \frac{1}{\alpha^{n_j}} \mu^*_{n_j} \qquad \text{for} \quad j \in \mathbb{N}$$

$$\nu_n = 0 \qquad \text{for} \quad n \neq n_j .$$

Then

$$\limsup_{n\in\mathbb{N}} \|\nu_n\|^{\frac{1}{n}} = \frac{1}{\alpha} < R$$

so that $\nu = (\nu_n) \in S(U;F)$ and

$$\langle \nu, \psi \rangle = \sum_{n=0}^{\infty} \langle \hat{\nu}_n, \psi_n \rangle = \sum_{j=0}^{\infty} \langle \nu_{n_j}, \psi_{n_j} \rangle .$$

But this is a contradiction since all the terms in the latter series are, grater than one.

Finally, for each $\mu = (\mu_n) \in S(U;F)$ we have

$$\langle \mu, \psi \rangle = \sum_{n=0}^{\infty} \langle \hat{\mu}_n, \psi \rangle = \sum_{n=0}^{\infty} \langle \mu_n, \psi_n \rangle .$$

This completes the proof.

<u>Corollary 3</u>. $(\mathcal{H}_b(U;F), \tau_b)$ is reflexive if and only if all $\mathcal{P}({}^nE;F)$, $n \in \mathbb{N}$, are reflexive.

<u>Proof</u>. From ([1], Proposition 5.3) we deduce that $\mathcal{H}_b(U;F)$ can be identified with the space of sequences $(P_n) \in \prod_{n=0}^{\infty} \mathcal{P}({}^nE;F)$ such that

$$\limsup_{n\in\mathbb{N}} \|P_n\|^{\frac{1}{n}} \leq \frac{1}{R}$$

associating to each $f \in \mathcal{H}_b(U;F)$ the sequence of its differential polynomials at the origin. Then, Proposition 4 completes the proof.

If $\dim E < \infty$ and F is reflexive, then $\mathcal{P}({}^nE;F)$ is reflexive for all n. Hence $(\mathcal{H}_b(U;F), \tau_b)$ is reflexive.

<u>Conjecture</u>. If $(\mathcal{H}_b(U;F), \tau_b)$ is reflexive, then $\dim E < \infty$.

REFERENCES

[1] Barroso, J.: Introducción a la holomorfía entre espacios normados, Publicaciones de la Universidad de Santiago de Compostela, Serie Cursos y Congresos, nº 7, 1976.

[2] Bierstedt, K.D. - Meise, R.: Bemerkungen über die Approximationseigenshaft lokalkonvexer Funktionenranme, Math. Ann. 209 (1974), 99-107.

[3] Chae, S.B.: Holomorphic germs on Banach spaces, Ann. Inst. Fourier Grenoble 21, 3 (1971), 107-141.

[4] Dineen, S.: Holomorphic functions on locally convex topological vector spaces I, Ann. Inst. Fourier, Grenoble, 23 (1973), 19-54.

[5] Grothendieck, A.: Sur les espaces ($\mathfrak{F}$) et ($\mathfrak{DF}$), Summa Brasiliensis Math. 3 (1954), 57-122.

[6] Isidro, J.M.: Topological duality on the space $(\mathcal{H}_b(U;F),\tau_b)$. Proc. Royal Irish Acad. 79, S, 12 (1979), 115-130.

[7] Mujica, J.: Gérmenes holomorfos y funciones holomorfas en espacios de Fréchet, Publicaciones del Departamento de Teoría de Funciones. Universidad de Santiago de Compostela, nº 1, 1978.

[8] Nachbin, L.: Topology on spaces of Holomorphic Mapping, Springer-Verlag, 1969.

THE LEVI PROBLEM
AND
THE RADIUS OF CONVERGENCE OF HOLOMORPHIC FUNCTIONS ON METRIC VECTOR SPACES

Aboubakr Bayoumi

Department of Mathematics
Uppsala University
Thunbergsvägen 3
S - 752 38 Uppsala, Sweden

1. INTRODUCTION

The Levi problem for domains Ω in (or over) many locally convex topological vector spaces E, i.e. to prove that if Ω is pseudoconvex in (or over) E, then it is a domain of holomorphy, has been attacked by many mathematicians. For example Gruman & Kiselman [5] have obtained a solution when E is a Banach space with basis. Schottenloher [17] has solved the Levi problem for a domain Ω in (or spread over) a Banach space with a finite-dimensional Schauder decomposition. As for negative results, Josefson [7] gave a counter-example: a pseudoconvex domain in $\ell^\infty(A)$, the Banach space of bounded functions on a non-countable set A with supremum norm, which is not a domain of holomorphy.

Here we study the Levi problem for a domain in (or over) a non-locally convex space E with a finite-dimensional Schauder decomposition. We shall give in Section 3.a a solution to the Levi problem when E is an infinite-dimensional complex metric vector with a finite-dimensional Schauder decomposition and a logarithmically plurisubharmonic metric d such that $-\log d_\Omega$, $d_\Omega(x) = \inf_{y\in\partial\Omega} d(x,y)$, is plurisubharmonic in Ω for every pseudoconvex domain Ω in E.

Moreover, we shall solve the Levi problem in any locally bounded topological vector space E with a finite-dimensional Schauder decomposition such that the equivalent p-homogeneous metric d defining the original topology of E is logarithmically plurisubharmonic (Corollary 3.a.4). An example of a non-locally convex space which admits no solution to the Levi problem is L^p, the space of all measurable functions on $[0,1]$ metrized by the metric $d(f,g) = \int_0^1 |f-g|^p \, d\mu$, $0 < p < 1$. In fact these results have been obtained as a consequence of studying the radius of convergence problem in non-locally convex spaces.

We have seen in [2] that every bounding subset of certain separable complex metric vector spaces E (i.e. the subsets of E on which every holomorphic function on E is bounded) is relatively compact. This means that for every non-compact closed set $A \subset E$ one can find an $f \in H(E)$ such that $\|f\|_A = \sup_{x\in A} |f(x)| = \infty$; i.e. there exists an $f \in H(E)$ with finite radius of convergence $R_f(x)$ at every point $x \in E$. It can even happen that $\inf(R_f(x), x \in A) = 0$, see Aron [1] and Theorem 3.1.a below.

The radius of convergence problem in the sense of Kiselman [8] is whether one can construct a holomorphic function f on a normed space E such that its radius of convergence satisfies $R_f = R$ at least approximately, where $R: E \to]0,\infty[$ is given and satisfies:

(*) $-\log R_f(x)$ is plurisubharmonic on E and $|R_f(x)-R_f(y)| \leq c\|x-y\|$, $x,y \in E$,

where c is some constant depending on f, $0 \leq c \leq 1$. Here we mean by $R_f(x)$ the supremum of all $r > 0$ such that the Taylor series of f at x converges uniformly in the ball $B(x,r)$.

C.O. Kiselman showed in [8] that the conditions (*) are not sufficient for a function $R: \ell^q \to]0,\infty[$, $1 < q < \infty$, to be the radius of convergence of some $f \in H(\ell^q)$. For $E = c_0$ the result is due to B. Josefson.

C.O. Kiselman [8] constructed for a normed space E under the hypothesis that R satisfies (*) and the restriction that R depends on finitely many variables, an $f \in H(E)$ with radius of convergence $R_f \leq R$. He removed in [10] this restriction on R for $E = \ell^q$, $1 \leq q < \infty$ or $E = c_o$, $q = \infty$, by constructing an $f \in H(E)$ such that $(1+c^p)^{-1/p} R \leq R_f \leq R$, $p = q/(q-1)$ (and thus $R_f = R$ if $q = 1$). For a polynomially convex set $\Omega \subset \ell^1$, Coeuré [3] proved that there exists an $f \in H(\Omega)$ with $R_f = F$, when R satisfies (*).

M. Schottenloher [15,16] completes and generalizes the results of Kiselman and Coeuré by studying this problem of constructing holomorphic functions in (or over) normed spaces with a Schauder decomposition.

In this paper we shall also study the radius of convergence problem, i.e. the problem of constructing holomorphic functions with a certain radius of convergence in (or over) infinite-dimensional non-locally convex spaces E, and we shall prove that Schottenloher's results can be generalized to infinite-dimensional complex metric vector spaces E with a finite-dimensional Schauder decomposition.

In Section 2 we shall give some general results on the radius of convergence R_f and the radius of boundedness R_f^b, i.e. the supremum of all $r > 0$ such that $f \in H(\Omega)$ is bounded in $B(x,r)$ with $B(x,r) \subset \Omega$, where Ω is a domain in E. We also prove that $-\log R_f^b$ is plurisubharmonic in a pseudoconvex domain $\Omega \subset E$ for every $f \in H(\Omega)$ assuming the metric is what we call a PB-metric, and we prove that this is the case if $t \to -\log d(e^t x, 0)$ is convex in $\mathbb{C}$ for every fixed $x \in E$.

In Section 3, Theorem 3.a.1, we prove that for an infinite-dimensional complex metric vector space E with a translation invariant metric d and a finite-dimensional Schauder decomposition (π_n) such that $\log d(x,0)$ is plurisubharmonic on E and $R: \Omega \to]0,\infty[$ is defined on a domain $\Omega \subset E$ with $-\log R$ plurisubharmonic and $R \leq d_\Omega$, there exists an $f \in H(\Omega)$ with radius of convergence $R_f \leq R$.

In Section 3.b we prove that if in addition to the above condition on R it is locally Lipschitz continuous, then there exists $f \in H(\Omega)$ such that $\varphi(KR) \leq R_f \leq R$, where K is a positive constant and φ is a continuous function on the real numbers $\mathbb{R}$ with $\varphi(0)=0$, $\varphi(t) > 0$ for $t > 0$, which depends on d and (π_n). In the particular case $E = \ell^{\{p_n\}}$ the metric vector space of all complex sequences $(x_j)_1^\infty = x$, $d(x,0) = \sum_{j=1}^\infty |x_j|^{p_j} < \infty$, $0 < p_n < 1$ there will always exist $f \in H(\Omega)$ with $R_f = R$.

In Section 3.c we shall see that $T_1(R) = \{f \in H(\Omega);\ R_f \leq R\}$ and $T_2(R) = \{f \in H(\Omega);\ KR \leq R_f \leq R\}$ are dense in the compact open topology on $H(\Omega)$ if R satisfies the hypotheses of Theorem 3.a.1 and Proposition 3.b.1 respectively.

In Section 3.d, finally, general results on Riemann domains are given.

ACKNOWLEDGEMENT

I want to express my deep gratitude to Professor Christer Kiselman for fruitful discussions, valuable suggestions and encouraging support. He has shown a never failing interest in my work.

2. PROPERTIES OF THE RADIUS OF CONVERGENCE

Let E be a complex metric vector space with a translation invariant metric d. Let $\Omega \subset E$ be open. Denote by $H(\Omega)$ the space of holomorphic functions $f: \Omega \to \mathbb{C}$, i.e. continuous and Gâteaux-analytic functions on Ω.

The <u>radius of convergence</u> $R_f(x)$ of a function $f \in H(\Omega)$ at a point $x \in \Omega$ is the least upper bound of all numbers $r > 0$ such that the Taylor series of f at x converges uniformly in $B(x,r)$, the closed ball of center x and radius r in E.

The <u>radius of boundedness</u> $R_u^b(x)$ of a numerical function

$u: \Omega \to [-\infty,\infty[$ at a point $x \in \Omega$ is the least upper bound of all numbers $r > 0$ such that u is bounded above in $B(x,r)$ with $B(x,r) \subset \Omega$. For a function $f \in H(\Omega)$,

$$R_f^b(x) = \sup(r > 0;\ \|f\|_{B(x,r)} \text{ is finite, } B(x,r) \subset \Omega), \quad x \in \Omega.$$

By these definitions of the radius of convergence R_f and the radius of boundedness R_f^b for a function $f \in H(\Omega)$, it is easy to prove that

$$(2.1) \qquad R_f^b = \inf(R_f, d_\Omega),$$

where $d_\Omega(x) = \inf_{y \in E\setminus\Omega} d(x,y)$, $x \in \Omega$, is the distance function on Ω defined by the metric d of E. It may happen that f can be continued analytically beyond the boundary of Ω and hence R_f^b will be less than R_f.

It is clear that the radius of boundedness R_u^b for any numerical function u is globally Lipschitz continuous. However, the radius of convergence is only locally Lipschitz continuous. We formulate these results as two lemmas.

Lemma 2.1. If u is any numerical function defined in a subset Ω of a metric space E we define $u(x) = +\infty$, $R_u^b(x) = 0$ for $x \in E\setminus\Omega$. With this convention we have

$$(2.2) \qquad |R_u^b(x) - R_u^b(y)| \le d(x,y) \quad \text{for all } x,y \in E.$$

Proof. If $R_u^b(x) = R_u^b(y)$ the conclusion is valid; let us assume that $R_u^b(x) > R_u^b(y) \ge 0$. Then u is bounded above in $B(y, r-d(x,y)) \subset$ $\subset B(x,r)$ for all $r < R_u^b(x)$. Hence $R_u^b(y) \ge r-d(x,y)$ and letting r tend to $R_u^b(x)$ we get

$$R_u^b(x) > R_u^b(y) \ge R_u^b(x) - d(x,y).$$

This proves (2.2) in this case, and by symmetry the estimate holds everywhere.

Lemma 2.2. If $f \in H(\Omega)$, Ω being an open subset of a metric vector space E with a translation invariant metric d satisfying $d(tx,0) \leq d(x,0)$ for all $x \in E$ and all $t \in \mathbb{C}$, $|t| \leq 1$, then

$$(2.3) \qquad |R_f(x)-R_f(y)| \leq d(x,y), \quad \text{if} \quad x,y \in \Omega \quad \text{and} \quad d(x,y) < \sup(d_\Omega(x),d_\Omega(y)).$$

Proof. By symmetry it is again enough to consider the case $R_f(x) > R_f(y) > 0$. Let ω denote the open ball $B(x,R_f(x))^\circ$ and denote by g the extension of $f|B(x,d_\Omega(x))^\circ$ to ω. Then $R_g(x) = R_f(x)$ and g is bounded in $B(x,r)$ for all $r < R_f(x)$. Now, if $R_f(x) \leq d(x,y)$ we are done; assume that $R_f(x) > d(x,y)$. Then $y \in \omega$ and if $d(x,y) < d_\Omega(x)$ or $d(x,y) < d_\Omega(y)$ the whole segment between x and y is contained in $\omega \cap \Omega$ so we must have $R_f(y) = R_g(y)$. By Lemma 2.1, $R_g^b(y) \geq R_g^b(x) - d(x,y) = R_f(x) - d(x,y)$. In conclusion we have

$$R_f(x) > R_f(y) = R_g(y) \geq R_g^b(x) - d(x,y) = R_f(x) - d(x,y),$$

i.e. (2.3) is proved.

The open set $\Omega \subset E$ is called <u>pseudoconvex</u> if for every finite-dimensional subspace $F \subset E$, $\Omega \cap F$ is pseudoconvex.

Since a metric d on a complex metric vector space E is not always homogeneous, i.e. we do not necessarily have $d(tx,0) = |t|\, d(x,0)$, $t \in \mathbb{C}$, $x \in E$, we have found it useful to introduce the following definition in the study of the Levi and radius of convergence problems in Sections 2 and 3:

A complex metric vector space E with a translation invariant metric d is said to have the <u>pseudoconvex-boundary-distance property</u> (in short: the PB-property) if, for every pseudoconvex domain Ω in E, the function $x \mapsto -\log d_\Omega(x)$ is plurisubharmonic in Ω. Such a space E is also called a PB-space.

That $\mathbb{C}^n$ is a PB-space for any norm is well known; see e.g. Hörmander [6], Theorem 2.5.4, where it is only assumed that the distance is measured by a 1-homogeneous function which is positive which is positive except at the origin.

The following result shows that R_f^b for $f \in H(\Omega)$ admits a certain geometric property which is a consequence of the PB-property.

Proposition 2.3. Let Ω be a pseudoconvex domain in a PB-space. Then $-\log R_u^b$ is plurisubharmonic in Ω if u is plurisubharmonic in Ω. Consequently $-\log R_f^b$ is plurisubharmonic in Ω for every function $f \in H(\Omega)$.

Proof. Let u be a plurisubharmonic function in Ω and let

$$\Omega_k = \{x \in \Omega\, ;\ u(x) < k\}, \quad k \in \mathbb{N}.$$

For every $k \in N$, the set Ω_k is pseudoconvex. This follows from the definition of pseudoconvexity and the fact that u is plurisubharmonic in Ω. Let

$$d_{\Omega_k}(x) = \inf_{y \in \partial\Omega_k} d(x,y), \quad x \in \Omega_k, \quad k \in \mathbb{N}.$$

Hence $R_u^b = \sup_k d_{\Omega_k} = \lim_{k\to\infty} d_{\Omega_k}$, i.e.

$$-\log R_u^b = \inf_k (-\log d_{\Omega_k}).$$

The functions $-\log d_{\Omega_k}$ are plurisubharmonic in Ω_k for every $k \in \mathbb{N}$, since Ω_k is pseudoconvex and E is a PB-space. Thus $-\log R_u^b$ is plurisubharmonic as a decreasing limit of a sequence of plurisubharmonic functions.

Since $\log|f|$ is plurisubharmonic in Ω for every function $f \in H(\Omega)$, we also get that $-\log R^b_{\log|f|} = -\log R_f^b$ is plurisubharmonic in Ω.

Remark. This proposition was proved by P. Lelong [11] for $\Omega = E$, a normed space. Of course, every normed space is a PB-space and $R_f^b = R_f$ if $f \in H(E)$. For other generalizations of R_f^b in normed spaces, see Kiselman [8].

The following proposition will give a sufficient condition for a metric to have the PB-property.

Proposition 2.4. Let d be a translation invariant metric on a complex metric vector space E such that the function

$$s \mapsto -\log d(e^s x, 0), \quad s \in \mathbb{C}, \tag{2.4}$$

is convex in $\mathbb{C}$ for each fixed $x \in E$. Then E is a PB-space. In particular, this is the case if d is given by a p-homogeneous norm $\|\cdot\|_p$, i.e. $d(x,y) = \|x-y\|_p$.

Proof. Assume that Ω is an arbitrary pseudoconvex domain in E. We claim that $-\log d_\Omega$ is plurisubharmonic in Ω. Since the function (2.4) is convex and decreasing, it has the Legendre transform ψ satisfying

$$-\log d(e^s x, 0) = \sup_{\tau \le 0} (\tau s - \psi(\tau, x)), \quad x \in E,\ s \in \mathbb{R},$$

$$\psi(\tau, x) = \sup_s (\tau s + \log d(e^s x, 0)), \quad x \in E, \quad \tau \in \mathbb{R}.$$

Changing notation by putting $s = \log|t|$, we have

$$d(tx, 0) = \inf_{\tau \ge 0} |t|^\tau \varphi(\tau, x), \quad x \in E,\ t \in \mathbb{C},$$

$$\varphi(\tau, x) = \sup_{t \in \mathbb{C}} |t|^{-\tau} d(tx, 0), \quad x \in E,\ \tau \in \mathbb{R}.$$

Note that if $d(x,0)$ is p-homogeneous, i.e. $d(tx,0) = |t|^p d(x,0)$, $t \in \mathbb{C}$, $x \in E$, $0 < p \le 1$, then we get $\varphi(p,x) = d(x,0)$ when $\tau = p$ and $\varphi(\tau,x) = +\infty$ otherwise.

Now

$$\begin{aligned} d_\Omega(x) &= \inf_{y \in \partial\Omega} d(x,y) = \inf_{x+y \in \partial\Omega} d(y,0) \\ &= \inf_{x+tz \in \partial\Omega} d(tz,0) = \inf_z \inf_t \inf_{\tau \ge 0} |t|^\tau \varphi(\tau,z) \\ &= \inf_z \inf_{\tau \ge 0} \inf_t |t|^\tau \varphi(\tau,z) = \inf_z \inf_{\tau \ge 0} \varphi(\tau,z)\, \delta_\Omega(x,z)^\tau \end{aligned}$$

for $\delta_\Omega(x,y) = \inf(|t|;\ x+ty \in \partial\Omega)$. Hence

$$-\log d_\Omega(x) = \sup_z \sup_{\tau \ge 0} [-\log \varphi(\tau,z) - \tau \log \delta_\Omega(x,z)]$$

and since this function is either $\equiv -\infty$ or else continuous, it is plurisubharmonic as a supremum of a family of plurisubharmonic functions. Here we have used the following property of a pseudoconvex domain $\Omega \subset E$: if $\Omega \subset E$ and E is a topological vector space, then Ω is pseudoconvex if and only if the function $(x,y) \mapsto -\log \delta_\Omega(x,y)$ is plurisubharmonic on $\Omega \times E$, where $\delta_\Omega(x,y) = \sup(r; x+r\,Dy \subset \Omega)$, $x \in \Omega$, $y \in E$, $D = \{t \in C; |t| \leq 1\}$. In fact, this is equivalent to our definition of pseudoconvexity, i.e. $\Omega \cap F$ is pseudoconvex in F for every finite-dimensional subspace F of E. (See Noverraz [13], Lemme 2.1.5).

Remark. According to the definition of a PB-space E, for every pseudoconvex domain Ω in E, $-\log d_\Omega$ is plurisubharmonic in Ω. The converse is also true. Indeed, assume $-\log d_\Omega$ is plurisubharmonic in Ω and F is an arbitrary finite-dimensional subspace of E. Then $\Omega \cap F$ is pseudoconvex in view of the classical properties of plurisubharmonic functions of finitely many variables. Hence Ω is pseudoconvex.

Let E_p, $0 < p \leq 1$, be a complex metric vector space with a p-homogeneous norm $\|\cdot\|_p$, i.e. $d(x,y) = \|x-y\|_p$ and $\|tx\|_p = |t|^p\, d(x,0)$, $x \in E$, $t \in C$. The following Lemma 2.5 gives a formula for the radius of convergence R_f for $f \in H(\Omega)$, $\Omega \subset E_p$ open, which is well known for normed spaces, i.e. the 1-homogeneous spaces.

Lemma 2.5. The radius of convergence R_f for $f \in H(\Omega)$, Ω open in E_p, $0 < p \leq 1$, is given by

$$R_f(x) = \liminf \|P_n\|^{-p/n}, \quad x \in \Omega, \tag{2.5}$$

where $\|P_n\| = \sup(|P_n(x)|; \|x\|_p < 1)$ and P_n is defined by the formula $f(x+y) = \Sigma P_n(y)$, $y \in E_p$, $\|y\|_p$ is small.

Proof. Similar to the method used by Nachbin [12].

Remark 1. It follows again from (2.5) that $-\log R_f$ is plurisubharmonic in Ω, which now may be any open set in the p-homogeneous space E_p, $0 < p \leq 1$. This is because $P_n = P_{n,x}$ is an n-homogeneous polynomial which depends analytically on x.

Remark 2. From the definition of R_f, $f \in H(\Omega)$, we have for a p-homogeneous space E_p, $0 < p \leq 1$, the formula

$$R_f(x) = \sup\,(r > 0;\ \Sigma\|P_n\|\ r^{n/p} < \infty), \qquad x \in \Omega.$$

3. CONSTRUCTION OF HOLOMORPHIC FUNCTIONS

Let E be an infinite-dimensional complex metric vector space with a metric d. We recall that for an open set $\Omega \subset E$, the boundary distance $d_\Omega(x) = \inf_{y\in\partial\Omega} d(x,y)$, $x \in \Omega$, and that the space E is called a PB-space if for every pseudoconvex domain Ω in E, the function $-\log d_\Omega$ is plurisubharmonic in Ω. By $B(x,r)$ we mean the closed ball of center $x \in E$ and radius r.

3.a. The Levi problem.

To solve the Levi problem in E means to show that every pseudoconvex domain Ω in E is a domain of holomorphy. The following Theorem 3.a.1 will give the solution to this problem for some infinite-dimensional complex metric vector spaces with a finite-dimensional Schauder decomposition (π_n), i.e. with an equicontinuous sequence (π_n) of linear projections $\pi_n: E \to \pi_n(E)$ such that $\dim \pi_n(E) < \infty$, $\pi_n \circ \pi_m = \pi_{\min\{n,m\}}$ for all $n,m \in N$ and $x = \lim \pi_n(x) = \Sigma(\pi_{n+1}(x) - \pi_n(x))$, $x \in E$; see Corollary 3.a.3 and 3.a.4 for the precise statement.

Theorem 3.a.1. Let E be an infinite-dimensional complex metric vector space, with a translation invariant metric d and with a finite-dimensional Schauder decomposition (π_n), such that $x \mapsto \log d(x,0)$ is plurisubharmonic on E. Let $R: \Omega \to \mathbb{R}_+$ be defined on a domain $\Omega \subset E$ with $-\log R$ plurisubharmonic and $R \leq d_\Omega$. Then there exists a holomorphic function $f \in H(\Omega)$ with radius of convergence $R_f \leq R$.

Proof. We shall construct, under the hypotheses of the theorem, a holomorphic function f on Ω such that for a suitably chosen sequence (x_n), which as we shall see, depends on the metric d, we have $|f(x_n)| \geq n$, $n \in \mathbb{N}$ and $R_f \leq R$. Here a covering (V_n) on Ω will be defined (cf. Schottenloher [16]) such that $\|f\|_{V_n} = \sup_{x \in V_n} |f(x)| < \infty$ for all $n \in \mathbb{N}$.

Let (z_n) be dense in Ω, $z_n \in \Omega_n = \Omega \cap \pi_n(E)$ such that

(3.a.1) $$d(z_n,0) + R(z_n) \leq n.$$

Let $e_n \in \pi_n(E)$ with $\pi_{n-1}(e_n) = 0$ and $d(e_n,0) = 1$ (this is possible if $\pi_n(E) \neq \pi_{n-1}(E)$ which we may assume without loss of generality). Put

$$\epsilon_1 = R(z_1)/2, \quad \epsilon_n = \min(R(z_n), \epsilon_{n-1}, R(x_{n-1}))/2,$$

where

$$x_n = z_n + \lambda_n e_{n+1}, \qquad n \in \mathbb{N},$$

and $\lambda_n \in \mathbb{C}$ is chosen such that $d(\lambda_n e_{n+1}, o) = R(z_n) - \epsilon_n/2$ i.e. λ_n is such that

(3.a.2) $$d(x_n, z_n) = R(z_n) - \epsilon_n/2, \qquad n \in \mathbb{N}.$$

(For a normed space it is obvious that $|\lambda_n| = R(z_n) - \epsilon_n/2$, $n \in \mathbb{N}$).

Assume first that (π_n) is a monotone Schauder decomposition, i.e. $d(\pi_n(x),0) \leq d(x,0)$, $x \in E$. Let

$$X_n = \{x \in \Omega;\ \pi_m(x) \in \Omega \text{ for all } m \geq n\} = \bigcap_{m \geq n} (\Omega \cap \pi_m^{-1}(\Omega_m)), \qquad n \geq 1,$$

$$\sigma_{n+1} = \Omega_{n+1} \cap \pi_n^{-1}(\Omega_n) = \Omega_{n+1} \cap X_n, \qquad n \geq 1,$$

$$L_{n+1} = \{x \in \sigma_{n+1};\ d(x,0) \leq n,\ d_{\sigma_{n+1}}(x) \geq \epsilon_{n+1}\}, \qquad n \geq 1,$$

$$\hat{L}_{n+1} = \text{the holomorphically convex hull of } L_{n+1}, \qquad n \geq 1,$$

$$V_n = \bigcap_{m \geq k \geq n} \{x \in X_n;\ R(\pi_k(x)) \geq \epsilon_n + d(\pi_k(x), \pi_m(x)) \text{ and } \pi_m(x) \in \hat{L}_m\},\ n \geq 2,$$

and $V_1 = \{z_1\}$. From these definitions it follows that:

(3.a.3) $$X_n \subset X_{n+1} \text{ and } \pi_m(X_n) = \Omega_m \cap X_n \text{ for } m \geq n,$$

(3.a.4) $$V_n \subset V_{n+1} \text{ and } \pi_m(V_n) = \Omega_m \cap V_n \text{ for } m \geq n.$$

Also the following statements will hold:

(3.a.5) For all $x \in \Omega$ and $r < d_\Omega(x)$, there exists $n \in N$ with $B(x,r) \subset X_n$ and $\pi_m(B(x,r)) \subset L_m$ for $m \geq n$.

To see this, let $x \in \Omega$ and $r < d_\Omega(x)$. There exists $\delta > 0$ such that $r + 3\delta < d_\Omega(x)$. We can find $n \in N$ with $\varepsilon_n < \delta$, $d(x,0) + r < n-1$ and $d(\pi_m(x),\pi_n(x)) < \delta$ for $m \geq n-1$. Since $\pi_m(B(\pi_n(x),r+2\delta)) \subset B(x,r+3\delta) \subset \Omega$ for $m \geq n-1$, we have $B(\pi_n(x), r+2\delta) \subset X_{n-1} \subset X_m$ and consequently $d_{\sigma_m}(\pi_n(x)) \geq r + 2\delta$, $m \geq n$. For $y \in B(x,r)$ and $m \geq n$ we have $d_{\sigma_m}(\pi_m(y)) \geq \delta \geq \varepsilon_n \geq \varepsilon_m$ and $d(\pi_m(y),0) \leq d(x,0) + r < m-1$. Hence $B(x,r) \subset X_{n-1}$ and $\pi_m(B(x,r)) \subset L_m$.

(3.a.6) For all $x \in \Omega$, there exist $\delta > 0$ and $n \in \mathbb{N}$ with $B(x,\delta) \subset V_n$.

For $x \in \Omega$, there exists $r > 0$ with $R(x) > 5r$. Since R is lower semi-continuous, $R(y) > 4r$ for $d(y,x) < 2\delta$, if $\delta > 0$ is suitably chosen. For $s = \min(r,\delta)$ there exists $n \in \mathbb{N}$ such that $\varepsilon_n < s$ and $d(\pi_m(x),\pi_k(x)) \leq s$ for all $m,k \geq n$. We choose n so big that $B(x,s) \subset X_n$ and $\pi_m(B(x,r)) \subset L_m$ for $m \geq n$ and hence $\pi_m(B(x,r)) \subset \hat{L}_m$ for $m \geq n$. For $y \in B(x,s)$ and $m,k \geq n$, it follows that $R(\pi_k(y)) > 4s$ since $d(\pi_k(y),x) \leq d(\pi_k(y-x),0) + d(\pi_k(x),x) < 2\delta$. Hence $R(\pi_k(y)) > \varepsilon_n + 3s > \varepsilon_n + d(\pi_k(y),\pi_m(y))$. This is because $d(\pi_k(y), \pi_m(y)) \leq d(\pi_k(y-x),0) + d(\pi_k(x), \pi_m(x)) + d(\pi_m(x-y),0) < 3s$. Thus $B(x,\delta) \subset V_n$.

(3.a.7) $K_{n+1} = \pi_{n+1}(V_n)$ is compact and holomorphically convex in σ_{n+1}, $n \geq 1$, and $\pi_n(K_{n+1}) \subset \sigma_n$.

Since $-\log R$ is plurisubharmonic in Ω, for every compact subset K in Ω,

$$\sup_{x \in \hat{K}_{p(\Omega)}} (-\log R(x)) = \sup_{x \in K} (-\log R(x)),$$

where $\hat{K}_{p(\Omega)}$ is the plurisubharmonic convex hull of K in Ω. Since $R \leq d_\Omega$,

$$\inf_{x\in\hat{K}_{p(\Omega)}} d_\Omega(x) \geq \inf_{x\in\hat{K}_{p(\Omega)}} R(x) = \inf_{x\in K} R(x) > 0.$$

Hence $p(\Omega)$ is relatively compact in $\Omega \cap F$ for every finite-dimensional subspace $F \subset E$, i.e. Ω is pseudoconvex. Consequently, $\Omega_{n+1} \cap \pi_n^{-1}(\Omega_n) = \sigma_{n+1}$ is pseudoconvex. The following expression of K_{n+1} will show that it is compact:

$$\begin{aligned} K_{n+1} &= \Omega_{n+1} \cap V_n = \\ &= \{x \in \hat{L}_{n+1} \cap \pi_n^{-1}(\hat{L}_n);\ R(x) \geq \epsilon_n \text{ and } R(\pi_n(x)) \geq \epsilon_n + d(\pi_n(x),x)\} \\ &= \{x \in \hat{L}_{n+1} \cap \pi_n^{-1}(\hat{L}_n);\ \log \epsilon_n - \log R(x) \leq 0 \text{ and} \\ &\qquad -\log R(\pi_n(x)) + \log(\epsilon_n + d(\pi_n(x),x)) \leq 0\}. \end{aligned}$$

The set K_{n+1} is holomorphically convex in $\hat{L}_{n+1}$, according to its above expression and assumption that $-\log R(x)$ and $\log d(x,0)$, $x \in \Omega$, are plurisubharmonic functions, and consequently K_{n+1} is holomorphically convex in σ_{n+1}. It is obvious that $\pi_n(K_{n+1}) \subset \hat{L}_n$ $\subset \sigma_n$.

For the proof we now need the following lemma (cf. Gruman & Kiselman [5] and Schottenloher [16]).

Lemma 3.a.2. For any holomorphic function f on $\sigma_n = \Omega_n \cap \pi_{n-1}^{-1}(\Omega_{n-1})$ and any $\epsilon > 0$ there exists a holomorphic function g on Ω with

$$\|g - f\circ\pi_n\|_{V_n} < \epsilon \quad \text{and} \quad \|g\|_{V_m} < \infty \quad \text{for all } m \in \mathbb{N}.$$

Proof. By induction there exists a sequence of functions $(f_j)_{j\geq n}$ with

(3.a.8) $$f_j \in H(\sigma_j) \text{ and } \|f_{j+1} - f_j\circ\pi_j\|_{K_{j+1}} < \epsilon\, 2^{-(j+1)}, \qquad j \geq n.$$

To see this let $f_n = f$, and assume that $f_n, f_{n+1}, \ldots, f_j$ satisfying (3.a.8) have already been found. By (3.a.7), K_{j+1} is a compact holomorphically convex subset of σ_{j+1} and $K_{j+1} \subset \pi_j^{-1}(\sigma_j)$, hence K_{j+1} lies in $\Omega_{j+1} \cap \pi_j^{-1}(\sigma_j)$. By the approximation theorem (see Hörmander [6], p.91) there exists for $f_j\circ\pi_j$ a holomorphic function

$f_{j+1} \in H(\sigma_{j+1})$ satisfying (3.a.8). Now for $\ell \geq k \geq m \geq n$ it follows that

$$\|f_\ell \circ \pi_\ell - f_k \circ \pi_k\|_{V_j} \leq \sum_{j=k}^{\ell-1} \|f_{j+1} \circ \pi_{j+1} - f_j \circ \pi_j\|_{V_j} \leq$$

$$\leq \sum_{j=k}^{\ell-1} \|f_{j+1} - f_j \circ \pi_j\|_{\pi_{j+1}(V_j)}$$

$$\leq \sum_{j=k}^{\ell-1} \|f_{j+1} - f_j \circ \pi_j\|_{K_{j+1}} < \epsilon 2^{-k}.$$

Thus $(f_k \circ \pi_k)_{k\geq m}$ converges uniformly on V_m, hence the limit is a holomorphic function g on V_m^o. Since m is arbitrary g is holomorphic on Ω according to (3.a.6). Letting $m=k$ and $\ell \to \infty$, we obtain

$$\|g - f_m \circ \pi_m\|_{V_m} < \epsilon 2^{-m}.$$

Thus $\|g - f \circ \pi_n\|_{V_n} < \epsilon 2^{-n} < \epsilon$ and $\|g\|_{V_m} \leq \|f_m \circ \pi_m\|_{V_m} + \|g - f_m \circ \pi_m\|_{V_m} \leq$ $\leq \|f_m\|_{K_{n+1}} + \epsilon < \infty$.

We are now going to construct the required holomorphic function. The choice of $x_n = z_n + \lambda_n e_{n+1}$, where $\lambda_n \in \mathbb{C}$ is such that $d(x_n, z_n) = R(z_n) - \epsilon_n/2$, and the construction of (V_n) will give $x_n \in V_{n+1} \setminus V_n$. By (3.a.2),

$$\epsilon_n + d(\pi_n(x_n), x_n) = \epsilon_n + R(z_n) - \epsilon_n/2 > R(z_n) = R(\pi_n(x_n)),$$

i.e. $x_n \notin V_n$. According to $x_n \in \sigma_{n+1} = \Omega_{n+1} \cap \pi_n^{-1}(\Omega_n)$ and the hypothesis $R \leq d_\Omega$ we get $d_{\sigma_{n+1}}(x_n) \geq d_{\sigma_{n+1}}(z_n) - d(z_n, x_n) \geq$ $\geq d_\Omega(z_n) - (R(z_n) - \epsilon_n/2) \geq \epsilon_n/2 \geq \epsilon_{n+1}$. Also $d(x_n, 0) \leq d(z_n, 0) +$ $+ d(x_n, z_n) = d(z_n, 0) + R(z_n) - \epsilon_n/2 \leq d(z_n, 0) + R(z_n) \leq n$ by (3.a.1). By the definition of (ϵ_n), $R(x_n) \geq \epsilon_{n+1}$. Thus $x_n \in V_{n+1}$.

By (3.a.7) and $x_n \notin K_{n+1} = V_n \cap \Omega_{n+1}$, there exists a holomorphic function f_n on σ_{n+1} with

$$|f_n(x_n)| > 1 > \|f_n\|_{K_{n+1}}.$$

By Lemma 3.a.2 one can find $g_n \in H(\Omega)$ with $\|g_n\|_{V_m} < \infty$ for all $m \in \mathbb{N}$ and

$$\|g_n - f_n \circ \pi_{n+1}\|_{V_{n+1}} < \min(|f_n(x_n)|-1,\ 1-\|f_n\|_{K_{n+1}}).$$

Hence

$$|g_n(x_n)| > 1 > \|g_n\|_{V_n}.$$

Let $h_n = c_n g_n^{\alpha_n}$, $\alpha_n \in N$, be a suitable power of g_n, $c_n > 0$. Then

$$\begin{aligned} &\|h_n\|_{V_m} < \infty \\ &\|h_n\|_{V_n} < 2^{-n} \end{aligned} \qquad \text{for all} \quad m,n \in \mathbb{N}$$

and

$$|h_n(x_n)| > n + 1 + \sum_{j=1}^{n-1} |h_j(x_n)| \quad \text{for all} \quad n \in \mathbb{N}.$$

Therefore, the series $\sum_{n=1}^{\infty} h_n$ converges uniformly on V_n and by (3.a.6) it converges to a holomorphic function $f = \sum_1^{\infty} h_n \in H(\Omega)$ with $|f(x_n)| \geq n$.

We claim that $R_f \leq R$. Let $y \in \Omega$ be arbitrary, hence there exists a subsequence $(z_{n_j})_{j\geq 1}$ of (z_n) with $z_{n_j} \to y$. The function f is unbounded on the ball $B(y,r)$ if $r > \liminf d(y,x_{n_j})$, since x_{n_j} will lie in $B(y,r)$ for j large enough and $|f(x_{n_j})| \geq n_j$. Consequently,

$$\begin{aligned} R_f(y) &\leq \liminf d(y,x_{n_j}) \\ &\leq \liminf d(y,z_{n_j}) + d(z_{n_j},x_{n_j}) \\ &\leq \liminf (R(z_{n_j}) - \epsilon_{n_j}/2) \\ &\leq R(y). \end{aligned}$$

The last inequality follows since as R is lower semi-continuous we have $\liminf_{j\to\infty} (R(z_{n_j})) \leq R(\lim_{j\to\infty} z_{n_j}) = R(y)$. The theorem is now proved for the monotone case.

The general case is obtained by defining a new metric $d^*(x,0) = \sup_n d(\pi_n(x),0)$, $x \in E$. Then

$$d(x,0) \leq d^*(x,0)$$

and (π_n) relative to d^* is monotone. Moreover, d and d^* define the same topology on E in view of the assumption that (π_n) is equicontinuous. For the corresponding boundary distance and the radius of convergence we get

$$d_\Omega \le d_\Omega^*$$
$$R_f \le R_f^*.$$

Hence, for every $R: \Omega \to \mathbb{R}_+$ with the hypotheses of the theorem, it follows that $R \le d_\Omega^*$ and consequently a holomorphic function f on Ω with $R_f^* \le R$ will exist. Thus $R_f \le R$. This completes the proof of Theorem 3.a.1.

The following result gives a solution to the Levi problem for a PB-space satisfying the hypotheses of Theorem 3.a.1.

Corollary 3.a.3. Let E be an infinite-dimensional PB-space with a finite-dimensional Schauder decomposition and such that the metric d is logarithmically plurisubharmonic, i.e. $x \mapsto \log d(x,0)$ is plurisubharmonic. Then every pseudoconvex domain Ω in E is a domain of holomorphy.

Proof. Take $R = d_\Omega$. According to hypothesis Ω is pseudoconvex, we get $-\log R$ is plurisubharmonic. Hence, applying Theorem 3.a.1, we find a holomorphic function $f \in H(\Omega)$ with $R_f \le R = d_\Omega$. This implies that Ω is the domain of existence of f and consequently a domain of holomorphy.

Corollary 3.a.4. Let E be a locally bounded topological vector space with a finite-dimensional Schauder decomposition and such that the p-homogeneous norm which defines the original topology is logarithmically plurisubharmonic. Then every pseudoconves domain Ω in E is a domain of holomorphy.

Proof. The topology of any locally bounded topological vector space E can be defined by a p-homogeneous norm $\|\cdot\|_p$ for some p, $0 < p \le 1$; see Rolewicz [14, p.61]. We now apply Corollary 3.a.3, noting that Proposition 2.4 shows that E, equipped with $\|\cdot\|_p$, is a PB-space.

Remark. The assumption that the space has a Schauder decomposition plays an important role in solving the Levi problem. To show this it suffices to give the following example of a non-locally convex space in which the Levi problem has no solution.

Let $E = L^p = \{f: \int_0^1 |f|^p d\mu < \infty\}$ be the space of measurable functions metrized by the metric $d(f,g) = \int_0^1 |f-g| d\mu$, $0 < p < 1$. The dual space $(L^p)' = \{0\}$ which means that every subset of L^p, $0 < p < 1$, is bounding. In fact, this is an example of a non-locally convex space which has non-compact bounding subsets. (See [2] where bounding subsets of topological vector spaces are discussed.) Hence a pseudoconvex domain $\Omega \neq L^p$, $0 < p < 1$, can never be a domain of holomorphy in any sense. Of course, the space L^p, $0<p<1$, has no Schauder decomposition (π_n). Otherwise, continuous linear functionals $T_n = f\circ\pi_n$ would exist, where $f: \pi_n(E) \to C$ is a continuous linear functional, contradicting that $(L^p)' = \{0\}$. Hence, the assumption that the locally bounded space has a Schauder decomposition cannot be removed in Corollary 3.a.4.

Examples. We shall now give some examples of metric vector spaces which satisfy the hypotheses of Theorem 3.a.1, but has a metric which is not a p-homogeneous norm.

Let E_j, $j \in \mathbb{N}$, be metric vector spaces, the metrics of which are given by pseudonorms $\|\cdot\|_{E_j}$, i.e. $d_{E_j}(x,y) = \|x-y\|_{E_j}$. If $\varphi: \mathbb{R}^{\mathbb{N}} \to [0,+\infty]$ is convex, homogeneous of degree 1 and $\varphi(t_1,t_2,\dots) = 0$ only if $t_1 = t_2 = \dots = 0$ (i.e. a norm on the subspace of $\mathbb{R}^{\mathbb{N}}$ where it is finite) we define

$$E = \{x \in \Pi E_j;\ \|x\|_E = \varphi(\|x\|_{E_1}, \|x\|_{E_2}, \dots) < +\infty\}.$$

Then E is a metric space, and $\|\cdot\|_E$ is p-homogeneous if all $\|\cdot\|_{E_j}$ are p-homogeneous. However, we may take $\|\cdot\|_{E_j}$ p_j-homogeneous with e.g. $p_j \to 0$ as $j \to \infty$ and $\varphi(\|x_1\|_{E_1},\dots,\|x_n\|_{E_n},\dots) = (\Sigma \|x_j\|^q_{E_j})^{1/q}$,

$1 \le q < +\infty$; E will not be p-pseudoconvex in the sense of Rolewicz for any $p > 0$.

Let $(\pi_{j,n})_{n\in\mathbb{N}}$ be a Schauder decomposition of E_j, let $\varphi(t_1,t_2,\ldots) = (\Sigma\, |t_j|^q)^{1/q}$, $1 \le q < \infty$. Then

$$\pi_n(x) = (\pi_{1,n}(x),\pi_{2,n}(x),\ldots,\pi_{n,n}(x)) \in E_1 \times\ldots\times E_n$$

defines a Schauder decomposition of E. (In general we need some kind of continuity of φ, roughly

$$\varphi(t_1^{(n)},\ldots,t_n^{(n)},\ 0,0\ldots) \to \varphi(t_1,t_2,\ldots)$$

if $t_k^{(n)} \to t_k$, $n \to \infty$, for every fixed k.) If every E_j is 1-dimensional and all $p_j = 1$, $j \in N$, we get $E \simeq \ell^q$ which has a 1-homogeneous norm. If $E_2 = E_3 = \ldots = 0$ we get just $E \simeq E_1 = \ell^q$, $0 < p < 1$, and we get $E \simeq \ell^{\{p_n\}}$ if $q = 1$, $\dim E_j = 1$ where $\ell^{\{p_j\}} = \{x = (x_j);\ \sum_1^\infty |x_j|^{p_j} < \infty\}$, $\|x\|_E = \sum_1^\infty |x_j|^{p_j}$. In general we therefore get a mixture of ℓ^p, $p < 1$ and ℓ^q, $q \ge 1$. Now, all these metric vector spaces E have the property that $\log \|x\|_E$ is plurisubharmonic in E, hence Theorem 3.a.1 can be applied to all of them.

3b. Holomorphic functions with prescribed radius of convergence.

By using the method of Section 3.a, we can construct a holomorphic function bounded on all V_m, $m \in \mathbb{N}$, and $|f(x_n)| \ge n$, $n \in \mathbb{N}$, and with $R_f(x) \le R(x)$, $x \in \Omega$.

Here we shall see that if, in addition to the hypotheses of Theorem 3.a.1, R is locally Lipschitz continuous in Ω, then there is an $f \in H(\Omega)$ with $R_f \le R$ and a lower bound for R_f which depends only on R, the metric d, and the Schauder decomposition (π_n). More precisely we have:

<u>Proposition 3.b.1</u>. Let Ω be a domain in an infinite-dimensional metric vector space E with a monotone Schauder decomposition (π_n)

and with a translation invariant metric d such that $\log d(x,0)$ is plurisubharmonic in E. Let $R: \Omega \to \mathbb{R}_+$ be such that $-\log R$ is plurisubharmonic in Ω, $R \leq d_\Omega$ and $|R(x)-R(y)| \leq cd(x,y)$ for $x,y \in \Omega$, $d(x,y) < d_\Omega(x)$ and some $c \in]0,1[$. Then there exists $f \in H(\Omega)$ and a constant $K = K(c) > 0$ depending on c with $KR \leq R_f \leq R$. Without assuming (π_n) to be monotone we get $\varphi(KR) \leq$ $\leq R_f \leq R$ where $\varphi: \mathbb{R} \to \mathbb{R}$ is a continuous function with $\varphi(0) = 0$, $\varphi(t) > 0$ for $t > 0$, which depends only on d and (π_n). In the particular case $E = \ell^{\{p_n\}}$, $0 < p_n < 1$, there always exists an $f \in H(\Omega)$ with $R_f = R$.

For similar results in Banach spaces see Schottenloher [16], Coeuré [3] and Kiselman [10].

<u>Proof</u>. Let first (π_n) be monotone with respect to d, i.e. $d\circ\pi_n \leq d$. According to the discussion in Section 3.a we can construct a holomorphic function f which is bounded on all V_m, $m \in \mathbb{N}$. Hence it suffices to show:

(3.b.1) For all $x \in \Omega$ and $s < \frac{1}{2+c} R(x)$, there exists $n \in \mathbb{N}$ with $B(x,s) \subset V_n$.

By (3.a.5) and since $s < \frac{1}{2+c} R(x) < R(x) < d_\Omega(x)$, there exists $n \in \mathbb{N}$ such that $B(x,s) \subset X_n$ and $\pi_m(B(x,s)) \subset L_m \subset \hat{L}_m$ and

(3.b.2) $$R(\pi_k(y)) \geq \epsilon_n + d(\pi_k(y),\pi_m(y))$$

for all $y \in B(x,s)$ and $m \geq k \geq n$. To show (3.b.2), let $\delta > 0$ and choose $n \in \mathbb{N}$ such that

$$(2+c)s + 2\delta + \epsilon_n < R(x) \quad \text{and} \quad d(\pi_k(x),\pi_m(x)) < \delta, \quad m,k \geq n.$$

For $y \in B(x,s)$ and $m,k \geq n$ we get $d(\pi_k(y),x) < s+\delta$. It follows by hypothesis that

(3.b.3) $$\begin{aligned} R(\pi_k(y)) &\geq R(x) - cd(\pi_k(y),x) \\ &\geq (2+c)s + 2\delta + \epsilon_n - c(s+\delta) \geq \epsilon_n + 2s + \delta \\ &\geq \epsilon_n + d(\pi_k(y),\pi_m(y)). \end{aligned}$$

(The last inequality follows since $d(\pi_k(y),\pi_m(y)) \le d(\pi_k(y),\pi_k(x)) + d(\pi_k(x),\pi_m(x)) + d(\pi_m(x),\pi_m(y)) \le s + \delta + s = 2s + \delta$.) Hence $B(x,s) \subset V_n$ for some n. Let $s \to \frac{1}{2+c} R(x)$; we get $R_f(x) \ge \frac{1}{2+c} R(x) = K(c)R(x)$.

In the general case, i.e. without assuming (π_n) to be monotone with respect to d, we first solve the problem with the metric

$$d^*(x,0) = \sup_n d(\pi_n(x),0).$$

Since d^* and d define the same topology we have

$$\varphi(d^*) \le d \le d^*$$

for some function φ with the properties mentioned in the statement of the theorem. We now get, by applying the result already proved, a function $f \in H(\Omega)$ with radius of convergence R_f^* measured by d^* satisfying

$$KR \le R_f^* \le R.$$

Since $\varphi(R_f^*) \le R_f \le R_f^*$ we get the desired conclusion

$$\varphi(KR) \le R_f \le R.$$

Finally for the spaces $\ell^{\{p_n\}} = \{x = (x_n);\ \sum_1^\infty |x_n|^{p_n} < \infty\}$, $0 < p_n < 1$ where the metric is defined by $d(x,y) = \sum_1^\infty |x_n - y_n|^{p_n}$, let $s < R(x)$, choose $\delta > 0$ and $n \in N$ such that

$$s + 2\delta + \epsilon_n < R(x) \quad \text{and} \quad d(\pi_k(x),\pi_m(x)) < \delta \quad \text{for } m,k \ge n.$$

For $y \in B(x,s)$ and $m,k \ge n$ and by hypothesis it follows that

$$R(\pi_k(y)) \ge R(x) - d(\pi_k(y),x)$$

$$\ge s + 2\delta + \epsilon_n - \delta - d(\pi_k(y),\pi_k(x)) = \epsilon_n + \delta + s - d(\pi_k(y),\pi_k(x)).$$

But for $m \ge k \ge n$ we have

$$d(\pi_m(y),\pi_k(y)) + d(\pi_k(y),\pi_k(x)) = d(\pi_m(y),\pi_k(x)) \le \delta + s.$$

Hence

$$\delta + s - d(\pi_k(y),\pi_k(x)) \ge d(\pi_m(y),\pi_k(y)).$$

Consequently,

$$R(\pi_k(y)) \geq \epsilon_n + d(\pi_m(y), \pi_k(y)).$$

This implies that $B(x,s) \subset V_n$. Hence $R_f(x) \geq s$. By letting $s \to R(x)$ we get $R_f \geq R(x)$.

3c. Richness of the space of holomorphic functions.

By using the results of Sections 3.a and 3.b we shall obtain the following result.

Proposition 3.c.1. Let E, Ω and R be as in Theorem 3.a.1. Then the set $T_1(\rho) = \{f \in H(\Omega);\ R_f \leq R\}$ is sequentially dense in $(H(\Omega), \tau_o)$. Moreover, if R in addition to this is locally Lipschitz continuous (i.e. satisfies the hypotheses of Proposition 3.b.1) and the Schauder decomposition is monotone, then $T_2(\rho) = \{f \in H(\Omega);\ KR \leq R_f \leq R\}$ is sequentially dense in $(H(\Omega), \tau_o)$. Here τ_o denotes the compact open topology on $H(\Omega)$.

For Banach spaces with Schauder decomposition the result is due to Schottenloher [16].

Proposition 3.c.1 can be proved along the lines of Schottenloher [16] provided we substitute Proposition 3.b.1 and Lemma 3.a.2 for the corresponding results in Banach spaces. In view of this, the proof will not be given here.

3d. Results for Riemann domains.

The previous results in 3.a, 3.b, and 3.c can be generalized to non-schlicht domains over a suitable space E. Let us first recall the following concepts which are analogous to those considered by Schottenloher [15,16] for locally convex spaces:

A Riemann domain spread over a metric vector space E is a pair (Ω, q) where Ω is a connected Hausdorff space and $q: \Omega \to E$ is a local homeomorphism (i.e., for every $x \in \Omega$ there exists a neighborhood ω of x such that $q|_{\omega}: \omega \to E$ is a homeomorphism of ω onto $q(\omega)$). If q is injective, the domain (Ω,q) is called a <u>Schlicht domain</u>, and can then be identified, via q, with a domain in E.

The boundary distance function d_Ω on a fixed domain (Ω,q) over E is defined by:

$$d_\Omega(x) = \sup\,(r;\ \text{there exists a neighborhood } U \text{ of } x \text{ such that } q|_U: U \to B(q(x),r) \text{ is a homeomorphism}),\quad x \in \Omega.$$

The ball $B(x,r)$ for $x \in \Omega$, $r < d_\Omega(x)$ is just the component of $q^{-1}(B(q(x),r))$ which contains x. The plurisubharmonic, the holomorphic, or for that matter any locally defined class of functions, can now be defined on (Ω,q) using restrictions $q|_\omega$ of the projection q.

We have proved the following result:

<u>Theorem 3.d.1</u>. Let E be an infinite-dimensional complex metric vector space with a translation invariant metric d such that $x \mapsto \log d(x,0)$ is plurisubharmonic in E and having a finite-dimensional Schauder decomposition (π_n). Let $R: \Omega \to]0,\infty[$ be defined on a domain (Ω,q) spread over E such that $-\log R$ is plurisubharmonic and $R \le d_\Omega$. Then there exists a holomorphic function $f \in H(\Omega)$ with radius of convergence $R_f \le R$.

The proof is analogous to that for a normed space E given by Schottenloher [16] and using Theorem 3.a.1 instead of the corresponding result in the normed case.

<u>Corollary 3.d.2</u>. Let E be an finite-dimensional metric vector space with a finite-dimensional Schauder decomposition and such that the metric d is logarithmically plurisubharmonic in E. Then every domain (Ω,q) spread over E such that $-\log d_\Omega$ is plurisubharmonic, is a domain of holomorphy.

<u>Proof</u>. Follows from Theorem 3.d.1 with the methods used in the proof

of Corollary 3.a.3.

The results of Propositions 3.b.1 and 3.c.1 can also be obtained for a Riemann domain (Ω,q) over the given space E if R is locally Lipschitz continuous in the following sense: There exists $c \in]0,1]$ with

$$|R(x)-R(y)| \leq cd(q(x),q(y)), \quad x,y \in \Omega, \quad y \in B(x,d_\Omega(x))^o.$$

REFERENCES

[1] Aron, R., Entire functions of unbounded type on Banach spaces. Boll. Un. Mat. Ital. (4)9, 28-31 (1974).

[2] Bayoumi, A., Bounding subsets of some metric vector spaces. To appear in Arkiv for Mat. (1980).

[3] Coeureé, G., Sur le rayon de bornologie des fonctions holomorphes. Sém. P. Lelong, Lecture Notes in Mathematics 578, 183-194. Springer-Verlag. 1977.

[4] Dineen, S., Unbounded holomorphic functions on a Banach space. J. London Math. Soc. (2)4, 461-465 (1972).

[5] Gruman, L., Kiselman, C.O., Le problème de Levi dans les espaces de Banach à base. C.R. Acad. Sci. Paris, A 274, 1296-1299 (1972).

[6] Hörmander, L., An introduction to Complex Analysis in Several Variables. Princeton, Van Nostrand (1966).

[7] Josefson, B., A counterexample to the Levi problem. In Proceedings on Infinite Dimensional Holomorphy. Lecture Notes in Mathematics 364, 168-177 (Springer 1974).

[8] Kiselman, C.O., On the radius of convergence of an entire function in a normed space. Ann. Polon. Math. 33, 39-55 (1976).

[9] Kiselman, C.O., Geometric aspects of the theory of bounds for entire functions in normed spaces. In Infinite Dimensional Holomorphy and Applications. Ed. M.C. Matos, North-Holland, Amsterdam (1977).

[10] Kiselman, C.O., Constructions de fonctions entières à rayon de convergence donné. Lecture Notes in Mathematics 578, 246-253. Springer-Verlag (1977).

[11] Lelong, P., Fonctions plurisousharmoniques dans les espaces vectoriels topologiques. Lecture Notes in Mathematics 71, 167-190. Springer-Verlag 1968.

[12] Nachbin, L., Topology on Spaces of Holomorphic Mappings. Springer-Verlag, Berlin, Heidelberg, New York (1969).

[13] Noverraz, P., Pseudoconvexité, convexité polynomiale et domaines d'holomorphie en dimension infinie. Amsterdam: North-Holland (1973).

[14] Rolewicz, S., Metric Linear Spaces. Instytut Matematyczny Polskiej Akademii Nauk. Monografie Matematyczne (1972).

[15] Schottenloher, M., Richness of the class of holomorphic functions on an infinite dimensional space. Functional Analysis, Results and Surveys. Conf. in Paderborn (1976). Amsterdam: North-Holland.

[16] Schottenloher, M., Holomorphe Funktionen auf Gebieten über Banachräumen zu vorgegebenen Konvergenzradien. Manuscripta Math. 21, 315-327 (1977).

[17] Schottenloher, M., The Levi problem for domains spread over locally convex spaces with a finite dimensional Schauder decomposition. Ann. Inst. Fourier, Grenoble 26, 207-237 (1976).

Added in proof

Recently, we have succeeded in improving the results of section 3 by avoiding the assumption $\log d(x,0)$ is plurisubharmonic in Theorem 3.a.1. Consequently we got the solutions of the Levi problem over locally bounded and PB-spaces with finite-dimensional Shauder decomposition.

EXTENDING NONARCHIMEDEAN NORMS ON ALGEBRAS

Edward Beckenstein

St. John's University

Staten Island, New York, U.S.A.

Lawrence Narici

St. John's University

Jamaica, New York, U.S.A.

Solving equations in ordinary Banach algebras has been considered in [1], [2] and [4], among other places. A very elementary level at which the problem must be confronted occurs in determinations of singularity: to ask if x is singular is to inquire is there a c for which $cx = e$; if there is no such x in the original algebra, is there one in an extension? This enlarges the question to include: Is there such an extension? Questions such as these for nonarchimedean Banach algebras have been considered in [3].

To solve polynomial equations whose coefficients come from a field F, one does not try to solve $f(x) = 0$ in F actually; instead one looks for solutions in an extension K of F. And by "extension" is meant that K contains a copy of F, an isomorphic image. To imitate this approach for Banach algebras the extension algebra must contain an image of the original Banach algebra which is a copy in the metric as well as the algebraic sense, and, naturally enough, the metric requirement makes for some more restrictive hypotheses than in the purely algebraic situation.

For the sake, ultimately, of solving equations with coefficients from nonarchimedean Banach algebras X (i.e. to be able to create extensions containing appropriate copies) we consider the problem of extending norms from X to extensions as the subject of this paper. Specifically we show that if $(X,\| \|) \subset (Y,\| \|')$ and $\alpha\| \| \le \| \|' \le \beta\| \|$ (for some real numbers α and β) then $\| \|$ may be extended to Y.

Let X be a commutative nonarchimedean Banach algebra with identity e over a complete nonarchimedean non-trivially valued field F. We consider the problem of how to extend a norm from a subalgebra of X to all of X (cf. [2] and [4]). To effect the solution we construct an algebra which contains a copy of X and our first results are concerned with mechanics of this construction.

Let K be an index set, let $Z = \{z_k : k \in K\}$ be a family of (commuting) indeterminates over X and let $T = \{t_k : k \in K\}$ be a family of positive real numbers. Let X(Z,T) denote the collection of all finite sums of the form

$$w = \Sigma x_{\mu_1,\ldots,\mu_n} z_{k_1}^{m_1} \cdots z_{k_n}^{m_n}$$

where $x_{\mu_1,\ldots,\mu_n} \in X$ and the m_i are each non-negative. We define a norm $\| \|_T$ on X(Z,T) by

$$(1) \qquad \|w\|_T = \max \|x_{\mu_1,\ldots,\mu_n}\| t_{k_1}^{m_1} \cdots t_{k_n}^{m_n} .$$

X(Z,T) is now a nonarchimedean normed extension of X in which, for convenience, we write sums such as w above as simply $\Sigma x_\mu z^\mu$.

Consider the special case of X(Z,T) where $Z = \{z\}$ and $T = \{t\}$, which we denote as X(z,t). It is straightforward to verify that the following characterization of the completion obtains.

<u>Proposition 1</u>. The completion Y of X(z,t) is the set of all series $y = \sum_{n=0}^{\infty} x_n z^n$ for which $\|x_n\| t^n \to 0$; $\|y\| = \max_n \|x_n\| t^n$.

A <u>Gelfand</u> <u>ideal</u> [5, p.118] M in X is one for which $X/M = F$. The collection of all Gelfand ideals is denoted by $\mathfrak{m}'$ and is assumed to carry the strong Gelfand topology, the weakest topology with respect to which the maps $M \to x+M = x(M)$ are continuous for each $x \in X$. We often identify $M \in \mathfrak{m}'$ with the homomorphism $f: X \to F$

sending x into $x(M)$.

<u>Proposition 2</u>. The space $\mathfrak{m}'_Y$ of Gelfand ideals of Y is homeomorphic to $\mathfrak{m}'_X \times \{\mu \in F : |\mu| \leq t\}$; X and Y are as in Proposition 1.

<u>Proof</u>. Let $U_t = \{\mu \in F : |\mu| \leq t\}$. With $(f,\mu) \in \mathfrak{m}'_X \times U_t$, we associate the homomorphism $(f,\mu)^* : X \to F$ taking $\Sigma x_n z^n$ into $\Sigma f(x_n)\mu^n$. Since

$$|\Sigma f(x_n)\mu^n| \leq \max|f(x_n)|\,|\mu|^n \leq \max\|x_n\|\,t^n = \|\Sigma x_n z^n\|,$$

$(f,\mu)^*$ is seen to be continuous.

Conversely, if $g: Y \to F$ is a continuous homomorphism, $|g(z)| \leq \|z\| = t$, so a pre-image for g under the above mapping would be $(g|_X, g(z))$, i.e., $g = (g|_X, g(z))^*$. It is routine to verify that this is a homeomorphism.

For ease in reference, we recall the following result.

<u>Proposition 3</u>. [5,p.124] If F is locally compact, then $\mathfrak{m}'_X$ and $\mathfrak{m}'_Y$ are compact.

<u>Proposition 4</u>. With $\sigma(w)$ denoting the spectrum of w, $\sigma(z) = U_t$. Thus $r_\sigma(z) = t$.

<u>Proof</u>. With notation as in the proof of Proposition 2, let $(f,\mu)^* \in \mathfrak{m}'_Y$. Then $(f,\mu)^*(z-\mu) = \mu-\mu = 0$ so $\mu \in \sigma(z)$ for all $\mu \in U_t$.

On the other hand if $|\mu| > \|z\| = t$, $\|e-(e-\mu^{-1}z)\| < 1$ so (by a standard Banach algebra result) $e-\mu^{-1}z$ is invertible. Therefore $\mu e-z$ is invertible and $\mu \notin \sigma(z)$.

Let $\mathfrak{m}_X$ denote the collection of all maximal ideals of X. The purpose of the hypothesis of Proposition 5 - namely $\bigcup\mathfrak{m}_X = \bigcup\mathfrak{m}'_X$ - is to guarantee that each singular element of X belongs to some Gelfand ideal.

<u>Proposition 5</u>. Suppose that the valuation on F is dense, $\bigcup\mathfrak{m}_X = \bigcup\mathfrak{m}'_X$, and that no positive integral power of t can be expressed

as a ratio $\|x\|/\|y\|$ for any $x,y \in X$. If $r_\sigma(x) = \|x\|$ for each $x \in X$ then $r_\sigma(w) = \|w\|$ for all $w \in Y$.

Proof. By Proposition 4, $r_\sigma(z) = \|z\| = t$. For $p = \sum_{r=0}^{\infty} x_n z^n$, $\|w\|_T = \max_n \|x_n\| t^n = \|x_j\| t^j$ for some j. (Since $\|x_n\| t^n \to 0$, beyond a certain point all $\|x_n\| t^n < (\max_n \|x_n\| t^n)/2$. Thus the maximum is actually the maximum of a finite set and is therefore assumed.) As it happens, j must be unique. In fact if $n \neq j$ then $\|x_n\| t^n < \|x_j\| t^j$. If these two terms were equal, then t^{n-j} would be equal to $\|x_j\|/\|x_n\|$ which is contradictory.

Since $r_\sigma(x) = \|x\|$, $\cup m_X = \cup m'_X$ and $|F^*|$ is dense in R^+ we may choose $f \in m'_X$ and $\mu \in F$ such that

$$\|x_j\| t^j - \varepsilon < |(f,\mu)^*(x_j z^j)| = |f(x_j)| |\mu|^j \leq \|x_j\| t^j,$$

given $0 < \varepsilon < \|x_j\| t^j - \|x_n\| t^n$ for all $n \neq j$. Thus for $n \neq j$

$$\|x_n\| t^n < |f(x_j)| |\mu|^j \leq \|x_j\| t^j.$$

Moreover, for all n

$$|(f,\mu)^* x_n z^n| = |f(x_n)| |\mu^n| \leq \|x_n\| t^n$$

so

$$|(f,\mu)^*(\Sigma x_n z^n)| = |f(x_j)| |\mu|^j.$$

Now $\|x_j\| t^j - \varepsilon < |f(x_j)| |\mu|^j \leq \|x_j\| t^j$ and it follows that $r_\sigma(w) = \|w\|$.

If $\|X^*\| = |F^*|$ (the nonzero values of X and F respectively), the condition on t could be altered to read that no power of t belongs to $|F^*|$. If F is algebraically closed as well, the condition would simplify to just $t \notin |F^*|$. These observations lead to the following corollary.

Corollary. Let F be algebraically closed, $\|X^*\| = |F^*|$, $t \notin |F^*|$ and let X be a closed subalgebra of the sup-normed algebra $C(T,F)$ of continuous functions mapping the compact 0-dimensional Hausdorff space T into F. Then, for all $w \in Y$, $r_\sigma(w) = \|w\|$.

It remains an open question whether Y is a Gelfand algebra if X is.

We may now present the main result, concerning the extendibility of a norm from a subalgebra of a nonarchimedean Banach algebra to the whole algebra. The main idea is to create a larger algebra (of the $X(Z,T)$ type) which contains a copy of the original one, but where the exact nature of the norm is better known, thus facilitating the extension.

<u>Theorem.</u> Let A be a subalgebra containing the identity of the commutative nonarchimedean normed algebra with identity $(B, \| \|')$. If $\| \|$ is a norm on A satisfying the condition (2) $\alpha\| \| \leq \| \|' \leq \beta\| \|$ on A for some real numbers α and β, then $\| \|$ can be extended in such a way that (3) $(\alpha/\beta)\| \| \leq \| \|' \leq \beta\| \|$ everywhere on B.

<u>Proof.</u> Since $\|e\| = \|e\|' = 1$, substituting e in (2) yields $\alpha \leq 1 \leq \beta$. Thus (3) is seen to be satisfied on A.

Let $Z = \{z_b : b \in B-A\}$ be a set of commuting indeterminates over A and let T be the set of norms $\|b\|' = t_b$ for $b \in A$. In the notation introduced at the beginning of the section, consider

$$A(Z,T) = \{\Sigma a_\mu z_b^\mu : a_\mu \in A\}.$$

The map $H: A(Z,T) \to B$ sending $\Sigma a_\mu z_b^\mu$ into $\Sigma a_\mu b^\mu$ is a homomorphism so, letting $J = \ker H$, B is isomorphic to $A(Z,T)/J$. The norm, $\| \|_T'$, on $A(Z,T)$ is as in (1): specifically

$$\|\Sigma a_\mu z_b^\mu\|_T' = \max\|a_{\mu_1,\ldots,\mu_n}\| (\|b_1\|')^{\mu_1} \cdots (\|b_n\|')^{\mu_n}.$$

Since (using the obvious abbreviation)

$$(4) \qquad \|\Sigma a_\mu b^\mu\|' \leq \max\|a_\mu\|' (\|b\|')^\mu = \|\Sigma a_\mu z_b^\mu\|_T' ,$$

H is norm-decreasing; J, therefore, is a closed ideal.

We now introduce the first quotient norm. For $b \in B-A$, choose $p \in A(Z,T)$ such that $H(p) = b$ (take $p = z_b$, for example) and define

(5) $$\|b\|'_Q = \inf_{j\in J} \|p+j\|'_T \,.$$

We show that $\| \|'_Q = \| \|'$. We consider elements in A and $B-A$ separately. If $a \in A$ then $H(a) = a$ so

(6) $$\|a\|'_Q = \inf \|a+J\|'_T \le \|a\|' .$$

Suppose $j = \Sigma a_{\mu_1,\ldots,\mu_n} z_{b_1}^{\mu_1} \ldots z_{b_n}^{\mu_n} \in J$. Since $H(j) = 0$ $a + \Sigma a_{\mu_1,\ldots,\mu_n} b_1^{\mu_1} \ldots b_n^{\mu_n} = a$. Hence, by (4),

(7) $$\|a + \Sigma a_{\mu_1,\ldots,\mu_n} b_1^{\mu_1} \ldots b_n^{\mu_n}\|' = \|a\|' \le$$
$$\le \|a + \Sigma a_{\mu_1,\ldots,\mu_n} z_{b_1}^{\mu_1} \ldots z_{b_n}^{\mu_n}\|'_T$$

and thus

(8) $$\|a\|' \le \|a\|'_Q \text{ and (therefore) } \|a\|' = \|a\|'_Q .$$

For $b \in B-A$

(9) $$\|b\|'_Q = \inf \|z_b+J\|'_T \le \|z_b\|'_T = \|b\|' .$$

Using an argument similar to that which led to (8), we get

(10) $$\|b\|' \le \|b\|'_Q \text{ and } \|b\|' = \|b\|'_Q .$$

The situation is now that $(B,\| \|')$ is isometrically isomorphic to $(A(Z,T)/J,\| \|'_Q)$. We next introduce another norm $\| \|''_Q$ to $A(Z,T)/J$ which will have the effect of making $(A/J, \| \|''_Q)$ isometrically isomorphic to $(A,\| \|)\cdot\| \|''_Q$ will then be shown to be the desired extension.

First we define a second T-norm $\| \|''_T$ on $A(Z,T)$:

(11) $$\|\Sigma a_{\mu_1,\ldots,\mu_n} z_{b_1}^{\mu_1} \ldots z_{b_n}^{\mu_n}\|''_T =$$
$$= \max(\|a_0\|, (\beta/\alpha)\max_{\mu\ne 0}\{\|a_{\mu_1,\ldots,\mu_n}\| (\|b_1\|')^{\mu_1}\ldots(\|b_n\|')^{\mu_n}\} .$$

Note that $\| \|$ and $\| \|'$ are used in defining $\| \|''_T$. Since $\beta/\alpha \ge 1$, $\| \|''_T$ is an algebra norm. Moreover $\| \|''_T$ is stronger than $\| \|'_T$ as the following inequalities show.

$$(12)\quad \|\Sigma a_\mu z_b^\mu\|'_T \le \max(\|a_o\|', \|\Sigma_{\mu\neq 0}\, a_\mu z_b^\mu\|'_T)$$

$$\le \max(\beta\|a_o\|, (\beta/\alpha)\|\Sigma a_\mu z_b^\mu\|'_T)$$

$$= \max(\beta\|a_o\|, (\beta/\alpha)\max\|a_{\mu_1,\ldots,\mu_n}\|'(\|b_1\|')^{\mu_1} \ldots (\|b_n\|')^{\mu_n}$$

$$\le \max(\beta\|a_o\|, (\beta/\alpha)\max(\beta\|a_{\mu_1,\ldots,\mu_n}\|(\|b_1\|')^{\mu_1} \ldots (\|b_n\|')^{\mu_n}$$

$$= \beta\|\Sigma a_\mu z_b^\mu\|''_T .$$

It now follows that J is also closed with respect to $\| \|''_T$.

Next, for any $b \in B$, consider

$$(13)\qquad \|b\|_Q = \inf \|p+J\|''_T \quad \text{where} \quad H(p) = b.$$

If $b \in B-A$, then $b = H(z_b)$ and, by (11),

$$(14)\qquad \|b\|_Q = \inf \|z_b+J\|''_T \le \|z_b\|''_T = (\beta/\alpha)\|b\|'.$$

Therefore

$$(15)\qquad ((\alpha/\beta)\|b\|_Q \le \|b\|'.$$

By (10) and (12) it follows that $\|b\|' \le \beta\|b\|_Q$, and therefore we have

$$(16)\qquad (\alpha/\beta)\|b\|_Q \le \|b\|' \le \|b\|_Q \quad \text{for} \quad b \in B-A.$$

By showing $\|a\|_Q = \|a\|$ for each $a \in A$ the proof will be complete. To this end let $a \in A$ and let $j \in J$ be as defined after (6). By the properties of $\| \|''_T$ and (8)

$$(17)\quad \|a+j\|''_T = \max(\|a+a_o\|, (\beta/\alpha)\max\|a_{\mu_1,\ldots,\mu_n}\|(\|b_1\|')^{\mu_1} \ldots (\|b_n\|')^{\mu_n}$$

$$\ge \max(\|a+a_o\|, (1/\alpha)\max\|a_{\mu_1,\ldots,\mu_n}\|'(\|b_1\|')^{\mu_1} \ldots (\|b_b\|')^{\mu_n}$$

$$= \max(\|a+a_o\|, (1/\alpha)\|j-a_o\|'_T)$$

$$\ge \max(\|a+a_o\|, (1/\alpha)\|a_o\|')$$

$$\ge \max(\|a+a_o\|, \|a_o\|) \ge \|a\|.$$

Thus $\|a\|_Q \ge \|a\|$. Again, however, since $\| \|_Q$ is a quotient norm, we must have $\|a\|_Q \le \|a\|$ for each $a \in A$. This completes the proof.

REFERENCES

[1] Arens, R. Extensions of Banach algebras, Pacific J. Math., vol. 10, 1960, 1-16.

[2] Arens, R. and Hoffman, K., Algebraic extensions of normed algebras, Proc. A.M.S., vol. 7, 1956, 203-210.

[3] Beckenstein, E., Narici, L. and Suffel, C., A notė on permanently singular elements in topological algebras, Coll. Math., vol. 31, 1974, 115-123.

[4] Lindberg, J., Extension of algebraic norms and applications, Studia Math. vol. XL, 1971, 35-39.

[5] Narici, L., Beckenstein, E. and Bachman, G., Functional analysis and valuation theory, Marcel Dekker, New York, 1971.

M-STRUCTURE IN TENSOR PRODUCTS OF BANACH SPACES

Ehrhard Behrends

I. Mathematisches Institut

der Freien Universität

Hüttenweg 9

D-1000 Berlin 33 Germany

ABSTRACT

We define the basic concepts of the theory of M-structure and investigate the M-structure properties of the ε-tensor product. Our main result generalizes a theorem due to author. It describes how the centralizer of the tensor product can be constructed from the centralizers of the factors. In the last sections we investigate some applications and indicate some open problems.

1. M-STRUCTURE

Let X be a Banach space. "M-structure of X" means the structure of the collection of M-ideals, M-summands, and the centralizer of X.

M-structure measures, in a sense, to what extent X behaves like an abstract M-space (those readers who are not familiar with the theory of Banach lattices may replace "M-space" by "space of continuous functions"). Most of the basic ideas are already contained in a paper of Cunningham ([8]), the development of the theory as well as the applications of M-structure methods to problems of geometric functional analysis are due to Alfsen and Effros ([1]). A number of consequences for different branches of functional analysis and approximation theory have been considered by several authors (see, f.ex., [4-7], [11], [13].

Note: As the most authors who investigated M-structure properties we will restrict ourselves for simplicity to real spaces. We refer the reader to the forthcoming Lecture Notes volume ("M-structure and the Banach-Stone theorem"; Springer Verlag, 1979) of the author where in part I M-structure is studied systematically for arbitrary Banach spaces.

1.1 Definition. Let J be a closed subspace of X.

(i) J is called an M-summand (resp. L-summand) if there is a closed subspace $J^{\perp}$ of X such that (algebraically) $X = J \oplus J^{\perp}$ and $\|x+x^{\perp}\| = \max\{\|x\|, \|x^{\perp}\|\}$ (resp. $\|x+x^{\perp}\| = \|x\| + \|x^{\perp}\|$) for $x \in J$, $x^{\perp} \in J^{\perp}$.

(ii) J is called an M-ideal if J^{π}, the annihilator of J in X', is an L-summand.

1.2 Examples:

a) Let K be a compact Hausdorff space. For any closed subset L of K we define J_L by $\{f \mid f \in CK,\ f|_L = 0\}$. Then

1. the M-ideals in CK are exactly the subspace J_L, $L \subset K$ closed
2. J_L is an M-summand iff L is clopen

b) Every M-summand is an M-ideal (a2. shows that the converse is not true in general)

c) Let A be a C*-algebra with unit, X the self-adjoint part of A. The M-ideals in X are exactly the self-adjoint parts of the closed two-sided ideals of A

d) Let K be a compact convex set. The M-ideals in AK are exactly the annihilators of the closed split-faces of K.

1.3 Definition. An operator $T: X \to X$ is called M-bounded if there is a $\lambda \geq 0$ such that Tx is contained in every ball which contains $\pm \lambda x$. $Z(X)$, the centralizer of X, means the collection of all M-bounded operators on X. $Z(X)$ is a commutative Banach algebra which can be represented as a space $X(K_X)$ (K_X a compact Hausdorff space).

1.4 Examples:

a) If L is a locally compact Hausdorff space, then the M-bounded operators on C_oL are exactly the multiplication operators M_h: $f \mapsto hf$, where $h: L \to \mathbb{R}$ is a bounded continuous function (this implies that $K_{C_oL} = \beta L$)

b) If X is smooth or strictly convex or if X contains a non-trivial L-summand, then $Z(X) = \mathbb{R}Id$. Spaces for which $Z(X)$ is "small" are interesting in connection with theorems of the Banach-Stone type ([5],[6]; cf. also 4.2 below)

c) $Z(X)$ is finite-dimensional for every reflexive space ([5])

d) For X as in 1.2c, $Z(X)$ consists precisely of the multiplication operators associated with the self-adjoint elements of the centre of A.

The following definition is essential for our considerations. It enables us to treat operators in the centralizer in a simple way:

1.5 Definition. Let K (the base space) be a compact Hausdorff space, $(X_k)_{k \in K}$ (the component spaces) a family of Banach space. A closed subspace X of $\Pi^\infty X_k$ is called a function module, if

(i) $k \mapsto \|x(k)\|$ is upper semicontinuous for $x \in X$

(ii) $hx \in X$ for $x \in X$ and $h \in CK$

(iii) $X_k = \{x(k) \mid x \in X\}$ for $k \in K$

(iv) $\{k \mid X_k \neq 0\}$ is dense in K

It is easy to see that $M_h \in Z(X)$ for $h \in CK$ (M_h: $x \mapsto hx$). In fact, we have the following important converse of this remark.

1.6 Theorem. ([8], cf. also [6]) Every Banach space X can be regarded as a function module such that, in addition, the operators in $Z(X)$ are exactly the operators M_h, $h \in CK$ (K and the X_k are uniquely determined in this case; one may take $K = K_X$).

1.7 Examples:

a) For every locally compact Hausdorff space L, C_oL can be regarded as a function module in the obvious way.

b) Every finite product $\pi^\infty \{X_i \mid i=1,\dots,n\}$ is a function module with base space $\{1,\dots,n\}$.

If X is any function module, the M-structure of X can be investigated by considering the M-structure properties of the component spaces X_k $(k \in K)$. We have the following localization results:

1.8 Theorem. K, $(X_k)_{k\in K}$, X as in 1.5.

(i) Let $T: X \to X$ be a linear continuous operator. Then the following are equivalent:

(a) T is M-bounded

(b) There is a bounded family $(T_k)_{k\in K}$ of operators, $T_k \in Z(X_k)$, such that $(Tx)(k) = T_k(x(kk)$ (all $x \in X$).

(ii) Let J be a closed subspace of X. Then J is an M-ideal iff $M_hJ \subset J$ for every $h \in CK$ (i.e. J is a CK-submodule) and $J_k = \{x(k) \mid x \in J\}$ is an M-ideal in X_k for every $k \in K$ (if J is an M-summand then the J_k are also M-summands).

Proof.

(i) "a ⇒ b" Let T be an M-bounded operator. For $h \in CK$ we have $M_hT = TM_h$ so that $h(k)(Tx)(k) = [T(hx)](k)$ for every $x \in X$. Since, for $k \in K$ and $x \in X$ such that $x(k) = 0$, we may choose for $\varepsilon > 0$ a function $h \in CK$ such that $h(k) = 1$ and $\|hx\| \leq \varepsilon$ (this is a consequence of 1.5(i)) it follows that $\|(Tx)(k)\| \leq$ $\leq \|T(hx)\| \leq \|T\|\|hx\| \leq \varepsilon\|T\|$ so that $(tx)(k) = 0$. Therefore $T_k: X_k \to X_k$, defined by $x(k)$ $(Tx)(k)$ is a well-defined linear operator (T_k is defined on all of X_k by 1.5(iii)). For $x(k)\in X_k$ and $\varepsilon > 0$ we may choose $h \in CK$ as in the first part of the proof such that $\|hx\| \leq \|x(k)\| + \varepsilon$ and $h(k) = 1$. Thus $\|T_k(x(k))\| =$ $= \|T(hx)(k)\| \leq \|T\|$ $(\|x(k)\| + \varepsilon)$, i.e. the T_k are continuous with

$\|T_k\| \le \|T\|$. It remains to prove that the T_k are M-bounded. We will show that they satisfy the condition of 1.3 for the same λ as T does. To this end, let $\|\pm\lambda x(k) - y(k)\| < r$, where $x(k), y(k)$ are arbitrary points in X_k and $r > 0$. Similarly to the beginning of the proof we choose $h \in CK$ such that $h(k) = 1$, $\|\pm\lambda hx - hy\| < r$. Since T is M-bounded we have $\|Thx - hy\| < r$ so that, in particular, $\|T_k(x(k)) - y(k)\| = \|h(k)[T_k(x(k)) - y(k)]\| = \|(Thx-hy)(k)\| \le$ $\le \|Thx - hy\| < r$.

"<u>b ⇒ a</u>" It has been pointed out in [5] that, for M-bounded operators S, λ in 1.3 can be taken to be any number greater than $2\|S\|$. This implies that we are able to choose a $\lambda_o > 0$ such that all T_k satisfy 1.3 for this particular λ_o. We will prove that λ_o works also for T. Let $x,y \in X$, $r > 0$ be given such that $\|\pm\lambda_o x - y\| < r$. For $k \in K$ it follows that $\|\pm\lambda_o x(k) - y(k)\| < r$ so that $\|(Tx-y)(k)\| = \|T_k x(k) - y(k)\| < r$. Thus $\|Tx-y\| < r$ which proves that T is M-bounded.

(ii) Let J be an M-ideal. Since $(M_h)'$ commutes with the L-projection onto J^{π} (both are operators in the Cunningham algebra of X'; cf. th. 4.8 in [1]) it follows that $M_h J \subset J$ (all $h \in CK$). Define J_k as in the theorem. We first prove that J_k is a closed subspace of X_k. It suffices to show that, for $x_k^n \in J_k$ such that $\sum_n x_k^n \in J_k$. We write $x_k^n = x_n(k)$ with $x_n \in J$. Replacing, if necessary, x_n by $h_n x_n$ for a suitable function h_n with $h_n(k) = 1$ we may assume that $\|x_n\| \le \|x_k^n\| + 1/2^n$. Thus $\sum_n \|x_n\| < \infty$ so that $x := \sum x_n \in J$. This yields $\sum_n x_k^n = x(k) \in J_k$. Finally, we show that J_k is an M-ideal. The direct verification of 1.1(ii) is complicated so that we prefer to use the characterization of M-ideals by intersection properties ([1], th. 5.9). Let $B(x_i(k), r_i) =: B_i$ (i=1,2,3) be three open balls in X_k such that $B_1 \cap B_2 \cap B_3 \neq \emptyset$, $B_i \cap J_k \neq \emptyset$ for i=1,2,3. We choose $x \in X$ and $y_1, y_2, y_3 \in J$ such that $\|(x_i - x)(k)\| < r_i$ and $\|(x_i - y_i)(k)\| < r_i$ for i=1,2,3. For a suitable function $h \in CK$ with $h(k) = 1$ we have $\|h(x_i - x)\| < r_i$ and $\|h(x_i - y_i)\| < r_i$, i.e. the balls $B_i^* := B(hx_i, r_i)$ satisfy $B_1^* \cap B_2^* \cap \cap B_3^* \neq \emptyset$, $B_i^* \cap J \neq \emptyset$ for $i=1,2,3$ so that $\bigcap_{i=1}^{3} B_i^* \cap J \neq \emptyset$. But this implies that $\bigcap_{i=1}^{3} B_i \cap J_k \neq \emptyset$ so that J_k is an M-ideal. If in addition J is an M-summand with $J^\perp$ as in 1.1.(i), a routine computation shows that J_k is also an M-summand with $(J_k)^\perp := (J^\perp)_k = \{y(k) \mid y \in J^\perp\}$.

Now let J be a closed subspace of X for which $M_h J \subset J$ (all $h \in CK$) and for which all J_k are M-ideals. We will show that J satisfies the three-ball intersection property of [1]. Let $B_i^* := B(x_i, r_i)$ be open balls such that $B_1^* \cap B_2^* \cap B_3^* \neq \emptyset$, $B_i^* \cap J \neq \emptyset$ for $i=1,2,3$. For arbitrary $k \in K$, the balls $B_i^k := B(x_i(k), r_i)$ satisfy $B_1^k \cap B_2^k \cap B_3^k \neq \emptyset$, $B_i^k \cap J_k \neq \emptyset$ $(i=1,2,3)$ so that there is a vector $y_k \in J$ such that $\|y_k(k) - x_i(k)\| < r_i$ for $i=1,2,3$. Let U_k be a neighbourhood of k such that the same inequalities are satisfied for all ℓ in U_k. We choose $U_{k_1} \ldots U_{k_n}$ in $(U_k)_{k \in K}$ such that $\bigcup_{j=1}^{n} U_{k_j} = K$. Further, let $h_1, \ldots, h_n$ be a partition of unity subordinated to the cover $U_{k_1}, \ldots, U_{k_n}$. Then $y := \Sigma h_k y_{k_i}$ is contained in J by hypothesis and it is easy to see that $\|y - x_i\| < r_i$ for $i=1,2,3$, i.e. $\bigcap_{i=1}^{3} B_i \cap J \neq \emptyset$. This proves that J is an M-ideal.

Theorem 1.6 states that, in a sense, the multiplication operators M_h $(h \in CK)$ on function modules are the "typical" M-bounded operators. It is often sufficient to know that, on a given function module, every operator in the centralizer is a multiplication operator associated with a (not necessarily continuous) scalar-valued function.

<u>1.9 Definition</u>. Let X be a function module in $\prod_{k \in K}^{\infty} X_k$.

(i) A bounded function $a: K \to \mathbb{R}$ is called a <u>multiplier</u> if $M_a X \subset X$ and $a(k) = 0$ for $X_k = 0$. It is easy to see that $M_a \in Z(X)$

for every multiplier α.

(ii) X is said to have a <u>scalar-function centralizer</u> if $Z(X) = \{M_\alpha \mid \alpha$ is a multiplier$\}$.

<u>1.10 Examples:</u>

a) If X is a function module as in 1.6, then X has a scalar-function centralizer (note that $M_h = M_\alpha$ when α is defined by $\alpha(k) = h(k)$ resp. $= 0$ if $X_k \neq 0$ resp. $= 0$).

b) If $\dim X_k \leq 1$ for every k, then X has a scalar-function centralizer. This example shows that the class of function modules described in 1.9(ii) is strictly larger than that of 1.6.

The following results prepare our investigations of Section 2.

<u>1.11 Proposition</u> ([4], 2.2): Every multiplier is contained in the strong operator closure of $\{M_h \mid h \in CK\}$. It follows that a function module has a scalar-function centralizer iff $\{M_h \mid h \in CK\}$ is dense with respect to the strong operator topology in $Z(X)$.

If X is a Banach space, we denote by E_X the extreme points in the unit ball of X' (usually, E_X will be provided with the structure topology for which the closed sets are the intersections of E_X with the polars of the M-ideals in X; cf. [1], p. 143). It is known ([9]) that for function modules X one has $E_X = \dot{\bigcup}\{E_{X_k} \mid k \in K,\ X_k \neq 0\}$ (more precisely: $p \in E_X$ iff there are a $k \in K$ with $X_k \neq 0$ and a $p_k \in E_{X_k}$ such that $p(x) = p_k(x(k))$ for every $x \in X$).

<u>1.12 Proposition.</u> Suppose that X is a function module with scalar-function centralizer and that $a: E_X \to \mathbb{R}$ is a bounded structurally continuous function. Then a is constant on every E_{X_k}.

<u>Proof.</u> By the Dauns-Hofmann type theorem of [1] (th. 4.9) there is an M-bounded operator $T: X \to X$ such that $p \circ T = a(p)p$ for $p \in E_X$. Since X has a scalar-function centralizer, there is a bounded func-

tion $\alpha : K \to R$ such that $Tx = \alpha x$ for $x \in X$ so that $\alpha(k)p = a(p)p$ for every $p \in E_{X_k}$. It follows that a has the constant value $\alpha(k)$ on E_{X_k}.

2. THE CENTRALIZER OF TENSOR PRODUCTS

Let X and Y be real Banach spaces, $X \hat{\otimes}_\varepsilon Y$ their usual ε-tensor product (we will restrict our investigations of M-structure to this type of tensor products; cf. the discussion in Section 5). For the definition of $X \hat{\otimes}_\varepsilon Y$ we refer the reader to any textbook of functional analysis. As examples we note that $C(K,X) \cong CK\hat{\otimes}_\varepsilon$ ($C(K,X)$ = the space of continuous X-valued functions on K; K a compact Hausdorff space) and that $C(K \times L) \cong CK\hat{\otimes}_\varepsilon CL$ (K, L compact Hausdorff spaces).

Let X, Y be function modules with base spaces K and L and component spaces $(X_k)_{k \in K}$ and $(Y_\ell)_{\ell \in L}$.

__2.1 Proposition__ ([4], 2.3). For $\sum_{i=1}^{r} x_i \otimes y_i \in X \otimes Y$, let $\sum x_i \tilde{\otimes} y_i$ in $\prod^\infty X_k \hat{\otimes}_\varepsilon Y_\ell$ be defined by $(\sum x_i(k) \otimes y_i(\ell))_{k,\ell}$. Then $\|\sum x_i \otimes y_i\| = \|\sum x_i \tilde{\otimes} y_i\|$ so that $X \hat{\otimes}_\varepsilon Y$ can be though of as a closed subspace of $\prod^\infty X_k \hat{\otimes}_\varepsilon Y_\ell$. With this identification, $X \hat{\otimes}_\varepsilon Y$ is a function module in $\prod^\infty X_k \hat{\otimes}_\varepsilon Y_\ell$, and the operators in $\{M_h \mid h \in CK\} \hat{\otimes}_\varepsilon \{M_g \mid g \in CL\}$ which (as can easily be shown) are contained in $Z(X \hat{\otimes}_\varepsilon Y)$ are precisely the operators M_h, $h \in C(K \times L)$.

Our main result is the following theorem.

__2.2 Theorem__. If X and Y are function modules with scalar-function centralizer, then the function module $X \hat{\otimes}_\varepsilon Y$ of 2.1 has the same property.

__Proof__. The proof is a refinement of the proof of Theorem 4.2 in [4]. Let T be an M-bounded operator on $X \hat{\otimes}_\varepsilon Y$. By [1], 4.8 every element of $E_{X \hat{\otimes}_\varepsilon Y}$ is an eigenvector of T'. It can be shown that this is also true for every $p\otimes q$, where $(p,q) \in E_X \times E_Y$. The proof of this fact depends on elementary properties of tensor product and weak*-topologies. We refer the reader to [13], p.506. Therefore there is a function $a: E_X \times E_Y \to R$ such that $(p\otimes q)\circ T = a(p,q)(p\otimes q)$ for $(p,q) \in E_X \times E_Y$. We claim that a is separately continuous. Let $p \in E_X$ be fixed and x a vector in X such that $p(x) = 1$. For $y \in Y$, the mapping $Y' \ni y' \mapsto (p\otimes y')(T(x\otimes y))$ is linear and weak*-continuous (by the Krein-Smulian theorem we have only to prove continuity on bounded sets, and this is obvious). So there is a vector $T_p y$ such that $y'(T_p y) = (p\otimes y')(T(x\otimes y))$ for every $y' \in Y'$. It is easy to see that $y \mapsto T_p y$ is linear and continuous. In fact we have $T_p \in Z(Y)$ since every $q \in E_Y$ is an eigenvector for T_p' (cf. [1], 4.8): $q\circ T_p(y) = (p\otimes q)(T(x\otimes y)) = a(p,q)(p\otimes q)(x\otimes y) = a(p,q)q(y)$. It follows that the corresponding eigenvalue for $q \in E_Y$ is $a(p,q)$ so that, by [1], 4.9, $q \quad a(p,q)$ must be structurally continuous. By symmetry, $p \quad a(p,q)$ has the same property for every $q \in E_Y$. By 1.12, a is constant on very subset $E_{X_k} \times E_{Y_\ell}$ (where $E_{X_k} \neq \{0\}$, $E_{Y_\ell} \neq \{0\}$); let $\alpha(k,\ell)$ denote this common value. This definition induces a mapping $\alpha: K\times L \to \mathbb{R}$ (we define $\alpha(k,\ell) := 0$ if $X_k = \{0\}$ or $Y_\ell = 0$). Since for $p \in E_{X_k}$ and $q \in E_{Y_\ell}$ we have $(p\otimes q)\circ T = \alpha(k,\ell)p\otimes q$ it follows that $(Tx)(k,\ell) = \alpha(k,\ell)x(k,\ell)$ so that $T = M_\alpha$.

<u>Note:</u> By 1.10a, the representation 1.6 is a special case of a scalar-function centralizer. Thus 4.2 in [4] is a corollary to 2.2.

<u>2.3 Corollary</u> ([13]). $Z(X\hat{\otimes}_\varepsilon Y)$ is the strong operator closure of $Z(X)\otimes Z(Y)$ (X and Y arbitrary Banach spaces).

<u>Proof.</u> We represent X and Y as in 1.6 so that $Z(X) \cong CK$ and $Z(Y) \cong CL$ (the base spaces are denoted by K and L, the component spaces by $(X_k)_{k\in K}$ and $(Y_\ell)_{\ell\in L}$). The natural embedding $Z(X)\otimes Z(Y) \hookrightarrow Z(X\hat{\otimes}_\varepsilon Y)$ maps $CK\otimes CL$ onto the subset $\{M_h \mid h \in XK\otimes CL \subset C(K\times L)\}$ which is uniformly dense in $\{M_h \mid h \in C(K\times L)\}$ (note that

$C(K\times L) = CK\hat{\otimes}_\varepsilon CL)$. 2.3 now follows from 2.2 and 1.11.

We already noted that function modules with scalar-function centralizer share much properties with function modules as in 1.6. However, if one is interested in invariants of the Banach space X, the $(X_k)_{k\in K}$ and K of 1.6 are of much more importance than the component spaces and the base space in the case of scalar-function centralizer: uniqueness is guaranteed only in the first case. It is therefore of interest to know sufficient conditions that, if X and Y are as in 1.6, the function module 2.1 has also the property that every M-bounded operator T is an operator M_h with $h \in C(K\times L)$. 2.2 only asserts that $T = M_\alpha$ where α is a multiplier so that we have to look for condtions such that multipliers are continuous. Such conditions have been given in [4], and we restate the main result in this connection.

2.4 Theorem ([4], th. 4.5). Let X and Y be Banach spaces such that the norm topology and the strong operator topology coincide on the centralizers of X and Y. Suppose that X and Y are represented as function modules in $\prod^\infty_{k\in K} X_k$ and $\prod^\infty_{\ell\in L} Y_\ell$ as in 1.6. Then every multiplier on $K\times L$ is continuous. This implies that $Z(X\hat{\otimes}_\varepsilon Y)$ is the norm closure of $Z(X)\otimes Z(Y)$.

Proof. By hypothesis there are $x_1,\dots,x_n \in X$, $y_1,\dots,y_m \in Y$, $r,r' > 0$ such that $r\|T\| \le \sup\{\|Tx_i\| \mid i=1,\dots,n\}$ and $r'\|S\| \le$ $\le \sup\{\|Sy_j\| \mid j=1,\dots,m\}$ for $T \in Z(X)$ and $S \in Z(Y)$. Now let α be a multiplier, $(k_o,\ell_o) \in K\times L$, and $\varepsilon > 0$. $(\alpha(k_o,\ell_o) - \alpha)x_i\otimes y_j$ beongs to $X\hat{\otimes}_\varepsilon Y$ and vanishes at (k_o,ℓ_o) so that there are neighbourhoods U of k_o and V of ℓ_o such that
$\|(\alpha(k_o,\ell_o) - \alpha(k,\ell))(x_i(k)\otimes y_j(\ell)\| \le \varepsilon$ for $k \in U$, $\ell \in V$. Since $\sup\{\|y_j(\ell)\| \mid j=1,\dots,m\} \ge r'$ for every $k \in K$, $\ell \in L$ (this is an easy consequence of the above conditions) it follows that
$|\alpha(k,\ell)| \le \varepsilon/r\,r'$ for $(k,\ell) \in U\times V$ so that α is continuous at

(k_o, ℓ_o).

By 2.2, every M-bounded operator is of the form $M_{\mathfrak{a}}$, $\mathfrak{a}$ a multiplier, so that we have $Z(X\hat{\otimes}_{\varepsilon}Y) = \{M_h \mid h \in C(K\times L)\} = \{M_h \mid h \in CK\otimes CL\}^{-} = (Z(X)\otimes Z(Y))^{-}$ (we used the fact that $(CK\otimes CL)^{-} = CK\hat{\otimes}_{\varepsilon}CL = C(K\times L)$).

3. SPACES FOR WHICH THE NORM TOPOLOGY AND THE STRONG OPERATOR TOPOLOGY COINCIDE ON THE CENTRALIZER

In view of Theorem 2.4 it is important to investigate the class of Banach spaces X for which the norm topology and the strong operator topology are equivalent on the centralizer of X.
It is obvious that this is the case iff there are vectors $x_1,\dots,x_n$ in X and a number $r > 0$ such that $\max\{\|Tx_i\| \mid i=1,\dots,n\} \geq r\|T\|$ for every $T \in Z(X)$. Such a set of vectors will be called a <u>centralizer-norming system</u> (abbr.: <u>cns</u>).

<u>3.1 Examples</u>:

a) Every Banach space for which $Z(X)$ is finite-dimensional has a cns.

b) C_oL has a cns iff L is compact. In this case $\{\underline{1}\}$ is a cns.

c) Let K be a compact convex set. Then $\{\underline{1}\}$ is a cns in AK.

d) If A is a C*-algebra with unit e, then $\{e\}$ is a cns in the self-adjoint part of A.

e) The number $\inf\{n \mid X$ has a cns consisting of n elements$\}$ may be arbitrarily large.

For details we refer the reader to Section 3 in [4].

4. APPLICATIONS

Let X and Y be real Banach spaces having a cns.

4.1 By 1.8 some M-structure properties of $X\hat{\otimes}_{\mathbb{C}}Y$ can be derived from the M-structure properties of X and Y. For example, the M-summands in $X\hat{\otimes}_{\mathbb{C}}Y$ correspond to the clopen subsets of $K_X \times K_Y$, and $K_{X\hat{\otimes}_{\mathbb{C}}Y} = K_X \times K_Y$ (cf. 1.6).

In particular, if X and Y have no nontrivial M-summands (resp. a trivial centralizer) then the same is true for $X\hat{\otimes}_{\mathbb{C}}Y$.

4.2 Let M be a compact Hausdorff space. Then $C(M,X) = CM\hat{\otimes}_{\mathbb{C}}X$ so that $Z(C(K,X)) = Z(CM\hat{\otimes}_{\mathbb{C}}X) = Z(CM)\otimes_{\mathbb{C}}Z(X) = CM\hat{\otimes}_{\mathbb{C}}Z(X) = C(M,Z(X)) = CM\hat{\otimes}_{\mathbb{C}}CK_X = C(M\times K_X)$.

Corollary. Suppose that M and N are compact Hausdorff spaces and that there exists an isometric isomorphism from $C(M,X)$ onto $C(N,Y)$. Then $M\times K_X$ and $N\times K_Y$ are homeomorphic. In particular, if $Z(X)$ is one-dimensional, then X has the Banach-Stone property, i.e. the existence of an isometric isomorphism between $C(M,X)$ and $C(N,X)$ implies that M and N are homeomorphic.

Note: This result has already been stated in [5].

4.3 Let K and L be compact convex sets. Then (by 3.1c) the centralizer of $AK\hat{\otimes}_{\mathbb{C}}AL$ is the tensor product of the centralizers of the components. This result is due to Vincent-Smith ([12]).

4.4 Let A and B be unital C^*-algebras. Then the centre of the C^*-algebra $A\hat{\otimes}_{\mathbb{C}}B$ consists exactly of the elements in $A_Z\hat{\otimes}_{\mathbb{C}}B_Z$ (where A_Z resp. B_Z denotes the centre of A resp. B). This result (which is due to Haydon and Wassermann ([10])) follows from 1.4d and 3.1d.

Note: It can be shown that this result concerning $A\hat{\otimes}_{\mathbb{C}}B$ implies a similar reslt for the completion of $A\otimes B$ in any C^*-algebra norm ([2],[3],[12]).

5. REMARKS/PROBLEMS

1. A more detailed investigation of the functions $\mathfrak{a} \in M(K\times L)$ shows that for all "familiar" classes of Banach spaces these functions are continuous at every point (k,ℓ) for which $X_k \hat{\otimes}_\varepsilon Y_\ell \neq 0$ (there are known only very pathological examples of spaces not having this property). Since $K = \beta\{k \mid X_k \neq 0\}$ ([4], prop. 2.5(i)) it follows that in all important cases the difference between $Z(X\hat{\otimes}_\varepsilon Y)$ and $Z(X)\hat{\otimes}_\varepsilon Z(Y)$ is just the difference between $\beta(\{k|X_k\neq 0\} \times \{l|Y_l\neq 0\})$ and $(\beta\{k|X_k\neq 0\}) \times (\beta\{l|Y_l\neq 0\})$.

2. It is well-known that a Banach space with trivial centralizer can have non-trivial M-ideals. There seems to be no way analogous to the methods of the present paper to decide the following <u>problem</u>:

Suppose that X and Y have no nontrivial M-ideals.
Is then the same true for $X\hat{\otimes}_\varepsilon Y$?

3. The foregoing discussion concerned questions of M-structure in the ε-tensor product. A systematic investigation of these problems should also contain a consideration of the projective $(\pi-)$ tensor product and the Cunningham algebra (the Cunningham algebra $C(X)$ is the Banach algebra generated by the set of projections associated with the L-summands of X). We have only partial results in this direction. There is some evidence that the following assertions (which have been proved under certain additional assumptions) are valid in general:

a) $Z(X\hat{\otimes}_\pi Y) = \mathbb{R}\mathrm{Id}$ whenever X and Y are at least two-dimensional.

b) $C(X\hat{\otimes}_\varepsilon Y) = \mathbb{R}\mathrm{Id}$ whenever X and Y are at least two-dimensional.

c) $C(X\hat{\otimes}_\pi Y)$ is the strong operator closure (and only in trivial cases the iniform closure) of $\overline{C(X)\otimes C(Y)}$.

REFERENCES

[1] Alfsen-Effros, E.M., M-structure in real Banach spaces I/II Ann. of Math. 96 (1972), 78-173.

[2] Archbold, R.J., On the centre of a tensor product of C*-algebras J. of the Ldn. Math. Soc. 10 (1975), 257-262.

[3] Batty, C.J., Tensor products of compact convex sets and Banach algebras, Math. Proc. Camb. Phil. Soc. 83 (1978), 419-427.

[4] Behrends, E., The centralizer of tensor products of Banach spaces, Pacific Journal of Math. (to appear, 1979).

[5] Behrends, E., An application of M-structure to theorems of the Banach-Stone type, in: Notas de Mathematica, Math. Studies 27 (1977), 29-49.

[6] Behrends, E. - Schmidt-Bichler, U., M-structure and the Banach-Stone theorem, Studia Math. 68 (1979) (to appear)

[7] Chui, C.K. et al., L-ideals and numerical range preservation, Ill. J. of Math. 21 (1977), 365-73.

[8] Cunningham, F., M-structure in Banach spaces, Proc. of the Ca Camb. Phil. Soc. 63 (1967), 613-629.

[9] Cunningham, F. - Roy, N.M., Extreme functionals on an upper semicontinuous function space, Proc. of the American Math. Soc. 42 (1974), 461-465.

[10] Haydon, R.G. - Wassermann, A.S., A commutation result for tensor products of C*-algebras, Bull. Ldn. Math. Soc. 5 (1973), 283-287.

[11] Holmes, R. et al., Best approximation by compact operators, Bull. of the AMS 80 (1974), 98-102.

[12] Vincent-Smith, G.F., The centre of the tensor product of AK-spaces, Quart. J. Math. Oxford 28 (1977), 87-91.

[13] Wickstead, A.W., The centralizer of $E\otimes_\lambda F$, Pac. J. of Math. 65 (1976), 563-571.

SILVA-HOLOMORPHY TYPES, BOREL TRANSFORMS and PARTIAL DIFFERENTIAL OPERATORS

Mauro Bianchini

Instituto de Matemática
UNICAMP - Brasil

ABSTRACT

Dineen in [2] described and studied various topological vector spaces of holomorphic functions and introduced the α-holomorphy, α-β-holomorphy and α-β-γ-holomorphy types solving questions about Borel transforms, convolution and partial differential operators. Matos & Nachbin in working with Silva-holomorphic functions between two complex locally spaces defined Silva-holomorphy types θ and obtained results about Borel transforms and Malgrange's theorem for convolution operators. In this work, using the techniques developed in [2] and using the study of the Silva-holomorphic functions in complex locally convex spaces, we generalize the results presented by Dineen in [2].

1. PRELIMINARIES

In this paper $\mathbb{N}$, $\mathbb{C}$ and E will denote, respectively, the set of positive integers and zero, the field of complex numbers and a complex locally convex space. $\mathcal{B}_E$ will denote the family of all closed absolutely convex bounded subsets of E. If $B \in \mathcal{B}_E$, E_B is the vector subspace of E generated by B and with the norm topology given by the Minkowsky functional determined by B. Hence if $x \in E_B$, $\|x\|_B = \inf\{\rho > 0;\ x \in \rho B\}$. For each $m \in \mathbb{N}$, $\mathcal{P}_b(^mE)$ will denote the vector space of all m-homogeneous polynomials from E to $\mathbb{C}$ which are bounded on the bounded subsets of E. On $\mathcal{P}_b(^mE)$ we consider the locally convex topology defined by the semi-norms:

$$\|P\|_B = \sup\{|P(x)|;\ x \in B\}, \quad \text{for each } B \in \mathcal{B}_E.$$

$\mathcal{P}_b(E)$ will denote the direct sum of the spaces $\mathcal{P}_b(^mE)$, $m \in \mathbb{N}$.

$\mathcal{H}_S(E)$ will indicate the vector space of all functions from E to $\mathbb{C}$ Silva-holomorphic at every point of E. For each $f \in \mathcal{H}_S(E)$ and $\xi \in E$, the Taylor series expansion of f, at ξ, is

$$f(x) = \sum_{m=0}^{\infty} \frac{1}{m!} \hat{\delta}^m f(\xi)(x-\xi),$$

for all $x \in E$ and the corresponding differential of order $m \in \mathbb{N}$ is $\hat{\delta}^m f(\xi) \in \mathcal{P}_b(^mE)$. As in [1] and [5] we use the notations $\mathcal{P}(^mE)$ with $m \in \mathbb{N}$, $\mathcal{P}(E)$ and $\mathcal{H}(E)$ to indicate, respectively, the vector space of all continuous m-homogeneous polynomials from E to $\mathbb{C}$, the vector space of all continuous polynomials from E to $\mathbb{C}$ and the vector space of all holomorphic functions from E to $\mathbb{C}$. E' will denote the vector space of all continuous linear functional from E to $\mathbb{C}$ and $E^* = \mathcal{P}_b(^1E)$. A subset K of E is strict compact if there exists $B \in \mathcal{B}_E$ with K a compact subset of E_B. $\mathcal{K}_e(E)$ will denote the set of all convex balanced strict compact subset of E. Finally C_o^+ will denote the set of all sequences of positive real number which tend to zero at infinity.

<u>(1.01) Definition</u>. A Silva holomorphy type θ (see [4]) is a sequence of complete complex locally convex spaces $\mathcal{P}_{b\theta}(^mE)$, $m \in \mathbb{N}$, the topology of each $\mathcal{P}_{b\theta}(^mE)$ being defined by a family of semi-norms $P \to \|P\|_{\theta,B}$, with $B \in \mathcal{B}_E$, denoted by $\Gamma_{\theta,\mathcal{B}}$. This sequence must satisfy the following conditions:

1) For each $m \in \mathbb{N}$, $\mathcal{P}_{b\theta}(^mE) \subset \mathcal{P}_b(^mE)$ as a vector space.

2) $\mathcal{P}_{b\theta}(^oE) = \mathcal{P}_b(^oE) = \mathbb{C}$ as a topological vector space.

3) There exists $\sigma_\theta \geq 1$ such that for each $B \in \mathcal{B}_E$, $P \in \mathcal{P}_{b\theta}(^mE)$, $x \in E_B$, $n \in \mathbb{N}$ with $n \leq m$ we have $\hat{\delta}^n P(x) \in \mathcal{P}_{b\theta}(^nE)$ and

$$\left\|\frac{1}{n!} \hat{\delta}^n P(x)\right\|_{\theta,B} \leq \sigma_\theta^m \|P\|_{\theta,B} \|x\|_B^{m-n} .$$

4) If $P \in \mathcal{P}_{b\theta}(^mE)$, B and $D \in \mathcal{B}_E$ with $B \subset D$, then $\|P\|_{\theta,B} \leq$ $\leq \|P\|_{\theta,D}$.

<u>(1.02) Examples</u>. We can verify, by definitions in [4] that the fol-

lowing sequences of spaces are examples of Silva holomorphy types.

a) For each $m \in \mathbb{N}$, $\mathcal{P}_{b\theta}(^mE) = \mathcal{P}_b(^mE)$. This is the current Silva holomorphy type.

b) For each $m \in \mathbb{N}$, $\mathcal{P}_{b\theta}(^mE) = \mathcal{P}_{bc}(^mE)$. This is the compact Silva holomorphy type.

c) For each $m \in \mathbb{N}$, $\mathcal{P}_{b\theta}(^mE) = \mathcal{P}_{bN}(^mE)$. This is the nuclear Silva holomorphy type.

<u>(1.03) Proposition</u>. The inclusion mapping $\mathcal{P}_{b\theta}(^mE) \subset \mathcal{P}_b(^mE)$ is a continuous mapping and $\|P\|_B \leq \sigma_\theta^m \|P\|_{\theta,B}$ for all $B \in \mathcal{B}_E$ and $P \in \mathcal{P}_{b\theta}(^mE)$.

<u>Proof</u>. (See [4]).

<u>(1.04) Definition</u>. $f \in \mathcal{H}_S(E)$ is said to be of Silva holomorphy type θ at $x \in E$ if:

1) $\hat{\delta}^m f(x) \in \mathcal{P}_{b\theta}(^mE)$, for all $m \in \mathbb{N}$.

2) For each $B \in \mathcal{B}_E$ there exists constants $c_1 \geq 0$ and $c_2 \geq 0$ such that

$$\left\|\frac{1}{m!}\hat{\delta}^m f(x)\right\|_{\theta,B} \leq c_1 c_2^m, \quad \text{for all } m \in \mathbb{N}.$$

A function f is said to be of Silva holomorphy type θ if f is of Silva holomorphy type θ at all points of E. $\mathcal{H}_{S\theta}(E)$ will denote the set of all these functions.

<u>(1.05) Definition</u>. Let K be a strict compact subset of E and let $B \in \mathcal{B}_E$ with K a compact subset of E_B. A semi-norm p on $\mathcal{H}_{S\theta}(E)$ is said to be ported by K and B if for all $\epsilon > 0$ there exists $c(\epsilon) > 0$ such that

$$p(f) \leq c(\epsilon) \sum_{m=0}^{\infty} \epsilon^m \sup_{x \in K} \left\|\frac{1}{m!}\hat{\delta}^m f(x)\right\|_{\theta,B}$$

for all $f \in \mathcal{H}_{S\theta}(E)$. $\mathcal{T}_{\omega\theta}$ will denote locally convex topology on $\mathcal{H}_\theta(E)$ which is generated by those semi-norms. When θ is the current type we denote $\mathcal{T}_{\omega\theta}$ by $\mathcal{T}_{\omega S}$ (see [4]).

2. α-SILVA HOLOMORPHY TYPES AND THE SPACE $(\mathcal{H}_{S\theta}(E), \mathcal{T}_{S\theta})$

<u>(2.01) Definition.</u> An α-Silva holomorphy type is a Silva holomorphy type θ which satisfies the following conditions:

1) $(\mathcal{P}_{b\theta}(^mE), \Gamma_{\theta,\mathcal{B}})$ and σ_θ depend only on the topological vector space structure of E.

2) If B_1 and $B_2 \in \mathcal{B}_E$ and if $c > 0$ with $cB_1 \subset B_2$, then $c^m\|P\|_{\theta,B_1} \leq \|P\|_{\theta,B_2}$ for all $P \in \mathcal{P}_{b\theta}(^mE)$ and for all $m \in \mathbb{N}$.

<u>(2.02) Example.</u> It is not difficult to verify (see [4]) that a), b) and c) of (1.02) are examples of α-Silva holomorphy types.

<u>(2.03) Definition.</u> Let θ be a α-Silva holomorphy type. $\mathcal{H}_{S\theta}(E)$ is the set of all functions on E which satisfy the conditions:

1) $f \in \mathcal{H}_S(E)$.

2) For all $m \in \mathbb{N}$, $\hat{\delta}^m f(0) \in \mathcal{P}_{b\theta}(^mE)$.

3) For all $K \in \hat{\mathcal{K}}_e(E)$ and all $B \in \mathcal{B}_E$ with K a compact subset of E_B, there exists $\varepsilon > 0$ such that

$$\sum_{m=0}^{\infty} \left\|\frac{1}{m!}\hat{\delta}^m f(0)\right\|_{\theta,K+\varepsilon B} < \infty.$$

<u>(2.04) Proposition.</u> Let θ be a α-Silva holomorphy type. If $f \in \mathcal{H}_S(E)$, with $\hat{\delta}^m f(0) \in \mathcal{P}_{b\theta}(^mE)$ for all $m \in \mathbb{N}$, then the following conditions are equivalent:

1) $f \in \mathcal{H}_{S\theta}(E)$.

2) For all $K \in \hat{\mathcal{K}}_e(E)$, for all $B \in \mathcal{B}_E$ with K is a compact subset of E_B, and for all sequence $(\alpha_n)_0^\infty \in C_o^+$ we have:

$$\sum_{m=0}^{\infty} \left\|\frac{1}{m!}\hat{\delta}^m f(0)\right\|_{\theta,K+\alpha_m B} < \infty.$$

3) For all $K \in \hat{\mathcal{K}}_e(E)$, for all $B \in \mathcal{B}_E$ with K is a compact subset of E_B, and for all sequence $(\alpha_n)_0^\infty \in C_o^+$ we have:

$$\lim_{m\to\infty} \left\|\frac{1}{m!}\hat{\delta}^m f(0)\right\|_{\theta,K+\alpha_m B}^{1/m} = 0.$$

Proof. Let $K \in \hat{\mathcal{K}}_e(E)$, $B \in \mathcal{B}_E$ with K a compact subset of E_B, and $(\alpha_n)_0^\infty \in c_o^+$. By 1) there exists $\varepsilon > 0$ such that

$$\sum_{n=0}^{\infty} \left\|\frac{1}{n!}\hat{\delta}^n f(0)\right\|_{\theta, K+\varepsilon B} < \infty .$$

Let $n_o \in \mathbb{N}$ such that $\alpha_n \leq \varepsilon$ if $n \geq n_o$. We have $K + \alpha_n B \subset K + \varepsilon B$ for all $n \geq n_o$ and then

$$\begin{aligned}\sum_{n=0}^{\infty} \left\|\frac{1}{n!}\hat{\delta}^n f(0)\right\|_{\theta, K+\alpha_n B} &= \\ = \sum_{n=0}^{n_o-1} \left\|\frac{1}{n!}\hat{\delta}^n f(0)\right\|_{\theta, K+\alpha_n B} &+ \\ + \sum_{n=n_o}^{\infty} \left\|\frac{1}{n!}\hat{\delta}^n f(0)\right\|_{\theta, K+\alpha_n B} &\leq \\ \leq \sum_{n=0}^{n_o-1} \left\|\frac{1}{n!}\hat{\delta}^n f(0)\right\|_{\theta, K+\alpha_n B} &+ \\ + \sum_{n=n_o}^{\infty} \left\|\frac{1}{n!}\hat{\delta}^n f(0)\right\|_{\theta, K+\varepsilon B} &< \infty .\end{aligned}$$

Thus we have proved 1) ⇒ 2). To prove that 2) ⇒ 3), let $K \in \hat{\mathcal{K}}_e(E)$, $B \in \mathcal{B}_E$ with K a compact subset of E_B, $(\alpha_n)_0^\infty \in c_o^+$ and for all $n \in \mathbb{N}$ $\beta_n = c^n$ with $c > 0$. We have $cK \in \hat{\mathcal{K}}_e(E)$ and $(\beta_n^{1/n}\alpha_n)_0^\infty \in c_o^+$. Let $B \in \mathcal{B}_E$ such that cK is a compact subset of E_B. By 2) we have

$$\sum_{n=0}^{\infty} \left\|\frac{1}{n!}\hat{\delta}^n f(0)\right\|_{\theta, cK+\beta_n^{1/n}\alpha_n B} < \infty .$$

Since θ is a α-Silva holomorphy type and since

$$(\beta_n)^{1/n}[K+\alpha_n B] = cK + \beta_n^{1/n}\alpha_n B$$

we have

$$\beta_n\left\|\frac{1}{n!}\hat{\delta}^n f(0)\right\|_{\theta, K+\alpha_n B} = \left\|\frac{1}{n!}\hat{\delta}^n f(0)\right\|_{\theta, cK+\beta_n^{1/n}\alpha_n B}$$

and then

$$\sum_{n=0}^{\infty} \beta_n\left\|\frac{1}{n!}\hat{\delta}^n f(0)\right\|_{\theta, K+\alpha_n B} < \infty$$

which implies

$$\limsup_{n\to\infty} \left(\beta_n\left\|\frac{1}{n!}\hat{\delta}^n f(0)\right\|_{\theta, K+\alpha_n B}\right)^{1/n} \leq 1,$$

that is,

$$\limsup_{n\to\infty} \left\|\frac{1}{n!}\hat{\delta}^n f(0)\right\|^{1/n}_{\theta,K+\alpha_n B} \le 1/c,$$

with c arbitrary. Since K is a compact set of E_B we get 3).

To prove that 3) ⇒ 1) we suppose that there exist $K \in \hat{\mathcal{K}}_e(E)$ and $B \in \mathcal{B}_E$ with K a compact subset of E_B and such that for every $\varepsilon > 0$ we have

$$\sum_{n=0}^{\infty} \left\|\frac{1}{n!}\hat{\delta}^n f(0)\right\|_{\theta,K+\varepsilon B} = \infty.$$

Then $\limsup_{n\to\infty} \left\|\frac{1}{n!}\hat{\delta}^n f(0)\right\|^{1/n}_{\theta,K+B} \ge 1$ and we can choose $n_1 \in \mathbb{N}$ such that

$$\left\|\frac{1}{n_1!}\hat{\delta}^{n_1} f(0)\right\|^{1/n_1}_{\theta,K+B} \ge 1/2.$$

Since also

$$\limsup_{n\to\infty} \left\|\frac{1}{n!}\hat{\delta}^n f(0)\right\|^{1/n}_{\theta,K+(1/2)B} \ge 1,$$

we can choose $n_2 \in \mathbb{N}$ such that $n_2 > n_1$ and

$$\left\|\frac{1}{n_2!}\hat{\delta}^{n_2} f(0)\right\|^{1/n_2}_{\theta,K+(1/2)B} \ge 1/2.$$

By induction we can choose $n_k > n_{k-1}$ such that

$$\left\|\frac{1}{n_k!}\hat{\delta}^{n_k} f(0)\right\|^{1/n_k}_{\theta,K+(1/k)B} \ge 1/2.$$

We define, for $n \in \mathbb{N}$,

$$\alpha_n = \begin{cases} 1 & \text{if} \quad n \le n_1 \\ 1/k & \text{if} \quad n_{k-1} < n \le n_k \end{cases}$$

We have that $(\alpha_n)_0^\infty \in c_o^+$ and

$$\limsup_{n\to\infty} \left\|\frac{1}{n!}\hat{\delta}^n f(0)\right\|^{1/n}_{\theta,K+\alpha_n B} \ge 1/2.$$

This contradicts 3). Then 3) ⇒ 1).

<u>(2.05) Definition.</u> A semi-norm p on $H_{S\theta}(E)$ is $S\theta$-ported by $K \in \hat{\mathcal{K}}_e(E)$ and $B \in \mathcal{B}_E$, with K a compact subset of E_B, if for every $\varepsilon > 0$ there exists $c(\varepsilon) > 0$ such that

$$p(f) \leq c(\delta) \sum_{n=0}^{\infty} \left\| \frac{1}{n!} \hat{\delta}^n f(0) \right\|_{\theta, K+\delta B}, \quad f \in H_{S\theta}(E).$$

(2.06) Definition. The topology $T_{S\theta}$ on $H_{S\theta}(E)$ is that one generated by all semi-norms which are $S\theta$-ported by some element in $\hat{\mathcal{K}}_e(E)$ and some element in $\mathcal{B}_E$. If θ is the current Silva holomorphy type we denote, respectively by $H_S(E)$ and $\tau_{\omega se}$ the space $H_{S\theta}(E)$ and the topology $T_{S\theta}$ (See [6]).

(2.07) Proposition. If $f \in H_{S\theta}(E)$, then the Taylor series of f at 0 converges to f in $(H_{S\theta}(E), T_{S\theta})$.

Proof. Let p be a semi-norm $S\theta$-ported by $K \in \hat{\mathcal{K}}_e(E)$ and $B \in \mathcal{B}_E$. Then, for each $\varepsilon > 0$, there exists $c(\varepsilon) > 0$ such that

$$p(f - \sum_{n=0}^{\infty} \frac{1}{n!} \hat{\delta}^n f(0)) \leq c(\varepsilon) \sum_{n=j+1}^{\infty} \left\| \frac{1}{n!} \hat{\delta}^n f(0) \right\|_{\theta, K+\varepsilon B}.$$

Since $f \in H_{S\theta}(E)$, for some $\varepsilon > 0$ the second member of the inequality tends to zero when j tends to zero. This proves the proposition.

The next proposition characterizes the topology $T_{S\theta}$.

(2.08) Proposition. The topology $T_{S\theta}$ on $H_{S\theta}(E)$ is generated by the semi-norms of the form:

$$p(f) = \sum_{n=0}^{\infty} \left\| \frac{1}{n!} \hat{\delta}^n f(0) \right\|_{\theta, K+\alpha_n B}$$

where $K \in \hat{\mathcal{K}}_e(E)$, $B \in \mathcal{B}_E$ with K a compact subset of E_B and $(\alpha_n)_0^{\infty} \in c_0^+$.

Proof. By Proposition (2.04) for each $K \in \hat{\mathcal{K}}_e(E)$, $B \in \mathcal{B}_E$ with K a compact subset of E_B and $(\alpha_n) \in c_0^+$ we have

$$p(f) = \sum_{n=0}^{\infty} \left\| \frac{1}{n!} \hat{\delta}^n f(0) \right\|_{\theta, K+\alpha_n B} < \infty,$$

for all $f \in H_{S\theta}(E)$. It is then obviously a semi-norm on $H_{S\theta}(E)$. Given $\varepsilon > 0$ choose $n_o \in N$ such that $\alpha_n \leq \varepsilon$ for all $n \geq n_o$. For $n = 0,1,\ldots,n_o-1$, we take $\rho \geq 1$ such that $\alpha_n \rho^{-1} \leq \varepsilon$ and we put $\delta = \rho^{-1}$. We have

$$\delta(K+\alpha_n B) = \delta K + \alpha_n \rho^{-1} B \subset K + \varepsilon B.$$

Then, for $n = 0,1,\ldots,n_o-1$ we have

$$K + \alpha_n B \subset \delta^{-1}(K+\varepsilon B)$$

and

$$\left\|\frac{1}{n!}\hat{\delta}^n f(0)\right\|_{\theta,K+\alpha_n B} \le \delta^{-n}\left\|\frac{1}{n!}\hat{\delta}^n f(0)\right\|_{\theta,K+\varepsilon B}.$$

If we take $c(\varepsilon) = \sup\{\delta^{-i};\ i = 0,1,\ldots,n_o-1\}$, we have

$$p(f) = \sum_{n=0}^{\infty}\left\|\frac{1}{n!}\hat{\delta}^n f(0)\right\|_{\theta,K+\alpha_n B} \le c(\varepsilon)\sum_{n=0}^{\infty}\left\|\frac{1}{n!}\hat{\delta}^n f(0)\right\|_{\theta,K+\varepsilon B}.$$

Hence p is continuous on $(H_{S\theta}(E),T_{S\theta})$.

Now let p_1 be a semi-norm on $H_{S\theta}(E)$ which is $S\theta$-ported by K and B. For every $\varepsilon > 0$ choose $c(\varepsilon) > 0$ such that

$$p_1(f) \le c(\varepsilon)\sum_{n=0}^{\infty}\left\|\frac{1}{n!}\hat{\delta}^n f(0)\right\|_{\theta,K+\varepsilon B}$$

for all $f \in H_{S\theta}(E)$. If $P_m \in \mathcal{P}_{b\theta}(^mE)$ we have

$$p_1(P_m) \le c(\varepsilon)\,\|P_m\|_{\theta,K+\varepsilon B}.$$

For all $m \in N$ and for each $\varepsilon > 0$ let $k_m(\varepsilon)$ be the smallest positive number or zero such that

$$p_1(P_m) \le k_m(\varepsilon)\,\|P_m\|_{\theta,K+\varepsilon B}, \quad \text{for all } P_m \in \mathcal{P}_{b\theta}(^mE).$$

The sequence $(k_m(\varepsilon))_o^{\infty}$ is bounded and we have

$$\limsup_{n\to\infty} k_m(\varepsilon)^{1/n} \le 1.$$

Let n_1 be a positive integer such that $k_n(1)^{1/n} \le 2$ for all $n \ge n_1$. By induction we take n_s such that $n_s > n_{s-1}$ and $k_n(1/s)^{1/n} \le 2$ for $n \ge n_s$.

We define

$$\alpha_n = \begin{cases} 1 & \text{for } n < n_2 \\ 1/s & \text{for } n_s \le n < n_{s+1} \end{cases}$$

Then $(\alpha_n)_0^\infty \in c_o^+$ and $k_n(\alpha_n)^{1/n} \leq 2$ for all $n \geq n_1$.

Then $k_n(\alpha_n) \leq 2^n$ for all $n \geq n_1$. Let $c > 0$ such that $k_n(\alpha_n) \leq c2^n$ for all $n \in \mathbb{N}$.

We have

$$p_1(f) = p_1\Big(\sum_{n=0}^{\infty} \frac{1}{n!} \hat{\delta}^n f(0)\Big) \leq c \sum_{n=0}^{\infty} 2^n \Big\|\frac{1}{n!} \hat{\delta}^n f(0)\Big\|_{\theta, K+\alpha_n B},$$

and then

$$p_1(f) \leq c \sum_{n=0}^{\infty} \Big\|\frac{1}{n!} \hat{\delta}^n f(0)\Big\|_{\theta, 2K+2\alpha_n B}.$$

for all $f \in H_{S\theta}(E)$, with $2K \in \hat{K}_e(E)$, $2B \in \mathcal{B}_E$ with $2K$ a compact set of E_{2B} and $(\alpha_n)_0^\infty \in c_o^+$. This completes the proof.

We now get a necessary and sufficient condition on the sequence $(P_m)_0^\infty$ with $P_m \in \mathcal{P}_{b\theta}(^mE)$ so that $\sum_{n=0} P_m$ is the Taylor series expansion of an element of $H_{S\theta}(E)$. For this we consider first the current type.

<u>(2.09) Proposition.</u> Let $P_m \in \mathcal{P}_{b\theta}(^mE)$ for $m \in \mathbb{N}$; then the following conditions are equivalent:

1) $\sum_{n=0}^{\infty} P_m$ is the Taylor series expansion of an element $f \in \mathcal{H}_S(E)$.

2) For each $K \in \hat{K}_e(E)$, $B \in \mathcal{B}_E$ with K a compact subset of E_B and $(\alpha_n)_0^\infty \in c_o^+$, we have $\lim_{m\to\infty} \|P_m\|_{K+\alpha_m B}^{1/m} = 0$.

3) For each $K \in \hat{K}_e(E)$, $B \in \mathcal{B}_E$ with K a compact subset of E_B and $(\alpha_n)_0^\infty \in c_o^+$, we have $\sum_{m=0}^{\infty} \|P_m\|_{K+\alpha_m B} < \infty$.

4) For each $K \in \hat{K}_e(E)$, $B \in \mathcal{B}_E$ with K a compact subset of E_B, there exists $\varepsilon > 0$ such that $\sum_{m=0}^{\infty} \|P_m\|_{K+\varepsilon B} < \infty$.

<u>Proof.</u> Since the current type is an α-Silva holomorphy type, by Proposition (2.04), we get that 2), 3) and 4) are equivalent. We therefore prove that 1) and 4) are equivalent. Let $K \in \hat{K}_e(E)$, $B \in \mathcal{B}_E$ with K compact subset of E_B and $(\alpha_n)_0^\infty \in c_o^+$.

Suppose that $\sum_{m=0}^{\infty} P_m$ is the Taylor series expansion of an element

$f \in \mathcal{H}_S(E)$. Since, (see [4]),

$$\|P_m\|_{K+\alpha_m B} = \sup \{|P_m(X)|;\ x \in K + \alpha_m B\} =$$

$$= \sup \{|\frac{1}{2\pi i}\int_{|\lambda|=\rho} \frac{f(\lambda x)}{\lambda^{m+1}} d\lambda|;\ x \in K + \alpha_m B\} \quad \text{for all} \quad \rho > 0,$$

we have

$$\|P_m\|_{K+\alpha_m B} \le \frac{2\pi\rho}{2\pi}\frac{1}{m!}\frac{1}{\rho^{m+1}} \sup \{|f(x)|;\ x \in \rho K + \rho\alpha_m B\}$$

with $\rho > 0$ arbitrary. We have that $\rho K \in \hat{\mathcal{K}}_e(E)$ and $\rho B \in \mathcal{B}_E$ with ρK a compact set of $E_{\rho B}$. Let V be a neighborhood of ρK on $E_{\rho B}$ which f is bounded. Choose $m_o \in \mathbb{N}$ such that $\rho K + \rho\alpha_m B \subset V$ for $m \ge m_o$. We have

$$\|P_m\|_{K+\alpha_m B} \le \frac{1}{\rho^m} \sup \{|f(x)|;\ x \in \rho K + \rho\alpha_m B\} \le \frac{M}{\rho^m}$$

for all $m \ge m_o$. Then

$$\limsup_{m\to\infty} \|P_m\|^{1/m}_{K+\alpha_m B} \le \frac{1}{\rho}.$$

Since $\rho > 0$ is arbitrary, we have

$$\limsup_{m\to\infty} \|P_m\|^{1/m}_{K+\alpha_m B} = 0.$$

This proves that 1) $\Rightarrow$ 2).

Now let $B \in \mathcal{B}_E$ and for each $x \in E_B$ we take $K_x = \{x\} \in \hat{\mathcal{K}}_e(E)$ the closed convex balanced hull of $\{x\}$. We have that K_x is a compact set of E_B and, by 4), there exists $\epsilon > 0$ such that

$$\sum_{m=0}^{\infty} \|P_m\|_{K_x+\epsilon B} < \infty.$$

Then

$$\limsup_{m\to\infty} \|P_m\|^{1/m}_{K_x+\epsilon B} \le 1$$

and then (see [4]) $\sum_{m=0}^{\infty} P_m|$ is the Taylor series expansion, at zero, of the function $f_B\colon E_B \to \mathbb{C}$ which is holomorphic on the interior of $K_x + \epsilon B$ and therefore on x. Since $E = \bigcup \{E_B;\ B \in \mathcal{B}_E\}$, by the uniqueness of the Taylor series expansion, we get that $\sum_{m=0}^{\infty} P_m$ is the Taylor series, at zero, of a function $f: E \to C$ such that $f/E_B \in \mathcal{H}(E_B)$, for all $B \in \mathcal{B}_E$, that is, of a function $f \in \mathcal{H}_S(E)$.

(2.10) Corollary. $H_S(E) = \mathcal{H}_S(E)$.

Proof. Since $H_{S\theta}(E) \subset \mathcal{H}_S(E)$ for all Silva holomorphy type, we get that $H_S(E) \subset \mathcal{H}_S(E)$. By other hand, since the current Silva holomorphy type is an α-Silva holomorphy type, if $f \in \mathcal{H}_S(E)$ we have that $\sum_{m=0}^{\infty} \frac{1}{n!} \hat{\delta}^n f(0)$ is the Taylor series expansion, at zero, of f and therefore, by Proposition (2.09), if we take $K \in \hat{\mathcal{K}}_e(E)$ and $B \in \mathcal{B}_E$ with K a compact set of E_B, there exists $\epsilon > 0$ such that

$$\sum_{m=0}^{\infty} \left\|\frac{1}{m!} \hat{\delta}^m f(0)\right\|_{K+\epsilon B} < \infty .$$

Hence $f \in H_{\theta S}(E)$.

(2.11) Corollary. Let $P_m \in \mathcal{P}_{b\theta}(^m E)$ for $m \in \mathbb{N}$. Then the following conditions are equivalent:

1) $\sum_{m=0}^{\infty} P_m$ is the Taylor series expansion of an element $f \in H_{S\theta}(E)$.

2) For each $K \in \hat{\mathcal{K}}_e(E)$, $B \in \mathcal{B}_E$ with K compact in E_B and $(\alpha_n)_0^{\infty} \in c_o^+$ we have $\lim_{m\to\infty} \|P_m\|_{\theta, K+\alpha_m B}^{1/m} = 0$.

3) For each $K \in \hat{\mathcal{K}}_e(E)$, $B \in \mathcal{B}_E$ with K compact in E_B and $(\alpha_n)_0^{\infty} \in c_o^+$ we have $\sum_{m=0}^{\infty} \|P_m\|_{\theta, K+\alpha_m B} < 0$.

4) For each $K \in \hat{\mathcal{K}}_e(E)$, $B \in \mathcal{B}_E$ with K compact in E_B there exists $\epsilon > 0$ such that $\sum_{m=0}^{\infty} \|P_m\|_{\theta, K+\epsilon B} < \infty$.

Proof. For any α-Silva holomorphy type θ, there exists $\sigma_\theta \geq 1$ such that, for all $m \in \mathbb{N}$, $\|P_m\|_B \leq \sigma_\theta^m \|P_m\|_{\theta, B}$. Suppose that 2) is true. If we take $K \in \hat{\mathcal{K}}_e(E)$, $B \in \mathcal{B}_E$ with K a compact set of E_B, and $(\alpha_n)_0^{\infty} \in c_o^+$ we get $\sigma_\theta K \in \hat{\mathcal{K}}_e(E)$, $\sigma_\theta B \in \mathcal{B}_E$ with $\sigma_\theta K$ a compact subset of $E_{\sigma\theta}B$. By 2) we have

$$\sum_{m=0}^{\infty} \|P_m\|_{\theta, \sigma_\theta K+\alpha_m \sigma_\theta B} < \infty .$$

Hence

$$\sum_{m=0}^{\infty} \|P_m\|_{K+\alpha_m B} \leq \sum_{m=0}^{\infty} \sigma_\theta^m \|P_m\|_{\theta, K+\alpha_m B} < \infty .$$

Then the condition 2) of the Proposition (2.09) hold and therefore the condition 1) of the same proposition hold. Then $\sum_{n=0}^{\infty} P_m$ is the Taylor series expansion, at zero, of an element $f \in \mathcal{H}_S(E)$. The Definition (2.03) and the Proposition (2.04) complete the proof.

<u>(2.12) Proposition</u>. $(H_{S\theta}(E), T_{S\theta})$ is complete.

<u>Proof</u>. Let $(f_\alpha)_{\alpha \in A}$ be a Cauchy net in $(H_{S\theta}(E), T_{S\theta})$. For each $m \in \mathbb{N}$,

$$\begin{array}{ccc} H_{S\theta}(E) & \rightarrow & \mathcal{P}_{b\theta}(^mE) \\ f & \longmapsto & \hat{\delta}^m f(0) \end{array}$$

is a linear and continuous mapping. Then $(\hat{\delta}^m f_\alpha(0))_{\alpha \in A}$ is a Cauchy net in the complete space $(\mathcal{P}_{b\theta}(^mE); \Gamma_{\theta,\beta})$. For each $m \in \mathbb{N}$, we denote by P_m the limit of $(\hat{\delta}^m f_\alpha(0))_{\alpha \in A}$. Let $K \in \hat{K}_e(E)$, $B \in \mathcal{B}_E$ with K a compact subset of E_B and $(\alpha_n) \in C_o^+$. Given $\epsilon > 0$ choose $\beta_o \in A$ such that for $\beta_1, \beta_2 \geq \beta_o$ we have

$$\sum_{m=0}^{\infty} \left\|\frac{1}{m!} \hat{\delta}^m (f_{\beta_1} - f_{\beta_2})(0)\right\|_{\theta, K+\alpha_m B} \leq \epsilon .$$

Hence for any $M \in N$ and $\beta_1, \beta_2 \geq 0$ we have

$$\sum_{m=0}^{M} \left\|\frac{1}{m!} \hat{\delta}^m f_{\beta_1}(0) - \frac{1}{m!} \hat{\delta}^m f_{\beta_2}(0)\right\|_{\theta, K+\alpha_m B} \leq \epsilon .$$

Then

$$\sum_{m=0}^{M} \left\|\frac{1}{m!} P_m - \frac{1}{m!} \hat{\delta}^m f_{\beta_2}(0)\right\|_{\theta, K+\alpha_m B} \leq \epsilon .$$

In particular,

$$\sum_{m=0}^{M} \left\|\frac{1}{m!} P_m\right\|_{\theta, K+\alpha_m B} \leq \sum_{m=0}^{M} \left\|\frac{1}{m!} \hat{\delta}^m f_{\beta_o}(0)\right\|_{\theta, K+\alpha_m B} + \epsilon \leq$$

$$\leq \sum_{m=0}^{\infty} \left\|\frac{1}{m!} \hat{\delta}^m f_{\beta_o}(0)\right\|_{\theta, K+\alpha_m B} + \epsilon < \infty .$$

Thus

$$\sum_{m=0}^{\infty} \left\|\frac{1}{m!} P_m\right\|_{\theta, K+\alpha_m B} < \infty .$$

By Proposition (2.04) we have

$$f = \sum_{m=0}^{\infty} \frac{1}{m!} P_m \in H_{S\theta}(E).$$

Since

$$\sum_{m=0}^{M} \left\| \frac{1}{m!} P_m - \frac{1}{m!} \hat{\delta}^m f_{\beta_2}(0) \right\|_{\theta, K+\alpha_m B} \leq \varepsilon$$

for all $M \in \mathbb{N}$ and $\beta_2 \geq \beta_0$ we have

$$p(f-f_{\beta_2}) = \sum_{m=0}^{\infty} \left\| \frac{1}{m!} P_m - \frac{1}{m!} \hat{\delta}^m f(0) \right\|_{\theta, K+\alpha_m B}$$

for $\beta_2 \geq \beta_0$. Hence the net $(f_\alpha)_{\alpha \in A}$ converges to f on the topology $T_{S\theta}$.

3. THE BORNOLOGICAL SPACE ASSOCIATED WITH $(H_{S\theta}(E), T_{S\theta})$

In this paragraph we study two problems. The first one is the study of the bornological topology $t_{S\theta}$ associated with the space $(H_{S\theta}(E), T_{S\theta})$. We characterize the family of semi-norms which define the topology $t_{S\theta}$. The second problem is the relationship between the spaces $(H_{S\theta}(E), T_{S\theta})$ and $(\mathcal{H}_{S\theta}(E), \mathcal{T}_{\omega\theta})$.

<u>(3.01) Definition</u>. $t_{S\theta}$ is the finest locally convex topology on $H_{S\theta}(E)$ having the same bounded sets as $T_{S\theta}$.

<u>(3.02) Proposition</u>. If $f \in H_{S\theta}(E)$, then the Taylor series expansion of f, at zero, converges to f in $(H_{S\theta}(E), t_{S\theta})$.

<u>Proof</u>. Let $K \in \hat{K}_e(E)$, $B \in \mathcal{B}_E$ with K compact in E_B and $(\alpha_n)_0^\infty \in c_0^+$. We consider the semi-norm

$$p(f) = \sum_{m=0}^{\infty} \left\| \frac{1}{m!} \hat{\delta}^m f(0) \right\|_{\theta, K+\alpha_m B} \quad \text{for all } f \in H_{S\theta}(E).$$

For $0 < \varepsilon < 1/2$ we can choose $m_0 \in \mathbb{N}$ such that if $m \geq m_0$ we have

$$\left\| \frac{1}{m!} \hat{\delta}^m f(0) \right\|_{\theta, K+\alpha_m B} \leq \varepsilon^m.$$

For $k \geq m_0$ we then have

$$p(2^k \sum_{m=k}^{\infty} \|\frac{1}{m!} \hat{\delta}^m f(0)\|_{\theta, K+\alpha_m B}) \le 2^k (\sum_{m=k}^{\infty} \|\frac{1}{m!} \hat{\delta}^m f(0)\|_{\theta, K+\alpha_m B}) \le$$

$$\le 2^k \sum_{m=k}^{\infty} \varepsilon^m = \frac{2^k \varepsilon^{k+1}}{1-\varepsilon} = \frac{(2\varepsilon)^k}{1-\varepsilon} \varepsilon .$$

But $\frac{(2\varepsilon)^k}{1-\varepsilon}\varepsilon$ tends to zero as $k \to \infty$. Thus the sequence

$$(2^k \sum_{m=k}^{\infty} \frac{1}{m!} \hat{\delta}^m f(0))_{k=0}^{\infty}$$

is bounded in $(H_{S\theta}(E), T_{S\theta})$. If q is a semi-norm which is continuous on $(H_{S\theta}(E), t_{S\theta})$, there exists $M > 0$ such that

$$q(2^k \sum_{m=k}^{\infty} \hat{\delta}^m f(0)) \le M$$

for all $k \in \mathbb{N}$. Then

$$q(\sum_{m=k}^{\infty} \frac{1}{m!} \hat{\delta}^m f(0))$$

converges to zero as $k \to \infty$. Since

$$f = \sum_{m=0}^{k-1} \frac{1}{m!} \hat{\delta}^m f(0) + \sum_{m=k}^{\infty} \frac{1}{m!} \hat{\delta}^m f(0)$$

we have that

$$q(f - \sum_{m=0}^{k-1} \frac{1}{m!} \hat{\delta}^m f(0))$$

converges to zero as $k \to \infty$. This proves the proposition.

<u>(3.03) Proposition</u>. Let p be a continuous semi-norm in $(H_{S\theta}(E), t_{S\theta})$. Then $\lim_{m\to\infty} p(\frac{1}{m!} \hat{\delta}^m f(0))^{1/m} = 0$ for all $f \in H_{S\theta}(E)$.

<u>Proof</u>. Let $c > 0$ and $f \in H_{S\theta}(E)$. For all $m \in \mathbb{N}$, we get $\frac{c^m}{m!} \hat{\delta}^m f(0) \in \mathcal{P}_{b\theta}(^mE;)$. Let $K \in \hat{\mathcal{K}}_e(E)$, $B \in \mathcal{B}_E$ with K compact in E_B and $(\alpha_m)_0^{\infty} \in c_o^+$. By Proposition (2.04) we have

$$\lim_{m\to\infty} \|\frac{c^m}{m!} \hat{\delta}^m f(0)\|^{1/m}_{\theta, K+\alpha_m B} \le c \lim_{m\to\infty} \|\frac{1}{m!} \hat{\delta}^m f(0)\|^{1/m}_{\theta, K+\alpha_m B} = 0.$$

Hence the sequence $(\frac{c^m}{m!} \hat{\delta}^m f(0))_0^{\infty}$ is bounded in $(H_{S\theta}(E), T_{S\theta})$ and the sequence $(p(\frac{c^m}{m!} \hat{\delta}^m f(0)))_0^{\infty}$ is bounded in $\mathbb{R}$. Let $M > 0$ $p(\frac{c^m}{m!} \hat{\delta}^m f(0)) \le \frac{M}{c^m}$ for all $m \in \mathbb{N}$. We have $\lim_{m\to\infty} (\frac{1}{m!} \hat{\delta}^m f(0))^{1/m} \le \frac{1}{c}$. Since $c > 0$ is arbitrary we have that $\lim_{m\to\infty} p(\frac{1}{m} \hat{\delta}^m f(0))^{1/m} = 0$.

<u>(3.04) Corollary.</u> Let p be a continuous semi-norm in $(H_{S\theta}(E), t_{S\theta})$ and $f = \sum_{m=0}^{\infty} \frac{1}{m!} \hat{\delta}^m f(0) \in H_{S\theta}(E)$. Then

$$\sum_{m=0}^{\infty} p(\frac{1}{m!} \hat{\delta}^m f(0)) < \infty.$$

<u>(3.05) Proposition.</u> $(H_{S\theta}(E), T_{S\theta})$ induces on $\mathcal{P}_{b\theta}(^mE)$ the same topology generated by $\Gamma_{\theta,\mathfrak{B}}$ on $\mathcal{P}_{b\theta}(^mE)$.

<u>Proof.</u> Since, for all $B \in \mathfrak{B}_E$, $p(f) = \|\frac{1}{m!} \hat{\delta}^m f(0)\|_{\theta,B}$ is a continuous semi-norm on $(H_{S\theta}(E), T_{S\theta})$ we have that $T_{S\theta}$ induces on $\mathcal{P}_{b\theta}(^mE)$ a topology stronger than or equal to the topology generated by $\Gamma_{\theta,\mathfrak{B}}$. On other hand, if we take $K = \{0\}$, $(\alpha_m)_0^{\infty}$ definited by $\alpha_m = 1$ and $\alpha_n = 0$ for all $n \neq m$, we have

$$\sum_{m=0}^{\infty} \|P_m\|_{\theta, K+\alpha_m B} = \|P_m\|_{\theta,B} .$$

Then $T_{S\theta}$ induces on $\mathcal{P}_{b\theta}(^mE)$ a topology weaker than or equal to the topology defined by $\Gamma_{\theta,\mathfrak{B}}$.

<u>(3.06) Corollary.</u> The topology $t_{S\theta}$ induces on $\mathcal{P}_{b\theta}(^mE)$ a topology stronger than or equal to the topology defined by $\Gamma_{\theta,\mathfrak{B}}$.

<u>(3.07) Proposition.</u> Let p a semi-norm on $H_{S\theta}(E)$ with the following properties:

1) For each $m \in \mathbb{N}$, the restriction of p to $\mathcal{P}_{b\theta}(^mE)$ is a continuous semi-norm on the topology induced by $t_{S\theta}$ on $\mathcal{P}_{b\theta}(^mE)$.

2) If $f = \sum_{m=0}^{\infty} \frac{1}{m!} \hat{\delta}^m f(0) \in H_{S\theta}(E)$, then

$$\sum_{m=0}^{\infty} p(\frac{1}{m!} \hat{\delta}^m f(0)) < \infty.$$

Then $p_1(f) = \sum_{m=0}^{\infty} p(\frac{1}{m!} \hat{\delta}^m f(0))$ defines a continuous semi-norm on $(H_{S\theta}(E), t_{S\theta})$.

<u>Proof.</u> Since $t_{S\theta}$ is a bornological topology, to prove that p_1 is continuous semi-norm is suffices to show that for each bounded set $\mathfrak{X}$ of $(H_{S\theta}(E), t_{S\theta})$ we have $\sup\{p_1(f);\ f \in \mathfrak{X}\} < \infty$. By condition 1),

for each $m \in \mathbb{N}$, $\sup\{p(\frac{1}{m!}\hat{\delta}^m f(0));\ f \in \mathfrak{X}\} < \infty$. Now suppose $\sup\{p_1(f);\ f \in \mathfrak{X}\} = \infty$. Then, for each $m_o \in \mathbb{N}$,

$$\sup\{\sum_{m=m_o}^{\infty} p(\frac{1}{m!}\hat{\delta}^m f(0));\ f \in \mathfrak{X}\} = \infty.$$

Choose $f_1 \in \mathfrak{X}$ and $m_1 \in \mathbb{N}$ such that

$$\sum_{m=0}^{\infty} p(\frac{1}{m!}\hat{\delta}^m f_1(0)) \geq 2 \quad \text{and} \quad \sum_{m=0}^{m_1} p(\frac{1}{m!}\hat{\delta}^m f_1(0)) \geq 1.$$

By induction choose $f_k \in \mathfrak{X}$ and $m_k \in \mathbb{N}$ such that

$$\sum_{m=m_{k-1}+1} p(\frac{1}{m!}\hat{\delta}^m f_k(0)) \geq 2 \quad \text{and} \quad \sum_{m=m_{k-1}+1}^{m_k} p(\frac{1}{m!}\hat{\delta}^m f_k(0)) \geq 1.$$

Let

$$g_m = \begin{cases} f_1 & \text{for } 0 \leq m \leq m_1 \\ f_k & \text{for } m_{k-1} < m \leq m_k \quad (k \geq 2). \end{cases}$$

The sequence $(g_m)_0^\infty$ is bounded on $(H_{S\theta}(E), T_{S\theta})$.
Let $K \in \hat{\mathcal{K}}_e(E)$, $B \in \mathcal{B}_E$ with K a compact set of E_B, $(\alpha_n)_0^\infty \in C_o^+$ and $c > 0$. We have $cK \in \hat{\mathcal{K}}_e(E)$, cK a compact set of E_B. Them

$$\sup\{\sum_{m=0}^{\infty} \|\frac{1}{m!}\hat{\delta}^m g_n(0)\|_{\theta, cK+c\alpha_m B};\ n \in \mathbb{N}\} < \infty.$$

Since θ is a α-Silva holomorphy type we have

$$\sup\{\sum_{m=0}^{\infty} c^m \|\frac{1}{m!}\hat{\delta}^m g_n(0)\|_{\theta, K+\alpha_m B};\ n \in \mathbb{N}\} < \infty.$$

In particular

$$\sup\{c^m \|\frac{1}{m!}\hat{\delta}^m g_m(0)\|_{\theta, K+\alpha_m B};\ m \in \mathbb{N}\} < \infty.$$

Then

$$\lim_{m\to\infty} \|\frac{1}{m!}\hat{\delta}^m g_m(0)\|_{\theta, K+\alpha_m B}^{1/m} = 0.$$

Since $\hat{\delta}^m g_m(0) \in \mathcal{P}_{b\theta}(^mE)$, by Corollary (2.11) we have

$$g = \sum_{m=0}^{\infty} \frac{1}{m!}\hat{\delta}^m g_m(0) \in H_{S\theta}(E).$$

By definition of g_m we have

$$p_1(g) = \sum_{m=0}^{\infty} p(\frac{1}{m!}\hat{\delta}^m g_m(0)) = \infty$$

which contradicts the condition 2). Then $\sup\{p_1(f);\ f \in \mathfrak{X}\} < \infty$ and p_1 is a continuous semi-norm on $(H_{S\theta}(E), t_{S\theta})$.

(3.08) Proposition. The topology $t_{S\theta}$ on $H_{S\theta}(E)$ is generated by all semi-norm p on $H_{S\theta}(E)$ which satisfy the following conditions:

1) $p(f) = \sum_{m=0}^{\infty} p(\frac{1}{m!}\hat{\delta}^m f(0))$ for all $f \in H_{S\theta}(E)$.

2) For each $m \in \mathbb{N}$, the restriction of p to $\mathcal{P}_{b\theta}(^mE)$ is a continuous semi-norm on the topology induced by $t_{S\theta}$ on $\mathcal{P}_{b\theta}(^mE)$.

Proof. By Proposition (3.07) all such semi-norm are $t_{S\theta}$-continuous. Let q be a continuous semi-norm on $(H_{S\theta}(E), t_{S\theta})$. By Corollary (3.04) we have

$$\sum_{m=0}^{\infty} q(\frac{1}{m!}\hat{\delta}^m f(0)) < \infty \quad \text{for all } f \in H_{S\theta}(E).$$

Since q is $t_{S\theta}$-continuous, the restriction of q on $\mathcal{P}_{b\theta}(^mE)$ is continuous on $(\mathcal{P}_{b\theta}(^mE), t_{S\theta})$. Hence the conditions 1) and 2) of the Proposition (3.07) are satisfied and

$$p_1(f) = \sum_{m=0}^{\infty} q(\frac{1}{m!}\hat{\delta}^m f(0)), \qquad f \in H_{S\theta}(E)$$

defines a continuous semi-norm on $(H_{S\theta}(E), t_{S\theta})$. By Proposition (2.07) the Taylor series expansion at zero of $f \in H_{S\theta}(E)$ converges to f on the topology $T_{S\theta}$. Then $q(f) \leq p_1(f)$. Hence every continuous semi-norm on $(H_{S\theta}(E), t_{S\theta})$ is dominated by a continuous semi-norm which satisfies the conditions 1) and 2). This proves the proposition.

(3.09) Proposition. The spaces $(H_{S\theta}(E), t_{S\theta})$ and $(H_{S\theta}(E), T_{S\theta})$ induce the same topology on all bounded set of $(H_{S\theta}(E), T_{S\theta})$.

Proof. Since $t_{S\theta}$ is a topology stronger than or equal to $T_{S\theta}$ it suffices to show if $(f_\alpha)_{\alpha\in A}$ is a bounded net in $(H_{S\theta}(E), T_{S\theta})$ which converges to zero in $(H_{S\theta}(E), T_{S\theta})$, then $(f_\alpha)_{\alpha\in A}$ converges to zero in $(H_{S\theta}(E), t_{S\theta})$. Suppose this is not true and let p be a semi-norm on $(H_{S\theta}(E), t_{S\theta})$ of the form described in Proposition (3.08),

$(f'_{\alpha'})_{\alpha' \in A'}$ and $\varepsilon > 0$ such that

$$\sum_{m=0}^{\infty} p(\frac{1}{m!} \hat{\delta}^m f_{\alpha'}(0)) \geq \varepsilon \quad \text{for all} \quad \alpha' \in A'.$$

Since $\lim_{\alpha' \in A} f_{\alpha'} = 0$ in $(H_{S\theta}(E), T_{S\theta})$ we get that for each $n \in N$,

$$\lim_{\alpha' \in A'} p(\frac{1}{m!} \hat{\delta}^m f_{\alpha'}(0)) = 0.$$

For each $k \in \mathbb{N}$, choose $f_k \in \{f_{\alpha'}; \alpha' \in A'\}$ and $m_k \in \mathbb{N}$ such that

(i) $m_k \geq k$ and (ii) $\sum_{m_{k-1} < m \leq m_k} p(\frac{1}{m!} \hat{\delta}^m f_k(0)) \geq \varepsilon/2.$

We define

$$g_m = \begin{cases} f_1 & \text{if } 0 \leq m \leq m_1 \\ f_k & \text{if } m_{k-1} < m \leq m_k \quad (k \geq 2). \end{cases}$$

Then

$$g = \sum_{m=0}^{\infty} \frac{1}{m!} \hat{\delta}^m g_m(0) \in H_{S\theta}(E)$$

and

$$p(g) = \sum_{m=0}^{\infty} p(\frac{1}{m!} \hat{\delta}^m g_m(0)) = \infty$$

which is a contradiction. Hence we get the required result.

<u>(3.10) Corollary</u>. $(H_{S\theta}(E), t_{S\theta})$ and $(H_{S\theta}(E), T_{S\theta})$ have the same compact sets.

<u>Proof</u>. Since $t_{S\theta}$ is a topology stronger than or equal to the topology $T_{S\theta}$ the $t_{S\theta}$-compact sets are $T_{S\theta}$-compact sets. Let $K \subset H_{S\theta}(E)$ and suppose that K is a $T_{S\theta}$-compact set. Since K is a bound-set on $(H_{S\theta}(E), T_{S\theta})$ by Proposition (3.09) the spaces $(K, t_{S\theta})$ and $(K, T_{S\theta})$ are topologically equivalent. Hence we conclude that K is a $t_{S\theta}$-compact set.

<u>(3.11) Remark</u>. The spaces $(H_{S\theta}(E), t_{S\theta})$ and $(H_{S\theta}(E), T_{S\theta})$ have the same strict compact subsets since they have the same bounded sets.

<u>(3.12) Proposition</u>. Let θ be an α-Silva holomorphic type. If $P_m \in \mathcal{P}_{b\theta}(^mE)$ $m \in \mathbb{N}$ and if for each continuous semi-norm p on $(H_{S\theta}(E), t_{S\theta})$ we have

$$\sum_{m=0}^{\infty} p(P_m) < \infty, \quad \text{then} \quad \sum_{m=0}^{\infty} P_m \in H_{S\theta}(E).$$

Proof. By Proposition (2.08) the topology $T_{S\theta}$ of $H_{S\theta}(E)$ is generated by the semi-norms

$$p(f) = \sum_{n=0}^{\infty} \left\|\frac{1}{n!}\, \hat{\delta}^n f(0)\right\|_{\theta, K+\alpha_n B}$$

where $K \in \hat{K}_e(E)$, $B \in \mathcal{B}_E$ with K a compact set of E_B and $(\alpha_n)_0^{\infty} \in c_o^+$. Since $t_{S\theta} \geq T_{S\theta}$, p is a $t_{S\theta}$-continuous semi-norm. We have

$$\sum_{m=0}^{\infty} p(P_m) < \infty.$$

But, for each $m \in \mathbb{N}$

$$p(P_m) = \sum_{n=0}^{\infty} \left\|\frac{1}{n!}\, \hat{\delta}^n P_m(0)\right\|_{\theta, K+\alpha_n B} = \|P_m\|_{\theta, K+\alpha_m B}.$$

Then

$$\sum_{m=0}^{\infty} \|P_m\|_{\theta, K+\alpha_m B} < \infty$$

and, by Corollary (2.11) we get

$$\sum_{m=0}^{\infty} P_m \in H_{S\theta}(E).$$

(3.13) Proposition. $(H_{S\theta}(E), t_{S\theta})$ is a complete space if, and only if, for each $m \in \mathbb{N}$, $(P_{b\theta}(^mE), t_{S\theta})$ is a complete space.

Proof. Suppose that $(P_{b\theta}(^mE), t_{S\theta})$ is a complete space and let $(f_\alpha)_{\alpha \in A}$ be a Cauchy net in $(H_{S\theta}(E), t_{S\theta})$. Hence, for $m \in \mathbb{N}$, $(\hat{\delta}^m f_\alpha(0))_{\alpha \in A}$ is a Cauchy net in the complete space $(P_{b\theta}(^mE), t_{S\theta})$. Let $P_m = \lim_{\alpha \in A} \hat{\delta}^m f_\alpha(0)$. Given $\varepsilon > 0$ choose $\alpha_o \in A$ such that if $\alpha_1, \alpha_2 \geq \alpha_o$ then

$$\sum_{m=0}^{\infty} p\left(\frac{1}{m!}\, \hat{\delta}^m f_{\alpha_1}(0) - \frac{1}{m!}\, \hat{\delta}^m f_{\alpha_2}(0)\right) \leq \varepsilon.$$

In particular, we get

$$\sum_{m=0}^{k} p\left(\frac{1}{m!}\, \hat{\delta}^m f_{\alpha_1}(0) - \frac{1}{m!}\, \hat{\delta}^m f_{\alpha_2}(0)\right) \leq \varepsilon$$

for each $k \in \mathbb{N}$. Hence

$$\sum_{m=0}^{k} p\left(\frac{1}{m!} P_m - \frac{1}{m!} \hat{\delta}^m f_{\alpha_2}(0)\right) =$$

$$= \lim_{\alpha_1 \in A} \sum_{m=0}^{k} p\left(\frac{1}{m!} \hat{\delta}^m f_{\alpha_1}(0) - \frac{1}{m!} \hat{\delta}^m f_{\alpha_2}(0)\right) \leq \varepsilon$$

for $\alpha_2 \geq \alpha_o$.

Then
$$\sum_{m=0}^{\infty} p\left(\frac{1}{m!} P_m\right) \leq \sum_{m=0}^{\infty} p\left(\frac{1}{m!} \hat{\delta}^m f_{\alpha_o}(0)\right) + \varepsilon < \infty.$$

By Proposition (3.12) we have

$$f = \sum_{m=0}^{\infty} \frac{1}{m!} P_m \in H_{S\theta}(E).$$

Since
$$\sum_{m=0}^{\infty} p\left(\frac{1}{m!} P_m - \frac{1}{m!} \hat{\delta}^m f_{\alpha_2}(0)\right) \leq \varepsilon,$$

we have that

$$p(f-f_{\alpha_2}) = \sum_{m=0}^{\infty} p\left(\frac{1}{m!} \hat{\delta}^m f(0) - \frac{1}{m!} \hat{\delta}^m f_{\alpha_2}(0)\right)$$

for $\alpha_2 \geq \alpha_o$. Then $\lim_{\alpha \in A} f_\alpha = f$ in $(H_{S\theta}(E), t_{S\theta})$. Reciprocally, suppose that $(H_{S\theta}(E), t_{S\theta})$ is a complete space. As $(\mathcal{P}_{b\theta}(^mE), t_{S\theta})$, for each $m \in \mathbb{N}$, is a closed subspace of a complete space it is a complete subspace.

<u>(3.12) Proposition</u>. The space $(H_{S\theta}(E), t_{S\theta})$ is a quasi-complete space.

<u>Proof</u>. Suppose $\mathfrak{X} \subset H_{S\theta}(E)$ is a $t_{S\theta}$-closed and $T_{S\theta}$-bounded subset. Let $f \in \bar{\mathfrak{X}}^{T_{S\theta}}$ and $(f_\alpha)_{\alpha \in A}$ be a net in which converges to f in $(H_{S\theta}(E), T_{S\theta})$. By Proposition (3.09) $(f_\alpha)_{\alpha \in A}$ converges to f in $(H_{S\theta}(E), t_{S\theta})$. Then $\mathfrak{X}$ is $T_{S\theta}$-closed in the complete space $(H_{S\theta}(E), T_{S\theta})$. Hence $\mathfrak{X}$ is $T_{S\theta}$-complete and by Proposition (3.09) again we have that $\mathfrak{X}$ is $t_{S\theta}$-complete.

<u>(3.15) Corollary</u>. The space $(H_{S\theta}(E), t_{S\theta})$ is a barrelled space.

<u>Proof</u>. $(H_{S\theta}(E), t_{S\theta})$ is a bornological and quasi-complete space.

(3.16) Proposition. The topology $t_{S\theta}$ induces on $\mathcal{P}_{b\theta}(^mE)$ the bornological topology associated with the usual topology on $\mathcal{P}_{b\theta}(^mE)$.

Proof. Let $\mathcal{T}$ be the bornological topology associated at the usual topology on $\mathcal{P}_{b\theta}(^mE)$. If p_m is a semi-norm $\mathcal{T}$-continuous on $\mathcal{P}_{b\theta}(^mE)$, then $p(f) = p_m(\frac{1}{m!}\hat{\delta}^m f(0))$ defines a semi-norm on $H_{S\theta}(E)$ which is bounded on the bounded sets of $(H_{S\theta}(E),T_{S\theta})$. Hence, if $\mathfrak{X}$ is a bounded subset of $(H_{S\theta}(E),T_{S\theta})$, we have that $\{\frac{1}{m!}\hat{\delta}^m f(0); f \in \mathfrak{X}\}$ is a bounded set of $(\mathcal{P}_{b\theta}(^mE), \Gamma_{\theta,\mathcal{B}})$ and then p_m is bounded on this set. Hence p is $t_{S\theta}$-continuous on $H_{S\theta}(E)$ and p_m is continuous on the topology induced by $t_{S\theta}$ on $\mathcal{P}_{b\theta}(E)$. On the other hand if q is a $t_{S\theta}$-continuous semi-norm on $H_{S\theta}(E)$, then q is bounded on the bounded sets of $\mathcal{P}_{b\theta}(^mE)$. Hence the restriction of q on $\mathcal{P}_{b\theta}(^mE)$ is $\mathcal{T}$-continuous. This completes the proof.

(3.17) Remark. There exist spaces which are metrizable but are not distinguished (see [3], p.435). In those spaces E' (which is equal to E^*) with the topology $\beta(E',E)$ is complete but it is not a bornological space. Then we have a example where $\mathcal{P}_b(^1E)$ with the usual topology is not bornological. We conclude, by a previous proposition that there exist spaces such that $t_{S\theta}$ do not induce on $\mathcal{P}_{b\theta}(^mE)$ the usual topology.

(3.18) Proposition. Let θ be an α-Silva holomorphy type. Then $(H_{S\theta}(E),T_{S\theta}) \subset (\mathcal{H}_{S\theta}(E), \mathcal{T}_{\omega\theta})$ continuously.

Proof. Since $H_{S\theta}(E) \subset \mathcal{H}_S(E)$, to prove that $H_{S\theta}(E) \subset \mathcal{H}_{S\theta}(E)$ it suffices to show that if $f \in H_{S\theta}(E)$ it is of Silva holomorphy type θ at each point x of E. Let $x \in E$, $\varepsilon > 0$ and $B \in \mathcal{B}_E$. We denote by X_ε the convex closed hull of the set $(1+\varepsilon)\,\sigma_o\{x\}$. Let $B_o \in \mathcal{B}_E$ such that $B_o \supset B$ and $B_o \supset X$. Take $\rho > 0$ such that

$$\sum_{m=0}^{\infty} \left\|\frac{1}{m!}\hat{\delta}^m f(0)\right\|_{\theta,X_\varepsilon+\rho B_o} < \infty.$$

Then

$$\limsup_{m\to\infty} \left\|\frac{1}{m!}\hat{\delta}^m f(0)\right\|^{1/m}_{\theta, X_\varepsilon + \rho B_o} \le 1$$

and then there exists $c > 0$ such that

$$\left\|\frac{1}{m!}\hat{\delta}^m f(0)\right\|_{\theta, X_\varepsilon + \rho B_o} \le c(1 + \varepsilon/2)^m$$

for each $m \in \mathbb{N}$. For all $k \in \mathbb{N}$ we have

$$\sum_{m=k}^{\infty} \left\|\hat{\delta}^k\left(\frac{1}{m!}\hat{\delta}^m f(0)\right)(x)\right\|_{\theta, X_\varepsilon + \rho B_o} \le$$

$$\le k! \sum_{m=k}^{\infty} \sigma_\theta^m \left\|\frac{1}{m!}\hat{\delta}^m f(0)\right\|_{\theta, X_\varepsilon + \rho B_o} \|x\|^{m-k}_{X_\varepsilon + \rho B_o} \le$$

$$\le k! \sum_{m=k}^{\infty} \sigma_\theta\, c(1 + \varepsilon/2)^m \|x\|^{m-k}_{X_\varepsilon + \rho B_o} =$$

$$= k!\, c\left[\sum_{m=k}^{\infty} \frac{(1+\varepsilon/2)^{m-k}}{(1+\varepsilon)^{m-k}}\right](1+\varepsilon/2)^k < \infty .$$

As $(1+\varepsilon)\sigma_\theta x \in X_\varepsilon + \rho B_o$ we have $\|x\|_{X_\varepsilon + \rho B_o} \le 1/(1+\varepsilon)\sigma_\theta$.

Since $\rho B_o \subset X_\varepsilon + \rho B_o$, we have now

$$\sum_{m=k}^{\infty} \left\|\hat{\delta}^k\left(\frac{1}{m!}\hat{\delta}^m f(0)\right)(x)\right\|_{\theta, \rho B_o} \le$$

$$\le \sum_{m=k}^{\infty} \left\|\hat{\delta}^k\left(\frac{1}{m!}\hat{\delta}^m f(0)(x)\right\|_{\theta, X_\varepsilon + \rho B_o} < \infty .$$

Hence

$$\sum_{m=k}^{\infty} \left\|\hat{\delta}^k\left(\frac{1}{m!}\hat{\delta}^m f(0)\right)(x)\right\|_{\theta, \rho B_o} =$$

$$= \rho^k \sum_{m=k}^{\infty} \left\|\hat{\delta}^k\left(\frac{1}{m!}\hat{\delta}^m f(0)\right)(x)\right\|_{\theta, B_o} < \infty .$$

Then

$$\sum_{m=k}^{\infty} \left\|\hat{\delta}^k\left(\frac{1}{m!}\hat{\delta}^m f(0)\right)(x)\right\|_{\theta, B_o} < \infty .$$

Since B is arbitrary we have that the serie

$$\sum_{m=k}^{\infty} \hat{\delta}^k\left(\frac{1}{m!}\hat{\delta}^m f(0)\right)(x)$$

converges absolutely in $\mathcal{P}_{b\theta}(^k E)$. Since this space is complete the limit exists. On the other hand (see [4])

$$\hat{\delta}^k f(x)(t) = \sum_{m=k}^{\infty} \hat{\delta}^k(\frac{1}{m!}\hat{\delta}^m f(0))(x)(t) \quad \text{for} \quad t \in E.$$

Furthermore, by the above results, we have

$$\|\frac{1}{k!}\hat{\delta}^k f(x)\|_{\theta,B} \le (1/\rho^k)\|\frac{1}{k!}\hat{\delta}^k f(x)\|_{\theta,\rho B_o} \le$$

$$\le (1/\rho^k)\sum_{m=k}^{\infty}\|\frac{1}{k!}\hat{\delta}^k(\frac{1}{m!}\hat{\delta}^m f(0))(x)\|_{\theta,\rho B_o} \le$$

$$\le (1/\rho^k)\sum_{m=k}^{\infty}\sigma_\theta^m c(1+\epsilon/2)^m (1/(1+\epsilon)\sigma_\theta)^{m-k} =$$

$$= (c/\rho^k)(1+\epsilon/2)^k \sum_{m=k}^{\infty}(\frac{1+\epsilon/2}{1+\epsilon})^{m-k} =$$

$$= 2c(1+\epsilon)(\frac{1+\epsilon/2}{\epsilon})^k \quad \text{for all} \quad k \in \mathbb{N}.$$

Therefore f is of type θ at x.

To prove that the inclusion is continuous it suffices to show that all semi-norm on $(\mathcal{H}_{S\theta}(E), \tau_{\omega\theta})$ which is ported by $K \in \hat{\mathcal{K}}_e(E)$ and $B \in \mathcal{B}_E$ also is $S\theta$-ported. Given a semi-norm p on $\mathcal{H}_{S\theta}(E)$ ported by $K \in \hat{\mathcal{K}}_e(E)$ and $B \in \mathcal{B}_E$ with K a compact subset of E_B and $\epsilon > 0$, let $c(\epsilon) > 0$ such that

$$p(f) \le c(\epsilon)\sum_{m=0}^{\infty}\epsilon^m \sup_{x\in K}\|\frac{1}{m!}\hat{\delta}^m f(x)\|_{\theta,B}$$

for all $f \in H_{S\theta}(E) \subset \mathcal{H}_{S\theta}(E)$. Let $V = 2\sigma_\theta K + 2\sigma_\theta \epsilon B$ and $\rho = \sup\{\|x\|_{2K+2\epsilon B};\ x \in K\}$. With this notations we have that if $x \in K$, then $x \in 1/2(2K+2\epsilon B)$ and $\rho \le 1/2$. We will prove

$$p(f) \le c(\epsilon)\sum_{m=0}^{\infty}\|\frac{1}{m!}\hat{\delta}^m f(0)\|_{\theta,2K+2\epsilon B}$$

for all $f \in H_{S\theta}(E)$. If

$$\sum_{m=0}^{\infty}\|\frac{1}{m!}\hat{\delta}^m f(0)\|_{\theta,2K+2\epsilon B} = \infty,$$

we have nothing to prove. Suppose that this is not true. In this case we have (see [4])

$$\|\frac{1}{k!}\hat{\delta}^k f(x)\|_{\theta,2K+2\epsilon B} \le \sum_{m=0}^{\infty}\sigma_\theta^m\|\frac{1}{m!}\hat{\delta}^m f(0)\|_{\theta,2K+2\epsilon B}\|x\|_{2K+2\epsilon B}^{m-k} \le$$

$$\le \sum_{m=0}^{\infty}\|\frac{1}{m!}\hat{\delta}^m f(0)\|_{\theta,V}\ \rho^{m-k}.$$

Then

$$\sum_{m=0}^{\infty} \epsilon^m \sup_{x\in K} \|\frac{1}{m!} \hat{\delta}^m f(x)\|_{\theta,B} \le$$

$$\le \sum_{m=0}^{\infty} (1/2)^m \sup_{x\in K} \|\frac{1}{m!} \hat{\delta}^m f(x)\|_{\theta,2\epsilon B} <$$

$$\le \sum_{m=0}^{\infty} (1/2)^m \sup_{x\in K} \|\frac{1}{m!} \hat{\delta}^m f(0)\|_{\theta,2K+2\epsilon B} \le$$

$$\le \sum_{k=0}^{\infty} (1/2)^k \sum_{m=k}^{\infty} \|\frac{1}{m!} \hat{\delta}^m f(0)\|_{\theta,V}\, \rho^{m-k} =$$

$$= \sum_{m=0}^{\infty} \|\frac{1}{m!} \hat{\delta}^m f(0)\|_{\theta,V}\, \rho^m \sum_{k=0}^{m} (1/2\rho)^k =$$

$$= \sum_{m=0}^{\infty} \frac{(1/2)^m - 2\rho^{m+1}}{1-2\rho} \|\frac{1}{m!} \hat{\delta}^m f(0)\|_{\theta,V} .$$

Since

$$c\cdot\sup \{\frac{(1/2)^m - 2\rho^{m+1}}{1-2\rho} ; m \in \mathbb{N}\} < \infty$$

we have

$$p(f) \le c(\epsilon) c \sum_{m=0}^{\infty} \|\frac{1}{m!} \hat{\delta}^m f(0)\|_{\theta,V} .$$

This completes the proof.

<u>(3.19) Proposition.</u> $(H_S(E), \tau_{\omega se}) = (\mathcal{H}_S(E), \tau_{\omega S})$.

<u>Proof.</u> By Corollary (2.11) we have $H_S(E) \subset \mathcal{H}_S(E)$. By Proposition (3.18) we have $(H_S(E), \tau_{\omega se}) \subset (\mathcal{H}_S(E), \tau_{\omega S})$ continuously. We have to prove that all semi-norm which is $\tau_{\omega se}$-continuous is also $\tau_{\omega S}$-continuous. Let $K \in \hat{K}_e(E)$, $B \in \mathcal{B}_E$ with K a compact set of E_B and $(\alpha_n)_0^\infty \in c_0^+$. We define

$$p(f) = \sum_{m=0}^{\infty} \|\frac{1}{m!} \hat{\delta}^m f(0)\|_{K+\alpha_m B}$$

for all $f \in \mathcal{H}_S(E)$. Let V a neighborhood of K in E_B and choose $\epsilon > 0$ such that $(\frac{1+\epsilon}{1+(1/2)\epsilon})K \; (1/2)\epsilon B \subset (\frac{1+\epsilon}{1+(1/2)\epsilon})K + \epsilon B \subset V$. Choose $m_o \in N$ such that $\alpha_m \le (1/2)$ for all $m \in \mathbb{N}$. We have $K + \alpha_m B \subset \frac{1+\epsilon}{1+(1/2)\epsilon} K + \alpha_m B \subset V$. Then for $\rho > 0$ and $m \ge m_o$

$$\left\|\frac{1}{m!}\hat{\delta}^m f(0)\right\|_{K+\alpha_m B} =$$

$$= \sup \{ |\frac{1}{m!}\hat{\delta}^m f(0)\cdot x| ; x \in K + \alpha_m B\} =$$

$$= \sup \{ |\frac{1}{2\pi i}\int_{|\lambda|=\rho} \frac{f(\lambda x)}{\lambda^{m+1}} d\lambda| ; x \in K + \alpha_m B\} \leq$$

$$\leq \frac{1}{\rho^m} \sup \{ |f(x)| ; x \in \rho K + \rho\alpha_m B\}.$$

If we choose $\rho > 0$ such that $1 < \rho < \frac{1+\varepsilon}{1+\frac{1}{2}\varepsilon}$ we have:

1) for $m \geq m_o$

$$\left\|\frac{1}{m!}\hat{\delta}^m f(0)\right\|_{K+\alpha_m B} \leq \frac{1}{\rho^m} \sup \{ |f(x)| ; x \in V\}$$

2) for $m = 0,1,\dots,m_o-1$,

$$\left\|\frac{1}{m!}\hat{\delta}^m f(0)\right\|_{K+\alpha_m B} \leq \frac{1}{\rho^m} \sup \{ |f(x)| ; x \in \rho_1 K + \rho_1\alpha_m B\}$$

where ρ_1 is such that $\rho_1(K+\alpha_m B) \subset V$ for $m = 0,1,\dots,m_o-1$.
Then if we take

$$c(V) = \sum_{i=0}^{m_o-1} \rho_1^{-i} + \sum_{i=m_o}^{\infty} \rho^{-i}$$

we have

$$p(f) = \sum_{m=0}^{\infty} \left\|\frac{1}{m!}\hat{\delta}^m f(0)\right\|_{K+\alpha_m B} \leq c(V) \sup \{ |f(x)| ; x \in V\}$$

for all $f \in \mathcal{H}_S(E)$.

4. α-β-SILVA HOLOMORPHY TYPES AND BOREL TRANSFORMS

In this paragraph we study the Borel transform of an element of $(H_{S\theta}(E), t_{S\theta})'$. For this we introduce the α-β-Silva holomorphy type. The main result is the characterization of the space $(H_{S\theta}(E), T_{S\theta})'$ using the Borel transform.

<u>(4.01) Definition</u>. An α-Silva holomorphy type θ is said to be an α-β-Silva holomorphy type if it satisfies the following conditions:

1) For all $m \in \mathbb{N}$, $\mathcal{P}_{b\theta}(^mE) \supset \mathcal{P}_{bN}(^mE)$ and for each $B \in \mathcal{B}_E$ $\|P_m\|_{\theta,B} \leq \|P_m\|_{N,B}$ for all $P_m \in \mathcal{P}_{bN}(^mE)$.

2) For all $m \in \mathbb{N}$, $\mathcal{P}_{bf}(^mE)$ is dense in $(\mathcal{P}_{b\theta}(^mE), \Gamma_{\theta,\mathcal{B}})$.

Remark. By condition 1) of Definition (4.01) we conclude that $(H_{SN}(E),T_{SN}) \subset (H_{S\theta}(E),T_{S\theta})$ continuously for all α-β-Silva holomorphy type θ.

(4.02) Definition. A function $f: E \to C$ is said to be an exponential function if there exists $\varphi \in E^*$ such that $f(x) = \exp(\varphi(x))$ for all $x \in E$. We denote by W the vector space spanned by all the exponential functions. We note that $W \subset H_{SN}(E) \subset H_{S\theta}(E)$ for all α-β-Silva holomorphy type.

(4.03) Lemma. If θ is an α-β-Silva holomorphy type, then the closure of W in $(H_{S\theta}(E),t_{S\theta})$ is $H_{S\theta}(E)$. Hence the closure of W in $(H_{S\theta}(E),T_{S\theta})$ is also $H_{S\theta}(E)$.

Proof. By Propositions (2.07) and (3.02) the Taylor series of f at zero converges to f for the topologies $T_{S\theta}$ and $t_{S\theta}$. By condition 1) of the Definition (4.01), since $\hat{\delta}^m[\exp(\varphi)](0) = \varphi^m$ for all $\varphi \in E^*$. We get the proof if we prove that φ^m belongs to the closure of W in the topology $t_{S\theta}$. We will prove by induction in $m \in \mathbb{N}$. Let p be a $t_{S\theta}$-continuous semi-norm of the form

$$p(f) = \sum_{m=0}^{\infty} p(\frac{1}{m!}\hat{\delta}^m f(0)).$$

If $\lambda \in C$ with $\lambda \neq 0$ we have

$$\exp(\lambda\varphi) = 1 + \lambda\varphi + \sum_{m=2}^{\infty} \frac{1}{m!}\lambda^m\varphi^m$$

and then

$$\frac{\exp(\lambda\varphi)-1}{\lambda} - \varphi = \sum_{m=2}^{\infty} \frac{1}{m!}\lambda^{m-1}\varphi^m$$

and

$$p(\frac{\exp(\lambda\varphi)-1}{\lambda} - \varphi) = \sum_{m=2}^{\infty} |\lambda|^{m-1} p(\frac{1}{m!}\varphi^m) \leq |\lambda| \sum_{m=2}^{\infty} |\lambda|^{m-2} p(\frac{1}{m!}\varphi^m).$$

Since $\exp(\varphi) \in H_{S\theta}(E)$ we have

$$\sum_{m=0}^{\infty} p(\frac{1}{m!}\varphi^m) = p(\exp(\varphi)) < \infty .$$

Then $p(\frac{\exp(\lambda\varphi)-1}{\lambda} - \varphi) \to 0$ with $\lambda \to 0$, and hence $\varphi \in \bar{W}^{t_{S\theta}}$.

Suppose that $\varphi^n \in \bar{W}^{t_{S\theta}}$ for $n \le k$.

$$\exp(\lambda\varphi) - \sum_{i=0}^{k} \frac{1}{i!}\lambda^i\varphi^i = \frac{1}{(k+1)}\lambda^{k+1}\varphi^{k+1} + \sum_{i=k+2}^{\infty} \frac{1}{i!}\lambda^i\varphi^i .$$

Then

$$p\left[\frac{\exp(\lambda\varphi) - \sum_{i=0}^{k}\frac{1}{i!}\lambda^i\varphi^i}{\lambda^{k+1}} - \varphi^{k+1}\right] \le \sum_{i=k+2}^{\infty} \frac{|\lambda|^{i-k-2}}{i!} p(\varphi^i).$$

Then we have $\varphi^{k+1} \in \bar{W}^{t_{S\theta}}$, this completes the proof.

(4.04) Lemma. The mapping $\beta\colon (\mathcal{P}_{bN}(^mE), \Gamma_{N;\mathcal{B}}) \to \mathcal{P}(^mE^*)$ defined by $(\beta(T))(\varphi) = T(\varphi^m)$ for all $\varphi \in E^*$ is a vector space isomorphism and

$$\sup\{\frac{|(\beta(T))(\varphi)|}{\|\varphi\|_B^m} ; \|\varphi\|_B \neq 0\} = \sup\{\frac{|T(P)|}{\|P\|_{N,B}} ; \|P\|_{N,B} \neq 0\}$$

for all $B \in \mathcal{B}_E$.

Proof. (See [4]).

We denote by $\|\beta(T)\|_B$ and $\|T\|_{N,B}$, respectively, the two members of expression in last lemma.

If θ is an α-β-Silva holomorphy type, then by condition 1) in definition (4.01) we have that

$$(\mathcal{P}_{bN}(^mE), \Gamma_{N,\mathcal{B}})' \supset (\mathcal{P}_{b\theta}(^mE), \Gamma_{\theta,\mathcal{B}})'$$

for all $m \in \mathbb{N}$. By Lemma (4.04) we can regard $(\mathcal{P}_{b\theta}(^mE), \Gamma_{\theta,\mathcal{B}})$ as a subset of $\mathcal{P}(^mE^*)$. We define, for each $m \in \mathbb{N}$,

$$\wedge\colon (\mathcal{P}_{b\theta}(^mE), \Gamma_{\theta,\mathcal{B}})' \to \mathcal{P}(^mE^*)$$

by $\hat{T}(\varphi) = \frac{1}{m!} T(\varphi^m)$ for all $\varphi \in E^*$. We denote by $\mathcal{P}_{b\theta^*}(^mE)$ the image of $(\mathcal{P}_{b\theta}(^mE), \Gamma_{\theta,\mathcal{B}})'$ under $\wedge$.

For $T_m \in (\mathcal{P}_{b\theta}(^mE), \Gamma_{\theta,\mathcal{B}})'$ and for $B \in \mathcal{B}_E$ we define

$\|\hat{T}\|^{\theta^*,B} = \frac{1}{m!} \|T\|_{\theta',B}$.

<u>(4.05) Definition</u>. The Borel transform of $T \in (H_{S\theta}(E), t_{S\theta})'$ is the mapping $T: E^* \to \mathbb{C}$ defined by $\hat{T}(\varphi) = T(\exp(\varphi))$.

The topology $t_{S\theta}$ induces on $\mathcal{P}_{b\theta}(^mE)$ a topology stronger than or equal to the usual topology $\mathcal{P}_{b\theta}(^mE)$, we have that if $T \in (\mathcal{P}_{b\theta}(^mE), \Gamma_{\theta,\mathcal{B}})'$, then $T \in (\mathcal{P}_{b\theta}(^mE), t_{S\theta})'$ and we can extend T to a continuous linear functional on $(H_{S\theta}(E), t_{S\theta})$ by the formula $T^*(f) = T(\frac{1}{m!}\hat{\delta}^m f(0))$ for all $f \in H_{S\theta}(E)$. Since, for all $m \in \mathbb{N}$ and all $\varphi \in E^*$, $\hat{\delta}^m[\exp(\varphi)](0) = \varphi^m$, we have $\hat{T}^*(\varphi) = T^*(\exp(\varphi)) = T(\frac{1}{m!}\varphi^m) = \hat{T}(\varphi)$. We conclude that the Borel transform is an extension of the mapping $\wedge$ defined previously.

<u>(4.06) Definition</u>. A function $f \in \mathcal{H}_S(E)$ is said to be a S-holomorphic nuclear function of the bounded type if it satisfies the following conditions:

1) For all $m \in \mathbb{N}$, $\hat{\delta}^m f(0) \in \mathcal{P}_{bN}(^mE)$.

2) For all $B \in \mathcal{B}_E$ $\lim\limits_{m\to\infty} (\frac{1}{m!}\|\hat{\delta}^m f(0)\|_{N,B})^{1/m} = 0$.

We denote by $\mathcal{H}_{SNb}(E)$ the vector space of all these functions. (See [4]).

<u>(4.07) Definition</u>. On $\mathcal{H}_{SNb}(E)$ we consider the following family of semi-norms:

$$\|f\|_{N,B,\rho} = \sum_{m=0}^{\infty} \frac{\rho^m}{m!}\|\hat{\delta}^m f(0)\|_{N,B}$$

for all $\rho > 0$ and all $B \in \mathcal{B}_E$. The topology on $\mathcal{H}_{SNb}(E)$ defined by this semi-norms will be denoted by $\mathcal{T}_{SNb}$. (See [4]).

<u>(4.08) Proposition</u>. $(\mathcal{H}_{SNb}(E), \mathcal{T}_{SNb}) \subset (H_{SN}(E), T_{SN})$ continuously.

<u>Proof</u>. By definition $\mathcal{H}_{SNb}(E) \subset H_{SN}(E)$. Let p a semi-norm on $H_{SN}(E)$ which is T_{SN}-continuous. Then, for all $f \in H_{SN}(E)$, where $K \in \hat{\mathcal{K}}_e(E)$, $B \in \mathcal{B}_E$ with K a compact set of E_B and $(\alpha_n)_0^\infty \in c_o^+$.

Since $K \subset E_B$, there exists $\gamma > 0$ such that $K \subset \gamma B$. Since $\lim_{m\to\infty} \alpha_m = 0$, there exists $\delta > 0$ such that $\alpha_m \leq \delta$ for all $m \in \mathbb{N}$. Then if $\rho = \gamma + \delta$ we have $K + \alpha_m B \subset \rho B$ and for all $m \in \mathbb{N}$,

$$P(f) \leq \sum_{m=0}^{\infty} \frac{1}{m!} \| \hat{\delta}^m f(0) \|_{N,\rho B} = \sum_{m=0}^{\infty} \frac{\rho^m}{m!} \| \hat{\delta}^m f(0) \|_{N,B} .$$

This completes the proof.

<u>(4.09) Proposition.</u> Let θ be an α-β-Silva holomorphy type. Then $(\mathcal{H}_{SNb}(E), \mathcal{T}_{SNb}) \subset (H_{S\theta}(E), T_{S\theta})$ continuously.

<u>Proof.</u> By condition 1) of the Definition (4.01), for all $m \in N$, $\mathcal{P}_{bN}(^mE) \subset \mathcal{P}_{b\theta}(^mE)$ continuously. Hence $(H_{SN}(E), T_{SN}) \subset (H_{S\theta}(E), T_{S\theta})$ continuously. The Proposition (4.08) completes the proof.

<u>(4.10) Definition.</u> A function $f \in \mathcal{H}(E^*)$ is said to be of exponential type if there exist $c > 0$ and $B \in \mathcal{B}_E$ such that

$$|f(\varphi)| \leq c \exp(\|\varphi\|_B) \quad \text{for all} \quad \varphi \in E^*.$$

We denote by $\mathrm{Exp}(E^*)$ the algebra of the exponential functions under the usual operation of vector space and pointwise multiplication.

<u>(4.11) Proposition.</u> The mapping

$$\wedge : (\mathcal{H}_{SNb}(E), \mathcal{T}_{SNb})' \to \mathrm{Exp}(E^*)$$

defined by $\hat{T}(\varphi) = T(\exp(\varphi))$, for all $\varphi \in E^*$, is a algebraic isomorphism.

<u>Proof.</u> (See [4]).

<u>(4.12) Proposition.</u> Let θ be an α-β-Silva holomorphy type. If $T \in (H_{S\theta}(E), T_{S\theta})'$, then $\hat{T} \in \exp(E^*)$ and for all $m \in \mathbb{N}$, $\hat{\delta}^m T(0) \in \mathcal{P}_{b\theta^*}(^mE^*)$.

<u>Proof.</u> The first part of the result is a consequence of the previous proposition. Since $\hat{T} \in \mathcal{H}(E^*)$, we have $\hat{\delta}^m T(0) \in \mathcal{P}(^mE^*)$. If we denote by T_m the restriction of T to the space $\mathcal{P}_{b\theta}(^mE)$ by Proposition (3.05) we have

$$T_m \in (\mathcal{P}_{b\theta}(^mE), T_{S\theta})' \subset (\mathcal{P}_{b\theta}(^mE), \tau_{\theta,\mathcal{B}})'.$$

Then $\hat{T}_m \in \mathcal{P}_{b\theta^*}(^mE^*) \subset \mathcal{P}(^mE^*)$ and $\hat{\delta}^m T_m(0) = m!\, \hat{T}_m \in \mathcal{P}_{b\theta^*}(^mE^*)$.

<u>(4.13) Definition.</u> Let $F \in \mathcal{H}(E^*)$ such that for each $m \in \mathbb{N}$, $F_m = \hat{\delta}^m F(0) \in \mathcal{P}_{b\theta^*}(E^*)$. If there exist $K \in \hat{\mathcal{K}}_e(E)$ and $B \in \mathcal{B}_E$ with K a compact subset of E_B and such that for all $\varepsilon > 0$ $\limsup_{m\to\infty} (\|F_m\|^{\theta^*, K+\varepsilon B})^{1/m} \leq 1$, we say F is of θ^*-compact exponential type in E.

<u>(4.14) Proposition.</u> There exists a one to one correspondence between the elements of θ^*-compact exponential type in E and the elements of $(H_{S\theta}(E), T_{S\theta})'$.

<u>Proof.</u> Let $T \in (H_{S\theta}(E), T_{S\theta})'$ and p be a semi-norm on $H_{S\theta}(E)$ which is $S\theta$-ported by $K \in \hat{\mathcal{K}}_e(E)$ and $B \in \mathcal{B}_E$. Suppose that $|T(f)| \leq p(f)$ for all $f \in H_{S\theta}(E)$. Then for all $\varepsilon > 0$ there exists $c(\varepsilon) > 0$ such that

$$|T(f)| \leq p(f) \leq c(\varepsilon) \sum_{m=0}^{\infty} \|\frac{1}{m!} \hat{\delta}^m f(0)\|_{\theta, K+\varepsilon B}$$

for all $f \in H_{S\theta}(E)$. For each $m \in \mathbb{N}$, let T_m be the restriction of T to $\mathcal{P}_{b\theta}(^mE)$. Then, if $P_m \in \mathcal{P}_{b\theta}(^mE)$

$$|T(P_m)| = |T_m(P_m)| \leq c(\varepsilon) \sum_{m=0}^{\infty} \|\frac{1}{m!} \hat{\delta}^m P_m(0)\|_{\theta, K+\varepsilon B} =$$

$$= c(\varepsilon) \|P_m\|_{\theta', K+\varepsilon B}.$$

Then

$$\|T_m\|_{\theta', K+\varepsilon B} = \sup \{\frac{|T_m(P_m)|}{\|P_m\|_{\theta, K+\varepsilon B}};\ \|P_m\|_{\theta, K+\varepsilon B} \neq 0\} \leq c(\varepsilon).$$

Let $F = \hat{T}$. We have $F_m = \hat{\delta}^m F(0) = \hat{\delta}^m T(0) = (m!)\hat{T}_m$.

Hence

$$\limsup_{m\to\infty} (\|F_m\|^{\theta^*, K+\varepsilon B})^{1/m} = \limsup_{m\to\infty} (m!\|T\|^{\theta^*, K+\varepsilon B})^{1/m} =$$

$$= \limsup_{m\to\infty} (\frac{m!}{m!} \|T_m\|_{\theta', K+\varepsilon B})^{1/m} \leq \limsup_{m\to\infty} c(\varepsilon)^{1/m} = 1.$$

Then F is of θ^*-compact exponential type in E.

Conversely, assume that $F \in (E^*)$, with $F_m = \hat{\delta}^m F(0)$ in $\mathcal{P}_{b\theta^* b}(^m E^*)$ and that there exist $K \in \hat{\mathcal{K}}_e(E)$ and $B \in \mathcal{B}_E$ with K a compact subset of E_B and such that for all $\epsilon > 0$ we have

$$\lim_{m\to\infty} \sup(\|F_m\|^{\theta^*, K+\epsilon B})^{1/m} \leq 1.$$

We define $T: H_{S\theta}(E) \to C$ by $T(f) = \sum_{m=0}^{\infty} \frac{1}{m!} T_m(\frac{1}{m!} \hat{\delta}^m f(0))$, where $T_m \in \mathcal{P}_{b\theta}(^m E)$ is such that $\hat{T}_m = F_m$. If $\epsilon > 0$ is given choose $c(\epsilon) > 0$ such that $\|F_m\|^{\theta^*, K+\epsilon B} \leq 2^m c(\epsilon)$ for all $m \in \mathbb{N}$. We have

$$|T(f)| \leq \sum_{m=0}^{\infty} |\frac{1}{m!} T_m(\frac{1}{m!} \hat{\delta}^m f(0))| \leq$$

$$\leq \sum_{m=0}^{\infty} \|\frac{1}{m!} T_m\|_{\theta', K+\epsilon B} \|\frac{1}{m!} \hat{\delta}^m f(0)\|_{\theta, K+\epsilon B} =$$

$$= \sum_{m=0}^{\infty} \|F_m\|^{\theta^*, K+\epsilon B} \|\frac{1}{m!} \hat{\delta}^m f(0)\|_{\theta, K+\epsilon B} \leq$$

$$\leq c(\epsilon) \sum_{m=0}^{\infty} \|\frac{1}{m!} \hat{\delta}^m f(0)\|_{\theta, 2K+2\epsilon B} .$$

Then $T \in (H_{S\theta}(E), T_{S\theta})'$.

Now

$$\hat{T}(\varphi) = T(\exp(\varphi)) = \sum_{m=0}^{\infty} \frac{1}{m!} T_m(\frac{1}{m!} \varphi^m) =$$

$$= \sum_{m=0}^{\infty} \frac{1}{m!} \hat{T}_m(\varphi) = \sum_{m=0}^{\infty} \frac{1}{m!} F_m(\varphi) = F(\varphi).$$

Then $\hat{T} = F$.

(4.15) Example. The nuclear Silva holomorphy type is an α-β-Silva holomorphy type. Then $f \in \mathcal{H}(E^*)$ is a N^*-compact exponential type if and only if there exist $K \in \hat{\mathcal{K}}_e(E)$ and $B \in \mathcal{B}_E$ with K a compac subset of E_B and such that for all $\epsilon > 0$,

$$\lim_{m\to\infty} \sup(\|\hat{\delta}^m f(0)\|^{N^*, K+\epsilon B})^{1/m} \leq 1.$$

5. α-β-γ-SILVA HOLOMORPHY TYPES AND PARTIAL DIFFERENTIAL OPERATOR

In this paragraph we introduce the α-β-γ-Silva holomorphy type, formal power series and partial differential operators.

For α-β-γ-Silva holomorphy types we show that the partial differential operator on formal power series are onto maps and the solutions can be approximated by exponential polynomial solutions.

<u>(5.01) Definition</u>. For each $T_m \in (\mathcal{P}_{b\theta}(^mE), \Gamma_{\theta,\beta})'$ and each $P_n \in \mathcal{P}_{b\theta}(^nE)$, we define the mapping $\gamma(T_m)(P_n): E \to \mathbb{C}$ by $\gamma(T_m)(P_n)(x) = T_m [\frac{1}{m!} \hat{\delta}^m(\tau_{-x}P_n)(0)]$ where $(\tau_{-x}P_n)(a) = P_n(a+x)$ for all $a \in E$.

<u>(5.02) Remark</u>. The mapping $\gamma(T_m)(P_n)$ is a (n-m)-homogeneous polynomial on E which is bounded on the bounded subset of E if $n \geq m$ and $\{0\}$ if $n < m$.

<u>(5.03) Definition</u>. An α-β-γ-Silva holomorphy type θ is a α-β-Silva holomorphy type which is an α-β-Silva holomorphy type satisfying the following conditions:

1) If $P \in \mathcal{P}_{b\theta^*}(E^*)$, $\varphi \in E^*$ and $k \in \mathbb{N}$, then
$$\hat{\delta}^k P(\varphi) \in \mathcal{P}_{b\theta^*}(^kE^*).$$

2) If $T_m \in (\mathcal{P}_{b\theta}(^mE), \Gamma_{\theta,\beta})'$, then $\gamma(T_m)$ maps $\mathcal{P}_{b\theta}(^nE)$ into $\mathcal{P}_{b\theta}(^{n-m}E)$ continuously.

<u>(5.04) Example</u>. $(\mathcal{P}_{bN}(^mE), \Gamma_{N,\beta})_{m=0}^{\infty}$ is an α-β-γ-Silva holomorphy type.

<u>(5.05) Definition</u>. Let θ be a α-β-γ-Silva holomorphy type. $\mathcal{F}_{b\theta}(E) = \prod_{m=0}^{\infty} \mathcal{P}_{b\theta}(^mE)$ is called the set of all $S\theta$-formal power series on E. We define scalar multiplication and addition coordinate-wise and give $\mathcal{F}\mathcal{P}_{b\theta}(E)$ the product topology. We identify $f \in H_{S\theta}(E)$ with an element of $\mathcal{F}\mathcal{P}_{b\theta}(E)$ by taking the Taylor series of f at zero.

<u>(5.06) Definition</u>. For each $m \in \mathbb{N}$, Q_m is an n-homogeneous partial differential operator on $\mathcal{F}\mathcal{P}_{b\theta}(E)$ if it satisfies the following conditions:

1) $Q_m: \mathcal{J}\mathcal{P}_{b\theta}(E) \to \mathcal{J}\mathcal{P}_{b\theta}(E)$ is a continuous linear operator.

2) For each $\xi \in E$ and each $P \in \mathcal{P}_{b\theta}(E)$, $Q_m(\tau_{-\xi}P) = \tau_{-\xi}Q_m(P)$.

3) For each $n \in \mathbb{N}$, $Q_m(\mathcal{P}_{b\theta}(^nE)) \subset \mathcal{P}_{b\theta}(^{n-m}E)$.

For each $m \in \mathbb{N}$, we denote by $PD_{S\theta}(E_m)$ the set of m-homogeneous partial differential operators on $\mathcal{J}\mathcal{P}_{b\theta}(E)$.

(5.07) Definition. Q is a partial differential operator if it is a finite sum of homogeneous partial differential operators. We denote by $PD_{S\theta}(E)$ the set of all partial differential operators on $\mathcal{J}\mathcal{P}_{b\theta}(E)$.

(5.08) Definition. For each $Q \in PD_{S\theta}(E)$, we define $\beta(Q): \mathcal{P}_{b\theta}(E) \to C$ by $\beta(Q)(P) = Q(P)(0)$, for all $P \in \mathcal{P}_{b\theta}(E)$.

(5.09) Proposition. For each $Q \in PD_{S\theta}(E_m)$ we have $\beta(Q) \in (\mathcal{P}_{b\theta}(E), t_{S\theta})'$.

Proof. Since Q is a operator $\beta(Q)$ is linear. Let $Q \in PD_{S\theta}(E_m)$ and $P = \sum_{i=0}^{n} P_i$ with $P_i \in \mathcal{P}_{b\theta}(^iE)$ $i = 0,1,\dots,n$. $\beta(Q)(P) =$
$= Q(P_o)(0) + \dots + Q(P_{m-1})(0) + Q(P_m)(0) + Q(P_{m+1})(0) + \dots + Q(P_n)(0) =$
$= Q(P_m)(0) = Q(\frac{1}{m!}\hat{\delta}^m P)(0)$.

Let $(P_\alpha)_{\alpha\in A} \subset \mathcal{P}_{b\theta}(E)$ with $\lim_{\alpha\in A} P_\alpha = 0$ in $(H_{S\theta}(E), T_{S\theta})$.
Let $B \in \mathcal{B}_E$ and $c \geq 0$ such that

$$|\beta(Q)(P_\alpha)| = |Q(\frac{1}{m!}\hat{\delta}^m P_\alpha)(0)| \leq c \left\|\frac{1}{m!}\hat{\delta}^m P_\alpha(0)\right\|_{\theta,B} .$$

Then $\lim_{\alpha\in A} \beta(Q)(P_\alpha) = 0$. Hence $\beta(Q) \in (\mathcal{P}_{b\theta}(E), T_{S\theta})'$.

(5.10) Proposition. The mapping $Q \in PD_{S\theta}(E_m) \to \widehat{\beta(Q)} \in \mathcal{P}_{b\theta^*}(^mE^*)$ is a one to one linear onto mapping.

Proof. By Proposition (5.09), $\widehat{\beta(Q)}$ is well defined.
$\widehat{\beta(Q)}(\varphi) = \beta(Q)(\exp\varphi) = Q(\frac{1}{m!}\hat{\delta}^m(\exp\varphi))(0) = Q(\frac{1}{m!}\varphi^m) = \frac{1}{m!}Q(\varphi^m)$ for all $\varphi \in E^*$. If $\lambda \in C$, we have $\widehat{\beta(Q)}(\lambda\varphi) = \frac{1}{m!}Q((\lambda\varphi)^m) =$
$= \lambda^m \frac{1}{m!} Q(\varphi^m) = \lambda^m \widehat{\beta(Q)}(\varphi)$. Then $\widehat{\beta(Q)}$ is an n-homogeneous function and since $\beta(Q) \in (\mathcal{P}_{b\theta}(E), T_{S\theta})'$, we have $\widehat{\beta(Q)} \in \mathcal{P}_{b\theta^*}(E^*)$.

If $Q_1, Q_2 \in PD_{S\theta}(E)$ with $\widehat{\beta(Q_1)} = \widehat{\beta(Q_2)}$, we have $\beta(Q_1) = \beta(Q_2)$ since the mapping is one to one mapping. For all $P \in \mathcal{P}_{b\theta}(E)$ and all $x \in E$ we have $\beta(Q_1)(\tau_{-x}P) = \beta(Q_2)(\tau_{-x}P)$, then $Q_1(P)(x) = Q_2(P)(x)$, then $Q_1(P) = Q_2(P)$ and then $Q_1 = Q_2$, which proves that the mapping is one to one. Let $P_m \in \mathcal{P}_{b\theta^*}(^mE^*)$. By definition there exists $R_m \in (\mathcal{P}_{b\theta}(^mE), \Gamma_{\theta,\beta})'$ such that $\hat{R}_m = P_m$. Since θ is a α-β-γ-Silva holomorphy type, $\gamma(R_m): \mathcal{P}_{b\theta}(^nE) \to \mathcal{P}_{b\theta}(^{n-m}E)$ is well-defined and continuous and has an extension as a continuous linear mapping from $\mathfrak{F}\mathcal{P}_{b\theta}(E)$ into $\mathfrak{F}\mathcal{P}_{b\theta}(E)$. If $x, y \in E$ and $P \in \mathcal{P}_{b\theta}(E)$, we have

$$(\tau_{-x}[\gamma(R_m)(P)](y) = \gamma(R_m)(P)(y+x) = R_m[\frac{1}{m!}\tau_{-x}(\hat{\delta}^m P(0)](y) =$$

$$= R_m[\frac{1}{m!}\hat{\delta}^m(\tau_{-x}P)(0)](y) = \gamma(R_m)(\tau_{-x}P)(y).$$

This proves the condition 2) of Definition (5.06). The conditions 1) and 3) are obvious. Hence $\gamma(R_m) \in PD_{S\theta}(E)$. Now, $\widehat{\beta[\gamma(R_m)]}(\varphi) = \beta[\gamma(R_m)](\exp\varphi) = \gamma(R_m)(\exp\varphi)(0) = R_m(\frac{1}{m!}\hat{\delta}^m(\exp\varphi)(0)) = R_m(\frac{1}{m!}\varphi^m) = \frac{1}{m!}R_m(\varphi^m) = \hat{R}_m(\varphi) = P_m(\varphi)$. This completes the proof.

<u>(5.11) Corollary</u>. There is a one to one correspondence between the elements of $PD_{S\theta}(E)$ and the elements of $\mathcal{P}_{b\theta^*}(E^*)$. This correspondence is given by the linear mapping

$$Q \in PD_{S\theta}(E) \to \widehat{\beta(Q)} \in \mathcal{P}_{b\theta^*}(E^*).$$

<u>(5.12) Definition</u>. If $Q_1, Q_2 \in PD_{S\theta}(E)$, then we denote by $Q_1 * Q_2$ the mapping from $\mathfrak{F}\mathcal{P}_{b\theta}(E)$ into $\mathfrak{F}\mathcal{P}_{b\theta}(E)$ defined by $(Q_1 * Q_2)(f) = Q_1(Q_2(f))$. $Q_1 * Q_2$ is called the convolution of Q_1 and Q_2.

<u>(5.13) Proposition</u>. If $Q_1, Q_2 \in PD_{S\theta}(E)$, then $Q_1 * Q_2 \in PD_{S\theta}(E)$.

<u>Proof</u>. It suffices to show homogeneous partial differential operator case. Let $Q_1 \in PD_{S\theta}(E_{m_1})$ and $Q_2 \in PD_{S\theta}(E_{m_2})$. The condition 1) of the Definition (5.06) is obvious. If $x \in E$ and $P \in \mathcal{P}_{b\theta}(^nE)$, we have $\tau_{-x}(Q_1 * Q_2)(P) = \tau_{-x}(Q_1(Q_2(P)) = Q_1(\tau_{-x}(Q_2(P)) = Q_1(Q_2(\tau_{-x}P) =$

$= Q_1 * Q_2(\tau_{-x}P)$. Since $Q_2(\mathcal{P}_{b\theta}({}^{n}E)) \subset \mathcal{P}_{b\theta}({}^{n-m_2}E)$, we have

$$Q_1 * Q_2(\mathcal{P}_{b\theta}({}^{n}E) \subset Q_1(\mathcal{P}_{b\theta}({}^{n-m_2}E) \subset \mathcal{P}_{b\theta}({}^{n-m_2-m_1}E).$$

Hence $Q_1 * Q_2$ is a (m_1+m_2)-homogeneous partial differential operator.

<u>(5.14) Proposition.</u> Let θ be an α-β-γ-Silva holomorphy type. Then:

1) $\mathcal{P}_{b\theta *}(E^*)$ is a commutative algebra under pointwise multiplication.

2) $PD_{S\theta}(E)$ is a commutative algebra under convolution.

3) The mapping $Q \in PD_{S\theta}(E) \to \widehat{\beta(Q)} \in \mathcal{P}_{b\theta *}(E^*)$ is a one to one onto linear and algebraic isomorphism.

<u>Proof.</u> We complete the proof of 3). For each $x \in E$ and $\varphi \in E^*$ we have for all $y \in E$, $(\tau_{-x} \exp \varphi)(y) = \exp \varphi(y+x) = \exp \varphi(x) \cdot$ $\cdot \exp \varphi(y)$. Hence $\tau_{-x} \exp \varphi = \exp \varphi(x) \exp \varphi$.

If $Q_1, Q_2 \in PD_{S\theta}(E)$, we have $Q_2(\exp \varphi)(x) = Q_2(\tau_{-x} \exp \varphi)(0) =$ $= Q_2(\exp \varphi(x) \exp \varphi)(0) = \exp \varphi(x) Q_2(\exp \varphi)(0) = \exp \varphi(x) \beta(Q_2)(\varphi)$.

Now,

$$[\widehat{\beta(Q_1 * Q_2)}](\varphi) = [\beta(Q_1 * Q_2)](\exp \varphi) = (Q_1 * Q_2)(\exp \varphi)(0)$$

$$[Q_1(Q_2(\exp \varphi))](0) = [Q_1(\exp \varphi(\cdot)\widehat{\beta(Q_2)}(\varphi))](0) =$$

$$= \widehat{\beta(Q_2)}(\varphi)[Q_1(\exp \varphi)](0) = \widehat{\beta(Q_2)}(\varphi)\widehat{\beta(Q_1)}(\varphi).$$

Then

$$\widehat{\beta(Q_1 * Q_2)} = \widehat{\beta(Q_1)}\widehat{\beta(Q_2)}.$$

<u>(5.15) Lemma.</u> If $P_1, P_2 \in \mathcal{P}_{b\theta *}(E^*)$, with $P_2 \neq 0$ and $P_3 = P_1/P_2 \in$ $\in \mathcal{H}(E^*)$, then $P_3 \in \mathcal{P}_{b\theta *}(E^*)$.

<u>Proof.</u> Since $P_1, P_2 \in \mathcal{P}_{b\theta *}(E^*) \subset \mathcal{P}(E^*)$ and $P_3 \in \mathcal{H}(E^*)$ with $P_3 = \sum_{m=0}^{\infty} \frac{1}{m!} \hat{\delta}^m P_3(0)$, we have $P_2 P_3 = \sum_{m=0}^{\infty} P_2 \frac{1}{m!} \hat{\delta}^m P_3(0) = P_1$ and then $P_3 \in \mathcal{P}(E^*)$. By condition 1) of the Definition (5.03) it suffices to show that for some $\xi \in E^*$ and all $m \in \mathbb{N}$ we have $\hat{\delta}^m P_3(\xi) \in \mathcal{P}_{b\theta *}(E^*)$. Let $\xi \in E^*$ such that $P_2(\xi) \neq 0$. Then

$$P_3(\xi) = P_1(\xi)/P_2(\xi) \in \mathcal{P}_{b\theta^*}(E^*).$$

For $i = 1,2,3$

$$P_i(x) = \sum_{j=0}^{\infty} \frac{1}{j!} \hat{\delta}^j P_i(\xi)(x-\xi).$$

By hypothesis

$$\hat{\delta}^j P_i(\xi) \in \mathcal{P}_{b\theta^*}({}^jE^*), \quad j \in \mathbb{N}, \quad i = 1,2.$$

$$\hat{\delta}^j P_3(\xi) \in \mathcal{P}({}^jE^*), \quad j \in \mathbb{N}.$$

Suppose $\hat{\delta}^j P_3(\xi) \in \mathcal{P}_{b\theta^*}(E^*)$ for $j \leq k$. We have

$$\frac{1}{(j+1)!} \hat{\delta}^{(j+1)} P_3(\xi) =$$

$$= \frac{\hat{\delta}^{(k+1)} P_1(\xi) - \sum_{i=1}^{k+1} \frac{1}{i!} \hat{\delta}^i P_2(\xi) \frac{1}{(k+1-i)} \hat{\delta}^{(k+1-i)} P_3(\xi)}{P_2(\xi)}$$

Since $\mathcal{P}_{b\theta^*}(E^*)$ is an algebra by using induction we get the required result.

<u>(5.16) Definition</u>. A function f on E is called an $S\theta$-exponential polynomial if there exists $\varphi \in E^*$, $P \in \mathcal{P}_{b\theta}(E)$ such that $f(x) = P(x) \exp \varphi(x)$ for all $x \in E$.

<u>(5.17) Proposition</u>. Let θ be an α-β-γ-Silva holomorphy type and let $Q \in PD_{S\theta}(E)$. Then every solution of $Q(f) = 0$ can be approximated in $\mathcal{F}\mathcal{P}_{b\theta}$ by $S\theta$-exponential polynomial solution of $Q(f) = 0$.

<u>Proof</u>. If $Q = 0$, since the Taylor series expansion of any function on E converges in $\mathcal{F}\mathcal{P}_{b\theta}(E)$, we have that $\{P \exp \varphi;\ P \in \mathcal{P}_{b\theta}(E),\ \varphi \in E^*\}$ is dense in $\mathcal{F}\mathcal{P}_{b\theta}(E) = \ker(Q)$. Suppose $Q \neq 0$. Let $v \in \mathcal{F}\mathcal{P}_{b\theta}(E)'$ such that if $Q(P \exp(\varphi)) = 0$ $v(P \exp(\varphi)) = 0$. Since $\mathcal{B}(Q) \in \mathcal{P}_{b\theta^*}(E^*) \subset \mathcal{P}_{bN^*}(E^*)$ we can associate with it in a unique way a convolution operator $\mathcal{O}_q = Q/H_{SN}(E)$ on $(H_{SN}(E), T_{SN})$. On the other hand, since $v \in \mathcal{F}\mathcal{P}_{b\theta}(E)'$, we have $\hat{v} \in \mathcal{P}_{b\theta^*}(E^*) \subset \mathcal{P}_{bN^*}(E^*)$ and we can associate with in a unique way a continuous linear functional $v_N = v/H_{SN}(E)$ on $(H_{SN}(E), T_{SN})$.

Thus, if $\Theta_Q(P\exp(\varphi)) = 0$, then $v_N(P\exp(\varphi)) = 0$ and by result in [4], we have $v_N/\widehat{\beta(\Theta_Q)} \in \mathcal{H}(E^*)$. Since $\hat{v} = \hat{v}_N$ and $\widehat{\beta(\Theta_Q)} = \widehat{\beta(Q)}$ we get $\hat{v}/\widehat{\beta(Q)} \in \mathcal{H}(E^*)$. By Lemma (5.15) there exists Q_1 and $Q_v \in PD_{S\theta}(E)$ such that $\widehat{\beta(Q_v)} = \hat{v}$ and $\widehat{\beta(Q_1)} = \hat{v}/\beta(Q)$. Then $\widehat{\beta(Q_v)} = \widehat{\beta(Q_1)}\widehat{\beta(Q)}$ and we conclude $Q_v = Q_1 * Q$. If $h \in \mathcal{F}\mathcal{P}_{b\theta}(E)$ with $Q(h) = 0$, we have $v(h) = [Q_v(h)](0) = (Q_1 * Q)(h)(0) = Q_1[Q(h)](0) = 0$. Let $\mathcal{L}$ be the subspace generated by $\{P\exp\varphi;\ P \in \mathcal{P}_{b\theta}(E),\ \varphi \in E^* \text{ and } Q(P\exp\varphi) = 0\}$. We have just proved that every continuous functional which is zero in $\mathcal{L}$ is also zero in the $\ker(Q)$. Using Hahn-Banach theorem we have that $\mathcal{L}$ is dense in $\ker(Q)$.

<u>(5.18) Proposition</u>. Let θ be an α-β-γ-Silva holomorphy type and let $Q \in PD_{S\theta}(E)$. Then, if $Q \neq 0$ we have that ${}^tQ$ is one to one and $\mathrm{Im}({}^tQ)$ is closed for the weak topology on $\mathcal{F}\mathcal{P}_{b\theta}(E)'$ defined by $\mathcal{F}\mathcal{P}_{b\theta}(E)$.

<u>Proof</u>. First all we prove that $\mathrm{Im}({}^tQ) = \{f \in \mathcal{F}\mathcal{P}_{b\theta}(E);\ Q(f) = 0\}^{\perp}$. Since $Q\colon \mathcal{F}\mathcal{P}_{b\theta}(E)' \to \mathcal{F}\mathcal{P}_{b\theta}(E)'$ is defined by ${}^tQ(w) \in v$ such that $v(f) = w(Q)(f)$ for all $f \in \mathcal{F}\mathcal{P}_{b\theta}(E)$ and all $w \in \mathcal{F}\mathcal{P}_{b\theta}(E)'$. If $h \in \{f \in \mathcal{F}\mathcal{P}_{b\theta}(E);\ Q(f) = 0\} \subset \mathcal{F}\mathcal{P}_{b\theta}(E)$, we have $Q(h) = 0$. If $v \in \mathrm{Im}({}^tQ) \subset \mathcal{F}\mathcal{P}_{b\theta}(E)'$ we have $v(h) = {}^tQ(w(h)) = w(Q(h) = w(0) = 0$. Hence $v \in (\ker Q)^{\perp}$. We have proved $\mathrm{Im}({}^tQ)$ $(\ker Q)^{\perp}$. Now let $v \in (\ker Q)^{\perp} \subset \mathcal{F}\mathcal{P}_{b\theta}(E)'$. If $Q(P\exp\varphi) = 0$, then $v(P\exp\varphi) = 0$. By the proof of the Proposition (5.17) there exists $Q_1, Q_2 \in PD_{S\theta}(E)$ such that $Q_1 * Q = Q_v$ and $\beta(Q_v) = v$. Therefore ${}^tQ\beta(Q_1)(f) = \beta(Q_1)(Q(f)) = Q_1[Q(f)](0) = [Q_1 * Q](f)(0) = Q_v(f)(0) = \beta(Q_v)(f) = v(f)$, which implies $v = {}^tQ\beta(Q_1)$. Hence $v \in \mathrm{Im}({}^tQ)$ and $\mathrm{Im}({}^tQ) = (\ker Q)^{\perp}$. We have prove

$$\mathrm{Im}({}^tQ) = \{v \in \mathcal{F}\mathcal{P}_{b\theta}(E)';\ v(f) = 0 \text{ if } Q(f) = 0\} =$$

$$= \bigcap_{Q(f)=0} \{v \in \mathcal{F}\mathcal{P}_{b\theta}(E)';\ v(f) = 0\}.$$

Hence $I_m({}^tQ)$ is the intersection of weakly closed sets and hence is weakly closed. Now suppose ${}^tQ(v) = 0$. Let $Q_v \in PD_{S\theta}(E)$ such that $\beta(Q_v) = v$. For each $x \in E$ and $P \in \mathcal{P}_{b\theta}(E)$ we have $[Q_v * Q](P)(x) =$ $= [(Q_v * Q)(\tau_{-x}P)](0) = [Q_v(Q(\tau_{-x}P))](0) = \beta(Q_v)[Q(\tau_{-x}P)] =$ $= v[Q(\tau_{-x}P)] = 0$. Hence $Q_v * Q = 0$ and then $\widehat{\beta(Q_v)}\widehat{\beta(Q)} = 0$. Since $Q \neq 0$ and $\widehat{\beta(Q_v)}$, $\widehat{\beta(Q)} \in \mathcal{H}(E^*)$ we have $\widehat{\beta(Q_v)} = \hat{v} = 0$. Hence ${}^tQ$ is one to one.

(5.19) Corollary. If E has a countable basis for the elements of $\mathfrak{B}_E$ for the boundeds sets and $Q \in PD_{S\theta}(E)$ with $Q \neq 0$, then $Q(\mathfrak{F}\mathcal{P}_{b\theta}(E)) \subset \mathfrak{F}\mathcal{P}_{b\theta}(E)$.

Proof. In the conditions of stated we have that $\mathfrak{F}\mathcal{P}_{b\theta}(E)$ is a Fréchet space. By the Dieudonne-Schwartz theorem, to show that $Q \neq 0$ is onto is suffices to show $Im({}^tQ)$ is closed for the weak topology on $\mathfrak{F}\mathcal{P}_{b\theta}(E)'$ defined by $\mathfrak{F}\mathcal{P}_{b\theta}(E)$. But this fact we have proved.

REFERENCES

[1] Barroso, J.A., Topologia nos espaços de aplicações holomorfas entre espaços localmente convexos, Anais da Academia Brasileira de Ciências, Vol. 43 (1971).

[2] Dineen, S., Holomorphy types on Banach space, Studia Mathematica, T. XXXIX. (1971).

[3] Kothe, G., Topological Vector Spaces I, Springer-Verlag, Berlin, Heidelberg, New York, 1969.

[4] Matos, M.C. & Nachbin, L., Silva-holomorphy types (to appear in these Proceedings).

[5] Nachbin, L., Holomorfia em dimensão infinita, Lectures Notes, Universidade Estadual de Campinas, 1976.

[6] Paques, O.T.W., Tensor Products of Silva-holomorphic Functions, Advances in Holomorphy, North-Holland, 1977.

THE APPROXIMATION-THEORETIC LOCALIZATION OF SCHWARTZ'S APPROXIMATION PROPERTY FOR WEIGHTED LOCALLY CONVEX FUNCTION SPACES AND SOME EXAMPLES

Klaus-D. Bierstedt

Gesamthochschule Paderborn
FB 17, D 2 - 228
Warburger Str. 100, Postfach 16 21
D-4790 Paderborn
Germany (Fed. Rep.)

INTRODUCTION

The method of an approximation-theoretic localization of the locally convex approximation property (a.p.) for weighted function spaces (or rather modules) was first explained in [4]. Its usefulness was then demonstrated in the last section of [11] in connection with investigations on spaces of functions of "mixed dependence" on subsets of products.

At the time when [11] was written the present author "computed" a number of concrete examples in order to get a better idea how far the applications of the fundamental localization theorem went and where it had its limitations. Only the examples which had direct connections with product sheaves could be included in [11].

In 1977, we came back to the remaining examples, and, based on the results of our paper [7], we have then been able to construct some new ones, too. When we looked once more at the localization theorem in the light of recent work in vector-valued approximation theory by Machado-Prolla [22] and in connection with Prolla's paper [25], it turned out that it was now possible to remove the completeness assumptions, which were needed in [11], completely. That no change of the general method was actually necessary, but that the research of Machado-Prolla on Nachbin spaces of vector fibrations in the non self-adjoint case, appropriately reinterpreted and used, enabled us to generalize our former results considerably, came as some surprise to this author's attention. It demonstrates that the method of dealing with with spaces of vector-valued functions by aid of L. Schwartz's ε-product is not really limited to complete spaces, but applies in full generality. So we have decided to give the main idea of the approximation-theoretic localization of the a.p. and the general version of the fundamental localization theorem in the first part of this article (sections 1. and 2.) while the examples and applications form the second part (sections 3. and 4.).

Let us now review the contents briefly: In section 1, we present <u>the two methods</u> how one can apply approximation-theoretic results to proofs of the a.p. of a function space: a "direct" one which sets out to interpret the space of continuous linear operators on a locally convex function space as a space of vector-valued functions (cf. Prolla [25]) and the one using the ε-product (cf. [4], [11]). It is rather obvious that the two methods are not essentially different, and we discuss their relation. For the rest of the paper we follow the second method because it takes the generality of the problem into account from the very start and because ε-product representations are readily available in the applications. - Since some technical problems and results complicate the general proof of the main localization theorem and since it is quite probable that the method applies in some other cases, too, section 1 does not go into details and may also serve as an introduction.

The first part of section 2 collects all the notation and all "ingredients" for the proof of <u>the fundamental localization Theorem</u> 16 of Schwartz's a.p. for subspaces of weighted spaces of type $CV_o(X)$ (i.e. Nachbin spaces) [which are modules over an algebra such that the weighted (Bernstein-Nachbin) approximation problem is localizable]. Lemma 11 is the crucial point which makes it possible to do _without_ the former completeness assumptions: Here we show that the space of (only weakly continuous) F-valued functions which corresponds with the ε-product $CV_o(X) \varepsilon F$ is still a Nachbin space $L\tilde{V}_o$ of cross-sections in the sense of (say) [22]. In fact, whenever the topology of the function space is stronger than the compact-open topology, it is enough to assume only _hypocontinuity_ of the functions here, and the localization theorem also applies in such a situation (as we point out in Theorem 18).

Section 3 is devoted to (more) examples for the localization of the a.p. among _weighted spaces with "mixed dependence"_ on subsets Λ of products. In [11], we had dealt with a (rather) simple case already; now we construct concrete examples for the "regularity assumptions" on Λ which are needed if the "localized spaces" are not subspaces of nuclear spaces. We are mainly interested in the two cases where the "slices" Λ_t of Λ are open or compact. In the last case we make use of the results of [7] to get interesting new examples (see e.g. 29 and 30); we also include (in 33 and 34) "density theorems" completely analogous to [11], 4.11 and 4.12.

-At the end of the section, we look at the general setting for all examples in section 3 and point out that they have _a natural interpretation as Nachbin spaces of cross-sections_. (Let us note that 11, 18, and the remarks at the end of section 3 yield a number of interesting new examples of Nachbin spaces of cross-sections which arise quite naturally in the applications and which apparently have not been mentioned in the literature so far.)

Section 4 concludes the article with _other examples_ (a different kind of "mixed dependence") and _some applications to vector-valued approximation theory_: In connection with the paper [26] of Prolla-Machado we consider Weierstrass-Stone, Kakutani-Stone, and Grothendieck subspaces of $CV_o(X,F)$ and generalize Blatter's method of [13] to weighted spaces.

ACKNOWLEDGEMENT

As I have mentioned before, part of the results in the last two sections of this article date back to the time when the joint publication [11] was prepared. I would like to thank B. Gramsch and R. Meise for many helpful conversations and remarks in this connection. - During the Campinas Conference on Approximation Theory in 1977 I had the opportunity to speak on the results (of [7] and) of (mainly the second part of) this article (part of which, in some sense, is a sequel to [7]). The author gratefully acknowledges support under the GMD/CNPq - agreement during his stay at UNICAMP July-September 1977 without which it would not have been possible to attend the Conference in Campinas. I would also like to thank J.B. Prolla for some conversations on his papers [22] - [26] which we had at this occasion.

1. THE GENERAL IDEA

Let E be an arbitrary (always Hausdorff) locally convex (l.c.) space over $\mathbb{K} = \mathbb{C}$ (or $\mathbb{R}$). The following definition is taken from Grothendieck [16] (resp. Schwartz [29]):

<u>1 Definition</u>. E is said to have the approximation property (a.p.) (resp. Schwartz's a.p.) if the identity id_E of E can be approximated uniformly on every precompact (resp. every absolutely convex and compact) subset of E by continuous linear operators from E into E of finite rank (i.e. with finite dimensional range).

For two l.c. spaces E and F, let $\mathcal{L}(E,F)$ denote the space of all continuous linear operators from E into F, and put $\mathcal{L}(E) := \mathcal{L}(E,E)$. If $\mathcal{L}(E,F)$ is endowed with the topology of uniform convergence on all precompact (resp. absolutely convex compact) subsets of E, we write $\mathcal{L}_c(E,F)$ (resp. $\mathcal{L}_{cc}(E,F)$); so $\mathcal{L}_c(E)$ and $\mathcal{L}_{cc}(E)$ denote $\mathcal{L}_c(E,E)$, and $\mathcal{L}_{cc}(E,E)$, respectively. As, under the natural identification, the space of all continuous linear operators of finite rank from E into F corresponds with the tensor product

$E' \otimes F$, we get easily:

<u>2 Proposition</u> (Grothendieck [16], Proposition 35, p. 164-165). The following assertions are equivalent:

(1) E has the a.p. (resp. Schwartz's a.p.),

(2) $id_E \in E' \otimes E^{\mathcal{L}_c(E)}$ (resp. $id_E \in E' \otimes E^{\mathcal{L}_{cc}(E)}$),

(3) $E' \otimes E$ is dense in $\mathcal{L}_c(E)$ (resp. $\mathcal{L}_{cc}(E)$),

(4) for each l.c. space F, $E' \otimes F$ is dense in $\mathcal{L}_c(E,F)$ (resp. $\mathcal{L}_{cc}(E,F)$),

(5) for each l.c. space F, $F' \otimes E$ is dense in $\mathcal{L}_c(F,E)$ (resp. $\mathcal{L}_{cc}(F,E)$).

A counterexample of Enflo (1972), with subsequent refinements due to (among others) Davie and Szankowski, shows that there are even closed subspaces of the sequence spaces l^p <u>without a.p.</u> for each p, $1 \leq p \leq \infty$, $p \neq 2$. And recently Szankowski [30] proved (surprisingly) that the C^*-algebra $\mathcal{L}(H)$ of all continuous linear operators on an infinite dimensional (separable) Hilbert space H (under its canonical operator norm) does <u>not</u> have the a.p.

So, in view of the many interesting applications of the a.p. in the theory of topological tensor products (cf. Grothendieck [16]), it is reasonable to ask for methods to establish the a.p. for "concrete" l.c. spaces. One such method which applies to <u>a general class</u> of l.c. function spaces is discussed here. - Usually, a proof of the a.p. of a function space immediately implies useful results on (the approximation of) vector-valued functions and on ("slice products" of) functions of several variables (cf. e.g. [2], [3], [4]), but our general idea here is, conversely, to apply known theorems on the approximation of vector-valued functions to prove the a.p. of spaces of scalar functions.

<u>3 Remark</u>. The a.p. of a complete space $\mathfrak{F}$ of scalar functions, an ε-tensor product representation $\mathfrak{F}_F = \mathfrak{F} \check{\otimes}_\varepsilon F$ of a space $\mathfrak{F}_F$ of

F-valued functions, and the known fact (cf. Schwartz [29], Proposition 11, Corollaire 2, p. 48) that the ε-tensor product of two complete l.c. spaces with a.p. again has the a.p. together imply that $\mathfrak{F}_F$ has the a.p. for each complete l.c. space F with a.p., too. - Hence we may restrict our attention to spaces of <u>scalar</u> functions here without any real loss of generality (for most purposes).

Now the most powerful tools of approximation theory in spaces (rather, algebras or, more generally, modules) $\mathfrak{F} = \mathfrak{F}(X)$ of continuous functions on a (Hausdorff) topological space X, like generalizations of the Stone-Weierstrass theorem or of Bishop's theorem on uniform algebras, yield a covering $X = \bigcup_{\alpha\in A} X_\alpha$ of X by (pairwise disjoint closed) sets X_α such that approximation can be "localized" to the sets X_α $(\alpha\in A)$ in a natural way. Correspondingly, one may try to prove that (under suitable conditions) the a.p. of $\mathfrak{F}$ can be "<u>localized</u>" to the sets X_α in the following sense: If, for each $\alpha \in A$, $\mathfrak{F}|_{X_\alpha}$ has the a.p., then $\mathfrak{F}$ must have the a.p., too. - As many examples of spaces of continuous functions with a.p. are known (since Grothendieck and Schwartz), such a localization principle for the a.p. will yield a number of new concrete examples (say, among spaces of functions with "mixed dependence" on subsets of products, where Schwartz's result on the a.p. of ε-tensor products of complete l.c. spaces with a.p. can only be applied on <u>full product sets</u>.) The theorem is, in some sense, an interesting and very useful approximation-theoretic "<u>permanence property</u>" of the a.p. in spaces of continuous functions.

However, the definition of the a.p. (see Proposition 2) does not allow a <u>direct</u> application of approximation theorems for vector-valued continuous functions on X to our situation of approximating $id_{\mathfrak{F}(X)}$ by elements of $\mathfrak{F}(X)'\otimes \mathfrak{F}(X)$ in $\mathfrak{L}_c(\mathfrak{F}(X))$.

There are essentially <u>two ways</u> (which, as a careful analysis shows, are closely related and, in fact, equivalent) to overcome this difficulty. - If $(X,(F_x)_{x\in X})$ is a "vector fibration" over a Hausdorff topological space X with "fibers" F_x $(x\in X)$ and if $L = LV_o$ is a "Nachbin space of cross-sections" (see below) with $L(x) := \{f(x);\ f \in L\} = F_x$ for each $x \in X$ such that L is a module over a selfadjoint and separating subalgebra of $CB(X)$ (= continuous and bounded scalar functions on X), Prolla [25] (modifying a previous idea of G. Gierz) has recently shown that the a.p. of all spaces F_x (under the semi-norms $(v(x))_{v\in V}$) implies the a.p. of L. In his proof he represents the space $\mathfrak{L}(L)$ as a Nachbin space of cross-sections over X with fibers $\mathfrak{L}(L,F_x)$ and then applies the solution of the "weighted approximation problem" (for Nachbin spaces of cross-sections) in the separating and selfadjoint bounded case. - <u>The corresponding representation</u> is given as follows: Let, for each $x \in X$, $\delta_x: L \to F_x$ be the point evaluation at x $(\delta_x(f) := f(x)$ for all $f \in L)$. For $T \in \mathfrak{L}(L)$, $\delta_x \circ T$ belongs to $\mathfrak{L}(L,F_x)$, and $R: T \to \hat{T} := (\delta_x \circ T)_{x\in X}$ represents $\mathfrak{L}(L)$ as a Nachbin space of cross-sections over X with fibers $\mathfrak{L}(L,F_x)$.

In our general case, it is clear that $\mathfrak{L}(\mathfrak{F}(X))$ can similarly be represented as a space of functions $\hat{T}$ on X with values in $\mathfrak{F}(X)'$, if the topology of $\mathfrak{F}(X)$ is stronger than pointwise convergence on X. As $\mathfrak{F}(X)$ consists of continuous functions, any such function $\hat{T}: x \to (f \to Tf(x))$ is continuous from X into $\mathfrak{F}(X)'\ [\sigma(\mathfrak{F}(X)', \mathfrak{F}(X))]$ and continuous into $\mathfrak{F}(X)'_c := \mathfrak{L}_c(\mathfrak{F}(X),\mathbb{K})$ (resp. $\mathfrak{F}(X)'_{cc} := \mathfrak{L}_{cc}(\mathfrak{F}(X),\mathbb{K})$) if and only if, for each precompact (resp. absolutely convex and compact) set K in $\mathfrak{F}(X)$, the image $T(K)$ under T is equicontinuous on X (which certainly holds for each $T \in \mathfrak{L}(\mathfrak{F}(X))$ if each precompact [resp. absolutely convex compact] subset of $\mathfrak{F}(X)$ is equicontinuous on X). - Now the topology of $\mathfrak{L}_c(\mathfrak{F}(X))$ [resp. $\mathfrak{L}_{cc}(\mathfrak{F}(X))$] is given by the set $\{q_{K,p};$ p continuous seminorm on $\mathfrak{F}(X)$, K precompact (resp. absolutely convex compact) in $\mathfrak{F}(X)\}$ of seminorms $q_{K,p}(T) = \sup_{f\in K} p(Tf)$ for all

$T \in \mathfrak{L}(\mathfrak{F}(X))$. For a large class of function spaces $\mathfrak{F}(X)$, it turns out that this topology corresponds with a (it is enough: topology weaker than a) "natural" topology on the space $\widehat{\mathfrak{L}(\mathfrak{F}(X))}$ of all functions $\hat{T}: X \to \mathfrak{F}(X)'_c$ (resp. $\mathfrak{F}(X)'_{cc}$) in the sense that vector-valued approximation theorems apply to approximation of the "evaluation mapping" $\widehat{id_{\mathfrak{F}(X)}}: x \to \delta_x$ by elements from $\widehat{\mathfrak{F}(X)' \otimes \mathfrak{F}(X)}$ (which are functions on X with values in finite dimensional subspaces of $\mathfrak{F}(X)'$ obviously).

In general, if there are vector-valued approximation theorems "only" for spaces of <u>continuous</u> functions, then this fact clearly restricts the class of spaces $\mathfrak{F}(X)$ to which our method applies. (Compare the equivalent condition for continuity of $\hat{T}: X \to \mathfrak{F}(X)'_c$ [resp. $\mathfrak{F}(X)'_{cc}$] which we have mentioned above; in the applications, this usually amounts to <u>a completeness type assumption</u> on $\mathfrak{F}(X)$.) In Prolla's case, however, we realize that the notion of Nachbin space of cross-sections (and the corresponding generalization of the Stone-Weierstrass theorem) is already flexible enough to conclude <u>without</u> any restriction. - Thus we can avoid completeness assumptions in our fundamental theorem below. On the other hand, such an assumption is quite natural, and most of the function spaces which occur in the applications <u>are</u> complete. So there is no great loss of generality e.g. in dealing mainly with <u>Schwartz's</u> a.p. throughout this paper.

Let us remark at this point (cf. Schwartz [28]) that the equivalence (2) ⇔ (5) of Proposition 2 implies, after a similar representation of $\mathfrak{L}(F,\mathfrak{F}(X))$ as a space of F'-valued functions on X for arbitrary l.c. F, that the approximation of the (single) evaluation mapping $\widehat{id_{\mathfrak{F}(X)}}$ by elements of $\widehat{\mathfrak{F}(X)' \otimes \mathfrak{F}(X)}$ in the space $\widehat{\mathfrak{L}(\mathfrak{F}(X))}$ (under a suitable topology) is really equivalent to approximation of <u>all</u> functions in a certain space $\widehat{\mathfrak{L}(F,\mathfrak{F}(X))}$ of F'-valued functions on X (under a suitable topology) by elements of

$\widehat{F' \otimes \mathfrak{F}(X)}$ (which are functions on X with values in finite dimensional subspaces of F') for arbitrary l.c. spaces F. - This remark leads us immediately to the other method to allow application of vector-valued approximation theorems to a proof of the a.p. for spaces of scalar functions. This (in some sense less direct) method which takes into account the generality of the problem right from the start was already presented in [4] and [11] and relies on a useful equivalence of the a.p. due to Schwartz [29] (cf. also [2] and [8]):

4 Definition. For a l.c. space F, let $F'_{cc} = \mathcal{L}_{cc}(F,\mathbb{K})$ denote the dual of F with the topology of uniform convergence on all absolutely convex compact subsets of F. Schwartz's ε-product of E and F is defined by $E\varepsilon F := \mathcal{L}_e(F'_{cc},E)$, where the subscript e indicates the topology of uniform convergence on the equicontinuous subsets of F'.

(Originally, Schwartz's ε-product was defined in [29] in a different way, but [29], Proposition 4, Corollaire 2, p.34 shows that our definition is equivalent up to topological isomorphism.) - Then $E\varepsilon F \cong F\varepsilon E$ holds, and the ε-product of two complete spaces is complete. Moreover, the tensor product $E \otimes F$ with the ε-topology of A. Grothendieck [16], i.e. $E \otimes_\varepsilon F$, is (canonically identified with) a topological subspace of $E\varepsilon F$, namely just the space of all (even $\sigma(F',F)$-) continuous linear operators of finite rank in $E\varepsilon F$. Let $E \check{\otimes}_\varepsilon F$ denote the ε-tensor product of E and F, i.e. the completion of $E \otimes_\varepsilon F$. Then Schwartz's criterion for the a.p. (with some refinements) can be formulated as follows:

5 Theorem (Schwartz [29], Proposition 11, p. 46-47). The following assertions for a l.c. space E are equivalent:

(1) E has Schwartz's a.p.,

(2) $id_E \in \overline{E \otimes E'}^{E \varepsilon E'_{cc}}$,

(3) $E \otimes E'$ is dense in $E \varepsilon E'_{cc}$,

(4) for each l.c. space F, $E \otimes F$ is dense in $E \varepsilon F$,

(5) for each Banach space F, $E \otimes F$ is dense in $E \varepsilon F$.

If, additionally, E is <u>complete</u>, the a.p. of E (i.e. (1)-(5) above) is also equivalent to:

(6) for each complete l.c. (or each Banach) space F, $E \check{\otimes}_\varepsilon F = E \varepsilon F$ holds.

Here the proof of (2) ⇔ (1) follows from the fact (Schwartz [29], Proposition 5, Corollaire, p. 36-37) that, for two l.c. spaces E and F, $\mathfrak{L}_{cc}(E,F)$ is always a topological linear subspace of $F \varepsilon E'_{cc}$.

Now, in many cases, for a space $\mathfrak{F}(X)$ of continuous functions on X the ε-product $F \varepsilon \mathfrak{F}(X)$ for any (say) complete l.c. space F is known to be (up to the topological isomorphism $u \to (x \to u(\delta_x))$ nothing but "<u>the space of</u> F-<u>valued functions of type</u> $\mathfrak{F}(X)$", and $F \otimes \mathfrak{F}(X)$ corresponds with the subspace of all functions which take their values in finite dimensional subspaces of F. In view of this, it is immediate (and much more natural than we may have realized up to this point) to apply vector-valued approximation theorems to a proof of the a.p. of $\mathfrak{F}(X)$ (in the form of equivalence (5) of Theorem 5). (And, conversely, the approximation of vector-valued functions [by functions with values in finite dimensional subspaces] which is implied by the a.p. of a function space is also more interesting than we may have been tempted to think after a short glance at the first method.)

Furthermore, equivalence (2) of Theorem 5 shows then that we have to deal with the approximation of the <u>single</u> $\mathfrak{F}(X)'_{cc}$-valued function (evaluation mapping) $x \to \delta_x$ <u>only</u> (if we prefer). This remark links the first method with this one, and, since $\mathfrak{L}_{cc}(E,F)$ is always a topological linear subspace of $F \varepsilon E'_{cc}$, it turns out that both methods are essentially <u>equivalent</u> (except for technical details). Undoubtedly, the first method is more direct, but, in each case, it requires (new) investigations on a topological isomorphism of $\mathfrak{L}_c(\mathfrak{F}(X))$ with the space $\widehat{\mathfrak{L}(\mathfrak{F}(X))}$ of $\mathfrak{F}(X)'_c$-valued functions on X

under a "natural" topology, whereas one advantage of the second method comes from the fact that <u>known</u> results (needed for other applications) can be used in a natural way. Also, while it is useful to know that the "test space" $F = E'_{cc}$ suffices in equivalence (4) of Theorem 5, it is sometimes much easier to work only with Banach spaces F as in equivalences (5) and (6) above, because the topological vector space structure of E'_{cc} may be quite complicated even for "good" E.

Thus, in accordance with our interest in l.c. function spaces (and not in Nachbin spaces of cross-sections, where similar results hold), we will only discuss the second method <u>in more detail</u> from now on. - As pointed out above, to derive a localization theorem for the a.p. (in the sense described before) by the second method, we need <u>two types of results</u>: ε-product representations $\mathfrak{F}(X)\varepsilon F = \mathfrak{F}_F(X)$ and localization theorems for (vector-valued) approximation in $\mathfrak{F}_F(X)$ by elements of $\mathfrak{F}(X) \otimes F$. So let us now list some of the known results in this direction:

(I) ε-<u>product-representations</u>:

If F is a (quasi-) complete l.c. space, a representation of $\mathfrak{F}(X)\varepsilon F$ as a "natural" space $\mathfrak{F}_F(X)$ of F-valued functions on X can be found for weighted Nachbin spaces (of continuous functions) $\mathfrak{F}(X) = CV_o(X)$ resp. $CV(X)$ (with $\mathfrak{F}_F(X) = CV_o(X,F)$ resp. $CV^p(X,F)$ if X is a $V_{\mathbb{R}}$-space [see below]) in [2], II, Theorem 4 or [4], for "weighted spaces of differentiable functions" in the thesis [1] of B. Baumgarten (cf. also L. Schwartz [28] and Garnir-de Wilde-Schmets [14]), for spaces of functions satisfying general Lipschitz conditions and spaces of continuously differentiable functions with Hölder conditions in W. Kaballo [18], 3 and 4 or [19], 3a). From these results and some general theorems on the ε-product of inductive limits, ε-product representations can also be derived for certain inductive limits of the spaces mentioned before. For inductive limits of weighted spaces e.g., this was done in [9]. (Recently some of the open problems here have been solved; see R. Hollstein [17], the author's paper [6] and [12], where it is shown that, in many interesting cases, inductive limits of weighted spaces are again [topologically] weighted spaces.) - At this point, it should be remarked that the ε-product obviously "preserves topological linear subspaces", and hence ε-product representations as above lead to analogous theorems for all topological subspaces, too. In fact, most of the spaces mentioned above are already known to have the a.p. (this usually follows from the ε-product representation theorem and results on the approximation of vector-valued functions, too), but not many results are available on the a.p. of topological subspaces, and here localization theorems might apply. (In the case of inductive limits, we still have the so-called "subspace problem" whether subspaces G of inductive limits $E = \operatorname{ind}_{\alpha \to} E_\alpha$ inherit the natural inductive limit topology $\operatorname{ind}_{\alpha \to}(G \cap E_\alpha)$, but in some applications this can be deduced from a lemma of A. Baernstein, cf. [9], 1 Satz 14).

(II) <u>Localization theorems for vector-valued approximation</u>:

Localization theorems for approximation by elements of modules W in weighted Nachbin spaces $CV_o(X,F)$ of continuous F-valued functions were first given in the selfadjoint case by Nachbin-Machado-Prolla [23]. For modules over non selfadjoint algebras, vector-valued localization was obtained in the complex (restricted) bounded case of the weighted (Bernstein-Nachbin) approximation problem by G. Kleinstück [21] (generalizing a previous results of J.B. Prolla). The general complex case has recently been treated by Machado-Prolla [22] (even in the setting of Nachbin spaces of cross-sections) with a reduction to the finite dimensional Bernstein approximation problem on $\mathbb{R}^n$ $(n \geq 1)$ and yielding, as corollaries, the analytic and quasi-analytic criteria; see also Prolla's book [24], Chapter 5, §2-3 for the case of vector-valued functions. Of course, for the spaces

$CV^p(X,F)$ instead of $CV_o(X,F)$, localization results can only hold when, for the functions f in $CV^p(X,F)$ and the weights $v \in V$, vf is considered (by "canonical" extension) as function on a compactification $\hat{X}$ of X (or even on $\beta\tilde{X}$, the Stone-Čech compactification of $\tilde{X} := X$ under an appropriate finer topology), and approximation is then localized to certain subsets of $\hat{X}$ (or $\beta\tilde{X}$) rather. Such localization theorems for approximation from modules W in $CV^p(X,F)$ were proved in the (complex) "bounded case" by G. Kleinstück [21], 2, Theorems 9 and 11. - No (general) localization theorems for approximation in spaces of continuously differentiable functions or spaces of Lipschitz functions are known. (But, by the above - mentioned theorem of [6] resp. [12] that, at least in the case of Banach space-valued functions, inductive limits of weighted spaces may again be weighted spaces [topologically], localization theorems are also available for most of the interesting "inductively weighted spaces". - In general, for a proof of the a.p. of a function space with an inductive limit topology, it is important that one may restrict the attention to <u>Banach</u> "test spaces" in equivalence (5) of Theorem 5. On the other hand, the a.p. is inherited, say, under quasi-complete "<u>compactly regular</u>" inductive limits, cf. [8], and hence, in many interesting cases, the a.p. of $\operatorname{ind}_{\alpha\to} E_\alpha$ follows from the a.p. for all the spaces E_α already.)

2. THE FUNDAMENTAL LOCALIZATION THEOREM FOR THE a.p.

Let us now turn to the statement (and proof) of the fundamental localization theorem for the a.p. of weighted function spaces, which follows from some known results listed in (I) and (II) at the end of section 1. First we need some notation (and a number of definitions):

<u>6 Definitions</u>. From now on, let X always denote a completely regular Hausdorff space. A non-negative upper semicontinuous function v on X is called a <u>weight</u> (on X). A family $V \neq \emptyset$ of weights on X which is directed in the sense that for all $v_1, v_2 \in V$ and any

$\lambda \geq 0$ there exists $v \in V$ with $\lambda v_1, \lambda v_2 \leq v$ (pointwise on X) is said to be a <u>Nachbin family</u> (on X). We will always assume that $V > 0$, i.e. that for each point $x \in X$ there is $v \in V$ with $v(x) > 0$.

Let F be an arbitrary l.c. space. Two <u>weighted (Nachbin) spaces</u> of continuous F-valued functions on X with respect to the Nachbin family V (on X) are introduced as follows:

$CV^p(X,F) := \{f: X \to F$ continuous; $(vf)(X) := \{v(x)f(x);\ x \in X\}$ precompact in F for each $v \in V\}$,

$CV_o(X,F) := \{f: X \to F$ continuous; $vf: x \to v(x)f(x)$ <u>vanishes at infinity</u> on X (i.e., for each continuous semi-norm p on F and each $\varepsilon > 0$ there is a compact subset K of X with $p((vf)(x)) < \varepsilon$ for all $x \in X\backslash K$) for each $v \in V\}$,

both endowed with the l.c. topology generated by the system $\{b_{v,p};\ v \in V,\ p$ continuous semi-norm on $F\}$ of semi-norms

$$b_{v,p}(f) := \sup_{x \in X} v(x)p(f(x)) \quad \text{for all} \quad f \in CV^p(X,F).$$

We put $CV(X) := CV^p(X,\mathbb{K}) = \{f: X \to \mathbb{K}$ continuous; vf bounded on X for each $v \in V\}$ and $CV_o(X) := CV_o(X,\mathbb{K})$. - Since all $v \in V$ are upper semicontinuous (and hence bounded on compact subsets of X), $CV_o(X,F)$ is a <u>closed</u> subspace of $CV^p(X,F)$, and the assumption $V > 0$ implies that the topology of $CV^p(X,F)$ (resp. $CV_o(X,F)$) is stronger than pointwise convergence on X (and hence Hausdorff).

The spaces $CV(X)$ and $CV_o(X)$ were introduced by L. Nachbin (in connection with the weighted approximation problem). For more information on the weighted spaces $CV^p(X,F)$ and $CV_o(X,F)$ and some examples see e.g. [2] and [4]. The following is a <u>sufficient</u> condition for completeness in weighted spaces ([4], Proposition 22, p.38):

<u>7 Proposition</u>. For a given Nachbin family $V > 0$ on X, let X be a $V_{\mathbb{R}}$-<u>space</u>, i.e.: A function $f: X \to \mathbb{R}$ (or, equivalently, $f: X \to Y$, Y <u>any</u> completely regular [Hausdorff] space) is continuous if (and

always only if) $f|_{\{x\in X;\ v(x)\geq 1\}}$ is continuous for each $v \in V$. Then $CV^p(X,F)$ and $CV_o(X,F)$ are complete for each complete l.c. space F.

As usual, a space X is called a $k_{\mathbb{R}}$-space, if a function $f: X \to \mathbb{R}$ is continuous if (and only if) $f|_K$ is continuous for each compact set $K \subset X$. (All locally compact or metrizable spaces, and more generally the k-spaces of Kelley, are $k_{\mathbb{R}}$-spaces.) Then, if $W \leq V$ holds, i.e. if for each compact subset K of X we can find a weight $v \in V$ with $\inf_{x\in X} v(x) > 0$ (which implies that the topology of $CV^p(X,F)$ resp. $CV_o(X,F)$ is stronger than uniform convergence on compact subsets of X), then any $k_{\mathbb{R}}$-space X is a fortiori a $V_{\mathbb{R}}$-space.

In our proof below, we need an ε-product representation theorem (see [2], II, 2.1. (4) and 3.1.(1) resp. [4], Theorem 24, p.39) which requires the following definition:

8 Definition. $CV^{\sigma,c}(X,F) := \{f: X \to F\,[\sigma(F,F')]$ continuous; $\Gamma((vf)(x))$ [:= absolutely convex hull of $(vf)(X)$] relatively compact in F for each $v \in V\}$,
$CV_o^{\sigma,c}(X,F) := \{f \in CV^{\sigma,c}(X,F);\ vf$ vanishes at infinity (as a function from X into F) for each $v \in V\}$;
on these spaces, the semi-norms $b_{v,p}$ as in 6 are still well-defined, and we equip the spaces with the corresponding l.c. topology (such that $CV^p(X,F) \subset CV^{\sigma,c}(X,F)$ and $CV_o(X,F) \subset CV_o^{\sigma,c}(X,F)$ topologically).

Since on a relatively compact subset of F the topology of F coincides with $\sigma(F,F')$, it is easy to see (cf.[4], Prop.23, p.39) that any function in $CV^{\sigma,c}(X,F)$ is already continuous from X into F if X is a $V_{\mathbb{R}}$-space, and hence we obtain $CV^{\sigma,c}(X,F) = CV^p(X,F)$ and $CV_o^{\sigma,c}(X,F) = CV_o(X,F)$ if X is a $V_{\mathbb{R}}$-space and F quasi-complete.

9 Theorem. (1) $F\varepsilon CV(X) \cong CV(X)\varepsilon F \cong CV^{\sigma,c}(X,F)$ and $F\varepsilon CV_o(X) \cong CV_o(X)\varepsilon F \cong CV_o^{\sigma,c}(X,F)$ (up to the following canonical topological isomorphisms: $u \to (x \to u(\delta_x))$ of $F\varepsilon CV(X)$ [resp.

$F\epsilon CV_o(X)]$ onto $CV^{\sigma,c}(X,F)$ [resp. $CV_o^{\sigma,c}(X,F)$] and $f \to (f' \to f' \circ f)$ of $CV^{\sigma,c}(X,F)$ [resp. $CV_o^{\sigma,c}(X,F)$] onto $CV(X)\epsilon F$ [resp. $CV_o(X)\epsilon F$]).

(2) Hence $CV(X)\epsilon F \cong CV^p(X,F)$ and $CV_o(X)\epsilon F \cong CV_o(X,F)$ hold if X is a $V_{\mathbb{R}}$-space and F quasi-complete.

(3) Moreover, for any topological linear subspace E of $CV(X)$ resp. $CV_o(X)$, we obtain:

$E\epsilon F \cong \{f \in CV^{\sigma,c}(X,F)$ (resp. $CV_o^{\sigma,c}(X,F)$); $f' \circ f: x \to f'(f(x))$ belongs to E for each $f' \in F'\}$ (with the induced topology),

which under the conditions of (2) becomes (more simply):

$E\epsilon F \cong \{f \in CV^p(X,F)$ (resp. $CV_o(X,F)$); $f' \circ f \in E$ for each $f' \in F'\}$.

Since we must make use of the solution of the weighted (Bernstein-Nachbin) approximation problem <u>for Nachbin spaces of cross-sections</u> later on, we recall the necessary definitions and prove an important lemma next.

<u>10 Definitions</u> (cf. [22], [23]). A <u>vector fibration</u> over X is a pair $(X,(F_x)_{x\in X})$, where each F_x is a vector space over the field $\mathbb{K}$. A <u>cross-section</u> is then any element of $\prod_{x\in X} F_x$, i.e., $f = (f(x))_{x\in X}$. A <u>"weight"</u> v on X is a function v on X such that $v(x)$ is a semi-norm on F_x for each $x \in X$. A <u>Nachbin space</u> LV_o is a vector space of cross-sections f such that the mapping $x \to v(x)[f(x)]$ is upper semicontinuous on X and vanishes at infinity for each "weight" $v \in V$, equipped with the l.c. topology defined by the family $\{\|\cdot\|_v\}_{v\in V}$ of semi-norms $\|f\|_v := \sup_{x\in X} v(x)[f(x)]$.

Of course, $CV_o(X,F)$ is certainly a Nachbin space $L\tilde{V}_o$ of cross-sections $f = (f(x))_{x\in X}$ with respect to the vector fibration $(X,(F_x)_{x\in X})$, where $F_x := F$ for each $x \in X$, and to the set $\tilde{V} := \{\tilde{v}_{v,p};\ v \in V,\ p$ continuous semi-norm on $F\}$ of "weights" on X, defined by

$$\tilde{v}_{v,p}(x)[e] := v(x)p(e) \quad \text{for each } x \in X \text{ and } e \in F.$$

However, we observe:

<u>11 Lemma.</u> (1) Let $v \in V$, p a continuous semi-norm on F, and $f \in CV_o^{\sigma,c}(X,F)$ be arbitrary. Then the function $vp\circ f\colon x \to v(x)p(f(x))$ is <u>upper semicontinuous</u> on X.

(2) In the same way as described above for $CV_o(X,F)$, $CV_o^{\sigma,c}(X,F)$ is also <u>a Nachbin space</u> $L\tilde{V}_o$ <u>of cross-sections.</u>

<u>Proof.</u> (1): Let $\epsilon > 0$ and fix an arbitrary point $x \in X$.

<u>First case</u>: $v(x) \neq 0$. Let $0 < \delta \leq 1$ satisfy $\delta p(f(x)) < \frac{\epsilon}{2}$ and put $\alpha := \frac{v(x)+1}{v(x)} > 1$.

Since $C := \alpha\Gamma((vf)(X))$ is relatively compact (and absolutely convex) in F, the (uniform structure resp.) topology of F coincides with $\sigma(F,F')$ on this set, and so p is uniformly continuous on C with respect to $\sigma(F,F')$. Hence there exists a balanced neighbourhood V of 0 in $\sigma(F,F')$ such that $e_1,e_2 \in C$ and $e_1-e_2 \in V$ imply $|p(e_1)-p(e_2)| < \frac{\epsilon}{2}$. As v is upper semicontinuous and $f\colon X \to F\,[\sigma(F,F')]$ continuous, we can find a neighbourhood $U(x)$ of x in S such that $v(y) < v(x) + \delta$ and $f(y) - f(x) \in \frac{1}{v(x)+1} V$ for all $y \in U(x)$. Then for any such $y \in U(x)$ we have:

$v(y)f(y) - v(y)f(x) \in \frac{v(y)}{v(x)+1} V \subset V$ with $v(y)f(y) \in (vf)(X) \subset C$ and $v(y)f(x) = \frac{v(y)}{v(x)} v(x)f(x) \in \alpha C((vf)(X)) \subset C$ (where C indicates the balanced hull), hence $|p(v(y)f(y)) - p(v(y)f(x))| < \frac{\epsilon}{2}$. It follows $p(v(y)f(y)) < p(v(y)f(x)) + \frac{\epsilon}{2} \leq p(v(x)f(x)) + \delta p(f(x)) + \frac{\epsilon}{2} < p(v(x)f(x)) + \epsilon$, that is, $vp\circ f$ is upper semicontinuous at x.

<u>Second case:</u> $v(x) = 0$. Since vf vanishes at infinity (as a function from X into F), there exists a compact subset K of X such that $p(v(y)f(y)) < \epsilon$ for all $y \in X\setminus K$. Since $f\colon X \to F\,[\sigma(F,F')]$ is continuous, $f(K)$ is $\sigma(F,F')$-compact and hence bounded in F; let $M > 0$ satisfy $p(f(y)) \leq M$ for all $y \in K$. Since v is upper semicontinuous and $v(x) = 0$, there exists a neighbourhood $U(x)$ of x in X such that $v(y) < \frac{\epsilon}{M+1}$ for any $y \in U(x)$. Then for any such $y \in U(x)$ we have $p(v(y)f(y)) < \epsilon = p(v(x)f(x)) + \epsilon$, since $y \notin K$ certainly implies $p(v(y)f(y)) < \epsilon$ while $p(v(y)f(y)) \leq \frac{\epsilon}{M+1}\, p(f(y)) < \epsilon$ for any $y \in U(x) \cap K$. So $vp\circ f$ is again upper semicontinuous at x.

(2): By definition, any function $vp\circ f\colon x \to \tilde{v}_{v,p}(x)[f(x)] = v(x)p(f(x))$ with $f \in CV_o^{\sigma,c}(X,F)$ vanishes at infinity, and by (1) this function is also upper semicontinuous on X (for arbitrary $v \in V$ and p = continuous semi-norm on F). Also, the topology of $CV_o^{\sigma,c}(X,F)$ is given by the directed family $\{\|\cdot\|_{\tilde{v}_{v,p}}\}_{v,p}$ of semi-norms

$$\|f\|_{\tilde{v}_{v,p}} = \sup_{x\in X} \tilde{v}_{v,p}(x)[f(x)] = \sup_{x\in X} v(x)\, p(f(x)). \quad \sqsupset$$

The following are the definitions and results we need from approximation theory (cf. [23]):

<u>12 Definitions</u>. Let A be a subalgebra of $C(X)$ (= continuous scalar functions on X), let LV_o be a Nachbin space of cross-sections over X and let Z denote a vector subspace of LV_o which is an A-module (with respect to pointwise multiplication, i.e., $a \in A$ and $z = (z(x))_{x\in X} \in Z$ imply $az = (a(x)z(x))_{x\in X} \in Z$, too). In this context, the <u>weighted (Bernstein-Nachbin) approximation problem</u> asks for a description of the closure of Z in LV_o.

Let $\mathcal{K}$ be a covering of X by pairwise disjoint closed subsets. Z is said to be <u>$\mathcal{K}$-localizable in</u> LV_o if $f \in LV_o$ belongs to $\bar{Z}^{LV_o}$ if (and always only if), given any $K \in \mathcal{K}$, any $v \in V$, and any $\epsilon > 0$, there is some $z \in Z$ such that $v(x)[f(x)-z(x)] < \epsilon$ for all $x \in K$.

$\mathcal{K}_A$ denotes the system of all maximal A-antisymmetric subsets of X: A subset K of X is called A-<u>antisymmetric</u> if $f \in A$, $f|_K$ real-valued always imply $f|_K$ constant. Each A-antisymmetric set is contained in a (with respect to inclusion) <u>maximal</u> A-antisymmetric subset, and hence the system $\mathcal{K}_A$ of all such sets is a covering of X by pairwise disjoint closed sets. - If A is <u>selfadjoint</u>, $\mathcal{K}_A$ coincides with the set of equivalence classes with respect to the equivalence relation $\underset{A}{\sim}$ on X: $x \underset{A}{\sim} y$ if and only if $a(x) = a(y)$ for all $a \in A$.

For simplicity, let us agree to say that Z is <u>localizable under</u> A in LV_o if Z is $\mathcal{K}_A$-localizable in LV_o.

Sufficient conditions for an A-module $Z \subset LV_o$ to be localizable under A in LV_o were derived by Machado-Prolla [22]. (Remark that "Z sharply localizable under A in LV_o" in the terminology of [22] implies "Z localizable under A in LV_o" in our notation.) In Theorem 14 resp. 15 of [22], Machado-Prolla reduce the search for sufficient conditions for localizability to the n-dimensional resp. one-dimensional Bernstein approximation problem (on fundamental weights on $\mathbb{R}^n$ resp. $\mathbb{R}$), generalizing previous results of Nachbin-Machado-Prolla [23] in the selfadjoint case. As corollaries (Theorems 16, 17, 18 of [22]), they derive the analytic resp. quasi-analytic criterion of localizability and localizability in the so-called "bounded case" of the weighted approximation problem. - It would take us too far to state all these results here, and so we confine ourselves to a specialization of Machado-Prolla's "bounded case" (which, however, is essentially enough for all the examples we have in mind).

13 Definition. Let $Z \subset LV_o$ be an A-module. We say that we are in the bounded case (of the weighted approximation problem for Z) if, given any $v \in V$, $a \in A$, and $z \in Z$, the function a is bounded on the support of $v \circ z: x \to v(x)[z(x)]$. This is certainly true if, given any $v \in V$ and $a \in A$, a is bounded on $\overline{\text{supp } v := \{x \in X; v(x) \neq 0\}}$. When the latter condition holds, we say that we are in the restricted bounded case.
So the restricted bounded case occurs for instance, if all the functions $a \in A$ are bounded or if each $v \in V$ has compact support.

14 Theorem (Machado-Prolla [22], Theorem 18). In the bounded case, Z is always localizable under A in LV_o.

In fact, as Machado-Prolla [22], Theorem 18 shows, it is sufficient that

$$(*) \qquad a\big|_{\text{supp}(v \circ z)} \text{ is bounded}$$

for all $v \in V$, $a \in A$, and $z \in G(Z)$, a "set of generators" for Z, that is, the A-module of Z generated by $G(Z)$ is dense in Z for the topology of LV_o. Similarly, the class of functions $a \in A$ for which condition (*) must be "tested" may be restricted to $G(A)$, a so-called "strong set of generators" for A. (If A has a set $G(A)$ of real-valued functions such that the subalgebra of A generated by $G(A)$ is dense in A for the topology of uniform convergence on the compact subsets of X, then $G(A)$ is a strong set of generators; and the whole algebra itself is always such a set.) - On the other hand, Kleinstück's previous result in [21] even assumed the restricted bounded case.

In (the proof of) our fundamental theorem, we assume that E is a topological vector subspace of $CV_o(X)$ which is an A-module for a subalgebra A of $C(X)$ and apply, for arbitrary l.c. space F, solutions of the weighted approximation problem to $Z := E \otimes F$ in $CV_o^{\sigma,c}(X,F) = L\tilde{V}_o$ (cf. 11 (2)). Now Z is clearly an A-module, too, and if $G(E)$ is a set of generators for E (in $CV_o(X)$), then $G(E) \otimes F$ is a set of generators for Z in $CV_o^{\sigma,c}(X,F)$, because $CV_o^{\sigma,c}(X,F)$ (or, equivalently, $CV_o(X,F)$) always induces the ϵ-topology on the tensor product $CV_o(X) \otimes F$ (and hence also on $E \otimes F = Z$), cf. Theorem 9 (1) above (and the remarks after Definition 4). - Moreover, any "reasonable" sufficient condition for localizability applies to $Z = E \otimes F$ (and the set of generators $G(E) \otimes F$) in $L\tilde{V}_o = CV_o^{\sigma,c}(X,F)$ if it applies to E (and the set of generators $G(E)$) in $CV_o(X)$. This is true e.g. for the conditions of Theorems 14 and 15 of [22] (reduction to the n - resp. one-dimensional Bernstein approximation problem) as well as for the analytic and quasi-analytic criterion (say, in the form of [22], Theorems 16 and 17). Again, we state only:

15 Remark. If the A-module $E \subset CV_o(X)$ satisfies (e.g. for a set of generators $G(E)$) the conditions of the bounded [resp. restricted

bounded] case, then the A-module $Z = E \otimes F$ satisfies (for the set $G(E) \otimes F$ of generators) the conditions of the bounded [resp. restricted bounded] case in $CV_o^{\sigma,c}(X,F) = L\tilde{V}_o$ (F any l.c. space).

After all these preparations, we can finally state the fundamental localization theorem for the a.p. of modules in weighted spaces $CV_o(X)$:

16 Theorem. Let X be a completely regular Hausdorff space and $V > 0$ a Nachbin family on X. Let A denote a subalgebra of $C(X)$ and E a topological linear subspace of $CV_o(X)$ which is an A-module. Let $\mathcal{K}$ be a covering of X by pairwise disjoint closed subsets such that, for an arbitrary l.c. space F, (the A-module) $Z := E \otimes F$ is $\mathcal{K}$-localizable in the space $L\tilde{V}_o = CV_o^{\sigma,c}(X,F)$. [E.g. let $\mathcal{K} = \mathcal{K}_A$ and assume that $E \otimes F$ is localizable under A in LV_o.] Then, if $E|_K = \{f|_K;\ f \in E\}$, as a topological linear subspace of $C(V|_K)_o(K)$, has Schwartz's a.p. for each $K \in \mathcal{K}$, the space E has Schwartz's a.p., too.

For instance, if under the assumptions of the first paragraph of the theorem E satisfies the condition of reduction to the n- resp. one-dimensional Bernstein approximation problem ([22], Theorems 14 and 15) or the condition in the analytic or quasi-analytic criterion ([22], Theorems 16 and 17), then by the remarks before 15 (and the results of Machado-Prolla [22] mentioned before 13) $E \otimes F$ is always localizable under A in $L\tilde{V}_o$, and 16 applies. Especially:

17 Corollary. Let X, V, A, and E be as in Theorem 16. Assume that we are in the bounded case of the weighted approximation problem for E in $CV_o(X)$. Then, if $E|_K$ (as a topological linear subspace of $C(V|_K)_o(K)$) has Schwartz's a.p. for each maximal A-antisymmetric subset K of X, the space E has Schwartz's a.p. (and hence the a.p., if E is quasi-complete).

For the restricted bounded case, this generalizes Satz 4.5 of [11], where a completeness assumption (X $V_{\mathbb{R}}$-space) for $CV_o(X)$ was necessary. - The method of proof of Theorem 16 (see below) is the same as for our previous result, but instead of using Kleinstück's solution of the weighted approximation problem for vector-valued continuous functions in the restricted bounded case, we have preferred here to formulate the result in full generality and to apply the recent approximation theorems of Machado-Prolla for Nachbin spaces of cross-sections. This not only allows to relax the approximation-theoretic conditions in the theorem, but, as we have already seen in section 1, the use of Nachbin spaces of cross-sections (inspired by the results of Prolla's paper [25]) in this context <u>makes the previous completeness assumptions superfluous</u>.

<u>Proof of Theorem 16</u>. By Theorem 5 (4), it a enough to show that, for an arbitrary l.c. space F, $E \otimes F$ is dense in $E\varepsilon F$. Since $E\varepsilon F \cong \{f \in CV_o^{\sigma,c}(X,F);\ f' \circ f \in E \text{ for all } f' \in F'\}$ by Theorem 9 (3), we have to prove: Each function $f \in CV_o^{\sigma,c}(X,F)$ with $f' \circ f \in E$ for each $f' \in F'$ belongs to the closure of $E \otimes F$ in $(E\varepsilon F$ or, equivalently) $CV_o^{\sigma,c}(X,F)$. AS $CV_o^{\sigma,c}(X,F)$ is a Nachbin space $L\tilde{V}_o$ of cross-sections (cf. 11 (2)) in which (by assumption) $Z := E \otimes F$ is $\mathcal{K}$-localizable, it suffices to verify that: Given any $K \in \mathcal{K}$, $f|_K$ belongs to the closure of $Z|_K = E|_K \otimes F$ in $L\tilde{V}_o|_K = CV_o^{\sigma,c}(X,F)|_K$ (under the weighted topology generated by the Nachbin family $V|_K$ on K).

But $f|_K \in CV_o^{\sigma,c}(X,F)|_K \subset C(V|_K)_o^{\sigma,c}(K,F)$ (as any $K \in \mathcal{K}$ is closed in X) satisfies $f' \circ (f|_K) = (f' \circ f)|_K \in E|_K \subset CV_o(X)|_K \subset C(V|_K)_o(K)$ for each $f' \in F'$, and by the description $(E|_K)\varepsilon F = \{g \in C(V|_K)_o^{\sigma,c}(K,F);\ f' \circ g \in E|_K \text{ for each } f' \in F'\}$ (which follows again from Theorem 9 (3)), we get $f|_K \in (E|_K)\varepsilon F$. By assumption $E|_K \subset C(V|_K)_o(K)$ has Schwartz's a.p., and so (once more) Theorem 5 (4) implies $f|_K \in \overline{E|_K \otimes F}^{(E|_K)\varepsilon F} \subset \overline{Z|_K}^{C(V|_K)_o^{\sigma,c}(K,F)}$ or $f|_K \in \overline{Z|_K}^{CV_o^{\sigma}(X,F)|_K}$, which is just what we had left to verify. □

<u>Remark</u>. It is not clear whether <u>a converse of 16</u> holds in general, that is, whether Schwartz's a.p. for E also implies Schwartz's a.p.

for the topological subspaces $E|_K$ of $C(V|_K)_o(K)$, $K \in \mathcal{K}$. - In our scheme, this question is of course related to the problem whether, for l.c. (or only Banach) spaces F, $(E\varepsilon F)|_K = (E|_K)\varepsilon F$ holds algebraically. (The inclusion $(E\varepsilon F)|_K \subset (E|_K)\varepsilon F$ is obvious, and the topologies agree. - Remark also that $(E\otimes F)|_K = (E|_K) \otimes F$ is certainly always true; hence $(E|_K)\varepsilon F \subset (E\varepsilon F)|_K$ holds whenever both $(E|_K) \otimes F$ is dense in $(E|_K)\varepsilon F$ and $(E\varepsilon F)|_K$ is a closed subspace.) In fact, if $(E\varepsilon F)|_K = (E|_K)\varepsilon F$ holds for all Banach spaces F, the a.p. of E (by density of $E \otimes F$ in $E\varepsilon F$) yields density of $(E|_K) \otimes F = (E \otimes F)|_K$ in $(E|_K)\varepsilon F = (E\varepsilon F)|_K$ such that $E|_K$ has the a.p. by 5 (5).

In the situation of 16 and for Banach spaces F, we are thus led to ask (and this may be of independent interest): Does a function $f \in C(V|_K)_o^{\sigma,c}(K,F)$ which "extends to E weakly", i.e. satisfies $f' \circ f \in E|_K$ for each $f' \in F'$, extend to an element $g \in E\varepsilon F$, i.e. satisfy $f = g|_K$ for some $g \in CV_o^{\sigma,c}(X,F)$ such that $f' \circ g \in E$ for each $f' \in F'$? - Here the methods of Gramsch [15] (above all, cf. 2.5.-2.8.) can be applied and yield (at least) an idea how one might proceed: Fix $K \in \mathcal{K}$ and a Banach space F. Let E_o be the (by $V > 0$) closed linear subspace $\{e \in E;\ e|_K = 0\}$ of E. Then any $f \in (E|_K)\varepsilon F$ induces a canonical linear mapping $\tilde{f}: F' \to E/E_o$ (say, $\tilde{f}(f') =$ "extension" of $f' \circ f \in E|_K$ to E, modulo E_o) which, as one can immediately verify, is closed for the weak topologies $\sigma(F',F)$ and $\sigma(E/E_o, (E/E_o)')$ [and hence for all stronger topologies]. In many cases, $\tilde{f}$ must then already be continuous from F'_b into E/E_o: This follows sometimes directly from (general) closed graph theorems. Furthermore, it may also turn out that E/E_o carries (it is enough: a topology weaker than) the projective topology of a system $(X_\alpha)_{\alpha \in A}$ of Banach spaces with respect to linear mappings $\pi_\alpha: E/E_o \to X_\alpha$ such that all the compositions $\pi_\alpha \circ \tilde{f}$ are closed linear mappings; it is then sufficient to apply the classical closed graph theorem to get continuity of $\pi_\alpha \circ \tilde{f}$ for all α and hence continuity of $\tilde{f}: F'_b \to E/E_o$

(cf. Gramsch [15], 2.13). Of course, we would like to prove continuity of $\tilde{f}: F'_{cc} \to E/E_o$ rather and then to get a continuous linear "lifting" $\tilde{\tilde{f}}: F'_{cc} \to E$ of $\tilde{f}$ (i.e., for the quotient map $\pi: E \to E/E_o$, $\tilde{f} = \pi \circ \tilde{\tilde{f}}$ holds). In this case (after the canonical identification of $\tilde{\tilde{f}}$ with an element of $CV_o^{\sigma,c}(X,F))$ clearly $\tilde{\tilde{f}}|_K = f$, i.e. $f \in (E \varepsilon F)|_K$. (For the existence of liftings in concrete cases see e.g. Kaballo [18] - [20].)

Naturally, from the point of view of applications, a converse of 16 is of secondary importance anyway: It is much more interesting to derive the a.p. of the "complicated" space E from the a.p. of the "simpler" spaces $E|_K$ than conversely. - In fact, the smaller the sets $K \in \mathcal{K}$ in 16 are, the simpler the spaces $E|_K$ will become. Then the chances are better that the a.p. of all $E|_K$ is already known and that 16 can be applied to prove the a.p. of E.

This is perhaps a good point to remark that, instead of restricting our attention to spaces E of continuous functions as in 16, we could also have started with weighted spaces E of scalar functions on X of which only the restrictions to certain "characteristic" subsets of X are continuous: For a given completely regular space X, a Nachbin family $V > 0$ on X and a l.c. space F let, as in [4], p.39, $\mathfrak{F}_V$ denote the system of all sets $F_v := \{x \in X; v(x) \geq 1\}$, $v \in V$, and $RV_o(X,F) := \{f: X \to F;$ $f|_S$ continuous for each $S \in \mathfrak{F}_V$, $(vf)(X)$ precompact in F and vf vanishes at infinity for each $v \in V\}$, equipped with the natural l.c. topology given by the system $\{b_{v,p}\}$ of semi-norms as defined in 6. Again put $RV_o(X) := RV_o(X,\mathbb{K})$. $RV_o(X,F)$ is complete whenever F is.

We will assume from now on that $W \leq V$. Then each function $f \in RV_o(X,F)$ is a fortiori hypocontinuous, i.e. the restriction of f to each compact subset of X is continuous, and so clearly $(vf)(X)$ is precompact in F whenever vf vanishes at infinity. This yields, more simply, $RV_o(X,F) = \{f: X \to F; f|_{\{x \in X; v(x) \geq 1\}}$

is continuous and vf vanishes at infinity for each $v \in V\}$.

From [4], Theorem 27, p. 40, we know that for quasi-complete F

$$RV_o(X)\varepsilon F = RV_o(X,F)$$

holds, and so for topological linear subspaces E of $RV_o(X)$ we get again: $E\varepsilon F \cong \{f \in RV_o(X,F);\ f'\circ f \in E \text{ for all } f' \in F'\}$.

Similarly as in Lemma 11 above, let us now prove that for arbitrary $v \in V$, p continuous semi-norm on F and $f: X \to F$ hypocontinuous with vf vanishing at infinity the function $vp\circ f: x \to v(x)\ p(f(x))$ is still <u>upper semicontinuous</u> on X: To do so, take $\varepsilon > 0$ and fix $x \in X$. Since vf vanishes at infinity there exists a compact subset K of X such that $p(v(y)f(y)) < p(v(x)f(x)) + \varepsilon$ for all $y \in X\setminus K$. If $x \notin K$, $X\setminus K$ is an open neighbourhood $U(x)$ of x in X such that $p(v(y)f(y)) < p(v(x)f(x)) + \varepsilon$ for all $y \in U(x)$. Now let $x \in K$. Since f is hypocontinuous, $f|_K$ is continuous, and so ($v \in V$ being upper semicontinuous) there exists a neighbourhood $U(x)$ of x in X such that, with $\delta := \min(1, \frac{\varepsilon}{3(p(f(x))+1)})$ and $\tilde{\delta} := \min(\frac{\varepsilon}{3}, \frac{\varepsilon}{3(v(x)+1)})$, $v(y) < v(x) + \delta$ for all $y \in U(x)$ and $p(f(y)-f(x)) < \tilde{\delta}$ for all $y \in U(x) \cap K$. Then we get again $p(v(y)f(y)) < p(v(x)f(x)) + \varepsilon$ for all $y \in U(x)$ since this is true whenever $y \in X\setminus K$ while for $y \in U(x) \cap K$:

$$v(y)p(f(y)) < (v(x)+\delta)(p(f(x))+\tilde{\delta}) = v(x)p(f(x)) + \delta p(f(x))+\tilde{\delta}v(x) + \delta\tilde{\delta}$$
$$< v(x)p(f(x)) + \frac{\varepsilon}{3} + \frac{\varepsilon}{3} + \frac{\varepsilon}{3} = v(x)p(f(x)) + \varepsilon.$$

So f is upper semicontinuous at x.

It follows that even $RV_o(X,F)$ is a Nachbin space $L\tilde{V}_o$ of cross-sections in the same way as before (cf. 11 (2))! An inspection of the proof of 16 (and use of, say, 5 (5) instead of 5 (4)) now shows:

<u>18 Theorem</u>. If $W \leq V$, then 16 and 17 hold also for subspaces E of $RV_o(X)$ (and $L\tilde{V}_o = RV_o(X,F)$, F quasi-complete or Banach, as

well as $R(V|_K)_o(K)$, $K \in \mathcal{K})$ [instead of $CV_o(X)$, $L\tilde{V}_o = CV_o^{\sigma,c}(X,F)$, and $C(V|_K)_o(K)$, respectively]. - Remark that E then even has <u>the a.p.</u> whenever it is <u>closed</u> in $RV_o(X)$.

Finally, let us note that our method of proof (as outlined in section 1) can also be applied, mutatis mutandis, to A-<u>modules</u> <u>$E \subset CV(X)$</u> (instead of $CV_o(X)$): Here the ε-product representation of $E\varepsilon F$ (F an arbitrary l.c. space) as a space of F-valued functions on X (in fact, a topological linear subspace of $CV^p(X,F)$ for $V_{\mathbb{R}}$-spaces X and quasi-complete F) is contained in Theorem 9, too, and, say, Theorem 9 (resp. 11) of section 2 of Kleinstück [21] yieldsa "localization" of approximation from A-modules in $CV^p(X,F)$ in the (restricted) "bounded ($\hat{X}$-) case" (if CV(X) satisfies condition (*) of [21], p.11). Since Kleinstück's approximation theorems work "only" in spaces of <u>continuous</u> functions, the (completeness type) assumption X $V_{\mathbb{R}}$-space is needed. Moreover, the results necessarily involve compactifications $\hat{X}$ of X (and extensions $(vf)^\wedge$ to $\hat{X}$ of the functions vf for arbitrary $v \in V$ and $f \in CV^p(X,F)$). As no "splitting" $(vf)^\wedge = v^\wedge f^\wedge$ is possible in general, we must rather suppose in this case that, for arbitrary complete l.c. (or Banach) space F and arbitrary $v \in V$, the F-valued functions $(vf)^\wedge$ with $f \in E\varepsilon F$ can be approximated, uniformly on maximal A_V-antisymmetric subsets K_{A_V} of $\hat{X}_V$, by functions $(vz)^\wedge$ with $z \in E \otimes F$ (instead of assuming Schwartz's a.p. for restrictions $E|_K$ of E to maximal A-antisymmetric subsets K of X as in 17). - The corresponding proposition is (necessarily) <u>technically involved</u>, and as we do not want to repeat the notation and the definitions of [21] here, we leave its exact formulation to the interested reader. Because of the technical difficulties, this proposition is not as useful as 17, but Kleinstück [21], section 3, Theorem 5 and Korollar 6, has still been able to derive at least the a.p. of CV(X) on $V_{\mathbb{R}}$-spaces X (and a slightly stronger result) in this way.

3. EXAMPLES FOR THE LOCALIZATION OF THE a.p. AMONG WEIGHTED SPACES WITH MIXED DEPENDENCE ON SUBSETS OF A PRODUCT

We turn to a number of examples for the localization of the a.p. which follow from Corollary 17 and which illustrate the application of the fundamental localization Theorem 16 to <u>one</u> concrete situation, namely to the case of weighted spaces "with mixed dependence" on subsets of a topological product (cf. [11]).

<u>The general setting</u> of these examples is as follows: Let Ω and X be completely regular (Hausdorff) spaces, $\Lambda \subset \Omega \times X$ a topological subspace, $V > 0$ a Nachbin family on Λ and Y a topological linear subspace of $CV_o(\Lambda)$. The canonical projection $\Omega \times X \to \Omega$ is denoted by π_1. For $t \in \pi_1(\Lambda)$ we identify $\Lambda_t := \{x \in X;\ (t,x) \in \Lambda\}$ with the "slice" $\{t\} \times \Lambda_t$. Correspondingly, we identify $Y_t := Y|_{\{t\} \times \Lambda_t}$ with a function space on $\Lambda_t \subset X$ and $V_t := V|_{\{t\} \times \Lambda_t}$ with a Nachbin family on Λ_t, too. Y_t is then a topological linear subspace of $C(V_t)_o(\Lambda_t)$.

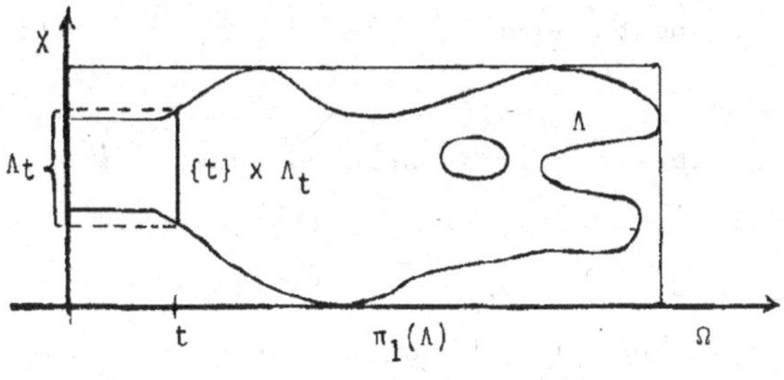

<u>19 Proposition</u>. Let Y be a module over a subalgebra A of $C(\Lambda)$ and assume that we are in the bounded case of the weighted approximation problem for Y in $CV_o(\Lambda)$ (e.g., let $a|_{\mathrm{supp}\, v}$ be bounded for each $a \in A$ and each $v \in V$). We suppose that, for an appropriate subset T of $\pi_1(\Lambda)$, $K_A = \{\{t\} \times \Lambda_t;\ t \in T\} \cup \{\{p\};\ p \in \Lambda \setminus (\bigcup_{t \in T} \{t\} \times \Lambda_t)\}$. Then if the completion $\hat{Y}_t$ of Y_t has (Schwartz's) a.p. for each $t \in T$, $\hat{Y}$ has Schwartz's a.p., too.

Proof. We have only to check that $Y|_K$ has Schwartz's a.p. for each $K \in \mathcal{K}_A$ and may then apply Corollary 17. But in case $K \in \mathcal{K}_\Lambda$ is the one point set $\{p\}$, $p \in \Lambda \setminus (\bigcup_{t\in T} \{t\}\times\Lambda_t)$, $Y|_K$ trivially has the a.p., whereas for each $t \in T$, Schwartz's a.p. for $Y|_{\{t\}\times\Lambda_t} \cong Y_t$ follows from the a.p. of the completion $\hat{Y}_t$ (cf. Séminaire Schwartz [27], Exposé nº 15, Théorème 7). □

To demonstrate the rôle of the set $T \subset \pi_1(\Lambda)$ in 19, we consider just one simple example:

20 Example. Let $X = \mathbb{C}^N$ $(N \geq 1)$ [or, more generally, a quasi-complete dual-nuclear locally convex space]. Let Λ be a subspace of $\Omega \times X$ and T a closed subset of $\pi_1(\Lambda) \subset \Omega$ such that Λ_t is open in X for each $t \in T$ (e.g. $\Lambda \subset \Omega\times X$ open). $C\mathcal{O}_T(\Lambda) \subset \Omega$ denotes the space of all continuous functions f on Λ such that $f(t,\cdot)$ is holomorphic [i.e. continuous and G-analytic] on Λ_t for each $t \in T$, endowed with the topology of uniform convergence on all compact subsets of Λ. Then $C\mathcal{O}_T(\Lambda)$ has Schwartz's a.p.

Proof. $Y := C\mathcal{O}_T(\Lambda)$ is a module over the algebra $A := \{f \in CB(\Lambda)$ (i.e. continuous and bounded on Λ); $f|_{\{t\}\times\Lambda_t}$ constant for each $t \in T\}$.

We are in the restricted bounded case of the weighted approximation problem for $C\mathcal{O}_T(\Lambda)$ in $CV_o(\Lambda)$, where $V = W = \{\lambda\chi_K;\ \lambda > 0,\ K \text{ compact in } X\}$ (and $\chi_K :=$ characteristic function of K). Since A is selfadjoint, $\mathcal{K}_A$ coincides with the system of maximal subsets of Λ on which all functions in A are constant. Now (real-valued) bounded continuous functions on $\pi_1(\Lambda)$ separate points, and by constant "extension" along X it follows that each set $K \in \mathcal{K}_A$ is contained in a slice $\{t\}\times\Lambda_t$, $t \in \pi_1(\Lambda)$. Since T is closed, each slice $\{t_o\} \times \Lambda_{t_o}$ with $t_o \notin T$ clearly "splits up" in one point sets. (There exists a [real-valued] $\varphi \in CB(\pi_1(\Lambda))$ with $\varphi(t_o) = 1$, $\varphi|_T = 0$; then the functions $\varphi \otimes g|_\Lambda : \lambda = (t,x) \to \varphi(t)g(x)$, g [real-valued] $\in CB(X)$, belong to A

and separate points of $\{t_o\} \times \Lambda_{t_o}$.) Hence $\mathcal{K}_A = \{\{t\}\times\Lambda_t;\ t \in T\} \cup \{\{p\};\ p \in \Lambda \setminus (\bigcup_{t\in T} \{t\} \times \Lambda_t)\}$, and 19 applies: For each $t \in T$, Y_t is a topological subspace of $(\mathcal{O}(\Lambda_t),co)$, the space of holomorphic functions on Λ_t, endowed with the topology of uniform convergence on all compact subsets of Λ_t. Since $(\mathcal{O}(\Lambda_t),co)$ is nuclear [in the infinite dimensional case, this follows from a recent theorem of Boland and Waelbroeck], Y_t (and, a fortiori, $\hat{Y}_t$) is nuclear, too, and hence has the a.p.. □

For all other examples from now on, we will no longer be interested in this (full) generality of 19, thus we will assume that $T = \pi_1(\Lambda)$, i.e. $\mathcal{K}_A = \{\{t\}\times\Lambda_t;\ t \in \pi_1(\Lambda)\}$, holds. This is certainly <u>the most interesting case</u>.

Let us introduce some notation at this point: $\mathfrak{F}$ will always denote a (topological) <u>sub-(pre-)sheaf</u> of the sheaf C_X of continuous functions on X, that is, for each open subset U of X, $\mathfrak{F}(U)$ denotes a (topological) linear subspace of $(C(U),co)$, the space of all continuous (scalar) functions on U (endowed with the compact-open topology co of uniform convergence on all compact subsets of U). - The first case in which we are interested is in some sense similar to Example 20 and requires $\Lambda \subset \Omega\times X$ to be a subspace with the property that Λ_t is <u>open</u> in X for each $t \in \pi_1(\Lambda)$. Then we can define:

<u>21 Definition</u>. $C\mathfrak{F}(\Lambda) := \{f$ continuous on $\Lambda;\ f(t,\cdot) \in \mathfrak{F}(\Lambda_t)$ for each $t \in \pi_1(\Lambda)\}$, endowed with the topology co of uniform convergence on all compact subsets of Λ; $C\mathfrak{F}V_o(\Lambda) := C\mathfrak{F}(\Lambda) \cap CV_o(\Lambda)$ with the topology induced by $CV_o(\Lambda)$, and, similarly, for $t \in \pi_1(\Lambda)$: $\mathfrak{F}(V_t)_o(\Lambda_t) := \mathfrak{F}(\Lambda_t) \cap C(V_t)_o(\Lambda_t)$ (with the weighted topology from the Nachbin family V_t).

<u>22 Proposition</u>. Let $Y := C\mathfrak{F}V_o(\Lambda)$ and assume that one of the following conditions (a) or (b) is satisfied for each $t \in \pi_1(\Lambda)$:

(a) $\mathfrak{F}(V_t)_o(\Lambda_t)$ is <u>nuclear</u>,

(b) Y_t is <u>dense</u> in $\mathfrak{F}(V_t)_o(\Lambda_t)$, and $\mathfrak{F}(V_t)_o(\Lambda_t)$ has Schwartz's a.p. Then $C\mathfrak{F}V_o(\Lambda)$ has Schwartz's a.p., too.

<u>Proof</u>. Take A as in Example 20 (with $T = \pi_1(\Lambda)$), so $K_A = \{\{t\}\times\Lambda_t, t \in \pi_1(\Lambda)\}$ holds. Again Y is quite obviously a module over this algebra. Then for each $t \in \pi_1(\Lambda)$ Schwartz's a.p. for $Y|_{\{t\}\times\Lambda_t} \cong Y_t$ follows from our conditions: In case (a), Y_t is a subspace of the nuclear space $\mathfrak{F}(V_t)_o(\Lambda_t)$ and hence nuclear, too, while in case (b), Y_t inherits Schwartz's a.p. from $\mathfrak{F}(V_t)_o(\Lambda_t)$ of which it is a dense topological linear subspace. ☐

Under additional completeness assumptions, Theorem 19 and Proposition 22 had already been given as 4.7 and 4.8 in [11]. See ([10] and) [11] for examples of concrete sheaves $\mathfrak{F}$, for some concrete examples following from 22 and for applications to vector-valued functions and "<u>density theorems</u>" (which we will not repeat here). - 4.9 and 4.10 of [11] considered only the compact-open topology, and 4.10 (2) is the only example for an application of case (b) in Proposition 22 which was mentioned there. (More examples were promised in [11], and we are now ready to keep this promise.)

<u>The density condition</u> in 22 (b) leads to <u>a "regularity" assumption</u> on the set Λ, as we shall see in a moment (cf. also [11], 2): We will state a <u>sufficient</u> condition which implies the required density in our next remark. - For the rest of our discussion of this first case, we will assume that Λ <u>is an open subset of</u> $\Omega\times X$; this is certainly satisfied in the most interesting examples based on 22.

<u>23 Remark</u>. Fix $t_o \in \pi_1(\Lambda)$. Let $\mathfrak{F}_{t_o}V$ denote the set of all functions g as follows:

We suppose that there is an open neighbourhood $U \subset \pi_1(\Lambda)$ of t_o in Ω and a function $\varphi \in CB(\Omega)$ with $\varphi(t_o) \neq 0$, but $\varphi \equiv 0$ outside of U. Let Λ_U be the open topological subspace $\bigcup_{t\in U} \Lambda_t$ of X.

Then $g \in \mathfrak{F}(\Lambda_U)$, and the function $\varphi_g \in C\mathfrak{F}(\Lambda)$, defined by

$$\varphi_g(t,x) = \begin{cases} \varphi(t)g(x), & t \in \bar{U} \text{ and } x \in \Lambda_t \\ 0 \quad , & \text{elsewhere on } \Lambda \end{cases},$$

belongs to $C\mathfrak{F}V_o(\Lambda)$ (i.e. satisfies the weight conditions induced by V). If $\mathfrak{F}_{t_o}V|_{\Lambda_{t_o}}$ is <u>dense</u> in $\mathfrak{F}(V_{t_o})_o(\Lambda_{t_o})$, then density of Y_{t_o} (for $Y = C\mathfrak{F}V_o(\Lambda)$) in $\mathfrak{F}(V_{t_o})_o(\Lambda_{t_o})$ holds, as required in 22 (b).

<u>Proof</u>. Let U, φ and g be arbitrary as in the definition of $\mathfrak{F}_{t_o}V$. Let $(t,x) \in \Lambda$ satisfy $t \in \bar{U}\setminus U$. Since Λ is open, there exist neighbourhoods U_1 of t in Ω and U_2 of x in X such that $U_1 \times U_2 \subset \Lambda$. Because of $t \in \bar{U}$ we can find $t_1 \in U_1 \cap U$. Then $(t_1,x) \in U_1 \times U_2 \subset \Lambda$, and hence $x \in \Lambda_{t_1} \subset \Lambda_U$ holds. It is now clear that φ_g is <u>well-defined and continuous</u> on Λ and so belongs to $C\mathfrak{F}(\Lambda)$.

After this remark, it suffices to point out that for any $g \in \mathfrak{F}_{t_o}V$ obviously

$$g|_{\Lambda_{t_o}} = \frac{1}{\varphi(t_o)}\varphi_g(t_o,\cdot) \in Y_{t_o}$$

holds and that therefore density of $\mathfrak{F}_{t_o}V|_{\Lambda_{t_o}}$ obviously implies $Y_{t_o} \subset \mathfrak{F}(V_{t_o})_o(\Lambda_{t_o})$ dense. ❐

We note that the definition of $\mathfrak{F}_{t_o}V$ in 23 involves <u>two different restrictions</u> (as compared with elements of $\mathfrak{F}(V_{t_o})_o(\Lambda_{t_o})$): The functions $g \in \mathfrak{F}_{t_o}V$ are defined and belong to the sheaf $\mathfrak{F}$ on some open set Λ_U which may be <u>strictly</u> larger than Λ_{t_o}, and their constant "extensions" along Ω, multiplied by a suitable "cut-off" function φ, must <u>satisfy the weight conditions</u> given by V on $\bigcup_{t\in U}\{t\} \times \Lambda_t \subset \Lambda$. <u>The cutting-off process</u> was introduced here in order to let the elements by which we approximate in $\mathfrak{F}(V_{t_o})_o(\Lambda_{t_o})$ belong to the sheaf $\mathfrak{F}$ only on some Λ_U (and not on all of $\bigcup_{t\in\pi_1(\Lambda)}\Lambda_t$), and, on the other hand, to take care of the growth conditions "in Ω-direction". (If Ω is locally compact, U may be chosen relatively compact, then φ has compact support in Ω.) -

The regularity assumption on Λ which we have mentioned before 23 comes - even if weight conditions do not exist, cf. [11], 4.10 (2) - from the approximation in $\mathfrak{F}(V_{t_o})_o(\Lambda_{t_o})$ by functions extending to elements of $\mathfrak{F}$ on some open set $\Lambda_U \supset \Lambda_{t_o}$.

In our next corollary, we will put much stronger regularity conditions on Λ than an application of 23 would really require. We will also restrict our attention to the case that $\mathfrak{F} = \mathfrak{O}$, the sheaf of holomorphic functions on $\mathbb{C}^N$ $(N \geq 1)$ here.

24 Corollary. Let $X = \mathbb{C}^N$ $(N \geq 1)$, and let Λ be an open subset of $\Omega \times X$. Then $C\mathfrak{O}V_o(\Lambda)$ has Schwartz's a.p. if for each $t \in \pi_1(\Lambda)$ the following conditions hold:

(i) $\mathfrak{O}(V_t)_o(\Lambda_t)$ has Schwartz's a.p.,

(ii) for each polynomial p (on $\mathbb{C}^N$), constant "extension" along Ω leads, after multiplication by a suitable cut-off function $\varphi \in CB(\Omega)$ with $\varphi(t) \neq 0$, to a function in $C\mathfrak{O}V_o(\Lambda)$ (i.e. the function φ_p, defined as in 23, satisfies the weight conditions given by V),

(iii) (restrictions of) polynomials are dense in $\mathfrak{O}(V_t)_o(\Lambda_t)$.

It should be clear by now that 24 is immediate from 22 (and 23). - Condition (ii) of 24 already implies that $Y_t \subset \mathfrak{O}(V_t)_o(\Lambda_t)$ (for $Y = C\mathfrak{O}V_o(\Lambda)$) contains all (restriction of) polynomials. Together with (iii), this is certainly a rather crude way to ensure that Y_t is dense in $\mathfrak{O}(V_t)_o(\Lambda_t)$! However, as on the question when (iii) is satisfied much information can be found in the literature, Corollary 24 makes it easy to construct concrete examples. - Roughly spoken, by condition (ii), V may impose rather arbitrary growth conditions in Ω-direction (above all if Ω is locally compact), but V (or rather the Nachbin families V_t on Λ_t for $t \in \pi_1(\Lambda)$) must allow polynomial growth in X-direction. (This is indeed fulfilled if the sets

Λ_t are relatively compact and if all the weights in V_t vanish at infinity on Λ_t.)

Not too many general results on the a.p. of the spaces $\mathfrak{G}V_o(G)$ on open sets $G \subset \mathbb{C}^N$ are known, however. All these results require a very special form of G and / or very special Nachbin families V. Let us mention the following:

(1) On the open unit disk $D \subset \mathbb{C}$, $\mathfrak{G}V_o(D)$ has the a.p. if V consists of "normal" weights v only (i.e. v is positive, continuous, and radial, and there exist $0 < \epsilon < k$ and $r_o < 1$ such that for all $r \geq r_o$ as $r \to 1-$: $\frac{v(r)}{(1-r)^{\epsilon}} \searrow 0$ and $\frac{v(r)}{(1-r)^k} \nearrow \infty$). - This follows from a theorem of Shields-Williams, see [8], §2, where we have also pointed out that then the a.p. of $\mathfrak{G}V_o(G)$ follows for (certain) simply connected regions G in $\mathbb{C}$ and some product regions $G \subset \mathbb{C}^N$ (if V is of a special form).

(2) Kaballo carried the methods of (1) over to more general domains in $\mathbb{C}^N$ (and developed new methods) to prove analogues of the Shields-Williams theorem, and to get the a.p., say, for $\mathfrak{G}v_o(G)$ ($= \mathfrak{G}(V_v)_o(G)$ with $V_v := \{\lambda v; \lambda > 0\}$)

- on the unit ball G (or a polydisc G) $\subset \mathbb{C}^N$ if the weight v satisfies only a weaker condition than normality ([19], 3.12), or

- on a bounded strictly pseudoconvex region G with C^∞-boundary if the weight is normal in a restricted sense ([20], 2.7).

- Moreover, $\mathfrak{G}V_o(G)$ also has the a.p. on a bounded region G which is "approximable from the exterior" if V is a countable Nachbin family of "admissible" weights which satisfies a certain "compactness" condition ([18], 6.6).

(3) Some work has also been done in the case $V = C_o^+(G)$ (= all non-negative continuous functions which vanish at infinity on G), when $\mathfrak{G}V_o(G) = (H^\infty(G), \beta)$ is the space of all bounded holomorphic functions with the strict topology β: In [2] and [3], we established the a.p.

of this space for arbitrary simply connected regions $G \subset \mathbb{C}$ and for products of such regions. Recently Kaballo ([18], 6.6 and 6.9) proved the a.p. of $(H^\infty(G),\beta)$ also for strictly pseudoconvex regions $G \subset \mathbb{C}^N$ with C^4-boundary and for bounded regions which are approximable from the exterior.

Normal weights clearly vanish at infinity. Remark also that (say, by the results of Shields-Williams) polynomials are dense in $\mathfrak{G}V_o(D)$ for Nachbin families V of normal weights on the open unit disk. Hence we can easily construct examples of sets Λ and of Nachbin families V, such that each V_t consists of normal weights on Λ_t $(t \in \pi_1(\Lambda))$, and such that $C\mathfrak{G}V_o(\Lambda)$ has Schwartz's a.p. by 24 (and the preceding remarks).

We leave the formulation of a general theorem of this type to the reader and note only the following (somewhat "curious") situation: Even on product sets, say, on $\Lambda = (0,1) \times D$, it is possible to exhibit Nachbin families V of continuous functions, with V_t consisting only of normal weights on D for each $t \in (0,1)$, such that $C\mathfrak{G}V_o(\Lambda)$ (is necessarily complete and) has the a.p. by 24, but such that no ε-tensor product "decomposition" of $C\mathfrak{G}V_o(\Lambda)$ holds (which would allow an easier proof of the a.p.)!

If the Nachbin family V is of the very special form $V_1 \otimes V_2$ (with Nachbin families V_1 on Ω and V_2 on X), however, general "slice product theorems" (cf. [3]) will usually give

$$C\mathfrak{F}(V_1 \times V_2)_o(\Omega \times X) = C(V_1)_o(\Omega) \check{\otimes}_\varepsilon \mathfrak{F}(V_2)_o(X),$$

and then the a.p. of $C\mathfrak{F}V_o(\Omega \times X)$ may also follow from the fact (Schwartz [29], p.48) that $E\varepsilon F$ inherits Schwartz's a.p. from the (quasi-complete) spaces E and F. (Because of this, we had only been interested in subsets of products and not in product sets in [11].)

Let us finish the first case with another example which is based on the results recalled in (3) above:

25 Example. Let $X = \mathbb{C}$, let Ω be locally compact and $\Lambda \subset \Omega \times \mathbb{C}$ open. We assume that V is a Nachbin family of continuous weights on Λ such that for each $t \in \pi_1(\Lambda)$:

(i) Λ_t is a bounded region in $\mathbb{C}$ with $\Lambda_t = \mathring{\overline{\Lambda}}_t$ and $\mathbb{C} \setminus \overline{\Lambda_t}$ connected,

(ii) $V_t = C_o^+(\Lambda_t)$ (up to "equivalence"),

(iii) each polynomial belongs to Y_t (for $Y = C\mathfrak{G}V_o(\Lambda)$).

Then $C\mathfrak{G}V_o(\Lambda)$ has the a.p.

Proof. Local compactness of Λ, continuity of the weights and (ii) combine to yield $CV_o(\Lambda)$ complete; then its closed subspace $C\mathfrak{G}V_o(\Lambda)$ is complete, too. - Now fix $t \in \pi_1(\Lambda)$. Since $\Lambda_t = \mathring{\overline{\Lambda}}_t$ holds, $\mathbb{C} \setminus \Lambda_t$ is the closure of the connected set $\mathbb{C} \setminus \overline{\Lambda}_t$ by (i) and hence itself connected so that Λ_t is simply connected. Clearly (ii) gives $\mathfrak{G}(V_t)_o(\Lambda_t) = (H^\infty(\Lambda_t), \beta)$ which then has the a.p. by (3) above. Moreover, (i) allows to apply the well-known theorem of Farrel on "pointwise bounded" approximation by polynomials to get density of polynomials in $(H^\infty(\Lambda_t), \beta)$. Hence the proof is finished by 24. □

Several different sufficient conditions which imply 25 (iii) are conceivable, but instead of discussing possible special cases of 25, we turn to the second general class of weighted spaces with mixed dependence in which we are interested and where the localization method can be applied to prove the a.p.

We start by introducing (resp. recalling, cf. [7]) some notation: Ω and X are like at the beginning of this paragraph, and $\mathfrak{F}$ again denotes a (pre-) sheaf (of continuous functions) on X. We assume now that the topological subspace $\Lambda \subset \Omega \times X$ has the property that Λ_t is compact in X for each $t \in \pi_1(\Lambda)$. Let V be a Nachbin family on Λ which satisfies $W \leq V$ (such that the weighted topology of $CV_o(\Lambda)$ is stronger than uniform convergence on compact subsets of Λ, and hence $CV_o(\Lambda)$ is complete whenever Λ is a

$k_{\mathbb{R}}$-space, see 7 above). Then for any topological linear subspace Y of $CV_o(\Lambda)$, the space Y_t is obviously (topologically isomorphic to) a <u>normed</u> subspace of $C(\Lambda_t)$ (= Banach space of all continuous functions on Λ_t under its canonical sup-norm) for each $t \in \pi_1(\Lambda)$, because upper semicontinuous functions are bounded on compact subsets. (<u>So there are no weight conditions whatsoever in</u> X-<u>direction</u> here!) Hence a condition like the one in 22(a) is of no use in this case, but under the assumptions of 19 the space Y has Schwartz's a.p. if for each $t \in T$ (here again $= \pi_1(\Lambda)$ for simplicity) the closure $\overline{Y_t}$ of Y_t in $C(\Lambda_t)$ has the a.p.

For a compact set K in X and a l.c. space F we define (with C(K,F) = space of all continuous F-valued functions on K under the topology of uniform convergence on K):

$A_{\mathfrak{F}}(K,F) := \{f \in C(K,F);\ f' \circ f|_{\overset{\circ}{K}} \in \mathfrak{F}(\overset{\circ}{K})$ for each $f' \in F'\}$, and

$H_{\mathfrak{F}}(K,F)$:= the closure in $C(K,F)$ of

$\{f \in C(K,F)$; there exists an open neighbourhood U of K (depending on f) and a function $g: U \to F$ continuous with $f' \circ g \in \mathfrak{F}(U)$ for any $f' \in F'$ such that $g|_K = f\}$.

If $\mathfrak{F}$ is a closed locally convex sub- (pre-) sheaf of C_X, i.e. if for each open subset U of X the topological linear subspace $\mathfrak{F}(U)$ of $(C(U),co)$ is closed - <u>which we will assume from now on</u> -, $A_{\mathfrak{F}}(K,F)$ is a <u>closed</u> subspace of $C(K,F)$ and hence $H_{\mathfrak{F}}(K,F) \subset A_{\mathfrak{F}}(K,F)$ holds. Both spaces are endowed with the topology induced by $C(K,F)$. If $F = \mathbb{K}$, we omit this symbol and so have introduced the Banach spaces $A_{\mathfrak{F}}(K)$ and $H_{\mathfrak{F}}(K)$.

<u>26 Definition</u>. $CA_{\mathfrak{F}}(\Lambda,F)$ [resp. $CH_{\mathfrak{F}}(\Lambda,F)$] := $\{f: \Lambda \to F$ continuous; $f(t,\cdot) \in A_{\mathfrak{F}}(\Lambda_t,F)$ [resp. $H_{\mathfrak{F}}(\Lambda_t,F)$] for each $t \in \pi_1(\Lambda)\}$, endowed with the topology co of uniform convergence on all compact subsets of Λ;

$$CA_{\mathfrak{F}}V_o(\Lambda,F) := CA_{\mathfrak{F}}(\Lambda,F) \cap CV_o(\Lambda,F)$$

and

$$CH_{\mathfrak{Z}}V_o(\Lambda,F) := CH_{\mathfrak{Z}}(\Lambda,F) \cap CV_o(\Lambda,F),$$

both endowed with the weighted topology induced by $CV_o(\Lambda,F)$. - If $F = \mathbb{K}$, we again omit this symbol in each case.

Under our general assumptions ($\mathfrak{Z}$ closed and $W \leq V$), $CH_{\mathfrak{Z}}V_o(\Lambda,F) \subset CA_{\mathfrak{Z}}V_o(\Lambda,F)$ are closed subspaces of $CV_o(\Lambda,F)$ and hence complete whenever Λ is a $k_{\mathbb{R}}$-space and F complete. - The following proposition is for the present case what 22(b) was for the first case:

<u>27 Proposition</u>. Let $Y := CA_{\mathfrak{Z}}V_o(\Lambda)$ [resp. $CH_{\mathfrak{Z}}V_o(\Lambda)$]. Assume that for each $t \in \pi_1(\Lambda)$ the following two conditions hold:

(a) Y_t is <u>dense</u> in $A_{\mathfrak{Z}}(\Lambda_t)$ [resp. $H_{\mathfrak{Z}}(\Lambda_t)$],

(b) $A_{\mathfrak{Z}}(\Lambda_t)$ [resp. $H_{\mathfrak{Z}}(\Lambda_t)$] has the a.p.

Then Y has Schwartz's a.p. (and hence the a.p. if Λ is a $k_{\mathbb{R}}$-space).

<u>Proof</u>. After our preceding remarks we have only to observe that Y is again quite obviously a module over the algebra A, defined as in Example 20 (with $T = \pi_1(\Lambda)$).

Conditions (a) and (b) imply that Y_t has the a.p. as dense subspace of a Banach space with a.p. for each $t \in \pi_1(\Lambda)$, so we may apply 19. □

As before, the density assumption in 27 (a) leads to regularity restrictions on Λ. In fact:

<u>28 Remark</u>. Fix $t_o \in \pi_1(\Lambda)$. With $Y = CA_{\mathfrak{Z}}V_o(\Lambda)$, the following are two examples of <u>sufficient</u> conditions for density of Y_{t_o} in $A_{\mathfrak{Z}}(\Lambda_{t_o})$, as required in 27 (a):

(i) Let $A_{\mathfrak{Z}}^{t_o}V$ denote the set of all functions g as follows: We suppose that there is a neighbourhood U of t_o in $\pi_1(\Lambda)$ with $\Lambda_U := \bigcup_{t\in U} \Lambda_t$ <u>relatively compact</u> in X and a function $\varphi \in CB(\pi_1(\Lambda))$ with $\varphi(t_o) \neq 0$, but $\varphi \equiv 0$ off U. Then $g \in A_{\mathfrak{Z}}(\overline{\Lambda_U})$, and the function $\varphi_g \in CA_{\mathfrak{Z}}(\Lambda)$, defined by:

$$\varphi_g(t,x) := \begin{cases} \varphi(t)g(x), & t \in U \text{ and } x \in \Lambda_t \\ 0, & \text{elsewhere on } \Lambda \end{cases},$$

belongs to $CA_{\mathfrak{F}}V_o(\Lambda)$ (i.e. satisfies the weight conditions induced by V). - Under these assumptions, our condition reads: $A_{\mathfrak{F}}^{t_o}V|\Lambda_{t_o}$ is <u>dense</u> in $A_{\mathfrak{F}}(\Lambda_{t_o})$.

(ii) There is a neighbourhood U of t_o in $\pi_1(\Lambda)$ such that $\overline{\Lambda_U} = \overline{\bigcup_{t\in U} \Lambda_t}$ is <u>compact in</u> X and $\overline{\Lambda \cap (U\times X)} = \overline{\bigcup_{t\in U} \{t\} \times \Lambda_t}$ compact in Λ and such that we have $A_{\mathfrak{F}}(\overline{\Lambda_U})|\Lambda_{t_o} \subset A_{\mathfrak{F}}(\Lambda_{t_o})$ <u>dense</u>.

<u>Proof</u>. Since any function $g \in A_{\mathfrak{F}}(\overline{\Lambda_U})$ is uniformly bounded, it is easy to see that φ_g (defined as above) is continuous on Λ and hence belongs to $CA_{\mathfrak{F}}(\Lambda)$, as claimed. Then the proof of (i) follows (cf. 23) from the fact that for any function $g \in A_{\mathfrak{F}}^{t_o}V$, $g|\Lambda_{t_o} \in Y_{t_o}$. - For (ii) remark that the weight conditions given by V are certainly satisfied by all functions which vanish off the compact subset $\overline{\Lambda \cap (U\times X)}$ of Λ. Hence we may apply the method of (i) even with the <u>fixed</u> neighbourhood U of t_o alone (and arbitrary cut-off function φ). □

Similarly, with $Y = CH_{\mathfrak{F}}V_o(\Lambda)$, the following is a <u>sufficient</u> condition for density of Y_{t_o} in $H_{\mathfrak{F}}(Y_{t_o})$: $H_{\mathfrak{F}}^{t_o}V|\Lambda_{t_o}$ <u>dense</u> in $H_{\mathfrak{F}}(\Lambda_{t_o})$, where $H_{\mathfrak{F}}^{t_o}V$ is the set of all functions g as follows: We suppose that there is a neighbourhood U of t_o in $\pi_1(\Lambda)$ and a function $\varphi \in CB(\pi_1(\Lambda))$ with $\varphi(t_o) \neq 0$, but $\varphi \equiv 0$ off U. Let N denote an open neighbourhood of $\Lambda_{\overline{U}} := \bigcup_{t\in\overline{U}} \Lambda_t$. Then $g \in \mathfrak{F}(N)$ and the function $\varphi_g \in CH_{\mathfrak{F}}(\Lambda)$, defined as in 28(i), belongs to $CH_{\mathfrak{F}}V_o(\Lambda)$ (i.e. satisfies the weight conditions induced by V).

However, if we are willing to assume <u>a stronger regularity condition for</u> Λ, the density assumption 27(a) becomes <u>superfluous</u> for $Y = CH_{\mathfrak{F}}V_o(\Lambda)$:

29 Proposition. Assume that Λ is regular in the following sense: For each $t_o \in \pi_1(\Lambda)$ and arbitrary open neighbourhood N of Λ_{t_o} in X, there exists a neighbourhood U of t_o in $\pi_1(\Lambda)$ such that $\overline{\Lambda_U} = \overline{\bigcup_{t\in U} \Lambda_t}$ is compact and contained in N and $\overline{\Lambda \cap (U\times X)} = \overline{\bigcup_{t\in U} \{t\} \times \Lambda_t}$ compact in Λ. Then Y_{t_o} (with $Y = CH_{\mathfrak{F}}V_o(\Lambda)$) is dense in $H_{\mathfrak{F}}(\Lambda_{t_o})$ for each $t_o \in \pi_1(\Lambda)$, and hence $CH_{\mathfrak{F}}V_o(\Lambda)$ has Schwartz's a.p. whenever $H_{\mathfrak{F}}(\Lambda_t)$ has the a.p. for each $t \in \pi_1(\Lambda)$.

Proof. Fix $f \in H_{\mathfrak{F}}(\Lambda_{t_o})$ and $\varepsilon > 0$. By definition there exists an open neighbourhood N of Λ_{t_o} and a function $g \in \mathfrak{F}(N)$ such that $\sup_{x\in\Lambda_{t_o}} |f(x)-g(x)| < \varepsilon$. By the assumed regularity of Λ, we can find a neighbourhood U of t_o in $\pi_1(\Lambda)$ such that $\overline{\Lambda_U}$ is compact $\subset N$ and $\overline{\Lambda \cap (U\times X)}$ compact in Λ. The function φ_g, defined as in 28(i), is continuous on Λ (since $g|_{\Lambda_U}$ must be bounded) and clearly belongs to $CH_{\mathfrak{F}}V_o(\Lambda)$ because it vanishes off the compact subset $\overline{\Lambda \cap (U\times X)}$ of Λ. But since $g|_{\Lambda_{t_o}} = \frac{1}{\varphi(t_o)} \varphi_g|_{\{t_o\}\times\Lambda_{t_o}}$, it turns out that $g|_{\Lambda_{t_o}} \in Y_{t_o}$, and as $\varepsilon > 0$ was arbitrary, f may be approximated by elements of Y_{t_o}, which proves our claim. □

We have treated the general problem of the a.p. for $A_{\mathfrak{F}}(K)$ and $H_{\mathfrak{F}}(K)$ in [7], and surveyed the known theorems in the cases $\mathfrak{F} = \mathcal{O}$, the sheaf of holomorphic functions on $\mathbb{C}^N$ $(N \geq 1)$, and, say, $\mathfrak{F} = \mathcal{H}$, the sheaf of harmonic functions on $\mathbb{R}^N$ $(N \geq 2)$ [resp. (some) sheaves of harmonic functions in axiomatic potential theory], in the last sections of that paper. We will not repeat the results of [7] here, but shall now note some of the examples of spaces $CH_{\mathfrak{F}}V_o(\Lambda)$ with a.p. which immediately follow from these results (and from 29 above):

30 Proposition. Assume that Λ is regular in the sense of 29. Then $Y = CH_{\mathfrak{F}}V_o(\Lambda)$ has Schwartz's a.p. in each of the following cases:

(a) $X = \mathbb{C}$, $\mathfrak{F} = \mathcal{O}$;

(b) $X = \mathbb{C}^N$ $(N > 1)$; $\mathfrak{F} = \mathcal{O}$, and for each $t \in \pi_1(\Lambda)$ the compact set Λ_t = the closure of a strictly pseudoconvex region with sufficiently smooth (say, C^3-) boundary or the closure of a regular Weil polyeder; more generally:

(c) $X = \mathbb{C}^N$ $(N>1)$, $\mathfrak{F} = \mathfrak{G}$, and for each $t \in \pi_1(\Lambda)$, Λ_t = a product $\Lambda_t^1 \times \ldots \times \Lambda_t^k$, where each compact set Λ_t^j $(j=1,\ldots,k)$ is either a subset of $\mathbb{C}$ or the closure of a strictly pseudoconvex region with sufficiently smooth boundary or the closure of a regular Weil polyeder;

(d) $X = \mathbb{R}^N$ $(N\geq 2)$ [or X the space of definition of a (suitable) harmonic sheaf of axiomatic potential theory], $\mathfrak{F} = \mathfrak{K}$ (and for each $t \in \pi_1(\Lambda)$ the compact set Λ_t = the closure of an open subset of X).

When $A_{\mathfrak{F}}(\Lambda_t) = H_{\mathfrak{F}}(\Lambda_t)$ holds for all $t \in \pi_1(\Lambda)$, $CH_{\mathfrak{F}}V_o(\Lambda)$ is equal to $CA_{\mathfrak{F}}V_o(\Lambda)$, and hence 30. gives some information on the a.p. of $CA_{\mathfrak{F}}V_o(\Lambda)$, too. It is interesting, however, to presente some simple concrete examples of spaces of type $CA_{\mathfrak{F}}V_o(\Lambda)$ with a.p. which follow from the results mentioned in [7] and from 27, because we will not require Λ to be regular in the sense of 29 here. For simplicity let us consider the case V = W (of the compact-open topology on Λ) only and so restrict our attention to $CA_{\mathfrak{F}}(\Lambda)$ (instead of $CA_{\mathfrak{F}}V_o(\Lambda)$ for general Nachbin families V on Λ).

31 Example. Let Ω be locally compact, $X = \mathbb{C}^N$, and Λ a closed subset of $\Omega \times X$ (with Λ_t compact for each $t \in \pi_1(\Lambda)$). Then $Y = CA_{\mathfrak{F}}(\Lambda)$ has the a.p. in each of the following cases:

(a) $N = 1$, $\mathfrak{F} = \mathfrak{G}$ or $\mathfrak{K}$, and for each $t \in \pi_1(\Lambda)$, the set $\mathbb{C}\setminus\Lambda_t$ is connected; (b) $N > 1$, $\mathfrak{F} = \mathfrak{G}$, and for each $t \in \pi_1(\Lambda)$, Λ_t is polynomially convex, has the so-called "segment property" and is a product $\Lambda_t^1 \times \ldots \times \Lambda_t^k$, where each compact set Λ_t^j $(j=1,\ldots,k)$ is either a "fat" compact subset of $\mathbb{C}$ with $\mathbb{C} \setminus \Lambda_t^j$ connected or the closure of a strictly pseudoconvex region with sufficiently smooth boundary.

Proof. We remark that Λ is locally compact as a closed subset of $\Omega \times \mathbb{C}^N$. Moreover, under the conditions above, polynomials [resp. real parts of complex polynomials] are dense in $A_{\mathfrak{G}}(\Lambda_t)$ [resp. $A_{\mathfrak{K}}(\Lambda_t)$]

for each $t \in \pi_1(\Lambda)$. (In case (a), this is Mergelyan's theorem resp. the Walsh-Lebesgue theorem, for (b) use e.g. [5], Theorem 5.2.) Therefore the density assumption 27(a) is certainly satisfied. For the a.p. of the spaces $A_{\mathfrak{J}}(\Lambda_t)$, $t \in \pi_1(\Lambda)$, which is needed in 27(b), we refer to [7]. □

Let us now have a look at the spaces of vector-valued functions introduced in 26 and derive a "density theorem" (similar to [11], 4.11 resp. 4.12 for $C\mathfrak{J}V_o(\Lambda,F)$ resp. $C\mathfrak{J}(\Lambda,F)$).

32 Proposition. Let F be quasi-complete and Λ a $k_{\mathbb{R}}$-space.

(1) Then $CA_{\mathfrak{J}}V_o(\Lambda,F) = CA_{\mathfrak{J}}V_o(\Lambda)\varepsilon F$ holds.

(2) Hence we have $CA_{\mathfrak{J}}V_o(\Lambda,F) = CA_{\mathfrak{J}}V_o(\Lambda)\check{\otimes}_\varepsilon F$ whenever F is even complete and $CA_{\mathfrak{J}}V_o(\Lambda)$ (or F) has the a.p.

Let now F be complete (and Λ a $k_{\mathbb{R}}$-space).

(3) Then we have the following inclusions of topological linear subspaces of $CV_o(\Lambda,F)$:

$$CH_{\mathfrak{J}}V_o(\Lambda)\check{\otimes}_\varepsilon F \subset CH_{\mathfrak{J}}V_o(\Lambda,F) \subset CH_{\mathfrak{J}}V_o(\Lambda)\varepsilon F.$$

(4) Hence we have $CH_{\mathfrak{J}}V_o(\Lambda,F) = CH_{\mathfrak{J}}V_o(\Lambda)\varepsilon F = CH_{\mathfrak{J}}V_o(\Lambda)\check{\otimes}_\varepsilon F$ whenever $CH_{\mathfrak{J}}V_o(\Lambda)$ (or F) has the a.p..

Proof. (1) follows directly from the ε-product representation Theorem 9(3) (with $E = CA_{\mathfrak{J}}V_o(\Lambda)$). [Remark that for each $t \in \pi_1(\Lambda)$ the space $A_{\mathfrak{J}}(\Lambda_t,F)$ is clearly nothing but $A_{\mathfrak{J}}(\Lambda_t)\varepsilon F$ by our definition and compare [7], 3(1).]

(2) is then an obvious consequence of Schwartz's Theorem 5. - Since one can easily show that $CH_{\mathfrak{J}}V_o(\Lambda)\otimes_\varepsilon F$ is a (topological linear) subspace of $CH_{\mathfrak{J}}V_o(\Lambda,F)$ and since $CH_{\mathfrak{J}}V_o(\Lambda,F)$ is complete under our assumptions, the first inclusion of (3) follows. But $CH_{\mathfrak{J}}V_o(\Lambda,F) \subset CH_{\mathfrak{J}}V_o(\Lambda)\varepsilon F$ can again be deduced from 9(3). (For each $t \in \pi_1(\Lambda)$ we have $H_{\mathfrak{J}}(\Lambda_t)\check{\otimes}_\varepsilon F \subset H_{\mathfrak{J}}(\Lambda_t,F) \subset H_{\mathfrak{J}}(\Lambda_t)\varepsilon F$, cf. [7], Theorem 4 and the following remark.) □

The next propositions are again formulated for the case of the compact-open topology (i.e. $V = W$) only although suitable analogues hold in the general weighted case, too.

33 Proposition. (1) If, for each $t \in \pi_1(\Lambda)$, $\mathfrak{F}(X)|_{\Lambda_t}$ is dense in $H_{\mathfrak{F}}(\Lambda_t)$, $C(\Omega) \otimes \mathfrak{F}(X)|_\Lambda$ is dense in $CH_{\mathfrak{F}}(\Lambda)$. (If we assume that $\mathfrak{F}(X)|_{\Lambda_t}$ is even dense in $A_{\mathfrak{F}}(\Lambda_t)$ for each $t \in \pi_1(\Lambda)$, this clearly implies $A_{\mathfrak{F}}(\Lambda_t) = H_{\mathfrak{F}}(\Lambda_t)$ for all t and hence $CA_{\mathfrak{F}}(\Lambda) = CH_{\mathfrak{F}}(\Lambda)$ such that then $C(\Omega) \otimes \mathfrak{F}(X)|_\Lambda$ is also dense in $CA_{\mathfrak{F}}(\Lambda)$.)

(2) Let $\Lambda_\Omega := \overline{\bigcup_{t\in\pi_1(\Lambda)} \Lambda_t}$ be compact in X. If, for each $t \in \pi_1(\Lambda)$, $H_{\mathfrak{F}}(\Lambda_\Omega)|_{\Lambda_t}$ [resp. $A_{\mathfrak{F}}(\Lambda_\Omega)|_{\Lambda_t}$] is dense in $H_{\mathfrak{F}}(\Lambda_t)$ [resp. $A_{\mathfrak{F}}(\Lambda_t)$], then $C(\Omega) \otimes H_{\mathfrak{F}}(\Lambda_\Omega)|_\Lambda$ [resp. $C(\Omega) \otimes A_{\mathfrak{F}}(\Lambda_\Omega)|_\Lambda$] is dense in $CH_{\mathfrak{F}}(\Lambda)$ [resp. $CA_{\mathfrak{F}}(\Lambda)$].

Proof. Apply (the scalar version of) 14 to the module $Z := C(\Omega) \otimes \mathfrak{F}(X)|_\Lambda$ (or $Z := C(\Omega) \otimes H_{\mathfrak{F}}(\Lambda_\Omega)|_\Lambda$ resp. $Z := C(\Omega) \otimes A_{\mathfrak{F}}(\Lambda_\Omega)|_{\Lambda_B}$) over the selfadjoint algebra $A := CB(\Omega) \otimes \mathbb{K}|_\Lambda = CB(\Omega) \otimes \{\text{constants on } X\}|_\Lambda$ (in the space $L\tilde{V}_o = (C(\Lambda), co)$): Since $\mathcal{K}_A = \{\{t\}\times\Lambda_t;\ t \in \pi_1(\Lambda)\}$, $f \in C(\Lambda)$ belongs to $\bar{Z}^{(C(\Lambda),co)}$ if (and only if) for each $t \in \pi_1(\Lambda)$ the restriction $f|_{\{t\}\times\Lambda_t}$ is an element of

$$\overline{Z|_{\{t\}\times\Lambda_t}}^{C(\{t\}\times\Lambda_t)}$$ or, equivalently, if, for each $t\in\pi_1(\Lambda)$,

$$f(t,\cdot) \in \overline{\mathfrak{F}(X)|_{\Lambda_t}}^{C(\Lambda_t)} \quad \text{(or } \overline{H_{\mathfrak{F}}(\Lambda_\Omega)|_{\Lambda_t}}^{C(\Lambda_t)} \text{ resp. } \overline{A_{\mathfrak{F}}(\Lambda_\Omega)|_{\Lambda_t}}^{C(\Lambda_t)}\text{)}.$$

But this is satisfied for $f \in CH_{\mathfrak{F}}(\Lambda)$ resp. $CA_{\mathfrak{F}}(\Lambda)$ under our respective assumptions. □

It is of course possible to combine 32 and 33 to derive density of $C(\Omega) \otimes \mathfrak{F}(X) \otimes F|_\Lambda$ (or $C(\Omega) \otimes H_{\mathfrak{F}}(\Lambda_\Omega) \otimes F|_\Lambda$ resp. $C(\Omega) \otimes A_{\mathfrak{F}}(\Lambda_\Omega) \otimes F|_\Lambda$) in $CH_{\mathfrak{F}}(\Lambda,F)$ resp. $CA_{\mathfrak{F}}(\Lambda,F)$ for a l.c. space F, but the corresponding result can also be obtained directly as follows:

By applying the vector-valued version of 14 to the module

$Z := C(\Omega) \otimes \mathfrak{Z}(X) \otimes F|_\Lambda$ (or $Z := C(\Omega) \otimes H_{\mathfrak{Z}}(\Lambda_\Omega) \otimes F|_\Lambda$ resp. $Z := C(\Omega) \otimes A_{\mathfrak{Z}}(\Lambda_\Omega) \otimes F|_\Lambda$) over the algebra $A := CB(\Lambda) \otimes \mathbb{K}|_\Lambda$ (in the space $L\tilde{V}_o = (C(\Lambda,F),co))$, we get similarly as in the proof of 33: Let F be an arbitrary l.c. space.

(1) If, for each $t \in \pi_1(\Lambda)$, $\mathfrak{Z}(X) \otimes F|_{\Lambda_t}$ is dense in $H_{\mathfrak{Z}}(\Lambda_t,F)$, $C(\Omega) \otimes \mathfrak{Z}(X) \otimes F|_\Lambda$ is dense in $CH_{\mathfrak{Z}}(\Lambda,F)$. (If we assume that $\mathfrak{Z}(X) \otimes F|_{\Lambda_t}$ is even dense in $A_{\mathfrak{Z}}(\Lambda_t,F)$ for each $t \in \pi_1(\Lambda)$, this clearly implies $A_{\mathfrak{Z}}(\Lambda_t,F) = H_{\mathfrak{Z}}(\Lambda_t,F)$ for all t and hence $CA_{\mathfrak{Z}}(\Lambda,F) = CH_{\mathfrak{Z}}(\Lambda,F)$ such that then $C(\Omega) \otimes \mathfrak{Z}(X) \otimes F|_\Lambda$ is also dense in $CA_{\mathfrak{Z}}(\Lambda,F)$.)

(2) Let $\Lambda_\Omega := \overline{\bigcup_{t\in\pi_1(\Lambda)} \Lambda_t}$ be <u>compact</u> in X. If, for each $t\in\pi_1(\Lambda)$, $H_{\mathfrak{Z}}(\Lambda_\Omega) \otimes F|_{\Lambda_t}$ [resp. $A_{\mathfrak{Z}}(\Lambda_\Omega) \otimes F|_{\Lambda_t}$] is dense in $H_{\mathfrak{Z}}(\Lambda_t,F)$ [resp. $A_{\mathfrak{Z}}(\Lambda_t,F)$], then $C(\Omega) \otimes H_{\mathfrak{Z}}(\Lambda_\Omega) \otimes F|_\Lambda$ [resp. $C(\Omega) \otimes A_{\mathfrak{Z}}(\Lambda_\Omega) \otimes F|_\Lambda$] is dense in $CH_{\mathfrak{Z}}(\Lambda,F)$ [resp. $CA_{\mathfrak{Z}}(\Lambda,F)$].

Let now $\hat{F}$ denote <u>the completion of</u> F and fix $t \in \pi_1(\Lambda)$. Since $A_{\mathfrak{Z}}(\Lambda_t,F)$ is a topological linear subspace of $A_{\mathfrak{Z}}(\Lambda_t,\hat{F}) = A_{\mathfrak{Z}}(\Lambda_t)\epsilon\hat{F}$ (cf. [7], 3.1) in which $A_{\mathfrak{Z}}(\Lambda_t) \otimes \hat{F}$ is dense if $A_{\mathfrak{Z}}(\Lambda_t)$ or $\hat{F}$ has the a.p. (cf. 5), we get a fortiori <u>density</u> of [$\mathfrak{Z}(X) \otimes F|_{\Lambda_t} = \mathfrak{Z}(X)|_{\Lambda_t} \otimes F$ resp.] $A_{\mathfrak{Z}}(\Lambda_\Omega) \otimes F|_{\Lambda_t} = A_{\mathfrak{Z}}(\Lambda_\Omega)|_{\Lambda_t} \otimes F$ in $A_{\mathfrak{Z}}(\Lambda_t,F)$ whenever [$\mathfrak{Z}(X)|_{\Lambda_t}$ resp.] $A_{\mathfrak{Z}}(\Lambda_\Omega)|_{\Lambda_t}$ is dense in $A_{\mathfrak{Z}}(\Lambda_t)$ and one of the spaces $A_{\mathfrak{Z}}(\Lambda_t)$ or $\hat{F}$ has the a.p.

Similarly $H_{\mathfrak{Z}}(\Lambda_t,F)$ is a topological linear subspace of $H_{\mathfrak{Z}}(\Lambda_t,\hat{F})$, and $H_{\mathfrak{Z}}(\Lambda_t,\hat{F})$ equals $H_{\mathfrak{Z}}(\Lambda_t) \check{\otimes}_\epsilon \hat{F}$ if $H_{\mathfrak{Z}}(\Lambda_t)$ of $\hat{F}$ has a.p. or if, for some basis $\mathfrak{G}$ of open neighbourhoods of Λ_t, $\mathfrak{Z}(U)$ has Schwartz's a.p. for each $U \in \mathfrak{G}$. (See [7], Theorem 4 (1) and the following remark. - Note also that $(C(U,\hat{F}),co)$ is <u>always</u> a topological linear subspace of $(C(U),co)\epsilon\hat{F}$ by 9 (1), and thus the assumption $X = k_{\mathbb{R}}$-space which was made in [7] is not needed under our present definitions.) Hence <u>density</u> of $\mathfrak{Z}(X) \otimes F|_{\Lambda_t}$ resp. $H_{\mathfrak{Z}}(\Lambda_\Omega) \otimes F|_{\Lambda_t}$ in $H_{\mathfrak{Z}}(\Lambda_t,F)$ follows whenever $\mathfrak{Z}(X)|_{\Lambda_t}$ resp.

$H_{\mathfrak{F}}(\Lambda_{\Omega})|_{\Lambda_t}$ is dense in $H_{\mathfrak{F}}(\Lambda_t)$ and (i) $H_{\mathfrak{F}}(\Lambda_t)$ or (ii) $\hat{F}$ has the a.p. or (iii) $\mathfrak{F}$ is a sheaf with Schwartz's a.p. (i.e. $\mathfrak{F}(U)$ has Schwartz's a.p. for <u>each</u> open $U \subset X$, e.g. $\mathfrak{F}$ nuclear).

So we have proved:

<u>34 Proposition</u>. Let F be an arbitrary l.c. space.

(1) Suppose that:

(a) $\mathfrak{F}(X)|_{\Lambda_t}$ is dense in $H_{\mathfrak{F}}(\Lambda_t)$ for each $t \in \pi_1(\Lambda)$, and:

(b) (i) $\hat{F}$ has the a.p. or (ii) $H_{\mathfrak{F}}(\Lambda_t)$ has the a.p. for each $t \in \pi_1(\Lambda)$ or (iii) $\mathfrak{F}$ is a sheaf with Schwartz's a.p.

Then $C(\Omega) \otimes \mathfrak{F}(X) \otimes F|_{\Lambda}$ is dense in $CH_{\mathfrak{F}}(\Lambda,F)$.

(If we assume instead of (a) that $\mathfrak{F}(X)|_{\Lambda_t}$ is even dense in $A_{\mathfrak{F}}(\Lambda_t)$ for each $t \in \pi_1(\Lambda)$, then $A_{\mathfrak{F}}(\Lambda_t) = H_{\mathfrak{F}}(\Lambda_t)$ for all t and hence the a.p. of $\hat{F}$ or the a.p. of $A_{\mathfrak{F}}(\Lambda_t)$ for each $t \in \pi_1(\Lambda)$ implies $A_{\mathfrak{F}}(\Lambda_t,F) = H_{\mathfrak{F}}(\Lambda_t,F)$ for all t such that also $CA_{\mathfrak{F}}(\Lambda,F) = CH_{\mathfrak{F}}(\Lambda,F)$, and $C(\Omega) \otimes \mathfrak{F}(X) \otimes F|_{\Lambda}$ is then dense in $CA_{\mathfrak{F}}(\Lambda,F)$.)

(2) Let $\Lambda_{\Omega} := \overline{\bigcup_{t\in\pi_1(\Lambda)} \Lambda_t}$ be <u>compact</u> in X. Suppose that:

(a) $H_{\mathfrak{F}}(\Lambda_{\Omega})|_{\Lambda_t}$ [resp. $A_{\mathfrak{F}}(\Lambda_{\Omega})|_{\Lambda_t}$] is dense in $H_{\mathfrak{F}}(\Lambda_t)$ [resp. $A_{\mathfrak{F}}(\Lambda_t)$] for each $t \in \pi_1(\Lambda)$, and

(b) (i) $\hat{F}$ has the a.p. or (ii) $H_{\mathfrak{F}}(\Lambda_t)$ has the a.p. for each $t \in \pi_1(\Lambda)$ or (iii) $\mathfrak{F}$ is a sheaf with Schwartz's a.p. [resp. (i) $\hat{F}$ has the a.p. or (ii) $A_{\mathfrak{F}}(\Lambda_t)$ has the a.p. for each $t \in \pi_1(\Lambda)$].

Then $C(\Omega) \otimes H_{\mathfrak{F}}(\Lambda_{\Omega}) \otimes F|_{\Lambda}$ [resp. $C(\Omega) \otimes A_{\mathfrak{F}}(\Lambda_{\Omega}) \otimes F|_{\Lambda}$] is <u>dense</u> in $CH_{\mathfrak{F}}(\Lambda,F)$ [resp. $CA_{\mathfrak{F}}(\Lambda,F)$].

To finish let us point out that <u>the abstract setting</u> for all the examples in this section, as outlined before 19, allows also <u>examples of a more general kind</u> than we have considered so far:

Let Λ be a topological subspace of the product $\Omega \times X$ (of completely regular spaces) and let $V > 0$ be a Nachbin family on Λ. For each $t \in \pi_1(\Lambda)$ identify $V|_{\{t\}\times\Lambda_t}$ with the Nachbin family V_t on Λ_t. Now take a topological linear subspace $\mathfrak{F}_t$ of the weighted space $C(V_t)_o(\Lambda_t)$ for each $t \in \pi_1(\Lambda)$ and put (topologically, with the induced weighted topology)

$$Y := \{f \in CV_o(\Lambda);\ f(t,\cdot) \in \mathfrak{F}_t \text{ for each } t \in \pi_1(\Lambda)\}.$$

Then Y is clearly a module over the selfadjoint algebra

$$A := \{f \in CB(\Lambda);\ f(t,\cdot) \text{ constant on } \Lambda_t \text{ for arbitrary } t \in \pi_1(\Lambda)\},$$

we are in the bounded case of the weighted approximation problem for Y in $CV_o(\Lambda)$, and $K_A = \{\{t\}\times\Lambda_t;\ t \in \pi_1(\Lambda)\}$. Hence by the localization theorem Y has Schwartz's a.p. whenever for each $t \in \pi_1(\Lambda)$ the space $\mathfrak{F}_t$ is <u>nuclear</u> or $Y_t = Y|_{\{t\}\times\Lambda_t}$, identified with a space of functions on Λ_t, is <u>dense</u> in $\mathfrak{F}_t$ and $\mathfrak{F}_t$ has Schwartz's a.p..

In all our examples, except 20, the spaces $\mathfrak{F}_t$ have been "of the same type" for each $t \in \pi_1(\Lambda)$, e.g. $\mathfrak{F}_t = \mathfrak{F}(V_t)_o(\Lambda_t)$ for Λ_t <u>open</u> and a <u>fixed</u> sheaf $\mathfrak{F}$ on X or $\mathfrak{F}_t = A_{\mathfrak{F}}(\Lambda_t)$ resp. $H_{\mathfrak{F}}(\Lambda_t)$ for Λ_t <u>compact</u>, $W \leq V$, and again a <u>fixed</u> sheaf $\mathfrak{F}$ on X. In Example 20 we took $\mathfrak{F}_t = \mathfrak{G}(\Lambda_t)$ for all $t \in T$ <u>closed</u> $\subset \pi_1(\Lambda)$ (with Λ_t <u>open</u> for each such t), but $\mathfrak{F}_t = (C(\Lambda_t), co)$ for each $t \in \pi_1(\Lambda)\setminus T$. More generally, it is of course possible to construct examples where the "type" of the spaces $\mathfrak{F}_t$ <u>changes</u> with $t \in \pi_1(\Lambda)$: For instance, let T_1 and T_2 be two <u>closed</u> disjoint subsets of $\pi_1(\Lambda)$, let Λ_t be <u>open</u> for all $t \in T_1 \cup T_2$ and take $\mathfrak{F}_t = \mathfrak{F}_i(\Lambda_t)$ for $t \in T_i$, $i = 1,2$, and $\mathfrak{F}_t = (C(\Lambda_t), co)$ for all $t \in \pi_1(\Lambda)\setminus(T_1 \cup T_2)$ with two <u>different</u> sheaves $\mathfrak{F}_1$ and $\mathfrak{F}_2$ on X. Or even, which is much more interesting, let all Λ_t be <u>open</u> [resp. <u>compact</u>] and put $\mathfrak{F}_t = \mathfrak{F}(\Lambda_t)$ [resp. $A_{\mathfrak{F}_t}(\Lambda_t)$ or $H_{\mathfrak{F}_t}(\Lambda_t)$] where the sheaves $\mathfrak{F}_t$ on X <u>depend on the parameter</u> $t \in \pi_1(\Lambda)$ (e.g. sheaves of [null-] solutions of hypoelliptic partial differential operators $P(x,D,t)$), etc.

Finally, we should perhaps point out that each space Y of scalar functions as above has <u>a natural interpretation as a vector space of cross-sections</u> with respect to the "vector-valued" vector fibration $(\pi_1(\Lambda), (\mathfrak{F}_t)_{t\in\pi_1(\Lambda)})$ by taking $f = (f(t,\cdot))_t$. The topology of Y is also given by the family $\tilde{V} = \{\tilde{v};\ v \in V\}$ of "weights"

$\tilde{v}$ on $\pi_1(\Lambda)$, defined by $\tilde{v}(t)[g] := \sup_{x\in\Lambda_t} v(t,x)|g(x)|$ for all $g \in \mathfrak{F}_t$, $t \in \pi_1(\Lambda)$ and $v \in V$. Moreover, for arbitrary $v \in V$ and $f \in Y$ the mapping $s: t \to \tilde{v}(t)[f(t,\cdot)] = \sup_{x\in\Lambda_t} v(t,x)|f(t,x)|$ is <u>upper semicontinuous on</u> $\pi_1(\Lambda)$ and <u>vanishes at infinity</u>:

To prove this, fix $\varepsilon > 0$. Since $f \in CV_o(\Lambda)$, there exists a compact subset K of Λ with $v(\lambda)\,|f(\lambda)| < \varepsilon$ for all $\lambda \in \Lambda\setminus K$. $\pi_1(K)$ is a compact subset of $\pi_1(\Lambda)$, and for each point $t \notin \pi_1(\Lambda) \setminus \pi_1(K)$ we get $(t,x) \notin K$ and hence $v(t,x)\,|f(t,x)| < \varepsilon$ for all $x \in \Lambda_t$ which implies $s(t) = \sup_{x\in\Lambda_t} v(t,x)\,|f(t,x)| \leq \varepsilon$, i.e. s vanishes at infinity. - Now we show upper semicontinuity of s at $t_o \in \pi_1(\Lambda)$: Let $S := s(t_o) = \sup_{x\in\Lambda_{t_o}} v(t_o,x)\,|f(t_o,x)|$. Since $v|f|: \lambda \to v(\lambda)|f(\lambda)|$ is upper semi continuous on Λ and vanishes at infinity, the set $K := \{\lambda \in \Lambda: v(\lambda)\,|f(\lambda)| \geq S + \frac{\varepsilon}{2}\}$ is compact. Let $\mathfrak{A}$ be the system of closed neighbourhoods of t_o in $\pi_1(\Lambda)$, and for each $U \in \mathfrak{A}$ let $F_U := \bigcup_{t\in U} \{t\} \times \Lambda_t$ $(= \{\lambda \in \Lambda;\ \pi_1(\lambda) \in U\})$. Then F_U is a <u>closed</u> subset of Λ and

$$\bigcap_{U\in\mathfrak{A}} (F_U \cap K) = (\bigcap_{U\in\mathfrak{A}} F_U) \cap K = (\{t_o\}\times\Lambda_{t_o}) \cap K = \emptyset$$

(Ω is [completely] regular). By the finite intersection property of compact sets we get a (closed) neighbourhood U of t_o in $\pi_1(\Lambda)$ such that $F_U \cap K = \emptyset$, i.e. $v(t,x)\,|f(t,x)| < S + \frac{\varepsilon}{2}$ for all $t \in U$ and $x \in \Lambda_t$ which implies $s(t) = \sup_{x\in\Lambda_t} v(t,x)|f(t,x)| \leq S + \frac{\varepsilon}{2} < s(t_o)+\varepsilon$ for all $t \in U$.

After what we have just proved, the canonical identification of Y with a vector space of cross-sections [with respect to the vector fibration $(\pi_1(\Lambda), (\mathfrak{F}_t)_{t\in\pi_1(\Lambda)})$] <u>yields a Nachbin space</u> $L\tilde{V}_o$. In fact, as such a vector space of cross-sections, Y is a module over $CB(\pi_1(\Lambda))$, and as we are in the selfadjoint case, the localization of the a.p. of Y to $Y|_{\{t\}\times\Lambda_t} = Y_t \subset \mathfrak{F}_t$ can also be deduced from Prolla's main theorem in [25]. - However, with this identification we cannot get localization to smaller sets than whole "slices"

$\{t\} \times \Lambda_t$ whereas considering Y as a space of functions on Λ has the advantage that a "finer" localization is possible whenever Y is a module over an "essentially larger" algebra than A above (cf. 20, but one can easily find more striking examples).

4. OTHER EXAMPLES

Another (obvious) case where the localization of the a.p. applies involves a different kind of "mixed dependence" (cf. already [4], Corollary 15, p. 13/14 for a very simple example):
Let X be an arbitrary completely regular space, $V > 0$ a Nachbin family on X, Λ an arbitrary topological subspace of X (with closure $\bar{\Lambda}$ in X), and $\mathfrak{F} = \mathfrak{F}(\Lambda)$ a topological linear subspace of $(C(\Lambda), co)$. Now define $CV_o(X;\mathfrak{F}) := \{f \in CV_o(X);\ f|_\Lambda \in \mathfrak{F}(\Lambda)\}$ with the induced weighted topology. (E.g. let Λ be open in X and $\mathfrak{F}$ a [pre-] sheaf of continuous functions on X as in the preceding section.) This space is a module over the algebra
$A := \{f \in CB(X);\ f|_\Lambda \text{ constant}\}$ with $K_A = \bar{\Lambda} \cup \{\{x\};\ x \in X \setminus \bar{\Lambda}\}$.
Since the restriction of $CV_o(X;\mathfrak{F})$ to each one point set certainly has the a.p., the localization Theorem 17 reduces Schwartz's a.p. of $CV_o(X;\mathfrak{F})$ to the question whether
$CV_o(X;\mathfrak{F})|_{\bar{\Lambda}} \subset C(V|_{\bar{\Lambda}})_o(\bar{\Lambda};\mathfrak{F}) = \{f \in C(V|_{\bar{\Lambda}})_o(\bar{\Lambda});\ f|_\Lambda \in \mathfrak{F}(\Lambda)\}$ (with the restricted weighted topology) has Schwartz's a.p..

35 Proposition. $CV_o(X;\mathfrak{F})$ has Schwartz's a.p. whenever $CV_o(X;\mathfrak{F})|_\Lambda$ has.

36 Remark. If X is a $V_{\mathbb{R}}$-space and F a quasi-complete l.c. space, 9 (3) yields: $CV_o(X;\mathfrak{F})\varepsilon F = \{f \in CV_o(X,F);\ f' \circ f|_\Lambda \in \mathfrak{F}(\Lambda)$ for each $f' \in F'\}$ with the weighted topology of $CV_o(X,F)$, and hence then $\{f \in CV_o(X,F);\ f' \circ f|_\Lambda \in \mathfrak{F}(\Lambda)$ for each $f' \in F'\} = CV_o(X;\mathfrak{F}) \check{\otimes}_\varepsilon F$ holds whenever F is even complete and $CV_o(X;\mathfrak{F})|_{\bar{\Lambda}}$ (or F) has Schwartz's a.p..

In Proposition 35 one would sometimes like to replace $CV_o(X;\mathfrak{F})|_{\bar{\Lambda}}$ by $C(V|_{\bar{\Lambda}})_o(\bar{\Lambda};\mathfrak{F})$. Similarly as before this is possible whenever density of $CV_o(X;\mathfrak{I})|_{\bar\Lambda}$ in $C(V|_{\bar\Lambda})_o(\bar\Lambda;\mathfrak{I})$ is known, that is, if the elements $f \in C(V|_{\bar\Lambda})_o(\bar\Lambda;\mathfrak{I})$ which extend to functions in $CV_o(X)$ form a dense subset.

37 Remark. (a) (Even) $CV_o(X;\mathfrak{I})|_{\bar\Lambda} = C(V|_{\bar\Lambda})_o(\bar\Lambda;\mathfrak{I})$ holds if we have $CV_o(X)|_{\bar\Lambda} \supset C(V|_{\bar\Lambda})_o(\bar\Lambda;\mathfrak{I})$, e.g. if (*) $CV_o(X)|_{\bar\Lambda}$ equals $C(V|_{\bar\Lambda})_o(\bar\Lambda)$.

(b) Condition (*) of (a) is satisfied for instance in the following cases:

(i) $V = W$ (if suffices that each function $v \in V$ has compact support which implies $CV_o(X) = C(X)$ algebraically) and $\Lambda \subset X$ relatively compact or X normal, or

(ii) V = positive constants on X (hence $CV_o(X) = C_o(X)$ and $C(V|_{\bar\Lambda})_o(\bar\Lambda) = C_o(\bar\Lambda)$) and X locally compact or

(iii) X locally compact and $\Lambda \subset X$ relatively compact (but $V > 0$ arbitrary).

In case (iii) of (b), each function $f \in C(\bar\Lambda)$ clearly extends to a function in $C(X)$ which has compact support and thus satisfies arbitrary weight conditions. - At this point we should perhaps also observe that, if X is a topological subspace of (the completely regular space) X_o, e.g. $\mathbb{C}^N$ or $\mathbb{R}^N$, the closure of $\Lambda \subset X$ with respect to X_o need not coincide with $\bar\Lambda \subset X$ in general, but $\bar\Lambda$ is the intersection of the closure in X_o with X.

We note some interesting examples that follow from 35, 37 (b) (iii) and from the results in [7] (already used in the last section) on the a.p. of $A_{\mathfrak{I}}(K)$:

38 Example. Let X be locally compact, V a Nachbin family on X with $W \le V$, $\mathfrak{I}$ a closed l.c. sub-(pre-) sheaf of C_X, and Λ an open relatively compact subset of X with $\Lambda = \mathring{\bar\Lambda}$. Then $CV_o(X;\mathfrak{I}(\Lambda))$ has the a.p. whenever (in the notation of the preceding section) the Banach space $A_{\mathfrak{I}}(\bar\Lambda)$ has.

E.g., let X be a locally compact subspace of $\mathbb{C}^N$ $(N \ge 1)$ resp.

$\mathbb{R}^N$ $(N \geq 2)$ [or of the space of definition of a (suitable) harmonic sheaf $\mathcal{K}$ of axiomatic potential theory], $W \leq V$, $\mathfrak{F} = \mathfrak{G}$ resp. $\mathcal{K}$, and Λ an open subset of $\mathbb{C}^N$ resp. $\mathbb{R}^N$ such that its closure $\bar{\Lambda}$ is a compact subset of X with $\mathring{\bar{\Lambda}} = \Lambda$.

Then $CV_o(X;\mathfrak{F}(\Lambda))$ has the a.p. provided $\mathfrak{F} = \mathcal{K}$ or $\mathfrak{F} = \mathfrak{G}$ and $N = 1$ or $\mathfrak{F} = \mathfrak{G}$, $N > 1$, and Λ is (a strictly pseudoconvex region with sufficiently smooth boundary or a regular Weil polyeder or) a <u>product</u> $\Lambda_1 \times \ldots \times \Lambda_k$ where each open (relatively compact) set Λ_j $(j=1,\ldots,k)$ is either contained in $\mathbb{C}$ or a strictly pseudoconvex region with sufficiently smooth boundary or a regular Weil polyeder.

<u>Proof</u>. Since X is locally compact and $W \leq V$, the topology of $CV_o(X)$ is <u>complete</u> and stronger than co. Since $\mathfrak{F}(\Lambda)$ is a <u>closed</u> topological subspace of $(C(\Lambda),co)$, $CV_o(X;\mathfrak{F}(\Lambda))$ is clearly a closed subspace of $CV_o(X)$ and hence complete, too. $\bar{\Lambda} \subset X$ compact and $\Lambda = \mathring{\bar{\Lambda}}$ imply $C(V|_{\bar{\Lambda}})_o(\bar{\Lambda};\mathfrak{F}) = A_{\mathfrak{F}}(\bar{\Lambda})$. Now the first part of 38 follows from 35 and 36 (b) (iii) while the second part is a consequence of the results surveyed in the third section of [7]. □

Let us now take X and V as before, but assume that there is a whole (finite or infinite) system $(\Lambda_\alpha)_\alpha$ of (disjoint) topological subspaces of X and a corresponding system $\mathfrak{F}_\alpha = \mathfrak{F}_\alpha(\Lambda_\alpha)$ of topological linear subspaces of $(C(\Lambda_\alpha),co)$. We look at the space $CV_o(X;(\mathfrak{F}_\alpha)) := \{f \in CV_o(X);\ f|_{\Lambda_\alpha} \in \mathfrak{F}_\alpha(\Lambda_\alpha) \text{ for each } \alpha\}$ with the induced weighted topology.

For instance, the sets Λ_α may be <u>open</u> in X, and $(\mathfrak{F}_\alpha)_\alpha$ may denote different (pre-)sheaves of continuous functions on X. It is also quite interesting to take $\mathfrak{F}_\alpha$ = the same sheaf $\mathfrak{F}$ for all α and to let $(\Lambda_\alpha)_\alpha$ denote the system of (connected) components of an open set $\Lambda = \bigcup_\alpha \Lambda_\alpha$ in a <u>locally connected</u> space X. (The sets Λ_α are then <u>open</u>, too.)

Clearly $CV_o(X;(\mathfrak{F}_\alpha))$ is a module over the selfadjoint algebra

$A := \{f \in CB(X); f|_{\Lambda_\alpha}$ constant for each $\alpha\}$, and all the sets $\bar{\Lambda}_\alpha$ are contained in maximal A-antisymmetric sets. But, even for only two sets Λ_1, Λ_2, it may happen that $\Lambda_1 \cap \Lambda_2 = \emptyset$, but that $\bar{\Lambda}_1 \cap \bar{\Lambda}_2$ is non-void; in this case $\bar{\Lambda}_1 \cup \bar{\Lambda}_2$ is a closed subset of X on which all functions in A are constant. So in general the localization of the a.p. of $CV_o(X;(\mathfrak{F}_\alpha))$ to $\mathcal{K}_A$, i.e. to maximal sets of constancy of A, will not lead to a complete "splitting" of the different spaces $\mathfrak{F}_\alpha$. In fact, for a space X on which the system $(\Lambda_\alpha)_\alpha$ induces a kind of "swiss cheese" structure, it may be quite complicated to "compute" $\mathcal{K}_A$ explicitly, and a number of different situations occur. - We will not deal with such a topological problem here any more, but mention only a very simple case in which the localization theorem is useful (and which fornunately arises sometimes in concrete applications):

39 Proposition. Assume that the algebra A (as above) separates the sets $(\overline{\Lambda_\alpha})_\alpha$ and points in $X \setminus \bigcup_\alpha \overline{\Lambda_\alpha}$ from $\bigcup_\alpha \overline{\Lambda_\alpha}$, i.e. $\mathcal{K}_A = \{\overline{\Lambda_\alpha}; \alpha\} \cup \{\{x\}; x \in X \setminus \bigcup_\alpha \overline{\Lambda_\alpha}\}$. (This is certainly the case if (*) X is normal, the sets $\overline{\Lambda_\alpha}$ are disjoint, and if $\bigcup_{\alpha \neq \alpha_o} \overline{\Lambda_\alpha}$ is closed for each α_o.)
Then $CV_o(X;(\mathfrak{F}_\alpha))$ has Schwartz's a.p. whenever all the spaces $CV_o(X;(\mathfrak{F}_\alpha))|_{\overline{\Lambda_\alpha}}$ have.

Here $CV_o(X;(\mathfrak{F}_\alpha))|_{\overline{\Lambda_\alpha}} \subset C(V|_{\overline{\Lambda_\alpha}})_o(\overline{\Lambda_\alpha};\mathfrak{F}_\alpha) =$
$= \{f \in C(V|_{\overline{\Lambda_\alpha}})_o(\overline{\Lambda_\alpha});\ f|_{\Lambda_\alpha} \in \mathfrak{F}_\alpha(\Lambda_\alpha)\}$ (with the restricted weighted topology), and similarly as in 37 one can find conditions such that (even) $CV_o(X;(\mathfrak{F}_\alpha))|_{\overline{\Lambda_\alpha}} = C(V|_{\overline{\Lambda_\alpha}})_o(\overline{\Lambda_\alpha};\mathfrak{F}_\alpha)$ holds: For instance this is the case if, additionally to the assumption (*) of 39, we have $V = W$ or: X locally compact and $\overline{\Lambda_\alpha} \subset X$ compact.

40 Example. Let X_o be a completely regular space and $\mathfrak{F}$ a closed l.c. subsheaf of C_{X_o}. Let then K denote a compact subset of X_o which is the union $\bigcup_\alpha K_\alpha$ of (disjoint) compact sets K_α such that

$\mathring{K} = \bigcup_\alpha \mathring{K}_\alpha$ holds and the algebra $A := \{f \in C(K);\ f|_{K_\alpha}$ constant for each $\alpha\}$ separates the sets K_α. Then $A_{\mathfrak{F}}(K)$ has the a.p. if $A_{\mathfrak{F}}(K)|_{K_\alpha} \subset A_{\mathfrak{F}}(K_\alpha)$ has the a.p. for each α.

Proof. Take $X = K$, $\Lambda_\alpha = K_\alpha$, and $\mathfrak{F}_\alpha(\Lambda_\alpha) = A_{\mathfrak{F}}(K_\alpha)$ for each α in 39. With, say, V = positive constants on K we get $CV_o(X;(\mathfrak{F}_\alpha)) = \{f \in C(K);\ f|_{\mathring{K}_\alpha} \in \mathfrak{F}(\mathring{K}_\alpha)$ for each $\alpha\}$ which by $\mathring{K} = \bigcup_\alpha \mathring{K}_\alpha$ and the sheaf property of $\mathfrak{F}$ equals $A_{\mathfrak{F}}(K)$. - Of course we need these facts also to show that $A_{\mathfrak{F}}(K)$ is a module over the algebra A above. □

It is easy to construct (non-trivial) examples of sets (say) $K = \bigcup_{n\in\mathbb{N}} K_n$ as in 40 and even to arrange these examples in order to get $A_{\mathfrak{F}}(K)|_{K_n} = A_{\mathfrak{F}}(K_n)$ for each $n \in \mathbb{N}$. - In many cases density of $A_{\mathfrak{F}}(K)|_{K_\alpha}$ in $A_{\mathfrak{F}}(K_\alpha)$ will hold anyway, and this already suffices to replace by $A_{\mathfrak{F}}(K_\alpha)$ in 40. - It is of course possible to combine the two kinds of "mixtures" and to use spaces of type $CV_o(X;(\mathfrak{F}_\alpha))$ as $\mathfrak{F}_t$ in the general scheme given at the end of the preceding section. - We prefer, however, to conclude with applications of the localization theorem (and of the ε-product) in the vector-valued weighted approximation theory of continuous functions. These applications were mentioned very briefly (and without proofs) in Remark 4.6 d) of [11]; they are connected with the paper [26] by Prolla-Machado and generalize B Blatter's method from [13], Theorem 1.10 to arbitrary weighted spaces.

Let us start by introducing resp. recalling (cf. [26]) some notation: Let X be completely regular, $V > 0$ a Nachbin family on $X, F \neq \{0\}$ a locally convex space, and $Y = Y_F$ a topological linear subspace of $CV_o(X,F)$. The set G_Y of all pairs $(x,y) \in X \times X$ such that (with $\alpha\delta_x\colon f \to \alpha f(x)$ for all $f \in CV_o(X,F)$, arbitrary $x \in X$ and $\alpha \in \mathbb{K}$) either $\delta_x|_Y = \delta_y|_Y = 0$ or there exists $t \in \mathbb{K}$, $t \neq 0$, with $\delta_x|_Y = t\,\delta_y|_Y \neq 0$ yields an equivalence relation on X. Define $g = g_Y\colon G_Y \to \mathbb{K}$ by $g(x,y) = 0$ if $0 = \delta_x|_Y = \delta_y|_Y$ and $g(x,y) = t$ if $0 \neq \delta_x|_Y = t\,\delta_y|_Y$.

Similarly the subsets $KS_Y := \{(x,y) \in G_Y;\ g(x,y) \geq 0\}$ and $WS_Y := \{(x,y) \in G_Y;\ g(x,y) \in \{0,1\}\}$ yield equivalence relations on X. Now consider a "symbol" $\Delta \in \{G, KS, WS\}$. The closed (topological linear) subspace $\Delta(Y) := \{f \in CV_o(X,F);\ f(x) = g(x,y)f(y)$ for all $(x,y) \in \Delta_Y\} \supset Y$ of $CV_o(X,F)$ is called the Δ-hull of Y in $CV_o(X,F)$. Y is said to bo a Δ-subspace of $CV_o(X,F)$ if $\Delta(Y)$ is just the closure of Y in $CV_o(X,F)$. (The letters G, KS, WS stand for Grothendieck, Kakutani-Stone, and Weierstrass-Stone, respectively.)

41 Proposition. Let Y $(= Y_{\mathbb{K}})$ be a topological linear subspace of $CV_o(X)$.

(a) Then $WS(Y)$ has Schwartz's a.p.

(b) Let $\Delta = G$ or KS. If $A(\Delta) := \{f \in CB(X);\ f$ is constant on each equivalence class modulo $\Delta_Y\}$ separates the equivalence classes mod Δ_Y, $\Delta(Y)$ has Schwartz's a.p..

Proof. Let $\Delta = G$, KS, or WS. Since $CV_o(X)$ is a module over $CB(X)$, $\Delta(Y)$ is a module over

$A(\Delta) = \{f \in CB(X);\ f$ is constant on all equivalence classes mod $\Delta_Y\}$. We are in the bounded case of the weighted approximation problem for $\Delta(Y)$ in $CV_o(X)$, and A is selfadjoint. - Let us prove that $A(\Delta)$ always separates the equivalence classes mod Δ_Y in the case $\Delta = WS$ (a fact which is mentioned [for real scalars] in [26], p. 248):

Take $x_1, x_2 \in X$ which belong to different equivalence classes mod WS_Y, i.e. there exists $h \in Y \subset WS(Y)$ such that $h(x_1) \neq h(x_2)$. $WS(Y)$ is clearly selfadjoint, hence we can find a real-valued $g \in WS(Y)$ such that $g(x_1) < g(x_2)$. But for real-valued functions $g_1, g_2 \in WS(Y)$ also $\sup(g_1,g_2) \in WS(Y)$. So without loss of generality we may assume $g \geq 0$. The function $f := \inf(g, g(x_2))$ is continuous, by $0 \leq f \leq g$ belongs to $CV_o(X)$, and it is easily checked that then even $f \in WS(Y)$, too. So f must be constant on the equivalence classes mod WS_Y, and we have constructed a real-valued function $f \in A(WS)$ which separates x_1 and x_2.

For the cases $\Delta = G$ resp. KS, we assume in (b) this separation property of $A(\Delta)$, and hence we have in all three cases that $\mathcal{K}_A$ is nothing but the system of equivalence classes mod Δ_Y. By the localization Theorem 17, $\Delta(Y)$ now has Schwartz's a.p. if $\Delta(Y)|_K$ has Schwartz's a.p. for each equivalence class K modulo Δ_Y.

In the case $\Delta = WS$ all functions $f \in \Delta(Y)$ are <u>constant</u> on each equivalence class K mod Δ_Y, that is $WS(Y)|_K = \{0\}$ or $\mathbb{K}$ which clearly has the a.p.. In the other cases $\Delta = KS$ or G it is easy to see that the values of an arbitrary function $f \in \Delta(Y)$ on K are completely determined by the value at one <u>single</u> point $x_o \in K$, and hence $\Delta(Y)|_K$ is again <u>at most one-dimensional</u>. □

For the assumption in 41 (b) see [26], 3.15. - From 41 we get by <u>density</u> of Δ-subspaces Y in $\Delta(Y)$, $\Delta = G$, KS or WS:

<u>42 Corollary</u>. Each WS-subspace $Y \subset CV_o(X)$ has Schwartz's a.p., and for $\Delta = G$ or KS any Δ-subspace Y of $CV_o(X)$ for which $A(\Delta)$ as in 41 <u>separates</u> the equivalence classes mod Δ_Y has Schwartz's a.p., too.

Let us now turn to the <u>vector-valued</u> case where (up to the a.p. of arbitrary closed KS- and G-subspaces of $CV_o(X)$ which followed in Blatter's case [X locally compact and V = positive constants on X] from <u>a theorem of Lindenstrauss</u>) we will <u>generalize</u> Theorem 1.10 of Blatter [13] below.

Let Y_F denote a topological linear subspace of $CV_o(X,F)$. Define $Y_o := \{f' \circ f;\ f' \in F',\ f \in Y_F\} \subset CV_o(X)$ with the weighted topology. We will always <u>assume</u> that

(*) $Y_o \times F := \{g \otimes e;\ g \in Y_o,\ e \in F\}$ is contained in Y_F.

Then obviously Y_o is <u>a linear subspace</u> of $CV_o(X)$, and Y_o is <u>closed</u> in $CV_o(X)$ whenever Y_F is closed in $CV_o(X,F)$. (See [2], II, 3.5 and compare [13], Remark 1.15 (i) as well as [26], Lemma 1.1.)

<u>43 Lemma</u>. Let $\Delta \in \{G, KS, WS\}$. Then $\Delta_{Y_F} = \Delta_{Y_o}$, and $g_{Y_F}(x,y) = g_{Y_o}(x,y)$ for all $(x,y) \in \Delta_{Y_o} = \Delta_{Y_F}$.

<u>Proof</u>. Using the Hahn-Banach theorem this is easily checked. □

<u>44 Proposition</u>. (a) For an arbitrary topological linear subspace $Y\ (= Y_{\mathbb{K}})$ of $CV_o(X)$ we have

$\Delta(Y)\varepsilon F = \{f \in CV_o^{\sigma,c}(X,F);\ f(x) = g_Y(x,y)f(y)$ for all $(x,y) \in \Delta_Y\}$.

(b) Let Y_F be a topological linear subspace of $CV_o(X,F)$ with (*). Then we get

$$\Delta(Y_F) = \Delta(Y_o)\varepsilon F$$

whenever F is quasi-complete and X a $V_{\mathbb{R}}$-space.

<u>Proof</u>. (a) is immediate from 9 (3) and the Hahn-Banach theorem. Under the assumptions of (b) we have $CV_o^{\sigma,c}(X,F) = CV_o(X,F)$, and hence (b) follows from (a) and Lemma 43. □

Without any assumptions on F and X we get in case (b): $\Delta(Y_F) = \Delta(Y_o)\varepsilon F \cap CV_o(X,F)$.

<u>45 Proposition</u>. An arbitrary <u>closed</u> linear subspace Y_F of $CV_o(X,F)$ is a Δ-<u>subspace</u> if and only if $Y_o := \{f'\circ f;\ f'\in F',\ f \in Y_F\}$ is a closed Δ-subspace of $CV_o(X)$ satisfying $Y_F = Y_o\varepsilon F \cap CV_o(X,F)$. (The intersection of $Y_o\varepsilon F$ with $CV_o(X,F)$ is <u>not necessary</u> if X is a $V_{\mathbb{R}}$-space and F quasi-complete.)

<u>Proof</u>. 1. Let Y_o be a closed Δ-subspace of $CV_o(X)$, i.e. $Y_o = \Delta(Y_o)$. Then $Y_F := Y_o\varepsilon F \cap CV_o(X,F)$ clearly contains $Y_o \times F$, and it is immediate that $\{f'\circ f;\ f'\in F',\ f \in Y_F\} = Y_o$ (F being $\neq \{0\}$). Hence we may apply 44 to get

$\Delta(Y_F) = \Delta(Y_o)\varepsilon F \cap CV_o(X,F) = Y_o\varepsilon F \cap CV_o(X,F) = Y_F$ which proves that Y_F is a closed Δ-subspace of $CV_o(X,F)$.

2. Let now Y_F be a closed Δ-subspace of $CV_o(X,F)$. Then one can verify <u>directly</u> that $Y_o := \{f'\circ f;\ f'\in F',\ f \in Y_F\}$ satisfies $Y_o \times F \subset \Delta(Y_F) = Y_F$. By 44 we get consequently:

$Y_F = \Delta(Y_F) = \Delta(Y_o)\varepsilon F \cap CV_o(X,F)$, and it remains to show $\Delta(Y_o) = Y_o$. But for $h \in \Delta(Y_o)$ and $e \in F$, $e \neq 0$, we have

$f := h \otimes e \in \Delta(Y_o)\times F \subset \Delta(Y_o) \otimes F \subset \Delta(Y_o)\varepsilon F \cap CV_o(X,F) = Y_F$. Choosing

$f' \in F'$ with $f'(e) = 1$ we finally obtain $f' \circ f = h \in Y_o$, i.e. $Y_o = \Delta(Y_o)$. □

__46 Corollary__. Let X be a $V_{\mathbb{R}}$-space, F a complete l.c. space, and Y_F a closed Δ-subspace of $CV_o(X,F)$. If $\Delta = WS$ or if $\Delta = G$ or KS and $A(\Delta)$ in 41 separates the equivalence classes mod Δ_Y^* (or if F has the a.p.), we obtain:

$$Y_F = Y_o \check{\otimes}_\varepsilon F.$$

__Proof__. Combine 41 and 45. □

The ε-product characterization 45 of closed Δ-subspaces Y_F of $CV_o(X,F)$ (in some sense) __reduces__ the study of Δ-subspaces in the vector-valued case to __scalar__ functions. (For some characterizations for Δ-subspaces $Y_o \subset CV_o(X)$ see [26].)

Finally let us note that the above results lead to new proofs of some propositions in [26] and yield better insight. We illustrate this with two examples (compare [26], 3.19):

In view of Lemma 2.2 and Remark 2.5 of [26], Theorem 2.9 of [26] can be __rephrased__ as follows:

Let Y_F be a topological linear subspace of $CV_o(X,F)$ such that Y_o satisfies $Y_o \times F \subset Y_F$ and is a WS-subspace of $CV_o(X)$. Then Y_F is a WS-subspace of $CV_o(X,F)$.

- A proof based on the preceding results runs as follows: We have $Y_F \supset Y_o \otimes_\varepsilon F$ and a fortiori: $\bar{Y}_F \supset \overline{Y_o \otimes_\varepsilon F}^{CV_o(X,F)}$ which clearly contains $\overline{\bar{Y}_o \otimes_\varepsilon F}^{CV_o(X,F)}$ and hence $\bar{Y}_o \varepsilon F \cap CV_o(X,F)$ since $\bar{Y}_o \otimes F$ is dense in $\bar{Y}_o \varepsilon F$ by 5 because $\bar{Y}_o = WS(Y_o)$ has Schwartz's a.p. by Proposition 41 (a). On the other hand 44 yields:

$$WS(Y_F) = WS(Y_o) \varepsilon F \cap CV_o(X,F) = \bar{Y}_o \varepsilon F \cap CV_o(X,F),$$

and it follows that $WS(Y_F) \subset \bar{Y}_F$ which implies $\bar{Y}_F = WS(Y_F)$, i.e. Y_F is a WS-subspace of $CV_o(X,F)$.

And in view of [26], Proposition 3.11, Theorem 3.14 of [26] reads:

Let Y_F be a topological linear subspace of $CV_o(X,F)$ such that Y_o satisfies $Y_o \times F \subset Y_F$ and is a Δ-subspace of $CV_o(X)$ for $\Delta = KS$ or G. Assume that $A(\Delta)$ separates the equivalence classes modulo Δ_{Y_F}. Then Y_F is a Δ-subspace of $CV_o(X,F)$.

- A proof can be given exactly as before, using Proposition 41 (b) (and Lemma 43) this time.

REFERENCES

[1] Baumgarten, B., Gewichtete Räume differenzierbarer Funktionen, Dissertation, Darmstadt 1976.

[2] Bierstedt, K.-D., Gewichtete Räume stetiger vektorwertiger Funktionen und das injektive Tensorprodukt I, II, J. reine angew. Math. 259, 186-210; 260, 133-146 (1973).

[3] ----------------, Injektive Tensorprodukte und Slice-Produkte gewichteter Räume stetiger Funktionen, J. reine angew. Math. 266, 121-131 (1974).

[4] ----------------, The approximation property for weighted function spaces; Tensor products of weighted spaces, in: Function Spaces and Dense Approximation, Proc. Conference, Bonn 1974, Bonner Math. Schriften 81, 3-25; 26-58 (1975).

[5] ----------------, Some generalizations of the Weierstrass and Stone-Weierstrass theorems, An. Acad. Brasil. Ciênc. 49, 507-523 (1977).

[6] ----------------, A question on inductive limits of weighted locally convex function spaces, in: Atas do 11º Colóquio Brasileiro de Matemática, Poços de Caldas 1977, Vol.I, 213-226, Rio de Janeiro (1978).

[7] ----------------, A remark on vector-valued approximation on compact sets, approximation on product sets, and the approximation property, Approximation Theory and Functional Analysis, Proc. Conference Approximation Theory, Campinas 1977, North-Holland Math. Studies 35, 37-62 (1979).

[8] Bierstedt, K.-D., Meise, R., Bemerkungen über die Approximationseigenschaft lokalkonvexer Funktionenräume, Math. Ann. 209, 99-107 (1974).

[9] ----------------, ---------, Induktive Limites gewichteter Räume stetiger und holomorpher Funktionen, J. reine angew. Math. 282, 186-220 (1976).

[10] Bierstedt, K.-D., Gramsch, B., Meise, R., Lokalkonvexe Garben und gewichtete induktive Limites $\mathfrak{F}$-morpher Funktionen, Function Spaces and Dense Approximation, Proc. Conference Bonn 1974, Bonner Math. Schriften 81, 59-72 (1975).

[11] -------, --------, --------, Approximationseigenschaft, Lifting und Kohomologie bei lokalkonvexen Produktgarben, Manuscripta math. 19, 319-364 (1976).

[12] Bierstedt, K.-D., Meise, R., Summers, W.H., A projective characterization of inductive limits of weighted spaces, to appear.

[13] Blatter, J.B., Grothendieck spaces in approximation theory, Mem. Amer. Math. Soc. 120 (1972).

[14] Garnir, H.G., de Wilde, M., Schmets, J., Analyse fonctionnelle, Tome III: Espace fonctionnels usuels, Birkhäuser 1973.

[15] Gramsch, B. Über eine Fortsetzungsmethode der Dualitätstheorie lokalkonvexer Räume, Ausarbeitung, Kaiserslautern WS 1975/76, partially published in: Ein Schwach-Stark-Prinzip der Dualitätstheorie lokalkonvexer Räume als Fortsetzungsmethode, Math. Z. 156, 217-230 (1977).
An extension method of the duality theory of locally convex spaces with applications to extension kernels and the operational calculus, Proc. Conference Paderborn 1976, Functional Analysis: Surveys and Recent Results, North-Holland Math. Studies 27, 131-147 (1977).

[16] Grothendieck, A., Produits tensoriels topologiques et espaces nucléaires, Mem. Amer. Math. Soc. 16 (1955).

[17] Hollstein, R., ε-Tensorprodukte von Homomorphismen, Habilitationsschrift, Paderborn 1978.

[18] Kaballo, W., Lifting-Sätze für Vektorfunktionen und das ε-Tensorprodukt, Habilitationsschrift, Kaiserslautern 1976.

[19] ----------, Lifting-Sätze für Vektorfunktionen und ($\mathcal{L}$L)-Tripel, to appear.

[20] ----------, Lifting-Probleme für holomorphe Funktionen mit Wachstumsbedingungen, to appear.

[21] Kleinstück, G., Der beschränkte Fall des gewichteten Approximationsproblems für vektorwertige Funktionen, Manuscripta math. 17, 123-149 (1975).

[22] Machado, S., Prolla, J.B., The general complex case of the Bernstein-Nachbin approximation problem, Ann. Inst. Fourier 28, 1, 193-206 (1978).

[23] Nachbin, L., Machado, S., Prolla, J.B., Weighted approximation, vector fibrations and algebras of operators, J. math. pures et appl. 50, 299-323 (1971).

[24] Prolla, J.B., Approximation of vector-valued functions, North-Holland Math. Studies 25 (1977).

[25] -----------, The approximation property for Nachbin spaces, Approximation Theory and Functional Analysis, Proc. Conference Approximation Theory, Campinas 1977, North-Holland Math. Studies 35, 371-382 (1979).

[26] Prolla, J.B., Machado, S., Weighted Grothendieck subspaces, Transact. Amer. Math. Soc. 186, 247-258 (1973).

[27] Séminaire Schwartz 1953/54: Produits tensoriels topologiques d'espaces vectoriels topologiques. Espaces vectoriels topologiques nucléaires. Applications, Paris 1954.

[28] Schwartz, L., Espaces de fonctions différentiables à valeurs vectorielles, J. d'analyse math. 4, 88-148 (1954).

[29] -----------, Théorie des distributions à valeurs vectorielles I, Ann. Inst. Fourier 7, 1-142 (1957).

[30] Szankowski, A., $\mathcal{B}$(H) does not have the approximation property, preprint 1978, to appear.

AN APPLICATION OF KOROVKIN'S THEOREM TO CERTAIN PARTIAL DIFFERENTIAL EQUATIONS

Bruno Brosowski
Johann Wolfgang Goethe Universität
Frankfurt/Main
Fachbereich Mathematik
Robert-Mayer-Str. 6 - 10
D-6000 Frankfurt

I. INTRODUCTION

In 1953 Korovkin [6] proved the following theorem:

If $L_n: C[0,1] \to C[0,1]$ is a sequence of monotonic linear operators such that

$$L_n(1) \to 1, \quad L_n(t) \to t, \quad L_n(t^2) \to t^2$$

converges in the sup-norm, then $L_n(x) \to x$ converges in the sup-norm for every continuous function x in $C[0,1]$.

In other words: If L_n converges to the identity I on the subset $K := \{1,t,t^2\} \subset C[0,1]$, then L_n converges on $C[0,1]$ to the identity. Every subset $K \subset C[0,1]$ with this property is called a Korovkin-set. With the aid of Korovkin's theorem one can give an easy proof of the approximation theorem of Weirstrass. For this, one constructs a sequence of monotonic linear operators which map $C[0,1]$ into the polynomials. As an example we mention the Bernstein-polynomials:

$$B_n(x)(t) := \sum_{\nu=0}^{n} x\left(\frac{\nu}{n}\right)\binom{n}{\nu} t^{n-\nu}(1-t)^{\nu}.$$

Since Korovkin published his result there have been many articles written that extend this theorem. If one, for instance, replaces the interval $[0,1]$ by a compact subset $T \subset R^m$, then one can prove that the set of functions

$$1, t_1, \ldots, t_m, \quad t_1^2 + \ldots + t_m^2$$

is a Korovkin-set in $C[T]$. For further generalizations we refer to the survey articles of Bauer [1] and Berens & Lorentz [2].

In [3] we gave a new approach to Korovkin's theorem which used the Dedekind-completion of a partially ordered vector space. Korovkin's theorem can then be considered as an extension of the convergence behaviour of an operator sequence on a subset $K \subset C[T]$ to its Dedekind-completion $\delta\,\mathrm{span}(K)$. The special Korovkin theorem is obtained if one takes into account that the Dedekind-completion of the polynomials of degree 2 (endowed with pointwise ordering) contains $C[0,1]$.

In this paper we continue the investigations begun in [3] and indicate how this theory can be applied to the case where the identity operator is replaced by certain mappings A of $C[T]$ into a Dedekind-complete vector space. Especially we consider the case, where A is the solution operator of the Dirichlet-problem for the potential equation. Finally we show that there are also Korovkin-type results for the approximation of the solution of the Dirichlet-problem by difference methods.

II. DEDEKIND-COMPLETION OF PARTIALLY ORDERED VECTOR SPACES

A partially ordered real vector space X is called Dedekind-complete iff every non-empty subset which is bounded from above (below) has a supremum (infimum) in X. The spaces $C[T]$ and $\mathbb{P}_n$, $n \geq 2$, are not Dedekind-complete. ($\mathbb{P}_n$ denotes the linear space of all polynomials of degree $\leq n$, the partial ordering is pointwise, i.e. $x \leq y$ iff $x(t) \leq y(t)$ for each t). Examples of Dedekind-complete spaces are: $\mathbb{R}$, $\mathbb{P}_0$, $\mathbb{P}_1$ and the space $\mathrm{HAR}(\Omega)$ which is defined as follows: Let Ω be a non-empty, open and bounded subset of $\mathbb{R}^m$. Then $\mathrm{HAR}(\Omega)$ is the real vector space of all functions $v = v_1 - v_2$, with v_1, v_2 harmonic and non-negative in Ω. With the pointwise ordering $HAR(\Omega)$ is Dedekind-complete (cf. Luxemburg & Zaanen [7]).

In Luxemburg & Zaanen [7] one can find the following

<u>Completion Theorem.</u> If X is an Archimedean [which is defined by $\forall_{u,v\in X} \forall_{n\in\mathbb{N}}\ nv \leq u \Rightarrow v \leq 0$] partially ordered real vector space, then there exists a uniquely determined Dedekind-complete partially ordered real vector space $\delta X \supset X$ such that every $x \in \delta X$ satisfies

$$x = \sup \{\ell \in X \mid \ell \leq x\}$$
$$= \inf \{u \in X \mid x \leq u\}.$$

δX is called the Dedekind-completion of X.

Like in the case of the rational numbers the elements of δX can be considered as cuts i.e. as pairs (L,U) of subsets of X with the following properties:

(α) $\forall_{\ell\in L} \forall_{u\in U}\ \ell \leq u$

(β) If $c \in X$ and $\forall_{\ell\in L}\ \ell \leq c$ then $c \in U$

(γ) If $c \in X$ and $\forall_{u\in U}\ c \leq u$ then $c \in L$

The vector space X is embedded into δX by the mapping

$$x \rightarrow (L_x, U_x),$$

where

$$L_x := \{\ell \in X \mid \ell \leq x\}$$

and

$$U_x := \{u \in X \mid x \leq u\}.$$

There is a simple sufficient condition that $\delta X \supset C[T]$ for a subspace $X \subset C[T]$:

<u>Theorem 1.</u> Let T be a compact Hausdorff-space and let X be a linear subspace of $C[T]$ such that

(α) $1 \in X$,

(β) $\forall_{t_o\in T} \exists_{p_{t_o}\in X} [p_{t_o}(t_o) = 0\ \&\ \forall_{t\neq t_o}\ p_{t_o}(t) > 0]$

then one has $\delta X \supset C[T]$.

<u>Proof</u>. Let $y \in C[T]$ be given. Choose an arbitrary point $t_o \in T$ and $0 < \varepsilon < \|y\|$.

There exists an open set U containing t_o such that

$$\underset{t \in U}{\forall} \; y(t) \leq y(t_o) + \varepsilon .$$

Now consider the function

$$p(t) := (y(t_o) + \varepsilon) + \beta \cdot p_{t_o}(t)$$

which is an element of X. If one chooses

$$\beta > \frac{2\|y\| - (y(t_o) + \varepsilon)}{\inf\limits_{t \in T \setminus U} p_{t_o}(t)} > 0$$

then we have

$$p(t) \geq y(t_o) + \varepsilon \geq y(t), \quad \text{if} \quad t \in U,$$

and

$$p(t) \geq y(t_o) + \varepsilon + 2\|y\| - (y(t_o) + \varepsilon)$$
$$= 2\|y\| \geq y(t), \quad \text{if} \quad t \notin U.$$

Consequently, $p \in U_y$ and

$$y(t_o) \leq p(t_o) \leq y(t_o) + \varepsilon .$$

Since $\varepsilon > 0$ and $t_o \in T$ are arbitrary we conclude

$$y(t) = \inf_{u \in U_y} u(t).$$

Similarly one can prove

$$y(t) = \sup_{\ell \in L_y} \ell(t).$$

<u>Corollary 2</u>. Let (T,d) be a compact metric space and let X be a subspace of $C[T]$ such that $1 \in X$ and for every $t_o \in T$ the function $d(\cdot,t_o) \in X$. Then $\delta X \supset C[T]$.

<u>Corollary 3</u>. Let T be a compact subset of a Hilbert-space $(H,(\cdot,\cdot))$. If X is a subspace of $C[T]$, which contains 1, the linear functionals, and the function $t \to (t,t)$ (restricted to T), then $\delta X \supset C[T]$.

<u>Corollary 4</u>. $\delta\mathbb{P}_2 \supset C[a,b]$.

III. AN EXTENSION THEOREM FOR MONOTONIC MAPPINGS

<u>Theorem 5</u>. Let X and Z be partially ordered Archimedean vector spaces with Z Dedekind-complete and let $A: X \to Z$ be a monotonic linear mapping. Then the following holds:

(1) There is a monotonic linear mapping

$$A^{\#}: \delta X \to Z$$

that extends A.

(2) For every y in

$$G := \{y \in \delta X \mid \sup A(L_y) = \inf A(U_y)\}$$

$A^{\#}(y)$ is uniquely determined.

(3) If A is inverse monotonic (i.e. $Ax \leq Ay \Rightarrow x \leq y$) then $A^{\#}$ is also inverse monotonic and $A^{\#}$ preserves suprema and infima.

<u>Proof</u>. (1) Since every element y in δX can be considered as a cut (L,U) in X, there exists an element u in X such that $y \leq u$. Hence X is cofinal in δX. Since Z is Dedekind-complete by a theorem in Jameson [5, p.65] there is a monotonic linear mapping $A^{\#}: \delta X \to Z$ that extends A.

(2) By the monotonicity of $A^{\#}$ we conclude for every y in δX

$$\sup A(L_y) \leq A^{\#}(y) \leq \inf A(U_y)$$

and hence $A^{\#}$ is uniquely determined on G.

(3) Now assume that $A: X \to Z$ is inverse monotonic. Let $A^{\#}(x) \leq$ $\leq A^{\#}(y)$. We have to show that

$$L_x \subset L_y \;\&\; U_x \supset U_y.$$

Let $\ell \in L_x$ and $u \in U_y$ arbitrary. Then we have

$$A(\ell) \leq A^{\#}(x) \leq A^{\#}(y) \leq A(u).$$

Since A is inverse monotonic it follows that $\ell \leq u$ for every

$u \in U_y$. Since (L_y, U_y) is a cut we have $\ell \in L_y$. Similarly one can show $U_x \supset U_y$.

Now let $W \subset \delta X$ and $s := \sup W$. Then we have $A^{\#}(s) \geq A^{\#}(W)$. Now let z be in $IM(A^{\#})$ such that $z \geq A^{\#}(W)$. Since $A^{\#}$ is inverse monotonic it follows that $A^{\#-1}(z) \geq W$ and hence $A^{\#-1}(z) \geq s$. Consequently $z = A^{\#}(A^{\#-1}(z)) \geq A^{\#}(s)$, i.e.

$$A^{\#}(s) = \sup A^{\#}(W). \quad \blacksquare$$

An application of this extension theorem to boundary value problems is as follows:

Let $\emptyset \neq \Omega \subset \mathbb{R}^m$ be an open bounded set and let $Z := HAR(\Omega)$ be the vector space of harmonic functions defined in Section II. Further let X be a linear subspace of $C[\delta X]$ such that $\delta X \supset C[\partial\Omega]$. Now consider the Dirichlet-problem

$$(*) \qquad \Delta v = 0 \text{ in } \Omega \quad \& \quad v|_{\partial\Omega} = x \quad \& \quad v \in C[\bar{\Omega}].$$

Let A be the operator which maps $x \in X$ to the solution of $(*)$. The operator A is a monotonic and inverse monotonic operator. Then we have the following

<u>Theorem 6</u>. If the Dirichlet-problem $(*)$ is solvable for a linear subspace $X \subset C[\partial\Omega]$ such that $\delta X \supset C[\partial\Omega]$ then the Dirichlet-problem $(*)$ is solvable for every $x \in C[\partial\Omega]$.

<u>Proof</u>. By the extension theorem we can extend A to the Dedekind-completion $\delta X \supset C[\partial\Omega]$. Since for every $y \in C[\partial\Omega]$ one has $A^{\#}(y) \in HAR(\Omega)$ it suffices to show

$$A^{\#}(y)|_{\partial\Omega} = y.$$

Since $\sup L_y = y = \inf U_y$ we conclude by using the maximum principle

$$\sup A(L_y) = A^{\#}(y) = \inf A(U_y).$$

Since $\sup A(L_y)|_{\partial\Omega} = y$ it follows that $A^{\#}(y)$ is the solution of the Dirichlet-problem $\Delta v = 0$ in Ω and $v|_{\partial\Omega} = y$. $\blacksquare$

For X we can choose for example the vector space

$$X := \operatorname{span}(1, t_1, \dots, t_m, t_1^2 + \dots + t_m^2).$$

Since the Dirichlet-problem is always solvable for the boundary values $1, t_1, \dots, t_m$ we conclude from Theorem 6 the

<u>Corollary 7</u>. The Dirichlet-problem

$$\Delta v = 0 \text{ in } \Omega \;\&\; v|_{\partial\Omega} = y \;\&\; v \in C[\bar{\Omega}]$$

has a solution for each $y \in C[\partial\Omega]$ iff it has a solution for the special boundary values

$$y(t_1, \dots, t_m) := t_1^2 + t_2^2 + \dots + t_m^2.$$

<u>Corollary 8</u>. Let the Dirichlet-problem (*) be solvable for a linear subspace $X \subset C[\partial\Omega]$ such that $\delta X \supset C[\partial\Omega]$. Let $A^{\#}$ be the extension of the solution operator $A: X \to HAR(\Omega)$, then

$$\{y \in \delta X \mid \sup A(L_y) = \inf A(U_y)\} \supset C[\partial\Omega].$$

IV. CONVERGENCE IN A DEDEKIND-COMPLETE VECTOR SPACE

For stating the generalized Korovkin-theorem we need a mode of convergence in the Dedekind-complete space Z. For this we use methods developed in [8]. Let Y be a linear subspace of Z such that $\delta Y = Z$. We assume that in the space Y, a mode of convergence is given, i.e. to certain sequences (y_n) in Y a limit point y in Y is assigned. In this case we write "$y_n \to y$". This mode of convergence can be extended to δY by the following

<u>Definition 9</u>. (1) A sequence of subsets $A_n \subset Y$ converges to a subset $A \subset Y$ iff

$$A = \lim A_n := \{y \in Y \mid \forall_{n \in N} \; \exists_{a_n \in A_n} \; a_n \to y\}.$$

(2) A sequence (L_n, U_n) of cuts in Y converges to a cut (L, U) in Y iff

$$L = \mathcal{L}(\lim U_n) \;\&\; U = \mathcal{U}(\lim L_n),$$

where we have defined

$$\mathcal{L}(W) := \{y \in Y \mid y \leq W\}$$

and

$$\mathcal{U}(W) := \{y \in Y \mid y \geq W\}$$

for any subset W in Y.

In general the mode of convergence for cuts is not an extension of the mode of convergence in Y. A necessary and sufficient condition is given by the following

<u>Theorem 10.</u> Definition 9 extends the mode of convergence to δY if and only if the mode of convergence in Y satisfies the condition

(*) If $x_n \to x$, $y_n \to y$, and $\forall_{n \in \mathbb{N}}\ x_n \leq y_n$, then $x \leq y$.

<u>Proof.</u> Let (*) be satisfied in Y. Then we have to show:

If $y_n \to y$ then

$$L_y = \mathcal{L}(\lim U_{y_n}) \ \& \ U_y = \mathcal{U}(\lim L_{y_n}).$$

If $a \in L_y$ then $a \leq y$. Since $\lim L_{y_n} \leq y \leq \lim U_{y_n}$ it follows that $a \leq \lim U_{y_n}$; hence $a \in \mathcal{L}(\lim U_{y_n})$.

If $a \in \mathcal{L}(\lim U_{y_n})$ then we have $a \leq b$ for each b in $\lim U_{y_n}$. Since $y \in \lim U_{y_n}$ we conclude $a \leq y$; hence $a \in L_y$.

Similarly one can prove $U_y = \mathcal{U}(\lim L_{y_n})$.

Now assume that by Definition 9 the mode of convergence is extended. Then we have to show that (*) is fulfilled. By assumption we have

$$(L_{x_n}, U_{x_n}) \to (L_x, U_x) \ \& \ (L_{y_n}, U_{y_n}) \to (L_y, U_y)$$
$$\& \ \forall_{n \in \mathbb{N}} \ (L_{x_n}, U_{x_n}) \leq (L_{y_n}, U_{y_n}).$$

From this we conclude

$$\forall_{n \in \mathbb{N}} \ L_{x_n} \subset L_{y_n} \ \& \ U_{x_n} \supset U_{y_n}$$

and hence $\lim L_{x_n} \subset \lim L_{y_n}$ which implies $\mathcal{U}(\lim L_{x_n}) \supset \mathcal{U}(\lim L_{y_n})$.

Since (L_x, U_x) and (L_y, U_y) are cuts we have

$$L_x = \mathcal{L}(\mathcal{U}(\lim L_{x_n})) \subset \mathcal{L}(\mathcal{U}(\lim L_{y_n})) = L_y \quad \text{and hence} \quad x \leq y. \quad ■$$

<u>Example 11</u>. Let T be a compact Hausdorff-space and let X be a linear subspace of $C[T]$ with $1 \in X$. The convergence in the sup-norm satisfies condition (*). By Theorem 10 we can extend this mode of convergence to δX. If we consider convergence in $\delta X \cap C[T]$ we have the following result:

A sequence (x_n) in $\delta X \cap C[T]$ converges in the sup-norm to $x \in \delta X \cap C[T]$ if and only if $L_x = \lim L_{x_n}$ and $U_x = \lim U_{x_n}$.

<u>Proof</u>. Assume x_n converges to x in the sup-norm. If $p \in \lim L_{x_n}$ then there exists a sequence $p_n \in L_{x_n}$ such that $p_n \to p$. Since $p_n \leq x_n$ and $x_n \to x$ it follows that $p \leq x$ and hence $p \in L_x$. Now let $p \in L_x$. For each $k \in \mathbb{N}$ choose an $n_k \in \mathbb{N}$ such that

$$\underset{n>n_k}{\forall} \|x-x_n\| < \frac{1}{k}.$$

We can assume that $n_1 < n_2 < \ldots$ The element $p(t) - \frac{1}{k}$ is contained in L_{x_n} for $n > n_k$. Now define a sequence (p_n) in X converging in the sup-norm to p by

$$p_n := \begin{cases} \text{any element in } L_{x_n}, & \text{if } n \leq n_1 \\ p - \frac{1}{k}, & \text{if } n_k < n \leq n_{k+1}. \end{cases}$$

Consequently we have $p \in \lim L_{x_n}$. Similarly one can prove $U_x = \lim U_{x_n}$.

Now assume $L_x = \lim L_{x_n}$ and $U_x = \lim U_{x_n}$. For $\varepsilon > 0$ choose elements $p_1,\ldots,p_k \in L_x$, $q_1,\ldots,q_k \in U_x$, and an open cover $B_1,B_2,\ldots,B_k$ of T such that

$$(\#) \qquad \underset{t\in B_\varkappa}{\forall} q_\varkappa(t) - \varepsilon < x(t) < p_\varkappa(t) + \varepsilon,$$

$\varkappa = 1,2,\ldots,k$. Since $p_1,\ldots,p_k \in \lim L_{x_n}$ and $q_1,\ldots,q_k \in \lim U_{x_n}$ we can find elements $p_\varkappa^n \in L_{x_n}$ and $q_\varkappa^n \in U_{x_n}$ such that $p_\varkappa^n \to p_\varkappa$ and $q_\varkappa^n \to q_\varkappa$. Now choose an $n_o \in \mathbb{N}$ such that

$$\forall_{n\geq n_o} \|p_\varkappa - p_\varkappa^n\| < \varepsilon \quad \& \quad \|q_\varkappa - q_\varkappa^n\| < \varepsilon ,$$

$\varkappa = 1,2,\ldots,k$. Then we have

$$\forall_{n\geq n_o} \quad p_\varkappa(t) - \varepsilon < x_n(t) < q_\varkappa(t) + \varepsilon$$

and with the aid of (#) we conclude

$$\forall_{n\geq n_o} \quad -2\varepsilon < x(t) - x_n(t) < 2\varepsilon . \quad \blacksquare$$

V. GENERALIZED KOROVKIN-THEOREM

Theorem 12. Let X be a linear subspace of $C[T]$ such that $\delta X \supset C[T]$. Further let Z be the space $\delta C[T]$ or $HAR(\Omega)$ endowed with the convergence which is an extension of the uniform convergence in $C[T]$ or in the subspace of all harmonic functions with continuous boundary values resp. Let $L_n: C[T] \to Z$ be monotonic operators and let $A: X \to Z$ be a monotonic and inverse monotonic mapping.

If for every $x \in X$ the sequence $L_n(x)$ converges uniformly to $A(x)$, then for every $y \in C[T]$ the sequence $L_n(y)$ converges also uniformly to the unique monotonic extension $A^{\#}(y)$.

Proof. For abbreviation we set

$$z := A^{\#}(y) \quad \& \quad z_n := L_n(y).$$

We prove

$$L_z = \lim L_{z_n} \quad \& \quad U_z = \lim U_{z_n}.$$

Let $\ell' \in L_z$ then we have that $\ell' \leq z = A^{\#}(y)$ and so $A^{-1}(\ell') \leq y$. By assumption

$$v_n := L_n(A^{-1}(\ell')) \to A(A^{-1}(\ell')) = \ell'$$

and $v_n \leq z_n$. Since the convergence is uniform it follows that

$$\lim L_{v_n} = L_{\ell'} .$$

Since $\ell' \in L_{\ell'}$ there exists a sequence $\ell'_n \in L_{v_n}$ such that $\ell'_n \to \ell'$. Since also $\ell'_n \leq v_n \leq z_n$ it follows that $\ell' \in \lim L_{z_n}$. Similarly one can show $U_z \subset \lim U_{z_n}$.

Now let $\ell' \in \lim L_{z_n}$. Then there exists a sequence $\ell'_n \leq z_n$ such that $\ell'_n \to \ell'$. Since $\ell'_n \leq z_n \leq L_n(U_y)$ it follows that $\ell' \leq A(U_y)$ and hence $\ell' \leq \inf A(U_y) = \Lambda^{\#}(y)$, i.e. $\ell' \in L_z$. Similarly one can show $\lim U_{z_n} \subset U_z$. ∎

We can apply this theorem to the Dirichlet-problem

$$(*) \qquad \Delta v = 0 \text{ in } \Omega \ \& \ v\big|_{\partial\Omega} = x \ \& \ v \in C[\bar{\Omega}].$$

Theorem 13. Let $L_n : C[\partial\Omega] \to HAR(\Omega)$ be a sequence of monotonic operators such that for each element x of the vector space

$$X := \operatorname{span}\{1, t_1, \ldots, t_m, t_1^2 + \ldots + t_m\} \subset C[\partial\Omega]$$

the sequence $L_n(x)$ converges uniformly to the solution of the Dirichlet-problem (*).

Then the sequence $L_n(y)$ converges uniformly to the solution of the Dirichlet-problem (*) for each $y \in C[\partial\Omega]$.

We consider now an application of Theorem 12 to the problem of convergence in the finite-difference method.

Let Ω be the set

$$\{(t_1, t_2) \in \mathbb{R}^2 \mid 0 < t_1 < 1 \ \& \ 0 < t_2 < 1\}$$

and consider the Dirichlet-problem

$$(*) \qquad \Delta v = 0 \text{ in } \Omega \ \& \ v\big|_{\partial\Omega} = x \ \& \ v \in C[\bar{\Omega}].$$

We introduce a mesh

$$t_1^{\nu} := \frac{\nu}{n} \ \& \ t_2^{\nu} := \frac{\nu}{n}, \qquad \nu = 0, 1, \ldots, n$$

and replace the Dirichlet-problem (*) by a system of difference equations

$$(\#) \qquad D_n v = 0 \ \& \ v\big|_{\partial\Omega} = x,$$

where for each point (t_1^{ν}, t_2^{μ}) in Ω, $D_n v(t_1^{\nu}, t_2^{\mu})$ is defined by

$$D_n\, v(t_1^\nu, t_2^\mu) := \frac{v(t_1^{\nu+1},\; t_2^\mu) - 2v(t_1^\nu, t_2^\mu) + v(t_1^{\nu-1},\; t_2^\mu)}{(1/n)^2} + \frac{v(t_1^\nu, t_2^{\mu+1}) - 2v(t_1^\nu, t_2^\mu) + v(t_1^\nu, t_2^{\mu-1})}{(1/n)^2}$$

Now let $\tilde{L}_n$ be the solution operator of the system (#). This linear mapping $\tilde{L}_n$ is monotonic (cf. Isaacson & Keller [4]). The function $\tilde{L}_n x$ is defined only in the mesh point. We extend the functions $\tilde{L}_n x$ linearly to a continuous function in $C[\Omega]$ by using the right directional grid illustrated in figure 1.

Figure 1.

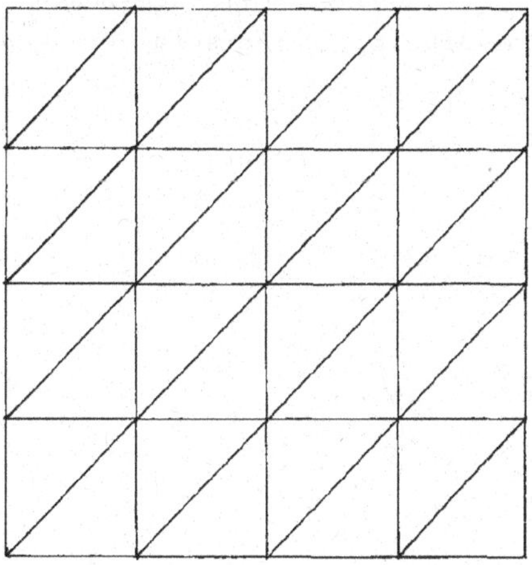

Thus we have defined a monotonic linear mapping

$$L_n\colon C[\,\Omega] \to C[\bar{\Omega}].$$

Then we have the following

<u>Theorem 14.</u> Assume that the difference-method converges for every

$$x \in \text{span } (1, t_1, t_2, t_1^2, +t_2^2) \subset C[\partial\,\Omega],$$

i.e. $L_n(x) \to A(x)$ in the sup-norm.

Then $L_n(y) \to A(y)$ in the sup-norm for each $y \in C[\partial\,\Omega]$.

<u>Proof.</u> Since by Theorem 1 $\delta X \supset C[\partial\,\Omega]$ Theorem 14 follows from Theorem 12.

REFERENCES

[1] Bauer, H.: Approximationssätze und abstrakte Ränder. Math. Phys. Semesterberichte 12 (1976) 141-173.

[2] Berens, H.; Lorentz, G.G.: Theorems of Korovkin Type for positive linear operators on Banach lattices. Proc. Int. Symp. Approximation Theory, Austin, Texas, 1973.

[3] Brosowski, B.: The completion of partially ordered vector spaces and Korovkin's theorem. Approximation Theory and Functional Analysis, North-Holland, 1979, p. 63-69.

[4] Isaacson, E.; Keller, H.B.: Analyse numerischer Verfahren. Verlag Harri Deutsch, Zürich, Frankfurt 1973.

[5] Jameson, G.: Ordered linear spaces. Springer-Verlag, Berlin, Heidelberg, New York 1970.

[6] Korovkin, P.P.: Über die Konvergenz positiver linearer Operatoren im Raum der stetigen Funktionen (Russisch). Doklady Akad. Nauk. SSSR (N.S.) 90, 961-964 (1953).

[7] Luxemburg, W.A.J; Zaanen, A.G.: Riesz spaces, Vol. I. North-Holland Publishing Company, Amsterdam-London 1971.

[8] Starke, P.: Diplom-Arbeit, Universität Frankfurt, 1978.

THE FOURIER-BOREL TRANSFORM IN INFINITELY MANY DIMENSIONS AND APPLICATIONS

J.F. Colombeau and B. Perrot

UER de Mathématiques et d'Informatique
Université de Bordeaux I
351, Cours de la Libération
33405 Talence - France

ABSTRACT

We study the Fourier-Borel transform in the case of infinite dimensional holomorphic functions. We first show (th.1) under a very general assumption on the space E that the image of $\mathcal{H}'(E)$ through the Fourier-Borel transform is the space $\mathfrak{Z}(E)$ introduced in [8], [9], [12]. An application of this result is a new proof of nuclearity of $\mathcal{H}(\Omega)$.

When the space E has some additional properties of nuclearity th. 1 is improved in th. 3 which generalizes a result of Boland [1]. Th. 3 is used in the section 6 of this paper where we obtain a general result (th. 4) on the approximation of solutions of some infinite dimensional convolution equations. This th. 4 unifies and improves some results of [1] and [3].

1. NOTATIONS AND TERMINOLOGY

The notations and terminology are those of [4] [15] [18]. All the vector spaces considered in this article are complex if the converse is not explicitely stated. A Hausdorff locally convex vector space is denoted by the letters "l.c.s.". Convex bornological vector spaces ([15], [18]) (denoted by the letters "b.v.s.") are used in this paper in order to state the results in their natural setting and are also used in some proofs. Since this article should be readable also for people not very familiar with them we shall recall most of the definitions and results which we need. We shall always consider complete b.v.s. which are separated by their dual: they are algebraic inductive limits of Banach spaces $(E_{B_i})_{i\in I}$ with an injective linear continuous mapping from E_{B_i} to E_{B_j} if $j > i$. By definition a subset of E is bounded iff it is contained and bounded in one of the Banach spaces E_{B_i}. A complete b.v.s. E is called a Schwartz b.v.s. iff for each index i there exists an index j such that B_i is relatively compact in the Banach space E_{B_j} (there are a lot of such spaces: for example the bornology of the compact sets in a Fréchet space [17] or the Von Neumann bornology of a quasi-complete l.c.s. the strong dual of which is a Schwartz l.c.s.). A complete b.v.s. is called nuclear [15] [16] iff for each i there exists j such that the injection $E_{B_i} \to E_{B_j}$ is a nuclear mapping (for example the Von Neumann bornology of a quasi complete l.c.s. the strong dual of which is a nuclear l.c.s.). A subset Ω of a b.v.s. E is said to be open for the topology τE of the Mackey closure iff: for each $i \in I$, $\Omega \cap E_{B_i}$ is open in the normed space E_{B_i}. A subset of E is said to be strictly compact iff it is contained and compact in one of the Banach spaces E_{B_i}.

A Silva holomorphic function ([5][6][22][24]) from Ω to $\mathbb{C}$ is a G-analytic function such that:

$\forall x \in \Omega, \forall B$ bounded set in E $\exists\, \varepsilon > 0$ such that $f(x+\varepsilon B)$ is bounded in $\mathbb{C}$.

We denote by $\mathcal{H}_S(\Omega)$ the space of the Silva holomorphic functions on Ω; this space is equipped with the topology of the uniform convergence on the strictly compact subsets of E contained in Ω.

Remark 1. If Ω is balanced it is proved in appendix 3 that the space of the G analytic and continuous functions is dense in $\mathcal{H}_S(\Omega)$ in the usual cases for the space E. Hence the results obtained on the Fourier-Borel transform in $\mathcal{H}'_S(\Omega)$ are interpretable in terms of the more usual concept of G-analytic and continuous functions.

Remark 2. The results of sections 2 and 3 may be generalized in the vector valued case (the results of section 2 are announced in [11] in the vector valued case).

Finally let us recall that if E is a Schwartz b.v.s. separated by its dual (which is always assumed in this paper), $E^{\tilde{\otimes}n}$ denotes the completion of the n-fold bornological tensor product of E (see [16], section 2 of [12] and the appendix 4; $E^{\tilde{\otimes}n}$ is an abbreviated notation for $E^{\tilde{\otimes}n}_{\pi_b^\circ}$).

The symbol □ indicates the end of a proof.

2. THE FOURIER-BOREL TRANSFORM IN THE GENERAL CASE

Let E be a b.v.s. and let $E^\times = L(E;\mathbb{C})$ denote the bornological dual of E (linear functions from E to $\mathbb{C}$ which are bounded on each bounded subset of E). $E^\times$ is naturally equipped with the topology of the uniform convergence on the bounded subsets of E and with the Von Neumann bornology of this topology. Let $\langle\ ,\ \rangle$ denote the duality brackets between $E^\times$ and E.

For T in $E^\times$ it is immediate to verify that the function

$$e^T: E \longrightarrow \mathbb{C} \qquad \alpha \to e^{\langle T,\alpha\rangle} \quad (\alpha \in E)$$

is in $\mathcal{H}_S(E)$.

Let now L be an element of $\mathcal{H}'_S(E)$, the dual of $\mathcal{H}_S(E)$. It is easy to prove that the function

$$\mathcal{F}L: E^\times \longrightarrow \mathbb{C} \qquad T \longrightarrow \mathcal{F}L(T) = L(e^T)$$

is in $\mathcal{H}_S(E^\times)$.

As usual, the mapping $\mathcal{F}: \underset{L}{\mathcal{H}'_S(E)} \to \underset{\mathcal{F}L}{\mathcal{H}_S(E^\times)}$ is called "the Fourier-Borel transform".

Let Ω be a convex balanced open subset for the topology τE.

Let us recall that a sequence (x_n) of elements of E is said to be Mackey convergent to 0 iff this sequence is contained in a normed space E_{B_i} and is a null sequence in this normed space. If M is a subset of E we denote by $\Gamma_{\ell_1}(M)$ the set of all the sums of the series $\Sigma\lambda_n x_n$ (where $x_n \in M$ and $\Sigma|\lambda_n| \leq 1$) which are convergent in some (variable) Banach space E_{B_i}. If b is a strictly compact subset contained in Ω there exists a strictly compact subset K contained in Ω, such that K contains b and a Mackey-null sequence (x_n) of points in Ω such that $K = \Gamma_{\ell_1}\{x_n\}$ (see appendix 1). From now on, all the sets K that we shall consider will be of this type.

Let $\mathcal{V}(K) = \{\varphi \in \mathcal{H}_S(\Omega)$ such that $\sup_{x\in K} |\varphi(x)| \leq 1\}$ where K is a strictly compact subset of Ω of the type described above and let $\mathcal{V}^o(K) \subset \mathcal{H}'_S(\Omega)$ be the polar of the set $\mathcal{V}(K)$.

Let $L_n((E^\times)^n;\mathbb{C})$ denote the space of the n-multilinear functions from $(E^\times)^n$ to $\mathbb{C}$ which are bounded on each bounded subset of $(E^\times)^n$.

For more simplicity in the formulation of the definitions and results let us assume that the canonical mapping:

$$E^{\otimes n} \to L_n((E^\times)^n;\mathbb{C})$$

admits an injective continuation in $E^{\tilde{\otimes} n}$, hence $E^{\tilde{\otimes} n}$ may be considered as a part of $L_n((E^\times)^n;\mathbb{C})$.

<u>Remark 3</u>. This last property is true if E admits a basis of bounded sets (B_i) such that E_{B_i} is a reflexive Banach space with the approximation property (see the prop. 2 of appendix 4), hence it is true in "usual" cases for E. In appendix 2 we indicate how one has to modify the formulation of the results of this section in the general case.

Let now:

$[K] = \{\phi \in \mathcal{H}_S(E^\times)$ such that, for each integer n, $\phi^{(n)}(0)\in\Gamma_{\ell_1}(K^{\otimes n})$ in $E^{\tilde{\otimes} n}\}$.

As usual let $L_n(E^n;\mathbb{C})$ denote the space of the n-multilinear bounded functions from E^n to $\mathbb{C}$ and let us equip it with its natural bornology of the equi-bounded sets. Then we have: $L_n(E^n;\mathbb{C}) = (E^{\tilde{\otimes} n})^\times$ bornologically ([16]) hence $L_n^\times(E^n;\mathbb{C}) = (E^{\tilde{\otimes} n})^{\times\times}$. Equip this space $L_n^\times(E^n;\mathbb{C})$ with the topology of the uniform convergence on the bounded subsets of $L_n(E^n;\mathbb{C})$.

Since $\tilde{E}^{\otimes n}$ is separated by its dual (because it is easy to prove that the completion of a polar b.v.s. is separated by its dual), it may be considered as a part of its bidual $(\tilde{E}^{\otimes n})^{\times\times} = L_n^{\times}(E^n;C)$. If we consider now that $\tilde{E}^{\otimes n}$ is contained in $L_n^{\times}(E^n;C)$:

<u>Lemma 1.</u> $\Gamma_{\ell_1}(K^{\otimes n})$ <u>is compact in the l.c.s.</u> $L_n^{\times}(E^n;C)$.

<u>Proof.</u> It is proved in the prop. 1 of appendix 4 that $\Gamma_{\ell_1}(K^{\otimes n})$ is strictly compact in the b.v.s. $\tilde{E}^{\otimes n}$. Now the result follows from the fact that the inclusion of $\tilde{E}^{\otimes n}$ in $L_n^{\times}(E^n;C)$ is bounded. □

<u>Proposition 1. The image through the Fourier-Borel transform of the set</u> $\mathcal{V}^o(K)$ <u>is contained in the set</u> $[K]$.

<u>Proof.</u> For the duality between $L_n(E^n;C)$ and $L_n^{\times}(E^n;C)$, by Lemma 1, $(\Gamma_{\ell_1} K^{\otimes n})^{oo} = \Gamma_{\ell_1} K^{\otimes n}$. If φ is in $(\Gamma_{\ell_1} K^{\otimes n})^o \subset L_n(E^n;C)$ and if we consider φ as an holomorphic function on E, $\sup_{x\in K} |\varphi(x)| \le 1$. Hence, if L is in $\mathcal{V}^o(K)$, $|L(\varphi)| \le 1$. Let $\mathcal{L}_n$ denote the restriction of L to $L_n(E^n,C) \subset \mathcal{H}_S(\Omega)$. Then we have just proved that $\mathcal{L}_n \in (\Gamma_{\ell_1} K^{\otimes n})^{oo} = \Gamma_{\ell_1} K^{\otimes n}$. But

$$\mathfrak{F}L(T) = L(e^T) = \sum_n \frac{1}{n!} \mathcal{L}_n(T^{\otimes n}).$$

Hence $\mathfrak{F}L^{(n)}(0) = \mathcal{L}_n \in \Gamma_{\ell_1} K^{\otimes n}$ hence $\mathfrak{F}L$ is in $[K]$. □

Let e denote the real number such that $\operatorname{Log} e = 1$ et us assume now $eK \subset \Omega$ and let $k > 1$ be such that $k\,e\,K \subset \Omega$.

<u>Proposition 2. If</u> ϕ $(\in \mathcal{H}_S(E^{\times}))$ <u>is in</u> $[K]$ <u>and</u> φ <u>is an element of</u> $\mathcal{H}_S(\Omega)$ <u>then the numerical series</u> $\sum_n \frac{1}{n!} |\varphi^{(n)}(0)\,\phi^{(n)}(0)|$ <u>are convergent; let</u> $(\check{\mathfrak{F}}\,\phi)(\varphi) = \sum_n \frac{1}{n!} \varphi^{(n)}(0)\,\phi^{(n)}(0)$. <u>Then the image of</u> $[K]$ <u>through the mapping</u> $\check{\mathfrak{F}}$ <u>is contained in</u> $\frac{k}{k-1}\mathcal{V}^o(k\,e\,K)$.

<u>Proof.</u>
$$\varphi^{(n)}(0)\,\alpha_1\cdot\ldots\cdot\alpha_n = \frac{1}{(2i\pi)^n}\int_{\substack{|\xi_1|=r_1\\ \vdots\\ |\xi_n|=r_n}} \frac{\varphi(\xi_1\alpha_1+\ldots+\xi_n\alpha_n)}{\xi_1^2\ldots\xi_n^2}\,d\xi_1\ldots d\xi_n.$$

choose $r_i = \frac{ke}{n}$ and $\alpha_i \in K$. Then if φ is in $\mathcal{V}$ $(k\,e\,K)$

$$\left| \frac{1}{n!} \varphi^{(n)}(0)\, \alpha_1 \ldots \alpha_n \right| \le \frac{n^n}{n!\, e^n k^n} \le \frac{1}{k^n} .$$

If ϕ is in $[K]$, $\phi^{(n)}(0) = \sum_i \mu_i \alpha_{i_1} \otimes \ldots \otimes \alpha_{i_n}$ with $\Sigma |\mu_i| \le 1$ and $\alpha_{i_j} \in K$ hence

$$\left| \frac{1}{n!} \varphi^{(n)}(0)\, \phi^{(n)}(0) \right| \le \frac{1}{k^n} \quad \square$$

Now let $\mathfrak{F}(E)$ denote the subspace of $\mathcal{H}_S(E^\times)$ spanned by the sets $[K]$ when K varies in a basis of subsets of E of the type $\Gamma_{\ell_1}\{x_n\}$ described above. Let us equip this space $\mathfrak{F}(E)$ with the bornology of the sets $[K]$ and with the bornological topology of this bornology (a basis of neighbourhoods of the origin for this last topology is the family of the convex and balanced bornivorous sets).

Let us equip $\mathcal{H}'_S(E)$ with the bornology of the equicontinuous sets and with its strong topology of dual.

<u>Theorem 1</u>. <u>Let</u> E <u>be a Schwartz b.v.s. separated by its dual</u>. <u>The Fourier-Borel transform</u> $\mathfrak{F}$ <u>is a bornological and topological isomorphism between</u> $\mathcal{H}'_S(E)$ <u>and</u> $\mathfrak{F}(E)$.

<u>Proof</u>. The algebraic and bornological isomorphisms are a consequence of prop. 1 and 2 in the case $\Omega = E$ because an easy computation shows that $\mathfrak{F}$ and $\check{\mathfrak{F}}$ are inverse mappings. The topological isomorphism is due to the fact that the strong topology on $\mathcal{H}'_S(E)$ is the bornological topology associated to the equicontinous bornology since $\mathcal{H}_S(E)$ is a "completely reflexive l.c.s." according to the terminology of Hogbé-Nlend ([18]). (If $T\mathcal{H}'_S(E)$ denotes the above bornological topology, the dual of $T\mathcal{H}'_S(E)$ is $\mathcal{H}_S(E)$. Hence, by Mackey's theorem [18], the strong topology on $\mathcal{H}'_S(E)$ is stronger than $T\mathcal{H}'_S(E)$. The converse is immediate). $\square$

3. A NEW PROOF OF THE NUCLEARITY OF $\mathcal{H}(\Omega)$

The nuclearity of $\mathcal{H}(\Omega)$ was proved by Boland [2] and Waelbroeck [26]. We give here a new proof as an immediate consequence of the results of the preceding section. We follow the classical terminology of Pietsch [21].

Let E be a nuclear complex b.v.s. and Ω a convex balanced open subset of τE. Let B be a strictly compact subset of Ω and let $(\mathcal{H}'_S(\Omega))_{\mathfrak{b}^{\circ}(B)}$ be the subspace of $\mathcal{H}'_S(\Omega)$ spanned by $\mathfrak{b}^{\circ}(B)$ and normed with the gauge of $\mathfrak{b}^{\circ}(B)$. We assume that all the B_i's we shall consider are convex balanced strictly compact subsets of Ω of the type considered in section 2. We denote more simply $\mathfrak{F}(E)$ by $\mathfrak{F}$ and $\mathcal{H}_S(\Omega)$ by $\mathcal{H}(\Omega)$.

Lemma 1. Let B_1 and B_2 be as above with $B_1 \subset B_2$ and such that the injection $i: E_{B_1} \to E_{B_2}$ is a nuclear map with: $i(x) = \sum_n \lambda_n f_n(x) y_n$ with $\sum |\lambda_n| \leq 1$, $f_n \in (E_{B_1})'$ and $|f_n|_{B_1} \leq 1$, $y_n \in B_2$. Then the injection $\mathfrak{F}_{[B_1]} \to \mathfrak{F}_{[B_2]}$ is nuclear.

Proof. As usual $\mathfrak{F}_{[B_i]}$ denotes the vector span of $[B_i] \subset \mathfrak{F} = \mathfrak{F}(E)$ normed with the gauge of $[B_i]$. If ϕ is in $[B_i]$

$$\phi^{(n)}(0) = \sum_r \mu_{n,r}\, x_{n,r_1} \otimes \ldots \otimes x_{n,r_n}$$

with $\sum_r |\mu_{n,r}| \leq 1$ and $x_{n,j} \in B_1$. Replace x_j by $i(x_j)$ and the proof is only an easy computation. □

Lemma 2. Same assumptions as in Lemma 1 and assume that $k e B_2 \subset \Omega$ for some $k > 1$. Then the canonical map:

$$(\mathcal{H}'(\Omega))_{\mathfrak{b}^{\circ}(B_1)} \xrightarrow{\;I'\;} (\mathcal{H}'(\Omega))_{\mathfrak{b}^{\circ}(ekB_2)}$$

is nuclear.

Proof.

$$\begin{array}{ccc}
\mathfrak{F}_{[B_1]} & \xrightarrow{\text{nuclear (Lemma 1)}} & \mathfrak{F}_{[B_2]} \\
\text{continuous (prop. 1)} \uparrow & & \downarrow \text{continuous (prop. 2)} \\
(\mathcal{H}'(\Omega))_{\mathfrak{b}^{\circ}(B_1)} & \xrightarrow{\;I'\;} & (\mathcal{H}'(\Omega))_{\mathfrak{b}^{\circ}(ekB_2)}
\end{array}$$

Let $(K(\Omega))_{U(B)}$ be the normed space canonically associated to the convex balanced o-neighbourhood $U(B)$ of $K(\Omega)$.

Lemma 3. Same assumptions as Lemma 2. Then the natural map

$$(K(\Omega))_{U(ekB_2)} \xrightarrow{\quad i \quad} (K(\Omega))_{U(B_1)}$$

is quasi nuclear.

Proof. Denoting by π_1, π_2 and i the canonical maps we have:

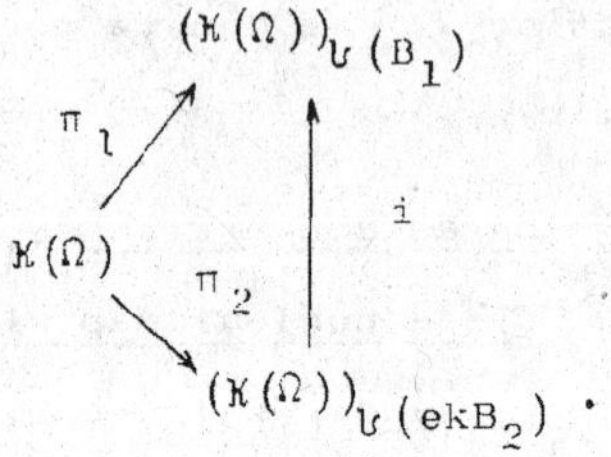

The transposed diagram is:

$$
\begin{array}{c}
(K'(\Omega))_{U^o(B_1)} \\
{}^{t}\pi_1 \swarrow \qquad \downarrow {}^{t}i \\
K'(\Omega) \\
\nwarrow {}^{t}\pi_2 \qquad \\
(K'(\Omega))_{U^o(ekB_2)}
\end{array}
$$

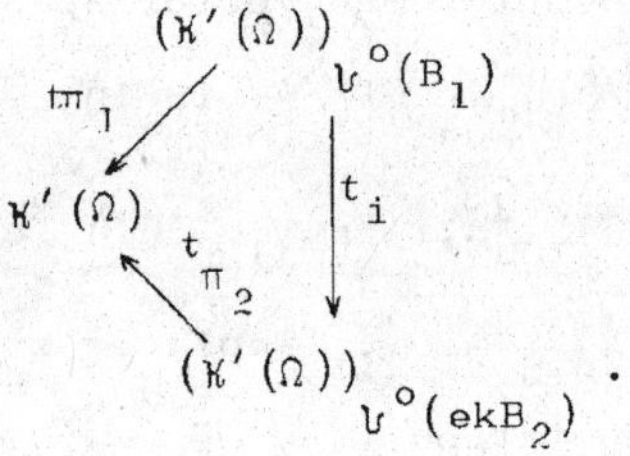

From Lemma 2 ${}^{t}i$ is nuclear. We have the diagram:

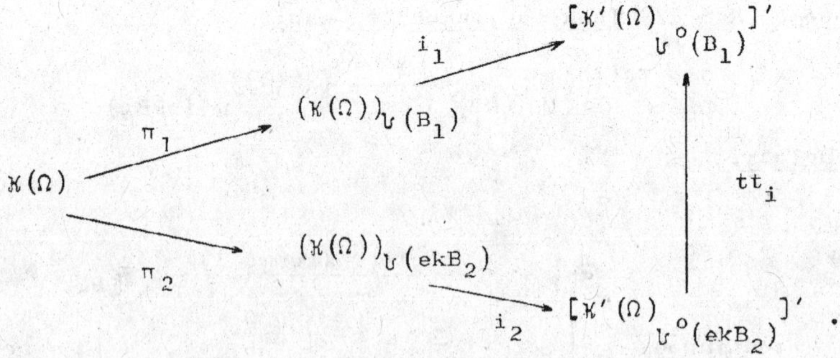

i_1 and i_2 are isometries (of a normed space into its bidual). tt_i is nuclear since t_i is nuclear. Hence i is quasi nuclear.

Now Ω is no longer necessarily convex balanced and we may prove the theorem, already stated in [10].

<u>Theorem 2</u>. Let E be a nuclear b.v.s. Then $\mathcal{H}_S(\Omega)$ is a nuclear l.c.s.

<u>Proof</u>. Let K be a strictly compact subset of Ω. There exists a convex balanced bounded subset B_1 of E such that K is compact in the normed space E_{B_1} and that there exists B_2 and B_3 (of the type above) such that the inclusion maps $E_{B_1} \to E_{B_2}$ and $E_{B_2} \to E_{B_3}$ are nuclear and satisfy the assumptions of Lemma 1. Let $k > 1$ be given. For every x in K there exists an $\epsilon_x > 0$ such that $x + \epsilon_x k^2 e^2 B_3 \subset \Omega$. From the compacity of K in E_{B_1}, $K \subset \bigcup_{1 \le i \le \ell} (x_i + \epsilon_{x_i} B_1)$. Let $K' = \bigcup_{1 \le i \le \ell} (x_i + \epsilon_{x_i} k^2 e^2 B_3)$. K' is strictly compact in Ω since B_3 is strictly compact in E.

From Lemma 3 the natural maps:

$$(\mathcal{H}(\Omega))_{\mathcal{V}(x_i + \epsilon_{x_i} e\, k\, B_2)} \longrightarrow (\mathcal{H}(\Omega))_{\mathcal{V}(x_i + \epsilon_{x_i} B_1)}$$

and

$$(\mathcal{H}(\Omega))_{\mathcal{V}(x_i + \epsilon_{x_i} e^2 k^2 B_3)} \longrightarrow (\mathcal{H}(\Omega))_{\mathcal{V}(x_i + \epsilon_{x_i} e\, k\, B_2)}$$

are quasi-nuclear. Hence the composed map is nuclear hence the product map:

$$\prod_{1 \le i \le \ell} (\mathcal{H}(\Omega))_{\mathcal{V}(x_i + \epsilon_{x_i} e^2 k^2 B_3)} \longrightarrow \prod_{1 \le i \le \ell} (\mathcal{H}(\Omega))_{\mathcal{V}(x_i + \epsilon_{x_i} B_1)}$$

is nuclear. From the following diagram the canonical map $I_{K,K'}$ is quasi-nuclear:

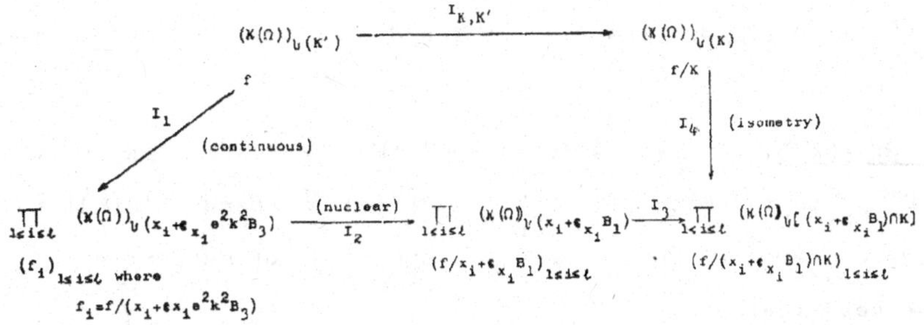

□

4. THE FOURIER-BOREL TRANSFORM AND SOME ASSUMPTIONS OF NUCLEARITY

In this section E is a nuclear b.v.s. separated by its dual then by [15], [16] E admits a basis of bounded sets (B_i) such that each E_{B_i} is a Hilbert space hence the technical assumption of the Remark 3 section 2 is true in this case. If B is a bounded subset of E and if T is in $E^{\times}$ we note:

$$|T|_B = \sup_{x \in B} |T(x)|.$$

Theorem 3. Let E be a nuclear b.v.s. separated by its dual. Then the image $\mathfrak{F}(E)$ of $\mathcal{H}'_S(E)$ via the Fourier-Borel transform is the space of the holomorphic functions ϕ on $E^{\times}$ such that: there exists a convex balanced bounded subset B of E and a number $c \geq 0$ such that:

$$|\phi(T)| \leq c\, e^{|T|_B} \quad \text{for each } T \text{ in } E^{\times}.$$

The bounded subsets of $\mathfrak{F}(E)$ are the families $\{\phi_i\}$ such that the above formula is true with the same c and B for all i.

Remark 1. Let E be a quasi complete dual nuclear l.c.s. and let $\mathcal{H}(E)$ denote the space of the G-analytic and continuous functions on E, equipped with the compact open topology. Let us consider the Von Neumann bornology of E: then E is a nuclear b.v.s.. $\mathcal{H}(E)$ is a dense subspace of $\mathcal{H}_S(E)$ (appendix 3) and each compact subset of E is also strictly compact hence $\mathcal{H}'(E) = \mathcal{H}'_S(E) \simeq \mathfrak{F}(E)$. This result was obtained in [1] (with the supplementary assumption that E is a nuclear l.c.s.) (in [1] E' is used instead of $E^{\times}$ but since $E' \subset E^{\times}$ it suffices to apply th. 2 and consider the restriction of ϕ to E').

Proof of th. 3. If ϕ is in $\mathfrak{F}(E)$, then ϕ is in $[K]$ for some K. If T is in $E^{\times}$, $\phi(T) = \sum_n \frac{1}{n!} \phi^{(n)}(0) T^{\otimes n}$ hence $|\phi(T)| \leq e^{|T|_K}$. For the converse we shall use the assumptions of nuclearity on E and we need the:

<u>Lemma</u>. <u>Let</u> X <u>be a semi-normed space</u>. <u>Let</u> H <u>be an entire function of exponential type from</u> X <u>to</u> $\mathbb{C}$ <u>(that is to say</u> H <u>is G-analytic and</u> $|H(x)| \le c\, e^{\rho\|x\|_X}$ <u>for some constants</u>, $c, \rho > 0$) <u>then</u>:

$$\sup_{\|x_i\|_X \le 1} |H^{(n)}(0)x_1,\dots,x_n| \le c(e\rho)^n \quad \text{for each } n.$$

<u>Proof of the lemma</u>. $H^{(n)}(0)x_1\dots x_n = \frac{1}{(2i\pi)^n}\int \frac{H(\xi_1 x_1+\dots+\xi_n x_n)}{\xi_1^2\dots\xi_n^2}\,d\xi_1\dots d\xi_n$.

From this formula we deduce that $|H^{(n)}(0)x_1\dots x_n| \le n^n\, c\,\frac{e^{\rho R}}{R^n}$ for every $R > 0$ if $\|x_i\|_X \le 1$. $R = \frac{n}{\rho}$ gives the result. □

Now we go on the proof of the Theorem 3. Let us assume that

(1) $$|\phi(T)| \le c\, e^{|T|_B}.$$

One may choose B such that, if i is the canonical injection $i: E_B \to E$ then $t_i(E^X)$ is dense in $(E_B)'$ [choose B such that E_B is a reflexive Banach space (since E is a nuclear b.v.s. one may choose for E_B a Hilbert space [15], [16]), then if $t_i(E^X)$ is not dense in $(E_B)'$ there exists x in $(E_B)'' = E_B$ such that $x \ne 0$ and $u(x) = 0$ for each u in E^X hence the b.v.s. E would not be separated by its dual]. Using the lemma above there exists c' and $\rho' > 0$ such that:

(2) $$\sup_{\substack{|X_i|_B \le 1 \\ X_i \in E^X}} |\phi^{(n)}(0)X_1\dots X_n| \le c'(\rho')^n \quad \text{for each } n.$$

Using this formula (2) and the aforementioned result that $t_i(E^X)$ is dense in the Banach space $(E_B)'$ there exists a unique map $\overline{\phi^{(n)}}(0)$ which is the continuation of the quotient map U, such that the following diagram is commutative:

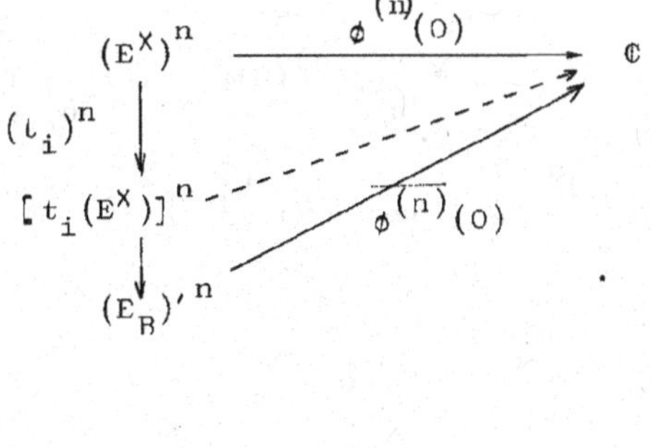

If y_i are in the unit ball of the Banach space $(E_B)'$

(2') $$|\overline{\phi^{(n)}}(0)y_1\ldots y_2| \leq c'(\rho')^n.$$

Let B_1 denote a convex balanced bounded subset of E such that the natural injection $j: E_B \to E_{B_1}$ is nuclear (such a B_1 exists because we have assumed that E is a nuclear b.v.s.) hence its transpose $t_j: (E_{B_1})' \to (E_B)'$ is nuclear hence, for x' in $(E_{B_1})'$ we may write:

$$t_j(x') = \sum_q \mu_q f_q(x')y_q$$

where $\sum_q |\mu_q| < +\infty$, the y_q are in the unit ball of $(E_B)'$ and the f_q are in B_1 (if we choose for E_{B_1} a reflexive Banach space which is possible as already remarked). Then the product mapping $(t_j)^n$ from $(E_{B_1})'^n$ to $(E_B)'^n$ is given by the formula:

$$(t_j)^n(x'_1,\ldots,x'_n) = \sum_{q_1,\ldots,q_n} \mu_{q_1}\ldots\mu_{q_n} f_{q_1}(x'_1)\ldots f_{q_n}(x'_n)(y_{q_1},\ldots,y_{q_n}).$$

If we denote by k the canonical injection from E_{B_1} to E we have the following commutative diagram:

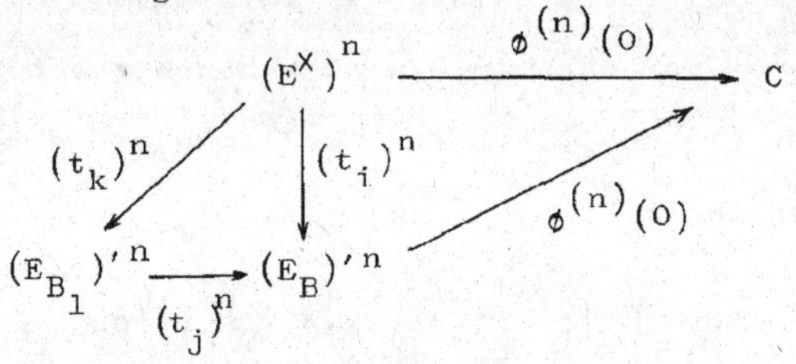

for x'_i in $E^{\times}$ $(1 \leq i \leq n)$. Then:

$$\phi^{(n)}(0)x'_1,\ldots,x'_n = \overline{\phi^{(n)}}(0) \circ (t_j)^n \circ (t_k)^n x'_1\ldots x'_n =$$

$$= \sum_{q_1,\ldots,q_n} \mu_{q_1}\ldots\mu_{q_n} f_{q_1}(t_k(x'_1))\ldots f_{q_n}(t_k(x'_n))\overline{\phi^{(n)}}(0)\, y_{q_1}\ldots y_{q_n}.$$

Hence

$$\phi^{(n)}(0) = \sum_{q_1,\ldots,q_n} \mu_{q_1}\ldots\mu_{q_n}(\overline{\phi^{(n)}}(0)y_{q_1}\ldots y_{q_n})\, f_{q_1}\otimes\ldots\otimes f_{q_n}$$

hence $\phi^{(n)}(0) \in c'(\rho')^n (\sum_q |\mu_q|)^n \Gamma_{\ell_1} B_1^{\otimes n}$ for each n which completes the proof. $\square$

Remark 2. The Theorem 3 in the general case is equivalent to the same th. 3 in the particular case when E is a nuclear Silva space (in which case the algebraic isomorphism is given by P.J. Boland in [1]). The proof is the following:

Let E be a nuclear b.v.s.: E may be naturally written as a bornological inductive limit of nuclear Silva spaces $(\mathcal{E}_i)_{i\in I}$. Let ϕ be a holomorphic function on E^x such that:

$$|\phi(T)| \leq e^{|T|_K} \quad \forall\, T \in E^x$$

where K is a strictly compact subset of E. Let $\mathcal{E}_{i_o}$ be one of these nuclear Silva spaces such that K is a strictly compact subset of $\mathcal{E}_{i_o}$. Let $r: E^x \to \mathcal{E}'_{i_o}$ be the restriction map. Analogously to a part of the proof of th. 3 one proves that r has a dense range. A standard proof based on the above majorization for ϕ shows that there exists a function $\dot{\phi}$ defined on $r(E^x)$ and such that $\phi = \dot{\phi}\cdot r$ and $\dot{\phi}$ may be continued as a holomorphic function $\tilde{\dot{\phi}}$ on $\mathcal{E}'_{i_o}$ that satisfies the same majorization above as ϕ, but with T in $\mathcal{E}'_{i_o}$. It suffices then to apply the th. 3 in the nuclear Silva space $\mathcal{E}_{i_o}$ to prove that ϕ is the Fourier-Borel transform of some element of $\mathcal{H}'_S(E)$. The converse is similar. ☐

5. THEOREM 3 AS A CONSEQUENCE OF A RESULT OF GUPTA [14]

If G is a Banach space we denote as usual by $\mathcal{H}_b(G)$ the space of the holomorphic functions on G of bounded type (i.e. which are bounded on each bounded subset of G) with the topology of uniform convergence on the bounded subsets of G. We denote by $\mathcal{H}_{N,b}(G)$ the space of the holomorphic functions on G of the nuclear bounded type (see [14]), $\mathcal{H}_{N,b}(G)$ is a Fréchet space.

Lemma 1. Let E_1 and E_2 be normed spaces and $i: E_1 \to E_2$ a linear continuous map which is nuclear. If f is in $\mathcal{H}_b(E_2)$, then $f\circ i$ is in $\mathcal{H}_{N,b}(E_1)$ and the map $f \to f\circ i$ is continuous from $\mathcal{H}_b(E_2)$ to $\mathcal{H}_{N,b}(E_1)$.

The proof of this lemma is a direct computation and is in [7]. Let now E be a nuclear b.v.s.; then one has

$$\mathcal{H}_S(E) = \varprojlim_{i\in I} \mathcal{H}_b(E_i) \qquad \text{(topologically)}$$

if $E = \varinjlim_{i\in E_i} E_i$ (bornologically) where the E_i's are Banach spaces. Applying Lemma 1 it follows immediately that

$$(1) \qquad \mathcal{H}_S(E) = \varprojlim_{i\in I} \mathcal{H}_{N,b}(E_i) \qquad \text{(topologically).}$$

<u>Lemma 2</u>. <u>The restriction map:</u>

$$\begin{array}{ccc} \mathcal{H}_S(E) & \to & \mathcal{H}_{N,b}(E_i) \\ f & & f/E_i \end{array}$$

<u>has a dense range</u>.

<u>Proof</u>. One may choose the E_i's as Hilbert spaces hence reflexive (because E is a nuclear b.v.s.). The polynomials of finite type are dense in $\mathcal{H}_{N,b}(E_i)$. The restriction map:

$$\begin{array}{ccc} E^\times & \to & E'_i \\ \ell & & \ell/E_i \end{array}$$

has a dense range in the Hilbert space E'_i (because $E''_i = E_i$ and because $E^\times$ separates E). Lemma 2 follows from these two remarks. □

From Lemma 2 and with the above choice of the E_i's it follows that if $E_i \subset E_j$ then the restriction map

$$\mathcal{H}_{N,b}(E_j) \to \mathcal{H}_{N,b}(E_i)$$

has a dense range hence its transposed

$$\mathcal{H}'_{N,b}(E_i) \to \mathcal{H}'_{N,b}(E_j)$$

is injective. Then algebraically one has (from (1)):

$$(2) \qquad \varinjlim_{i\in I} \mathcal{H}'_{N,b}(E_i) = \mathcal{H}'_S(E)$$

and the equicontinuous subsets of $\mathcal{H}'_S(E)$ are the subsets contained and equicontinuous in some $\mathcal{H}'_{N,b}(E_i)$.

If G is a complex Banach space let us define $\mathcal{E}xp\, G'$ as the space of the entire functions on G' of exponential type:

$f \in \mathcal{E}xp\, G'$ if $\exists\, c, \rho > 0$ such that $\forall\, x \in G'$ we have $|f(x)| \leq c\, e^{\rho \| x \|_{G'}}$.

Let us define the bounded subsets of $\mathcal{E}xp\, G'$ as the families $(f_i)_{i \in I}$ such that the above inequality is valid with the same constants c and ρ for all the f_i's.

If G is a Banach space with approximation property, Gupta proves in [14] that $\mathcal{H}'_{N,b}(G)$ is isomorphic algebraically to $\mathcal{E}xp\, G'$ via the Fourier-Borel transform. If we equip $\mathcal{H}'_{N,b}(G)$ with its equicontinuous bornology this algebraic isomorphism may be easily improved in a bornological isomorphism, using the closed graph theorem as it is stated in Hogbé Nlend [15] p. 44 or [18].

Since the E_i's are Hilbert spaces and that $\mathfrak{F}$ is injective (2) becomes:

$$\mathfrak{F}\mathcal{H}'_S(E) = \varinjlim_{i \in I} \mathcal{H}'_{N,b}(E_i) = \varinjlim_{i \in I} \mathcal{E}xp(E'_i)$$

hence the bornological equality:

$$\mathfrak{F}\mathcal{H}'_S(E) = \varinjlim_{i \in I} \mathcal{E}xp(E'_i) \tag{3}$$

(in the sense of the bornological inductive limits).

Now if E is a nuclear complex b.v.s. let us denote by $\mathcal{E}xp(E^\times)$ the b.v.s. defined in th. 3 (th. 3 asserts then that $\mathfrak{F}(E) = \mathcal{E}xp(E^\times)$).

Then th. 3 will follow immediately from the following:

Lemma 3. $\mathcal{E}xp(E^\times) = \varinjlim_{i \in I} \mathcal{E}xp\, E'_i$ bornologically (in the sense of the bornological inductive limits).

Proof. Since we recall that for each $i \in I$ the restriction map

$$r : E^\times \to (E_i)'$$

has a dense range, the proof is classical (more details are in Remark 2 of section 4). □

6. AN APPLICATION TO CONVOLUTION EQUATIONS

In this section E is still a nuclear b.v.s. separated by its dual; $\mathcal{H}$ denotes a l.c.s. of holomorphic functions on E; we assume that $\mathcal{H}$ is translation invariant. As usual, a convolution operator on $\mathcal{H}$ is a linear continuous mapping from $\mathcal{H}$ to $\mathcal{H}$ which commutes with the translations. We assume that $\mathcal{H}$ is a dense subspace of $\mathcal{H}_S(E)$ and we equip $\mathcal{H}$ with the topology induced by the topology of $\mathcal{H}_S(E)$.

Let F be a vector space injectively contained in the dual E^X of E and which separates E. We denote by "F-exponential-polynomial" a holomorphic function on E which can be written $\sum_{i=1}^{n} P_i e^{\sigma_i}$ with P_i in $F^{\otimes n}$ and α_i in F. Let us assume that the vector space of the F-exponential-polynomials is contained in $\mathcal{H}$.

Let $\mathfrak{G}$ be a convolution operator on $\mathcal{H}$.

<u>Theorem 4</u>. <u>Under the above hypothesis: Every solution</u> u <u>in</u> $\mathcal{H}$ <u>of the homogeneous equation</u> $\mathfrak{G}u = 0$ <u>is limit</u> (<u>for the topology of</u> $\mathcal{H}$) <u>of</u> F-<u>exponential-polynomial solutions</u>.

Before the proof let us state a few consequences:

<u>Consequence 1</u>: Let E be a complex vector space equipped with its finite dimensional bornology and let $\mathcal{H} = \mathcal{H}_S(E)$ be the space of the G-analytic functions on E; in this case th. 4 gives a result of [3].

<u>Consequence 2</u>: Let E be a quasi complete l.c.s. and let E'_c be its dual equipped with the compact open topology. Let us assume that E'_c is a nuclear l.c.s.: it follows from this last assumption that the compact bornology of E, denoted by $c(E)$, is a nuclear bornology. Let $\mathcal{H}$ be the space of the G-analytic and continuous functions on E. It is proved in appendix 3 that $\mathcal{H}$ is dense in $\mathcal{H}_S(c(E))$. Apply th. 4 in the nuclear b.v.s. $c(E)$ and with $F = E'$. In this case th. 4 improves the th. 2.1 of [1] (in which it is assumed that E is a quasi complete nuclear and dual nuclear l.c.s.).

<u>Proof of th. 4</u>. Let $\tilde{\mathfrak{G}}$ be the continuation of $\mathfrak{G}$ from $\mathcal{H}_S(E)$ to $\mathcal{H}_S(E)$. $\tilde{\mathfrak{G}}$ is a convolution operator on $\mathcal{H}_S(E)$ hence it suffices to

prove the theorem in the case $\mathcal{K} = \mathcal{K}_S(E)$. The proof follows now the same steps as the th. 1 of [14] hence we shall only sketch it.

Let us remark that F is dense in $E^\times$ ($E^\times$ is equipped with the topology of the uniform convergence on the bounded sets of E). Since E is a nuclear b.v.s. separated by its dual it is a reflexive b.v.s. [18] hence $E^{\times\prime} = E$ (with the notations of [18]). Let $y \in E^{\times\prime} = E$ such that y is null on F; since F separates E (by hypothesis) then $y = 0$ hence F is dense in $E^\times$.

Let us equip F with the bornology induced by $E^\times$ and let us consider the vector subspace of $\mathcal{K}_S(F)$ made of the functions ϕ such that: there exists a convex balanced bounded subset B of E and a real number $c > 0$ such that

$$|\phi(T)| \leq c\ e^{|T|_B} \quad \text{for every } T \text{ in } F \subset E^\times.$$

<u>Remark.</u> From the Theorem 2 it follows immediately that this vector subspace of $\mathcal{K}_S(F)$ is $\mathfrak{J}(E)$ (using the density of F in $E^\times$ and the lemma in the proof of the th. 2). A useful lemma is:

<u>Lemma.</u> <u>Let</u> f_1, f_2, f_3 <u>in</u> $\mathcal{K}_S(F)$ <u>such that</u> $f_1, f_2 \in \mathfrak{J}(E)$, $f_1 = f_2 f_3$ <u>and</u> $f_3 \not\equiv 0$; <u>then</u> f_3 <u>is in</u> $\mathfrak{J}(E)$.

<u>Proof.</u> By hypothesis there exists a convex balanced bounded subset B of E such that for each T in F:

$$|f_i(T)| \leq e^{|T|_B} \qquad i = 1,2 .$$

Applying the prop. 3 of [19] p. 305 there exists c and $p > 0$ such that

$$|f_3(T)| \leq c\ e^{p|T|_B}$$

for each T in F. The above remark gives the result (this kind of argument is detailed in [14]. End of the proof of the lemma. □

For each T in $\mathcal{K}_S'(E)$ let T^* be the convolution operator on $\mathcal{K}_S(E)$ defined by $\phi \to T*\phi$ if ϕ is in $\mathcal{K}_S(E)$), with $T*\phi(x) = T(\phi_x)$ (where

ϕ_x is the map $\phi_x: y \to \phi(x+y)$).

The mapping $T \to T*$ is bijective between $\mathcal{H}'_S(E)$ and the space of the convolution operators on $\mathcal{H}_S(E)$. The inverse map is: $\mathcal{O} \to (\phi \to \mathcal{O}\phi(0))$ if $\mathcal{O}$ is a convolution operator on $\mathcal{H}_S(E)$. The subspace of $\mathcal{H}_S(E)$ spanned by the functions $x \to e^{\langle T,x\rangle}$ (with T in E^X) is dense in $\mathcal{H}_S(E)$ and thus we may define the convolution product of two element X, Y of $\mathcal{H}'_S(E)$ be the formula:

$$\langle X*Y, e^T\rangle = \langle X, e^T\rangle\langle Y, e^T\rangle.$$

$\mathcal{H}'_S(E)$ is then an algebra isomorphic, via the Fourier-Borel transform, to the algebra $\mathfrak{F}(E)$ with the ordinary product of functions.

Lemma. Let X and T be in $\mathcal{H}'_S(E)$ with $T \neq 0$ such that if P is a F-exponential-polynomial: $T*P = 0 \Rightarrow \langle X,P\rangle = 0$. Then $\mathfrak{F}X$ is divisible by $\mathfrak{F}T$ and the quotient is in $\mathfrak{F}(E)$.

Proof. Using the proof of the prop. 14 of [14] there exists a holomorphic function h on F such that $\mathfrak{F}X = \mathfrak{F}T\cdot h$; hence, by Lemma 2, h is in $\mathfrak{F}(E)$. □

End of the proof of the theorem. Let X be an element of $\mathcal{H}'_S(E)$ which is null on the solutions of $\mathcal{O}u = 0$ which are F-exponential-polynomials. Let T be in $\mathcal{H}'_S(E)$ such that $T* = \mathcal{O}$ then $T*P = 0 \Rightarrow$ $\Rightarrow \langle X,P\rangle = 0$ with P a F-exponential-polynomial. Then by the last lemma there exists Y in $\mathcal{H}'_S(E)$ such that $\mathfrak{F}X = \mathfrak{F}T.\ \mathfrak{F}Y$, hence $X = T*Y$. Let f be a solution in $\mathcal{H}_S(E)$ of the equation $\mathcal{O}u = 0$. Then $X(f) = X*f(0) = ((T*Y)*f)_{(0)} = ((Y*T)*f)_{(0)} = (Y*(T*f))_{(0)} = 0$ because $T*f = \mathcal{O}f = 0$ (see [14] p. 45).

Appendix 1. Let E be a Banach space and Ω a convex balanced open subset of E. Let b be a compact subset of E contained in Ω. Then there exists a null sequence (x_n) of elements of Ω such that $b \subset \Gamma_{\ell_1}\{x_n\}$. The proof is an easy modification of the proof of [23] Chap. VII §2 lemma 2 given in this last book in the case $\Omega = E$.

<u>Appendix 2.</u> If the canonical mapping: $E^{\tilde{\otimes}n} \to L_n((E^{\times});C)$ is not injective then one cannot consider that $E^{\tilde{\otimes}n}$ is a part of $L_n((E^{\times})^n;C)$ and one cannot consider that $\mathfrak{F}(E)$ is contained in $\mathcal{H}_S(E^{\times})$; we have to state more general definitions and consider that $\mathfrak{F}(E)$ is contained in the product $\prod_n E^{\tilde{\otimes}n}$. We define

$$[K] = \{\phi = (\phi_n)_{n\in N} \text{ such that, for each } n, \ \phi_n \in \Gamma_{\ell_1} K^{\otimes n}\}.$$

Let us recall that $E^{\tilde{\otimes}n} \subset (E^{\tilde{\otimes}n})^{\times\times} = L_n^{\times}(E^n;C)$. Now the Fourier-Borel transform $\mathfrak{F}$ becomes the mapping:

$$\mathcal{H}'_S(E) \to \prod_{n=0}^{+\infty} L_n^{\times}(E^n;C)$$

$$\ell \qquad \mathfrak{F}\ell = [\ell/L_n(E^n;C)] \ n\in N$$

where $\ell/L_n(E^n;C)$ is the restriction of ℓ to $L_n(E^n;C) \subset \mathcal{H}_S(E)$. Obviously the proofs of the prop. 1 and 2 and of the th. 1 are not modified.

<u>Appendix 3.</u> Let E be a complex l.c.s., U a balanced open subset of E, B a convex balanced bounded subset of E such that E_B is a Banach space, $\mathcal{H}(\Omega)$ the vector space of the G analytic and continuous functions on Ω and $\mathcal{H}(\Omega\cap E_B)$ the space of the holomorphic functions on $\Omega \cap E_B$ (considered in the Banach space E_B). Let us equip $\mathcal{H}(\Omega\cap E_B)$ with the topology of the uniform convergence on the compact subsets of $\Omega \cap E_B$. Then:

<u>Proposition.</u> <u>If E_B has the approximation property then the restriction map $\mathcal{H}(\Omega) \to \mathcal{H}(\Omega\cap E_B)$ has a dense range.</u>

<u>Proof.</u> If f is in $\mathcal{H}(\Omega\cap E_B)$, approximate f by its Taylor expansion at the origin. Since E_B has the approximation property each polynomial on E_B can be approximated by finite linear combinations of finite proudct of continuous linear forms, and a continuous linear form on E_B can be uniformly approximated on compact subsets of E_B by restrictions of continuous linear forms on E. □

Proposition. Let Ω be a balanced open subset of a l.c.s. E. Let us assume that the Von-Neumann bornology of E has the property that each strictly compact subset is contained in a convex balanced strictly compact subset B such that E_B is a Banach space with the approximation property. Then the space of the G-analytic and continuous functions on Ω is dense in $\mathcal{K}_S(\Omega)$.

It is an immediate consequence of the above result.

Appendix 4. Let G be a Schwartz b.v.s. separated by its dual. Let $m \geq 1$ be an integer. We denote by $G^{\otimes m}$ ($G^{\otimes m}_{\pi_b}$ in [15],[16]) the m-fold tensor product of G equipped with the bornology generated by the convex hulls of the sets $B^{\otimes n}$ (where B is a bounded set in G). In general ([25]) a b.v.s. cannot be injectively embedded in its bornological completion. So it has been introduced in [16] a new bornology; this bornology denoted here by $G_o^{\otimes m}$ ($G^{\otimes m}_{\pi_{b\,o}}$ in [15],[16]) is generated by the weak closures of the bounded sets of $G^{\otimes m}$, for the duality between $G^{\otimes m}$ and $(G^{\otimes m})^{\times}$. We denote by $G^{\tilde{\otimes} m}$ ($G^{\tilde{\otimes} m}_{\pi_{b^o}}$ in [15], [16]) the bornological completion of $G_o^{\otimes m}$.

In fact we show further that $G^{\tilde{\otimes} m}$ is the bornological completion of $G^{\otimes m}$ in the case when G is a Schwartz b.v.s. with a basis b_i of bounded sets such that G_{b_i} is a reflexive Banach space with the approximation property (for example when G is a nuclear b.v.s.).

Proposition 1. Let G be a Schartz b.v.s. separated by its dual. Let m be an integer ≥ 1. Let (x_n) be a Mackey-null sequence of elements of G. Let $K[(x_n)_n] = \Gamma_{\ell_1}[\{x_{i_1} \otimes \ldots \otimes x_{i_m}\}_{i_j \in \mathbb{N}}]$ (in $G^{\tilde{\otimes} m}$). Then

i) $K[(x_n)_{n\in\mathbb{N}}]$ is strictly compact in $G^{\tilde{\otimes} m}$

ii) each bounded set in $G^{\tilde{\otimes} m}$ is contained in a set $[K(x_n)_{n\in\mathbb{N}}]$ (for some suitable sequence (x_n)).

Proof. (i) The product set $(\{x_n\}_{n\in\mathbb{N}})^m$ is countable and if $(x_{p_1},\ldots,x_{p_m})$ is in this set, let $y_p = x_{p_1} \otimes \ldots \otimes x_{p_m} \in G^{\otimes m} \subset G^{\tilde{\otimes} m}$. The sequence (x_n) is Mackey convergent to o in G hence there exists a bounded set B of G and a sequence (ϵ_n) of real positive numbers convergent to o such that for every n:

$$x_n \in \epsilon_n B.$$

ence $y_p \in \epsilon_{p_1} \dots \epsilon_{p_m} B^{\otimes m}$. $B^{\otimes m}$ is a bounded set of $G^{\otimes m}$, hence the sequence (y_p) is Mackey-convergent to o in $G^{\otimes m}$ and hence also in the complete b.v.s. $G^{\tilde{\otimes} m}$. Hence there exists a convex and balanced bounded set K in $G^{\tilde{\otimes} m}$ such that $(G^{\tilde{\otimes} m})_K$ is a Banach space which has the property that $(y_p)_{p\in\mathbb{N}}$ tends to o in this Banach space. Hence $\Gamma_{\ell_1}(\{y_p\}_{p\in\mathbb{N}})$ (in $(G^{\otimes m})_K$) is compact in $(G^{\tilde{\otimes} m})_K$ hence strictly compact in $G^{\tilde{\otimes} m}$ and we have $K[(x_n)_{n\in\mathbb{N}}] = \Gamma_{\ell_1}(\{y_p\}_{p\in\mathbb{N}})$.

(ii) By (i) $K[(x_n)_{n\in\mathbb{N}}]$ is a bounded subset of $G^{\tilde{\otimes} m}$. Conversely: since G is a Schwartz b.v.s. the subsets of G of the type $\Gamma_{\ell_1}\{x_n\}$ (where $\{x_n\}$ is a sequence of points of G which is Mackey convergent to o in G) are a basis of the bounde sets of G (see [17]). Hence the subsets of $G^{\otimes m}$ of the type $\Gamma[(\Gamma_{\ell_1}\{x_n\})^{\otimes m}]$ are a basis of the bounded sets of $G^{\otimes m}$. If H is a b.v.s. denote by $H^{\times}$ its bornological dual and let σ be the weak topology $\sigma(G^{\otimes m},(G^{\otimes m})^{\times})$.

$\overline{\Gamma[(\Gamma_{\ell_1}\{x_n\})^{\otimes m}]}^{\sigma} \subset \overline{K[(x_n)_{n\in\mathbb{N}}] \cap G^{\otimes m}}^{\sigma} = K[(x_n)_{n\in\mathbb{N}}] \cap G^{\otimes m}$ (this last equality is due to the fact that by (i) $K[(x_n)_{n\in\mathbb{N}}]$ is compact for the topology $\sigma(G^{\tilde{\otimes} m},(G^{\otimes m})^{\times})$. Hence the subsets $K[(x_n)_{n\in\mathbb{N}}] \cap G^{\otimes m}$ form a basis of the bounded sets of $Go^{\otimes m}$. Since these subsets are closed for the topology $\sigma(G^{\otimes m},(G^{\otimes m})^{\times})$ the b.v.s. $G^{\tilde{\otimes} m}$ is the bornological inductive limit of the Banach spaces $\widehat{(G^{\otimes m})_{K[(x_n)_{n\in\mathbb{N}}] \cap G^{\otimes m}}}$ with injective canonical mappings between these Banach spaces. The unit ball of this last Banach space may be considered as contained in $K[(x_n)_{n\in\mathbb{N}}]$ hence each bounded set in $G^{\tilde{\otimes} m}$ is contained in a suitable $K[(x_n)_{n\in\mathbb{N}}]$. □

Proposition 2. Let G be a Schwartz b.v.s. separated by its dual. Let m be an integer ≥ 1. Let us assume that G has a basis (b_i) of convex balanced bounded sets such that G_{b_i} is a reflexive Banach space with the approximation property. Then the canonical map: $G^{\tilde{\otimes} m} \to L_m((G^\times)^m;\mathbb{C})$ is injective.

Proof. The image of $G^\times$ through the restriction map: $G^\times \xrightarrow{i} (G_{b_i})'$ is a dense subspace of $(G_{b_i})'$ equipped with its strong topology: if x in $(G_{b_i})'' = G_{b_i}$ is null on $G^\times$ then $x = o$ in G hence $x = o$ in G_{b_i}. Hence the restriction mapping $L_m((G_{b_i})'^m;\mathbb{C}) \to L_m((G^\times)^m;\mathbb{C})$ is injective. By a lemma due to Grothendieck ([16] p. 263) the completion of $G^{\otimes m}$ is the inductive limit of the Banach spaces $G_{b_i}{}^{\hat{\otimes} m}_{\pi}$ where the canonical maps of this inductive limit are injective. Since G_{b_i} is a Banach space with the approximation property the canonical map: $G_{b_i}{}^{\hat{\otimes} m}_{\pi} \to L_m((G_{b_i})'^m;\mathbb{C})$ is injective (Grothendieck [13] Corollary 3 of prop. 35, 36 p.168). Hence the canonical map: $G_{b_i}{}^{\hat{\otimes} m}_{\pi} \to L_m((G^\times)^m;\mathbb{C})$ is injective et us remark now that the canonical mapping: $G_o^{\otimes m} \to L_m(G^\times)^m;\mathbb{C})$ is bounded (where $G^\times$ and $L_m((G^\times)^m;\mathbb{C})$ are equipped with their natural bornologies): let σ be the weak topology $\sigma(G^{\otimes m},(G^{\otimes m})^\times)$ and let $\overline{\Gamma\, b^{\otimes m}}^{\sigma}$ be a bounded set in $G_o^{\otimes m}$, let $C^\times$ be a bounded set in $G^\times$; we have to show that $\overline{\Gamma\, b^{\otimes m}}^{\sigma}$ is equi-bounded on $(C^\times)^m$: $(C^\times)^m$ is contained in $(G^{\otimes m})^\times = (G_o^{\otimes m})^\times$; since there exists $\alpha > 0$ such that $\sup\limits_{\substack{X\in\Gamma(b^{\otimes m})\\ Y\in(C^\times)^m}} |Y(x)| \le \alpha$ (because $C^\times$ is a bounded set in $G^\times$) this inequality remains valid if x is in $\overline{\Gamma\, b^{\otimes m}}^{\sigma}$. Hence there exists a bounded canonical mapping: $G^{\tilde{\otimes} m} \xrightarrow{i_3} L_m[(G^\times)^m;\mathbb{C}]$ and the following diagram is commutative:

$$\begin{array}{ccc} (\text{completion of } G^{\otimes m}) & \xrightarrow{\quad i_1 \quad} & L_m[(G^\times)^m;\mathbb{C}] \\ & {\scriptstyle i_2}\searrow \qquad \nearrow{\scriptstyle i_3} & \\ & G^{\tilde{\otimes} m} & \end{array}$$

Since we have proved that i_1 is injective, i_2 and i_3 are injective. □

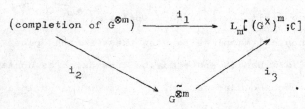

Remark. Under the above hypothesis the completion of $G^{\otimes m}$ is $G^{\tilde{\otimes} m}$: from (ii) of prop. 1, i_2 is surjective hence bijective; i_2 is obviously bounded; the inverse of i_2 is bounded ((ii) of prop. 1).

REFERENCES

[1] Boland, P.J., Malgrange theorem for entire functions on nuclear spaces, Proceedings on Infinite Dimensional Holomorphy - Lecture Notes in Math. nº 364 (Springer) p. 135-144.

[2] Boland, P.J., An example of a nuclear space in Infinite Dimensional Holomorphy, Arkiv fur Math. 15 (1977) p. 87-91.

[3] Boland, P.J. and Dineen, S., Convolution operators... Trans. of the AMS, Vol. 190 (1974) p. 313-323.

[4] Bourbaki, N., Espaces vectoriels topologiques, Hermann, Paris.

[5] Colombeau, J.F., Sur les applications G-analytiques et analytiques... Seminaire P. Lelong 1972, Lecture Notes in Math. Nº 332 (Springer) p. 48-58.

[6] Colombeau, J.F., On some various notions of Infinite Dimensional Holomorphy, Proceedings on Infinite Dimensional Holomorphy, Lecture Notes in Math. nº 364 (Springer) p. 145-149.

[7] Colombeau, J.F. and Matos, M.C., Convolution equations in spaces of infinite dimensional entire functions (preprint).

[8] Colombeau, J.F. and Perrot, B., Theorèmes de Najaux Analytiques en dimension infinie, Comptes Rendus Acad. des Sci., Paris t. 284 serie A (1977) p. 759-762.

[9] Colombeau, J.F. and Perrot, B., Transformation de Fourier-Borel.. Comptes Rendus (Paris) t. 284 série A (1977) p. 963-966.

[10] Colombeau, J.F. and Perrot, B., Une caractérisation de la nucléarité... Comptes Rendus (Paris) t 284 série A (1977) p. 1275-1278.

[11] Colombeau, J.F. and Perrot, B., Transformation de Fourier-Borel... Comptes Rendus (Paris) t. 285 série A (1977) p. 19-21.

[12] Colombeau, J.F. and Perrot, B., Infinite dimensional holomorphy normal forms of operators on the Fock spaces of Boson fields and an extension of the concept of Wick product, in "Advances in Holomorphy" (editor J.A. Barroso) North Holland Math. Studies (1979), p. 249-274.

[13] Grothendieck, A., Produits tensoriels topologiques et espaces nucléaires, Memoirs of the AMS nº 16 (1966).

[14] Gupta, C., Malgrange theorem... Notas de Matemática nº 37, IMPA, Rio de Janeiro, 1968.

[15] Hogbé-Nlend, H., Théorie des Bornologies et Applications Lecture Notes in Math. nº 213 (1971) (Springer).

[16] Hogbé-Nlend, H., Complétion, tenseurs et nucléarité en Bornologie, Journal de Math, pures et appliquées 49, (1970), p. 193-288.

[17] Hogbé-Nlend, H., Les espaces de Fréchet-Schwartz et la propriéte d'approximation, Comptes Rendus (Paris) t. 275 série A (1972) p. 1073-1075.

[18] Hogbé-Nlend, H., Bornologies and Functional Analysis, North Holland Math. Studies nº 26 (1977).

[19] Lazet, D., Applications analytiques... Séminaire P. Lelong 1972, Lecture Notes in Math. nº 332 (Springer) p. 1-47.

[20] Malgrange, B., Existence et approximation des solutions... Annales de l'Institut Fourier VI (1955-56) p. 271-355.

[21] Pietsch, A., Nuclear locally convex spaces, Ergelnisse der Math. 66, Springer 1972.

[22] Pisanelli, D., Sur les applications analytiques... Bulletin des Sciences Math., 2eme série, 96 (1972) p. 181-191.

[23] Robertson, A.P. and W., Topological vector spaces, Cambridge University Press, 1973.

[24] Silva, J.S. e, Le Calcul differentiel et intégral... Atti Acad. Naz. Lincei vol. 20 (1956) p. 743-750, vol. 21 (1956) p.40-46.

[25] Waelbroeck, L., Le complété et le dual d'un espace à bornés, Comptes Rendus Acad. des Sci. Paris t. 253 série A (1961) p. 2827-2828.

[26] Waelbroeck, L., The nuclearity of G(U) . In "Infinite Dimensional Holomorphy and Applications" (M.C. Matos editor), North Holland, Math. Studies 12 (1977), p.425-435.

ON THE SOLVABILITY OF DIFFERENTIAL EQUATIONS OF INFINITE ORDER IN NON-METRIZABLE SPACES

J.F. Colombeau, and B. Perrot

U.E.R. d'Informatique
Université de Bordeaux I
33405 Talence, France

and

T.A. W. Dwyer, III(*)

Department of Mathematical Sciences
Northern Illinois University
DeKalb, IL 60115

1. INTRODUCTION

In [9] an existence theorem was given for differential equations of infinite order in the "Fock spaces of order q" $\mathfrak{F}_N^q(E)$, $1 \leq q < \infty$ of analytic functions Φ, on a large class of Fréchet spaces E, such that

$$(1) \qquad \|\Phi\|_{r,N,q}^{q} := \sum_{n=0}^{\infty} (1/n!) \, \|\hat{d}^n\Phi(0)\|_{r,N}^{q} < \infty$$

for some continuous seminorm $\|\cdot\|_r$, where $\|\hat{d}^n\Phi(0)\|_{r,N}$ are the nuclear norms of the polynomial derivatives of Φ. The results of [9] yield extensions of the existence and approximation theorems in [16], [1] (namely $\mathcal{H}_{Nb}(E)$ with $p = 1$, and $\mathrm{Exp}_N(E)$ with $p = \infty$, although with a different norm from (1)), from nuclear Fréchet spaces or their duals to more general Fréchet spaces or DF-spaces, including Fréchet-Schwartz and Silva spaces. In [2] the approximation and existence theorems are extend to operators on the space $\mathfrak{F}\mathcal{H}'(\Omega)$ of Fourier-Borel transforms of analytic functionals carried by an absolutely convex domain Ω in a Schwartz bornological vector space E, not necessarily Fréchet-

---

(*) Research supported in part by NSF Grant MCS 77-03900-A01.

Schwartz nor a Silva space (but with the approximation property in the sense of [9]). If $\Omega = E$ one gets the analogue of $\mathcal{F}_N^q(E)$ with $q = \infty$. The existence theorem in [2] is the first result of its kind in a setting general enough to apply to $\mathcal{D}$ and $\mathcal{D}'$ (distributions). The technique consists of constructing a Silva space $\mathcal{E}$ imbedded in the b.v.s. E, chosen so that $\mathcal{F}\mathcal{H}'(\Omega \cap \mathcal{E})$ contains the second member of the equation as well as intersecting the kernel of the operator, then solving the equation first over $\Omega \cap \mathcal{E}$. It turns out that an analogous technique can be applied to $\mathcal{F}_N^q(E)$ when E is a Schwartz topological vector space, or more generally an uncountable projective limit of Hilbert spaces: a Fréchet space $\tilde{E}$ extending the t.v.s. E replaces the Silva subspace $\mathcal{E}$ of a b.v.s. E, but still depends on the second member of the equation and on the kernel of the operator. The proof presented here is carried out for a differential operator of infinite order $g(d)$ with its Fourier-Borel symbol g in the space $\mathcal{F}^p(E')$, $1/p + 1/q = 1$, of functions $g: E' \to C$ such that

$$\|g\|_{r,p}^p := \sum_{n=0}^{\infty} (1/n!) \|\hat{d}^n g(0)\|_r^p < \infty$$

for every continuous seminorm $\|\cdot\|_r$ on E, where $\|\hat{d}^n g(0)\|_r$ is the current norm on E_r' (Banach space dual of E with respect to $\|\cdot\|_r$). The problem is reduced to constructing the Fourier-Borel symbol $\tilde{g}$ of the "restriction" of $g(d)$ to $\mathcal{F}_N^q(\tilde{E})$, solving the equation for $\tilde{g}(d)$ then returning to E by a purely "algebraic" process. This technique is applicable to other situations, such as the Hilbert-Schmidt case, [6], [13], [15] and the representation $g(d)$ opens the way to the study of operators with variable coefficients, such as products of creation and annihilation operators [3], [4], [5], [14], or hyperdifferential operators arising in the study of infinite-dimensional dynamical systems [10], [11], [12], not yet considered over $\mathcal{D}$ or $\mathcal{D}'$, to be treated elsewhere.

2. EXISTENCE THEOREM

Theorem 2.6.2 of [9] without metrizability takes the form below.

Theorem. Let E be a complete locally convex complex Hausdorff space, with its topology determined by a family of seminorms with properties (1) and (2) below.

(1) For each seminorm r there is a seminorm s such that the associated norms satisfy the following inequality:

$$\|\cdot\|_r^q \leq (1/2)\,\|\cdot\|_s^q \,.$$

(2) Either each seminorm r is dominated by a seminorm s such that the completion E_s of E for $\|\cdot\|_s$ is a Hilbert space, or each seminorm r is dominated by a seminorm s such that the extension $\iota_{rs}: E_s \to E_r$ of the natural mapping $\iota_r: E \to E_r$ is a strong limit of operators of finite rank.

Then for each $g \in \mathfrak{F}^p(E')$ with $1/p + 1/q = 1$, which is not identically zero, and each $\Phi \in \mathfrak{F}_N^q(E)$, the equation $g(d)\psi = \Phi$ has a solution ψ in $\mathfrak{F}_N^q(E)$.

Remark. It was shown in [7], sec. 1.0 that by taking "sufficiently many" seminorms r multiplied by appropriate scale factors in the construction of $\mathfrak{F}_N^q(E)$ and $\mathfrak{F}^p(E')$, property (1) can always be satisfied. Property (2) guarantees that the strong dual of $\mathfrak{F}_N^q(E)$ is a Fréchet space whenever E is metrizable; cfr [9], p. 293 and [7] Prop. 2.5.2.

The proof of the theorem depends on the representation of E and $\mathfrak{F}^p(E')$ as projective limits of the Banach spaces E_r and $\mathfrak{F}^p(E'_r) = \{f: E'_r \to \mathbb{C} : \|f\|_{r,p} < \infty\}$ respectively, as well as the representation of $\mathfrak{F}_N^q(E)$ as the locally convex inductive limit of the Banach spaces $\mathfrak{F}_N^q(E_r) := \{\Phi: E_r \to \mathbb{C} : \|\Phi\|_{r,N,q} < \infty\}$, via the lemmas below.

Let E be the projective limit of a projective family $(E_r, \iota_{rs})_{r<s}$ of Banach spaces $(E_r, \|\cdot\|_r)$ and intertwining continuous linear maps $\iota_{rs}: E_s \to E_r$, with respect to continuous linear projecting maps $\iota_r: E \to E_r$, that is: E has the coarsest topology for which the maps ι_r are continuous; if $r < s$ for the ordering of the directed index set then $\iota_{rs} \circ \iota_s = \iota_r$; and for every family $(x_r)_r$ of $x_r \in E_r$ such that $\iota_{rs}x_s = x_r$ when $r < s$, there is a <u>unique</u> $x \in E$ such that $\iota_r x = x_r$ for each r. We shall also suppose that $\iota_r E$ is dense in E_r for each r. (This is always verified if the E_r are completions of quotient spaces $E/r^{-1}(0)$ of a locally convex Hausdorff space with seminorms r).

Let $(r_n)_n$ be a countable increasing subfamily of indices from the index set of E. Let $\tilde{E}$ be the projective limit of the projective system $(E_{r_n}, \iota_{r_n r_m})_{n<m}$ relative to a family of projecting maps $j_n: \tilde{E} \to E_{r_n}$ (e.g., let $\tilde{E} = \{(x_n)_n \in \Pi_n E_{r_n} : \iota_{r_n r_m} x_m = x_n$ for $n < m\}$ and $j_m((x_n)_n) := x_m$). We have:

<u>Lemma 1</u>. There is a continuous linear map $\lambda: E \to \tilde{E}$ such that $j_n \circ \lambda = \iota_{r_n}$ for every n.

<u>Proof</u>. Given $x \in E$ let $x_n := \iota_{r_n} x \in E_{r_n}$: then $\iota_{r_n r_m} x_m = x_n$ for $n < m$, hence there is a unique $\tilde{x} \in \tilde{E}$ such that $j_n x = x_n$ for each n, so by letting $\lambda x := \tilde{x}$ we get $(j_n \circ \lambda)x = j_n \tilde{x} = x_n = \iota_{r_n} x$.

Let $\iota^{rs}: \mathfrak{F}^p(E'_s) \to \mathfrak{F}^p(E'_r)$ for $r < s$ be given by $\iota^{rs} f := f \circ {}^T\iota_{rs}$, and $\iota^r: \mathfrak{F}^p(E') \to \mathfrak{F}^p(E'_r)$ by $\iota^r f := f \circ {}^T\iota_r$: in [7], Sec. 2.2 it is shown that if $\iota_r E$ is dense in E then $\mathfrak{F}^p(E')$ (with the topology generated by the maps ι^r) is the projective limit of the projective system $(\mathfrak{F}^p(E'_r), \iota^{rs})_{r<s}$ relative to the projecting maps ι^r. To transfer this representation to $\mathfrak{F}^p(\tilde{E}')$ we need the next lemma.

<u>Lemma 2</u>. If $\iota_r E$ is dense in E_r for each r then $j_n \tilde{E}$ is dense in E_{r_n} for each n.

<u>Proof</u>. Given $x \in E_{r_n}$ there is a sequence of $x_k \in E$ such that $\lim_k \iota_{r_n} x_k = x$. Letting $\tilde{x}_k := \lambda x_k \in E$ we have $j_n \tilde{x}_k = \iota_{r_k} x_k$ by Lemma 1, hence $\lim_k j_n \tilde{x}_k = x$.

It follows that $\mathfrak{F}^p(\tilde{E})$ is the projective limit of the projective system $(\mathfrak{F}^p(E_{r_n}), \iota_{r_n r_m})_{n<m}$ relative to the projecting maps

$j^n: \mathfrak{F}^p(\tilde{E}') \to \mathfrak{F}^p(E'_{r_n})$ given by $j^n\tilde{f} := \tilde{f}\circ {}^T j_n$.

Lemma 3. There is a continuous linear map $\Lambda: \mathfrak{F}^p(E') \to \mathfrak{F}^p(\tilde{E}')$ such that $j^n\circ\Lambda = \iota^{r_n}$ for all n.

Proof. Let $\Lambda f := f\circ {}^T\lambda$: then $(j^n\circ\Lambda)f = f\circ {}^T\lambda\circ {}^T j_n = f\circ {}^T\iota_{r_n} = \iota^{r_n} f$ for $f \in \mathfrak{F}^p(E')$ by Lemma 1.

Let $(g\cdot f)(x') := g(x')f(x')$ over E', the same for $\tilde{g}\cdot\tilde{f}$ over $\tilde{E}'$. In [8], Coroll. to Prop. 1.8.1 it was shown that the multiplication $f \mapsto g\cdot f$ maps $\mathfrak{F}^p(E')$ continuously into itself, provided the hypothesis (1) in the theorem is satisfied. The next lemma shows that Λ is a homomorphism for multiplication.

Lemma 4. $\Lambda(g\cdot f) = \Lambda g\cdot\Lambda f$ in $\mathfrak{F}^p(\tilde{E}')$ for f and g in $\mathfrak{F}^p(E')$.

Proof. $\Lambda(g\cdot f)(\tilde{x}') := (g\cdot f)({}^T\lambda\tilde{x}') := g({}^T\lambda\tilde{x}')f({}^T\lambda\tilde{x}') =: \Lambda g(\tilde{x}')\Lambda f(\tilde{x}')$ $=: (\Lambda g\cdot\Lambda f)(\tilde{x}')$ for every $\tilde{x}' \in \tilde{E}'$.

Let ${}^{rs}\iota: \mathfrak{F}_N^q(E_s) \to \mathfrak{F}_N^q(E_r)$ for $r < s$ be given by ${}^{rs}\iota\Phi := \Phi\circ\iota_{rs}$ and ${}^r\iota: \mathfrak{F}_N^q(E_r) \to \mathfrak{F}_N^q(E)$ by ${}^r\iota\Phi := \Phi\circ\iota_r$: in [7], Sec. 2.3 it is shown that $\mathfrak{F}_N^q(E)$ is the locally convex inductive limit of the inductive system $(\mathfrak{F}_N^q(E_r), {}^{rs}\iota)_{r<s}$ relative to the injective maps ${}^r\iota$. The same holds in particular for $\mathfrak{F}_N^q(E)$ and the maps ${}^n j: \mathfrak{F}_N^q(E_{r_n}) \to$ $\to \mathfrak{F}_N^q(E)$ given by ${}^n j\tilde{\Phi} := \tilde{\Phi}\circ j_n$.

Lemma 5. There is a continuous linear map $\Lambda': \mathfrak{F}_N^q(\tilde{E}) \to \mathfrak{F}_N^q(E)$ such that $\Lambda'\circ {}^n j = {}^{r_n}\iota$ for all n.

Proof. Let $\Lambda'\tilde{\Phi} := \tilde{\Phi}\circ\lambda$: then $(\Lambda'\circ {}^n j)\tilde{\Phi} = \tilde{\Phi}\circ j_n\circ\lambda = \tilde{\Phi}\circ\iota_{r_n} = {}^{r_n}\iota$ for $\tilde{\Phi} \in \mathfrak{F}_N^q(\tilde{E})$ by Lemma 1.

If $1/p + 1/q = 1$ we let $\langle\langle\cdot,\cdot\rangle\rangle$ be the Fourier-Borel bilinear form on $\mathfrak{F}^p(E') \times \mathfrak{F}_N^q(E)$: i.e., given $T \in \mathfrak{F}_N^q(E)'$ and $f(x') = BT(x') := T(\exp\circ x')$ then $\langle\langle f,\Phi\rangle\rangle := T(\Phi)$, same for $\langle\langle\cdot,\cdot\rangle\rangle^\sim$ on $\tilde{E}$.

Lemma 6. $\langle\langle\Lambda f,\tilde{\Phi}\rangle\rangle^\sim = \langle\langle f,\Lambda'\tilde{\Phi}\rangle\rangle$ for every f in $\mathfrak{F}^p(E')$ and $\tilde{\Phi}$ in $\mathfrak{F}_N^q(\tilde{E})$.

Proof. From [7], Prop. 2.5.1' we have $\langle\langle\Lambda f,\tilde{\Phi}\rangle\rangle^\sim = \langle\langle j^n(f\circ {}^T\lambda),\Phi_n\rangle\rangle_n$

if $\Phi = {}^{n}j\Phi_n$ with $\Phi_n \in \mathfrak{F}_N^q(E_{r_n})$ where $\langle\langle\cdot,\cdot\rangle\rangle_n$ is the pairing of $\mathfrak{F}^p(E'_{r_n})$ with $\mathfrak{F}_N^q(E_{r_n})$. But $j^n(f\circ {}^T\lambda) = f\circ {}^T\lambda\circ {}^Tj_n = f\circ {}^T(j_n\circ\lambda) = f\circ {}^T\iota_{r_n} = \iota^{r_n} f$ by Lemma 1, hence $\langle\langle \Lambda f, \tilde{\Phi}\rangle\rangle^\sim = \langle\langle \iota^{r_n} f, \Phi_n\rangle\rangle_n$. On the other hand, $\Lambda'\tilde{\Phi} = {}^{n}j\Phi_n\circ\lambda = \Phi_n\circ j_n\circ\lambda = \Phi_n\circ\iota_{r_n} = {}^{r_n}\iota\Phi_n$, hence we also have $\langle\langle f,\Lambda'\Phi\rangle\rangle = \langle\langle j^n f,\Phi_n\rangle\rangle_n$, again by [4], Prop. 2.5.1'.

It was shown in [8], Prop. 1.7.1 and sequel under the hypothesis (1) of the theorem that $g(d)$ maps $\mathfrak{F}_N^q(E)$ continuously into itself, and in [8], Prop. 1.9.1, that $\langle\langle f,g(d)\psi\rangle\rangle = \langle\langle g\cdot f,\psi\rangle\rangle$ for Φ in $\mathfrak{F}_N^q(E)$, with f and g in $\mathfrak{F}^p(E')$. We then also have $\langle\langle\tilde{f},\tilde{g}(d)\tilde{\psi}\rangle\rangle^\sim = \langle\langle\tilde{g}\cdot\tilde{f},\tilde{\psi}\rangle\rangle^\sim$ in $\mathfrak{F}_N^q(\tilde{E})$ and $\mathfrak{F}^p(\tilde{E}')$.

<u>Lemma 7</u>. Given $g \in \mathfrak{F}^p(E')$ and $\tilde{\psi} \in \mathfrak{F}_N^q(\tilde{E})$ let $\tilde{g} := \Lambda g \in \mathfrak{F}^p(\tilde{E}')$ and $\psi := \Lambda'\tilde{\psi} \in \mathfrak{F}_N^q(E)$: then $\Lambda'(\tilde{g}(d)\tilde{\psi}) = g(d)\psi$.

<u>Proof</u>. Given $f \in \mathfrak{F}^p(E')$, by Lemmas 6 and 4 we have

$\langle\langle f,\Lambda'(\tilde{g}(d)\tilde{\psi})\rangle\rangle = \langle\langle\Lambda f,\tilde{g}(d)\tilde{\psi}\rangle\rangle^\sim = \langle\langle\tilde{g}\cdot\Lambda f,\tilde{\psi}\rangle\rangle^\sim = \langle\langle\Lambda g\cdot\Lambda f,\tilde{\psi}\rangle\rangle^\sim =$
$= \langle\langle\Lambda(g\cdot f),\tilde{\psi}\rangle\rangle^\sim = \langle\langle g\cdot f,\Lambda'\tilde{\psi}\rangle\rangle = \langle\langle g\cdot f,\psi\rangle\rangle = \langle\langle f,g(d)\psi\rangle\rangle$, hence $\Lambda'(\tilde{g}(d)\tilde{\psi}) = g(d)\Phi$, because by [7], Prop. 2.5.1 every continuous linear form on $\mathfrak{F}_N^q(E)$ is of the form $\langle\langle f,\cdot\rangle\rangle$.

<u>Proof of the existence theorem</u>. E is the projective limit of the projective system $(E_r, \iota_{rs})_{r<s}$, with E_r and ι_{rs} as in the statement of the theorem. We assume the norms modified if necessary so that the hypothesis (1) of the theorem is verified (cfr Remark after the statement of the theorem). Given $\Phi \in \mathfrak{F}_N^q(E)$ we have $\Phi = {}^{r}\iota\Phi_r$ for some r and some $\Phi_r \in \mathfrak{F}_N^q(E_r)$. Given $g \in \mathfrak{F}^p(E')$, if g is not identically zero on E' then $g(x') \neq 0$ for some $x' \in E'$. We also have $x' = {}^T\iota_s x'_s$ for some s and some $x'_s \in E'_s$. We may choose an index (seminorm) r_1 such that $r_1 > r$ and $r_1 > s$. For each integer $n \geq 1$, given a seminorm r_n we may by hypothesis (1) choose a seminorm r_{n+1} such that

$$\|\cdot\|^q_{r_n} \leq (1/2)\,\|\cdot\|^q_{r_{n+1}}\,.$$

Let $\tilde{E}$ be the projective limit of the projective system $(E_{r_n},\iota_{r_n r_m})_{n<m}$

relative to projecting maps $j_n : \tilde{E} \to E_{r_n}$. We set $\tilde{g} := \Lambda g \in \mathfrak{F}^p(\tilde{E}')$ from Lemma 3. Letting $x_1' := {}^T\iota_{sr_1} x_s' \in E_{r_1}'$ and $\tilde{x}' := {}^T j_1 x_1' \in \tilde{E}$, by Lemmas 1 and 3 we get $\tilde{g}(\tilde{x}') = g({}^T\lambda\, {}^T j_1 x_1') = g({}^T\iota_{r_1} x_1') =$ $= g({}^T\iota_{r_1}\, {}^T\iota_{sr_1} x_s') = g({}^T\iota_s x_s') = g(x') \neq 0$, hence $\tilde{g}$ is not identically zero on $\tilde{E}'$. Let $\Phi_1 := {}^{rr_1}\iota\,\Phi_r \in \mathfrak{F}_N^q(E_{r_1})$ and set $\tilde{\Phi} := {}^1 j\,\Phi_1 \in$ $\in \mathfrak{F}_N^q(\tilde{E})$. Under either form of hypothesis (2) of the theorem, the strong dual of $\mathfrak{F}_N^q(\tilde{E})$ is a Fréchet space (cfr. Remarks following the statement of the theorem). Hence, by [7], Theorem 2.6.2 there is a solution $\tilde{\psi}$ of the equation $\tilde{g}(d)\tilde{\psi} = \tilde{\Phi}$ in $\mathfrak{F}_N^q(\tilde{E})$. Let now $\psi := \Lambda'\tilde{\psi} \in \mathfrak{F}_N^q(E)$ from Lemma 5: by Lemma 7 we get $\Lambda'\tilde{\Phi} = \Lambda'(\tilde{g}(d)\tilde{\psi}) =$ $= g(d)\psi$. Finally, from Lemmas 1 and 5 we got $\Lambda'\tilde{\Phi} = \Lambda'\,{}^1 j\,\Phi_1 =$ $= \Phi_1 \circ j_1 \circ \lambda = \Phi_1 \circ \iota_{r_1} = {}^{rr_1}\iota\,\Phi_r \circ \iota_{r_1} = \Phi_r \circ \iota_{rr_1} \circ \iota_{r_1} = {}^r\iota\,\Phi_r = \Phi$, i.e., $g(d)\psi = \Phi$.

REFERENCES

[1] Boland, Ph.J., Holomorphic functions on DFN-spaces, Sem. Pierre Lelong 1973/74. Lecture notes in Math. nº 474, Springer-Verlag, N.Y. 1975, p. 109-113.

[2] Colombeau, J.F. and Perrot, B., Convolution equations in spaces of infinite dimensional entire functions of exponential type, to appear in Trans. Amer. Math. Soc.

[3] Colombeau, J.F. and Perrot, B., Théorèms de noyaux analytiques en dimension infinie, C.R. Acad. Sci. Paris, t. 284 (1977), Série A, 759-762.

[4] Colombeau, J.F. and Perrot, B., Transformation de Fourier-Borel et noyaux en dimension infinie, C.R. Acad. Sci. Paris, t. 284 (1977), Série A, 963-966.

[5] Colombeau, J.F. and Perrot, B., Infinite-dimensional holomorphic "normal forms" of operators on the Fock spaces of Boson fields and an extension of the concept of Wick product, to appear in Advances in Holomorphy ed. by J.A. Barroso, North-Holland, 1979.

[6] Dwyer, III, T.A.W., Holomorphic Fock representations and partial differential equations on countably Hilbert spaces, Bull. Amer. Math. Soc. 79 (1973), 1045-1050.

[7] Dwyer, III, T.A.W., Dualité des espaces de fonctions entières en dimension infinie, Ann. Inst. Fourier (Grenoble, 26 (1976), 151-195.

[8] Dwyer, III, T.A.W., Differential operators of infinite order on locally convex spaces I, Rendiconti di Matematica 10, nº 1 (1977), 149-179.

[9] Dwyer, III, T.A.W., Differential operators of infinite order on locally convex spaces II, Rendiconti di Matematica 10, nº2-3 (1978), 273-293.

[10] Dwyer, III, T.A.W., Fourier-Borel duality and bilinear realizations of control systems, 1976 Ames Research Center Conf. on Geometric Methods in Control Theory, ed. by R. Hermann and C. Martin, Math. Sci. Press, 53 Jordan Rd, Brookline, MA 02146, 1977, p. 405-438.

[11] Dwyer, III, T.A.W., Analytic evolution equations in Banach spaces, Vector Space Measures and Applications II, ed. by R. Aron and S. Dineen, Lecture Notes in Math. nº 645 (1978), p. 48-61.

[12] Dwyer, III, T.A.W., Infinite-dimensional analytic systems, 1977 Conf. on Decision and Control Proceddings, IEEE Press, 1978, p. 285-290.

[13] Krée, P., Solutions faibles d'équations aux dérivées fonctionnelles, I, Sém. Pierre Lelong 1972/73, Lecture Notes in Math. nº 410, Springer-Verlag, N.Y., 1974.

[14] Krée, P. and Raczka, R., Kernels and symbols of operators in quantum field theory, Ann. Inst. Henri Poincaré (nº 1) 28 (1978), 41-73.

[15] Lascar, B., Équations aux dérivées partielles en dimension infinie, Vector Space Measures and Applications I, ed. by R. Aron and S. Dineen, Lecture Notes in Math. nº 644, Springer Verlag, N.Y., 1978, p. 286-313.

[16] Matos, M. de, Sur le Théorème d'approximation et d'existence de Malgrange-Gupta, C.R. Acad. Sci. Paris, t. 271 (1970), Série A, 1258-1259.

C^∞-FUNCTIONS ON LOCALLY CONVEX AND ON BORNOLOGICAL VECTOR SPACES

Jean-François Colombeau(*) and Reinhold Meise

U.E.R. de Mathématiques et d'Informatique
Université de Bordeaux I
F-33405 Talence - France

Mathematisches Institut der Universität
D-4000 Düsseldorf
Federal Republic of Germany

PREFACE

The spaces of continuously γ-differentiable functions on (real) locally convex spaces were studied by Keller [12] and under a different point of view by Meise [13], [14]. Independently the spaces of Silva differentiable functions on Schwartz bornological vector spaces have been investigated in the papers of Colombeau [6], [7]. Both approaches led to results of great similarity. Hence the authors became interested in the relations between the various definitions and this finally led to the present article, which has two aims: Methods used in the locally convex, respectively the bornological, setting are combined in order to clarify the relations between various definitions of differentiability and to improve several known results. In doing this, a survey on recent developments in this area is given.

Let us now sketch the content of the article. In the first section we introduce some notation and basic definitions which we shall use later. In the second section we introduce two fundamental concepts of differentiability: the (continuously) γ-differentiable

(*) The work of this author was done at the State University of Campinas (UNICAMP) São Paulo, Brazil in the (local) winter term 1978, under financial support of FAPESP and FINEP.

functions and the Silva-γ-differentiable functions. Then we give various results of comparison of these concepts. In the third section we prove results on the completeness, the Schwartz property and the (non-)nuclearity of $C^{\infty}_{S,\gamma}(\Omega,F)$, the space of infinitely Silva-γ-differentiable functions on Ω with values in F. We give also a description of spaces of this type in terms of the ε-product of Schwartz and a kernel representation for C^{∞}-functions on product sets. In the fourth and last section we prove a density result in spaces of n times differentiable functions and characterize the completion of $C^{n}_{co}(\Omega,F)$ for open subsets Ω in certain l.c. spaces E.

1. PRELIMINARIES

For our notation from the theory of locally convex (l.c.) spaces we refer to Horváth [11] and Schaefer [20], while we shall refer to Hogbe Nlend [10] for notation from the theory of bornological (b.v.) spaces. Throughout this article all l.c. spaces are assumed to be Hausdorff. They are also assumed to be real vector spaces as far as the converse is not explicitly stated. A bornological vector (b.v.) space will alwyas denote a real, convex, and separated bornological vector space in the terminology of Hogbe Nlend [10]. If E is a b.v. space we define $\mathcal{L} = \mathcal{L}(E)$ as the set of all the convex balanced and bounded subsets of E.

Let us now recall or give some definitions and results used in the sequel. If E is a b.v. space and B is in $\mathcal{L}$ then E_B denotes the linear span of B in E, normed by the gauge of B. A b.v. space E is called <u>Schwartz</u> if for any $A \in \mathcal{L}$ there exists a $B \in \mathcal{L}$ such that A is relatively compact in the normed space E_B. A subset Ω of a b.v. space E is called <u>M-open</u>, if for every $x \in \Omega$ and every bounded set B in E there is $\varepsilon > 0$ such that $x+\varepsilon B \subset \Omega$, or equivalently, if for any $B \in \mathcal{L}$ the set $\Omega \cap E_B$ is open in the normed space E_B. A subset K of a b.v. space E is called <u>strictly compact</u> if there exists B in $\mathcal{L}$ such that K is compact in E_B.

If Ω is a M-open subset of E then a subset K of Ω is said to be strictly compact in Ω if it is contained in Ω and strictly compact in E. Remark that any bounded set in a Schwartz b.v. space is contained in a strictly compact set. A set $B \in \mathcal{L}$ is called a <u>Banach disc</u> if E_B is a Banach space. We say that a b.v. space E has property (a) if for any strictly compact set K in E there is a Banach disc $B \in \mathcal{L}$ such that K is compact in E_B and such that the identity on E_B can be approximated uniformly on K by continuous linear operators on E_B, which have finite rank. Obviously a sufficient condition in order to ensure that E has property (a) is that, for any strictly compact set K in E there is a Banach disc $B \in \mathcal{L}$ such that K is compact in E_B and such that E_B has the approximation property (the above property is true in any nuclear b.v. space). The vector space of all the bounded linear forms on E is denoted by $E^{\times}$. E is called t-separated or regular or separated by its (bornological) dual if $E^{\times}$ separates the points of E. It is an easy consequence of the Hahn-Banach theorem that E is t-separated iff the locally convex inductive limit $\underrightarrow{\text{ind}}\, E_B$ is Hausdorff or equivalently iff $\sigma(E,E^{\times})$ is Hausdorff.

Let E and F be l.c. spaces and let γ be a system of bounded subsets of E which cover E. In the sequel we shall furthermore assume that γ is directed by inclusion. Then on the space $L(E,F)$ of all continuous linear maps from E into F one can introduce the corresponding γ-topology of uniform convergence on the sets in γ. The resulting l.c. space is denoted by $L_\gamma(E,F)$. By γ_σ, γ_{co}, γ_c and γ_b we denote the systems of all finite dimensional bounded, all compact, all precompact and all bounded subsets of E. The corresponding spaces $L_\gamma(E,F)$ are denoted by $L_\sigma(E,F)$, $L_{co}(E,F)$, $L_c(E,F)$ and $L_b(E,F)$. We write E'_γ instead of $L_\gamma(E,R)$.

We shall use the ε-product of Schwartz [21] for two l.c. spaces E and F with the following definition:

$$E \,\varepsilon\, F := L_e(E'_c, F),$$

where the subscript e stands for the topology of uniform convergence on the equicontinuous subsets of E'.

2. VARIOUS SPACES OF DIFFERENTIABLE FUNCTIONS

We begin by recalling the definition of an n times continuously γ-differentiable function, as it was given in Meise [13], 2.4. Before doing this we define for l.c. spaces E and F and a system γ of bounded subsets of E (with the properties required above) the l.c. spaces $L^n_\gamma(E,F)$ by induction on n as

$$L^o_\gamma(E,F) := F \quad \text{and} \quad L^{n+1}_\gamma(E,F) := L_\gamma(E,L^n_\gamma(E,F)).$$

Obviously $L^n_\gamma(E,F)$ may be considered as a space of n linear maps from E^n into F.

<u>2.1 Definition</u>. Let E and F be l.c. spaces, $\Omega \neq \emptyset$ an open subset of E and γ a system of bounded subsets of E with the properties mentioned above. For $n \in \mathbb{N}$ we define the space $C^n_\gamma(\Omega,F)$ of n times continuously γ-differentiable functions on Ω with values in F by induction as

$$C^1_\gamma(\Omega,F) := \{f \mid f: \Omega \to F \text{ is continuous and Gâteaux-differentiable on } \Omega, \text{ for any } x \in \Omega \text{ the Gâteaux-derivative } f'(x) \text{ is in } L(E,F) \text{ and } f': \Omega \to L_\gamma(E,F) \text{ is continuous}\}$$

and

$$C^{n+1}_\gamma(\Omega,F) := \{f \mid f: \Omega \to F \text{ is in } C^n(\Omega,F) \text{ and } f^{(n)} \text{ is in } C^1_\gamma(\Omega,L^n_\gamma(E,F))\}$$

$C^\infty_\gamma(\Omega,F)$ is defined as $\bigcap_{n\in\mathbb{N}} C^n_\gamma(\Omega,F)$. For any $n \in \mathbb{N}_\infty$ $(:= \mathbb{N} \cup \{\infty\})$ the space $C^n_\gamma(\Omega,F)$ is endowed with the topology of uniform convergence of all derivatives (up to order n, if n is finite) on the compact subsets of Ω.

<u>Remark</u>. An application of the generalized mean value theorem shows that the functions in $C^n_\gamma(\Omega,F)$ are not only Gâteaux differentiable but even γ-differentiable in a natural sense (Meise [13], 2.3).

The following proposition indicates how C^n_γ-functions behave under composition with continuous linear mappings.

2.2 Proposition. For $j = 1,2$ let E_j be a l.c. space and γ_j a system of bounded subsets of E_j having the properties mentioned at the beginning of this section. Let $A: E_1 \to E_2$ be a continuous linear map such that for any $S_1 \in \gamma_1$ there exists $S_2 \in \gamma_2$ with $A(S_1) \subset S_2$. Let $\Omega_2 \neq \emptyset$ be an open set in E_2 and define $\Omega_1 := A^{-1}(\Omega_2)$.

Then the mapping $A^*: f \to f \circ A$ is a continuous linear mapping from $C^n_{\gamma_2}(\Omega_2,F)$ into $C^n_{\gamma_1}(\Omega_1,F)$ for any $n \in \mathbb{N}_\infty$ and any l.c. space F.

Proof. Trivially $f \circ A$ is continuous on Ω_2 for any $f \in C^n_{\gamma_2}(\Omega_2,F)$. Then one shows by induction that for $1 \leq j < n+1$, any $x \in \Omega_2$ and any $y \in (E_2)^j$ one has

$$(f \circ A)^{(j)}(x)[y] = f^{(j)}(Ax)[A^j y] .$$

From this identity it follows that $f \circ A$ is in $C^n_{\gamma_1}(\Omega_1,F)$, and it follows also by the definition of the topologies that A^* is continuous.

2.3 Corollary. Let E be a l.c. space and let γ_1 and γ_2 be appropriate systems of bounded subsets of E. Assume that for any $S_1 \in \gamma_1$ there exists $S_2 \in \gamma_2$ with $S_1 \subset S_2$. Then, for any l.c. space F, any $n \in \mathbb{N}_\infty$ and any open subset Ω of E, $C^n_{\gamma_2}(\Omega,F)$ is a linear subspace of $C^n_{\gamma_1}(\Omega,F)$ and has a topology which is finer than the induced one.

For certain l.c. spaces E topological properties of E and the continuity of the derivatives imply that any C^n_σ-function is even a C^n_{co}-function for any $n \in \mathbb{N}_\infty$. This fact was shown in Keller [12], 1.0.2., where also limit structures are considered. Here we give a result which is slightly more general because we use the notion of a $k_{\mathbb{R}}$-space: A topological Hausdorff space X is called a $k_{\mathbb{R}}$-space, if it has the property that a function $f: X \to \mathbb{R}$ is continuous, iff the restriction of f to any compact subset of X is continuous. It is easy to see that a topological Hausdorff space X is a $k_{\mathbb{R}}$-space iff for any completely regular topological space Y a function $f: X \to Y$ is continuous iff its restriction to any compact subset of X is continuous. Metrizable topological spaces are examples of $k_{\mathbb{R}}$-spaces and also E'_c for any metrizable l.c. space (see e.g. Horváth [11], chap. 3, §10).

<u>2.4 Proposition</u>. Let E be a barrelled l.c. space, which is a $k_{\mathbb{R}}$-space. Then for any open set $\Omega \neq \emptyset$ in E, any $n \in \mathbb{N}_\infty$ and any l.c. space F we have algebraically

$$C^n_\sigma(\Omega,F) = C^n_{co}(\Omega,F).$$

<u>Proof</u>. By Corollary 2.3 $C^n_{co}(\Omega,F)$ is a linear subspace of $C^n_\sigma(\Omega,F)$. In order to show the converse inclusion we prove that $C(\omega,L^n_\sigma(E,F)) = C(\omega,L^n_{co}(E,F))$ (as sets) for any $n \in \mathbb{N}$ and any $k_{\mathbb{R}}$-space ω. Then we observe that by a result of Blasco [4], Ω is a $k_{\mathbb{R}}$-space, since E is a $k_{\mathbb{R}}$-space.

Now let ω be any $k_{\mathbb{R}}$-space, and let $f \in C(\omega,L^1_\sigma(E,F))$ be given. Then $f(K)$ is compact in $L_\sigma(E,F)$ for any compact set K in ω. But then $f(K)$ is pointwise bounded and hence equicontinuous, since E is barrelled. It is known (see e.g. Schaefer [20], Chap. III, 4.5) that the topologies of $L_c(E,F)$ and $L_\sigma(E,F)$ coincide on equicontinuous subsets, hence they coincide on $f(K)$. Since ω is a $k_{\mathbb{R}}$-space this implies $f \in C(\omega,L^1_{co}(E,F))$. Tho general result now follows by induction.

Now we want to show that on certain l.c. spaces C^∞_c- and C^∞_b-functions coincide. This was proved in Keller [12], 2.9.1-2.9.4; we give a proof using only topological arguments.

<u>2.5 Lemma</u>. Let E and F be l.c. spaces, let $\Omega \neq \emptyset$ be an open subset of E and let $n \in \mathbb{N}$ be given. Assume that E^{n+2} is a $k_{\mathbb{R}}$-space. Then $C^{n+1}_{co}(\Omega,F)$ is a subset of $C^n_b(\Omega,F)$.

Proof. We shall give the proof for $n=1$, then the result will follow by induction (remember that for a $k_{\mathbb{R}}$-space $X\times Y$ any factor is a $k_{\mathbb{R}}$-space). First we observe that by a result of Blasco [4] an open subset of a $k_{\mathbb{R}}$-space is a $k_{\mathbb{R}}$-space again, hence $\Omega\times E\times E$ is a $k_{\mathbb{R}}$-space by hypothesis. Now let $f \in C^2_{co}(\Omega,F)$ be given. We show that the mapping $g: \Omega\times E\times E \to F$, $g(x,y,z) := f''(x)[y,z]$, is continuous. As we know that $\Omega\times E\times E$ is a $k_{\mathbb{R}}$-space it suffices to show that $g|K\times Q\times Q$ is continuous for any compact set K in Ω and any compact set Q in E. This, however, follows from the continuity of $f'': \Omega \to L^2_{co}(E,F)$ and the continuity of the restriction map $R_Q: L^2_{co}(E,F) \to C(Q\times Q,F)$, $R_Q(u)[y,z] = u(y,z)$, together with the observation that $C(K,C(Q\times Q,F)) \cong C(K\times Q\times Q,F)$.

Now take any continuous semi-norm q on F and any $x \in \Omega$. By the continuity of g there is a balanced convex neighbourhood U of zero such that for any $h,y,z \in U$ we have $q(g(x+h,y,z)) \leq 1$. Since $f''(x+h)$ is bilinear it follows from this by standard arguments that for any $\varepsilon > 0$ and any bounded set B in E there is a neighborhood V of zero in E such that for any $h,y \in V$ and any $b \in B$

$$q(f''(x+h)[y,b]) \leq \varepsilon.$$

By the generalized mean value theorem applied to f' this implies that for any $h \in V$ and any $b \in B$

$$q((f'(x+h) - f'(x))[b]) \leq \sup_{0\leq t\leq 1} q(f''(x+th)[h,b]) \leq \varepsilon.$$

Since $x \in \Omega$ was arbitrary we have shown that $f: \Omega \to L^1_b(E,F)$ is continuous.

The following result is now an immediate consequence of Lemma 2.5 and Proposition 2.4.

2.6 Proposition. Let E and F be l.c. spaces and let $\Omega \neq \emptyset$ be an open subset of E. Assume that E^j is a $k_{\mathbb{R}}$-space for any $j \in \mathbb{N}$. Then $C^\infty_{co}(\Omega,F) = C^\infty_b(\Omega,F)$ as sets. If moreover E is barrelled, then $C^\infty_\sigma(\Omega,F) = C^\infty_b(\Omega,F)$ as sets.

<u>Remark</u>. $C^\infty_{co}(\Omega,F)$ and $C^\infty_b(\Omega,F)$ have the same topology iff any bounded set of E relatively compact, i.e. iff E is a Semi-Montel space. This shows that on Banach spaces E and on Fréchet spaces E which are not Semi-Montel, $C^\infty_{co}(\Omega,F)$ and $C^\infty_b(\Omega,F)$ have different topologies though they are equal as sets (by 2.6). On (FM)- and (DFM)-spaces, however, we have by 2.6 the topological identity $C^\infty_\sigma(\Omega,F) = C^\infty_b(\Omega,F)$.

Now we turn to differentiable functions on b.v. spaces. One reason for doing this is the fact that the requirement of continuity in Definition 2.1 leads to the disadvantage that even many bilinear functions occuring in applications are not of type C^1 in the sense of 2.1, since they are hypocontinuous but not continuous. Knowing the theory of calculus according to Sebastião e Silva (see [22], [23] and [16]) one can overcome this difficulty by working on b.v. spaces and using the following definition.

<u>2.7 Definition</u>. a) Let E be a b.v. space and $\Omega \neq \emptyset$ a M-open subset of E. For any $B \in \mathfrak{L} = \mathfrak{L}(E)$ define $\Omega_B := \Omega \cap E_B$, endowed with the topology of the normed space E_B. Let $\gamma = (\gamma_B)_{B\in\mathfrak{L}}$ denote a family of systems of bounded sets in E, where γ_B is a system of bounded sets in E_B having the usual properties. We assume that for any $B_1,B_2 \in \mathfrak{L}$ with $B_1 \subset B_2$ and any $S_1 \in \gamma_{B_1}$ there is a set $S_2 \in \gamma_{B_2}$ with $S_1 \subset S_2$. Furthermore let F be a l.c. space and $n \in \mathbb{N}_\infty$. Then a map $f: \Omega \to F$ is called n times continuously Silva-γ-differentiable if its restriction to Ω_B is in $C^n_{\gamma_B}(\Omega_B,F)$ for any $B \in \mathfrak{L}$. The space of all these functions is denoted by $C^n_{S,\gamma}(\Omega,F)$. Using 2.2 it is easy to see that algebraically $C^n_{S,\gamma}(\Omega,F)$ is identical with $\underleftarrow{\text{proj}}_{B\in\mathfrak{L}}\, C^n_{\gamma_B}(\Omega_B,F)$. Hence we may endow $C^n_{S,\gamma}(\Omega,F)$ with the topology given by this projective limit. Obviously this topology is the topology of uniform convergence on the strictly compact subsets of Ω of the derivatives (relative to any E_B) up to order n.

b) If in a) for any $B \in \mathfrak{L}$ the system $\gamma_B = \gamma_\sigma$, γ_{co} or γ_c, then we write $C^n_{S,\sigma}(\Omega,F)$, $C^n_{S,co}(\Omega,F)$ or $C^n_{S,c}(\Omega,F)$. If $\gamma_B = \gamma_b$ for any $B \in \mathfrak{L}$ then we write $C^n_S(\Omega,F)$ only.

2.8 Remark. a) It is easy to see that $C^n_S(\Omega,F)$ is the set of all $f: \Omega \to F$ which are "Silva differentiable in the large sense" according to the definitions given by Colombeau [6], i.e. for any $B \in \mathfrak{L}$ and any balanced convex zero neighbourhood V in F the function $f_{B,V}: \Omega_B \to F_V$ naturally induced by f is a n times continuously Fréchet-differentiable function (in the usual sense) between normed spaces.

b) If E is a complete b.v. space then it is an immediate consequence of Proposition 2.4 that for any M-open subset Ω of E, any l.c. space F and any $n \in \mathbb{N}_\infty$ the algebraic identity $C^n_{S,\sigma}(\Omega,F) = C^n_{S,co}(\Omega,F) = C^n_{S,c}(\Omega,F)$ holds. Moreover it follows from Proposition 2.6 that $C^\infty_{S,\sigma}(\Omega,F) = C^\infty_S(\Omega,F)$ as sets.

c) If E is a Schwartz b.v. space then it follows easily from Proposition 2.2 that for any M-open subset Ω of E, any l.c. space F and any $n \in \mathbb{N}_\infty$ the topological identity $C^n_{S,co}(\Omega,F) = C^n_S(\Omega,F)$ holds.

If (E,τ) is a l.c. space for which the closed convex hull of any compact subset of (E,τ) is compact, then one can introduce in the vector space E the compact bornology of (E,τ). This b.v. space is denoted by E_c and its bounded sets are just the relatively compact subsets of (E,τ). Using a result which is related with the theorem of Banach-Dieudonné one shows that E_c is a Schwartz b.v. space iff any compact set in (E,τ) is very compact. Recall that a compact set K in E is called very compact if there exists a Banach disc B in (E,τ) such that K is contained and compact in E_B. If (E,τ) is quasi-complete then E_c is a Schwartz b.v. space iff $(E,\tau)'_c$ is a Schwartz space (for the standard proof see e.g. Meise [13], 2.12). Using these remarks we can show:

2.9 Proposition. Let (E,τ) be a l.c. space in which any compact set is very compact and denote by E_c the space E endowed with the compact bornology of (E,τ). Let $\Omega \neq \emptyset$ be an open subset of (E,τ) and denote by Ω_c the M-open subset Ω of E_c. Let $n \in \mathbb{N}_\infty$ and any l.c. space F be given.

a) $C^n_{co}(\Omega,F)$ is a topological subspace of $C^n_S(\Omega_c,F) = C^n_{S,co}(\Omega_c,F)$.

b) If E^j is a $k_{\mathbb{R}}$-space for $1 \leq j < n+2$, then the topological identity $C^n_{co}(\Omega,F) = C^n_{S,co}(\Omega_c,F) = C^n_S(\Omega_c,F)$ holds.

Proof. a) For any $B \in \gamma_{co}$ the canonical inclusion $E_B \to E$ is a continuous linear map and hence maps compact sets into compact sets. This implies by Proposition 2.2 that the restriction from Ω to Ω_B gives a continuous linear map from $C^n_{co}(\Omega,F)$ into $C^n_{co}(\Omega_B,F)$. Hence $C^n_{co}(\Omega,F)$ is a continuously embedded subspace of $C^n_{S,co}(\Omega_c,F)$ by Definition 2.7. Since any compact set in E is very compact it is easy to see that $C^n_{co}(\Omega,F)$ is a topological subspace of $C^n_{S,co}(\Omega_c,F)$. This completes the proof by Remark 2.8 c) and the recall above.

b) Let $f \in C^n_{S,co}(\Omega_c,F)$ be given. If K is any compact set in Ω, by hypothesis there exists a Banach disc B in E such that K is compact in E_B. It is a consequence of the theorem of Banach-Dieudonné that we may assume that B is even compact in E, i.e., $B \in \gamma_{co}$. Since K is compact in Ω_B, the topologies induced on K by E and E_B coincide. Hence $f|K$ is continuous, since $f \in C^n_{S,co}(\Omega_c,F)$. By a result of Blasco [4] it follows from our hypothesis that Ω is a k_R-space. Hence f is a continuous function on Ω.

Now we observe that $L_{co}(E,F) = \underleftarrow{\text{proj}}_{B \in \gamma_{co}} L_{co}(E_B,F)$ since E is a k_R-space and since any compact set in E is very compact. From this observation it follows that for any $x \in \Omega$ the Gâteaux-derivative $f'(x)$ of f at x is in $L_{co}(E,F)$. Moreover it implies that $f' : \Omega \to L_{co}(E,F)$ is continuous on any compact subset of Ω. Hence f' is continuous. An inductive application of these arguments completes the proof.

The hypotheses of Proposition 2.9 b) are satisfied for Fréchet spaces and Silva spaces (i.e. locally convex compact inductive limits of an increasing sequence of Banach spaces). Hence the following corollary holds.

<u>2.10 Corollary</u>. Let E be a Fréchet space or a Silva space, F a l.c. space and $\Omega \neq \emptyset$ an open subset of E. Then we have for any $n \in \mathbb{N}_\infty$ the topological identity

$$C^n_{co}(\Omega,F) = C^n_{S,co}(\Omega_c,F) = C^n_S(\Omega_c,F).$$

If E is a complete b.v. space one defines the Schwartz b.v. space E_s associated to E in the following way: The bounded subsets of E_s are the subsets of E which are contained in some strictly compact subset of E. From the theorem of Banach-Dieudonné it follows that E_s is a Schwartz b.v. space. The following proposition shows that the functions of type $C^n_{S,co}$ do not change if one goes from E to E_s.

<u>2.11 Proposition</u>. Let E be a complete b.v. space, F a l.c. space and $\Omega \neq \emptyset$ an M-open set in E. Denote by Ω_s the set Ω regarded as a subset of E_s, the Schwartz b.v. space associated to E. Then the following topological identity holds for any $n \in \mathbb{N}_\infty$

$$C^n_{S,co}(\Omega,F) = C^n_{S,co}(\Omega_s,F) = C^n_S(\Omega_s,F).$$

<u>Proof</u>. It is easy to see that any $f \in C^n_{S,co}(\Omega,F)$ belongs to $C^n_{S,co}(\Omega_s,F)$ and that $C^n_{S,co}(\Omega,F)$ is continuously embedded into $C^n_{S,co}(\Omega_s,F)$.

In order to prove the converse inclusion let $f \in C^n_{S,co}(\Omega_s,F)$ be given. If B is a Banach disc in E, let $(E_B)_c$ denote the space E_B endowed with its compact bornology and let $(\Omega_B)_c$ denote the set $\Omega_B = \Omega \cap E_B$ as a subset of $(E_B)_c$. Then it follows from $f \in C^n_{S,co}(\Omega_s,F)$ that $f|\Omega_B$ is in $C^n_{S,co}((\Omega_B)_c,F)$, which coincides by 2.1o with $C^n_{co}(\Omega_B,F)$. It is also easy to see that the mapping $f \to f|\Omega_B$ is continuous from $C^n_{S,co}(\Omega_s,F)$ into $C^n_{co}(\Omega_B,F)$. Since the b.v. space E is complete by hypothesis, we have shown that

$C^n_{S,co}(\Omega_s,F)$ is continuously imbedded into $C^n_{S,co}(\Omega,F)$. Because of Remark 2.8 c) the proof is complete.

3. TOPOLOGICAL PROPERTIES OF THE SPACES OF C^n-FUNCTIONS

In Meise [13], 2.5 a sufficient condition for the completeness of $C^n_\gamma(\Omega,F)$ was given. Using this and Proposition 2.11 it is easy to prove the following proposition.

<u>3.1 Proposition</u>. Let E be a b.v. space, $\Omega \neq \emptyset$ an M-open subset of E, F a l.c. space and $n \in \mathbb{N}_\infty$:

a) $C^n_S(\Omega,F)$ is complete iff F is complete.

b) If E is a complete b.v. space then $C^n_{S,co}(\Omega,F)$ is complete iff F is complete.

From this and 2.9 we get as corollary:

<u>3.2 Corollary</u>. Let (E,τ) be a l.c. space in which any compact set is very compact and let $\Omega \neq \emptyset$ be an open subset of (E,τ). Then for any complete l.c. space F and any $n \in \mathbb{N}_\infty$ the completion of $C^n_{co}(\Omega,F)$ is a topological linear subspace of $C^n_{S,co}(\Omega_c,F)$.

Concerning the Schwartz property of $C^\infty_{co}(\Omega)$ a sufficient condition (which is close to being necessary; see [13], 2.12) was given in [13], 2.11. This result can be extended to b.v. spaces as it was done in [7].

<u>3.3 Proposition</u>. Let E be a complete b.v. space, F a l.c. space and $\Omega \neq \emptyset$ an M-open subset of E. Then $C^\infty_{S,co}(\Omega,F)$ is a Schwartz l.c. space iff F is Schwartz. Remark that $C^\infty_{S,co}(\Omega,F) = C^\infty_S(\Omega,F)$ if E is a Schwartz b.v. space.

<u>Proof</u>. By [13], Proposition 2.11, $C^\infty_{co}(\Omega_B)$ is a Schwartz l.c. space for any Banach disc B in E. By the definition of $C^\infty_{S,co}(\Omega)$ it follows from this that $C^\infty_{S,co}(\Omega)$ is a Schwartz l.c. space. In 3.5 below we shall show that $C^\infty_{S,co}(\Omega,F) = C^\infty_{S,co}(\Omega)\varepsilon F$. Hence the result follows from the observation that the ε-product of two l.c. spaces is

Schwartz iff both factors are Schwartz. The rest is obvious by Proposition 2.11.

If E is a b.v. space having the finite dimensional bornology (i.e. its bounded sets are just the finite dimensional bounded sets) then it follows from the definition and from a classical result on the nuclearity of $C^\infty(\omega)$, $\omega \subset \mathbb{R}^n$ open, that $C_S^\infty(\Omega) = C_{S,co}^\infty(\Omega)$ is nuclear for any non-empty M-open subset Ω of E. We want to show now that $C_S^\infty(\Omega)$ and $C_{S,co}^\infty(\Omega)$ are not nuclear in general. This can be done exactly in the same way as it was shown in Meise [14] that $C_\gamma^\infty(\Omega)$ is not nuclear, if $\Omega \neq \emptyset$ is any open subset of a l.c. space E which contains a balanced convex compact subset of infinite dimension, and if the system γ contains the compact subsets of E.

<u>3.4 Theorem</u>. Let E be a regular b.v. space which contains a balanced convex strictly compact set K of infinite dimension. Then $C_S^\infty(\Omega)$ and $C_{S,co}^\infty(\Omega)$ are not nuclear for any M-open subset $\Omega \neq \emptyset$ of E.

<u>Proof</u>. a) First we observe that it suffices to show that $C_S^\infty(E)$ and $C_{S,co}^\infty(E)$ are not nuclear. This is a consequence of the following considerations (given only for C_S^∞): For any M-open set $\Omega \neq \emptyset$ in E and any $x \in E$ $C_S^\infty(\Omega)$ and $C_S^\infty(\Omega+x)$ are isomorphic and $C_S^\infty(E)$ can be regarded as a topological linear subspace of $\prod_{x\in E} C_S^\infty(\Omega+x)$ by means of the mapping $f \to (f|(\Omega+x))_{x\in E}$. Hence the nuclearity of $C_S^\infty(\Omega)$ for one M-open set $\Omega \neq \emptyset$ implies the nuclearity of $C_S^\infty(E)$.

b) Now we show that the nuclearity of $C_S^\infty(E)$ (resp. $C_{S,co}^\infty(E)$) implies the following assertion:

(*) There exists $m \in \mathbb{N}$ such that for any $N \in \mathbb{N}$ the canonical embedding B_N of the space $C_{2\pi}^m(N)$ (of $[0,2\pi]^N$-periodic m times continuously differentiable functions) in the space $L^2([0,2\pi]^N)$ is of type $l^{1/2}$ (cf. Pietsch [18], 8.2.1).

In order to prove this, let K be a balanced convex strictly compact set in E which is of infinite dimension. Then the semi-norm

$q\colon f \to \sup_{x\in K} |f(x)|$ is continuous on $C^\infty_S(E)$ and on $C^\infty_{S,co}(E)$. Assuming that $C^\infty_S(E)$ is nuclear (the case of $C^\infty_{S,co}(E)$ is treated similarly), we find a continuous semi-norm p on $C^\infty_S(E)$ such that the canonical map

$$\varkappa\colon (C^\infty_S(E))_p \to (C^\infty_S(E))_q$$

between the associated normed spaces is of type $l^{1/2}$. W.l.o.g. we may assume that p is of the form

$$p(f) = \sup_{0\le j\le m} \sup_{x\in Q} \sup_{y\in B^j} |f^{(j)}(x)[y]|,$$

where $m \in \mathbb{N}$, $B \in \mathfrak{L}(E)$ and where Q is a strictly compact subset of E. It is easy to see that the restriction map $\rho_K\colon C^\infty_S(E) \to C(K)$, $\rho_K(f) := f|K$, is a continuous linear map, hence it induces a continuous linear map $\tilde{\rho}_K\colon (C^\infty_S(E))_q \to C(K)$. Since mappings of type l^p have the ideal property (Pietsch [18], 8.2.8), it follows that $\varkappa_K := \tilde{\rho}_K\circ\varkappa$ is of type $l^{1/2}$ too. Since $C(K)$ is complete one can extend $\varkappa_K$ continuously to the completion $\widehat{(C^\infty_S(E))_p}$ of $(C^\infty_S(E))_p$. Then it follows from the definition of the approximation numbers (Pietsch [18], 8.1.1), that this extension $\hat{\varkappa}_K$ is again of type $l^{1/2}$.

Now we choose a linearly independent sequence $(x_j)_{j\in\mathbb{N}}$ in K and denote for $N \in \mathbb{N}$ the span of $\{x_1,\dots,x_n\}$ by E_N. Since E is a regular b.v. space by hypothesis we can extend linear functionals on E_N to bounded linear functionals on E. Hence for any $N\in\mathbb{N}$ we can find a bounded linear projection π_N on E which projects E onto E_N. Since π_N is bounded, $\pi_N|E_A$ is continuous for any $A \in \mathfrak{L}(E)$, hence it follows from the definition of $C^\infty_S(E)$ and Proposition 2.2 that

$$\tilde{\pi}_N\colon C^\infty_S(E_N) \to C^\infty_S(E),\quad \tilde{\pi}_N(f) := f\circ\pi_N,$$

is a continuous linear map. This implies that the composition of $\tilde{\pi}_N$ and the canonical map $\sigma\colon C^\infty_S(E) \to (C^\infty_S(E))_p$ gives a continuous linear map $\sigma_N := \sigma\circ\tilde{\pi}_N$. Obviously this map is continuous on $C^\infty(E_N)$ for the topology induced by $C^m(E_N)$. Since $C^\infty(E_N)$ is dense in $C^m(E_N)$ we can extend σ_N continuously and get a continuous linear map

$$\hat{\sigma}_N: C^m(E_N) \to (C_S^\infty(E))_p.$$

Consequently the mapping

$$\lambda_N := \hat{\kappa}_K \circ \hat{\sigma}_N: C^m(E_N) \to C(K)$$

is of type $l^{1/2}$.

Now we denote by I_N the interval $[0,\frac{1}{2^N}]$ and define the mapping $j_N: I_N^N \to E$ by $j_N(t_1,\ldots,t_N) := \sum_{n=1}^{N} t_n x_n$. Obviously j_N is continuous and has values in $K \cap E_N$ since K is balanced and convex. Hence the mapping

$$\tilde{j}_N: C(K) \to L^2(I_N^N), \quad \tilde{j}_N(f) := f \circ j_N,$$

is continuous and linear. Now we observe that the space $C^m_{I_N^N}(\mathbb{R}^N)$ of I_N^N - periodic functions in $C^m(\mathbb{R}^N)$ is mapped continuously into $C^\infty(E_N)$ by $i_N(f)[\sum_{n=1}^{N} t_n x_n] = f(t_1,\ldots,t_N)$. Hence the mapping

$$A_N := \tilde{j}_N \circ \lambda_N \circ i_N: C^m_{I_N^N}(\mathbb{R}^N) \to L^2(I_N^N)$$

is of type $l^{1/2}$. However, A_N is nothing else than the canonical inclusion. Hence (*) follows by applying a dilatation.

c) The proof is completed by showing that (*) does not hold. From this it follows that the assumption of the nuclearity of $C_S^\infty(E)$ leads to a contradiction.

In order to show that B_N is not of type $l^{1/2}$ for N sufficiently large, we use a modification of the idea of proof in Gramsch [9], Bemerkung after Lemma 1.

We define the Hilbert space H_N as

$$H_N := \{(a_n) \in \mathbb{R}^{\mathbb{N}^N} \mid \sum_{n\in\mathbb{N}^N} (\prod_{j=1}^{N} n_j^2)(\sum_{\substack{\alpha\in\mathbb{N}_0^N \\ |\alpha|\leq 1}} n^{2\alpha}|a_n|^2) < \infty\}.$$

Then the mapping $C_N: H_N \to C^1_{2\pi}(N)$, defined by

$$C_N((a_n)_{n\in\mathbb{N}^N}): (x_1,\ldots,x_N) \to \sum_{n\in\mathbb{N}^N} a_n \prod_{j=1}^{N} \sin n_j x_j,$$

is continuous, and hence $D_N := B_N \circ C_N$ is of type $l^{1/2}$ because of (*). Observing that the functions $f_n : x \to \prod_{j=1}^{N} \sin n_j x_j$, $n \in \mathbb{N}^N$, form an orthogonal system in $L^2([0,2\pi]^N)$, one gets the approximation numbers of D_N, similarly as in the proof of Lemma 1 in [9], as the decreasing rearrangement of the family

$$((2^N \prod_{j=1}^{N} n_j (\sum_{|\alpha|} n^{2\alpha})^{1/2})^{-1})_{n \in \mathbb{N}^N}.$$

A simple estimation shows that this family is not in $l^{1/2}$ for $N \geq 1+2$.

As a generalization of results of Aron [1] and Bombal Gordón and González Llavona [13] theorem 3.1 in Meise [13] gives (under appropriate hypotheses) a characterization of the ε-product of $C^n(\Omega)$ and any quasi-complete l.c. space F as topological linear subspace of $C^n(\Omega,F)$, if the system γ contains γ_{co}. This characterization leads to the relation $C^n_{co}(\Omega)\varepsilon F = C^n_{co}(\Omega,F)$ for any quasi-complete l.c. space F, any $n \in \mathbb{N}_\infty$ and any open subset $\Omega \neq \emptyset$ of an (F)- or a (DFM)-space E (see [13], 3.2).

Now we want to use this result in order to generalize it to $C^n_{S,co}$-functions on b.v. spaces. This and the kernel representation given below are easy corollaries of the corresponding results on Banach spaces and the inheritance properties of the ε-product.

<u>3.5 Theorem</u>. Let E be a complete b.v. space, F a quasi-complete l.c. space and $\Omega \neq \emptyset$ an M-open subset of E. Then we have for any $n \in \mathbb{N}_\infty$:

$$C^n_{S,co}(\Omega)\varepsilon F = C^n_{S,co}(\Omega,F).$$

Proof. Let $\bar{\mathfrak{L}}$ denote the subsystem of all Banach discs in $\mathfrak{L}(E)$. $\bar{\mathfrak{L}}$ is cofinal in $\mathfrak{L}$ since E is complete. Hence we have $C^n_{S,co}(\Omega,G) = \underleftarrow{\mathrm{proj}}_{B\in\bar{\mathfrak{L}}}\, C^n_{co}(\Omega_B,G)$ for any l.c. space G. By Bierstedt and Meise [3], 4.4 and Meise [13], 3.2 this implies

$$C^n_{S,co}(\Omega)\varepsilon F = L_e(F'_c, C^n_{S,co}(\Omega)) = L_e(F'_c, \underleftarrow{\mathrm{proj}}_{B\in\bar{\mathfrak{L}}}\, C^n_{co}(\Omega_B))$$

$$= \underleftarrow{\mathrm{proj}}_{B\in\bar{\mathfrak{L}}}\, L_e(F'_c, C^n_{co}(\Omega_B)) = \underleftarrow{\mathrm{proj}}_{B\in\bar{\mathfrak{L}}}\, C^n_{co}(\Omega_B)\varepsilon F$$

$$= \underleftarrow{\mathrm{proj}}_{B\in\bar{\mathfrak{L}}}\, C^n_{co}(\Omega_B,F) = C^n_{S,co}(\Omega,F).$$

3.6 Theorem. For $j = 1,2$ let E_j be a complete b.v. space and Ω_j an M-open subset of E_j. Then the following holds true:

$$C^\infty_{S,co}(\Omega_1\times\Omega_2) = C^\infty_{S,co}(\Omega_1, C^\infty_{S,co}(\Omega_2)) = C^\infty_{S,co}(\Omega_1)\varepsilon C^\infty_{S,co}(\Omega_2).$$

Proof. Since E_1 and E_2 are complete, their bornological product is complete. Denoting the system of Banach discs in E_1, resp. E_2, by $\bar{\mathfrak{L}}_1$, resp. $\bar{\mathfrak{L}}_2$, we have that $\mathfrak{L}_o = \{B_1\times B_2 \mid B_1 \in \bar{\mathfrak{L}}_1, B_2 \in \bar{\mathfrak{L}}_2\}$ is cofinal in $\mathfrak{L}(E_1\times E_2)$. Hence we get from Bierstedt and Meise [3], 4.4, Meise [13], 4.6, theorem 3.5 and proposition 3.1 b)

$$C^\infty_{S,co}(\Omega_1\times\Omega_2) = \underleftarrow{\mathrm{proj}}_{B\subset\mathfrak{L}_o}\, C^\infty_{co}(\Omega_1\times\Omega_2)_B)$$

$$= \underleftarrow{\mathrm{proj}}_{B_1\in\mathfrak{L}_1}\, \underleftarrow{\mathrm{proj}}_{B_2\in\mathfrak{L}_2}\, C^\infty_{co}(C^\infty_{co}((\Omega_1)_{B_1} \times (\Omega_2)_{B_2})$$

$$= \underleftarrow{\mathrm{proj}}_{B_1\in\mathfrak{L}_1}\, \underleftarrow{\mathrm{proj}}_{B_2\in\mathfrak{L}_2}\, C^\infty_{co}(C^\infty_{co}((\Omega_1)_{B_1}) \varepsilon C^\infty_{co}((\Omega_2)_{B_2})$$

$$= \underleftarrow{\mathrm{proj}}_{B_1\in\mathfrak{L}_1}\, C^\infty_{co}((\Omega_1)_{B_1}, C^\infty_{S,co}(\Omega_2))$$

$$= C^\infty_{S,co}(\Omega_1, C^\infty_{S,co}(\Omega_2)) = C^\infty_{S,co}(\Omega_1)\varepsilon C^\infty_{S,co}(\Omega_2).$$

4. A DENSITY RESULT

A modification of an idea of Aron and Schottenloher [2] in the complex case was used by Bombal Gordón and González Llavona [5] as well as by Prolla and Guerreiro [19] to show that $C^n_{co}(\Omega)$ has the approximation property (a.p.) for any open subset Ω of a Banach space E iff E has a.p. This result was presented in a more general form in Meise [13], 3.5 where also other density statements were given. We shall show now that the ideas used in [13] can also be applied to describe the completion of $C^n_{co}(\Omega,F)$ for open subsets Ω of certain l.c. spaces. A similar result in the complex case was given in [8]. Before proving the main lemma we recall the definition of a finite polynomial.

<u>4.1 Definition</u>. Let E be a l.c. space. A function $p: E \to \mathbb{R}$ is called a finite (continuous) monomial if it is a constant or if there exist $n \in \mathbb{N}$ and $y_1,\dots,y_n \in E'$ such that $p(x) = \prod_{j=1}^{n} \langle y_j,x\rangle$ for any $x \in E$. A finite sum of finite monomials is called a finite (continuous) polynomial on E. The vector space of all finite polynomials is denoted by $\mathcal{P}_f(E)$.

Remember that it was remarked in the preliminaries that on a regular b.v. space E there exists a l.c. topology τ with the property that any bounded set in the b.v. space E is bounded in the l.c. space (E,τ).

<u>4.2 Lemma</u>. Let E be a regular b.v. space with property (a) (see section 1), and let τ be a l.c. topology on E for which any bounded set in the b.v. space E is bounded in (E,τ). Then $\mathcal{P}_f(E,\tau)\otimes F$ is dense in $C^n_{S,co}(\Omega,F)$ for any M-open subset $\Omega \neq \emptyset$ in E, any $n \in \mathbb{N}_\infty$ and any l.c. space F.

<u>Proof</u>. The proof is along the lines of the proofs of 3.3 and 3.9 in [13]. Because of a standard argument it is enough to prove the lemma only for Banach spaces F. Let $f \in C^n_{S,co}(\Omega,F)$ and a contin-

uous semi-norm p on $C^n_{S,co}(\Omega,F)$ be given. We may assume that there exist $l \in \mathbb{N}$ with $0 \le l < n+1$ and strictly compact subsets $K_o \subset \Omega$ and $Q_\emptyset \subset E$ such that p is the semi-norm

$$p(g) = \sup_{0 \le j \le l} \sup_{x \in K_o} \sup_{y \in Q_o^j} \| f^{(j)}(x)[y] \| .$$

Then $K := K_o + Q_o$ is a strictly compact subset of E and by property (a) we can find a Banach disc B in E such that K is compact in E_B and such that the id_{E_B} can be approximated uniformly on K by continuous linear operators on E_B which have finite rank. Hence we can find a sequence $(u_m)_{m \in \mathbb{N}}$ in $E'_B \otimes E_B$ such that

$$\sup_{x \in K} \| u_m(x) - x \|_{E_B} \le \frac{1}{m} .$$

Then the proof of Lemma 3.3 in Meise [13] shows that for a given $\varepsilon > 0$ there is $s \in \mathbb{N}$ such that for $u := u_s$ the following holds:

(1) $u(K_o) \subset \Omega_B$.

(2) There exists an open neighbourhood ω of K_o in E_B such that $(f|\Omega_B) \circ u$ is in $C^n_{co}(\omega,F)$ and such that

$$\sup_{0 \le j \le l} \sup_{x \in K_o} \sup_{y \in Q_o^j} \| f^{(j)}(x)[y] - ((f|\Omega_B) \circ u)^{(j)}(x)[y] \| \le \varepsilon .$$

Now we define $E_o := \operatorname{Im} u$, $\Omega_o := \Omega \cap E_o = \Omega_B \cap E_o$ and $f_o := f|\Omega_o$. Then f_o belongs to $C^n_{co}(\Omega_o,F) = C^n_{co}(\Omega_o) \check{\otimes}_\varepsilon F$. Since the polynomials are dense in $C^n_{co}(\Omega_o)$ we can find because of (1) and $h_o \in \mathcal{P}(E_o) \otimes F = \mathcal{P}_f(E_o) \otimes F$ such that

$$\sup_{0 \le j \le l} \sup_{x \in u(K_o)} \sup_{y \in u(Q_o)^j} \| f_o(x)[y] - h_o(x)[y] \| \le \varepsilon .$$

From this and (2) it follows that $h := h_o \circ u \in \mathcal{P}_f(E_B) \otimes F$ has the property

(3) $$\sup_{0 \le j \le l} \sup_{x \in K_o} \sup_{y \in Q_o^j} \| f^{(j)}(x)[y] - h^{(j)}(x)[y] \| \le 2\varepsilon .$$

Now we remark that the transpose of $j_B : E_B \to (E,\tau)$ has $\lambda(E'_B,E_B)$-dense range. Hence it follows from the definition of a finite polynomial that there is $k \in \mathcal{P}_f(E,\tau) \otimes F$ such that

$$\sup_{0\le j\le 1}\ \sup_{x\in K_o}\ \sup_{y\in Q_o^j} \|h^{(j)}(x)[y] - k^{(j)}(x)[y]\| \le \epsilon .$$

From this and (3) we finally get $p(f-k) \le 3\epsilon$, which finishes the proof of the lemma.

4.3 Theorem. Let (E,τ) be a l.c. space in which any compact set is very compact and denote by E_c the space E endowed with the compact bornology of (E,τ). Let $\Omega \neq \emptyset$ be an open subset of (E,τ) and denote by Ω_c the M-open subset Ω of E_c. Then $P_f(E) \otimes F$ and hence $C_b^\infty(\Omega,F)$, $C_{co}^\infty(\Omega,F)$ are dense in $C_{S,co}^n(\Omega_c,F)$ for any $n \in \mathbb{N}_\infty$ and any l.c. space F, provided E_c has property (a).

Proof. It is obvious from the definition of E_c that E_c is regular and that the l.c. topology τ has the properties required in Lemma 4.2. Hence it follows from this lemma that $P_f(E,\tau)\otimes F$ is dense in $C_{S,co}^n(\Omega_c,F)$. Since $P_f(E,\tau)\otimes F$ is contained in $C_b^\infty(\Omega,F)$ the proof of the theorem is complete.

4.4 Corollary. Let (E,τ) and E_c be as in 4.3 and assume that E_c has property (a). Let $\Omega \neq \emptyset$ be an open subset of E, F a l.c. space and $n \in \mathbb{N}_\infty$. Denote the completion of F by $\hat{F}$. Then $C_{S,co}^n(\Omega_c,\hat{F})$ is the completion of $C_{co}^n(\Omega,F)$.

Proof. In 2.9 a) we have shown that $C_{co}^n(\Omega,F)$ is a topological subspace of $C_{S,co}^n(\Omega_c,F)$ which is of course a topological subspace of $C_{S,co}^n(\Omega_c,\hat{F})$. By 4.3 $P_f(E) \otimes \hat{F}$ is dense in $C_{S,co}^n(\Omega_c,\hat{F})$, and it easy to see that $P_f(E) \otimes F$ is dense in $P_f(E) \otimes \hat{F}$ for the topology induced by $C_{S,co}^n(\Omega_c,\hat{F})$. Hence we have shown that $C_{co}^n(\Omega,F)$ is a dense topological subspace of $C_{S,co}^n(\Omega_c,F)$ which is complete by 3.1.

Remark. If E is a quasi-complete l.c. space for which E_c' is nuclear, then E satisfies the hypotheses of 4.3 and 4.4.

4.5 Remark. a) Assume that the complete b.v. space E satisfies the hypotheses of Lemma 4.2. Then $C_{S,co}^n(\Omega)$ has the approximation property (a.p.). This is a consequence of 4.2, 3.5 and a characterization of the a.p. by means of the ϵ-product (cf. Schwartz [21]). By similar arguments one can improve this result by using the notion of Silva approximation property (introduced by Paques [17]) instead of property (a).

b) Further density results for polynomial algebras in (respectively submoduls of) $C^n_{co}(\Omega,F)$ are given in the articles of Nachbin [15] and [16], which also contain further references.

REFERENCES

[1] Aron, R., Compact polynomials and compact differentiable mappings between Banach spaces, in "Seminaire Pierre Lelong (Analyse) Année 1974/75", Springer Lectures Notes Math. 524 (1976), p. 213-222.

[2] Aron, R., Schottenloher, M., Compact holomorphic mappings on Banach spaces and the approximation property, J. Functional Analysis 21 (1976), 7-30.

[3] Bierstedt, K.-D., Meise, R., Lokalkonvexe Unterräume in topologischen Vektorräumen und das ε-Produkt, manuscripta math. 8 (1973), 143-172.

[4] Blasco, J.L., Two problems on k_R-spaces, Acta Math. Sci. Hung. 32 (1978), 27-30.

[5] Gordón Bombal, F., Llavona González J.L., La propriedad de aproximación en espacios de funciones diferenciables, Revista Acad. Ci. Madrid 70 (1976), 727-741.

[6] Colombeau, J.F., Sur les applications differentiables et analytiques au sens de J. Sebastião e Silva, to appear in Port. Math.

[7] Colombeau, J.F., Spaces of C^∞-mappings in infinitely many dimensions and applications, preprint Bordeaux 1977.

[8] Colombeau, J.F., Meise, R., Perrot, B., A density result in spaces of Silva holomorphic mappings, Pacific J. Math., 84 (1979), 35-42.

[9] Gramsch, B., Zum Einbettungssatz von Rellich bei Sobolevräumen, Math. Z. 106, 81-87 (1968).

[10] Hogbe-Nlend, H., Bornologies and functional analysis, North-Holland Mathematics Studies 26 (1977).

[11] Horváth, J., Topological vector spaces and distributions I, Reading, Mass., Addison Wesley 1965.

[12] Keller, H.H., Differential calculus in locally convex spaces, Springer Lecture Notes in Math. 417 (1974).

[13] Meise, R., Spaces of differentiable functions and the approximation property, p. 263-307 in "Approximation Theory and Functional Analysis" J.B. Prolla (Editor), North Holland Mathematics Studies (1979).

[14] Meise, R., Nicht-Nuklearität von Räumen beliebig oft differenzierbarer Funktionen, to appear in Arch. Math. (1980).

[15] Nachbin, L., Sur la densité des sous-algèbres polynomiales d'applications continument différentiables, in Seminaire P. Lelong, H. Skoda (Analyse) Année 1976/77, Springer Lect. Math. 694, 196-203.

[16] Nachbin, L., On the closure of modules of continuously differentiable mappings, preprint, to appear in Rendiconti del Seminário Matemático dell'Università di Padova 59 (1979).

[17] Paques, O.W., The approximation property for certain spaces of holomorphic mappings, p.351-370 in "Approximation Theory and Functional Analysis", J.B. Prolla (Editor), North Holland Mathematics Studies (1979).

[18] Pietsch, A., Nuclear locally convex spaces, Ergebnisse der Math. 66, Springer (1972).

[19] Prolla, J.B., Guerreiro, C.S., An extension of Nachbin's theorem to differentiable functions on Banach spaces with the approximation property, Ark. Mat. 14, 251-258 (1976).

[20] Schaefer, H.H., Topological Vector Spaces, Springer 1970.

[21] Schwartz, L., Théorie des distributions à valeurs vectorielles I, Ann. Inst. Fourier 7, 1-142 (1957).

[22] Sebastião e Silva, J., Le calcul différentiel et intégral dans les espaces localement convexes réels ou complexes, Att. Acad. Naz. Lincei 20, 743-750 (1956) and 21, 40-46 (1956).

[23] Sebastião e Silva, J., Les espaces à bornés et la notion de fonction différentiable, p. 57-61 in "Colloque d'Analyse Fonctionelle CBRM" (1961).

UNIFORM MEASURES AND COSAKS SPACES

J.B. Cooper and W. Schachermayer

Institut für Mathematik
Johannes Kepler-Universität
A-4045 Linz-Auhof Austria

INTRODUCTION

In this note we present a survey of results on uniform measures, based on the theory of Saks spaces and the dual concept - that of CoSaks spaces. The main result is that in the space of uniform measures, weak compactness is equivalent to compactness with respect to a natural strong topology. In the two extreme cases where the uniform space is uniformly discrete resp. where it is compact this theorem reduces to two familiar facts, namely Schur's Lemma on compactness in ℓ^1, respectively the fact that on the dual of a Banach space E, the $\sigma(E',E)$-compact sets coincide with those sets which are compact for $\tau_c(E',E)$, the topology of uniform convergence on the compact sets of E, and in fact our proofs use essentially the fact that the general case may be regarded as a combination of these two extreme cases. The theory of uniform measures has been developed mainly bei Fedorova [16] Berezanskii [3], Deaibes [11], Azzam [2], Frolik [19] and Pachl [34] and the main result on compactness is due to Pachl whose proof is complicated and technical. In the topological case analogous results were obtained by Berruyer and Ivol [4], Buchwalter [6], Dudley [14], Leger and Soury [29], Rome [38] and Wheeler [42]. Here the compactness result was obtained by Rome and Wheeler using partition of unity techniques. In the uniform case no such technique is available. Using our approach, a proof based on a gliding hump technique which reduces the result to Shur's Lemma presents itself very naturally. Here the notion of "uniform Lipschitz-tightness" is of central importance. It replaces the classical notion of uniform tightness for topological measure theory. The proof given is considerably simpler and more natural than that of Pachl.

We have taken the opportunity of presenting a sketch of a systematic development of the theory of uniform measures from the point of view of CoSaks spaces which we believe to be the natural and correct framework.

The general line of attack in this paper is to do the analysis in metric spaces and then lift to uniform spaces using formal manipulations with projective and inductive limits. For this reason we have been unable to resist the temptation of showing in the first two sections how the basic concepts of Lipschitz functions and uniform measures arise naturally from categorical considerations and how this serves to simplify and illuminate them. Readers with a distaste for category theory can skip these two sections after familiarising themselves with the notation introduced there.

The idea for writing this paper came from a very stimulating talk on the work of the Prague uniform space group given by Z. Frolik in Linz, Summer 1978. A crucial role is played by the concept of a "compactology" which was introduced by Waelbroeck and systematically studied by Buchwalter and his coworkers.

§1. LIPSCHITZ SPACES

We begin with a brief exposition of results on Lipschitz functions. They are essentially due to Arens and Eells (who showed that every metric space can be isometrically embedded into a Banach space) resp. de Leeuw (who instigated the theory of Banach spaces of Lipschitz functions). In the following we shall combine these two treatments by showing that the construction of Arens and Eells can be made in a categorical fashion and that then many of the results of de Leeuw (and in particular, those that we shall require) follow from very general considerations.

We consider the category M_o whose objects are metric spaces (X,d) with base point x_o and radius ≤ 1 (i.e. $\sup\{d(x,x_o): x \in X\} \leq 1$). The morphisms $f: (X,x_o) \to (Y,y_o)$ are contractions which respect the base point. If f is a Lipschitz mapping between metric spaces, $\mathrm{Lip}(f)$ denotes the Lipschitz constant of f (i.e. $\sup\{d_1(f(x),f(y))/d(x,y): x,y \in X,\ x \neq y\}$).

BAN_1 denotes the category of real Banach spaces with linear contractions as morphisms. There is a natural forgetful functor

$$O: BAN_1 \to M_o$$

where if E is an object of BAN_1, OE is the unit ball of E with O as base point.

Now it follows from Freyd's adjoint functor theorem (see, e.g. Mac Lane [30]) that O has a left adjoint which we shall denote by Λ_{BAN_1}. This means that if (X,x_o) is an object of M_o, there is a Banach space $\Lambda_{BAN_1}(X)$ and a contraction from (X,x_o) into $\Lambda_{BAN_1}(X)$ with the following universal property:

> if $f: (X,x_o) \to E$ (E a Banach space) is a Lipschitz function with $\mathrm{Lip}(f) \leq 1$ and $f(x_o) = 0$, then there is a unique linear contraction $\tilde{f}: \Lambda_{BAN_1}(X) \to E$ so that the following diagram commutes:

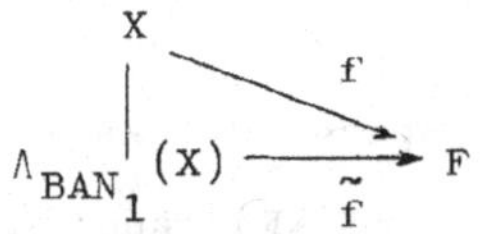

$\Lambda_{BAN_1}(X)$ is called the <u>free Banach space over</u> X. One can deduce from the above universal property that the mapping from X into $\Lambda_{BAN_1}(X)$ is isometric so that we can (and do) regard X as a subspace of $\Lambda_{BAN_1}(X)$. For if x, x' are points in X with $x \neq x'$, then the function

$$f: y \to d(y,x) - d(x,x_o)$$

is Lipschitz with $f(x_o) = 0$, $\mathrm{Lip}(f) \leq 1$. The extension $\tilde{f}$ of f

is an element of $\Lambda_{BAN_1}(X)'$ with $\|\tilde{f}\| \leq 1$. Also

$$\tilde{f}(x'-x) = f(x') - f(x) = d(x',x) - d(x,x_o) - d(x,x) + d(x,x_o) = d(x',x)$$

and so $\|x'-x\| \geq d(x',x)$.

Now if X is an object of M_o we construct a Banach space $Lip_o(X)$ as follows: the elements of $Lip_o(X)$ are Lipschitz functions from X into $\mathbb{R}$ with the property that $f(x_o) = 0$. The norm is the Lipschitz norm

$$\|f\|_{L_o} : \sup\{\frac{|f(x)-f(y)|}{d(x,y)} : x,y \in X, x \neq y\}$$

i.e. $\|f\|_{L_o} = Lip(f)$ - which is in this case a norm).

If we apply the above universal property to the Banach space $\mathbb{R}$ as target we see that $Lip_o(X)$ is naturally isometric to $\Lambda_{BAN_1}(X)'$, the dual of $\Lambda_{BAN_1}(X)$, the isomorphism being the mapping $f \to \tilde{f}$ (the extension of f to $\Lambda_{BAN_1}(X)$). For the mapping is a linear isomorphism. On the other hand, it is norm decreasing since the extension does not increase the Lipschitz constant. Conversely its inverse - the restriction operator - is trivially norm-decreasing. With this hindsight it is then easy to see how to construct $\Lambda_{BAN_1}(X)$ without having recourse to the Freyd theorem. We can embed X in $Lip_o(X)'$ in the usual manner and the above consideration whows that this embedding is isometric. Then $\Lambda_{BAN_1}(X)$ has the closed linear span of X in $Lip(X)'$ as a model.

More generally, we can define in the obvious way a Banach space $Lip_o(X;F)$ for each Banach space F. As above it follows from the universal property that $L(\Lambda_{BAN_1}(X),F)$ and $Lip_o(X;F)$ are naturally isometric for any Banach space. This implies the result that $Lip_o(X;F)$ is a dual space if F is (cf. de Leeuw [28], Johnson [26]). Now we reduce the case of metric space without base point to the above by the following trick. First we consider metric spaces with diameter ≤ 2 (of course every metric space is uniformly equivalent to such a space). Then we add a base point x_o and extend the metric by defining $d(x_o,x) = 1$ $(x \in X)$. We denote this object of M_o by $\tilde{X}$. Then if $f: X \to R$ is Lipschitz with constant K the extension of f to $\tilde{X}$ obtained by putting $f(x_o) = 0$ is Lipschitz with constant K if and only if $|f| \leq K$ on X. Hence if $\mathrm{Lip}(X)$ denotes the Banach space of Lipschitz functions $f: X \to \mathbb{R}$ with the norm

$$\| \ \|_L : f \to \max\{\sup\{|f(x)|: x \in X\}, \mathrm{Lip}(f)\}$$

then $\mathrm{Lip}(X)$ and $\mathrm{Lip}_o(\tilde{X})$ are isometric. Similarly, if F is a Banach space, $\mathrm{Lip}(X;F)$ (obvious definition) and $\mathrm{Lip}_o(\tilde{X};F)$ are isometric. Hence we obtain from the above that each metric space X with diameter ≤ 2 can be embedded isometrically and functorially in a Banach space $\Lambda_{BAN_1}(X)$ so that each Lipschitz function $f: X \to F$ (F a Banach space) with $\mathrm{Lip}(f) \leq 1$ and $\|f\|_\infty \leq 1$ i.e. $\|f\|_L \leq 1$ has a unique extension to a linear contraction $\tilde{f}: \Lambda_{BAN_1}(X) \to F$. Then we have naturalisometries $L(\Lambda_{BAN_1}(X);F) = \mathrm{Lip}(X;F)$.

§2. FREE TOPOLOGICAL VECTOR SPACES

Before constructing the space of uniform measures by duality in the next section, we show here how its existence is ensured by the Freyd adjoint functor since it is the solution of a universal problem, that of linearisation of bounded uniformly continuous functions (a fact which has been observed by several authors).

Again once one is assured of the existence of such a solution, it follows easily how one can construct it by duality. One sees thus how this construction fits into a general scheme of linearisation which has attracted some attention recently (see Dostal [13], John [24], Ptak [35], Raikov [37], Tomasek [41] and references there) and which include the construction of spaces of distributions, analytic functionals and various spaces of measures.

Consider the following categories:

a1) CREG - the completely regular spaces;

a2) UNIF - uniform spaces;

b1) LCS - locally convex spaces;

b2) WLCS - locally convex spaces with the weak topology;

b3) SS - Saks spaces (see Cooper [9] or §3 below);

b4) W - Waelbroeck spaces (see Buchwalter [5] or Cigler [8]).

Then there are natural forgetful functors from each of the categories in the b) list into the categories in the a) list (in the case of b3) and b4) we restrict to the unit ball before forgetting the linear structure). Each of these functors has (once more by Freyd's theorem) a left adjoint. We denote them by

$$\Lambda_{LCS}, \Lambda_{WLCS}, \Lambda_{SS}, \Lambda_{W} \quad \text{(for the functors on CREG)}$$

resp.

$$\Lambda^U_{LCS}, \Lambda^U_{WLCS}, \Lambda^U_{SS}, \Lambda^U_{W} \quad \text{(for the functors on UNIF).}$$

A hat on the functor symbol (e.g. $\hat{\Lambda}_{LCS}$) will denote the composition of $\Lambda_?$ with the corresponding completion functor (in the b) list). For each of these Λ-functors, there are natural mappings $X \to \Lambda(X)$. It follows from the universal property that this mapping is an isomorphic embedding onto a closed subspace for the functors Λ_{LCS}, Λ_{SS}, Λ_{WLCS}, Λ_W, Λ^U_{LCS}, Λ^U_{SS} as we shall now prove:

We begin with the topological case. The mapping is in each case injective since the bounded continuous functions separate the points of X. Hence we can regard X as a (set theoretical) subspace of $\Lambda_{LCS}(X)$ etc.

We now show that if A is closed in X then $A = \tilde{A} \cap X$ there $\tilde{A}$ is closed in $\Lambda_{LCS}(X)$. From this it follows that $X \to \Lambda_{LCS}(X)$ is a topological embedding. Since X is completely regular, there is a family $M \subseteq C^b(X)$ so that $A = \bigcap_{f \in M} f^{-1}(0)$. Then $\tilde{A} = \bigcap_{f \in M} \tilde{f}^{-1}(0)$ is the required set.

To show that X is closed in $\Lambda_{LCS}(X)$ we first note the following consequences of the universal property:

1) X is linearly independent in $\Lambda_{LCS}(X)$ (for if $\{x_1,\ldots,x_n\}$ is a finite sequences of distinct elements of X, there is a continuous

$f: X \to \mathbb{R}$ with $f(x_1) = 1$, $f(x_i) = 0$ $(i=2,\dots,n)$. Then $\tilde{f}$ is an element of $\Lambda_{LCS}(X)'$ with $\tilde{f}(x_1) = 1$, $\tilde{f}(x_i) = 0$ $(i=2,\dots,n)$ q.e.d.

2) $\Lambda_{LCS}(X)$ is the span of X. For by the uniqueness part of the universal property, the span $L(X)$ is dense in $\Lambda_{LCS}(X)$. On the other hand, the extension $\tilde{I}$ of the natural injection $I: X \to L(X)$ ($L(X)$ with the topology induced from $\Lambda_{LCS}(X)$) is a continuous linear mapping from $\Lambda_{LCS}(X)$ onto $L(X)$ which is the identity when restricted to the dense subspace $L(X)$. Hence it is the identity on $\Lambda_{LCS}(X)$ i.e. $L(X) = \Lambda_{LCS}(X)$.

We now show X is closed in $\Lambda_{LCS}(X)$. The injection $\bar{X} \to \Lambda_{LCS}(X)$ ($\bar{X}$ the closure of X in $\Lambda_{LCS}(X)$) satisfies the universal property for $\bar{X}$ and so we have $\Lambda_{LCS}(X) = \Lambda_{LCS}(\bar{X})$. But we have seen that X is a basis for $\Lambda_{LCS}(X)$ and of course, $\bar{X}$ is also a basis. This implies $X = \bar{X}$.

Almost exactly the same proof shows that X is embedded as a closed subspace of $\Lambda_{WLCS}(X)$ and $\Lambda_{SS}(X)$. On the other hand, X is embedded as a topological subspace of $\Lambda_W(X)$, which will be closed only if X is compact.

We now consider the Λ^U functors. Here the proofs that $X \to \Lambda^U_{LCS}(X)$ is an injection and that X is a basis for $\Lambda^U_{LCS}(X)$ are exactly as above. We now show that the mapping is a uniform isomorphism. For this it suffices to show that if a uniformly bounded family $\{f_\alpha\}_{\alpha \in A}$ from X into $\mathbb{R}$ is uniformly equicontinuous for the original structure on X then it is uniformly equicontinuous for the structure induced from $\Lambda^U_{LCS}(X)$. But $\{f_\alpha\}$ induces in a natural way a mapping

$$f: X \to (f_\alpha(x))_{\alpha \in A}$$

from X into $\ell^\infty(A)$ and the mapping is uniformly continuous if and only if $\{f_\alpha\}_{\alpha \in A}$ is uniformly equicontinuous. Hence if this is the case, f lifts to a continuous linear mapping $\tilde{f}$ from $\Lambda^U_{LCS}(X)$ into $\ell^\infty(A)$ and so the family $\{\tilde{f}_\alpha\}$ of its components is equicontinuous on $\Lambda^U_{LCS}(X)$. Hence the restrictions to X are uniformly equicontinuous q.e.d.

Exactly the same proof shows that X is uniformly isomorphic to a closed subspace of $\Lambda^U_{SS}(X)$. On the other hand the mappings $X \to \Lambda^U_{WLCS}(X)$, $X \to \Lambda^U_W(X)$ are not isomorphisms since $\Lambda^U_{WLCS}(X)$ induces the weak uniformity on X (i.e. that defined by the uniformly continuous R-valued functions on X) and $\Lambda^U_X(X)$ induces the precompact uniformity.

In particular, we can regard a completely regular space X as a subspace of $\hat{\Lambda}_{LCS}(X)$, $\hat{\Lambda}_{WLCS}(X)$, $\hat{\Lambda}_{SS}(X)$, $\Lambda_W(X)$.

Note that the closure of X in these spaces is

θX (the c-repletion - cf. Buchwalter [5]);

υX (the realcompletion or real compactification - cf. e.g. Buchwalter [5]);

cX (the topological completion c.f. Engelking [15]);

βX (the Stone-Cech compactification c.f. Engelking [15]); respectively.

For readers who are unhappy at the use of Freyd's theorem we sketch briefly how these adjoints can be constructed directly. To be concrete, we construct Λ_{LCS}. If X is a completely regular space, we consider $\Lambda(X)$, the free vector space over X (i.e. the set of formal linear combinations of elements of X) and give it one of the following two (equivalent) structures:

a) the finest locally convex structure so that $X \to \Lambda(X)$ is continuous;

b) the projective structure induced by all mapping

$$\tilde{f}: \Lambda(X) \to E$$

where $\tilde{f}$ is the canonical extension of a continuous $f: X \to E$ (E a locally convex space).

The equivalence of these two structures is ensured by the fact that both by their very definitions, have the required universal property and this uniquely determines the topology of $\Lambda_{LCS}(X)$.

Now there exists a duality theory (for example, the duality between locally convex spaces and spaces with convex compactologies - see Buchwalter [5]) for each of the categories in the b) list and it follows once again from the universal properties that we have:

$$\hat{\Lambda}_{LCS}(X)' = C(X)$$
$$\hat{\Lambda}_{WLCS}(X)' = C(X)$$
$$\hat{\Lambda}_{SS}(X)' = C^b(X)$$
$$\Lambda_W(X)' = C^b(X)$$
$$\hat{\Lambda}^U_{LCS}(X)' = U(X)$$
$$\hat{\Lambda}^U_{WLCS}(X)' = U(X)$$
$$\hat{\Lambda}^U_{SS}(X)' = U^b(X)$$
$$\Lambda^U_W(X)' = U^b(X)$$

where $C(X)$ resp. $U(X)$) denotes the space of continuous (resp. uniformly continuous) real valued functions and a superscript b beans "bounded". From this it easy to deduce the (more usual) definitions of the Λ-spaces as the duals of spaces of (uniformly) continuous functions with suitable structure. In particular it is now clear how we must define the space $\hat{\Lambda}^U_{SS}(X)$ of uniform measures on a uniform space X by duality and this is what we shall do in the next section.

§3. COSAKS SPACES

In this section we recall some definitions and constructions from Cooper [9]. A Saks space is a triple $(E, \| \|, \tau)$ where $(E, \| \|)$ is a normed space and τ is a locally convex topology on E so that $\bigcirc E$, the unit ball, is τ-closed and bounded. The Saks spaces form a category when we define morphisms from $(E, \| \|, \tau)$ into $(F, \| \|_1, \tau_1)$ to be linear norm contraction T so that $T|_{\bigcirc E}$ is τ-τ_1 continuous. The category of Saks spaces is complete and cocomplete i.e. possesses products, sums, subspaces and quotients. If $(E, \| \|, \tau)$ is a Saks space, we define a new locally convex topology, the mixed topology

$\gamma(\|\ \|,\tau)$ (or γ for short) on E to be the finest locally convex topology on E which agrees with τ on $\bigcirc E$. Then $(E,\|\cdot\|,\gamma)$ is also a Saks-space and we call it the <u>fine Saks space</u> associated with $(E,\|\cdot\|,\tau)$. A Saks-space $(E,\|\cdot\|,\tau)$ is called <u>fine</u> if $\tau = \gamma(\|\cdot\|,\tau)$. Note that every Saks space is isomorphic to the fine Saks-space associated to it, whence in every isomorphism-class of Saks spaces there is exactly one fine Saks-space. Also a linear map $T:(E,\|\cdot\|,\tau) \to (F,\|\cdot\|,\tau_1)$ is a Saks-morphism iff it is a norm-contraction and $\gamma(\|\cdot\|,\tau)$ - $\gamma(\|\cdot\|_1,\tau_1)$-continuous.

A Saks-space $(E,\|\cdot\|,\tau)$ is complete if $\bigcirc E$ is τ-complete or equivalently if (E,γ) is a complete locally convex space.

We define the dual space E'_γ of a Saks-space $(E,\|\cdot\|,\tau)$ to be $(E,\gamma)'$, which is a Banach-space with respect to the dual norm $\|\cdot\|'$ of E. In fact we have an additional structure on E'_γ, namely the bornology $\mathfrak{B}$ of τ-equicontinuous sets and these sets are relatively compact with respect to $\sigma(E'_\gamma,E)$. $\tilde{\mathfrak{B}}$, the bornology of $\gamma(\|\cdot\|,\tau)$-equi-continuous sets is what we shall call the <u>$\|\cdot\|$-saturation</u> of $\mathfrak{B}$, namely those balls C in E'_γ such that for every $\epsilon > 0$ there is $B \in \mathfrak{B}$ with $C \subseteq B + \epsilon\bigcirc E'$.

Now if $(E,\|\cdot\|,\gamma)$ is a complete fine Saks space then by Grothendieck's completeness theorem, c.f. Schaefer [40], we can recover $(E,\|\cdot\|,\gamma)$ from $(E'_\gamma,\|\cdot\|',\tilde{\mathfrak{B}},\sigma(E',E))$ as the set of linear functionals on E'_γ such that the restriction to every member of $\tilde{\mathfrak{B}}$ is $\sigma(E'_\gamma,E)$-continuous, equipped with the norm dual to $\|\cdot\|'$ and the topology of uniform convergence on $\tilde{\mathfrak{B}}$. We define the topology $\tilde{\sigma}$ on E' to be the finest locally convex topology that agrees with $\sigma(E'_\gamma,E)$ on the members of $\tilde{\mathfrak{B}}$. Then $E = (E',\tilde{\sigma})'$. This motivates the

<u>Definition</u>. A quadruple $(\mathcal{E},\|\cdot\|,\tilde{\mathfrak{B}},\tilde{\sigma})$ is called a <u>CoSaks space</u> if $(\mathcal{E},\|\cdot\|)$ is a Banach-sapce, $\tilde{\mathfrak{B}}$ is a bornology of $\|\cdot\|$-bounded sets that is $\|\cdot\|$-saturated (i.e. if $C \subseteq \mathcal{E}$ is a ball such that $\forall\, \epsilon > 0$ $C \subseteq B + \epsilon\bigcirc\mathcal{E}$ for some $B \in \tilde{\mathfrak{B}}$, then $C \in \tilde{\mathfrak{B}}$) and $\tilde{\sigma}$ is a locally

convex Hausdorff-topology for which OE is closed and for which the members of $\tilde{\mathfrak{B}}$ are relatively compact in $\mathcal{E}$, and which is the finest locally convex topology whose traces on the members of $\tilde{\mathfrak{B}}$ coincide with the traces of $\tilde{\sigma}$

$$T: (\mathcal{E},\|\cdot\|,\tilde{\mathfrak{B}},\tilde{\sigma}) \to (\mathfrak{F},\|\cdot\|_1,\tilde{\mathfrak{B}}_1,\tilde{\sigma}_1)$$

is a CoSaks morphism if it is a linear norm-contraction, a bornological morphism with respect to $\tilde{\mathfrak{B}}$ and $\tilde{\mathfrak{B}}_1$ and $\tilde{\sigma}$ - $\tilde{\sigma}_1$ continuous: It is clear from the above discussion that the definition is chosen so that the following proposition holds.

<u>Proposition</u>. The category of complete fine Saks spaces is dual to the category of CoSaks spaces.

<u>Remark</u>. The reader will perhaps have been irritated by the lack of symmetry in the definitions of Saks and CoSaks spaces. The reason for this lies in the fact that, in order to conform with the notation of [9], we have been forced to distinguish between a Saks space and its associated fine space althrough they are indistinguishable from the categorial point of view whereas in the definition of CoSaks spaces we have singled out one particular member of each isomorphism class.

We also note the following simple result:

<u>Lemma</u>. OE is σ-bounded, and so $\tilde{\sigma}$ is coarser than the $\|\cdot\|$-topology on $\mathcal{E}$.

<u>Proof</u>. If there exists a $\tilde{\sigma}$-bounded squence $\{x_n\}_{n=1}^{\infty}$ in OE then it is easy to see that for some increasing subsequence $\{n_k\}_{k=1}^{\infty}$ the sequence $\{k^{-1}x_{n_k}\}_{k=1}^{\infty}$ is not $\tilde{\sigma}$-bounded either. But as $\{k^{-1}x_{n_k}\}_{k=1}^{\infty}$ is $\|\cdot\|$-precompact and is therefore contained in some closed ball B of saturated bornology $\tilde{\mathfrak{B}}$ this contradicts the assumption that $\tilde{\sigma}$ induces a compact topology on every closed $B \in \tilde{\mathfrak{B}}$.

In the following proposition we examine further the relation between the bornologies of τ-equicontinuous sets and γ-equicontinuous sets.

<u>Proposition</u>. Let $(E,\|\cdot\|,\tau)$ be a complete Saks-space, $\gamma = \gamma[\|\cdot\|,\tau]$ the associated mixed topology. Let $(\mathcal{E},\|\cdot\|,\tilde{\mathfrak{B}},\tilde{\sigma})$ be the dual CoSaks-space and denote by $\mathfrak{B}$ the bornology of τ-equicontinuous sets in $\mathcal{E}$.

Then for a subset H of E the following are equivalent:

(i) H is $\tilde{\sigma}$-equicontinuous;

(ii) H is $\tilde{\sigma}$-equicontinuous on every member of $\mathfrak{B}$;

(iii) H is $\|\cdot\|$-bounded and $\tilde{\sigma}$-equicontinuous on every member of $\mathfrak{B}$;

(iv) H is relatively γ-compact;

(iv)' H is relatively countably γ-compact;

(iv)" H is γ-precompact;

(v) H is $\|\cdot\|$-bounded and relatively τ-compact;

(v)' H is $\|\cdot\|$-bounded and relatively countably τ-compact;

(v)" H is $\|\cdot\|$-bounded and τ-precompact.

<u>Proof</u>. (i) $\Leftrightarrow$ (ii): follows from the definition of $\tilde{\sigma}$ and (iv) $\Leftrightarrow$ (iv)' $\Leftrightarrow$ (iv)" $\Leftrightarrow$ (v) $\Leftrightarrow$ (v)' $\Leftrightarrow$ (v)" from the fact that (E,γ) is complete, γ-bounded sets are norm-bounded and that, on bounded subsets of E, γ coincides with τ.

(ii) $\Rightarrow$ (iii): We only have to show that H is $\|\cdot\|$-bounded and this follows from the uniform boundedness theorem since $\tilde{\mathfrak{B}}$ covers $\mathcal{E}$.

(iii) $\Rightarrow$ (v)": Fix a $\tilde{\sigma}$-compact $B \in \mathfrak{B}$. If we consider H as a subset of $C(B)$, the Banach space of continuous functions on B, then by Ascoli's theorem H is relatively compact in $C(B)$. Since this holds for each $B \in \mathfrak{B}$, H is τ-precompact.

(iv) $\Rightarrow$ (ii): If we now fix a $\tilde{\sigma}$-compact $B \in \tilde{\mathfrak{B}}$, then H is relatively compact in $C(B)$ whence equicontinuous on B again by Ascoli's theorem.

Let us illustrate the situation:

<u>Examples</u>

1) Let (Ω,Σ,μ) be a finite measure-space. $(L^\infty(\mu),\|\cdot\|_\infty,\|\cdot\|_1)$ is a

complete Saks-space and γ is the Mackey-topology with respect to the duality of $L^\infty(\mu)$ and $L^1(\mu)$.

The dual fine CoSaks space $(\mathcal{E},\|\cdot\|,\tilde{\mathfrak{B}},\tilde{\sigma})$ is the Banach-space $(L^1(\mu),\|\cdot\|_1)$ equipped with the bornology of relatively σ (L^1,L^∞)-compact balls and $\tilde{\sigma}$ is the topology of uniform convergence on the $\|\cdot\|_\infty$-bounded and $\|\cdot\|_1$-compact subsets of $L^\infty(\mu)$. Finally the bornology of τ-equicontinuous sets consists of the $\|\cdot\|_\infty$-bounded balls in $L^1(\mu)$.

2) If $E = F'$ is a dual Banach-space, then $(E,\|\cdot\|,\sigma(F',F))$ is a complete Saks-space and $\gamma(\|\cdot\|,\sigma(F',F))$ is the topology of compact convergence. Whence the dual CoSaks space is the Banach-space $(F,\|\cdot\|)$ with the bornology of $\|\cdot\|$-relatively compact balls and σ is just the norm topology on F.

3) If S is a locally compact paracompact space, then $(C^b(S),\|\cdot\|_\infty,\beta)$ is a complete fine Saks-space. The dual CoSaks space consists of the Banach space $M^R(S)$ of Radon-measures on S equipped with the bornology of bounded, uniformly tight balls and σ is the topology of uniform convergence on the $\|\cdot\|_\infty$-bounded equicontinuous subsets of $C^b(S)$.

In this paper we will, in contrast to the last example, consider spaces of measures as Saks-spaces, defined as the dual of the space of bounded uniformly continuous functions with a suitable CoSaks structure.

More preciesely let X be a uniform space. $(\mathcal{U}^b(X),\|\ \|_\infty)$ denotes the Banach-space (even algebra) of real-valued bounded, uniformly continuous functions on X. We consider the family $\mathcal{H}$ of all absolutely convex, uniformly bounded uniformly equicontinuous (abbreviated ueb) subsets of $\mathcal{U}^b(X)$. If $\mathfrak{D}$ denotes the family of uniformly continuous pseudometrics on X then a bounded absolutely convex set H is in $\mathcal{H}$ if and only if H is pointwise dominated by some $d \in \mathfrak{D}$ (in this the sense that for some $K > 0$),

$$|f(x)-f(y)| \leq Kd(x,y)(x,y \in X,\ f \in H).$$

Nothe that this means that H factorises over the appropriate metric space X_d and forms a bounded subset of $Lip(X_d)$ there. Then $(U^b(X), \|\ \|_\infty, \mathcal{H}, \tilde{\sigma})$ is a CoSaks space where $\tilde{\sigma}$ is the finest locally convex topology that agrees on $\mathcal{H}$ with that of pointwise convergence on X.

For historical reasons, this topology is denoted by β_∞ (c.f. Rome [38], Wheeler [42]).

Now we define the space $M_u(X)$ of uniform measures on X to be the dual of this CoSaks space (so that it can also be regarded as the dual of the locally convex space $U^b(X), \beta_\infty)$).

By the above $M_u(X)$ has a natural fine Saks space structure $(M_u(X), \|\ \|, \gamma)$ where $\|\ \|$ is the dual norm to that of $U^b(X)$ and γ is the topology of uniform convergence on the sets of $\mathcal{H}$.

By the Gelfand Naimark theorem we can embedd X (topologically) in a compact space X (the Samuel compactification of X) so that $U^b(X)$ and $C(X)$ are naturally isometric by extension). Then every uniform measure can regarded as a Radon measure μ on X, and this identification is isometric i.e. the norm of μ in $M_u(X)$ is just the variation norm of μ.

§4. APPROXIMATION OF UNIFORMLY CONTINUOUS FUNCTIONS BY LIPSCHITZ FUNCTIONS

In this section (X,d) is a metric space which we assume for convenience to have diameter ≤ 2. $(U^b(X), \|\ \|_\infty)$ is the Banach space of bounded, uniformly continuous functions on X. It contains the space $Lip(X)$ of course. If $\alpha \in \mathbb{R}_+$ we put

$$L_\alpha(X) := \{f \in U^b(X) : \|f\|_\infty \leq 1 \text{ and } Lip\ f \leq \alpha\}$$

in particular $L_1(X) = O\ Lip\ (X)$.

<u>Lemma</u>. Let Y be a subset of X, g a function from Y into $[-1,1]$

with $\mathrm{Lip}(g) \leq \alpha$. Then there is an extension of g to a function $\tilde{g}: X \to [-1,1]$ with $\mathrm{Lip}\, \tilde{g} \leq \alpha$. Hence the restriction operator maps $\mathrm{O\,Lip}(X)$ onto $\mathrm{O\,Lip}(Y)$.

<u>Proof</u>. For $y \in Y$ define the function f_y on X

$$f_y(x) := g(y) - \alpha d(x,y)$$

and let $f(x) = -1 \vee \sup\{f_y(x): y \in Y\}$.

<u>Proposition</u>. Let H be a ueb subset of $\mathcal{U}^b(X)$. Then for each $\varepsilon > 0$ there is an $n \in N$ so that

$$H \subseteq nL_1(X) + \varepsilon O\mathcal{U}^b(X)$$

i.e. in the language of §3, H is in the saturation of $\{n\mathrm{O\,Lip}(X)\}$ in $\mathcal{U}^b(X)$.

<u>Proof</u>. We may and do suppose $H \subseteq O\mathcal{U}^b(X)$. Suppose that $0 < \varepsilon \leq 1$ and choose $\delta > 0$ so that $d(x,y) < \delta$ implies that $|f(x)-f(y)| < \varepsilon$ for each $f \in H$. Let $\{x_\alpha\}_{\alpha \in A}$ be a maximal family of points in X so that $d(x_\alpha, x_\beta) \geq \delta \cdot \varepsilon$ for $\alpha \neq \beta$. If $f \in H$, let g be the restriction of f to $Y = \{x_\alpha\}$. Then $\mathrm{Lip}(g) \leq 2/\delta$ since

$$\frac{|g(x_\alpha) - g(x_\beta)|}{d(x_\alpha, x_\beta)} \leq 2 \cdot \delta^{-1}.$$

Indeed if $d(x_\alpha, x_\beta) \geq \delta$ then the inequality is implied by $|g(x_\alpha)-g(x_\beta)| \leq 2$, while if $d(x_\alpha, x_\beta) < \delta$ we have $|g(x_\alpha)-g(x_\beta)| < \varepsilon$ and $d(x_\alpha, x_\beta) \geq \varepsilon \cdot \delta$.

By the preceding lemma, there is an extension $\tilde{g}$ of g to X with Lipschitz constant $\leq 2 \cdot \delta^{-1}$. Then $\|\tilde{g}-f\|_\infty < 3\varepsilon$. Indeed, for $x \in X$ there is, by the maximality of $\{x_\alpha\}$, an x_{α_0} so that $d(x_{\alpha_0}, x) < \varepsilon \cdot \delta$. Then

$$|\tilde{g}(x)-f(x)| \leq |\tilde{g}(x)-\tilde{g}(x_{\alpha_0})| + |\tilde{g}(x_{\alpha_0})-f(x_{\alpha_0})|$$
$$+ |f(x_{\alpha_0})-f(x)| < \varepsilon \cdot \delta \cdot 2\delta^{-1} + 0 + \varepsilon = 3\varepsilon.$$

<u>Corollary</u>. The sets $\{nL_1(X)\}_{n \in N}$ generate the ueb compactology of $\mathcal{U}^b(X)$ in the sense of the saturation introduced in §3.

<u>Corollary</u>. Let Y be the topology on $M(X)$ of uniform convergence

on the ueb sets. Then the restriction of Υ to $OM_u(X)$ coincides with the topology of uniform convergence on $L_1(X)$ and is therefore induced by a norm on $M_u(X)$.

§5. UNIFORM MEASURES ON METRIC SPACES

In this section we show that if X is a complete metric space then the uniform measures on X are just the Radon (or tight) measures. We begin with the remark that if X is a uniform space and we regard $M_u(X)$ as a subspace of $M(\check{X})$ (cf. §3) then $M_u(X)$ inherits the order structure of $M(\check{X})$. In fact, $M_u(X)$ is a band in $M(\check{X})$ as the next result shows. Note that since the topology of $M_u(X)$ is not nicely related to the order structure (it is not locally solid) this result is not as evident as one might expect and in fact the corresponding result for "free uniform measures" (cf. Pachl [33]) is false. The result is well-known but we are unable to indicate the original reference. Our proof is essentially that of Deaibes [11] (but we correct an error in his proof):

<u>Proposition</u>. Let $\mu \in M(\check{X})$. Then

$$\mu \in M_u(X) \text{ iff } \mu^+, \mu^- \in M_u(X).$$

<u>Proof</u>. We begin with the remark that if $\nu \in M(\check{X})$ then to show that $\nu \in M_u(X)$ it suffices to show that for every net $(f_\alpha)_{\alpha\in A}$ such that $\{f_\alpha\}$ is a ueb set and (f_α) tends to zero pointwise, $\lim_\alpha \nu(f_\alpha)=0$. Furthermore, by considering (f_α^+) and (f_α^-) one sees that it suffices to suppose that $f_\alpha \geq 0$. Hence consider such a net (f_α) with $0 \leq f_\alpha \leq 1/2$ so that $f_\alpha \to 0$ pointwise. We show that if $\mu \in M_u(X)$ then $\mu^+(f) \to 0$.

Since $U^b(X) = C(\check{X})$ we have the formula

$$\mu^+(1) = \sup \{\mu(g): g \in U^b(X),\ 0 \leq g \leq 1\}.$$

For $\varepsilon > 0$ choose $g \in U^b(X)$ with $0 \leq g \leq 1$ and

$$\mu^+(1) \geq \mu(g)-\varepsilon.$$

Note that this implies that

$$\mu^+(1-g) \le \epsilon \quad \text{and} \quad \mu^+(f) - |\mu(f)| \le \epsilon$$

for $0 \le f \le g$.

Now consider the family $(f_\alpha \wedge g)$. By assumption $\mu(f_\alpha \wedge g) \to 0$. We have the estimate $f_\alpha \le (f_\alpha \wedge g) + (1-g)$ (consider the cases $g \le 1/2$, $g \ge 1/2$). Hence

$$\mu^+(f_\alpha) \le \mu^+(f_\alpha \wedge g) + \mu^+(1-g) \le |\mu(f_\alpha \wedge g)| + \epsilon + \epsilon,$$

which is less than 3ϵ for large α. Thus $\mu^+(f_\alpha) \to 0$ q.e.d.

We now show that if X is a complete metric space then the uniform measures on X coincide with the Radon measures. This result is well known but again we have been unable to trace the original source. Again our proof is essentially that of Deaibes but we reproduce it since we encounter here in a natural way the concept of "Lipschitz-tightness" which will be essential in the sequel.

First some notation: if $K \subseteq X$ and $\alpha \in \mathbb{R}_+$, $L_{\alpha,K}$ will be the set of $f \in \mathcal{U}^b(X)$, with $\|f\|_\infty \le 1$, $\mathrm{Lip}(f) \le \alpha$ and $f \equiv 0$ on K. Also if $K \subseteq X$, $\eta > 0$ then $B(K,\eta) = \{y \in X: d(y,K) \le \eta\}$.

<u>Definition</u>. a) A measure $\mu \in M(\check{X})$ is called <u>Lipschitz-tight</u> (or <u>L-tight</u> for short) if

$$\lim_{K \in \mathcal{K}(X)} \sup \{ |\mu(f)| : f \in L_{1,K} \} = 0$$

where $\mathcal{K}(X)$ denotes the family of compact subsets of X directed by inclusion.

b) A subset $H \subseteq M(X)$ is called <u>uniformly Lipschitz-tigh</u> if the above limit holds uniformly in $\mu \in H$.

<u>Proposition</u>. For $\mu \in M(X)$ the following are equivalent:

1) $\mu \in M_u(X)$;

1)' $\mu^+, \mu^- \in M_u(X)$;

2) μ is L-tight;

2)' μ^+, μ^- are L-tight;

3) μ is tight i.e. for each $\epsilon > 0$ there is a $K \in \mathcal{K}(X)$ so that if $f \in \mathcal{U}^b(X)$ and $f = 0$ on K then $|\mu(f)| < \epsilon$;

3)' μ^+ and μ^- are tight.

Proof. 1)' ⇒ 2)': The family $\{L_{1,K}: K \in \mathcal{K}(X)\}$ is ueb and tends to zero pointwise when K tends to X. Hence, by the definition of a uniform measure, if $\mu^+ \in M_u(X)$ then μ^+ is L-tight.

2)' ⇒ 3)': Note that if $\mu^+ \in M(\check{X})$ is L-tight then for any $\alpha \in \mathbb{R}_+$,

$$\lim_{K \in \mathcal{K}(X)} \{\sup |\mu^+(f)| : f \in L_{\alpha,K}\} = 0$$

(since $L_{\alpha,K}$ is contained in a multiple of $L_{1,K}$).
Then, if $\epsilon > 0$, the L-tightness of $|\mu^+|$ implies that for each $n \in N$, there is a $K_n \in \mathcal{K}(X)$ so that

$$|\mu^+|(X \setminus \check{B}(K_n, 1/n)) \leq \epsilon/2^n$$

where $B(K,\eta) := \overline{B(K,\eta)}^{\check{X}}$ $(K \subseteq X, \eta > 0)$.
Indeed there is a function in $L_{n,K}$ which is 1 outside of $\check{B}(K_n,1/n)$ (when extended to $\check{X}$). For we can take the function

$$f: x \to n \cdot d(K_n, X) \wedge 1.$$

Then if $K := \bigcap_{n=1}^{\infty} B(K_n,1/n)$, K is compact in X (see the lemma below which is well known but which we prove for completeness) and satisfies the required condition.

3)' ⇒ 1)': Evident since the pointwise convergence of a ueb net implies compact convergence.

The equivalence of 1) and 1)' was proved in the previous Proposition, that of 3) and 3)' is evident as is the implication 2)' ⇒ 2). 2) implies 1) is a corollary of the result of §4.

Lemma. Let (K_n) be a sequence in $\mathcal{K}(X)$. Then, with the notation of the above proof, $\bigcap_{n=1}^{\infty} \check{B}(K_n,1/n) \subseteq X$.

Proof. (cf. Engelking [15], Th. 3.8.2.): Let $\check{x} \in \bigcap_{n=1}^{\infty} \check{B}(K_n,1/n)$. $B(K_n,1/n)$ can be covered by finitely many balls of radius $2/n$.

Hence for each $n \in \mathbb{N}$, there is an $x_n \in X$ with $\check{x} \in \check{B}(x_n,2/n)$. The function

$$f_n: x \to d(x_n,x) \wedge 1$$

can be continued to $\check{X}$ and $U_n := \check{f}_n^{-1}([0,3/n[)$ is an open neighbourhood of $\check{x}$ in $\check{X}$ whose trace in X has diameter at most $6/n$. Let $\mathcal{V}$ be the filter of closed neighbourhoods of $\check{x}$ in $\check{X}$. As $\mathcal{V}$ is finally contained in every U_n, the family $\{V \cap X\}_{V\in\mathcal{V}}$ forms a filter of closed subsets of X whose diameters tend to zero. By the completeness of (X,d), $\bigcap_{V\in\mathcal{V}} (V\cap X) \neq \emptyset$ and this latter set is of course exactly the point $\check{x}$. Hence $\check{x} \in X$.

Note that condition 3) above implies that μ defines a functional on $\mathcal{U}^b(X)$ which is continuous on the unit ball with respect to the topology of compact convergence. As $\mathcal{U}^b(X)$ is a dense subspace of $(C^b(X),\beta)$, the space of bounded continuous functions with the strict topology, by the Stone-Weierstrass theorem (cdf. Cooper [9], p.84) μ extends to a unique continuous functional on $(C^b(X),\beta)$ i.e. to a Radon measure on the associated topological space. Hence the terminology "tight" is in agreement with the usage of topological measure theory.

§6. THE COMPACTNESS THEOREM FOR METRIC SPACES

We are now ready to prove the main result on compactness in $M_u(X)$ for the case where X is a complete metric space. The introduction of the concept of uniform Lipschitz tightness allows us to give a much shorter and more intuitive proof than the original one of Pachl. Of course, we use Schur's Lemma (i.e. the special case of a discrete space) in our proof.

Theorem (metric case). Let H be a subset of $M_u(X)$ (X a complete metric space). Then the following are equivalent:

1) H is relatively $\sigma(M_u,\mathcal{U}^b)$-compact;

1)' H is relatively countably $\sigma(M_u,\mathcal{U}^b)$-compact;

2) H is relatively γ-compact;

3) H is bounded and uniformly Lipschitz-tight.

Remark. Note that 2) is equivalent to all the conditions which are listed in the proposition of Chapter 3.

Thus the following conditions (for ecample) are all equivalent to those of the above list:

- H is β_∞-equicontinuous
- H is equicontinuous on every ueb-set (with respect to the pointwise topology)
- H is $\|\cdot\|$-bounded and equicontinuous on $L_1(X)$.

In particular, β_∞ is the Mackey topology for the duality $(U^b(X), M_u(X))$. We also remark that, contrary to the result of the previous paragraph, uniform L-tightness may not be replaced by uniform tightness (obvious definition): for if we take $X = R$ with its usual metric and

$$\mu_n = \delta_n - \delta_{n+\frac{1}{n}}$$

then $\{\mu_n\}$ is uniformly Lipschitz tight but not uniformly tight. Also the sequence $\{\sqrt{n}\mu_n\}$ of measures shows that the boundedness condition in 3) is indispensable.

Proof. 2) $\Rightarrow$ 1) and 1) $\Rightarrow$ 1)' are evident.

1)' $\Rightarrow$ 3): If 1)' holds, then H is bounded and so we can suppose that $H \subseteq OM_u(X)$. If H is not L-tight we shall show how to construct $\eta > 0$, a sequence (μ_n) in H, a sequence (K_n) of compact sets in X so that

$$B(K_n,\eta) \cap B(K_m,\eta) = \emptyset \qquad (n \neq m)$$

and a sequence (f_n) in $L_{\eta^{-1}}(X)$ so that $\mathrm{supp}(f_n) \subseteq B(K_n,\eta)$ and $|\mu_n(f_n)| \geq \eta$. Once this is done, we complete the proof as follows: for any sequence (λ_n) in $O\ell^\infty$, the unit ball of ℓ^∞, $\Sigma\lambda_n f_n$ is in $L_{\eta^{-1}}(X)$. Hence

$$T: (\lambda_n) \to \Sigma\lambda_n f_n$$

is a CoSaks morphism from ℓ^∞ in $u^b(X)$ (we regard ℓ^∞ as $u^b(\mathbb{N})$, $\mathbb{N}$ with the discrete metric). Then the transposed operator

$$T' : M_u(X) \to \ell^1$$

sends H into a relatively countably $\sigma(\ell^1,\ell^\infty)$-compact set and so, by Schur's Lemma, into a relatively norm compact set in ℓ^1. But this contradicts the fact that $|T'\mu_n(e_n)| \geq \eta$, e_n dending the n-th unitvector in e^∞.

We now show how to construct the above sequences inductively. By assumption there is an η $(0 < \eta \leq 1)$ so that for every compact set K there is an $f_K \in L_{1,K}$ and $\mu_K \in H$ with $|\mu_K(f_K)| \geq 4\eta$. Define $\tilde{f}_K$ by

$$\tilde{f}_K : x \longrightarrow \begin{cases} 0 & \text{if } |f_K(x)| \leq 2\eta \\ f_K - 2\eta & \text{if } f_K(x) \geq 2\eta \\ f_K + 2\eta & \text{if } f_K(x) \leq -2\eta \end{cases}$$

Then $f_K \in L_{1,B(K,2\eta)}$ is such that $|\mu_K(\tilde{f}_K)| \geq 2\eta$. We can now proceed with the construction. First we find a $g_1 \in L_1(X)$ and $\mu_1 \in H$ with $|\mu_1(g_1)| \geq 2\eta$. Since μ_1 is a Radon measure on X we can find a compact set K_1 so that $|\mu_1|(X\setminus K_1) \leq \eta$. Hence

$$f_1 : x \to [g_1(x) \wedge (1-\eta^{-1}\, d(x,K_1))] \vee [-1+\eta^{-1}\, d(x,K_1)]$$

is a member of $L_{\eta^{-1}}(x)$ whose support is contained in $B(K_1,\eta)$ and is such that $|\mu_1(f_1)| \geq \eta$.

At the n-the step let $g_n \in L_{1,B_n}$ $(B_n := B(K_1 \cup \ldots \cup K_{n-1}, 2\eta))$ and $\mu_n \in H$ be such that $|\mu_n(g_n)| \geq 2\eta$.

Choose a compact $K_n \subseteq X\setminus B_n$ so that

$$|\mu_n|(X\setminus B_n \cup K_n)) \leq \eta.$$

Note that $d(K_n,K_m) \geq 2\eta$ for $m < n$.

Again define

$$f_n : x \to [g_n(x) \wedge (1-\eta^{-1}\, d(x,K_n))] \vee [-1+\eta^{-1}\, d(x,K_n)].$$

Then this is a member of $L_{\eta^{-1}}(X)$ whose support is contained in

$B(K_n,\eta)$ and which is such that $|\mu_n(f_n)| \geq \eta$.

This completes the induction step and so the proof of 1) ⇒ 3).

3) ⇒ 2): By the remark following the theorem, it suffices to show that H is equicontinuous on $L_1(X)$. If $(f_\alpha)_{\alpha\in I}$ is a net in $L_1(X)$ tending pointwise to $f \in L_1(X)$, then $(f_\alpha - f)_{\alpha\in I}$ is a net in $2L_1(X)$ which tends to zero uniformly on compact sets. Thus for each $K \in \mathcal{K}(X)$ and $\epsilon > 0$ there is α_0 so that for $\alpha \geq \alpha_0$

$$\{f_\alpha - f\} \in 2\cdot L_{1,K} + \epsilon\cdot O\mathcal{U}^b(X),$$

whence, by the $\|\cdot\|$-boundedness of H, $\mu(f_\alpha) \to \mu(f)$ uniformly in $\mu \in H$.

Using the above proof, we also get:

__Corollary__. Let (μ_n) be a weak Cauchy sequence in $M_u(X)$. Then (μ_n) is γ-Cauchy and so γ-convergent.

__Proof__. If $\{\mu_n\}_{n=1}^\infty$ is not relatively γ-compact, then one constructs as above $\eta > 0$, a sequence $\{f_k\}_{k=1}^\infty \in L_{\eta^{-1}}(X)$ with disjoint support and a subsequence $\{\mu_{n_k}\}_{k=1}^\infty$ such that

$$|\mu_{n_k}(f_k)| \geq \eta.$$

Defining again an operator T from ℓ^∞ to $\mathcal{U}^b(X)$ by

$$T: (\lambda_k) \to \sum_{k=1}^\infty f_k$$

we obtain a CoSaks-morphism. The transposed operator

$$T': M_u(X) \to \ell^1$$

sends $\{\mu_{n_k}\}_{k=1}^\infty$ to a weak Cauchy-sequence and $|T'\mu_{n_k}(e_k)| \geq \eta$ where e_k denotes the k-th unit-vector in ℓ^∞. Again this is contradictary to Schur's Lemma, since for a relatively $\|\cdot\|$-compact set K in ℓ^1

$$\lim_{k\to\infty} \sup \{ |\langle e_k, x\rangle| : x \in K\} = 0.$$

__Corollary__. Let T be a linear mapping from $\mathcal{U}^b(X)$ into a weakly compactly generated (in particular, separable or reflexive) Fréchet space F. Then T is β_∞-continuous if and only if it has a β_∞-closed graph.

Remark. In particular, if T is continuous for any locally convex Hausdorff topology on F which is coarser than the original topology (e.g. a weak topology defined by a total subset of F'), then T is continuous.

Proof of the Corollary. Note that we now know that $(U^b(X), \beta_\infty)$ is a Mackey space whose dual is weakly sequentially complete. The result now follows from a closed graph theorem of Kalton and Marquina (see e.g. Cooper [9], p.60).

Despite the remark after the statement of the theorem we do have equivalence of tightness and uniform tightness for positive measures.

Proposition. If H $M_u^+(X)$, then the conditions of the theorem are equivalent to

4) H is bounded and uniformly tight.

Proof. 4) implies 3) is evident (even without the positivity assumption).

3) implies 4): since $L_{1,K}$ contains a function which is 1 on $X\backslash B(K,1)$, 3) implies that for $\varepsilon > 0$ there is a $K_1 \in \mathcal{K}(X)$ so that $\check{\mu}(\check{X}\backslash\check{B}(K,1)) \leq \varepsilon/2$ for $\mu \in H$ ($\check{X}$ the Samuel compactification of X and $\check{\mu}$ the Radon measure on X corresponding to μ).

Similarly for $n \in \mathbb{N}$ there is a $K_n \in \mathcal{K}(X)$ so that

$$\mu(X\backslash B(K_n, 1/n)) \leq \varepsilon/2^n$$

for $\mu \in H$ since $L_{n,K} \subseteq nL_{1,K}$).

Putting $K := \bigcap B(K_n, \frac{1}{n})$ we obtain a compact subset of X so that

$$\mu(X\backslash K) = \check{\mu}(\check{X}\backslash K) \leq \varepsilon \qquad (\mu \in H).$$

§7. LIFTING TO UNIFORM SPACES

Let X be a uniform space. As in §3 we denote by $\mathfrak{D}$ the family of all uniformly continuous pseudometries on X which are bounded by 2. Then we have the natural representation

$$\hat{X} = \varprojlim \{\hat{X}_d : d \in \mathcal{D}\}$$

where, for $d \in \mathcal{D}$, $\hat{X}_d$ represents the associated complete metric space and $\hat{X}$ is the completion of X. More generally, if $\mathcal{D}_1$ is a subset of $\mathcal{D}$ which is closed under pointwise suprema and generates the uniformity of X, then once again $\{\hat{X}_d : d \in \mathcal{D}_1\}$ forms a projective system and $\hat{X}$ is its projective limit.

As noted above the bornology of ueb sets in $\mathcal{U}^b(X)$ consists of those balls which factor through $\mathcal{U}^b(X_d)$ for some $d \in \mathcal{D}$ and form a Lipschitz bounded set there. The set $\mathfrak{B}$ of those balls which factor in this way over some $\mathcal{U}^b(X_d)$ $(d \in \mathcal{D}_1)$ is, in general, a proper subfamily of the ueb sets but they are linked in the sense that the ueb bornology is exactly the $\| \|_\infty$-saturation $\tilde{\mathfrak{B}}$ of $\mathfrak{B}$ (the proof of this assertion is an easy adaptation of the arguments of §4).

If $\mathcal{D}_p$ denotes the family of all uniformly continuous precompact pseudometrics in $\mathcal{D}$ then, for $d \in \mathcal{D}_p$, $\hat{X}_d$ is compact. The projective limit $X = \varprojlim \{\hat{X}_d : d \in \mathcal{D}_p\}$ is compact and X embeds homeomorphically (as a topological space) into X. It is easily checked that X is just the Samuel-compactification of X i.e. the spectrum of the Banach-algebra $\mathcal{U}^b(X)$.

Now if X, $\mathcal{D}_1$ are as above we can obtain projective and inductive representations of $M_u(X)$ and $\mathcal{U}^b(X)$ as follows:

$$\{\mathcal{U}^b(X_d) \to \mathcal{U}^b(X_{d_1}) : d \leq d_1, \quad d, d_1 \in \mathcal{D}_1\}$$

forms an inductive system of CoSaks spaces while

$$\{M_u(X_{d_1}) \to M_u(X_d) : d \leq d_1, \quad d, d_1 \in \mathcal{D}_1\}$$

forms a projective system of Saks-spaces.

If $\mathcal{U}^b(X_d)$ is considered as subspace of $\mathcal{U}^b(X)$, the bornologies of Lipschitz-bounded sets in $\mathcal{U}^b(X_d)$ (or equivalently of ueb-sets in $\mathcal{U}^b(X_d))$, as d ranges through $\mathfrak{D}_1$ generate the bornology of ueb-sets in $\mathcal{U}^b(X)$ (in the sense of $\|\cdot\|$-saturation). Thus it is clear that

$$\mathcal{U}^b(X) = \varinjlim \{\mathcal{U}^b(X_d): d \in \mathfrak{D}_1\} = \varinjlim \{\mathcal{U}^b(X_d): d \in \mathfrak{D}\}$$

the injective limit being taken in the category of CoSaks spaces and

$$M_u(X) = \varprojlim \{M_u(X_d): d \in \mathfrak{D}_1\} = \varprojlim \{M_u(X_d): d \in \mathfrak{D}\}$$

this time the projective limit in the category of Saks spaces.

We remark that if we form $\varinjlim (\mathcal{U}^b(X_d): d \in \mathfrak{D}_p)$, we again get the Banach space $(\mathcal{U}^b(X), \|\cdot\|_\infty)$ while the bornology now consists of the relatively $\|\cdot\|_\infty$-compact balls in $\mathcal{U}^b(X)$.

Using this formalism it is now easy to lift results of §5, §6 to uniform spaces simply by observing that the appropriate statements hold in every component of the projective limit. We gather together the most important results.

<u>Proposition.</u> Let X be a uniform space. A functional $\mu \in (\mathcal{U}^b(X), \|\,\|_\infty)'$ is a member of $M_u(X)$ iff its image on every $\hat{X}_d$ ($d \in \mathfrak{D}_1$ or equivalent $d \in \mathfrak{D}$) is a uniform measure, i.e. a Radon-measure by the results of §6.

<u>Theorem</u> (<u>uniform case</u>). If X is a uniform space, H a subset of $M_u(X)$ then the following are equivalent:

1) H is relatively $\sigma(M_u, \mathcal{U}^b)$-compact;

1)' H is relatively countably $\sigma(M_u, \mathcal{U}^b)$-compact;

2) H is relatively γ-compact;

3) the image of H in every $M_u(\hat{X}_d)$, $d \in \mathfrak{D}$ is relatively γ-compact;

3)' H is $\|\cdot\|$-bounded and its image in every $M_u(\hat{X}_d)$, $d \in \mathfrak{D}_1$ is relatively γ-compact;

4) H is bounded and its image in every $M_u(\hat{X}_d)$, $d \in \mathfrak{D}$ is uniformly L-tight;

4)' H is bounded and its image in every $M_u(\hat{X}_d)$, $d \in \mathfrak{D}_1$, is uniformly L-tight.

Proof. A subset H in a projective limit of Saks-spaces is γ-compact (respectively weakly compact) iff it is $\|\cdot\|$-bounded and all its projections into the component spaces are γ-compact (respectively weakly compact), c.f. Cooper [9], pp.10,16. From this remark and the corresponding theorem for the metric case the theorem is easily deduced.

Corollary. Let X be a uniform space. Every $\sigma(M_u(X),U^b(X))$ Cauchy-sequence (μ_n) converges in the γ-topology.

Proof. For every $d \in \mathcal{D}$ the image of $\{\mu_n\}_{n=1}^{\infty}$ is weakly Cauchy in $M_u(\hat{X}_d)$, whence γ-convergent in $M_u(\hat{X}_d)$ by the result for the metric case. So $\{\mu_n\}_{n=1}^{\infty}$ is γ-convergent in $M_u(X)$.

Exactly as in the metric case one also derives the two following results for the uniform case.

Proposition. Let X be a uniform space and T be a linear mapping from $U^b(X)$ into a weakly compactly generated Fréchet space. Then T is β_∞-continuous iff it has a β_∞-closed graph.

Proposition. A subset $H \subseteq M_u^+(X)$ is relatively γ-compact iff it is $\|\cdot\|$-bounded and its image in every $M_u^+(X_d)$ ($d \in \mathcal{D}$ or, equivalently, $d \in \mathcal{D}_1$) is uniformly tight.

Following Pachl [34], we now indicate briefly how the concept of uniform measure embraces many important classes of measures:

I. Separable measures:

If X is a completely regular space, we can regard it as a uniform space with the fine uniformity i.e. the finest uniformity compatible with its topology. Then $U^b(X) = C^b(X)$ and the corresponding CoSaks structure on $C^b(X)$ is that of the bounded, equicontinuous subsets of $C^b(X)$. As mentioned in the introduction the corresponding topology on $C^b(X)$ has been studied by Wheeler and the corresponding space of measures by various authors from various points of view.

II. σ-additive abstract measures:

Now let (Ω,Σ) be a measure space i.e. Σ is a σ-algebra on Ω. In addition, we assume that Σ separates the points of Ω. For each countable partition $\mathfrak{A} = (A_n)$ of Ω by sets of Σ, we define a pseudometric $d_{\mathfrak{A}}$ by putting

$$d_{\mathfrak{A}} = \begin{cases} 1 & \text{if } x,y \text{ do not belong to the same } A_n \\ 0 & \text{otherwise} \end{cases}$$

The associated metric space is just N with the discrete metric. We can regard Ω as a uniform space with the structure induced by these pseudometrics (such uniform spaces have been studied by Hager under the name "measurable uniform spaces" in [20]). Then $U^b(\Omega)$ is the space $B(\Omega)$ of bounded measurable functions on Ω (for every bounded measurable function can be uniformly approximated by countably (even finitely) valued functions). From the equation $M_u(\Omega) = S \varprojlim_{\mathfrak{A}} M_u(\Omega_{d_{\mathfrak{A}}})$ we see that an element of $M_u(\Omega)$ is a bounded set function on Σ with the property that for each measurable partition (A_n) $\mu(\bigcup_{n=1}^{\infty} A_n) = \sum_{n=1}^{\infty} \mu(A_n)$ i.e. it is a σ-additive measure.

III. Cylindrical measures:

Let E, F be vector spaces in duality. We denote by $\mathcal{G}(F)$ the family of finite dimensional subspaces of F. They form an inductive system (ordered by inclusion) and so by duality we get a projective system of finite dimensional spaces (which we can regard as quotients of E) whose projective limit is F^*. We can then give E a uniform structure as a subset of this projective limit (i.e. this is just a complicated way of taling about the $\sigma(E,F)$-uniformity on E). The corresponding space of uniform measures is denoted by $M_{CYL}(E)$ - the space of cylindrical measures on E. Now the above complications begin to pay off because we can write

$$M_{CYL}(E) = S \varprojlim \{M_u(E/G^o): G \in \mathcal{G}(F)\}$$

and so an element of $M_{CYL}(E)$ can be regarded as a projective limit of Radon measures (in the category of Saks spaces) on finite dimensional quotients of E and this is the normal definition of a cylindrical measure.

§8. VECTOR-VALUED MEASURES AND ORLICZ-PETTIS TYPE THEOREMS

In this section we indicate briefly how the classical Orlicz-Pettis theorem can be interpreted as the statement that the dual $(\mathcal{U}^b(\Omega,\beta_\infty)$ of a space of measures is a Mackey space.

As in Example II of §7 let Σ be a σ-algebra of subsets of Ω and let

$$m: \Sigma \to E$$

be a weakly σ-additive set-function from Σ to a locally convex space E. By the Dieudonné-Grothendieck-theorem (c.f. Diestel-Uhl [12], th. I.3.3), m has bounded range and so we may, if E is quasicomplete, define the integration operator

$$T_m: \mathcal{U}^b(\Omega) \to E$$

$$T_m: (\chi_A) \to m(A)$$

where T_m is a continuous operator from $(\mathcal{U}^b(\Omega),\beta_\infty)$ into $(E,\sigma(E,E'))$. As $(\mathcal{U}^b(\Omega),\beta_\infty)$ is a Mackey space T_m is continuous with respect to the original topology of E. If $\{A_n\}_{n=1}^\infty$ decreases to $\emptyset$ in Σ, then $\{\chi_{A_n}\}_{n=1}^\infty$ is a ueb-set tending pointwise to zero, whence $\{T(\chi_{A_n})\}_{n=1}^\infty$ tends to zero in the original topology of E. Thus we have proved the Orlicz-Pettis-theorem:

"<u>A weakly σ-additive measure on a σ-algebra is strongly σ-additive</u>".

Now assume E is a weakly compactly generated Fréchet-sapce, F is an E-total subset of E′ and

$$m: \Sigma \to E$$

is a $\sigma(E,F)$ countably additive measure. Again by Dieudonné-Grothendieck m has bounded range and we may define an integration operator

$$T_m: \mathcal{U}^b(\Omega) \to E$$

$$T_m: \chi_A \to m(A)$$

where by the closed-graph theorem proved in §7, T_m is β_∞-continuous, which implies as above that m is strongly σ-additive.

REFERENCES

[1] Arens, R.F., Eells, J., On embedding uniform and topological spaces Pac, J. Math. 6 (1956) 397-403.

[2] Azzam, N., Meusures sur les espaces uniformes, Prépublications Université St.-Etienne 1974.

[3] Berezanskii, I.A., Measures on uniform spaces, Trans. Moscow Math. Soc. 19 (1968) 1-40.

[4] Berruyer, J., Ivol, B., Espaces de mesures et compactologies, Publ. Dép. Math. Lyon 9.1 (1974) 1-36.

[5] Buchwalter, H., Topologies, bornologies et compactologies (Lyon, undated)

[6] Buchwalter, H., Quelques curieuses topologies sur $M_\sigma(T)$ et $M_\beta(T)$ Ann. Inst. Fourier (to appear).

[7] Buchwalter, H., Pupier, R., Complétion d'un espace uniforme et formes linéaires, C.R. Acad. Sc. Paris A 273 (1971) 96-98.

[8] Cigler, J., Funktoren auf Kategorien von Banachräumen (Lecture Notes, Vienna 1974).

[9] Cooper, J.B., Saks spaces and applications to functional analysis (Amsterdam, 1978).

[10] Cooper, J.B., Schachermayer, W., Saks spaces and vector valued measures (Institutsbericht 98, Linz, 1978).

[11] Deaibes, A., Espaces uniformes et espaces de mesures, Publ. Dép. Math. Lyon 12.4 (1975) 1-166.

[12] Diestel, J., Uhl, J.J., The theory of vector measures (Providence 1977).

[13] Dostal, M.A., Some recent results on topological vector spaces (Springer Lecture Notes 384 (1974) 20-91).

[14] Dudley, R.M., Convergence of Baire measures, Studia Math. 27 (1966) 252-268.

[15] Engelking, R., Outline of general topology (Amsterdam, 1968).

[16] Fedorova, V.L., Linear functionals and the Daniell integral on spaces of uniformly continuous functions, Usp. Math. Nauk 133.1 (1967) 172-173 (Russian).

[17] Frolik, Z., Mesures uniformes, C.R. Acad. Sc. Paris A 277 (1973) 105-108.

[18] Frolik, Z., Représentation de Riesz des mesures uniformes, C.R. Acad. Sc. Paris, A 277 (1973) 163-166.

[19] Frolik, Z., Metric-fine uniform spaces (Seminar Abstract Analysis - Prague, 1974).

[20] Hager, A., Measurable uniform spaces, Fund. Math. 77 (1972) 51-73.

[21] Haydon, R.G., Sur les espaces $M(T)$ et $M^\infty(T)$, C.R. Acad. Sc. Paris 275 (1972) A 989-991.

[22] Isbell, J.R., Algebras of uniformly continuous functions, Ann. Math. 68 (1958) 96-125.

[23] Isbell, J.R., Uniform spaces (Providence, 1964).

[24] John, K., Differentiable manifolds as topological linear spaces, Math. Ann. 186 (1970) 177-190.

[25] Johnson, J., Lipschitz function spaces for arbitrary metrics, Bull. Amer. Math. Soc. 78 (1972) 702-706.

[26] Johnson, J., Banach spaces of Lipschitz functions and vector valued Lipschitz functions, Bull. Amer. Math. Soc. 75 (1969) 1334-1338.

[27] Kirk, R.B., Topologies on spaces of Baire measures, Trans. Amer. Math. Soc. 184 (1973) 1-29.

[28] Leeuw, K. de, Banach spaces of Lipschitz functions, Studia Math. 21 (1961) 55-66.

[29] Leger, C., Soury, P., Le convexe topologique des probabilités sur un espace topologique, J. Math. Pure et Appl. 50 (1971) 363-425.

[30] Mac Lane, S., Categories for the working mathematican (Berlin, 1971).

[31] Manes, E.G., A characterisation of dual spaces, Func. Anal. 17 (1974) 292-295.

[32] Michael, E., A short proof of the Arens-Eells embedding theorem Proc. Amer. Math. Soc. 15 (1964) 415-416.

[33] Pachl, J., Free uniform measures, Comm. Math. Univ. Carol. 15 (1974) 541-553.

[34] Pachl, J., Compactness in spaces of uniform measures, Trans. Amer. Math. Soc. (to appear).

[35] Ptak, V., Algebraic extensions of topological spaces, Proc. of a Symposium, Berlin 1967 - VEB Berlin (1969, 179-188).

[36] Pupier, R., Méthodes fonctorielles en topologies générale, Thèse, Univ. Lyon I (1971).

[37] Raikov, D.A., Free locally convex spaces, Math. Sbornik 63 (1964) 582-590.

[38] Rome, M., L'espace $M^{\infty}(T)$, Publ. Dép. Math. Lyon 9-1 (1972) 37-60.

[39] Roy, A.K., Extreme points and linear isometries of the Banach space of Lipschitz functions, Can. J. Math. 20 (1968), 1150-1164.

[40] Schaefer, H.H., Topological vector spaces (New York, 1966).

[41] Tomasek, S., On certain classes of Λ-structures I, II, Czech. Math. J. 20 (1970) 1-18, 19-33.

[42] Wheeler, R.F., The strict topology, separable measures and paracompactness, Pac. J. Math. 47 (1973) 287-302.

HOLOMORPHIC GERMS ON COMPACT SUBSETS OF LOCALLY CONVEX SPACES

Seán Dineen

Department of Mathematics
University College Dublin
Belfield, Dublin 4, Ireland

If K is a compact subset of a locally convex space E then the space of holomorphic germs about $K, H(K)$, is given the inductive limit topology $\varinjlim_{U} (H^{\infty}(U), \| \ \|_{U})$ where U ranges over all open subsets of E which contain K. It is known ([1],[2],[3],[12]) that $H(K)$ is a complete regular inductive limit whenever E is a quasi-normable metrizable space. In this paper, we show that $H(K)$ is a complete locally convex space whenever K is a compact subset of a metrizable locally convex space. We also give examples of non-metrizable locally convex spaces in which the same conclusion holds. In obtaining these results we prove some general results concerning locally convex spaces which may be of interest in their own right - for instance if (E,τ) is a complete locally convex space and τ_t is the barrelled topology associated with τ then we show that (E,τ_t) is also complete.

For background information and open problems concerning the space $H(K)$ we refer to the comprehensive and excellent survey article [3].

1. SOME RESULTS ON ARBITRARY LOCALLY CONVEX SPACES

Let E denote a locally convex space - over C or R with topology τ. The barrelled topology associated with τ, τ_t, is the infimum of all barrelled topologies on E finer than τ. Hence

(E,τ_t) is an inductive limit of barrelled topologies and since the finest locally convex topology on E is barrelled it is well defined and thus (E,τ_t) is a barrelled locally convex space. An alternate description of (E,τ_t) - by means of transfinite induction - is provided in the proof of the following theorem.

Theorem 1. If (E,τ) is a complete locally convex space then (E,τ_t) is also a complete locally convex space.

Proof. We first construct an ordered family of locally convex topologies on E indexed by the ordinals. Let $\tau_1 = \tau$. Now suppose α is an ordinal number and τ_β has been defined for all ordinals $\beta < \alpha$. Let $(E,T_\alpha) = \varprojlim_{\beta<\alpha} (E,\tau_\beta)$ - i.e. T_α is the projective limit topology defined by all τ_β, $\beta < \alpha$. We define τ_α as the topology which has a neighbourhood basis at zero consisting of all T_α closed convex balanced absorbing subsets of E. By transfinite induction τ_α is defined for every ordinal α. Since the cardinality of the set of all convex balanced absorbing subsets of E is less than or equal to $2^{|E|}$ it follows that there exists an ordinal number ω such that $\tau_\omega = \tau_{\omega_1}$ for all $\omega_1 \geq \omega$. By our construction (E,τ_ω) is a barrelled locally convex space and since it is the weakest barrelled topology finer than τ it follows that $\tau_\omega = \tau_t$. We now show that (E,τ_α) is complete for every ordinal number α. This will complete the proof. If this were not true, then there would exist a smallest ordinal α_1 such that (E,τ_{α_1}) was not complete. Hence (E,τ_α) is complete for all $\alpha < \alpha_1$ and since the projective limit of complete spaces is complete it follows that (E,T_{α_1}) is a complete locally convex space. Now let $(x_\beta)_{\beta\in B}$ denote a Cauchy set in (E,τ_{α_1}). Since $\tau_{\alpha_1} \geq T_{\alpha_1}$, $(x_\beta)_{\beta\in B}$ is also a Cauchy set in (E,T_{α_1}) and hence converges to some x in (E,T_{α_1}). Now let V denote a neighbourhood of zero in τ_{α_1}. Without loss of generality we may suppose that V is T_{α_1}-closed. Hence there exists $\beta_o \in B$ such that $x_{\beta_1} - x_{\beta_2} \in V$ for all $\beta_1,\beta_2 \geq \beta_o$.

Hence $x - x_{\beta_2} \in V$ for all $\beta_2 \geq \beta_0$ since V is T_{α_1}-closed and $x_{\beta_1} \to x$ in (E, T_{α_1}) as $\beta_1 \to \infty$. This implies that $x_\beta \to x$ as $\beta \to \infty$ in (E, τ_{α_1}) and completes the proof.

By modifying the above proof one easily shows the following.

<u>Proposition 2</u>. Let (E,τ) denote a locally convex space and let τ_t and τ_i denote respectively the barrelled and the infrabarrelled topology associated with τ;

(a) if (E,τ) is complete then (E,τ_i) is complete;

(b) if (E,τ) is quasi-complete then (E,τ_t) and (E,τ_i) are both quasi-complete;

(c) if (E,τ) is sequentially complete then (E,τ_t) and (E,τ_i) are both sequentially complete.

We now turn to a different type of result which might be described as a sort of open mapping theorem.

<u>Proposition 3</u>. Let τ_1, τ_2 and τ_3 denote three Hausdorff locally convex topologies on a vector space E such that

(1) $\tau_1 \geq \tau_2 \geq \tau_3$;

(2) (E,τ_1) is a bornological DF-space (or equivalently a countable inductive limit of normed linear spaces) with countable fundamental system of convex balanced bounded sets $(B_n)_n$;

(3) (E,τ_2) is a barrelled locally convex space;

(4) B_n is τ_3 compact for all n.

Then $\tau_1 = \tau_2$.

<u>Proof</u>. (See [10] p.72, Lemma 4). A fundamental system of neighbourhoods of zero in (E,τ_1) is given by sets of the form $\sum_{n=1}^{\infty} \lambda_n B_n$; $\lambda_n > 0$ all n. Let $V = \sum_{n=1}^{\infty} \lambda_n B_n$, $\lambda_n > 0$, and let $\tilde{V}$ denote the algebraic closure of V in E, i.e. $\tilde{V} = \{x \in E;\ \lambda x \in V \text{ for } 0 \leq \lambda < 1\}$. Since B_n is a compact subset of (E,τ_3) it follows that $\sum_{n=1}^{k} \lambda_n B_n$ is, for each k, also a compact subset of (E,τ_3) and hence it is a

closed subset of (E,τ_2). Now let $x \in C\tilde{V}$. Then there exists $\lambda > 1$ such that $x \in C(\lambda\tilde{V})$ and hence $x \in C(\lambda \cdot \sum_{n=1}^{k} \lambda_n B_n)$ for every integer k. Hence for each integer k we can choose $\phi_k \in (E,\tau_2)'$ such that $\phi_k(x) = \lambda$ and $|\phi_k(\sum_{n=1}^{k} \lambda_n B_n)| \leq 1$. Since $\sum_{n=1}^{\infty} \lambda_n B_n$ is an absorbing subset of (E,τ_2) it follows that $\{\phi_k\}_k$ is a pointwise bounded and hence a relatively weakly compact subset of $(E,\tau_2)'$ because of hypothesis (3). If ϕ belongs to the weak closure of $\{\phi_k\}_k$ then $\phi(x) = \lambda$ and $|\phi(V)| \leq 1$. Hence $\bar{V}^{\tau_2} = \tilde{V}$ and so $\bar{V}^{\tau_2} \subset (1+\epsilon)V$ for every $\epsilon > 0$. Since $\bar{V}^{\tau_2}$ is convex balanced absorbing and τ_2 closed it is a neighbourhood of zero in (E,τ_2) and hence every τ_1-neighbourhood of zero contains a τ_2-neighbourhood. This shows that $\tau_1 = \tau_2$ and completes the proof.

2. HOLOMORPHIC FUNCTIONS ON METRIZABLE LOCALLY CONVEX SPACES

For the remainder of this paper, E will denote a locally convex space over the field of complex numbers $\mathbb{C}$. $\mathcal{P}(^nE)$ will denote the space of continuous n-homogeneous polynomials on E. τ_o will denote the topology of uniform convergence on compact sets and β will denote the topology of uniform convergence on the bounded subsets of E. The barrelled topology associated with τ_o is also the barrelled topology associated with β. A semi-norm p on $\mathcal{P}(^nE)$ is τ_ω-continuous if for every neighbourhood V of zero there exists $C(V) > 0$ such that $p(P) \leq C(V)\|P\|_V$ for all $P \in \mathcal{P}(^nE)$. τ_ω is the locally convex topology generated by all τ_ω continuous semi-norms. $(\mathcal{P}(^nE),\tau_\omega) = \varinjlim_{V\ni 0} (\mathcal{P}_V(^nE),\| \|_V)$ where $\mathcal{P}_V(^nE) = \{P \in \mathcal{P}(^nE);\ \|P\|_V < \infty\}$. Hence $(\mathcal{P}(^nE),\tau_\omega)$ is an inductive limit of Banach spaces and hence is an ultrabornological space and in particular it is barrelled and bornological. We always have $\tau_o \leq \beta \leq \tau_t \leq \tau_\omega$ and τ_ω is the bornological topology associated with τ_o if the τ_o bounded subsets of $\mathcal{P}(^nE)$ are locally bounded.

Proposition 4. If E is a metrizable locally convex space then $\tau_t = \tau_\omega$ on $\mathcal{P}(^nE)$.

Proof. $(\mathcal{P}(^nE),\tau_\omega)$ is a bornological DF-space with fundamental system of bounded sets $B_n = \{P \in \mathcal{P}(^nE); \|P\|_{V_n} \leq 1\}$ as V_n ranges over the neighbourhoods of zero in E. Since B_n is a closed bounded subset of the semi-Montel space $(\mathcal{P}(^nE),\tau_o)$ ([1]) we may apply Proposition 3 to complete the proof.

Corollary 5. If E is a metrizable locally convex spaces then $(\mathcal{P}(^nE),\tau_\omega)$ is a complete locally convex space.

Proof. $(\mathcal{P}(^nE),\tau_o)$ is a complete locally convex space and hence, by Theorem 1, $(\mathcal{P}(^nE),\tau_t)$ is also a complete locally convex space. By Proposition 4, $\tau_t = \tau_\omega$ and this completes the proof.

Corollary 6. If E is a metrizable locally convex space then the following are equivalent:

(a) $(\mathcal{P}(^nE),\tau_o)$ (resp. $(\mathcal{P}(^nE),\beta)$) is a barrelled locally convex space;

(b) $(\mathcal{P}(^nE),\tau_o)$ (resp. $(\mathcal{P}(^nE),\beta)$) is a bornological locally convex space;

(c) $(\mathcal{P}(^nE),\tau_o)$ (resp. $(\mathcal{P}(^nE),\beta)$) is an ultrabornological locally convex space.

For $n = 1$ this gives A. Grothedieck's characterization of distinguished Fréchet Spaces ([10], [13]).

Corollary 7. If U is a balanced open subset of a Fréchet space, then $(H(U),\tau_\omega)$ and $(H(U),\tau_\delta)$ are complete locally convex spaces.

Proof. By [8], $(H(U),\tau_\omega)$ is complete since $(\mathcal{P}(^nE),\tau_\omega)$ is complete for all n and every Fréchet space is T.S.τ_ω complete. Since τ_δ is the barrelled topology associated with τ_ω ([13]), it follows, by Theorem 1, that $(H(U),\tau_\delta)$ is also a complete locally convex space.

We now look at holomorphic germs on a compact subset of a metrizable locally convex space. The main part of our proof consists of an application of Theorem 1, a use of the method of Proposition 3 and the construction of a quasi-complete locally convex topology on $H(K)$ which is weaker than the inductive limit topology. This last construction is similar to one found in [11].

Theorem 8. Let K denote a compact subset of a metrizable locally convex space. Then $H(K)$ is a complete locally convex space.

Proof. Let τ_1 denote the inductive limit topology on $H(K)$. Let τ_2 denote the locally convex topology on $H(K)$ generated by all semi-norms which have either of the following forms:

$$p_1(f) = \sum_{n=0}^{\infty} \sup_{x\in K} p\left(\frac{\hat{d}^n f(x)}{n!}\right) \quad \text{where } p \text{ is}$$

$$\text{a } \tau_1 \text{ continuous semi-norm on } H(0) \qquad (*)$$

$$p_2(f) = \sum_{n=1}^{\infty} \left(\frac{\lambda_n}{2}\right)^{k_n} \cdot \left| \sum_{j=0}^{k_n} \frac{\hat{d}^j f(x_n)}{j!}(y_n) - \sum_{j=0}^{k_n} \frac{\hat{d}^j f(x'_n)}{j!}(y'_n) \right| \qquad (**)$$

where $(x_n)_n$, $(x'_n)_n$ are two sequences in K, $(y_n)_n$ and $(y'_n)_n$ are null sequences in E, $x_n + y_n = x'_n + y'_n = x'_n + y'_n$ for all n, $(\lambda_n)_n$ is a sequence of complex numbers, $|\lambda_n| \to +\infty$ $(\lambda_n y_n)_n$ and $(\lambda_n y'_n)_n$ are null sequences in E and $(k_n)_n$ is a strictly increasing sequence of positive integers.

Since $(H(K),\tau_1)$ is a barrelled space and $H(K)$ induces the τ_ω topology on $\mathcal{P}(^nE)$ for all n it follows that $\tau_1 \geq \tau_2$ if each of the above semi-norms is finite for every f in $H(K)$.

If p is a τ_1 continuous semi-norm on $H(0)$ then there exists for every neighbourhood V of zero, $C(V) > 0$, such that $p(f) \leq C(V)\|f\|_V$ for every $f \in H(0)$. If $f \in H(K)$ then there exists a neighbourhood W of zero such that f is defined and bounded on $K + 2W$ (by C say). Hence $\|\frac{\hat{d}^n f(x)}{n!}\|_W \leq \frac{C}{2^n}$ for every x in K and every non-negative integer n. Hence

$$p_1(f) \leq \sum_{n=0}^{\infty} \sup_{x\in K} p(\frac{\hat{d}^n f(x)}{n!}) \leq C(W) \cdot \sum_{n=0}^{\infty} C/2^n < \infty .$$

Thus, each semi-norm of the form (*) is τ_1-continuous. Now suppose p_2 is a semi-norm which has the form (**).

If $f \in H(K)$ then f is defined and bounded on $K+\{y;\tilde{p}(y)<2\}$ by M where $\tilde{p}$ is a continuous semi-norm on E. Choose N, a positive integer, such that $\tilde{p}(y_n) + \tilde{p}(y'_n) \leq \frac{1}{|\lambda_n|} \leq \frac{1}{2}$ for all $n \geq N$. Hence, for $n \geq N$,

$$\sum_{j=0}^{\infty} \frac{\hat{d}^j f(x_n)}{j!}(y_n) = f(x_n+y_n) = f(x'_n+y'_n) = \sum_{j=0}^{\infty} \frac{\hat{d}^j f(x'_n)}{j!}(y'_n)$$

and
$$\left|\frac{\hat{d}^j f(x_n)}{j!}(y_n)\right| \leq M(p(y_n))^j \leq M(\frac{1}{\lambda_n})^j$$

and
$$\left|\frac{\hat{d}^j f(x'_n)}{j!}(y'_n)\right| \leq M(p(y'_n))^j \leq M(\frac{1}{\lambda_n})^j .$$

It now follows that

$$\sum_{n\geq N} (\frac{\lambda_n}{2})^{k_n} \left| \sum_{j=0}^{k_n} \frac{\hat{d}^j f(x_n)}{j!}(y_n) - \sum_{j=0}^{k_n} \frac{\hat{d}^j f(x'_n)}{j!}(y'_n) \right|$$

$$\leq \sum_{n\geq N} (\frac{\lambda_n}{2})^{k_n} \cdot 2M \cdot \sum_{j=k+1}^{\infty} (\frac{1}{\lambda_n})^j$$

$$\leq 4M \cdot \sum_{n\geq N} (\frac{1}{2})^{k_n} < \infty .$$

Hence p_2 is τ_1 continuous and $\tau_1 \geq \tau_2$.

We now show that τ_1 and τ_2 define the same bounded subsets of $H(K)$. Let $(f_\alpha)_{\alpha\in\Gamma}$ denote a τ_2 bounded subset of $H(K)$. by using the semi-norms of the form (*) it follows that $\left(\frac{\hat{d}^n f_\alpha(x)}{n!}\right)_{x\in K, \alpha\in\Gamma,\ n \text{ arbitrary}}$ is a bounded subset of $H(0)$ with the inductive limit topology. Hence there exists a convex balanced neighbourhood of zero in E, W, such that $\left\|\frac{\hat{d}^n f_\alpha(x)}{n!}\right\|_W \leq \frac{M}{2^n}$ for all α in Γ, all x in K and all n.

If $x \in K$ and $y \in W$ we let

$$\tilde{f}_\alpha(x)(y) = \sum_{n=0}^{\infty} \frac{\hat{d}^n f_\alpha(x)}{n!}(y).$$

To complete the proof, it suffices to show that there exists a neighbourhood of zero, $V, V \subset W$, such that $\tilde{f}_\alpha(x)(y) = \tilde{f}_\alpha(x')(y')$ for all $x, x' \in K$, $y, y' \in V$ such that $x+y = x'+y'$ and any α in Γ.

If not, there exists two sequences in K, $(x_n)_n$, $(x'_n)_n$, two null sequences in E, $(y_n)_n$ and $(y'_n)_n$ and $(\alpha_n)_n$ a sequence in Γ such that

$$x_n + y_n = x'_n + y'_n \quad \text{and} \quad \tilde{f}_{\alpha_n}(x_n)(y_n) \neq \tilde{f}_{\alpha_n}(x'_n)(y'_n).$$

Let $\delta_n = |\tilde{f}_{\alpha_n}(x_n)(y_n) - \tilde{f}_{\alpha_n}(x'_n)(y'_n)|$ and choose $(\lambda_n)_n$ a sequence of scalars such that $|\lambda_n| \to \infty$ as $n \to \infty$ and $\lambda_n y_n$ and $\lambda_n y'_n \to 0$ as $n \to \infty$. Without loss of generality we may suppose $\lambda_n y_n$ and $\lambda_n y'_n$ lie in W for all n. Now we choose inductively an increasing sequence of positive integers, $(k_n)_n$, such that

$$\left| \sum_{j=0}^{k_n} \frac{\hat{d}^j f_{\alpha_n}(x_n)}{j!}(y_n) - \sum_{j=0}^{k_n} \frac{\hat{d}^j f_{\alpha_n}(x'_n)}{j!}(y'_n) \right| > \frac{\delta_n}{2}$$

and $\left(\frac{\lambda_n}{2}\right)^{k_n} \cdot \frac{\delta_n}{2} > n$ for all n.

If p_2 is the semi-norm (**) then $p_2(f_{\alpha_n}) > n$ and this contradicts the fact that $(f_\alpha)_{\alpha \in \Gamma}$ is τ_2 bounded. Hence τ_1 and τ_2 define the same bounded subsets of $H(K)$.

We now show that $(H(K), \tau_2)$ is quasi-complete. Let $(f_\alpha)_{\alpha \in \Gamma}$ denote a τ_2-bounded Cauchy set. If $x \in K$ and n is a positive integer then $\left(\frac{\hat{d}^n f_\alpha(x)}{n!}\right)_{\alpha \in \Gamma}$ is a Cauchy net in $(P(^nE), \tau_\omega)$ and hence by Corollary 5, it converges to an element of $P(^nE)$, which we shall denote by $P_{n,x}$. Since the net is bounded, there exists a neighbourhood W_1 of zero and $M > 0$ such that $\|P_{n,x}\|_{W_1} \leq \frac{M}{2^n}$ for every x in K and every positive integer n. By the above, there exists a neighbourhood V of zero in E, $V \subset W_1$, such that

$$\tilde{f}_\alpha(x)(y) = \tilde{f}_\alpha(x')(y') \quad \text{if } \alpha \in \Gamma, \quad x,x' \in K, \quad y,y' \in V$$

and $x+y = x'+y'$.

Since $\sum_{j=0}^{N} \frac{\hat{d}^j f_\alpha(x)}{j!}(y) \to \sum_{j=0}^{N} P_{j,x}(y)$ for all x in K and $y \in E$ as $\alpha \to \infty$ it follows that

$$\sum_{n=0}^{\infty} P_{n,x}(y) = \sum_{n=0}^{\infty} P_{n,x'}(y') \quad \text{for all } x,x' \in K, \quad y,y' \in V$$

such that $x+y = x'+y'$.

Hence the Taylor series expansions defined by $(P_{n,x})_{n=0}^{\infty}$ as x ranges over K are coherent and define a holomorphic function on a neighbourhood of K and hence an element of $H(K)$ which we shall denote by f. Because of the form of the semi-norms (*) and (**) one easily shows that $f_\alpha \to f$ as $\alpha \to \infty$ in $(H(K),\tau_2)$ in the same manner as one shows that the normed linear spaces ℓ_1 and ℓ_∞ are complete.

Let τ_3 denote the barrelled topology associated with τ_2 on $H(K)$. Since $(H(K),\tau_1)$ is a barrelled space and $\tau_1 \geq \tau_3 \geq \tau_2$ Proposition 2(b) implies that $(H(K),\tau_3)$ is a quasicomplete locally convex space. Since τ_1 and τ_2 have the same bounded sets and $(H(K),\tau_1)$ is a DF-space it follows that $(H(K),\tau_3)$ is a quasicomplete DF-space and hence it is complete. We now show that $\tau_1 = \tau_3$.

Let $(V_n)_n$ denote a fundamental system of open neighbourhood of K and for each n let $B_n = \{f \in H(V_n); \|f\|_{V_n} \leq 1\}$. $(B_n)_n$ forms a fundamental system of convex balanced bounded subsets of $(H(K),\tau_1)$. We may complete the proof as in Proposition 3 if we first show that $\sum_{j=1}^{n} \lambda_j B_j$ is a τ_3-closed subset of $H(K)$ for any integer n and any finite sequence of sclars $(\lambda_j)_{j=1}^{n}$. Let $(f_\alpha)_{\alpha\in T} \in \sum_{j=1}^{n} \lambda_j B_j$ and suppose $f_\alpha \to f$ in $(H(K),\tau_3)$ as $\alpha \to \infty$. Let $f_\alpha = \sum_{j=1}^{n} \lambda_j f_{\alpha_j}$ for every α in T where $f_{\alpha_j} \in B_j$ for all α and j. Now for each j, $(f_{\alpha_j})_{\alpha\in T}$ is a bounded set in the semi-Montel space $(H(V_j),\tau_o)$ ([7]) and hence there exists $(f_j)_{j=1}^{n}$, $f_j \in B_j$, and a subset of $(f_\alpha)_{\alpha\in T}$, $(f_\beta)_{\beta\in B}$, such that $f_{\beta_j} \to f_j$ in $(H(V_j),\tau_o)$ as $\beta \to \infty$ for each integer j, $1 \leq j \leq n$. Let $g = \sum_{j=1}^{n} \lambda_j f_j$. Then $f_\beta \to g$ uniformly on the compact subsets of V_n and hence, since $f_\alpha \to f$ in $(H(K),\tau_3)$, $\hat{d}^n f(x) = \hat{d}^n g(x)$ for every x in K and every non-negative integer n. Thus $f = g$ in $H(K)$. Since $g \in \sum_{j=1}^{n} \lambda_j B_j$ we have completed the proof.

3. HOLOMORPHIC GERMS ON CERTAIN NON-METRIZABLE LOCALLY CONVEX SPACES

By using recent results concerning holomorphic functions on nuclear spaces ([5],[6],[9]) and by extending the method used in the previous section, we show that there exist non metrizable locally convex spaces such that $H(K)$ is complete and regular whenever K is a compact subset of one of these spaces. We frist describe the conditions which our spaces must satisfy and give some examples.

Our first and most important requirement is the following

(a) $H(0)$, the space of germs at 0 in E, is a regular inductive limit.

If E is a fully nuclear space with a basis ([4]) (i.e. E and E'_β are both reflexive nuclear spaces with a basis) then $H(0)$ is a regular inductive limit if and only if $(H(E'),\tau_o)$ is infra-barrelled ([5],[9]). Examples of non-metrizable locally convex spaces with this property are given in [6] and [9]. For example, this occurs if E is the strong dual of a B-nuclear space (e.g.

$E'_{\beta} = H(\mathbb{C})$ or $\mathcal{S}$) or if $E = \mathcal{D}'(\Omega)$. If E is fully nuclear with a basis and $H(0)$ is regular, then E'_{β} admits a continuous norm. This condition is not however sufficient ([9]).

For our second condition, we require the following definition.

<u>Definition 9</u>. A subset B of a locally convex space E is said to be a determining set for holomorphic functions if for every connected open set U which contains the origin and every f in $H(U)$,

$$f\big|_{U\cap B} = 0 \text{ implies } f \equiv 0.$$

(b) E contains a compact balanced determining set for holomorphic functions.

It is easily seen, by using a Taylor series expansion, that a balanced set is determining for holomorphic functions if and only if it is a determining set for continuous polynomials and hence, by induction (see [5]) if and only if it is a determining set for continuous linear functionals. Hence E contains a compact balanced determining set if and only if (E',τ_0) admits a continuous norm. In particular, if E is a semi-Montel space, then E has a compact balanced determining set if and only if E'_{β} admits a continuous norm. By our remarks concerning condition (a) we see that if E is a fully nuclear space with a basis and $H(0)$ is a regular inductive limit then E has a compact balanced determining set for holomorphic functions.

Our third requirement is:

(c) the compact subsets of E are metrizable.

This condition is satisfied if E is a reflexive dual nuclear space, in particular if E is a fully nuclear space, or if $(E',\sigma(E',E))$ is separable.

Our final condition is:

(d) every null sequence in E is a Mackey null sequence i.e. if $x_n \to 0$ as $n \to \infty$ in E then there exists a sequence of scalars $(\lambda_n)_n$, $|\lambda_n| \to \infty$ such that $\lambda_n x_n \to 0$ as $n \to \infty$.

If E is the strong dual of an infrabarrelled Schwartz space, then this condition is satisfied and in particular this holds if E is a fully nuclear space.

Proposition 10. If E is a locally convex space and

(a) $H(0)$, the space of germs at 0 in E, is regular;

(b) E contains a compact balanced determining set;

(c) the compact subsets of E are metrizable;

(d) every null sequence in E is a Mackey null sequence;

then $H(K)$ is a regular inductive limit for every compact subset K of E.

Proof. Let B denote a bounded subset of $H(K)$ and let p denote a continuous semi-norm on $H(0)$. Let $\tilde{p}(f) = \sum_{n=0}^{\infty} \sup_{x \in K} p(\frac{\hat{d}^n f(x)}{n!})$ for every f in $H(K)$. If $f \in H(K)$ then f is defined and bounded on some neighbourhood of K and hence there exists a neighbourhood of zero, W, such that $\|\frac{\hat{d}^n f(x)}{n!}\|_W \leq \frac{M}{2^n}$ for every x in K and every non negative integer n. Since p is continuous on $H(0)$ there exists $C(W) > 0$ such that $p(P) \leq C(W)\|P\|_W$ for every continuous polynomial on E, P.

Hence $\tilde{p}(f) \leq \sum_{n=0}^{\infty} C(W) \cdot \frac{M}{2^n} < \infty$, and since $H(K)$ is barrelled it follows that $\tilde{p}(f)$ is a continuous semi-norm on $H(K)$. Hence $\{\frac{\hat{d}^n f(x)}{n!}\}_{f \in B, x \in K}$ is a bounded subset of $H(0)$ and condition (a) implies that there exists a neighbourhood of zero V and $M > 0$ such that $\|\frac{\hat{d}^n f(x)}{n!}\|_V \leq \frac{M}{2^n}$ for every f in B and all n.

To complete the proof, we must show that the Taylor series

expansions about different points of K are coherent for all $f \in B$ in some neighbourhood of K.

To prove this, we consider all semi-norms on $H(K)$ which have the following from

$$p(f) = \sum_{1=i}^{\infty} \left(\frac{\lambda_i}{2}\right)^{n_i} \cdot \left| \sum_{n=0}^{n_i} \frac{\hat{d}^n f(X_i)}{n!}(w_i) - \sum_{n=0}^{n_i} \frac{d^n f(X'_i)}{n!}(X_i - X'_i + w_i) \right|$$

where $(n_i)_i$ is a strictly increasing sequence of positive integers, $(X_i)_i$ and $(X'_i)_i$ are two sequences in K which converge to the same point, $w_i \in L$ (the compact determining set) for all i, $\lambda_i \in C$ all i, $|\lambda_i| \to +\infty$, $\lambda_i w_i \to 0$ and $\lambda_i(X_i - X'_i) \to 0$ as $i \to \infty$.

We first show that any such semi-norm is continuous on $H(K)$. Since $H(K)$ is barrelled and the finite summands of p define a continuous semi-norm on $H(K)$ it suffices to show $p(f) < \infty$ for every f in $H(K)$.

Let $f \in H(B)$. Choose W_1 a neighbourhood of zero such that f is defined on $K + W_1$ and $\left\|\frac{\hat{d}^n f(x)}{n!}\right\|_{W_1} \leq \frac{M'}{2^n}$ for all x in K and all n.

Now choose N a positive integer such that $w_i \in W_1$, $X_i - X'_i + w_i \in W_1$, $\lambda_i w_i \in W_1$ and $\lambda_i(X_i - X'_i + w_i) \in W_1$ for all $i \geq N$.

For $i \geq N$

$$\sum_{n=0}^{\infty} \frac{\hat{d}^n f(X_i)}{n!}(w_i) = f(X_i + w_i) = f(X'_i + X_i - X'_i + w_i)$$
$$= \sum_{n=0}^{\infty} \frac{\hat{d}^n f(X'_i)}{n!}(X_i - X'_i + w_i)$$

and hence

$$\left| \sum_{n=0}^{n_i} \frac{\hat{d}^n f(X_i)}{n!}(w_i) - \sum_{n=0}^{n_i} \frac{\hat{d}^n f(X'_i)}{n!}(X_i - X'_i + w_i) \right|$$

$$\leq \sum_{n=n_i+1}^{\infty} \left| \frac{\hat{d}^n f(X_i)}{n!}(w_i) \right| + \sum_{n=n_i+1}^{\infty} \left| \frac{\hat{d}^n f(X'_i)}{n!}(X_i - X'_i + w_i) \right| \leq$$

$$\leq 2 \cdot \sum_{n=n_i+1} M' \cdot \left(\frac{1}{2\lambda_i}\right)^n = \frac{2M' \cdot (\frac{1}{2}\lambda_i)^{n_i}}{1 - \frac{1}{2\lambda_i}} .$$

Thus

$$\sum_{i\geq N} \left(\frac{\lambda_i}{2}\right)^{n_i} \left| \sum_{n=0}^{n_i} \frac{\hat{d}^n f(\chi_i)}{n!}(w_i) - \sum_{n=0}^{n_i} \frac{\hat{d}^n f(\chi'_i)}{n!}(\chi_i - \chi'_i + w_i) \right|$$

$$\leq \sum_{i\geq N} \left(\frac{\lambda_i}{2}\right)^{n_i} \cdot \frac{2M'}{1 - \frac{1}{2\alpha_i}} \left(\frac{1}{2\lambda_i}\right)^{n_i} = \sum_{i\geq N} \frac{2M'}{4^{n_i}\left(1 - \frac{1}{2\lambda_i}\right)} < \infty .$$

Hence p is a continuous semi-norm on $H(K)$. We now return to the bounded subset B of $H(K)$. To complete the proof, we must show that the elements of B are well defined on some neighbourhood of K. If not then for every convex balanced open subset of E, V_α, $V_\alpha \subset \frac{1}{2}V$, there exists $f_\alpha \in B$, $x_\alpha \in K$, $x'_\alpha \in K$, $y_\alpha, y'_\alpha \in V$ such that $x_\alpha + y_\alpha = x'_\alpha + y'_\alpha$ and

$$\tilde{f}_\alpha(x_\alpha)(y_\alpha) = \sum_{n=0}^{\infty} \frac{\hat{d}^n f(x_\alpha)}{n!}(y_\alpha) \neq \sum_{n=0}^{\infty} \frac{\hat{d}^n f_\alpha(x'_\alpha)}{n!}(y'_\alpha) = \tilde{f}(x'_\alpha)(y'_\alpha).$$

For each α let

$$g_\alpha(x) = \tilde{f}_\alpha(x_\alpha)(x) \quad \text{and} \quad h_\alpha(x) = \tilde{f}_\alpha(x'_\alpha)(x_\alpha - x'_\alpha + x).$$

If we order the sets $(V_\alpha)_\alpha$ by set inclusion, it follows that $y_\alpha, y'_\alpha \to 0$ as $\alpha \to \infty$ and $x_\alpha - x'_\alpha = y_\alpha - y'_\alpha \to 0$ as $\alpha \to \infty$. For all α g_α and h_α are both defined in some neighbourhood of zero. Since $g_\alpha(y_\alpha) = \tilde{f}_\alpha(x_\alpha)(y_\alpha)$ and $h_\alpha(y_\alpha) = \tilde{f}_\alpha(x'_\alpha)(x_\alpha - x'_\alpha + y_\alpha) = \tilde{f}_\alpha(x'_\alpha)(y_\alpha)$ it follows that $g_\alpha \neq h_\alpha$ and hence there exists $w_\alpha \in L \cap V_\alpha$ such that $g(w_\alpha) \neq h_\alpha(w_\alpha)$.

Since K and L are compact metrizable spaces, we can choose a sequence $(f_n)_n$ in B, $(x_n)_n$, $(x'_n)_n$ in K such that $(x_n)_n$ and $(x'_n)_n$ converge to the same element of K and $(w_n)_n$ a null sequence in L so that $\tilde{f}_n(x_n)(w_n) \neq \tilde{f}_n(x'_n)(x_n - x'_n + w_n)$ for all n. Now choose λ_n, $|\lambda_n| \to +\infty$ as $n \to \infty$ such that $\lambda_n w_n \to 0$ and $\lambda_n(x_n - x'_n) \to 0$ as $n \to \infty$. We may also assume that $\lambda_n w_n$ and

$\lambda_n(x_n - x'_n + w_n) \in V$ for all n.

Let $\delta_n = \tilde{f}_n(x_n)(w_n) - \tilde{f}_n(x'_n)(x_n - x'_n + w_n)|$ for all n. Now choose $(n_i)_i$ a strictly increasing sequence of positive integers such that

$$\left(\frac{\lambda_i}{2}\right)^{n_i} \cdot \left(\frac{\delta_i}{2}\right) > i$$

and

$$\left|\sum_{n=0}^{n_i} \frac{\hat{d}^n f_i(X_i)}{n!}(w_i) - \sum_{n=0}^{n_i} \frac{\hat{d}^n f_i(X'_i)}{n!}(X_i - X'_i + w_i)\right| > \frac{\delta_i}{2}.$$

If $p(f) = \sum_{i=1}^{\infty} \left(\frac{\lambda_i}{2}\right)^{n_i} \cdot \left|\sum_{n=0}^{n_i} \frac{\hat{d}^n f(X_i)}{n!}(w_i) - \sum_{n=0}^{n_i} \frac{\hat{d}^n f(X'_i)}{n!}(X_i - X'_i + w_i)\right|$

then $p(f_n) \geq n$ for all n and since p is a continuous semi-norm on $H(K)$ this contradicts the fact that $(f_n)_n \subset B$ is a bounded subset of $H(K)$ and completes the proof.

<u>Corollary 11</u>. If E is a fully nuclear space with a basis then the following are equivalent:

(1) $(H(E'_\beta), \tau_o)$ is an infrabarrelled locally convex space;

(2) $H(0)$ is a regular inductive limit;

(3) $H(K)$ is a complete locally convex space and a regular inductive limit for every compact subset K of E.

<u>Proof</u>. We have already noted the equivalence of (1) and (2). If (2) holds then all the conditions of Proposition 10 are satisfied and hence $H(K)$ is a regular inductive limit for any compact subset K of E. By [3] (Remark 13) this implies also that $H(K)$ is complete since every fully nuclear space is quasinormable. Hence (2) ⇒ (3) and (3) ⇒ (2) trivially.

We remark that it is also possible to show that $H(K)$ is Montel and boundedly retractive and to use the above to obtain properties of the τ_π topology on $H(U)$, U open. We refer to [3] for further details.

ADDED-IN PROOF

Since writing this paper we have been informed by a number of people that Theorem 1 and Proposition 2 are known results.

Theorem 1 is due to

Y. Komura, On linear topological spaces. Kumamoto J. of Sc., 5A, 148-157, 1962;

and Proposition 2 is due to

K. Noureddine and J. Schmets, Espaces Associes a un espace localement convexe et espaces de fonctions continues, B.S.R.Sc., Liege, 42, 109-117, 1973.

Further details can also be found in J. Schmets, Espaces de fonctions continues, Lecture Notes in Mathematics, 519, Springer-Verlag, 1976.

We decided not to rewrite the paper as the inclusion of the proofs of Theorem 1 and Proposition 2 makes the paper much more self-contained.

BIBLIOGRAPHY

[1] Aviles, P., Mujica, J., Holomorphic germs and homogeneous polynomials on quasi-normable metrizable spaces. Rend. Mat., 10, 117-127, 1977.

[2] Bierstedt, K.D., Meise, R., Nuclearity and the Schwartz property in the theory of holomorphic functions on metrizable locally convex spaces. Infinite Dimensional Holomorphy and Applications. Ed. M.C. Matos, North Holland Mathematical Studies, 12, 1977, p. 93-129.

[3] Bierstedt, K.D., Meise, R. Aspects of inductive limits in spaces of germs of holomorphic functions on locally convex spaces and applications to the study of $(H(U),\tau_\omega)$. Advances in Holomorphy Ed. J.A. Barroso, North Holland Mathematical Studies, 34, 1979, p. 111-178.

[4] Boland, P.J., Dineen, S., Holomorphic functions on fully nuclear spaces. B.S.M.Fr., 106, 311-336, 1978.

[5] Boland, P.J., Dineen, S., Duality theory for spaces of germs and holomorphic functions on nulcear spaces. Advances in Holomorphy. Ed. J.A. Barroso, North Holland Mathematical Studies, 34, 1979, p. 179-207.

[6] Boland, P.J., Dineen, S. Holomorphy on Spaces of Distributions. (to appear in Pac. Jour. Math.)

[7] Colombeau, J.F., Lazet, D., Sur les théorèmes de Vitali et de Montel en dimension infinie, C.R. Acad. Sc., Paris, t.274, p. 185-187, 1972.

[8] Dineen, S. Holomorphic functions on locally convex topological vector spaces I. Locally convex topologies on H(U). Ann. Inst. Fourier, 23, 1, 19-54, 1973.

[9] Dineen, S. Analytic functionals on fully nuclear spaces (preprint).

[10] Grothendieck, A. Sur les espaces (F) et (DF). Summa Brasil. Math. 3, 57-123, 1954.

[11] Hirschowitz, A. Bornologie des espaces de fonctions analytiques. Seminaire P. Lelong, 1969-1970, Springer Verlag Lecture Notes in Mathematics, 205, p. 21-33, 1970, (including typed corrections).

[12] Mujica, J. Spaces of germs of holomorphic functions, Studies in Analysis, Advances in Mathematics, Sup.Studies,4,p.1-41,1979.

[13] Noverraz, Ph. On topologies associated with the Nachbin Topology. Proc. Roy. Irish Acad., 77, p. 85-95, 1977.

[14] Pomes, R. Ouverts pseudo-convexes et domaines d'holomorphie en dimension infinie. These de 3e cycle, Université de Bordeaux I, 1977.

SOME MATHEMATICAL PROBLEMS IN NON-EQUILIBRIUM STATISTICAL MECHANICS

Gérard G. Emch
Departments of Mathematics and Physics
University of Rochester, NY (USA)

The purpose of this lecture is to review, for a mathematical audience, some of the connections recently explored between non-equilibrium statistical mechanics and several techniques in functional analysis. In doing so, we hope to stimulate further enquiries into some of the mathematical structures suggested by the process of building our physical understanding on reliable mathematical foundations.

In line with a time-honored tradition in applied mathematics, the first section is a heuristic presentation of the physical question to be addressed. In the second section we use the Nagy dilation technique for semi-groups of contractive operators to extract some positive information from a no-go theorem. We show in the third section the relevance of the concept of spectral concentration for the decay problem. In the fourth section we direct our attention to some analogy between the theory of the asymptotic behavior of solutions of certain stochastic differential equations and the van Hove limit in non-equilibrium statistical mechanics. In the fifth section we turn to certain aspects of the theory of completely positive maps on von Neumann algebras to propose a definition of dynamical systems which extends the classical notions to the quantum realm; as an illustration we discuss the Bloch equation governing spin-relaxation. Finally in the sixth section we briefly indicate some recent advances in the non-commutative extensions of classical ergodic theory.

1. HEURISTICS

The fundamental problem of non-equilibrium statistical mechanics is to reconcile two physical descriptions of nature: non-equilibrium thermodynamics and mechanics.

The termodynamical description is a macroscopic, phenomenological theory epitomized by dissipative transport equations. An important class of such equations present themselves as parabolic differential equations with time-independent coefficients, the so-called "transport coefficient". A typical example is provided by the diffusion equation:

(1) $$\partial_t \rho(x,t) = D\, \partial_x^2\, \rho(x,t)$$

with $t \in R^+$ and $x \in R^s$ ($s = 1,2$, or 3, the dimension of space). The solution of such equations can be written in the form

(2) $$\rho_t = \gamma(t)[\rho] \quad \text{with} \quad \gamma(t) = \exp(-\Lambda t)$$

and $\{\gamma(t) \mid t \in R^+\}$ is a contractive (or "dissipative") semi-group acting on an appropriate Banach space, the space of solutions of the given differential equation.

The microscopic description on the other hand assumes that one has a "large" collection of particles the motion of which is governed by the laws of classical or quantum mechanics. In the latter case, the corresponding equation is the Schroedinger equation:

(3) $$\partial_t\, \Psi(x,t) = -i\, H\, \Psi(x,t)$$

with $t \in R$ and $x \in R^{sN}$, N denoting the number of particles; and where the Hamiltonian H is a differential operator acting on $\mathfrak{H} = \mathcal{L}^2(R^{sN},dx)$ and typically breaking into the sum of a kinetic energy part:

(4) $$H_o = \frac{1}{2m} \sum_{n=1}^{N} \partial_{x_n}^2$$

and a potential energy part:

(5) $$V(x_1,\ldots,x_N) = \sum_n V_1(x_n) + \frac{1}{2}\sum_{n\neq m} V_2(|x_n - x_m|)$$

including there the effect of a common outside potential V_1 and a two-body interaction V_2. The solution of the Schroedinger equation can be written in the form:

(6) $$\Psi_t = U(t)\Psi \quad \text{with} \quad U(t) = \exp(-i\,H\,t) \quad \text{and} \quad \|\Psi\| = 1$$

so that $\{U(t) \mid t \in R\}$ is a one-parameter group of unitary operators acting on $\mathfrak{H}$. For the purpose of statistical mechanics it is necessary [65] to consider more general states than those corresponding to such normalized vector Ψ, i.e. states which are more general than the pure (or "extremal") positive normalized linear functionals:

(7) $$\psi: A \in B(\mathfrak{H}) \mapsto \langle \psi ; A\rangle = (A\Psi,\Psi) \in C .$$

In particular the ultraweakly continuous, normalized, positive, linear functionals:

(8) $$\psi: A \in B(\mathfrak{H}) \mapsto \langle \psi ; A\rangle = \mathrm{Tr}\, \rho A$$

with ρ self-adjoint, positive, trace-class and $\mathrm{Tr}\,\rho = 1$, have traditionally played an important role. Note that the pure states (7) appear as particular cases, namely those states corresponding to a density operator ρ of the form:

(9) $$\rho: f \in \mathfrak{H} \mapsto (f,\Psi)\Psi \in \mathfrak{H}.$$

The evolution of the states (8) is then given, in conformity with (6), by the von Neumann equation:

(10) $$\rho_t = U(t)\,\rho\, U(-t)$$

i.e.

(11) $$\rho_t = \alpha(t)[\rho] \quad \text{with} \quad \alpha(t) = \exp(-i\delta t)$$

corresponding to the <u>conservative</u> differential equation

(12) $$\partial_t \rho_t = -i\delta[\rho_t] \quad \text{with} \quad \delta[\rho] = H\rho - \rho H = [H,\rho]$$

which thus appear as a generalization of the Schroedinger equation (3) more suitable for the purpose of statistical mechanics, where the information available intially on the complex system of N particles is far from being sufficient to specify the state to the extend where it would be a pure state.

We can now reformulate our fundamental problem as one in which we ask whether the two equations (2) and (11) are compatible. This problem presents some similarity with the classical (and fully solved) problem of reconstructing a continuous Markov process from a continuous Markov semi-group. There, as here, the problem appears to be that of embedding the dissipative system of interest (thermodynamical description) into a larger conservative system (mechanistic description), and then to project the evolution in the large system down to the system of interest. We thus will first want to ask whether, given $\{Y(t) \mid t \in R^+\}$, there exist an embedding i, a conditional expectation E, and a conservative evolution $\{\alpha(t) \mid t \in R\}$ such that:

$$E \circ \alpha(t) \circ i = Y(t) \quad \text{for all} \quad t \in R^+. \tag{13}$$

The problem of the foundations of non-equilibrium statistical mechanics is moreover complicated by two facts. Firstly, we want the generator δ of $\{\alpha(t) \mid t \in R\}$ to have some connection with a realistic Hamiltonian H, say of the form (3-5). Secondly, we will have to account for the scale, spacial and temporal, in which macroscopic processes manifest themselves; this scale is quite different from that characteristic of most microscopic processes. It is in this connection that the notions of thermodynamical limit, and of asymptotic dynamics will have to be introduced.

2. THE NAGY MINIMAL DILATION

Although the well-known theory of minimal dilations ([73], [64] see also [58]) does not by itself provide a large enough frame for an answer to our problem, we nevertheless want to show in this section how this theory can be used to obtain some preliminary indication that some kind of "large system" limit is a necessary ingredient for a satisfactory solution to our problem.

Let us first recall the central result of the theory. If $\{S(t) \mid t \in R^+\}$ is a continuous one-parameter semi-group of contractive operators acting on a Hilbert space $\mathfrak{H}_o$, then there exist a continuous one-parameter group $\{U(t) \mid t \in R\}$ of unitary operators acting on a Hilbert space $\mathfrak{H}$, and an isometric embedding X of $\mathfrak{H}_o$ into $\mathfrak{H}$ such that:

(14) $$X^*U(t)X = S(t) \text{ for all } t \in R^+.$$

Moreover the dilation $\{\mathfrak{H}, U(R), X\}$ is completely determined, up to unitary equivalence of course, by $\{\mathfrak{H}_o, S(R^+)\}$ if one imposes the condition:

(15) $$\overline{\text{Span}}\, \{U(t)f \mid t \in R,\ f \in \mathfrak{H}_o\} = \mathfrak{H}.$$

This dilation is called minimal, and it is clearly contained in any dilation $\{\mathfrak{H}', U'(R), X'\}$ of $\{\mathfrak{H}_o, S(R^+)\}$. Together with the spectral theorem for the generator of a semi-group, this universality of the minimal Nagy dilation is responsible for the general character of the conclusion that we are about to derive from the elementary example which follows.

Consider indeed the most simple example where $S(t) = \exp(-at)$ (with $a > 0$) acting in the one-dimensional Hilbert space C. It is easy to check that the minimal Nagy dilation is:

(16) $$\begin{cases} \mathfrak{H} = \mathfrak{L}^2(R,dx) & \\ [U(t)f](x) = \exp(-ixt)f(x) & \text{for all } f \in \mathfrak{H} \\ X: z \in C \mapsto zf_o \in \mathfrak{H} & \text{for all } z \in C \\ f_o(x) = [a/\pi(a^2+x^2)]^{1/2} & \end{cases}$$

One has indeed, for all $z_1, z_2 \in C$ and $t \in R^+$

$$(17)\quad \begin{cases} (X^*U(t)Xz_1,z_2) = \frac{1}{\pi} z_1 z_2^* \int dx \, \exp(ixt)[a/(a^2+x^2)] = \\ z_1 z_2^* \exp(-at) = (S(t)z_1,z_2). \end{cases}$$

This example already shows an interesting feature, namely that the spectrum of the generator L of our unitary group $\{U(t) \mid t \in R\}$ extends over the entire real axis, and is absolutely continuous with respect to Lebesgue measure. This is no mere accident (see also [17], [88], and [58]). These properties of Sp(L) indicate two things. First, since L is unbounded below (as well as above), it can not be interpreted as the Hamiltonian of a stable quantum mechanical system in the usual sense. This however is no objection to the eventual interpretation of L as coming from an Hamiltonian in a way akin to the way δ is linked to H in (12), since the spectrum of such an operator is symmetric around the origin. On the other hand, even if L were to be interpreted in this manner, the absolute continuity of its spectrum does rule out the possibility that H itself have a discrete spectrum. Hence the Nagy dilation (and thus its non minimal extensions) cannot describe such things as a finite assembly of harmonic oscillators with discrete collective modes, nor can it describe an ordinary conservative quantum mechanical systems restricted to a compact subset of R^s. It thus appears that more general systems will have to be considered for a complete microscopic description of any system exhibiting the kind of dissipative behaviour alluded to in section 1.

Whereas the above argument is not to be taken as a general proof for the necessity of the thermodynamical limit (in which the number N of particles, and the volume V available to the system, both approach infinity, while the density N/V remains finite), it is nevertheless one of the simplest mathematical indication pointing in that direction, in conformity with the traditional wisdom of physicists. We shall later on in this lecture discuss further information on this account, as it becomes available from the study of some particular models.

3. ASYMPTOTIC DYNAMICS AND SPECTRAL CONCENTRATION

The argument of the preceding section involves no priviledged time-scales: it ignores the fact that the characteristic times of microscopic and macroscopic events are quite different (such as, for instance, the collision-time and the mean-free-time in a dilute gas). When one wishes to take the microscopic description as fundamental, one should then expect the macroscopic description to appear as some kind of asymptotic dynamics. This aspect of our problem is approached from two different points of view in this section and in the next section.

The relevant mathematical concept for this section is the notion of "spectral concentration". A beautiful mathematical presentation can be found in [51], together with a brief heuristic introduction as well as some indications of the physical motivation of this notion as the mathematical formulation of the idea of "weak quantization" ([52], [57]). Rather than repeating here Kato's presentation, we will review some of the principal features of the theory with the help of a particular model; moreover we use the information obtained from this specific model to illustrate some important consequences of the theory.

In $\mathfrak{H}^I = \mathcal{L}^2([0,\pi],dx)$ let $H^I = -\Delta$ be defined as a self-adjoint operator with domain $\mathfrak{D}^I = \{\Phi \in \mathfrak{H}^I \mid \Phi, \Phi' \text{a.c.}; \Phi'' \in \mathfrak{H}^I;$ $\Phi(0) = \Phi(\pi) = 0\}$. Clearly

$$\{\psi_n(x) = \sin(nx) \mid n = 1,2,\ldots\} \tag{18}$$

is an orthonormal basis of eigenvectors of H^I in $\mathfrak{H}^I$ satisfying:

$$H^I\psi_n = n^2\psi_n. \tag{19}$$

The physical interpretation of these objects is that H^I is the

Hamiltonian of a particle moving in the interval $[0,\pi]$ with rigid boundary conditions at 0 and π; ψ_n are the eigenmodes of this particle. We now approximate this situation by replacing the hard impenetrable barrier at $x = \pi$ by a weaker one, of appropriately large height a, and extending from $x = \pi$ to $x = \pi + b$. Mathematically, this new situation is described by defining in $\mathfrak{H} = \mathcal{L}^2(R^+, dx)$: $H_a = H_o + aV$ on $\mathfrak{D}_o = \{\Phi \in \mathfrak{H} \mid \Phi, \Phi' \text{ a.c.}; \Phi'' \in \mathfrak{H}; \Phi(0) = 0\}$ where $H_o = -\Delta$ and V is given by

(20) $$(Vf)(x) = \chi_{[\pi,\pi+b]}(x)f(x) \quad \text{for all} \quad f \in \mathfrak{H}.$$

For every $a > 0$ H_a is a positive, self-adjoint operator; in contradistinction with H^I, which has purely discrete spectrum, the spectrum of H_a is absolutely continuous with respect to Lebesgue spectrum. Physically, this is the simplest model for the "tunnel effect" (see for instance [61], or in a lighter vein [35]), first invoked to explain radio-active decay in nuclear physics. In this crude caricature of the nucleus, a low-energy particle initially confined in the well $[0,\pi]$ is expected to "leak" to the outside world $[\pi+b,\infty)$ through the semi-transparent region $[\pi,\pi+b]$. There is good experimental evidence that on a wide observable time-scale this leaking is described by an exponential decay. We now concentrate on the mathematical description of this phenomenon, through the use of the notion of spectral concentration. To this effect we first introduce $\mathfrak{H}_\infty = \mathfrak{H}^I + \mathfrak{H}^{III}$ with $\mathfrak{H}^{III} = \mathcal{L}^2([\pi+b,\infty), dx)$; $\mathfrak{D}^{III} = \{\Phi \in \mathfrak{H}^{III} \mid \Phi, \Phi' \text{ a.c.}; \Phi'' \in \mathfrak{H}^{III}; \Phi(\pi+b) = 0\}$; $H^{III} = -\Delta$ defined as a self-adjoint operator with domain $\mathfrak{D}^{III}$; and $H_\infty = H^I + H^{III}$. The notion of spectral concentration involves a family of self-adjoint operators, here $\{H_a \mid a \in R^+\}$, with H_∞ being the limit of H_a $(a < \infty)$ in some sense to be properly defined. Here, one can prove that for every $\Phi \in \mathfrak{D}$, a domain of essential self-adjointness of H_∞, there exists $\{\Phi_a \mid a \in R^+\} \subset \mathfrak{D}_o$, the domain of self-adjointness of H_a, such that we have, as $a \to \infty$:

$$(21)\qquad \begin{cases} \|\Phi_a - \Phi\| = O(a^{-1/2}) \\ \|H_a\Phi_a - H_\infty\Phi\| = O(a^{-1/2}) \end{cases}$$

This result has the following dynamical consequence: one can prove, using Kurtz [56] criterion of extended operator convergence, that for every $\Phi \in \mathfrak{H}_\infty$ and $T \in R^+$

$$(22)\qquad \lim_{a\to\infty} \sup_{0\le t\le T} \|[U_a(t) - U_\infty(t)]\Phi\| = 0$$

where $U_a(R)$ [resp. $U_\infty(R)$] is the continuous one-parameter group of unitary operators acting on $\mathfrak{H}$ [resp. $\mathfrak{H}_\infty$] generated by H_a [resp. H_∞].

The point of the notion of spectral concentration is that this convergence results can be significantly sharpened. Let indeed $\{E_a(S) \mid S \in B(R)\}$ denote, for each $a \in R^+$, the spectral family of the self-adjoint operator H_a. One can then prove that for each $n = 1,2,\ldots$ there exist positive-valued functions $\lambda(n,\cdot)$ and $\Gamma(n,\cdot)$ such that for every ordered pair $h_2 \geq h_1$ of real numbers:

$$(23)\qquad \begin{cases} \lim_{a\to\infty} (E_a([\lambda(n,a) + \Gamma(n,a)h_1,\ \lambda(n,a) + \Gamma(n,a)h_2])\psi_n,\psi_n) \\ \qquad = \frac{1}{\pi}\int_{h_1}^{h_2} dh(1+h^2)^{-1}. \end{cases}$$

To avoid the speculation that there might ([16], [12]) be a contradiction between (21) and (23), it is perhaps well to point out here that whereas, for large a

$$(24)\qquad \lambda(n,a) - n^2 = O(a^{-1/2})$$

$\Gamma(n,a)^{-1}$ depends in a non-polynomial manner on the coupling a; one has indeed:

$$(25)\qquad \Gamma(n,a)^{-1} \cong a\exp(a^{1/2}\, b).$$

The mathematical interpretation of (23) is clear: the absolutely continuous spectrum of H_a concentrates asymptotically (as $a \to \infty$) around the eigenvalues of H_∞. This is the essence of the mathematical notion of spectral concentration. Correspondingly, the physical interpretation of (23) is that, as the height a of the barrier increases, a sharp Lorentzian resonance of width $\Gamma(n,a)$ builds up in the region where an eigenvalue of H_∞ is located. Furthermore, an important dynamical interpretation of (23) can be obtained rigorously, namely:

$$\lim_{a\to\infty} (\exp\{-i[H_a - \lambda(n,a)]\, \Gamma(n,a)^{-1}\, \tau\}\psi_n,\psi_n) = \exp(-\tau) \tag{26}$$

the convergence being uniform in $\tau \in R^+$. Physically, the interpretation of (26) is that for large values of a, one thus has:

$$|(\exp[-iH_a t]\ \psi_n,\psi_n| \cong \exp(-2\,\Gamma(n,a)t). \tag{27}$$

The LHS of (27) is the quantum transition probability of interest; the RHS of (27) is the exponential decay we were looking for: $\Gamma(n,a)^{-1}$ is the half-life time of the eigenmode ψ_n, and as such is the characteristic time of the decay process; as the height of the barrier becomes large, this time becomes extremely large, see (25); hence the notion of spectral concentration, expressed here by (23), provides a powerful mathematical tool to focus on the asymptotic dynamics, here (26), of the process under investigation.

From both the mathematical and the physical points of view, a more thourough discussion of the phenomenon at hand (together with some generalizations of the above model) was presented in [30], where all the proofs of the above mathematical statements were given in details.

4. ASYMPTOTIC DYNAMICS AND STOCHASTIC DIFFERENTIAL EQUATIONS

A method to study the reduced dynamics of many-body open dynamical systems is sketched in this section. We first propose an abstract Hilbert space formulation of the problem. We next show how a certain type of stochastic differential equations does appear in this way. We then specify things further to construct a stochastic model for the evolution of a quantum particle (say an electron) interacting with a random static field (such as that created by random impurities in a crystal). We describe the asumptotic dynamics resulting from the stochastic differential equation corresponding to this model; as an application, we give the form of the collision kernel for the master equation governing the evolution of the observables. We finally indicate some of the generalizations recently studied in the literature.

Starting with the abstract Hilbert space formulation, we let $\mathfrak{H}^{(1)}$ be a Hilbert space; $\{U^{(1)}(t) \mid t \in R\}$ be a continuous one-parameter group of unitary operators acting on $\mathfrak{H}^{(1)}$; and $H^{(1)}$ be its generator. These ingredients are supposed to describe the evolution of the system of interest. The "outside world" with which this system interacts is similarly described by $\mathfrak{H}^{(2)}$, $\{U^{(2)}(t) \mid t \in R\}$, and $H^{(2)}$. In the absence of interaction the composite system is described by

(28)
$$\begin{cases} \mathfrak{H} = \mathfrak{H}^{(1)} \otimes \mathfrak{H}^{(2)} & \\ U_o(t) = U^{(1)}(t) \otimes U^{(2)}(t) & t \in R \\ H_o = H^{(1)} \otimes I + I \otimes H^{(2)} & \end{cases}$$

An interaction of the form

(29)
$$V = \sum_n A_n^{(1)} \otimes A_n^{(2)}$$

is then introduced, coupling the system of interest with the outside world. The joint evolution of the composite system is thus now given by the continuous one-parameter group of unitary operators

(30)
$$\{U_\lambda(t) \mid t \in R\} \quad \text{with generator} \quad H_\lambda = H_o + \lambda V$$

λ is a coupling constant which will ultimately determine the asymptotic time-scale. We finally suppose that there is a priviledged vector $\Phi^{(2)} \in \mathfrak{H}^{(2)}$ with

(31)
$$\|\Phi^{(2)}\| = 1 \quad \text{and} \quad H^{(2)}\Phi^{(2)} = 0$$

from which we construct the isometric embedding

$$X: f^{(1)} \in \mathfrak{H}^{(1)} \mapsto f^{(1)} \otimes \Phi^{(2)} \in \mathfrak{H} \tag{32}$$

and its adjoint

$$X^*: \sum_n f_n^{(1)} \otimes f_n^{(2)} \in \mathfrak{H} \mapsto \sum_n (f_n^{(2)}, \Phi^{(2)})\, f_n^{(1)} \in \mathfrak{H}^{(1)}. \tag{33}$$

We now ask the central question of this section, namely whether there exist a monotonically increasing function ν of the coupling $\lambda > 0$, and a contractive semigroup $\{S_\nu(\tau) \mid \tau \in R^+\}$ acting on $\mathfrak{H}^{(1)}$ such that:

$$S_\nu(\tau) = \lim_{\substack{t\to\infty \\ \nu(\lambda)t=\tau}} X^* U_o(-t)\, U_\lambda(t) X \tag{34}$$

in the appropriate topology. The case $\nu(\lambda) = \lambda^2$ seems to be of singular physical significance, and this particular case of (34) shall be referred to here as the van Hove limit, in the honor of L. van Hove who first proposed [42] conditions on V and H_o under which this limit could reasonably be expected to exist and to be non-trivial. It should nevertheless also be pointed out here that his pionnering effort has, to this day, not yet been cast into a general, mathematically reliable theory. We will review below a particular model [60] for which the mathematics of the situation have first been brought under complete control.

We now proceed to show how the question asked around equ.(34) above is very much akin to a well-known ([53], [66], [40], [11], [68], [67], [49]) problem in the theory of stochastic differential equations. Let indeed $\{\Omega, \Sigma, \mu\}$ be a probability space, and $\{T(t) \mid t \in R\}$ be a continuous one-parameter group of measurable, measure preserving transformations of Ω. We then identify:

$$\left\{\begin{array}{l} \mathfrak{H}^{(2)} = \mathcal{L}^2(\Omega,\Sigma,\mathbb{C},\mu) = \{f: \Omega \to \mathbb{C} \mid \int_\Omega d\mu(\omega)\,|f(\omega)|^2 < \infty\} \\ \left.\begin{array}{l} [U^{(2)}(t) f^{(2)}](\omega) = f^{(2)}(T(t)[\omega]) \\ \Phi^{(2)}(\omega) = 1 \end{array}\right\} \text{for all } \omega \in \Omega \end{array}\right. \tag{35}$$

and we further assume that

$$(36)\quad \begin{cases} V = \sum_n A_n \otimes v_n \quad \text{with} \\ [v_n f^{(2)}](\omega) = v_n(\omega) f^{(2)}(\omega). \end{cases}$$

We thus have now

$$(37)\quad \begin{cases} \mathfrak{H} = \mathcal{L}^2(\Omega, \Sigma, \mathfrak{H}^{(1)}, \mu) \\ \quad = \{f : \Omega \to \mathfrak{H}^{(1)} \mid \int_\Omega d\mu(\omega) \, \|f(\omega)\|^2 < \infty \end{cases}$$

Upon defining for every $f \in \mathfrak{H}^{(1)}$

$$(38)\quad f_\lambda(t,\omega) = (U_o(-t) U_\lambda(t) X f)(\omega)$$

we obtain a time-dependent $\mathfrak{H}^{(1)}$-valued random variable $f_\lambda(t,\cdot)$ satisfying the stochastic differential equation

$$(39)\quad \partial_t f_\lambda(t,\cdot) = -i \, \lambda V_t^{(1)}(\cdot) \, f_\lambda(t,\cdot)$$

which, in the interaction picture, represents the Schroedinger equation corresponding to the time-dependent random Hamiltonian $H^{(1)}(t,\cdot)$ acting in $\mathfrak{H}^{(1)}$ and defined by

$$(40)\quad \begin{cases} H^{(1)}(t,\cdot) = H^{(1)} + \lambda V^{(1)}(t,\cdot) \quad \text{with} \\ V^{(1)}(t,\cdot) = \sum_n v_n(T(t)[\cdot]) A_n \end{cases}$$

so that

$$(41)\quad V_t^{(1)}(\cdot) = U^{(1)}(-t) V^{(1)}(t,\cdot) U^{(1)}(t).$$

Note that the random operator $V_t^{(1)}(\cdot)$ governing (39) and defined by (40)-(41) depends on time in two ways: explicitly through $U^{(1)}(t)$, and hence $H^{(1)}$, since we work in the interaction picture; and implicity through $V^{(1)}(t, \;)$, and hence $H^{(2)}$, which reflects the fact that we reduced the description to $\mathfrak{H}^{(1)}$. Our principal question now becomes to prove, instead of (34), the existence of an operator Λ_ν, acting in $\mathfrak{H}^{(1)}$, such that:

$$(42)\quad (\exp[-\Lambda_\nu \tau] f, g) = \lim_{\substack{t \to \infty \\ \nu(\lambda)_t = \tau}} E\{(f_\lambda(t,\cdot), g)\}$$

for all $f,g \in \mathfrak{H}^{(1)}$ and all $\tau \in R^+$; E is the expectation with respect to μ. This reformulation of (34) exhibits the equivalence of this former equation with the well-known problem of determining the asymptotic behaviour [here (39)], a problem which has extensively been studied in the references mentioned above. As the methods of the latter papers do however not directly apply to some of the models we want to study in statistical mechanics, we shall outline here an alternate method ([60], see also [77]) which turns out to vindicate the scheme proposed by van Hove in the original context of our problem.

We now specify the model by defining

$$(43)\quad \begin{cases} \mathfrak{H}^{(1)} = \mathcal{L}^2(B,d\theta) \quad \text{with} \quad B = [-\pi,\pi]^3 \\ [H^{(1)}f](\theta) = \omega(\theta)\, f(\theta) \quad \text{with} \quad \omega(\theta) = \theta^2 \end{cases}$$

Upon using the three-dimensional Fourier transform

$$(44)\quad \tilde{f}(n) = (2\pi)^{-3/2} \int_B d\theta \quad \exp(-i\theta n) f(\theta)$$

we establish an isometric isomorphism between $\mathcal{L}^2(B,d\theta)$ and $\ell^2(Z^3)$. Let now $\{\xi_n \mid n \in Z^3\}$ be the orthonormal basis in $\ell^2(Z^3)$ defined by $\xi_n(k) = \delta_{n,k}$; for each $n \in Z^3$, let P_n be the projector

$$(45)\quad P_n: f \in \mathfrak{H}^{(1)} \mapsto (f,\xi_n)\xi_n \in \mathfrak{H}^{(1)}.$$

We further specify our probability space $\{\Omega,\Sigma,\mu\}$ as that sustaining the gaussian random distribution $\{v_n \mid n \in Z^3\}$ with mean zero and covariance

$$(46)\quad E\{v_n(\cdot)v_m(\cdot)\} = g_{|n-m|}$$

on which we impose the conditions

$$(47)\quad \begin{cases} \|g\|_1 = \sum_{n\in Z^3} |g_n| < \infty \\ g(\theta) = (2\pi)^{-3/2} \sum_{n\in Z^3} \exp(i\theta n)\, g_n > 0 \text{ for all } \theta \in B \end{cases}$$

For every finite subset $M \subset Z^3$ we define the random operator

$$(48)\quad V^M(\cdot) = \sum_{m\in Z^3} v_m(\cdot) P_m .$$

Note that in this simplified model $H^{(2)} = 0$. This corresponds physically to the recoilless approximation in which the impurities, randomly distributed on a cubic lattice, are supposed to be infinitely heavy. This very simple model has already the advantage to show that an infinite volume limit [here $M \to Z^3$] is indispensible to get a non-trivial result for the van Hove limit in (42). Indeed we can use the fact that $\mathcal{L}^2(B,d\theta)$ is an invariant subspace of $\mathcal{L}^2(R^3,dk)$ with respect to $\{\hat{U}(s) \mid s \in R\}$ defined by

(49) $$[\hat{U}(s)f](k) = \exp(-ik^2 s)f(k).$$

We have then

$$(\hat{U}(s)f,\xi_m) = (2\pi)^{-3/2} \int dk\, [\hat{U}(s)f](k) \exp(imk)$$
$$= (4\pi is)^{-3/2} \int dy\, \exp(i|x-y|/4s)\tilde{f}(y).$$

Hence for all $f \in \mathcal{L}^2(B,d\theta) \cap C_o^\infty(R^3)$ we have:

(50) $$|(\hat{U}(s)f,\xi_m| \le (4\pi s)^{-3/2} \|f\|_1.$$

Upon dropping the variable ω in the forthcoming estimate, we write:

$$\|[U_o(-t)U_\lambda^M(t) - I]f\| = \|[U_\lambda^M(-t)U_o(t) - I]f\| =$$
$$\|\int_0^t ds \frac{d}{ds} U_\lambda^M(-t)U_o(t)f\| \le \lambda \int_0^t ds\|V^M U_o(s)f\| \le$$
$$\sum_{m\in M} |v_m| \int_0^t ds\, |(U_o(s)f,\xi_m)|$$

so that for all $\tau \in R^+$ we have, upon using (50):

(51) $$\underset{\substack{t\to\infty \\ \lambda^2 t=\tau}}{\text{s-lim}}\; U_o(-t)U^M(t) = I$$

and thus $\Lambda = 0$ whenever M is finite. We therefore must take the limit $M \to Z^3$ before we take the van Hove limit. Even then to prove that the result is non-trivial involves some detailled computations, the idea of which we now briefly summarize. From the second resolvant equation, one first derive the well-known Dyson expansion:

(52)
$$\begin{cases} U_o(-t)U_\lambda^M(t) = \sum_{n=0} U_{\lambda,n}^M(t) \quad \text{with} \\ U_{\lambda,n}^M(t) = (-i\lambda)^n \int\limits_{t\geq t_n} dt_n \cdots \int\limits_{\geq\ldots\geq t_1\geq 0} dt_1 \quad V_{t_n}^M \cdots V_{t_1}^M \end{cases}$$

and $V_s^M = U_o(-s)V^M U_o(s)$.

One then prove, without much trouble, that for every $t \in R$ there exists a function $T_t: B \to C$, essentially bounded by 1 such that for every $f,g \in \mathfrak{H}^{(1)}$:

(53)
$$\lim_{M\to Z^3} E\{(U_o(-t)U_\lambda^M(t)f,g)\} = \int_B d\theta\ T_t(\theta)f(\theta)g(\theta)^*.$$

One then uses the fact that $\{v_n\}$ is a gaussian distribution with mean zero, and more specifically that

(54)
$$\begin{aligned} &E\{v_1(\cdot) \ldots v_{2k+1}(\cdot)\} = 0 \\ &|E\{v_1(\cdot) \ldots v_{2k}(\cdot)\}| = \frac{(2k)!}{k!2^k} \|g\|_1 \end{aligned}$$

to control separately each term of the Dyson expansion (52). Upon summing up these contributions in the van Hove limit, one obtains the central result of this section: there exists $0 < \tau_o < \infty$ such that for each fixed $\tau \in [0,\tau_o]$ and $f,g \in \mathfrak{H}^{(1)}$

(55)
$$\lim_{\substack{t\to\infty \\ \lambda^2 t=\tau}} \lim_{M\to Z^3} E\{(U_o(-t)U_\lambda^M(t)f,g)\} = \int_B d\theta\ S_\tau(\theta)f(\theta)g(\theta)^*$$

with

(56)
$$S_\tau(\theta) = \exp\{-[\Gamma(\theta) + i\Delta(\theta)]\}$$

where

(57)
$$\begin{cases} \Gamma(\theta) = \pi \int_B d\theta'\ W(\theta',\theta)\delta(\omega(\theta') - \omega(\theta)) \\ \Delta(\theta) = P\!\!\int_B d\theta'\ W(\theta',\theta)\ (\omega(\theta') - \omega(\theta))^{-1} \\ W(\theta',\theta) = (2\pi)^{-3/2}\ g(\theta'-\theta) \end{cases}$$

This result rests on three principal assumptions: the form of the energy dispersion $\omega(\theta) = \theta^2$ in (43); the space dimension $s \geq 3$;

and the gaussian character of the distribution (46)-(47). A closer examination of the detailled computations shows that the former two assumptions are central in explaining the emergence of the dissipative behaviour as being rooted in the "spreading of the wave-packet" (the physicist's way of referring to eqn. (50)).

The interest of this simple model is threefold. Firstly, from a mathematical point of view, it indicates a promising novel way to attack the problem of determining the asymptotic behavior of certain stochastic differential equations. In this respect, it is worth pointing out here that some generalizations of the model have already been obtained: one in the very paper [60] where the method was first carried out; and a further extension in an interesting paper by Spohn [77], the solution again going through the same essential steps of the original method outlined here. Secondly, a diagramatic representation of the various terms appearing in the Dyson expansion can be laid down; it parallels closely (and this fact actually served as a strong guide in the original investigation) the diagramatic description proposed by van Hove. In fact, van Hove's criterion for dissipative behavior is a "diagonal singularity" which can here be expressed as follows. Whereas $E\{V(\cdot)\} = 0$, one has for every operator A on $\mathfrak{H}^{(1)}$ of the form

$$(Af)(\theta) = A(\theta)f(\theta) \tag{58}$$

that the kernel of $E\{V(\cdot)AV(\cdot)\}$ has the singularity:

$$E\{V(\cdot)AV(\cdot)\}(\theta',\theta) = \delta(\theta'-\theta)\int_B d\theta''\ A(\theta'')W(\theta'',\theta) \tag{59}$$

with W given in (57). Thirdly, and still in connection with van Hove's scheme, the method used here to arrive at (55)-(57) can be used, in its essential features, to compute for every observable of the form (58) the van Hove limit of the expectation of the random operator:

$$A_t(\cdot) = U(-t,\cdot)AU(t,\cdot). \tag{60}$$

One then finds this limit to be given by

$$(61)\quad \begin{cases} A_\tau = \gamma(\tau)[A] \quad \text{with} \quad \gamma(\tau) = \exp(\delta\tau) \quad \text{and} \\ \delta[A](\theta) = 2\pi \int_B d\theta' \, \delta(\omega(\theta')-\omega(\theta)) W(\theta',\theta)(A(\theta')-A(\theta)) \end{cases}$$

which is to say that in the van Hove limit the evolution of the lattice-translation invariant observables is described by the momentum distribution function $\rho_\tau(\theta)$ which satisfies the "markovian master equation":

$$(62)\quad \begin{cases} \partial_\tau \rho_\tau = \delta^*[\rho_\tau] \quad \text{with} \\ \delta^*[\rho](\theta) = \int_B d\theta' [K(\theta,\theta')\rho(\theta') - K(\theta',\theta)\rho(\theta)] \quad \text{and} \\ K(\theta,\theta') = 2\pi\delta(\omega(\theta') - \omega(\theta)) \, W(\theta',\theta) \end{cases}$$

In line with the original argument given by Pauli [70] the symmetric kernel K in (62) plays the role of the collision operator appearing in the usual transport equations; again this argument, together with the computations associated to it, does generalize to the models referred to earlier [60], [77]; and in particular, the method can be used for models where the recoilless approximation $H^{(2)} = 0$ is removed. One short-coming of the method however is the appearance of $\tau_0 < \infty$; this seems to be intimately linked to the power expansion (52) used throughout; it should still either be removed or be properly understood.

5. CP MAPS ON VON NEUMANN ALGEBRAS

In this section we somewhat change perspective to concentrate on the observables and their evolution. To this effect we use a generalized version of the traditional framework of quantum mechanics. This section should however not be construed as a rigid - or dogmatic - axiomatization of non-equilibrium statistical mechanics, although it does seem to encompass some of its important aspects. Moreover, this choice seems also to lead to rich mathematical structures with their own intrinsic interest.

Pursuing the line opened by Kadison [46], we first introduce the concept of general dynamical system and we show, with the help of typical examples, that quantum as well as classical dynamical systems appear as particular cases. We will then generalize this notion somewhat further to encompass an interesting class of dissipative dynamical systems; we show that the resulting structure is nevertheless still tight enough to impose useful restrictions which give the theory some predictive value. We finally discuss briefly the formulation of a dilation problem pertinent to the extension of the notion of continuous Markov process in the quantum realm.

We now define, for the purpose of this section, a conservative dynamical system as a triple $\{\mathfrak{N},\Phi,\mathfrak{a}(R)\}$ consisting of the following elements. (1) A von Neumann algebra $\mathfrak{N}$, i.e. (see e.g. [18], [19], [74]; for their relevance to physical theories, see also [21]) an algebra of bounded operators, acting on a (separable) Hilbert space $\mathfrak{H}$, and satisfying the following conditions: (i) it is closed under the operation of taking the hermitian adjoint; and (ii) it is equal to its double commutant $\mathfrak{N}''$ [for any subset $\mathfrak{X} \subseteq B(\mathfrak{H})$, $\mathfrak{X}' = \{A \in B(\mathfrak{H}) \mid [A,X] = 0 \;\forall\; X \in \mathfrak{X}\}$; and $\mathfrak{X}'' = (\mathfrak{X}')'$]. The added generality where $\mathfrak{N}$ is defined abstractly as a C*-algebra dual to a Banach space is not needed at this point. What is important however is to note that every von Neumann algebra is complete with respect to each of the following five topologies: the uniform topology induced by the operator norm; the strong and the weak operator topologies; the ultrastrong and the ultraweak topologies. (2) The second ingredient in our definition of a dynamical system is a faithful normal state Φ on $\mathfrak{N}$, i.e. a positive linear functional $\Phi: A \in \mathfrak{N} \to \langle\Phi;A\rangle \in \mathbb{C}$ satisfying the following conditions: (i) Φ is a state: $\langle\Phi;I\rangle = 1$; (ii) Φ is normal (or equivalently is ultraweakly continuous): let $\mathfrak{N}^+$ denote the positive cone of $\mathfrak{N}$, equipped with the partial ordering $A \leq B$ whenever $(B-A)$ is a positive operator; we require that for any directed set $\mathfrak{F} \subseteq \mathfrak{N}^+$, bounded above in $\mathfrak{N}^+$:

$\sup_{\mathfrak{F}} \langle \Phi ; F \rangle = \langle \Phi ; \sup_{\mathfrak{F}} F \rangle$; (iii) Φ is faithful, i.e. $\{A \in \mathfrak{N}^+ \mid \langle \Phi ; A \rangle = 0\} = 0$. (3) The third, and last, element in our definition is a *-weakly continuous, one-parameter group $\alpha(R)$ of automorphisms of $\mathfrak{N}$, leaving Φ invariant, i.e. $\Phi \cdot \alpha(t) = \Phi$ for all t in R.

We give two simple examples.

Let $\{\Omega, \Sigma, \mu\}$ be a probability space and $T(R)$ be a continuous, one-parameter group of measurable, measure-preserving transformations of Ω. Construct then:

$$(63) \qquad \begin{cases} \mathfrak{H} = \mathcal{L}^2(\Omega, \Sigma, \mu) & \mathfrak{N} = \mathcal{L}^\infty(\Omega, \Sigma) \\ \Phi : f \in \mathfrak{N} \mapsto \int_\Omega d\mu(\omega) f(\omega) & \\ \alpha(t)[f](\omega) = f(T_t[\omega]) & \forall \text{ a.a. } \omega \in \Omega \end{cases}$$

One checks immediately that the triple $\{\mathfrak{N}, \Phi, \alpha(R)\}$ is indeed a conservative dynamical system in the above sense. Moreover $\mathfrak{N} = \mathfrak{N}'$, i.e. $\mathfrak{N}$ is maximally abelian. It is interesting to note that this case actually exhausts the class of all conservative dynamical systems where $\mathfrak{N}$ is an abelian von Neumann algebra. Classical conservative dynamical systems [7] are therefore completely characterized within our general conservative dynamical system. We now construct a non-classical conservative dynamical system on the algebra $\mathfrak{M}(2,C)$ of all two-by-two matrices with complex entries:

$$(64) \qquad \begin{cases} \mathfrak{N} = \mathfrak{M}(2,C) \\ \Phi : A \mapsto \operatorname{Tr} \rho A \quad \text{with } \rho = \begin{pmatrix} \lambda & 0 \\ 0 & 1-\lambda \end{pmatrix} ; \quad \lambda \in (0,1) \\ \alpha(t)[A] = U(t) A U(-t) \quad \text{with } U(t) = \exp(-iHt) \quad \text{and} \\ \qquad H = \epsilon \begin{pmatrix} 1 & 0 \\ 0 & -1 \end{pmatrix} ; \quad \epsilon \in R \end{cases}$$

Actually it turns out that the above triple $\{\mathfrak{N}, \Phi, \alpha(R)\}$ is (up to unitary equivalence of course) the most general conservative dynamical system on can build over $\mathfrak{M}(2,C)$. The theory however is not limited to such simple von Neumann algebras; from a physical point of view, the theory is rich enough to allow the description of infinite dynamical systems in the thermodynamical limit; an example of the latter will be given towards the end of this section (for other examples, see e.g. the early prototypes studied in [6]; and for further references [17], [21] or [27].

From a mathematical point of view, the existence of a faithful normal state on $\mathfrak{N}$ is a very powerful tool, exploited by Takesaki [81] in his theory of Tomita modular algebras; for an interesting reformulation see [47], [48] and references quoted there. The Tomita-Takesaki theory has received spectacular applications to which we can only allude here. Firstly, it provides a mathematical chacterization [38] (see also [39], [50], [3], [4], [72], [86]) of canonical equilibrium states of infinite systems via the KMS condition; as a further application along this line we should mention a mathematical theory of the chemical potential [5] (see also [50]); we might also mention here that this theory has provided a systematic approach to spontaneous symmetry breaking in phase transitions (for a review, see [27]). Closer to pure mathematics, the Tomita-Takesaki theory has provided considerable insights and advances in our understanding of the classification of type III factors [13]; in turn these factors [63] play an important role in the theory of the representations associated to certain states [44]; and in quantum field theory [2], [45], [20] as well as in equilibrium statistical mechanics [43], [79] (see also [1] and [21] for further references).

We now return to our study of general dynamical systems, and we enlarge the notion of conservative dynamical systems so as to include as well some interesting dissipative systems. We thus define along this line a general dynamical system as a triple $\{\mathfrak{N},\Phi,\gamma(R^+)\}$ consisting of the following elements: a von Neumann algebra $\mathfrak{N}$ and a faithful normal state Φ on $\mathfrak{N}$, as previously; the existence of $\alpha(R)$ is however replaced by a weaker conditions which we shall now state and then discuss. We indeed only assume that there exists a *weakly continuous one-parameter semigroup $\gamma(R^+)$ of linear completely positive maps of $\mathfrak{N}$ into itself such that: (i) $\Phi \circ \gamma(t) = \Phi$ for all $t \in R^+$; (ii) $\gamma(t)[I] = I$ for all $t \in R^+$; and (iii) there exists a one-parameter semigroup $\nu(R^+)$ of linear completely positive maps of $\mathfrak{N}$ into itself such that for every pair A, B of elements of $\mathfrak{N}$, and every $t \in R^+$:

$$(65) \qquad \langle \Phi ; \gamma(t)[A]B \rangle = \langle \Phi ; A\nu(t)[B] \rangle .$$

To make this presentation self-contained, we recall that a linear map γ from a von Neumann algebra $\mathfrak{N}_1$ into a von Neumann algebra $\mathfrak{N}_2$ is said to be positive if $A \in \mathfrak{N}_1^+$ implies $\gamma[A] \in \mathfrak{N}_2^+$; the map is said to be completely positive if for every positive integer n the map $\gamma \otimes id: \mathfrak{N}_1 \otimes \mathfrak{M}(n,C) \to \mathfrak{N}_2 \otimes \mathfrak{M}(n,C)$ is positive. These maps have been studied first by Stinespring [78] (see also [8], [10] and [80]), and their relevance to physics has been pointed out in [55] and [59]. Let it be sufficient here to recall that the transposition map on $\mathfrak{M}(2,C)$ is positive without being completely positive, so that the assumption of complete positivity is a genuine one; the reason why it does not appear in the classical context of abelian von Neumann algebras is that in this case positivity and complete positivity coincide; the distinction between these two concepts hence only appear in a non-commutative context. For orientation purposes, we give a few elementary examples. Every state on a von Neumann algebra is completely positive. Every automorphism of a von Neumann algebra is completely positive. Our next two examples are a little bit more involved. Let $\mathfrak{N}_1$ [resp. $\mathfrak{N}_2$] be a von Neuamnn algebra; Φ_1 [resp. Φ_2] be a state on $\mathfrak{N}_1$ [resp. $\mathfrak{N}_2$]. Suppose further that there exist an injective *-homomorphism i of $\mathfrak{N}_2$ into $\mathfrak{N}_1$ such that $\Phi_1 \circ i = \Phi_2$; and a conditional expectation E (i.e. a projector of norm 1, see [62], [85] and [83]) from $\mathfrak{N}_1$ onto $\mathfrak{N}_2$ which is faithful, normal and such that $E \circ i = id$, $\Phi_1 \circ i \circ E = \Phi_1$ (for the conditions necessary and sufficient to the existence of these objects, see [82]). Then E is completely positive. Furthermore, if $\alpha(t)$ is an automorphism of $\mathfrak{M}_1$ leaving Φ_1 invariant, then $\gamma(t) = E \circ \alpha(t) \circ i$ is again a faithful normal completely positive map of $\mathfrak{M}_2$ into itself, leaving Φ_2 invariant.

Clearly, if $\{\mathfrak{M}, \Phi, \alpha(R)\}$ is a conservative dynamical system, then $\{\mathfrak{M}, \Phi, \gamma(R^+)\}$ with $\gamma(t) = \alpha(t)$ for all $t \geq 0$ (and $\nu(-t) = \alpha(t)$ for all $t \leq 0$) is a dynamical system in the extended sense just introduced. We now give an example of a non-conservative dynamical system. Let $\{\mathfrak{M}, \Phi\}$ be as in (64), and define:

$$(66) \qquad a = \begin{pmatrix} 0 & 0 \\ 1 & 0 \end{pmatrix} \in \mathfrak{M}(2, C).$$

For every $A \in \mathfrak{M}$ let

$$(67) \qquad \begin{cases} L[A] = i[H, A] + \sum_k L_k A \quad \text{with} \\ L_k[A] = V_k^* A V_k - (V_k^* V_k A + A V_k^* V_k)/2 \end{cases}$$

$$(68) \qquad \begin{cases} H = \omega a^* a; \quad V_1 = \nu_1 a; \quad V_2 = \nu_2 a^* \\ V_3 = \nu_3 a^* a; \quad \omega \in R; \quad \nu_k \in C; \quad k = 1, 2, 3 \end{cases}$$

$$(69) \qquad \gamma(t)[A] = \sum_{n=0} \frac{t^n}{n!} L^n[A] \quad \text{for all} \quad t \geq 0.$$

Then $\{\mathfrak{M}, \Phi, \gamma(R^+)\}$ is a dynamical system ([59], [31]) which is dissipative (i.e. non-conservative) as soon as at least one of the ν_k is non-zero. In fact, up to unitary equivalence, this is the most general dynamical system one can build on $\mathfrak{M}(2, C)$ (when Φ is not a trace). For a proof, see [31] where an application is also given to the Bloch equation governing spin-relaxation. In particular, it is shown there that the restriction imposed here on a dynamical system, namely that $\gamma(t)$ be completely positive and not merely positive, implies that this differential equation contains only two relaxation times

$$(70) \qquad \begin{cases} T_1 = (\nu_1^* \nu_1 + \nu_2^* \nu_2)^{-1} \\ T_2 = [(\nu_1^* \nu_1 + \nu_2^* \nu_2 + \nu_3^* \nu_3)/2]^{-1} \end{cases}$$

which therefore are linked by

$$(71) \qquad 0 \leq T_1 \leq 2\, T_2$$

a remarkable relation which seems to be confirmed by experiment, and was already discovered, on less general premices, by Favre and Martin [33] (see also [36], [87]).

We now return to the general theory and we reformulate in this context the dilation problem touched upon in section 2. We ask the general question: Given a dynamical system $\{\mathfrak{N},\Phi,\gamma(R^+)\}$ does there exist a conservative dynamical system $\{\underline{\mathfrak{N}},\underline{\Phi},\alpha(R)\}$, an injective *-homomorphism i of $\mathfrak{N}$ onto $\underline{\mathfrak{N}}$, and a faithful normal conditional expectation E from $\underline{\mathfrak{N}}$ onto $\mathfrak{N}$ such that

$$(72) \qquad \begin{cases} \underline{\Phi}\circ i = \Phi; \quad E\circ i = id; \quad \underline{\Phi}\circ i\circ E = \underline{\Phi}; \\ \text{and} \quad E\circ\alpha(t)\circ i = \gamma(t) \quad \text{for all} \quad t \in R^+ \end{cases}$$

If such a dilation exists, it is said to be minimal whenever

$$(73) \qquad \{\alpha(t)\circ i[N] \mid N \in \mathfrak{N}, \quad t \in R\}'' = \underline{\mathfrak{N}}.$$

A similar, although not quite as sharp, question was already asked in [26], [31], [32] (and references quoted in these three papers); notice in particular the requirement imposed here in the definition of $\{\mathfrak{N},\Phi,\gamma(R^+)\}$ namely that there exists a "mirror" semi-group $\nu(R^+)$ of CP maps satisfying (65); this condition is necessary [31] to the existence of any dilation $\{\underline{\mathfrak{N}},\underline{\Phi},\alpha(R)\}$ of $\{\mathfrak{N},\Phi,\gamma(R^+)\}$ in the sense of (72). In the classical case where $\{\mathfrak{N},\Phi\}$ is as in (63), $\gamma(R^+)$ induces a Markov semigroup $S(R^+)$ on $\mathfrak{H}$:

$$(74) \qquad \begin{cases} S(t)N\Phi = \gamma(t)[N]\Phi, \quad t \in R^+, \quad N \in \mathfrak{N} \\ \text{with} \quad \Phi : \omega \in \Omega \mapsto 1 \in C \end{cases}$$

so that our problem admits then a positive answer which can readily be obtained from the Komogorov-Daniell reconstruction theorem (see e.g. [9] or [84]) of the Markov process, indexed by R, with state space $\{\Omega,\Sigma,\mu\}$ and underlying probability space $\{\underline{\Omega},\underline{\Sigma},\underline{\mu}\}$; we write indeed $\underline{\mathfrak{N}} = \mathcal{L}^\infty(\underline{\Omega},\underline{\Sigma})$; $\langle \underline{\Phi};\underline{N}\rangle = \underline{\mu}(N)$; and $\alpha(R)$ is then induced by the shift on $\underline{\Omega} = \Omega^R$; the injection is

(75)
$$\begin{cases} i: N \in \mathfrak{N} \mapsto N\circ\pi_o \text{ where} \\ \pi_o: \underline{\omega} \in \underline{\Omega} \mapsto \underline{\omega}(0) \in \Omega \end{cases}$$

and $\mathbb{E}$ is the conditional expectation $E(\cdot|\Sigma_o)$ by the present $(t=0)$. Note that $\{\underline{\mathfrak{H}},\underline{U}(R),X\}$ defined now by

(76)
$$\begin{cases} \underline{\mathfrak{H}} = \mathcal{L}^2(\underline{\Omega},\underline{\Sigma},\underline{\mu}) \\ \underline{U}(t)\underline{N}\underline{\Phi} = \alpha(t)[\underline{N}]\underline{\Phi}, \quad \underline{N} \in \underline{\mathfrak{N}} \\ \text{with } \underline{\Phi}: \underline{\omega} \in \Omega \mapsto 1 \in C \\ X: N\Phi \in \mathfrak{H} \mapsto i(N)\underline{\Phi} \in \underline{\mathfrak{H}} \end{cases}$$

is a unitary dilation of $\{\mathfrak{H}, S(R^+)\}$, but in general it is well-known [37] not to be a minimal one, although $\{\underline{\mathfrak{N}},\underline{\Phi},\alpha(R)\}$ is a minimal dilation of $\{\mathfrak{N},\Phi,\gamma(R^+)\}$ in the sense of (73).

Coming back to the case where $\mathfrak{N}$ is not abelian, we want to mention that one knows that the dissipative system corresponding to the Bloch equation (67)-(69) with $\nu_3 = 0$ admits a minimal dilation [71], [31]. Similarly, one knows [24] that the dissipative system corresponding to the diffusion of a quantum particle in a harmonic well [34], [15] does also admit a minimal dilation; this provides a non-commutative generalization of the classical flow of Brownian motion [41]. Some more general quasi-free quantum dissipative dynamical systems which admit a minimal dilation have further been studied in [28]. A complete answer to our problem for non quasi-free, non abelian dissipative dynamical systems however is still missing.

Since this will provide an example of the structures studied in both this section and the next one, we now outline a very simple particular case. With $1 < \Theta < \infty$ and $\mathfrak{J}$ a Hilbert space, let

(77)
$$\hat{\Phi}: f \in \mathfrak{J} \mapsto \exp\{-\Theta\|f\|^2/4\}$$

One knows (for the general theory, see for instance [21]; for an explicit construction and its physical interpretation, see [24]) that there exist: a Hilbert space $\mathfrak{H}$, a vector $\Phi \in \mathfrak{H}$ of unit norm, and a family $\{W(f) \mid f \in \mathfrak{I}\}$ of unitary operators acting in $\mathfrak{H}$ such that for every $f,g \in \mathfrak{I}$:

$$(78) \qquad \begin{cases} W(f)W(g) = W(f+g)\exp\{-i\,\mathrm{Im}(f,g)/2\} \\ (W(f)\Phi,\Phi) = \hat{\Phi}(f) \end{cases}$$

and Φ cyclic and separating in $\mathfrak{H}$ for the von Neumann algebra

$$(79) \qquad \mathfrak{N} = \{W(f) \mid f \in \mathfrak{I}\}''$$

so that

$$(80) \qquad \Phi: N \in \mathfrak{N} \mapsto (N\Phi,\Phi)$$

is a faithful normal state on $\mathfrak{N}$. We consider two particular cases. We first let $\mathfrak{I} = C$. With $\mathfrak{N}$ and Φ as above, and $a > 0$, let $\gamma(R^+)$ be the semigroup of transformations of $\mathfrak{N}$ determined by:

$$(81) \qquad \gamma(t)[W(z)] = W(e^{-at}z)\exp\{-\theta|z|^2(1-e^{-2at})/4\}.$$

It is easy to check that the triple $\{\mathfrak{N},\Phi,\gamma(R^+)\}$ so defined is a non conservative dynamical system. Let now $\mathfrak{I} = \mathfrak{L}^2(R,dx)$ and let $\{\underline{\mathfrak{N}},\underline{\Phi}\}$ be the pair associated to (77) by (78)-(80). With $U(R)$ as in (16), let $\alpha(R)$ be the group of automorphisms of $\underline{\mathfrak{N}}$ determined by:

$$(82) \qquad \alpha(t)[\underline{W}(f)] = \underline{W}(U(t)f).$$

It is then easy to check that the triple $\{\underline{\mathfrak{N}},\underline{\Phi},\alpha(R)\}$ so defined is a conservative dynamical system. Moreover, with X and f_o as in (16), let

$$(83) \qquad \begin{cases} i: W(z) \in \mathfrak{N} \mapsto \underline{W}(Xz) \in \underline{\mathfrak{N}} \\ E: \underline{W}(f) \in \underline{\mathfrak{N}} \mapsto \hat{\Phi}([I-XX^*]f)\,W(X^*f) \in \mathfrak{N}. \end{cases}$$

One then verifies that $\{\underline{\mathfrak{N}},\underline{\Phi},\alpha(R),i,E\}$ is a minimal dilation of $\{\mathfrak{N},\Phi,\gamma(R^+)\}$.

6. NON-COMMUTATIVE ERGODIC THEORY

In the same way that classical statistical mechanics has provided the traditional motivation for classical ergodic theory, the scheme proposed in the preceding section has prompted some non-commutative extensions of ergodic theory. As an illustration, we want to mention here a notion which generalizes the classical structure of Komogorov-Sinai systems and flows (for an exposition of these classical notions, see [7] or [69]; the original papers are [54] and [75]).

A conservative dynamical system $\{\underline{\mathfrak{N}},\underline{\Phi},\alpha(R)\}$ is said to be a generalized K-flow if there exists a von Neumann subalgebra $\mathfrak{A}$ of $\underline{\mathfrak{N}}$ with the following properties:

$$(84)\qquad \begin{cases} \mathfrak{A} \subseteq \alpha(t)[\mathfrak{A}] \quad \text{for all} \quad t \geq 0 \\ \bigcap_{t\in R} \alpha(t)[\mathfrak{A}] = \mathbb{C}I; \quad \bigvee_{t\in R} \alpha(t)[\mathfrak{A}] = \underline{\mathfrak{N}} \\ \exists \text{ a faithful normal cond.expect. } E: \underline{\mathfrak{N}} \to \mathfrak{A} \text{ w.r.t. } \underline{\Phi} \end{cases}$$

For instance, with $\{\underline{\mathfrak{N}},\underline{\Phi},\alpha(R)\}$ and $\mathfrak{N}$ as in the end of the preceding section, $\mathfrak{A} = \{\alpha(t)\circ i\,[\mathfrak{N}] \mid t \leq 0\}''$ equips $\{\underline{\mathfrak{N}},\underline{\Phi},\alpha(R)\}$ with the structure of a generalized K-flow. This and several other examples have been constructed in [23] (see also [22] and [28]) on factors $\underline{\mathfrak{N}}$ of type III_λ $(0 < \lambda < 1)$ (as is the case in the above example), but also on factors of type II_1 and III_1. The ergodic properties of generalized K-flows have been studied in [23]. For instance, knowing that we can assume without loss of generality that there exists a unit vector Φ which is cyclic and separating for $\mathfrak{N}$ and such that $\langle\underline{\Phi};\underline{N}\rangle = (\underline{N}\underline{\Phi},\underline{\Phi})$ for all $\underline{N} \in \underline{\mathfrak{N}}$, let in analogy with (76), $\underline{U}(R)$ be the unitary group determined by

$$(85)\qquad \underline{U}(t)\underline{N}\underline{\Phi} = \alpha(t)[\underline{N}]\underline{\Phi}, \qquad \underline{N} \in \underline{\mathfrak{N}}$$

and H be its generator. One can then prove ([23]; compare with [76]) among other things:

(86) $$Sp_d(H) = \{0\} \text{ and it is simple}$$

(87) $$\{\underline{N} \in \underline{\mathfrak{N}} \mid \alpha(t)[\underline{N}] = \underline{N} \text{ for all } t \in R\} = CI$$

(88) H has homogeneous Lebesgue spectrum on the orthocomplement of Φ in $\underline{\mathfrak{H}}$.

Strong mixing (or clustering) properties can be proven as well. Furthermore non commutative versions of the classical Kolmogorov-Sinai entropy (see [7] or [69]) have been defined and studied in [14] and in [22]-[25]. The version used in [22]-[25] is shown to be infinite for a class of dynamical systems including the above explicit example of a generalized K-flow, thus generalizing to the quantum domain the result that the entropy of the flow of Brownian motion [41] is infinite. Whereas the entropy defined in [14] appears to be a powerful mathematical tool, the entropy defined in [22]-[25] takes more closely into account the physics of the quantum measuring process by which information is actually acquired. Both are dynamical invariants, and both agree with the classical notion of Kolmogorov-Sinai when $\underline{\mathfrak{N}}$ is abelian. However the extend to which those two invariants are distinct is not known.

For lack of space, we shall not enter here into a discussion of the notion of asymptotic abelianness which nevertheless plays an important role in linking various notions in non commutative ergodic theory. For a general discussion, see [21]; for a review of its relevance to several aspects of equilibrium statistical mechanics, see [27]; for an application to the understanding of KMS states as states which are stable under local perturbations, see [50] (and references therein); for a particular consequence of this property for the relaxation of local deviations from equilibrium, see [29].

7. CONCLUSION

We have presented here some of the different mathematical concepts and techniques in functional analysis and probability theory that have been brought to bear on one central physical problem. All these share something in common; it is strongly suggested that this commonness of purpose is indicative of an underlying general mathematical structure that further investigations will uncover.

ACKNOWLEDGEMENTS

It is a pleasure to thank here Prof. S. Machado for his invitation to address this rich and stimulating seminar, and Prof. L. Nachbin for his friendly, but unyielding, insistence that these notes be written for the proceedings.

REFERENCES

[1] Dell'Antonio, G.F., Commun. math. Phys. 9 (1969) 81.

[2] Araki, H., J. Math. Phys. 4 (1963) 1343; 5 (1964) 1; Progr. Theor. Phys. 32 (1964) 844 and 956.

[3] ---------, in Mathematical Problems in Theoretical Physics, G. dell'Antonio, S. Doplicher & G. Jona-Lasino, eds., Springer, Heidelberg, 1978.

[4] ---------, in C*-Algebras and Applications to Physics, H. Araki & R.V. Kadison, eds., Springer, Heidelberg, 1978.

[5] Araki, H., Haag, R., Kastler, D. & Takesaki, M., Commun. math. Phys. 53 (1977) 97.

[6] Araki, H. & Woods, E.J., J. Math. Phys. 4 (1963) 637; --- & W. Wyss, Helv. Phys. Acta 37 (1964) 136.

[7] Arnold, V.I., & Avez, A., Ergodic Problems of Classical Mechanics, Benjamin, New York, 1968.

[8] Arveson, W., Acta Math. 123 (1969) 141.

[9] Breiman, L., Probability, Addison-Wesley, Reading, Mass., 1968.

[10] Choi, M.D., Canad. J. Math. 24 (1972) 520.

[11] Cogburn, R. & R. Hersh, Indiana Univ. Math. Journ. 22 (1973) 1067.

[12] Conley, C.C. & Reijto, P.A., in Perturbation Theory and its Applications in Quantum Mechanics, C.H. Wilcox, ed., Wiley, New York, 1966.

[13] Connes, A., Ann. Scient. Ec. Norm. Sup. (4e s.) 6 (1973) 123.

[14] Connes, A. & Stormer, E., Acta Math. 134 (1975) 289.

[15] Davies, E.B., Commun. math. Phys. 27 (1972) 309.

[16] ------------, Lett. Math. Phys. 1 (1975) 31.

[17] ------------, Quantum Theory of Open Systems, Academic Press, London, 1976.

[18] Dixmier, J., Les Algèbres d'Opérateurs dans l'espace Hilbertien, Gauthier-Villars, Paris, 1957.

[19] Dixmier, J., Les C*-Algèbres et leurs représentations, Gauthier-Villars, Paris, 1964.

[20] Driessler, W., Commun. math. Phys. 44 (1975) 133.

[21] Emch, G.G., Algebraic Methods in Statistical Mechanics and Quantum Field Theory, Wiley-Interscience, New York, 1972.

[22] ----------, J. Funct. Analysis 19 (1975) 1; in Mathematical Problems in Theoretical Physics, H. Araki, ed., Springer, New York, 1975; Asterisque 40 (1976) 315.

[23] ----------, Commun. math. Phys. 49 (1976) 191.

[24] ----------, Acta Physica Austriaca, Suppl. XV (1976) 79; or in Current Problems in Mathematical Physics, P. Urban, ed., Springer, New York, 1976.

[25] ----------, in Mathematical Problems in Theoretical Physics, G. dell'Antonio, S. Doplicher & G. Jona-Lasino, eds., Springer, Heidelberg, 1978.

[26] ----------, in C*-Algebras and Applications to Physics, H. Araki & R.V. Kadison, eds., Springer, Heidelberg, 1978.

[27] ----------, in Groups, Systems and Many-Body Physics, P. Kramer & C. Dal Cin, eds., Vieweg, Braunschweig, 1979.

[28] Emch, G.G., Albeverio, S. & Eckmann, J.P., Rep. Math. Phys. 13 (1978) 73.

[29] Emch, G.G. & Radin, C., J. Math. Phys. 12 (1971) 2013.

[30] Emch, G.G. & Sinha, K.B., J. Math. Phys. 20 (1979) 1336.

[31] Emch, G.G. & Varilly, J.C., Lett. Math. Phys. 3 (1979) 113.

[32] ---------- & -------------, Rep. Math. Phys. (to appear)

[33] Favre, C. & Martin, Ph., Helv. Phys. Acta 41 (1968) 333.

[34] Ford, G.W., Kac M. & Mazur, P., J. Math. Phys. 6 (1965) 504.

[35] Gamov, G., Mr. Tompkins in Wonderland, Cambridge University Press, London, 1953.

[36] Gorini, V., Kossakowski, A. & Sudarshan, E.C.G., J. Math. Phys. 17 (1976) 821.

[37] Guerra, F., Rosen, L. & Simon, B., Ann. Math. 101 (1975) 111.

[38] Haag, R., Hugenholtz, N. & Winnink, M., Commun. math. Phys. 5 (1967) 215.

[39] Haag, R., Kastler, D. & Trych-Pohlmeyer, Commun. math. Phys. 38 (1974) 173.

[40] Hersh, R. & Papanicolaou, G., Comm. Pure & Applied Math. 25 (1972) 337.

[41] Hida, T., Stationary Stochastic Processes, Princeton University Press, Princeton, NJ, 1970.

[42] Hove, L. van, Physica 21 (1955) 517; 23 (1957) 441.

[43] Hugenholtz, N., Commun. math. Phys. 6 (1967) 189.

[44] Kadison, R.V., Trans. Amer. Math. Soc. 103 (1962) 304.

[45] -------------, J. Math. Phys. 4 (1963) 1511.

[46] -------------, Topology 3, Suppl. 2 (1965) 177.

[47] Kadison, R.V., in Algebras of Operators and their Applications to Mathematical Physics (Marseille, 1977), D. Kastler & D.W. Robinson, eds. (to appear)

[48] -------------, Acta Math. 141 (1978) 147.

[49] Kampen, N.G. van, Physics Reports 24 (1976) 171.

[50] Kastler, D., in Mathematical Problems in Theoretical Physics, G. dell'Antonio, S. Doplicher & G. Jona-Lasino, eds., Springer, Heidelberg, 1978.

[51] Kato, T., Perturbation Theory for Linear Operators, Springer, New York, 1966.

[52] Kemble, E.C., Fundamental Principles of Quantum Mechanics, McGraw-Hill, New York, 1937.

[53] Khas'minskii, R.Z., Theory of Prob. & Appl. 11 (1966) 211 and 390.

[54] Kolmogorov, A.N., Dokl. Akad. Nauk SSSR 119 (1958) 861; 124 (1959) 754.

[55] Kraus, K., Ann. Phys. 64 (1970) 311.

[56] Kurtz, T.G., J. Funct. Analysis 3 (1969) 354; 12 (1973) 55.

[57] Landau, L.D. & Lifschitz, E.M., Quantum Mechanics, Pergamon Press, London, 1958.

[58] Lax, P.D. & Phillips, R.S., Scattering Theory, Academic Press, New York, 1967.

[59] Lindblad, G., Commun. math. Phys. 48 (1976) 119.

[60] Martin, Ph. & Emch, G.G., Helv. Phys. Acta 48 (1975) 59.

[61] Messiah, A., Mécanique Quantique, Dunod, Paris, 1959.

[62] Moy, S.-T.C., Pac. J. Math. 4 (1954) 47.

[63] Murray, F.J. & Neumann, J. von, Ann. Math. 37 (1936) 116.

[64] Nagy, B.Sz. & Foias, C., Harmonic Analysis..., North-Holland, Amsterdam, 1970.

[65] Neumann, J.von, Grundlagen der Quantenmechanik, Springer,Berlin, 1932.

[66] Papanicolaou, G. & Keller, J.B., J. Appl. Math. 21 (1971) 287.

[67] Papanicolaou, G. & Kohler, W., Comm. Pure & Appl. Math. 27 (1974) 641.

[68] Papanicolaou, G. & Varadhan, S., Comm. Pure & Appl. Math. 26 (1973) 497.

[69] Parry, W., Entropy and Generators in Ergodic Theory, Benjamin, New York, 1969.

[70] Pauli, W., in Festschrift zum 60. Geburtstage Sommerfelds, Hirzel, Berlin, 1928.

[71] Pulé, J.V., Commun. math. Phys. 38 (1974) 241.

[72] Pusz, W. & Woronowicz, S.L., in Mathematical Problems in Theoretical Physcis, G. dell'Antonio, S. Doplicher & G. Jona-Lasino, eds., Springer, Heidelberg, 1978; Commun. math. Phys. 58 (1978) 273.

[73] Riesz, F. & Nagy, B.Sz., Leçons d'analyse fonctionnelle, Gauthier-Villars, Paris, 1955.

[74] Sakai, S., C*-Algebras and W*-Algebras, Springer, New York, 1971.

[75] Sinai, Ya., Dokl. Akad. Nauk SSSR 124 (1959) 768.

[76] ----------, Amer. Math. Soc. Transl. (2) 39 (1961) 83.

[77] Spohn, H., J. Stat. Mech. 17 (1977) 385.

[78] Stinespring, W.F., Proc. Amer. Math. Soc. 6 (1955) 211.

[79] Stormer, E., Commun. math. Phys. 6 (1967) 194.

[80] -----------, in Foundations of Quantum Mechanics and Ordered Linear Spaces, A. Hartkaemper & H. Neumann, eds., Springer, Heidelberg, 1974.

[81] Takesaki, M., Tomita's Theory of Modular Hilbert Algebras and its Applications, Springer, New York, 1970.

[82] ------------, J. Funct. Analysis 9 (1972) 306.

[83] Tomiyama, J., Proc. Japan Acad. 33 (1957) 608.

[84] Tucker, H.G., A Graduate Course in Probability, Academic Press, New York, 1967.

[85] Umegaki, H., Tohoku Math. J. 6 (1954) 177.

[86] Verbeure, A., in Algebras of Operators and their Applications to Mathematical Physics (Marseille, 1977), D. Kastler & D.W. Robinson, eds. (to appear)

see also B. Demoen, P. Vanheuverzwijn and A. Verbeure, J.Math. Phys. 19 (1978) 2256.

[87] Verri, M. & Gorini, V., J. Math. Phys. 19 (1978) 1803.

[88] Williams, N.D., Commun. math. Phys. 21 (1971) 314.

NOTES ADDED IN PROOF

(i) Concerning the appearance of a finite τ_o, as discussed at the end of section 4, a way to bypass this difficulty is proposed in A. Nogueira, Ph.D. thesis, University of Rochester, Spring 1980.

(ii) The isomorphism between certain von Neumann algebras and their commutant, mentionned in section 5 as a consequence of the faithfulness of the state Φ, has been reinvestigated from the abstract point of view of the underlying JBW-structure and its dual, in P.King, Ph.D. thesis, University of Rochester, Spring 1980.

(iii) A dilation for a non quasi-free, quantum dissipative system (see section 5) has been found by J. Varilly, Ph.D. thesis, University of Rochester, Spring 1980.

GENERALIZED HEWITT-NACHBIN SPACES
ARISING IN STATE-SPACE COMPLETIONS

Benno Fuchssteiner
Gesamthochschule, D 4790 Paderborn

The paper deals with the question whether or not the well known notions of Stone-Czech compactification and realcompactification (Hewitt-Nachbin completion) can be transferred from the classical situation (CB(X), X completely regular) to a more general situation where X is replaced by an arbitrary set and where a convex cone of bounded functions plays the role of the continuous functions.

It turns out that this is in fact possible and that this procedure leads in a very satisfactory way to a simple and transparent theory which comprises the classical situation as well as the Choquet theory of the state space. Furthermore it is possible to adopt most of the fundamental results known for the classical situation with slight modifications for the general situation.

In the first part (Chapters I to IV) we define our basic notions (generalized compactness, realcompactness and pseudocompactness) and we show that - roughly spoken - these objects can be characterized by filter properties as well as geometric or lattice properties.

In the second part we transfer some basic results from the classical case to our situation. For example, it is demonstrated that a suitable generalization of Glicksberg's integral representation theorem contains the Choquet - theorem as a special case. This happens since the extreme points of a compact convex set are pseudocompact (in the generalized sense) with respect to the continuous affine functions.

The rest of the paper is devoted to the investigation of real-compact spaces.

I. BASIC DEFINITIONS

Let X be a nonempty set. We consider a convex cone F of bounded real functions on X such that F separates the points and contains all constant real-valued functions. A functional $\mu: F \to \mathbb{R}$ is said to be linear if it is additive and positive-homogeneous (i.e. $\mu(\lambda f) = \lambda\mu(f)$ for all $\lambda \geq 0$, $f \in F$). The functional μ is called order-preserving if $\mu(g) \geq \mu(f)$ whenever $g \geq f$.

At this point it should be remarked that every $\mathbb{R} \cup \{-\infty\}$ - valued order-preserving linear functional μ is automatically $\mathbb{R}$-valued. This is easily seen by

$$\mu(f) \geq \mu(\alpha) \geq -\mu(-\alpha) \in \mathbb{R}$$

where α is the constant function equal to $\inf_{x\in X} f(x)$.

A state is defined to be an order-preserving linear functional μ with

$$\mu(f) \leq \sup_{x\in X} f(x) \quad \text{for all} \quad f \in F.$$

The set of all states of F is called the state space and is denoted by SX_F. Identification of $x \in X$ with the point evaluation $f \to f(x)$ leads to an embedding of X into SX_F. The state space is made into a topological space by endowing it with the coarsest topology such that all the functions $\mu \to \mu(f)$, $f \in F$, are continuous. SX_F is a compact Hausdorff space (every ultrafilter converges). A state μ is defined to be Dini-continuous if we have

$$\inf_{n\in\mathbb{N}} \mu(f_n) \leq \sup_{x\in X} \inf_{n\in\mathbb{N}} f_n(x)$$

for all pointwise decreasing sequences f_n in F.

1 Remark. Let F consist of upper-semicontinuous functions on a compact (not necessarily Hausdorff) space X, then because of Dini's lemma all states are Dini-continuous.

As usual a state μ is called <u>maximal</u> if whenever ν is a state with $\mu \leq \nu$ then $\mu = \nu$ (there $\mu \leq \nu$ stands for $\mu(f) \leq \nu(f)$ $\forall f \in F$).

<u>2 Remark.</u> (i) States on vector spaces are always maximal.
(ii) By Zorn's lemma every state μ is dominated by a maximal state ρ (i.e. $\mu \leq \rho$).

Let $Y \subset X$ be a nonempty subset then we denote by $\sup_Y$ the sublinear functional given by

$$f \to \sup \{f(y) \mid y \in Y\}.$$

A convex cone G of functions on X is defined to be <u>max-stable</u> if for $g_1, g_2 \in G$ the function $x \to (g_1 \vee g_2)(x) = \max(g_1(x), g_2(x))$ is always in G. By VF we denote the smallest max-stable convex cone containing F, that is the cone

(1) $$VF = \{f_1 \vee f_2 \vee \ldots \vee f_n \mid n \in N, \; f_1, \ldots, f_n \in F\}.$$

A state $\hat{\mu}$ of VF is termed <u>dominated extension</u> of the F-state μ if $\hat{\mu}(f) \geq \mu(f)$ $\forall f \in F$. If we have equality for all $f \in F$ then, of course, $\hat{\mu}$ is called an <u>extension</u>. $\mu \in SX_F$ can always be extended to a state of VF [2, Lemma 2].

<u>3 Definition</u>. A state μ of F is called F-<u>character</u> if it has a unique dominated extension to VF and if for every finite cover $Y_1, \ldots, Y_n$ of X there is some $k \leq n$ such that $\mu \leq \sup_{Y_k}$.

The set of F-characters will be denoted by βX_F. Those characters which are Dini-continuous are called <u>Dini-characters</u>, and υX_F stands for the set of Dini-characters.

<u>4 Definition</u>. (1) βX_F and υX_F are called the F-<u>compactification</u> of X and the F-<u>realcompactification</u> (or F-<u>Hewitt-Nachbin-completion</u>) respectively.
(2) The set X is defined to be
 (i) F-<u>compact</u> if $\beta X_F \subset X$
 (ii) F-<u>realcompact</u> (or F-<u>Hewitt-Nachbin space</u>) if $\upsilon X_F \subset X$.

5 Definition. X is defined to be F-pseudocompact if every element of the sup-norm closure of VF attains its maximum on X.

II. THE CLASSICAL SITUATION

When X is a completely regular Hausdorff space and F is equal to $CB(X) = \{f \in C(X) \mid f \text{ bounded}\}$ ($C(X)$ being the space of continuous real-valued functions on X) then we call this the "classical situation".

6 Proposition. (i) $\beta X_{CB(X)}$ is the Stone-Czech compactification of X.
(ii) $\upsilon X_{CB(X)}$ is the set of multiplicative linear functionals on $C(X)$ (restricted to $CB(X)$). Hence it is the usual realcompactification of X.

Proof. (i) is left as an exercise.
(ii) Let $\mu \in \upsilon X_{CB(X)}$. Then μ is multiplicative by Proposition 6 (i) and we prove that $\mu(g) \in g(X)$ $\forall g \in CB(X)$, which is a well-known criterion for μ being extendable to a multiplicative linear functional of $C(X)$. For this purpose we consider for an arbitrary $g \in CB(X)$ the decreasing sequence $f_n = -n(g-\mu(g))^2 \leq 0$. We must then have $\sup_{x\in X} \inf_n f_n(x) = 0$ because of $\mu(f_n) = 0$ and the Dini-continuity of μ. This implies that $(g-\mu(g))$ is equal to zero at some $x_o \in X$. Hence $\mu(g) = g(x_o) \in g(X)$.

It remains to prove that whenever $\nu \in \beta X_{CB(X)}$ is not Dini-continuous then there is some $g \in CB(X)$ with $\nu(g) \notin g(X)$.

By definition there is a decreasing sequence f_n in $CB(X)$ such that

(3) $$\inf_n \nu(f_n) = \alpha > \delta = \sup_{x\in X} \inf_n f_n(x).$$

We consider the following σ-compact subset of the Stone-Czech compactification βX

$$S = \bigcup_n K_n, \quad \text{where} \quad K_n = \{z \in \beta X \mid 2f_n(z) \leq \alpha + \delta\}.$$

Then S contains X (consequence of (3)) and by a suitable Urysohn argument we find some $g \in C(\beta X)$ with $\nu(g_{|X}) = g(\nu) = \alpha$ and $g(z) < \alpha \ \forall z \in S$, where $g_{|X}$ denotes the restriction to X. Hence, because of $S \supset X$, we have $\nu(g_{|X}) \notin g_{|X}(X)$. □

Thus we have shown that the notions of "CB(X) - realcompactification" and "CB(X)-compactification" coincide with the usual notions of "realcompactification" and "Stone-Czech compactification" respectively. The observation that CB(X) - pseudocompact means pseudocompact in the usual sense is quite obvious.

So, after having seen that the notions we have defined so far are generalizing a well-known concept, we can state that the aim of this paper is to investigate if those results which do hold for the classical situation can be adopted for the general situation.

We show in the sequel that this is in fact possible.

III. THE PRINCIPAL TOOLS

We gather here those results from [2] to [5] which we need for our investigation. Although they are formulated for the rather special cone F we would like to mention that they are valid in more general situations.

A very useful application of the sandwich theorem (our beloved form of the Hahn-Banach theorem, see [4, p.152]) is the following:

<u>Finite Sum Theorem</u>. Let μ be a linear functional on F and let $p_1,\dots,p_n$ be sublinear order-preserving functionals on F with $\mu \leq \sum_{i=1}^{n} p_i$. Then there are order-preserving linear functionals $\mu_i \leq p_i$ such that $\mu \leq \sum_{i=1}^{n} \mu_i$.
(This is a special case of the sum theorem from [4]).

<u>7 Lemma</u>. Let $\mu \leq \sup_Z$ (where $Z \subset X$) be a state of F and let $\emptyset \neq Y_k$ (k=1,...,n) be a finite cover of Z then there are $\lambda_k \geq 0$ and states μ_k with $\mu \leq \sum_{k=1}^{n} \lambda_k\mu_k$ and $\sum_{k=1}^{n} \lambda_k = 1$ such that the μ_k are Y_k-order-preserving, i.e. $\mu_k(f) \geq \mu_k(g)$ whenever $f(y) \geq g(y)$

$\forall\, y \in Y_k$.

<u>Proof</u>. [4, finite decomposition theorem] gives us the λ_k and states $\tilde{\mu}_k \le \sup_{Y_k}$ with $\mu \le \sum_{k=1}^{n} \lambda_k \tilde{\mu}_k$. And from the sandwich theorem (applied with respect to the preorder given by pointwise order on Y_k) we get the desired Y_k-order-preserving states μ_k with $\tilde{\mu}_k \le \mu_k \le \sup_{Y_k}$.

□

The next result is of a much deeper nature. First, some notation. By Σ_F we denote the σ-algebra generated in X by F (that is the smallest σ-algebra such that the elements of F are measureable). We call a Σ_F-probability measure m a <u>representing measure</u> for an F-state μ if

$$\mu(g) \le \int_X g \, dm \quad \text{for all} \quad g \in F,$$

in case that we have equality then we speak of a <u>strict</u> representing measure.

<u>8 Theorem</u>. Every maximal Dini-continuous state μ of VF has a strict representing measure on X.

<u>Proof</u>. From [4, thm 1] we obtain that μ has the decomposition property, which means that whenever $\emptyset \ne Y_n \subset X$ are such that $\bigcup_{n\in\mathbb{N}} Y_n = X$ then there are states $\mu_n \le \sup_{Y_n}$ and $\lambda_n \ge 0$ with $\mu = \sum_{n=1} \lambda_n \mu_n$. We have to keep in mind that a state ν of VF has a unique extension to the vector lattice $E = VF - VF$, which we denote by $\hat{\nu}$. Furthermore, that when ν is maximal and $\le \sup_Y$ then we also have $\hat{\nu} \le \sup_Y$ (sandwich theorem, compare proof of the Main theorem in [3]).

Now, since μ is maximal μ_n must be maximal when $\lambda_n > 0$. Hence, $\hat{\mu} = \sum_{n=1}^{\infty} \lambda_n \hat{\mu}_n$ has the decomposition property, because we can drop all those $\hat{\mu}_n$ where $\lambda_n = 0$. And this was the condition required in [3, thm 1] for the existence of a representing measure m for $\hat{\mu}$. Obviously m is then also a representing measure for μ. □

IV. THE GENERAL SITUATION

IV. 1 THE F-COMPACTIFICATION

Our aim is to find many useful characterizations of the F-compactification. We begin with some definitions:

9 Definition. (i) A state ρ of VF is said to be a lattice state if $\rho(g_1 \vee \ldots \vee g_n) = \max\{\rho(g_1),\ldots,\rho(g_n)\}$ for all $g_1,\ldots,g_n \in F$.
(ii) $\mu \in SX_F$ is called an extreme point of SX_F if whenever $0 < \lambda < 1$ and $\nu_1,\nu_2 \in SX_F$ with $\mu \le \lambda\nu_1 + (1-\lambda)\nu_2$ then $\mu = \nu_1$.
(iii) By Face (μ) we denote the set of those states ν such that there are $0 < \lambda \le 1$ and a state ρ with $\mu \le \lambda\nu + (1-\lambda)\rho$.
(iv) $Z(\mu)$ stands for the family of those subsets $Z \subset X$ having the property that for arbitrary $\alpha < \beta < 0$ there is always an $f \in F$ with $f \le 0$, $\mu(f) \ge \beta$ and $\sup_{(X\setminus Z)}(f) \le \alpha$. $Z(\mu)$ is termed the set of strong domination.
(v) $\mathcal{D}(\mu) = \{Y \subset X \mid \mu \le \sup_Y\}$ and $h(\mu) = \{Y \subset X \mid \mu \not\le \sup_{(X\setminus Y)}\}$ are called the set of domination and the complementary set respectively.

We first gather some technical details:

10 Lemma. (i) μ is an extreme point of $SX_F \Leftrightarrow \mathrm{Face}(\mu) = \{\mu\}$.
(ii) $\nu \in \mathrm{Face}(\mu) \Rightarrow \mathrm{Face}(\nu) \subset \mathrm{Face}(\mu)$.
(iii) If two states $\mu,\nu \in SX_F$ have the property that for $f,g \in F$ the inequality (*) $\mu(f) \ge \mu(g)$ always implies $\nu(f) \ge \nu(g)$ then $\mu = \nu$.
(iv) $Z(\mu) \subset h(\mu) \cap \mathcal{D}(\mu) \quad \forall \mu \in SX_F$
(v) $\nu \in \mathrm{Face}(\mu) \Rightarrow Z(\mu) \subset Z(\nu)$
(vi) $Z_1, Z_2 \in Z(\mu) \Rightarrow Z_1 \cap Z_2 \in h(\mu)$.

Proof. (i) and (ii) are trivial.
(iii): The number $\mu(f+g)$ must have the sign of $\mu(f)$ or $\mu(g)$, say that of $\mu(g)$. Then from $\mu(|\mu(f+g)|g) = \mu(|\mu(g)|(f+g))$ we obtain with (*) that $\nu(|\mu(f+g)|g) = \nu(|\mu(g)|(f+g))$. Hence we have

$\mu(f+g)\nu(g) = \nu(f+g)\mu(g)$, or $\mu(f)\nu(g) = \nu(f)\mu(g)$. For $g = 1$ we obtain with $\mu(1) = \nu(1) = 1$ the equality $\mu(f) = \nu(f)$.

(iv): $Z(\mu) \subset \mathcal{N}(\mu)$ is obvious. Let $Z \in Z(\mu)$, we want to prove $\mu \leq \sup_Z$. According to Lemma 7 we have $\mu \leq \lambda\mu_1 + (1-\lambda)\mu_2$ with $0 \leq \lambda \leq 1$ and $\mu_1 \leq \sup_{(X\setminus Z)}$, $\mu_2 \leq \sup_Z$. For $\alpha = -n < \beta = -1 < 0$ there has to be some $f \leq 0$ with $\mu(f) \geq -1$ and $\sup_{(X\setminus Z)}(f) \leq -n$. Thus $-1 \leq \mu(f) \leq \lambda \sup_{(X\setminus Z)}(f) \leq -\lambda n$, or $\lambda \leq \frac{1}{n}$ for all n. Hence $\lambda = 0$ or $\mu = \mu_2 \leq \sup_Z$.

(v): Let $\nu \in \mathrm{Face}(\mu)$, i.e. $\mu \leq \lambda\nu + (1-\lambda)\rho$, $0 < \lambda \leq 1$, $\rho \in SX_F$ and let Z be an arbitrarily chosen element of $Z(\mu)$.

Then for arbitrary $\alpha < \beta < 0$ there is some $f \leq 0$ in F with $\mu(f) \geq \lambda\beta$ and $\sup_{(X\setminus Z)}(f) \leq \alpha$.

Hence $\nu(f) \geq \beta$, which shows that $Z \in Z(\nu)$.

(vi): Let $Z_1, Z_2 \in Z(\mu)$ and put $Y_i = X\setminus Z_i$, $i=1,2$. We have to show that $\mu \not\leq \sup_{(Y_1 \cup Y_2)}$. Assume therefore $\mu \leq \sup_{(Y_1\cup Y_2)}$. With the help of Lemma 7 we may write $\mu + \lambda\mu_1 + (1-\lambda)\mu_2$, where $0 \leq \lambda \leq 1$ and $\mu_i \leq \sup_{Y_i}$, $i=1,2$. Take $\alpha < \beta < 0$ such that $2\beta > \alpha$ then there are $f_1, f_2 \leq 0$ in F with $\mu(f_i) \geq \beta$ and $\sup_{Y_i}(f_i) \leq \alpha$, $k=1,2$. Because of $\mu_2(f_1) \leq 0$ and $\mu_1(f_2) \leq 0$ this leads to the contradiction:

$$\begin{aligned} 2\beta \leq \mu(f_1+f_2) &\leq \lambda\mu_1(f_1) + (1-\lambda)\mu_2(f_2) \\ &\leq \lambda \sup_{Y_1}(f_1) + (1-\lambda)\sup_{Y_2}(f_2) \leq \alpha. \end{aligned}$$

Hence $\mu \not\leq \sup_{(Y_1\cup Y_2)}$. □

Now, we are in the position to prove the main theorem of this chapter.

<u>11 Theorem.</u> Let μ be a state of F, then the following are equivalent:

(i) μ is a character of F

(ii) μ can be extended to a character of VF

(iii) μ has an extension $\hat{\mu}$ to VF such that $\hat{\mu}$ is maximal and is a lattice state

(iv) μ is maximal and $Z(\mu) = \cap(\mu)$

(v) $Z(\mu)$ is a filter on X which converges in the state space to μ

(vi) μ is an extreme point of SX_F

(vii) μ can be extended to an extreme point $\hat{\mu}$ of SX_{VF}.

<u>Proof</u>. We proceed in the following way:

(i) ⇒ (ii) ⇒ (iii) ⇒ (vii) ⇒ (vi) ⇒ (iv) ⇒ (v) ⇒ (vi) and (v)+(vi)⇒(i).

(i) ⇒ (ii): Let $\hat{\mu}$ be the unique dominated extension of μ to VF, and consider a finite nonempty cover $Y_1,\dots,Y_n$ of X. Then by definition there is some $k \leq n$ such that $\mu \leq \sup_{Y_k}$. By the sandwich theorem we find a dominated extension $\tilde{\mu} \leq \sup_{Y_k}$. Because of $\tilde{\mu} = \hat{\mu}$ (uniqueness) we have $\hat{\mu} \leq \sup_{Y_k}$.

(ii) ⇒ (iii): We have to show that a character $\hat{\mu}$ of VF is a lattice state. Consider for $f_1,\dots,f_n \in VF$ the sets $Y_i = \{x \mid f_i(x) \geq (f_1 \vee\dots\vee f_n)(x)\}$, $i = 1,\dots,n$. Then by definition there is some $k \leq n$ with $\hat{\mu} \leq \sup_{Y_k}$. Since $\hat{\mu}$ is maximal it must be Y_k-order-preserving (otherwise it is according to the sandwich theorem dominated by some Y_k-order-preserving state). So we have by definition of Y_k that $\hat{\mu}(f_k) \geq \hat{\mu}(f_1 \vee\dots\vee f_n)$, and everything follows from the trivial inequality $\hat{\mu}(f_1 \vee\dots\vee f_n) \geq \hat{\mu}(f_i)$, $i=1,\dots,n$ (which is a consequence of the fact that $\hat{\mu}$ is order-preserving).

(iii) ⇒ (vii): We have to show that if $\hat{\mu}$ is a maximal lattice state then it is an extreme point of SX_{VF}. So, let $\hat{\mu} \leq \lambda\mu_1 + (1-\lambda)\mu_2$ $0 < \lambda < 1$, $\mu_1,\mu_2 \in SX_{VF}$. We claim that

(*) $\hat{\mu}(f) \geq \lambda\mu_1(f_1) + (1-\lambda)\mu_2(f_2)$ $\forall\, f_1,f_2 \in VF$ with $f_1 \vee f_2 \leq f$.

Now, assume that for arbitrary f, g we have $\hat{\mu}(f) \geq \hat{\mu}(g)$. Then from (*) and the fact that $\hat{\mu}$ is a lattice state we get $\hat{\mu}(f \vee g) = \hat{\mu}(f) \geq \lambda\mu_1(f) + (1-\lambda)\mu_2(g)$. And because of $\lambda\mu_1(f) + (1-\lambda)\mu_2(f) \geq \hat{\mu}(f)$ we obtain $\mu_2(f) \geq \mu_2(g)$. Application of Lemma 10 (iii) gives $\hat{\mu} = \mu_2$.

Proof of the claim:

$\delta(f) = \sup\{\lambda\mu_1(f_1) + (1-\lambda)\mu_2(f_2) \mid f_1, f_2 \in VF \text{ with } f_1 \vee f_2 \leq f\}$ defines a superlinear functional on VF with $\hat{\mu} \leq \delta \leq \sup_X$. By the sandwich theorem there is some state $\tilde{\mu}$ such that $\delta \leq \tilde{\mu} \leq \sup_X$. Hence $\tilde{\mu} = \hat{\mu}$ and $\delta = \hat{\mu}$ since $\hat{\mu}$ was maximal.

(vii) ⇒ (vi): Assume that $\mu \in SX_F$ has an extension to an extreme point $\hat{\mu}$ of SX_{VF}. Define $\delta(g) = \sup\{\mu(f) \mid f \in F, f \leq g\}$. Obviously we must have $\delta \leq \tilde{\mu}$ for every dominated extension $\tilde{\mu}$ of μ to VF. We claim $\delta = \hat{\mu}$ which proves that $\hat{\mu}$ is the only dominated extension of μ since every extreme point is a maximal state. From this we easily deduce that μ is extreme via the following argument: Assume $\mu = \lambda\mu_1 + (1-\lambda)\mu_2$, with $0 < \lambda < 1$ and $\mu_1, \mu_2 \in SX_F$. Take dominated extensions $\tilde{\mu}_1, \tilde{\mu}_2$ of μ_1, μ_2 and consider $\tilde{\mu} = \lambda\tilde{\mu}_1 + (1-\lambda)\tilde{\mu}_2$. This is a dominated extension of μ, hence equal to $\hat{\mu}$. Thus $\tilde{\mu}_1 = \tilde{\mu}_2 = \hat{\mu}$ because $\hat{\mu}$ was extreme. Restriction to F gives $\mu_1 = \mu_2 = \mu$.

Proof of the claim:

Let $f_1, \ldots, f_n$ be arbitrary elements of F then it suffices to prove $\hat{\mu}(f_1 \vee \ldots \vee f_n) = \max\{\mu(f_1), \ldots, \mu(f_n)\}$. For this purpose we consider the cover $Y_i = \{x \mid f_i(x) \geq (f_1 \vee \ldots \vee f_n)(x)\}$, $i=1,\ldots,n$, of X. Lemma 7 gives us the decomposition $\hat{\mu} \leq \sum_{i=1}^{n} \lambda_i\hat{\mu}_i$ with Y_i-order-preserving states $\hat{\mu}_i$. Since $\hat{\mu}$ was extreme we have $\hat{\mu} = \hat{\mu}_k$ for some k. So $\hat{\mu}$ must be Y_k-order-preserving. This gives $\hat{\mu}(f_1 \vee \ldots \vee f_n) = \hat{\mu}(f_k) = \mu(f_k)$.

(vi) ⇒ (iv): $Z(\mu) \subset h(\mu)$ (Lemma 10 iv) and the maximality of μ are obvious. So, we have for arbitrary $Z \in h(\mu)$ to demonstrate that $Z \in Z(\mu)$. Consider arbitrary $\alpha < \beta < 0$ and put $\lambda = \beta\alpha^{-1}$, $Y = X \setminus Z$. If we had $\mu \leq \lambda \sup_Y + (1-\lambda)\sup_X$ then the sum theorem would give us states μ_2 and $\mu_1 \leq \sup_Y$ such that $\mu \leq \lambda\mu_1 + (1-\lambda)\mu_2$. This is in contradiction to $\mu \not\leq \sup_Y$ and the fact that μ is extreme. Hence there must be some $g \in F$ with $\mu(g) > \lambda \sup_Y(g) +$

$+ (1-\lambda)\ \sup_X(g)$.

Then $f = \alpha(\sup_Y(g) - \sup_X(g))^{-1}\ (g-\sup_X(g))$ has the required properties, namely $f \le 0$, $\mu(f) \ge \beta$ and $\sup_Y(f) \le \alpha$.

(iv) ⇒ (v): From the filter properties only $Z_1, Z_2 \in Z(\mu) \rightarrow Z_1 \cap Z_2 \in Z(\mu)$ is nontrivial, but this is an immediate consequence of (iv) and Lemma 10 (vi). So it remains to demonstrate that $Z(\mu)$ converges to μ. For that purpose it suffices to show $\lim \mathcal{U} = \mu$ for every ultrafilter $\mathcal{U} \supset Z(\mu)$. Let $Y \in \mathcal{U}$ then $\mu \not\le \sup_Y$ is impossible because (iv) then implies $X \setminus Y \in Z(\mu)$ which contradicts $\mathcal{U} \supset Z(\mu)$. Hence $\mu \le \sup_Y\ \forall\ Y \in \mathcal{U}$. This leads to $\mu \le \tilde{\mu} = \lim \mathcal{U}$, where $\tilde{\mu}$ is a state and must be equal to μ since μ is maximal.

(v) ⇒ (vi): Let $\nu \in \mathrm{Face}(\mu)$. Then from Lemma 10 (v) we obtain $Z(\nu) \supset Z(\mu)$. Because of $Z(\nu) \subset \emptyset(\nu)$ (Lemma 10 iv) this gives $\nu \le \sup_Y\ \forall\ Y \in Z(\mu)$. Hence $\nu \le \lim Z(\mu) = \mu$. Since this must hold for every element in Face(μ) we get $\nu = \mu$ and $\mathrm{Face}(\mu) = \{\mu\}$.

(v)+(vi) ⇒ (i): We first demonstrate that μ has a unique dominated extension to VF. For this purpose we consider arbitrary $f_1,\dots,f_n \in F$ and put $\delta = \max(\mu(f_1),\dots,\mu(f_n))$. We show that $\tilde{\mu}(f_1 \vee\dots\vee f_n) = \delta$ for every dominated extension $\tilde{\mu}$ of μ. $\tilde{\mu}(f_1 \vee\dots\vee f_n) \ge \delta$ is obvious. For the other inequality we consider $Y_i = \{x \mid f_i(x) \ge (f_1 \vee\dots\vee f_n)(x)\}$ and obtain with Lemma 7 $\tilde{\mu} \le \Sigma\, \lambda_i \tilde{\mu}_i$, where $\tilde{\mu}_i$ is Y_i-order-preserving (for those i with $Y_i \ne \emptyset$, otherwise we put $\lambda_i = 0$).

Hence $\tilde{\mu}(f_1 \vee\dots\vee f_n) \le \sum_{i=1}^{n} \lambda_i\, \tilde{\mu}_i(f_i)$. But we have $\tilde{\mu}_i(f_i) = \mu(f_i)$ if $\lambda_i \ne 0$ since μ is an extreme point. This immediatly leads to $\tilde{\mu}(f_1 \vee\dots\vee f_n) \le \delta$.

It remains to show that for any cover $Y_1,\dots,Y_n$ of X we find some $k \le n$ with $\mu \le \sup_{Y_k}$. For this we consider some ultrafilter $\mathcal{U} \supset Z(\mu)$ on X. Then $\lim \mathcal{U} = \mu$, which has $\mu \le \sup_Y\ \forall\ Y \in \mathcal{U}$ as consequence. But as an ultrafilter $\mathcal{U}$ has to contain one of the sets $Y_1,\dots,Y_n$. ⊐

IV 2. THE F-REALCOMPACTIFICATION

Let us turn our attention to the Dini-characters of F. First some remarks. By $\overline{VF}$ and LF we denote the sup-norm closures of VF and VF - VF respectively. LF is a vector lattice with respect to pointwise structure. Since states are sup-norm continuous they have unique extensions from VF to $\overline{VF}$ and LF. Hence every character $\mu \in \beta X_F$ must have unique dominated extensions to $\overline{VF}$ and LF. These extensions are also denoted by μ since no confusion can arise.

<u>12 Remark</u>. Consider $\mu \in \beta X_F$ and $Y \subset X$. Then $\mu(f) \leq \sup_Y(f)$ $\forall f \in F$ if and only if $\mu(h) \leq \sup_Y(h)$ $\forall h \in LF$.

<u>Proof</u>. The "if" is trivial. From the sandwich theorem we get a dominated extension $\tilde{\mu} \leq \sup_Y$ of μ to LF and by uniqueness we have $\tilde{\mu} = \mu$ (on LF). □

By $Z_F(\mu)$, $Z_{VF}(\mu)$, $Z_{LF}(\mu)$ we denote the sets of strong domination for μ with respect to the cones F, VF and LF respectively. The same notation is adopted for $\hbar(\mu)$.

<u>13 Consequence</u>: Let μ be a character of F then $Z_F(\mu) = Z_{VF}(\mu) = Z_{LF}(\mu)$.

<u>Proof</u>. From Remark 12 we obtain $\hbar_F(\mu) = \hbar_{VF}(\mu) = \hbar_{LF}(\mu)$ and Theorem 11 (iv) gives the desired result, since μ is obviously a character for all the cones under consideration. □

<u>14 Theorem</u>. Let μ be a character of F, then the following are equivalent:

(i) μ is Dini-continuous on F

(ii) μ is Dini-continuous on LF

(iii) Let $Z_n \in Z(\mu)$, $n \in \mathbb{N}$, then $\bigcap \{Z_n \mid n \in \mathbb{N}\}$ is not empty.

(iv) μ has a representing measure m which is a $\{0,1\}$-measure, i.e. $m(A) = 1$ or 0 $\forall A \in \Sigma_F$.

(v) μ has a representing measure

(vi) For every countable cover Y_n, $n \in \mathbb{N}$, of X there is some n_o with $\mu \leq \sup_{Y_{n_o}}$.

(vii) For every $h \in LF$ there is an $x \in X$ such that $\mu(h) \leq h(x)$.

(viii) For every sequence $f_n \leq 0$ in F with $\sum_{n\in\mathbb{N}} \mu(f_n) > -\infty$ there is an x with $\sum_{n\in\mathbb{N}} f_n(x) > -\infty$.

Proof. We proced as follows

(iii) ⇒ (iv) ⇒ (v) ⇒ (viii) ⇒ (iii), (v) ⇒ (ii) ⇒ (i) ⇒ (viii) and (vi) ⇔ (iii). The equivalence of (vii) is proved separately. Let us begin:

(iii) ⇒ (iv): Let $\bar{Z}(\mu)$ be the family of countable intersections of elements of $Z(\mu)$. Then $\emptyset \notin \bar{Z}(\mu)$. Hence $\bar{Z}(\mu)$ is a filter which is stable against countable intersections. Now, we define

$$m(A) = \begin{cases} 1 & \text{if } A \in \bar{Z}(\mu) \\ 0 & \text{if } X\setminus A \in \bar{Z}(\mu). \end{cases}$$

Then m is clearly a σ-additive $\{0,1\}$-measure on the σ-algebra $\Sigma = \{Y \subset X \mid Y \in \bar{Z}(\mu)$ or $X\setminus Y \in \bar{Z}(\mu)\}$. It remains to prove that every $f \in F$ is Σ-measurable and that $\int f dm = \mu(f)$. For this purpose it is certainly sufficient to show that an arbitrary $h \in LF$ is Σ-measurable and that for $A = \{x \in X \mid h(x) = \mu(h)\}$ we have $A \in \bar{Z}(\mu)$ (i.e. $m(A) = 1$). Let $\delta < \mu(h)$ then by definition we have $X\setminus h^{-1}(]-\infty,\delta]) \in h_{LF}(\mu)$, thus this must be an element of $Z(\mu) = Z_{LF}(\mu) = h_{LF}(\mu)$ (Theorem 11 (iv) and consequence 13). Replacing h by $-h$ we see in the same way that for $\gamma > \mu(h)$ $X\setminus h^{-1}([\gamma,+\infty[) \in Z(\mu)$. Using the σ-additivity of Σ we then find that h is Σ-measurable, and by application of the fact that $\bar{Z}(\mu) \supset Z(\mu)$ is stable against countable intersections we immediately get $A \in \bar{Z}(\mu)$.

(iv) ⇒ (v) is obvious and (v) ⇒ (viii) follows from the Lebesgue dominated convergence theorem.

(viii) ⇒ (iii): Consider $Z_n \in Z(\mu)$ and assume $\bigcap \{Z_n \mid n \in \mathbb{N}\} = \emptyset$.

Since $Z(\mu)$ is a filter we can - without loss of generality - restrict our considerations to the case where $Z_{n+1} \subset Z_n$ for all $n \in \mathbb{N}$. By definition we know that there are $f_n \leq 0$ in F with $\mu(f_n) \geq \geq -\frac{1}{n^2}$ and $\sup_{Y_n}(f_n) \leq -2$, where $Y_n = X \setminus Z_n$. Now, we have

$$\sum_{n \in N} \mu(f_n) \geq -\sum \frac{1}{n^2} = -\frac{\pi^2}{6} > -\infty$$

and

$$\sum_{n=1}^{\infty} f_n(x) \leq \sum_{n=m}^{\infty} (-2) = -\infty \quad \text{for} \quad x \in Y_m.$$

Because of our assumption the Y_m cover X. Hence the sequence f_n is in contradiction to (viii).

(v) $\Rightarrow$ (ii): Let m be a representing measure for $\mu \in SX_F$. LF consists of Σ_F-measurable functions, since the Σ_F-measurable functions are a σ-complete vector lattice. So

$$\nu(h) := \int_X h \, dm \quad \forall\, h \in LF$$

defines a dominated extension of μ to LF and must therefore be equal to μ on LF. Hence m is also with respect to LF a representing measure for μ and (ii) is a consequence of Lebesgues dominated convergence theorem.

(ii) $\Rightarrow$ (i) $\Rightarrow$ (viii) is trivial.

(vi) $\Leftrightarrow$ (iii) Because of Theorem 11 (iv) the converse of (vi) must be equivalent to the existence of $Z_n \in Z(\mu)$, $n \in \mathbb{N}$, such that the $Y_n = X \setminus Z_n$ with $\mu \not\leq \sup_{Y_n}$, $n \in \mathbb{N}$, are a cover of X.

The equivalence of (vii): Because of (i) $\Leftrightarrow$ (ii) it suffices to prove: (v) for LF $\Rightarrow$ (vii) $\Rightarrow$ (vi) for LF. But: (v) for LF $\Rightarrow$ (vii) is completely obvious. Now, let (vii) be fulfilled and assume that there is a countable cover Y_n, $n \in \mathbb{N}$, of X with $\mu \not\leq \sup_{Y_n}$ (for LF). Then by Theorem 11 (iv) we find $f_n \leq 0$ in LF with $\mu(f_n) = \beta_n > > \alpha_n = \sup_{Y_n}(f_n)$, where $\beta_n \leq 0$ and $\alpha_n < 0$.

Define $h_n = \lambda_n((f_n \wedge \beta_n) \vee \alpha_n)$ where $\lambda_n > 0$ is chosen such that

$\sup_{x\in X} |h_n(x)| \le \frac{1}{n^2}$. Then $h = \sum_{n=1}^{\infty} h_n$ is an element of LF with $\mu(h) = \sup_{x\in X} h(x) > h(x)$ for all $x \in \bigcup\{Y_n \mid n \in \mathbb{N}\}$. This contradicts (vii). □

We would like to conclude this chapter with some words of warning. Although we have $\beta X_F = \beta X_{VF}$ we do not have $\beta X_{LF} = \beta X_{VF}$. Indeed, a state of LF is uniquely determined by its restriction to VF, but not every state on VF is maximal whereas every state on LF is automatically maximal since LF is a vector space. So in general we have $\beta X_{LF} \supset \beta X_{VF}$ and this has $\upsilon X_{LF} \supset \upsilon X_{VF}$ as a consequence.

Another warning should be given with respect to the comparison of the usual realcompactification υX (of X as a subspace of the topological space SX_{VF}) with υX_{LF}. Here we have in general $\upsilon X \subset \upsilon X_{LF}$. The reason for this is the fact that although LF is isometrically lattice-isomorphic to $C(\Omega)$ (Ω = (closure of X in SX_{VF}) $= \beta X_{LF}$) it is in general only isometrically lattice-isomorphic to a subspace of $CB(X)$. This inequality $\upsilon X \subset \upsilon X_{LF}$ has some advantages with respect to products.

Another remark which seems appropriate is that we actually proved in (ii) ⇒ (iv) (of Theorem 14) a little bit more than we claimed, namely

<u>15 Corollary</u>. $\mu \in \beta X_F$ is a Dini-character if and only if $\mu(h) \in h(X)$ for all $h \in LF$.

V CONSEQUENCES

Those who are acquainted with the theory of Hewitt-Nachbin spaces have certainly realized that Theorems 11 and 14 already generalize very many classical results of Hewitt, Gillman, Jerison and others. We shall not elaborate this in great detail, but we shall present some more results along these lines. These results show that our theory comprises large parts of Choquet-theory as well as the classical theory of continuous functions. Of course, in view of Theorem 11 (vi) and (vii) this is not very surprising. In constrast to the first part of this paper, where we have insisted on giving the full details of the proofs of the fundamental Theorems 11 and 14 we are only going to sketch the proofs of the coming results.

V 1. F-PSEUDOCOMPACTNESS

We recall that X is defined to be F-pseudocompact if every element in $\overline{VF}$ (sup-norm closure) attains its maximum on X.

16 Theorem. The following are equivalent:

(i) X is F-pseudocompact

(ii) $\beta X_F = \upsilon X_F$

(iii) F is a Dini cone, that means that for every pointwise decreasing sequence (f_n), $n \in \mathbb{N}$, we have

$$\sup_{x\in X} \inf_n f_n(x) = \inf_n \sup_{x\in X} f_n(x)$$

(iv) Every state of F has a representing measure.

Proof. We prove the theorem via the following implications: (i) $\Rightarrow$ VF is a Dini cone $\Rightarrow$ (iv) $\Rightarrow$ (ii) $\Rightarrow$ (i) and (iii) $\Rightarrow$ (ii) (which is obvious since (iii) implies that every state is Dini-continuous).

Let us start:

(i) $\Rightarrow$ VF is a Dini cone (compare[4 Theorem 2]): Let h_n be a decreasing sequence in VF with $\alpha = \inf_n \sup_{x\in X} h_n(x) > -\infty$ and consider $\delta < \alpha$. Then $g_n = h_n \vee \delta$ is a decreasing and uniformly bounded sequence and the series $\sum_{n\in\mathbb{N}} \lambda_n g_n$ converges uniformly on X for $\lambda_n \geq 0$ with $\sum_{n\in\mathbb{N}} \lambda_n = 1$. So with the help of Simon's convergence lemma [7, p.104] we see that F-pseudocompactness implies:

$$\inf\{\sup_{x\in X} \Big(\sum_{n=1}^{m} \lambda_n g_n(x)\Big) \mid \lambda_1,\dots,\lambda_m \geq 0,\ \sum_{n=1}^{m} \lambda_n = 1\} \leq \sup_{x\in X} \limsup_{n\to\infty} g_n(x).$$

The left-hand side is equal to $\inf_n \sup_{x\in X} g_n(x) = \alpha$ (since the g_n

are decreasing and $\delta < \alpha$), and the right-hand side is equal to $\sup_{x\in X} \inf_n g_n(x)$ which again is equal to the maximum of δ and $\sup_{x\in X} \inf_n h_n(x)$. Hence

$$\sup_{x\in X} \inf_n h_n(x) \geq \inf_n \sup_{x\in X} h_n(x).$$

The other inequality is trivial. So in view of the arbitraryness of the sequence VF must be a Dini cone.

VF is A Dini cone $\Rightarrow$ (iv): First of all, we observe that every state of VF must be Dini continuous (VF is a Dini cone). Every state of VF is dominated by a maximal state thus Theorem 8 tells us, that every state of VF has a representing measure. Hence every state of F must have a representing measure.

(iv) $\Rightarrow$ (ii) follows immediately from Theorem 14.

(ii) $\Rightarrow$ (i): An elementary exercise leads to $\beta X_{VF} = \beta X_{\overline{VF}}$.

So with Theorems 11 and 14 we get that (ii) is equivalent to $\beta X_{VF} = \upsilon X_{VF}$. From Bauer's maximum principle [1] we know that for every $h \in VF$ there is an extreme point μ of the state space, (hence an element of $\beta X_{\overline{VF}}$) such that

$$\mu(h) \geq \sup_{x\in X} h(x).$$

Now, because of $\mu \in \upsilon X_{\overline{VF}}$ we find with Theorem 14 (vii) an $x_o \in X$ such that $\mu(h) \leq h(x_o)$. So h must attain its maximum on X. □

This theorem already generalizes some well known results of Hewitt and Glicksberg. It contains for example the well known Alexandrov-Glicksberg theorem [8, thm 21 or 23] which says that every state on CB(X) has an integral representation if and only if X is pseudocompact.

Another consequence is the integral representation theorem of [3].

V 2. THE GEOMETRICAL SITUATION

Consider a compact convex subset K of a Hausdorff locally

convex topological vector space and denote by $A(K)$ the affine continuous functions on K and by ∂K the extreme points of K. Then by Bauer's maximum principle the spaces $A(K)|_{\partial K}$ (restrictions of the $f \in A(K)$ to ∂K) and $A(K)$ are isometrically (with respect to the sup-norm) isomorphic and lattice isomorphic.

Hence they have the same state spaces (being equal to the point evaluations given by all $x \in K$).

__17 Lemma__. ∂K is $A(K)|_{\partial K}$-pseudocompact.

__Proof__. Take a decreasing sequence $f_n \in A(K)|_{\partial K}$ and denote by φ_n the unique extensions of f_n to elements in $A(K)$. Then the φ_n are also decreasing. From Bauer's maximum principle (for upper semicontinuous affine functions) and from Dini's lemma we know that there is some $x_o \in \partial K$ such that

$$\inf_n \sup_{x \in X} f_n(x) \leq \inf_n \sup_{x \in X} \varphi_n(x) = \inf_n \varphi_n(x_o) = \inf_n f_n(x_o).$$

Since f_n was chosen arbitrarily this proves the assertion (Theorem 16 (iii)). $\square$

So, Theorem 16 contains the celebrated Choquet-Bishop-de Leeuw theorem as special case.

__18 Corollary__. For every $x \in K$ there is a probability measure m on ∂K (with respect to the σ-algebra generated by $A(K)|_{\partial K}$) such that

$$\varphi(x) = \int_{\partial K} \varphi \, dm \quad \forall \varphi \in A(K).$$

V 3. EPIMORPHISMS AND ADMISSIBLE SUBSETS

In this subsection we turn our attention to subsets of the state space, and here especially to the admissible subsets. Roughly spoken, these are subsets coming out of epimorphisms of cones.

Let $\emptyset \neq \Omega \subset SX_F$. Then if not otherwise mentioned Ω will always carry the topology inherited from the state space. By (Ω, F)

we denote the convex cone of functions on Ω given by $\omega \to \omega(f)$, $f \in F$. Since X is embedded in the state space we can identify (X,F) with F. Cones of the form (Ω,F) are called <u>concrete order unit cones</u>.

Now, consider two concrete order unit cones (Ω,F) and (Z,G) and a map $\varphi: \Omega \to Z$. φ is said to be an <u>epimorphism</u> from (Ω,F) to (Z,G) if

(4) $$\sup_{\Omega} (g\circ\varphi) = \sup_Z(g) \quad \forall\, g \in G$$

(5) $$\{g\circ\varphi \mid g \in G\} = F_{|\Omega}.$$

Where, of course, $g\circ\varphi$ stands for the function

$$\Omega \ni \omega \to \varphi(\omega)(g),$$

and where $\sup_Z(g) = \sup\{y(g) \mid y \in Z\}$, and $F_{|\Omega}$ denotes the cone of functions given by $\Omega \ni \omega \to \omega(f)$, $f \in F$.

<u>19 Examples</u>. (i) Let $\tilde{\Omega} \subset \Omega$, then the embedding of $\tilde{\Omega}$ in Ω is an epimorphism $(\tilde{\Omega},F) \to (\Omega,F)$ if and only if $\tilde{\Omega}$ is a sup-boundary of Ω, i.e. $\sup_{\tilde{\Omega}}(f) = \sup_{\Omega}(f) \quad \forall\, f \in F$.
Such a subset will be called <u>admissible</u>.

(ii) In the geometrical situation the embedding ∂K in K gives an epimorphism $(\partial K, A(K)) \to (K, A(K))$.

<u>20 Observation</u>. Let φ be an epimorphism from (Ω,F) to (Z,G).
(i) $\varphi: \Omega \to Z$ is automatically injective and continuous with respect to the topology of the state space.

(ii) $\varphi^*: S\Omega_F \to SZ_G$ given by $\varphi^*(\mu)(g) = \mu(g\circ\varphi) \ \forall\, g \in G$ is an affine continuous injective map.

(iii) For every $\nu \in SZ_G$ there is a $\mu \in S\,\Omega_F$ such that $\varphi^*(\mu) \geq \nu$.

<u>Proof</u>. Only (iii) is less obvious. For this, we define a superlinear $\delta \leq \sup_{\Omega}$ (because of (4)) on (Ω,F) by

$$\delta(f) = \sup\{\nu(g) \mid g \in G,\ g\circ\varphi \leq f\}.$$

Then from the sandwich theorem we obtain the desired state μ with $\delta \leq \mu \leq \sup_\Omega$. □

Assertions 20 (i) and (ii) have the interesting consequence that φ^* restricted to $\{\mu \in S\,\Omega_F \mid \varphi^*(\nu)$ is maximal in $SZ_G\}$ is invertible. This immediately leads to

<u>21 Corollary</u>. (i) φ^* maps $\beta\Omega_F$ onto βZ_G

(ii) φ^* maps $\upsilon\Omega_F$ into υZ_G.

This together with the fact that Ω is dense in $\Omega \cup \beta\Omega_F$ (Theorem 11 (v)) leads to the following important universal properties (which are well known for the classical situation).

<u>22 Theorem</u>. Let φ be an epimorphism from (Ω,F) to (Z,G) then φ extends uniquely to epimorphisms from $(\Omega \cup \beta\Omega_F, F)$ to $(Z \cup \beta Z_G, G)$ and from $(\Omega \cup \upsilon\Omega_F, F)$ to $(Z \cup \upsilon Z_G, G)$.

This result is useful for formal proofs of structural properties. As an exercise the reader may use it for the proof of the idempotence of β and υ, i.e.

$$\beta(X \cup \beta X_F)_F = \beta X_F \quad \text{and} \quad \upsilon(X \cup \upsilon X_F)_F = \upsilon X_F.$$

After this detour let us get back to admissible subsets Ω of the state space SX_F, i.e. subsets with $\sup_\Omega(f) = \sup_X(f)\ \forall f \in F$.

<u>23 Theorem</u>. Let Y be an admissible subset of X.

(i) If Y is closed in X and if X is F-compact then Y is again F-compact.

(ii) If X is F-realcompact and if Y is an F_σ-subset of X (i.e. countable union of closed subsets of X) then Y is F-realcompact.

<u>Proof</u>. (i): The embedding $\varphi: Y \to X$ can be uniquely extended (thm. 22) to an epimorphism $\varphi^*: (Y \cup \beta Y_F, F) \to (X \cup \beta X_F, F) = (X,F)$ (since X is F-compact). Hence $\varphi^*: Y \cup \beta Y_F \to X$ and the dense subset $Y = \varphi^{*-1}(\varphi(Y))$ must be closed in $Y \cup \beta Y_F$. Thus $Y \supset \beta Y_F$.

(ii): First we remark that according to Theorem 14 (iii) or (vi) every F_σ-subset of $Y \cup \upsilon Y_F$ which contains Y must be equal to $Y \cup \upsilon Y_F$. Now, as in (i), the embedding $\varphi: Y \to X$ extends uniquely to an injective continuous $\varphi^*: Y \cup \upsilon Y_F \to X$ (realcompactness of X). Hence, Y must be an F_σ-subset of $Y \cup \upsilon Y_F$ which gives $Y \supset \upsilon Y_F$ with the help of our introductory remark. □

By a similar argument one proves:

<u>24 Theorem</u>. The following are equivalent:

(i) $\mu \in \beta X_F \backslash \upsilon X_F$.

(ii) There is an F_σ-subset Ω of $X \cup \beta X_F$ with $\Omega \supset X$ and $\mu \notin \Omega$.

V 4. REALCOMPACT SPACES

We like to look a little bit closer on the case when X is F-realcompact, i.e. $\upsilon X_F \subset X$. First we observe that we have already proved the following characterization of F-compactness, which is well known in the classical case [9, p 34]:

<u>25 Observation</u>. The following are equivalent:

(i) X is F-realcompact and F-pseudocompact

(ii) X is F-compact.

<u>Proof</u>. (i) ⇒ (ii): We have $\upsilon X_F \subset X$ (X is F-realcompact) and $\upsilon X_F = \beta X_F$ (X is F-pseudocompact), hence $\beta X_F \subset X$.

(ii) ⇒ (i): Obviously F-compact implies F-realcompact. And from $\beta X_F \subset X$ and the fact that all point evaluations are Dini-continuous follows $\beta X_F \subset \upsilon X_F$. □

This is a rather useful criterion for F-compact spaces. Because very many spaces are automatically F-realcompact and for them F-psoudocompactness is then a necessary and sufficient condition for the assertion that X contains already all extreme points of the state space. If one reformulates the Theorems 23 and 24 one easily finds examples for such situations:

26 Theorem. The following are equivalent:

(i) X is F-realcompact

(ii) X is G_δ-closed in $X \cup \beta X_F$, i.e. for every $\mu \in \beta X_F \setminus X$ there is an F_δ-subset Ω of $X \cup \beta X_F$ with $\mu \notin \Omega$ and $\Omega \supset X$.

(iii) X is the intersection of F_σ-subsets of $X \cup \beta X_F$.

This theorem generalizes well known results of Mrówka [10,p 80] and Wenjen [10, p 81].

Other examples for such X which are automatically F-realcompact are occuring in topological situations (compare [4]):

27 Theorem. Assume that there is a topology τ on X such that all $f \in F$ are τ-upper-semicontinuous and such that (X,τ) is a Lindelöf space. Then X is F-realcompact.

Proof. Take $\mu \in \upsilon X_F$ and consider the filter $Z(\mu)$. By $\bar{Y}$ we denote the τ-closure of $Y \subset X$. By the definitions of $h(\mu)$ and $Z(\mu)$ one easily shows that $\mathfrak{F} = \{\bar{Y} \mid Y \in Z(\mu)\}$ is a filter basis of $Z(\mu)$. Hence $\mu \notin X$ implies $\bigcap \{\bar{Y} \mid \bar{Y} \in \mathfrak{F}\} = \emptyset$ (since $\mathfrak{F} \to \mu$), or in other words $\{X \setminus \bar{Y} \mid \bar{Y} \in \mathfrak{F}\}$ is an τ-open cover of X. Thus there must be a countable subcover $X \setminus \bar{Y}_n$, $n \in \mathbb{N}$, since (X,τ) is Lindelöf. This gives $\bigcap \{\bar{Y}_n \mid n \in \mathbb{N}\} = \emptyset$ in contradiction to Theorem 14 (iii). □

REFERENCES

[1] Alfsen, E.M., Compact convex sets and Boundary Integrals, Springer Verlag, Berlin-Heidelberg-New York 1971.

[2] Fuchssteiner, B., W. Hackenbroch, Maximumpunkte, Arch. Math. 23, 415-421 (1972).

[3] Fuchssteiner, B., When does the Riesz Representation theorem hold? Arch. Math. 28, 173-181 (1977).

[4] Fuchssteiner, B., Decomposition theorems, Manuscripta Mathematica 22, 151-164 (1977).

[5] Fuchssteiner, B., Neumann, M., Small boundaries, Arch. Math. 30, 613-621 (1978).

[6] Gillman, L., M. Jerison, Rings of continuous functions, Van Nostrand, New York-Toronto-London-Melbourne, 1960.

[7] Simons, S., A convergence theorem with boundary, Pacific J. Math. 40, 703-708 (1972).

[8] Varadarajan, V.S., Measures on topological spaces, Amer. Math. Soc. Translations, 48, 161-228 (1965).

[9] Walker, R.C., The Stone-Czech-compactification, Springer Verlag, Berlin-Heidelberg-New York 1974.

[10] Weir, M.D., Hewitt-Nachbin Spaces, North-Holland Publ. Co., Amsterdam 1975.

ON THE TOPOLOGY OF COMPACT COMPLEX SURFACES

Ludger Kaup

Universität Konstanz
Fakultät für Mathematik
Postfach 5560
D-7750 Konstanz

In this lecture we outline some recent results concerning the topology of compact complex spaces of (complex) dimension two.

Let us briefly recall the classical case of dimension one, namely the topological classification of compact <u>Riemann</u> "<u>surfaces</u>" R (compact connected nonsingular complex curves), which is one of the most beautiful examples of mathematical classification: Up to hoemomorphism (even diffeomorphism) any such R can be obtained by attaching g handles to the sphere S^2. The number g is called the genus of R; it completely determines the topology of R.

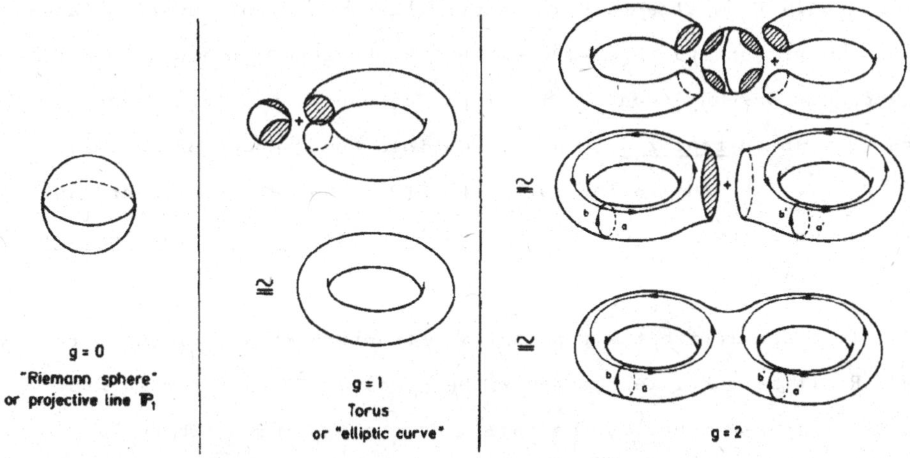

g = 0 "Riemann sphere" or projective line $\mathbb{P}_1$ — g = 1 Torus or "elliptic curve" — g = 2

g can be described in terms of a classical invariant of algebraic topology: If $H_j(R) = H_j(R,\mathbb{Z})$ denotes the j-th integral homology group of R, then it has a finite number $b_j(R)$ of free generators, called the j-th Betti number of R. It is not hard to see that

$b_o(R) = 1$ (since R is connected)

$b_2(R) = b_o(R)$ (by Poincaré duality, since R is an oriented compact manifold of topological dimension two)

$b_1(R) = 2g$ (since every handle contributes two free generators; cf. a, b and a', b' in the picture for $g = 2$).

The natural generalization of Riemann "surfaces" to higher dimensions are the complex spaces, i.e. Hausdorff spaces X with a complex structure given by a sheaf ${}_X O$ of holomorphic functions on X. We want to discuss the classification of the underlying topological space in the case of complex surfaces, where surface X means simply connected compact irreducible complex space of dimension two.*)

Simple examples are the projective plane $\mathbb{P}_2$ and $\mathbb{P}_1 \times \mathbb{P}_1$; in particular we will be concerned with a large class of examples arising naturally from algebraic geometry: the irreducible hypersurfaces $H \hookrightarrow \mathbb{P}_3$ in projective three-space. By the Lefschetz theorem on hypersurface sections any such H is simply connected and hence a surface. Since H can be realized by an irreducible homogeneous polynomial $P \in \mathbb{C}[z_o,\dots,z_3]$ (H is the zero set X(P)), we have to answer the following question: Given P, how can we determine the topological type of X(P)?

To characterize the topological structure of a Riemann "surface" R it is sufficient to know the homology of R. For surfaces this is no longer true, as the following example shows (where $\bar{\mathbb{P}}_2$ denotes $\mathbb{P}_2$ with the "wrong" orientation):

---

*) X admits the structure of a finite oriented polyhedron of real dimension four, cf. [Gi]. The hypothesis "simply connected" is necessary for the application of [Wh, thm. 2] and related results.

<u>1 Example</u>. The manifolds $\mathbb{P}_1 \times \mathbb{P}_1$ and $\mathbb{P}_2 \# \bar{\mathbb{P}}_2$ *) have isomorphic homology groups, but they are not of the same topological type.

By standard methods of algebraic topology one finds that $\mathbb{P}_1 \times \mathbb{P}_1$ and $\mathbb{P}_2 \# \bar{\mathbb{P}}_2$ have free homology groups, and one calculates the Betti numbers $b_o = b_4 = 1$, $b_2 = 2$ and $b_j = 0$ otherwise. On H_2 there exists an additional algebraic structure: For every compact oriented topological manifold M of real dimension four, there is a well defined intersection pairing

$$S(M): H_2(M) \times H_2(M) \to \mathbb{Z}$$
$$(\bar{a}, \bar{b}) \mapsto \bar{a} \cdot b \ .$$

Represent $\bar{a}$ and $\bar{b}$ by cycles a and b in "general" position, then a and b intersect transversally in a finite number of points:

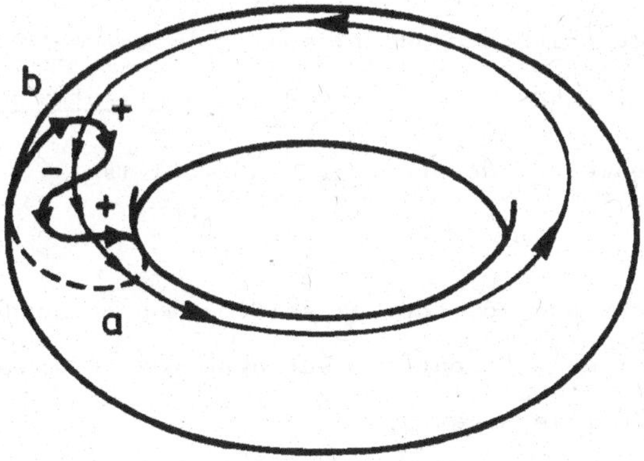

This number, counted with signs according to the induced orientation, is called the <u>intersection number</u> $\bar{a} \cdot \bar{b}$. Since M is of real dimension four, $S(M)$ is a <u>symmetric bilinear form</u>. - In our particular examples in appropriate bases of H_2 there are matrix representations

$$S(\mathbb{P}_1 \times \mathbb{P}_1) = \begin{pmatrix} 0 & 1 \\ 1 & 0 \end{pmatrix}, \qquad S(\mathbb{P}_2 \# \bar{\mathbb{P}}_2) = \begin{pmatrix} 1 & 0 \\ 0 & -1 \end{pmatrix} .$$

---

*) For two four-manifolds M and M', $M \# M'$ denotes the <u>connected sum</u>: Remove a small disc in M and M' and join the remainders along the boundaries, cf. the picture of the Riemann "surface" $g=2$. For example, $\mathbb{P}_2 \# \bar{\mathbb{P}}_2$ is diffeomorphic to a $\mathbb{P}_2$ blown up in a point.

Homeomorphic manifolds M and M' have <u>congruent matrices</u> $S(M)$ and $S(M')$, i.e. there is an invertible matrix $U \in GL(n,\mathbb{Z})$ such that

$$S(M') = U^t S(M) U.$$

But $\begin{pmatrix} 0 & 1 \\ 1 & 0 \end{pmatrix}$ and $\begin{pmatrix} 1 & 0 \\ 0 & -1 \end{pmatrix}$ are not congruent, hence $\mathbb{P}_1 \times \mathbb{P}_1$ and $\mathbb{P}_2 \# \bar{\mathbb{P}}_2$ cannot be homeomorphic. □

Actually the congruence type of the intersection matrix is even preserved under the following coarser equivalence relation of topological spaces which is basic in algebraic topology: Call two topological spaces S and T <u>homotopy equivalent</u> or <u>of the same homotopy type</u>, if there are continuous maps $S \overset{f}{\underset{g}{\rightleftarrows}} T$ such that

$$g \circ f \simeq Id_S, \qquad f \circ g \simeq Id_T$$

($\simeq$ means equal up to continuous deformation). We then write $S \simeq T$. If S and T are oriented and f and g respect the orientation, then we say that S and T are of the same <u>oriented homotopy type</u>.

A consequence of the proof of Example 1 is

$$\mathbb{P}_1 \times \mathbb{P}_1 \not\simeq \mathbb{P}_2 \# \bar{\mathbb{P}}_2.$$

Whereas homeomorphy and homotopy classification coincide for Riemann "surfaces", they are different for surfaces (cf. Example 21). Both aspects have to be investigated.

NONSINGULAR SURFACES

In this section we assume in addition that the surface X is nonsingular. Then the Poincaré duality homomorphisms

$$P_j(X): H^j(X) \to H_{4-j}(X)$$

are bijective and this implies:

i) $H_*(X)$ is a free abelian group,

ii) $b_0(X) = 1 = b_4(X)$ since X is connected,

iii) $b_1(X) = 0 = b_3(X)$ since X is simply connected,

iv) the intersection form $S(X)$ provides $H_2(X)$ with an

<u>inner product</u> structure ($S(X)$ has determinant ± 1).

The algebraic structure on $H_2(X)$ reflects much of the geometry of X:

<u>2 Theorem</u>. [MiHu V.1.5] Two nonsingular surfaces X and X' are of the same oriented homotopy type if and only if $H_2(X)$ and $H_2(X')$ are isomorphic as inner product spaces.

This solves the homotopy classification problem. The same criterion applies in the following aspect of a homeomorphy classification:

<u>3 Theorem</u>. [Wa] Two nonsingular surfaces X and X' are h-cobordant*) if and only if $H_2(X)$ and $H_2(X')$ are isomorphic as inner product spaces.

This result comes very close to a classification up to diffeomorphism: The h-cobordism theorem states that for compact simply connected C^∞-manifolds of real dimension at least five h-cobordism coincides with diffeomorphy [Sm]. Hence for nonsingular surfaces X and X' with congruent intersection forms, the complex threefolds $X \times \mathbb{P}_1$ and $X' \times \mathbb{P}_1$ are diffeomorphic. It is unknown whether the h-cobordism theorem holds in real dimension four, but the following "stable" version is true:

<u>4. Theorem</u>. [Wa] Two nonsingular surfaces X and X' have congruent intersection forms if and only if there is an integer m such that $X \# m(\mathbb{P}_1 \times \mathbb{P}_1)$ and $X' \# m(\mathbb{P}_1 \times \mathbb{P}_1)$ are diffeomorphic.

Here $m(\mathbb{P}_1 \times \mathbb{P}_1)$ denotes the connected sum of m copies of $\mathbb{P}_1 \times \mathbb{P}_1$ (for $m \geq 2$ we consider $m(\mathbb{P}_1 \times \mathbb{P}_1)$ as a C^∞-manifold). Actually, Theorems 2, 3 and 4 hold for compact oriented simply connected C^∞-manifolds of real dimension four.

---

*) Two compact oriented manifolds M and M' are called h-<u>cobordant</u>, if there is a compact oriented C^∞-manifold W with oriented boundary $M \cup \bar{M}'$, such that both M and M' are deformation retracts of W.

Since the geometric classification problem is now reduced to the investigation of the congruence type of $S(X)$, we consider the following invariants of symmetric bilinear forms: $S(X)$ interpreted as a real-valued matrix has well defined

i) rank, we have $\operatorname{rk} S(X) = b_2(X)$

ii) signature $\tau(X) = b_+ - b_-$ (number of positive eigenvalues minus number of negative eigenvalues counted with multiplicities, thus $b_+ = \frac{b_2(X)+\tau(X)}{2}$, $b_- = \frac{b_2(X)-\tau(X)}{2}$).

As an integer-valued matrix, $S(X)$ has a

iii) type which by definition is I(odd) if some diagonal entry is odd, II (even) if all diagonal elements are even.

Let us look at some examples:

X	$\mathbb{P}_2$	$\mathbb{P}_1 \times \mathbb{P}_1$	$\mathbb{P}_2 \# \bar{\mathbb{P}}_2$	$X(z_0^d+z_1^d+z_2^d+z_3^d)$
$\operatorname{rk} S(X)$	1	2	2	$1-\sum_{j=1}^{3}(1-d)^j$
$\tau(X)$	1	0	0	$d(2+d)(2-d)/3$
type	I	II	I	$\equiv d \bmod 2$

For surfaces X with $\operatorname{rk} S(X) \neq \pm\tau(X)$ there is the following

5 Proposition. [MiHu II.5.3] The congruence type of an indefinite inner product space is completely determined by rank, type and signature.

This result also holds for definite forms of rank at most eight, hence it covers the only case of a nonsingular surface with definite form known to us, namely $\mathbb{P}_2$.

6 Corollary. If $S(X)$ is indefinite and of type I, there is a homotopy equivalence

$$X \simeq b_+\mathbb{P}_2 \# b_-\bar{\mathbb{P}}_2 .$$

Proof. Proposition 5 implies the congruence

$$S(X) \sim \begin{pmatrix} I_{b_+} & 0 \\ 0 & I_{b_-} \end{pmatrix}.$$

From $S(M\#M') \sim S(M) \oplus S(M')$ we infer

$$S(b_+\mathbb{P}_2 \# b_-\bar{\mathbb{P}}_2) \sim \begin{pmatrix} I_{b_+} & 0 \\ 0 & I_{b_-} \end{pmatrix}.$$

Now Theorem 2 gives the result. ☐

For intersection forms of type II the situation is more complicated. Using the symmetric bilinear form

$$E_8 = \begin{pmatrix} 2 & 1 & & & & & & \\ 1 & 2 & 1 & & & & \bigcirc & \\ & 1 & 2 & 1 & & & & \\ & & 1 & 2 & 1 & & & \\ & & & 1 & 2 & 1 & 0 & 1 \\ & \bigcirc & & & 1 & 2 & 1 & 0 \\ & & & & 0 & 1 & 2 & 0 \\ & & & & 1 & 0 & 0 & 2 \end{pmatrix}$$

of rank 8, signature 8, determinant 1 and type II one has

<u>7. Theorem</u> [Ro]. For indefinite $S(X)$ of type II there is a congruence

$$S(X) \sim \frac{\tau(X)}{16}(E_8 \oplus E_8) \oplus b_- \begin{pmatrix} 0 & 1 \\ 1 & 0 \end{pmatrix};$$

in particular $\tau(X)$ is a multiple of 16.

But since no surface X with $S(X) \sim E_8 \oplus E_8$ is known (and there does not even seem to be much hope to find a C^∞-manifold of dimension four, representing $E_8 \oplus E_8$), Theorem 7 does not give us a geometric decomposition analogous to that of Corollary 6. However, there is a method to pass from type II to type I: Blowing up one point changes X to $X \# \bar{\mathbb{P}}_2$ and $S(X)$ to $S(X) \oplus (-1)$. Hence if $S(M)$ is indefinite, the surface $X \# \bar{\mathbb{P}}_2$ is homotopy-decomposable. Obviously this decomposibility is also true for a connected sum $X \# \mathbb{P}_2$ (which corresponds to blowing up a point with the wrong complex structure; $X \# \mathbb{P}_2$ still is a real analytic manifold). But there is a much nicer result:

8 Theorem [MaMo]. For every nonsingular hypersurface $X_d \hookrightarrow \mathbb{P}_3$ of degree d there is a diffeomorphism

$$X_d \# \mathbb{P}_2 = \frac{d}{3}(d^2-6d+11)\mathbb{P}_2 \# \frac{d-1}{3}(2d^2-4d+3)\bar{\mathbb{P}}_2.$$

This result also holds for twodimensional nonsingular complete intersections in $\mathbb{P}_m$ and for nonsingular hypersurface sections of sufficiently high degree in a nonsingular projective threefold $[MaMo]_2$.

By Kodaira's classification, every X is diffeomorphic to a surface which is

i) rational, or ii) elliptic, or iii) of general type (cf. [Ue 1.2.7]).

One has

i) Every rational surface is diffeomorphic to $\mathbb{P}_1 \times \mathbb{P}_1$ or to $\mathbb{P}_2 \# (b_2-1)\bar{\mathbb{P}}_2$.

In fact any such surface is obtained by a sequence of blowing up points from either $\mathbb{P}_2$ or from a $\mathbb{P}_1$-bundle over $\mathbb{P}_1$ [GrHa theorem on page 520]. Up to diffeomorphism there are precisely two such bundles: The trivial bundle $\mathbb{P}_1 \times \mathbb{P}_1$ and a nontrivial bundle B. Both have an intersection form of rank 2 and of signature 0, but their type is different. In fact B is diffeomorphic to $\mathbb{P}_2 \# \bar{\mathbb{P}}_2$. Finally $(\mathbb{P}_1 \times \mathbb{P}_1) \# \bar{\mathbb{P}}_2$ and $\mathbb{P}_2 \# 2\bar{\mathbb{P}}_2$ are diffeomorphic.

ii) If X is nonsingular and elliptic, then there is a diffeomorphism [Ma]

$$X \# \mathbb{P}_2 \cong (b_+(X)\mathbb{P}_2 \# b_-(X)\bar{\mathbb{P}}_2) \# \mathbb{P}_2.$$

iii) If X is nonsingular and projective algebraic, but not contained in $\mathbb{P}_4$, then there is a diffeomorphism [Mo]

$$X_d \# \mathbb{P}_2 \cong X \# [b_+(X_d)-b_+(X)]\mathbb{P}_2 \# [b_-(X_d)-b_-(X)]\bar{\mathbb{P}}_2 \# \mathbb{P}_2$$

where d = deg X and X_d a nonsingular hypersurface of degree d in $\mathbb{P}_3$ (cf. 9).

Let us finish this section with the remark that for nonsingular hypersurfaces in $\mathbb{P}_3$, the topological classification is analogous to that of Riemann "surfaces": Homotopy, homeomorphy and diffeomorphy classification coincide by the classical

<u>9 Proposition</u>. Any two nonsingular hypersurfaces in $\mathbb{P}_3$ of degree d are diffeomorphic. Moreover, d is uniquely determined by the second Betti number $1 - \sum_{j=1}^{3} (1-d)^j$.

We sketch the proof since its idea is basic for results like 14 and 20: The $t = \binom{d+3}{3}$-dimensional complex vector space of homogeneous polynomials $P \in \mathbb{C}[z_0,\ldots,z_3]$ of degree d has a basis of monic monomials $m_1,\ldots,m_t$, and every nonzero $(a_1,\ldots,a_t) \in \mathbb{C}^t$ defines $P_a = \sum_{j=1}^{t} a_j m_j \in \mathbb{C}[z_0,\ldots,z_3]$ and thus a (nonnecessarily irreducible) "surface" $X(P_a) \hookrightarrow \mathbb{P}_3$.

In the union $Z = \{(z,a) \in \mathbb{P}_3 \times \mathbb{C}^t;\ P_a(z) = 0\}$ of all $X(P_a)$, the set of the singular varieties $X(P_a)$ is given by $A := Z \cap \{\frac{\partial P_a}{\partial z_j} = 0;\ j=0,1,2,3\}$. Since the projection $pr: Z \to \mathbb{C}^t$ is a proper holomorphic map, the image $pr(A)$ is an analytic subset of $\mathbb{C}^t$ (Remmert's mapping theorem). The parameter space $U: = \mathbb{C}^t \setminus pr\,A$ of all nonsingular surfaces $X(P_a)$ is therefore connected. pr induces a proper holomorphic map

$$\pi : Z \setminus A \to U$$

which is even a submersion. By Ehresmann's theorem π is locally trivial as a C^∞-map, consequently all fibres $X(P_a)$ of π are diffeomorphic. □

THE SINGULAR CASE

For a surface X (as defined in the introduction) with singular points the investigation of the topology becomes considerably more complicated, since the singular Poincaré homomorphisms $P_j(X): H^j(X) \to H_{4-j}(X)$ are no longer bijective (for details cf [Ka]). Hence we encounter the following new phenomena:

a) $H_2(X,\mathbb{Z})$ <u>may have torsion</u> (cf. 18). We note that $H_2(X,\mathbb{Z})$ as a finitely generated abelian group has a decomposition

$$H_2(X,\mathbb{Z}) = \mathbb{Z}^{b_2(X)} \oplus \Sigma\mathbb{Z}/p^j$$

with uniquely determined prime powers p^j.

b) $H_3(X,\mathbb{Z})$ <u>may be different from zero</u> (cf. 12; it is nevertheless torsionfree [Ka, Kor. 3.6]).

c) <u>The classical intersection theory does not apply.</u>

In the nonsingular case, there is a close connection between the intersection form $S(X)$ and $P_2(X)$: The intersection of $a,b \in H_2(X,Z)$ satisfies

$$a \cdot b = P_4(X)(P_2^{-1}(X)(a) \cup P_2^{-1}(X)(b)),$$

where $H_o(X,\mathbb{Z})$ has been indentified with $\mathbb{Z}$. This implies that for a fixed basis of $H_2(X,\mathbb{Z})$ and the dual basis of $H^2(X,\mathbb{Z}) = \mathrm{Hom}(H_2(X,Z),\mathbb{Z})$, the matrices of $P_2(X,\mathbb{Z})$ and $S(X)$ are inverse to each other. One can formulate Theorem 2 by means of $P_2(X)$ instead of $S(X)$, which leads to the following generalization:

<u>10. Homotopy classification theorem</u> [BaKa]. Two surfaces X and X' are of the same oriented homotopy type if and only if:

i) $b_3(X) = b_3(X')$,

ii) there is an isomorphism

$$\vartheta: H_2(X,\mathbb{Z}) \to H_2(X',\mathbb{Z})$$

of abelian groups,

iii) ϑ is compatible with Poincaré homomorphisms, i.e. the diagrams

$$\begin{array}{ccc} H^2(X,\mathbb{Z}/m) & \xleftarrow{\vartheta^2} & H^2(X',\mathbb{Z}/m) \\ {\scriptstyle P_2(X,\mathbb{Z}/m)}\downarrow & & \downarrow{\scriptstyle P_2(X',\mathbb{Z}/m)} \\ H_2(X,\mathbb{Z}/m) & \xrightarrow{\vartheta_2} & H_2(X',\mathbb{Z}/m) \end{array}$$

which are naturally induced by ϑ commute for $m = 0$ and for all prime powers $m = p^j$ that occur in the decomposition a) of $H_2(X,\mathbb{Z})$,

iv) if $H_2(X,\mathbb{Z})$ has 2-torsion, then ϑ is compatible with the Pontrjagin squares $\mathfrak{p}_{2^j}$ for every $m = 2^j$ in iii): The diagram

$$\begin{array}{ccc} H^2(X,\mathbb{Z}/2^j) & \xleftarrow{\vartheta^2} & H^2(X',\mathbb{Z}/2^j) \\ \Big\downarrow \mathfrak{p}_{2^j}(X) & & \Big\downarrow \mathfrak{p}_{2^j}(X') \\ H^4(X,\mathbb{Z}/2^{j+1}) & \xleftarrow{\vartheta^4} & H^4(X',\mathbb{Z}/2^{j+1}) \end{array}$$

commutes, where ϑ^4 is induced by identifying the orientation classes $[X]$ and $[X']$.

The Pontrjagin square $\mathfrak{p}_{2^j}$ is an elementary cohomology operation refining the cup-product:

$$\mathfrak{p}_{2^j}(a) = a \cup a \bmod 2^j.$$

Conditions iii) and iv) can be considerably simplified for locally irreducible X (these are homeomorphic to a normal surface): Let $f: Y \to X$ denote a desingularization of X *). Then $H_1(Y,\mathbb{Z})$ is independent of the choice of Y, as there exists a minimal resolution $g: \tilde{Y} \to X$ [La 5.12] and every desingularization is obtained from g by blowing up points which does not change H_1.

<u>11 Remark</u>. i) If p is a prime number such that $H_1(Y,\mathbb{Z})$ has no p-torsion, then

$$P_2(X,\mathbb{Z}/p^j) = P_2(X,\mathbb{Z}) \otimes Id_{\mathbb{Z}/p^j}.$$

Hence $P_2(X,\mathbb{Z})$ completely determines $P_2(X,\mathbb{Z}/p^j)$ [Ba 3.3.1].

ii) If $H_1(Y,\mathbb{Z})$ has no 2-torsion, then all $\mathfrak{p}_{2^j}(X)$ are determined by the desingularization:

---

*) i.e. Y is a connected complex manifold and f a proper holomorphic map such that the restriction $Y\setminus f^{-1}(S(X)) \to X\setminus S(X)$ is biholomorphic and $Y = \overline{Y\setminus f^{-1}(S(X))}$. Y need not be simply connected (e.g. if $b_3(X) \neq 0$).

There is a commutative diagram

$$\begin{array}{ccccc}
H^2(X,\mathbb{Z}) & & \xrightarrow{\alpha\mapsto\alpha\cup\alpha \bmod 2^{j+1}} & & H^4(X,\mathbb{Z}/2^{j+1}) \\
 & \searrow^{\rho_X} & & \nearrow^{p(X)} & \\
\Big\downarrow f^2 & & H^2(X,\mathbb{Z}/2^j) & & \cong \Big\downarrow f^4 \\
 & & \Big\downarrow f^2 & & \\
 & & H^2(Y,\mathbb{Z}/2^j) & & \\
 & \nearrow^{\rho_Y} & & \searrow^{p(X)} & \\
H^2(Y,\mathbb{Z}) & & \xrightarrow{\beta\mapsto\beta\cup\beta \bmod 2^{j+1}} & & H^4(Y,\mathbb{Z}/2^{j+1})
\end{array}$$

where ρ is induced by the reduction $\mathbb{Z} \to \mathbb{Z}/2^j$. The homomorphism ρ_Y is surjective by a universal coefficient formula as $H_1(Y,\mathbb{Z})$ has no 2-torsion. For

$$\gamma \in H^2(X,\mathbb{Z}/2^j),$$

there is a $\beta \in H^2(Y,\mathbb{Z})$ such that $\rho_Y(\beta) = f^2(\gamma)$.

Then

$$f^4\, p(X)(\gamma) = \beta \cup \beta \bmod 2^{j+1}.$$

Let us discuss an explicit example:

<u>12 Example</u>. For every integer $d \geq 2$ the homogeneous polynomial $P = z_1^d+z_2^d+z_3^d \in \mathbb{C}[z_o,\ldots,z_3]$ defines a normal surface $X = X(P) \subset \mathbb{P}_3$ with the affine origin $0 = [1,0,0,0]$ as the only singular point. The part at infinity

$$X_\infty := V(X,z_o) = V(\mathbb{P}_2, z_1^d+z_2^d+z_3^d) \hookrightarrow \mathbb{P}_2$$

is a Riemann "surface" of genus $(d-1)(d-2)/2$. As to the smooth part $X\setminus 0$, there is a commutative diagram

$$\begin{array}{ccc}
X\setminus 0 & \longrightarrow & X_\infty \\
\downarrow & & \downarrow \\
\mathbb{P}_3\setminus 0 & \longrightarrow & \mathbb{P}_2
\end{array}$$

exhibiting $X\setminus 0$ as the total space of a complex line bundle on X_∞ induced by the normal bundle of $\mathbb{P}_2 \hookrightarrow \mathbb{P}_3$. - The affine part $X\setminus X_\infty =$

$= V(\mathbb{C}^3, P)$ is a complex cone and hence a contractible topological space. By inspection of the big commutative diagram (P) for (X, X_∞) (see [BaKa 2.1] or [Ba (3.3)]) one easily obtains that $H_*(X,\mathbb{Z})$ is a free abelian group with Betti numbers

$$b_2(X) = 1, \quad b_3(X) = (d-1)(d-2).$$

(For $d \geq 3$ this provides us with an example of a normal complex surface X (with just one singular point) such that $b_3(X) \neq 0$, hence the Poincaré homomorphisms $P_1(X)$ and $P_3(X)$ cannot be bijective). To apply Theorem 10 we still have to determine the congruence type of $P_2(X,\mathbb{Z})$. This follows from

13 Lemma [BaKa 3.2, 3.3]. Let $j\colon Y \hookrightarrow \mathbb{P}_3$ denote a surface and let $\omega \in H^2(\mathbb{P}_3,\mathbb{Z})$ be the canonical generator. Then $j^2\omega$ generates a direct factor of $H^2(Y,\mathbb{Z})$ and in $H_2(\mathbb{P}_3,\mathbb{Z})$ we have

$$(j_2 \circ P_2(Y,\mathbb{Z}) \circ j^2)(\omega) = (\deg Y)[\mathbb{P}_1].$$

In particular if $b_2(Y) = 1$, then $P_2(Y,\mathbb{Z})$ is multiplication by $\deg(Y)$ modulo torsion in $H_2(Y,\mathbb{Z})$.

We now intend to generalize this example: For any (non-constant) homogeneous polynomial $P \in \mathbb{C}[z_1,z_2,z_3]$, the surface $X(P) \hookrightarrow \mathbb{P}_3$ is normal if and only if the curve at infinity

$$X_\infty(P) = V(\mathbb{P}_2, P) \hookrightarrow \mathbb{P}_2$$

is nonsingular. In this case $X_\infty(P)$ and $X_\infty(z_1^d + z_2^d + z_3^d)$ are diffeomorphic for $d = \deg P$ (cf. the proof of Proposition 9). Such a diffeomorphism extends to a diffeomorphism of the total spaces of the normal bundles

$$X(P)\setminus 0 \to X_\infty(P),$$

which extends to a homeomorphism of the one-point compactifications, that is to the whole surfaces. Thus we have established

14 Proposition. If a normal surface $X \hookrightarrow \mathbb{P}_3$ can be realized by a homogeneous polynomial $P \in \mathbb{C}[z_1,z_2,z_3]$, then homeomorphy and homo-

topy type of $X = X(P)$ are completely determined by $d = \deg P$. The homotopy invariants are

$$H_2(X,\mathbb{Z}) = \mathbb{Z}, \quad b_3(X) = (d-1)(d-2), \quad P_2(X) = (d).$$

For a homogeneous polynomial $P \in \mathbb{C}[z_1,z_2,z_3]$, $X(P)$ is the union of all projective lines passing through the plane curve $X_\infty(P)$ and through the point $O = [1,0,0,0]$, i.e. $X(P)$ is a projective cone over $X_\infty(P)$ with vertex O. This structure is compatible with the natural $\mathbb{C}^*$-action on $\mathbb{P}_3$:

$$\lambda\cdot[z_0,z_1,z_2,z_3] = [z_0,\lambda z_1,\lambda z_2,\lambda z_3].$$

Obviously $X(P)$ is invariant under this action. If λ tends to zero, one obtains a contraction of the affine cone to the vertex O.

This aspect of a $\mathbb{C}^*$-action leads to an important extension of the class of homogeneous polynomials: A polynomial $P \in \mathbb{C}[z_1,z_2,z_3]$ is called weighted homogeneous (or quasihomogeneous) if there are positive integers q_1, q_2, q_3 and r *) such that

$$P(\lambda^{q_1}z_1, \lambda^{q_2}z_2, \lambda^{q_3}z_3) = \lambda^r P(z_1,z_2,z_3) \quad \text{for all } \lambda, z_j \in \mathbb{C}^*.$$

Then the weighted homogeneous hypersurface $X(P) \overset{**)}{=} V(\mathbb{P}_3,P) \subset \mathbb{P}_3$ is invariant under the $\mathbb{C}^*$-<u>action</u>

$$(\lambda,[z_0,z_1,z_2,z_3]) \mapsto [z_0, \lambda^{q_1}z_1, \lambda^{q_2}z_2, \lambda^{q_3}z_3]$$

on $\mathbb{P}_3$. Conversely for every surface X in $\mathbb{P}_3$, invariant under this action, there is (in appropriate affine coordinates) such a polynomial P that realizes X [OrWa 1.1.2]. - If λ tends to zero, then the affine part $V(\mathbb{C}^3,P)$ of $X(P)$ is contracted to the origin.

Introducing the <u>weights</u> $w_j := r/q_j$ and the <u>weight vector</u>

---

*) without loss of generality we may assume $\gcd(q_1,q_2,q_3) = 1$.

**) For non-homogeneous P, the hypersurface $X(P)$ is by definition $X(z_0^{\deg P}\cdot P(z_1/z_0, z_2/z_0, z_3/z_0))$.

$\underline{\underline{w}} = (w_1, w_2, w_3)$, we note that a polynomial $P = \Sigma a_{\nu_1\nu_2\nu_3} z_1^{\nu_1} z_2^{\nu_2} z_3^{\nu_3}$ is $\underline{\underline{w}}$-homogeneous if and only if

$$\nu_1/w_1 + \nu_2/w_2 + \nu_3/w_3 = 1$$

for every monomial that actually occurs in P. Here is a list of $\underline{\underline{w}}$-homogeneous polynomials which define an affine variety V such that the origin is nonsingular or an isolated singularity (in this case, $V \setminus 0$ is smooth):

class	equation	conditions	w_1	w_2	w_3
I	$z_1^{a_1}+z_2^{a_2}+z_3^{a_3}$		a_1	a_2	a_3
II	$z_1^{a_1}+z_2^{a_2}+z_2z_3^{a_3}$	$a_2 \geq 2$	a_1	a_2	$\frac{a_2a_3}{a_2-1}$
III	$z_1^{a_1}+z_2^{a_2}z_3+z_2z_3^{a_3}$	$a_2 \geq 2,\ a_3 \geq 2$	a_1	$\frac{a_2a_3-1}{a_3-1}$	$\frac{a_2a_3-1}{a_2-1}$
IV	$z_1^{a_1}+z_1z_2^{a_2}+z_2z_3^{a_3}$	$a_1 \geq 2$	a_1	$\frac{a_1a_2}{a_1-1}$	$\frac{a_1a_2a_3}{a_1(a_2-1)+1}$
V	$z_1^{a_1}z_2+z_2^{a_2}z_3+z_3^{a_3}z_1$		$\frac{a_1a_2a_3+1}{a_3(a_2-1)+1}$	$\frac{a_1a_2a_3+1}{a_1(a_3-1)+1}$	$\frac{a_1a_2a_3+1}{a_2(a_1-1)+1}$
VI	$z_1^{a_1}+z_1z_2^{a_2}+z_1z_3^{a_3}+z_2^{c_2}z_3^{c_3}$	$a_2 \geq 2$, $c_2/w_2+c_3/w_3=1$	a_1	$\frac{a_1a_2}{a_1-1}$	$\frac{a_1a_3}{a_1-1}$
VII	$z_1^{a_1}z_2+z_1z_2^{a_2}+z_1z_3^{a_3}+z_2^{c_2}z_3^{c_3}$	$a_1 \geq 2$, $a_2 \geq 2$ $c_2/w_2+c_3/w_3=1$	$\frac{a_1a_2-1}{a_2-1}$	$\frac{a_1a_2-1}{a_1-1}$	$\frac{a_3(a_1a_2-1)}{a_2(a_1-1)}$
VIII	$z_1z_2+z_3^{a_3}$		any $w_1 > 1$	$\frac{w_1}{w_1-1}$	a_3

Polynomials of class VI (resp. VII) exist if and only if (a_1-1) divides $\operatorname{lcm}(a_2,a_3)$ (resp. $(a_1-1)\cdot\gcd(a_2,a_3)$ divides $(a_2-1)a_3$) [Ar, Prop. 11.2].

These examples are of particular importance for the topological investigation of $X(P)$ in the isolated singularity $0 = [1,0,0,0]$: The boundary of a good neighbourhood of 0 in $X(P)$ is given by

$$K(P) := V(\mathbb{C}^3, P) \cap \{z \in \mathbb{C}^3;\ \|z\| = 1\},$$

which evidently is invariant under the action of $S^1 = U(1) \subset \mathbb{C}^*$. The affine part $V(\mathbb{C}^3, P)$ topologically is an open real cone over $K(P)$, thus the topology of $X(P)$ in 0 is determined by $K(P)$ and hence by $\underline{\underline{w}}(P)$:

<u>15 Remark</u> [OrWa 3.1.4]. For each $\underline{\underline{w}}$-homogeneous polynomial $P \in \mathbb{C}[z_1, z_2, z_3]$ with nonsingular $V(\mathbb{C}^3, P)\setminus 0$, there is a decomposi-

tion*) $P = F+G$ with $\underline{\underline{w}}$-homogeneous polynomials F and G that have no monomial in common such that

i) F occurs in one of the classes I-VIII (possibly after multiplication with nonzero complex numbers and a permutation of indices),

ii) $K(P)$ and $K(F)$ are equivariantly diffeomorphic.

But it is not true that the whole space $X(P)$ is determined by $\underline{\underline{w}}(P)$:

<u>16 Example</u>. The (2,3,4)-homogeneous polynomials

$$P_3 := z_1^2+z_2^3+z_1z_3^2, \qquad P_4 := z_1^2+z_2^3+z_3^4$$

define normal surfaces $X_j \hookrightarrow \mathbb{P}_3$ which are not homeomorphic, not even of the same homotopy type. In fact, $b_2(X_j) = 1$; consequently $P_2(X_j,\mathbb{Z}) = (j)$ for $j = 3,4$ by Lemma 13.

Two surfaces of the same weight and degree need not be homeomorphic:

<u>17 Example</u>. The (2,4,4)-homogeneous polynomials

$$Q_3 = z_1^2+z_1z_2^2+z_3^4, \qquad Q_4 = z_1^2+z_1z_2^2+z_2z_3^3$$

define normal surfaces $X_j \hookrightarrow \mathbb{P}_3$ with $b_2(X_j) = j$; consequently X_3 and X_4 are not of the same homotopy type. - In fact they differ in the number $b_2(X_{j\infty})$ of the irreducible components of $X_{j\infty}$. For certain weights, even the coincidence of these numbers is not sufficient:

<u>18 Example</u> [Ba 3.6.2, 3.6.3] The (2,4,4)-homogeneous polynomials of degree 4

$$P_1 = z_1^2+z_1z_2^2+z_2z_3^3, \qquad P_2 = z_1^2+z_1z_2^2+(z_2+z_3)^2z_3^2$$

define normal surfaces $X_j \hookrightarrow \mathbb{P}_3$ with

$$H_2(X_1,\mathbb{Z}) = \mathbb{Z}^4, \qquad H_2(X_2,\mathbb{Z}) = \mathbb{Z}^4 \oplus \mathbb{Z}/2.$$

Thus $b_2(X_{1\infty}) = 2 = b_2(X_{2\infty})$ but X_1 and X_2 are not of the same

---

*) non necessarily unique, e.g. $P = z_1^2 + z_1z_2^2 + z_2^3z_3 + z_3^4$.

homotopy type.

These examples lead to the following:

<u>19 Definition</u>. The <u>numerical type</u> of a weighted homogeneous polynomial $P \in \mathbb{C}[z_1,z_2,z_3]$ is given by

i) $\underline{\underline{w}}(P)$,

ii) $d := \deg P$,

iii) in case of $\underline{\underline{w}}(P) = (d/2,d,d)$: the non-ordered system of multiplicities m_j that occur in the factorization of the "leading form" (that is the homogeneous part of degree d)

$$P_d = \prod_{j=1}^{t} (\alpha_j z_2 + \beta_j z_3)^{m_j}$$

of P.

The multiplicities m_j can be interpreted as the multiplicities of the irreducible components of X_∞ with respect to their natural non-reduced complex structures.

Now the homeomorphy classification of normal weighted homogeneous surfaces is achieved by

<u>20 Theorem</u> [Ba]$_2$. Two normal weighted homogeneous surfaces in $\mathbb{P}_3$ are homeomorphic if and only if they can be realized (in appropriate affine coordinates) by weighted homogeneous polynomials of the same numerical type.

In order to prove that surfaces of the same numerical type are homeomorphic, one has to collect all these surfaces in a proper holomorphic family $\pi: \mathfrak{X} \to U$. This is done by a subtle modification of the proof of Proposition 9. The parameter space U turns out to be a connected complex manifold. There is a natural stratification $\{\mathfrak{X}\setminus S(\mathfrak{X}), S(\mathfrak{X})\}$; by verifying explicitly Teissier's "μ^* constant" criterion [Te, p.334] one obtains that Whitney's conditions a and b are satisfied. Finally the first Thom-Mather isotopy lemma implies that π is topologically locally trivail. - To prove the converse one has to reconstruct the numerical type from the topological data. This is mainly achieved by analyzing the local topological invariants at the singular points. The principal tool is Orlik's result [Or, §3] relating the local fundamental group of an isolated weighted homogeneous singularity to the weights.

It is easy to see that homeomorphy and homotopy classification do <u>not</u> coincide for normal weighted homogeneous surfaces:

<u>21 Example</u>. The normal weighted homogeneous surfaces $X(z_1+z_2^4+z_3^5)$ and $X(z_1^3+z_2^4+z_3^5)$ are <u>homotopy equivalent</u>, but <u>not homeomorphic</u>. In fact we have (cf. 23) $H_2(X,\mathbb{Z}) = \mathbb{Z}$, $b_3(X) = 0$ and $P_2(X,\mathbb{Z}) = (5)$ for both surfaces, thus they are of the same homotopy type by 10. On the other hand, the singular sets are $\{[0,1,0,0]\}$ and $\{[1,0,0,0], [0,1,0,0]\}$, respectively. In the case of normal surfaces, analytic singularities are always topological singularities; consequently the surfaces under consideration are not homeomorphic.

For a homotopy classification one has to calculate $H_*(X,\mathbb{Z})$ and the "congruence type" of $P_2(X,\mathbb{Z})$ (by 22.i)). Now $P_2(X,\mathbb{Z})$ in general has a non trivial kernel (e.g. 23). Hence the interesting part of $P_2(X,\mathbb{Z})$ is given by the induced homomorphism

$$\bar{P}_2(X)\colon H^2(X,\mathbb{Z})/\ker P_2(X,\mathbb{Z}) \to [\ker P_2(X,\mathbb{Z})]^{\perp}/\text{torsion}$$

where $[\ker P_2(X,\mathbb{Z})]^{\perp} \subset H_2(X,\mathbb{Z})$ refers to the pairing $H^2(X,\mathbb{Z}) \times H_2(X,\mathbb{Z}) \to \mathbb{Z}$.

We now exclude the case of homogeneous affine equations since it has been discussed in 14:

<u>22 Theorem</u> [Ba 3.5.4, 3.5.6]. Let $X \hookrightarrow \mathbb{P}_3$ be a normal weighted homogeneous surface with nonhomogeneous affine equation. Then the following holds:

i) $b_2(X) = b_3(X) + b_2(X_\infty)$,

ii) $P_2(X,\mathbb{Z}/n) = P_2(X,\mathbb{Z}) \otimes \mathrm{Id}_{\mathbb{Z}/n}$,

iii) rank $P_2(X,\mathbb{Z}) = b_2(X_\infty)$,

iv) $\ker P_2(X) = \mathrm{im}[H^2(X,X_\infty;\mathbb{Z}) \to H^2(X,\mathbb{Z})]$ is a direct factor of $H^2(X,\mathbb{Z})$,

v) $\det \bar{P}_2(X) = \pm|\mathrm{Tors}\,\Gamma(X,\mathcal{H}_2(\mathbb{Z}))|/|\mathrm{Tors}\,H_2(X,\mathbb{Z})|^2$,

vi) $\tau(\bar{P}_2(X)) = 2-b_2(X_\infty)$.

In v), the torsion group of the space of global sections of the second local homology sheaf $\mathcal{H}_2(X,\mathbb{Z})$ is the direct sum of the torsion groups of the second local homology groups $\mathcal{H}_2(X,\mathbb{Z})_a$ in the (fintely many) singular points a of X.

<u>23 Corollary</u> [KaBa 6]. If the part at infinity of the surface X in Theorem 22 is homeomorphic to a projective line, then

i) $H_2(X,\mathbb{Z}) \cong \mathbb{Z} \oplus \mathcal{H}_2(X,\mathbb{Z})_0$ and $b_3(X) = b_2(X)-1$,

ii) $H^2(X,\mathbb{Z}) \cong \mathbb{Z} \oplus \ker P_2(X,\mathbb{Z})$,

iii) $P_2(X,\mathbb{Z}) = (\deg X) \oplus 0$.

The local homology group $\mathcal{H}_2(X,\mathbb{Z})_0$ of the affine singularity 0 can be explicitly calculated (cf. [BaKa 9] for a complete list). For instance in the case of a Brieskorn polynomial $z_1^{a_1} + z_2^{a_2} + z_3^{a_3}$ the local homology $\mathcal{H}_2(X,\mathbb{Z})_0 = \mathbb{Z}^{b_{2,0}} \oplus T_{2,0}$ is given by

$$b_{2,0} = c^2 c_1 c_2 c_3 - c(c_1+c_2+c_3) + 2$$

$$T_{2,0} = \begin{cases} \mathbb{Z}/cc_1c_2c_3 \oplus \bigoplus_{k=1}^{3} [\mathbb{Z}/c_{ij}]^{cc_k-2} & \text{if all } \gcd(a_i,a_j)>1 \\ [\mathbb{Z}/c_{jk}]^{cc_i-1} \oplus [\mathbb{Z}/c_{ik}]^{cc_j-1} & \text{if } \gcd(a_i,a_j) = 1, \end{cases}$$

where $c = \gcd(a_1,a_2,a_3)$; $c_i = \gcd(a_j,a_k)/c$; $c_{ij} = a_k/cc_ic_j$ for $\{i,j,k\} = \{1,2,3\}$.

While $P_2(X)$ in 23 has rank 1, in the following example it is of rank 2:

<u>24 Corollary</u>. For $a \geq 2$ the polynomial $z_1^a + z_2^a + z_2 z_3^a$ of the numerical type $\underline{\underline{w}} = (a,a,a^2/(a-1))$ and degree $a+1$ define normal sur-

faces $X \subset \mathbb{P}_3$ with

$$b_3 = 0, \quad H_2(X) = \mathbb{Z}^2 \oplus [\mathbb{Z}/a]^{a-1} \quad \text{and} \quad P_2(X) = \begin{pmatrix} a+1 & 1 \\ 1 & a-1 \end{pmatrix}.$$

As a last case we mention the polynomials of degree 3:

<u>25 Corollary</u> [Ba] $_3$. Any normal weighted homogeneous cubic surfaces up to homeomorphy is given by one of the following equations

P	$\underline{\underline{w}}(P)$	T_2	$P_2(X)$
$z_1+z_2^2+z_3^3$	(1,2,3)	0	(3)
$z_1^3+z_2^3+z_3^3$	(3,3,3)	0	(3)
$z_1z_3+z_2^2+z_3^3$	(3/2,2,3)	$\mathbb{Z}/2$	(3)
$z_1^2+z_2^2+z_3^3$	(2,2,3)	$\mathbb{Z}/3$	(3)
$z_1+z_2^2+z_2z_3^2$	(1,2,4)	0	$\begin{pmatrix} 3 & 2 \\ 2 & 0 \end{pmatrix}$
$z_1z_3+z_2^2+z_2z_3^2$	(4/3,2,4)	0	$\begin{pmatrix} 3 & 2 \\ 2 & -2 \end{pmatrix}$
$z_1z_2+z_1z_3+z_2z_3^2$	(3/2,3,3)	0	$\begin{pmatrix} 3 & 0 \\ 0 & -6 \end{pmatrix}$
$z_1^2+z_2^2+z_2z_3^2$	(2,2,4)	$\mathbb{Z}/2$	$\begin{pmatrix} 3 & 1 \\ 1 & -1 \end{pmatrix}$
$z_1+z_2^3+z_3^3$	(1,3,3)	0	$\begin{pmatrix} 3 & 2 & -1 \\ 2 & 0 & 0 \\ -1 & 0 & -1 \end{pmatrix}$
$z_1z_3+z_2^3+z_3^3$	(3/2,3,3)	0	$\begin{pmatrix} 3 & 0 & 0 \\ 0 & -2 & 1 \\ 0 & 1 & -2 \end{pmatrix}$

Moreover $b_2 = \operatorname{rk} P_2(X)$ and, except for $z_1^3+z_2^3+z_3^3$ one has $b_3 = 0$.

REFERENCES

[Ar] Arnold, V.I., Normal Forms of Functions in Neighbourhoods of Degenerate Critical Points. Usp. Math. Nauk, Russian Math. Surveys 29:2, 10-50 (1974).

[Ba] Barthel, G., Topologie normaler gewichtet homogener Flächen. In P. Holm (editor): Real and Complex Singularities, Oslo 1976, Sijthoff & Noordhoff Intern. Publ. Groningen-Leyden; 99-126 (1977).

$[Ba]_2$ Barthel, G., Homöomorphieklassifikation normaler gewichtet homogener Flächen (in preparation).

$[Ba]_3$ Barthel, G., Zur Topologie normaler kubischer Flächen mit C^*-aktion. Preprint.

[BaKa] Barthel, G., Kaup L., Homotopieklassifikation einfach zusammenhängender normaler kompakter komplexer Flächen. Math. Ann. 212, 113-144 (1974).

[Gi] Giesecke, B., Simpliziale Zerlegung abzählbarer analytischer Räume. Math. Zeitschr. 83, 177-213 (1964).

[GrHa] Griffiths, P., Harris, J., Principles of Algebraic Geometry. Wiley-Interscience, (1978).

[Ka] Kaup, L., Poincaré-Dualität für Räume mit Normalisierung. Ann. Sc. Norm. Sup. Pisa XXVI, 1-31 (1972).

[KaBa] Kaup, L., Barthel, G., On the Homotopy Type of Weighted Homogeneous Normal Complex Surfaces, in "Several Complex Variables", Proceedings of Symposia in Pure Mathematics 30:1, 263-271 (1977).

[La] Laufer, H., Normal two-dimensional Singularities. Ann. of Math. Studies 71, Princeton Univ. Press (1971).

[Ma] Mandelbaum, R., On the Topology of Elliptic Surfaces. Preprint.

[MaMo] Mandelbaum, R., Moishezon, B., On the Topological Structure of Nonsingular Algebraic Surfaces in CP^3. Topology 15, 23-40 (1976).

$[MaMo]_2$ Mandelbaum, R., Moishezon, B., On the Topology of Simply-connected Algebraic Surfaces. Topology

[MiHu] Milnor, J., Husemoller, D., Symmetric Bilinear Forms. Erg. Math. 73, Springer Verlag, Berlin-Heidelberg-New York (1973).

[Mo] Moishezon, B., Complex Surfaces and Connected Sums of Complex Projective Planes. LNM 603 Springer (1977).

[Or] Orlik, P., Weighted Homogeneous Polynomials and Fundamental Groups. Topology 9, 267-273 (1970).

[OrWa] Orlik, P., Wagreich, P., Isolated Singularities of Algebraic Surfaces with C^*-Action. Ann. of Math. 93, 205-228 (1971).

[Ro] Rohlin, V.A., A New Result in the Theory of 4-dimensional Manifolds (russian). Dokl. Akad. Nauk. SSSR 84, 221-224 (1952).

[Sm] Smale, S., On the Structure of Manifolds. Amer. J. of Math. 84, 387-399 (1962).

[Te] Teissier, B., Cycles évanescents, sections planes et conditions de Whitney. In: Singularités à Cargèse, Astérisque 7/8, 285-362 (1973).

[Ue] Ueno, K., Introduction to Classification Theory of Algebraic Varieties and Compact Complex Spaces. In H. Popp (editor): Classification of Algebraic Varieties and Compact Complex Manifolds, LNM 412, Springer (1974).

[Wa] Wall, C.T.C., On Simply-connected 4-Manifolds. J. London Math. Soc. 39, 141-149 (1964).

[Wh] Whitehead, J.H.C., On Simply-connected 4-dimensional Polyhedra. Comm. Math. Helv. 22, 48-92 (1949).

JORDAN ALGEBRAS AND HOLOMORPHY

Wilhelm Kaup

Mathematisches Institut
der Universität Tübingen
Auf der Morgenstelle 10
7400 Tübingen Germany

Among the non-associative algebras there are essentially two important types - the <u>Lie algebras</u> and the <u>Jordan algebras</u>. The importance of Lie algebras in many contexts is widely known - for Jordan algebras this is not the case to the same extent. In the following I will try to explain the role of Jordan algebras for the description of bounded symmetric domains.

The now classical back ground is the relationship between formally real Jordan algebras, self-dual homogeneous cones and symmetric upper half planes due to Koecher in finite dimensions. The infinite-dimensional case is much more complicated - the objects here are the Jordan C*-algebras and their real analogues - the so-called JB-algebras introduced by Alfsen, Shultz and Størmer. In a joint paper with Braun and Upmeier we have shown that the symmetric tube domains in complex Banach spaces are in a one-one correspondence to JB-algebras.

The present note is expository - no proof is given and no result is really new except the classification of all bounded symmetric domains of tube type in reflexive complex Banach spaces (Theorem 3.15). A proof of this result will be given in a forthcoming paper in a more general setup.

1. JORDAN ALGEBRAS

Suppose A is an associative algebra over a field k. Then A becomes a new algebra A^L in the commutator product (also called

Lie product)

(1.1) $$[x,y] = xy - yx.$$

A^L satisfies

(1.2) $$[x,x] = 0$$

(1.3) $$[x,[y,z]] = [[x,y],z] + [y,[x,z]]$$

for all $x,y,z \in A^L$. (1.3) says that left multiplication is a derivation with respect to the product $[\,,\,]$ and is called the <u>Jacobi identity</u>. Every algebra with product $[\,,\,]$ satisfying (1.2) and (1.3) is called a <u>Lie algebra</u>. Note that a linear subspace B of A might be a Lie subalgebra without being an associative subalgebra - consider for instance a complex Hilbert space H and the associative algebra $A = \mathcal{L}(H)$ of all bounded linear operators on H. Then $\mathcal{S}(H) = \{\lambda \in A : \lambda^* = -\lambda\}$ is a real Lie subalgebra of A^L but not a subalgebra of A.

A^L is obtained from A by making the product anticommutative (and loosing associativity). Making the product commutative leads to Jordan algebras. Instead of putting $x \circ y = xy + yx$ it is convenient to introduce the <u>Jordan product</u>

(1.4) $$x \circ y = \frac{xy+yx}{2}$$

which has the same squares $x \circ x = x^2$ as the associative product. But we have to require that the characteristic of k is not 2 - which is not too restrictive for us since in the following we are mainly interested in the real and the complex case. The characteristic 2 case is much more complicated; for a nice review of the whole algebraic theory see McCrimmon [23]. A with the product (1.4) is an algebra A^J satisfying

(1.5) $$x \circ y = y \circ x$$

(1.6) $$x \circ (x^2 \circ y) = x^2 \circ (x \circ y).$$

The last identity says that the left multiplications by x and x^2 commute and is called the Jordan identity. Again a (not necessarily associative) algebra B over k, $\mathrm{char}(k) \neq 2$, with product $x \circ y$ is called a Jordan algebra if (1.5) and (1.6) are satisfied. For instance the bounded self-adjoint operators on a complex Hilbert space H, $\dim H > 1$, form a real Jordan subalgebra

$$\mathcal{H}(H) := \{\lambda \in \mathcal{L}(H) : \lambda^* = \lambda\}$$

of A^J, $A = \mathcal{L}(H)$. $\mathcal{H}(H)$ is not an associative algebra.

1.7 Example. Let $\mathbb{K}$ be one of the following real algebras $\mathbb{R}, \mathbb{C}, \mathbb{H}, \mathbb{O}$. $\mathbb{H}$ is the field of quaternions and $\mathbb{O}$ is the standard Cayley algebra (= octonion numbers). $\mathbb{K}$ has a natural conjugation $z \mapsto \bar{z}$ with $z = \bar{z} \Leftrightarrow z \in \mathbb{R} \subset \mathbb{K}$. Let $p \geq 1$ be an integer and denote by

$$V = \mathcal{H}_p(\mathbb{K})$$

the set of all $p \times p$-matrices $z = (z_{ij})$ over $\mathbb{K}$ which are hermitian, i.e. $z^* = z$. Then V is a real Jordan algebra with respect to the Jordan product

$$z \circ w = \frac{zw+wz}{2}$$

provided $p \leq 3$ in case $\mathbb{K} = \mathbb{O}$.

1.8 Example. Let Y be a real Hilbert space with inner product $(x|y)$. Put

$$V := \mathbb{R} \oplus Y$$

and consider the unique commutative real algebra structure on V with unit $e = (1,0)$ satisfying

$$x \circ y = (x|y)e \quad \text{for every} \quad x,y \in Y$$

(i.e. $(s,x) \circ (t,y) = (st+(x|y),\ sy+tx)$ for all $(s,x),(t,y) \in V$). V is a Jordan subalgebra of A^J, where A is the Clifford algebra of Y corresponding to the quadratic form $(x|y)$ on Y. With $p = 1 + \dim Y$ (= cardinality of an orthonormal basis of Y) V is called the spin factor of dimension p and denoted by V_p.

All the algebras V in (1.7) and (1.8) are formally real in the sense that

$$(1.9) \qquad x^2 + y^2 = 0 \quad \text{implies} \quad x = y = 0.$$

Algebras of this type were introduced and completely classified in the finite dimensional case by Jordan, von Neumann and Wigner in 1934 [13]:

<u>1.10 Theorem</u>. Every formally real Jordan algebra of finite dimension is a unique direct product of algebras in the following list ($p < \infty$):

V	$n = \dim V$	
$\mathcal{H}_p(\mathbb{C})$	p^2	$p > 0$
$\mathcal{H}_p(\mathbb{R})$	$\frac{p(p+1)}{2}$	$p > 1$
$\mathcal{H}_p(\mathbb{H})$	$\frac{2p(2p-1)}{2}$	$p > 2$
V_p	p	$p > 4$
$\mathcal{H}_3(\mathbb{O})$	27	

Note that $\mathcal{H}_1(\mathbb{K}) = \mathbb{R}$ and $\mathcal{H}_2(\mathbb{K}) \approx V_p$ for $\mathbb{K} = \mathbb{R}, \mathbb{C}, \mathbb{H}, \mathbb{O}$ and $p = 2 + \dim \mathbb{K}$. Furthermore $V_2 \approx V_1 \oplus V_1$ is not irreducible. For $p > 3$ the algebra $\mathcal{H}_p(\mathbb{O})$ is not a Jordan algebra. $\mathcal{H}_3(\mathbb{O})$ is the only simple formally real Jordan algebra which is not isomorphic to a subalgebra of some A^J with A associative. Therefore it is called an <u>exceptional algebra</u>. It occurs in a natural way together with some exceptional situations in analysis (e.g. some exceptional Lie groups and also exceptional bounded symmetric domains).

Now suppose V is an arbitrary Jordan algebra. Since the Jordan product in general is not associative for concrete computations it is often convenient to use operators (the product of which is associative). So denote for every $x \in V$ by $L(x) \in \mathrm{End}(V)$ the <u>left multiplication</u> by x, i.e. $L(x)y = x \circ y$. By the Jordan identity for every fixed $x \in V$ the subalgebra Σ of $\mathrm{End}(V)$ generated by $L(x)$ and $L(x^2)$ is commutative. It can be shown that Σ contains all operators $L(x^n)$, $n \geq 1$. The operator

$$(1.11) \qquad P(x) := 2L^2(x) - L(x^2) \in \Sigma$$

is of particular interest. It depends quadratically on x and is called the quadratic representation of x. In the special case $V = A^J$, A associative,

$$P(x)y = xyx$$

is easily checked. In A the identity

$$(xyx)z(xyx) = x(y(xzx)y)x$$

is true since the product in A is associative. The important fact is that it remains true in every Jordan algebra, i.e. the so called fundamental formula

$$P(P(x)y) = P(x)P(y)P(x)$$

is valid in every Jordan algebra.

Now suppose that the Jordan algebra V has a unit e. Then an element x is called invertible in V if there is a $y \in V$ with $x \circ y = e$ and $x^2 \circ y = x$. The element $y \in V$ is unique if it exists and is denoted by x^{-1}. It can be shown, that x is invertible if and only if the operator $P(x)$ on V is invertible (and then $x^{-1} = P(x)^{-1} x$). The subalgebra of V generated by x, x^{-1} and e is associative.

2. JORDAN C*-ALGEBRAS

In finite dimensions the formally real Jordan algebras V behave like the algebra $\mathbb{R}$ - for instance the multiplication operator $L(x)$ on V has always real spectrum. In infinite dimensions the purely algebraic condition (1.9) is no longer appropriate. Let us first study an example: Denote by V the set of all bounded holomorphic functions f on $\Delta = \{z \in \mathbb{C} : |z| < 1\}$ with $f(\Delta \cap \mathbb{R}) \subset \mathbb{R}$. Then V is a commutative real Banach algebra with respect to t e norm

$\|f\| := \sup|f(\Delta)|$. V is formally real in the sense (1.9), but L(f) has real spectrum only when f is a constant function. Also the cone $V^2 := \{f^2 : f \in V\}$ of squares in V does not have nice properties: The function $f_t(z) := z-t$, $t \in \mathbb{R}$, is in V^2 if and only if $|t| > 1$, i.e. $2f_o = f_t + f_{-t}$ shows that the cone V^2 is not convex. For $|t| > 1$ the function

$$g_t := (f_t/(1+|t|))^{1/2} \in V$$

has norm 1 and

$$\lim_{t\to\infty} \|g_t^2 + g_{-t}^2\| = 0,$$

i.e. the condition (1.9) is not satisfied approximately.

The appropriate generalizations of formally real Jordan algebras to infinite dimensions were introduced and studied by Alfsen-Shultz-Størmer [1]:

<u>2.1 Definition.</u> A JB-algebra is a real Jordan algebra V with unit e together with a complete norm such that

(i) $\|xy\| \leq \|x\| \, \|y\|$ [1)]

(ii) $\|x^2\| = \|x\|^2$

(iii) $\|x^2+y^2\| \geq \|x^2\|$

is true for all $x,y \in V$.

It can be shown that the norm on V is completely determined by the Jordan product on V: The cone $V^2 = \{x^2 : x \in V\}$ of all squares is closed, convex and defines a partial order on V by $x \leq y \Leftrightarrow y - x \in V^2$. The norm is just the order unit norm defined by

$$\|x\| = \inf\{t > 0 : -te \leq x \leq te\},$$

i.e. the closed unit ball in V is given by $(e-V^2) \cap (V^2-e)$. Therefore it makes sense to say, a given real Jordan algebra is a JB-algebra or not. The axioms (ii) and (iii) in 2.1 imply immediately that every JB-algebra is formally real in the sense (1.9). On the other

---

1) For simplicity we write xy instead of $x \circ y$ in the following.

hand it is known

<u>2.2 Theorem.</u> A real Jordan algebra V of finite dimension is a JB-algebra if and only if V is formally real.

In particular all the algebras in Examples 1.7 and 1.8 are JB-algebras. The spin factor $V_p = \mathbb{R} \oplus Y$ was already defined for arbitrary cardinal p. In fact V_p always is a JB-algebra even if p is infinite. The cone of squares in this case is given by

$V^2 = \{(t,y) : t \geq \|y\|\}$:

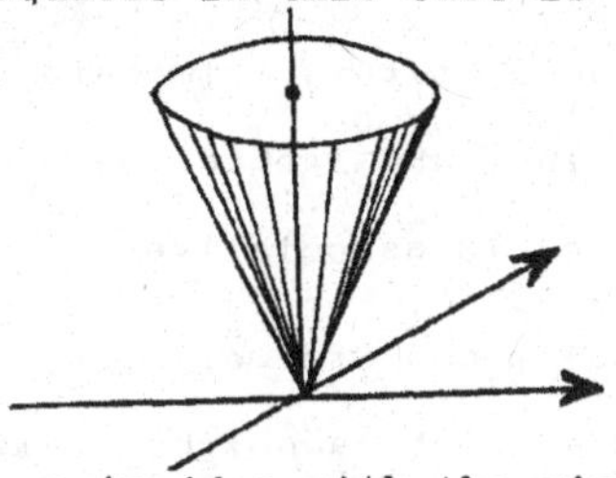

where on Y the JB-norm $\| \ \|$ coincides with the given Hilbert norm on Y. The JB-norm on $V_p = \mathbb{R} \oplus Y$ is given by

$$\|(t,y)\| = |t| + \|y\|, \quad \text{i.e.}$$

the closed unit ball in V_p is the double cone over the unit ball in Y with vertices in $\pm e$:

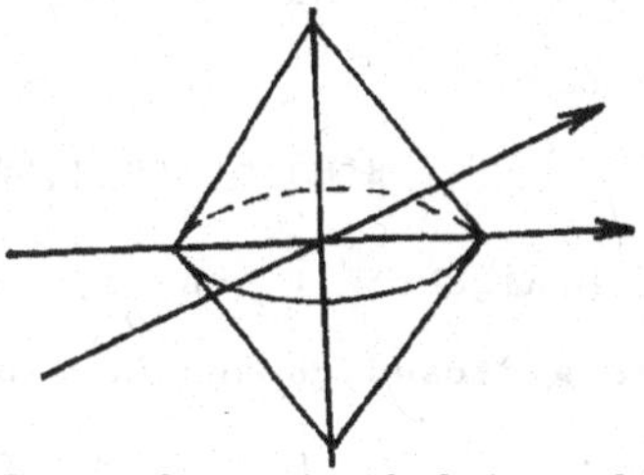

Also the Example in 1.7 can be extended to arbitrary cardinal p: Suppose $\mathbb{K} = \mathbb{R}, \mathbb{C}$ or $\mathbb{H}$ and H is a (right) $\mathbb{K}$-Hilbert space of dimension p over $\mathbb{K}$. Denote by $\mathcal{L}(H)$ the algebra of all bounded $\mathbb{K}$-linear operators on H. Then there is a natural involution (the adjoint $*$) on $\mathcal{L}(H)$ and

$$\mathcal{H}_p(\mathbb{K}) := \mathcal{H}(H) := \{\lambda \in \mathcal{L}(H) : \lambda^* = \lambda\}$$

is a JB-algebra with respect to the operator norm. The squares are precisely the positive operators in this case (i.e. operators with spectrum ≥ 0). Since every unital closed subalgebra of a JB-algebra again is a JB-algebra for every unital C^*-algebra X, the self-adjoint part $\{z \in X : z^* = z\}$ is a JB-algebra. For every compact topological space S and every JB-algebra V also the algebra $\mathbb{C}(S,V)$ of all continuous functions $S \to V$ is a JB-algebra. In particular $\mathbb{C}(S,\mathbb{R})$ is an associative JB-algebra. The converse is also true:

<u>2.3 Lemma</u>. Every associative JB-algebra is isometrically isomorphic to $\mathbb{C}(S,\mathbb{R})$ for some compact topological space S.

The importance of this Lemma stems from the fact that for every element x in a JB-algebra the closed subalgebra $C(x)$ generated by x and e is associative.

Alfsen, Shultz and Størmer showed that $\mathcal{H}_3(\mathbb{O})$ is essentially the only JB-algebra which cannot be realized as algebra of selfadjoint operators on a complex Hilbert space [1]:

<u>2.4 Gelfand-Neumark Theorem for JB-algebras</u>. Suppose V is a JB-algebra. Then there exists a complex Hilbert space H and a compact topological space S such that V is isometrically isomorphic to a closed subalgebra of

$$\mathcal{H}(H) \oplus \mathbb{C}(S,\mathcal{H}_3(\mathbb{O})).$$

Now fix a JB-algebra V and denote by $C = V^2$ the positive cone in V. C is a closed convex cone and $x \mapsto x^2$ is a homeomorphism of C. The interior Ω of C is an open convex cone which is not empty. Actually it can be shown that Ω is the connected unit-component of $V^{-1} := \{x \in V : x \text{ invertible}\}$. Also $\Omega = V^{-2}$ and $\Omega = \exp(V)$. The elements of Ω are called <u>positive definite</u> and we define $x < y$ for $x,y \in V$ by $y-x \in \Omega$. Denote by $\mathcal{L}(V)$ the algebra of all bounded operators in V and put for every total subset $A \subset V$

$$GL(A) := \{g \in \mathcal{L}(V) : g \text{ invertible and } g(A) = A\}.$$

For every $t \in \Omega$ the operator $P(t)$ is in $GL(\Omega)$ and sends e to t^2, i.e. $P(\Omega) \subset GL(\Omega)$ generates a transitive linear transformation group on Ω - in particular Ω is a (linearly) homogeneous cone. The isotropy subgroup of $GL(\Omega)$ at $e \in \Omega$ (i.e. the subgroup of all $g \in GL(\Omega)$ with $g(e) = e$) is the group $Aut(V)$ of all algebra automorphisms of V. Note that because of the uniqueness of the JB-norm on V every (algebra) automorphism of V is an isometry and in particular a homeomorphism. Therefore

$$GL(\Omega) = P(\Omega)\, Aut(V). \tag{2.5}$$

Now consider the complexification $U := V \oplus iV$ of V. U is a complex Jordan algebra with involution

$$(x+iy)^* = x-iy.$$

Consider furthermore the tube domain (<u>generalized upper half plane</u>)

$$D := D(\Omega) := \{z \in U : Im(z) \in \Omega\}$$

associated with the cone Ω. We identify $GL(V)$ in a natural way with a subgroup of $GL(U)$. Then the group of complex-affine transformations

$$Aff(D) := \{z \mapsto \lambda z + t : \lambda \in GL(\Omega),\ t \in V\}$$

is transitive on D, in particular D is (holomorphically) homogeneous. In the special case $V = \mathbb{R}$ we have $U = \mathbb{C}$ and $D = \{z \in \mathbb{C} : Im(z) > 0\}$ is the classical upper half plane. The Cayley transformation

$$C(z) = i\,\frac{z-i}{z+i}$$

maps D biholomorphically onto the open unit disc

$$\Delta = \{z \in \mathbb{C} : |z| < 1\}.$$

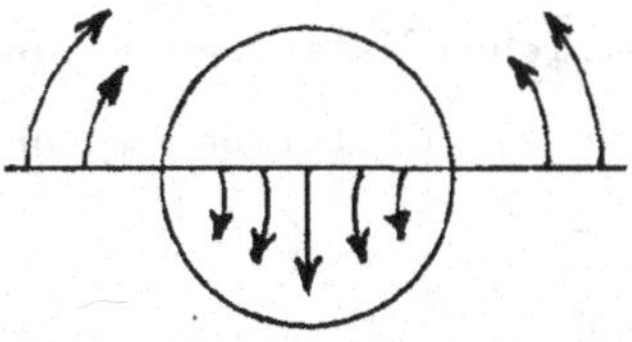

Actually Δ and D can be considered as hemispheres of the Riemann sphere $M := \mathbb{C} \cup \{\infty\}$ and C as a rotation of M by the angle $\frac{\pi}{2}$

around the points ± 1. Also in the general case every $z \in D$ is invertible and

$$C(z) := i(z-ie)(z+ie)^{-1}$$

defines a biholomorphic mapping from D onto a bounded balanced and convex domain $\Delta \subset U$. Actually

$$\Delta = \{z \in U : |P(z)P(z^*)|_\sigma < 1\}$$

where $|\ |_\sigma$ denotes the spectral radius. Therefore

(2.6) $$\|z\| := \|z\|_\infty := |P(z)P(z^*)|_\sigma^{1/4}$$

defines a complex norm on U such that Δ is the open unit ball of U. This norm extends the JB-norm on V. We call Δ the <u>generalized unit disc</u> and

$$\Sigma := \{z \in U^{-1} : z^* = z^{-1}\}$$

the <u>generalized unit circle</u>. The elements of Σ are called <u>unitary elements</u> in U. $z \in U$ is in Σ if and only if $P(z)$ is a surjective isometry of U. It can be shown that the extreme boundary $\check{S}$ of $\bar{\Delta}$ is a closed real-analytic submanifold of U and that Σ is a union of certain connected components of $\check{S}$. Actually $s \in \check{S}$ is in Σ if and only if the tangent space to $\check{S}$ in s is a real form of U. $\exp(iV)$ is a connected subset of Σ and for every connected component T of Σ (or more generally of $\check{S}$) $\bar{\Delta}$ is the closed convex hull $\overline{\text{co}T}$ of T in U. For the norm on U with closed unit ball $\overline{\text{co}(\exp(iV))}$ Wright has shown that U is a JB*-algebra (= Jordan C*-algebra, see definition below). But our norm (2.6) is the same as the JB*-norm of Wright:

<u>2.7 Definition</u>. A JB*-algebra is a complex Jordan algebra U with unit e, (conjugate linear) involution $*$ and complete norm such that

(i) $\|xy\| \leq \|x\|\ \|y\|$

(ii) $\|P(z)z^*\| = \|z\|^3$

for all $x,y,z \in U$.

For every unital C^*-algebra A the Jordanification $U = A^J$ is a JB^*-algebra. On the other hand every closed associative *-subalgebra of a JB^*-algebra U is a commutative C^*-algebra. Furthermore the self-adjoint part

$$V = \{x \in U : x^* = x\}$$

of a JB^*-algebra U is a JB-algebra. The converse is due to Wright:

<u>2.8 Theorem.</u> For every JB-algebra V there is a unique complex norm on $U := V \oplus iV$ such that U is a JB^*-algebra with self-adjoint part V. $V \leftrightarrow U$ defines an equivalence of the category of JB-algebras onto the category of JB^*-algebras.

Note that the morphisms in the JB^*-case are the *-homomorphisms.

A bounded domain B in a complex Banach space is called <u>symmetric</u> if to every $a \in B$ there is a holomorphic mapping $s_a : B \to B$ with $s_a^2 = id_B$ and a as isolated fixed point (s_a is uniquely determined if it exists and is called the <u>symmetry at</u> a). The generalized unit disc Δ is (holomorphically) homogeneous and $s_o(z) = -z$ is the symmetry at 0, i.e. Δ is symmetric. For every open cone P in a real Banach space X the domain $T = \{z \in X \oplus iX : Im(z) \in P\}$ is called a <u>symmetric tube domain</u> if T is biholomorphically equivalent to a bounded symmetric domain.

<u>2.9 Theorem.</u> Let V be a JB-algebra and $U = V \oplus iV$ the corresponding JB^*-algebra. Then

$$D = \{z \in U : Im(z) \in \Omega\}$$

is a symmetric tube domain. The symmetry at the point $ie \in D$ is given by $s(z) = -z^{-1}$ and

$$C(z) := i(z-ie)(z+ie)^{-1}$$

maps D biholomorphically on the open unit ball Δ of U. In particular Δ is a homogeneous domain.

Actually a complex Banach manifold M (generalized Riemann sphere) can be constructed such that

(i) U is a domain in M

(ii) Every (holomorphic) automorphism of D as well of Δ extends to an automorphism of M.

(iii) C extends to an automorphism of M.

C has order 4 and the set of fixed points is the set

$$\Sigma \cap V = \{x \in V : x^2 = e\}$$

of symmetries in V. Again C may be considered as a rotation by the angle $\frac{\pi}{2}$ around $\Sigma \cap V$. $\Sigma \cap V$ is the extreme boundary of $\bar{\Delta} \cap V$. Since $x \mapsto \frac{1}{2}(x+e)$ defines a linear homeomorphism of $\Delta \cap V$ onto $\Delta \cap V^2$ the extreme boundary of $\bar{\Delta} \cap V^2$ is the set of all idempotents in V. For instance for the spin factor $V_p = \mathbb{R} \oplus Y$ the set $\Sigma \cap V$ has three connected components, $\pm e$ and the unit sphere in Y.

The converse of 2.8 is also true but much harder to prove (compare [3]):

<u>2.10 Theorem</u>. Suppose V is a real Banach space and D is a symmetric tube domain in $U = V \oplus iV$. Then to every $e \in \Omega$ there exists a unique Jordan product on V such that V is a JB-algebra with unit e and upper half plane D.

Therefore the JB-algebras as well as the JB*-algebras are in a one-to-one correspondence to symmetric tube domains.

3. VARIOUS FINITENESS CONDITIONS

In the following V is always a JB-algebra and $U = V \oplus iV$ is the corresponding JB*-algebra. $\Delta \subset U$ is the open unit ball in U and $D = D(V) = \{z \in U : Im(z) > 0\}$ the upper half-plane in U - where $x > 0$ for $x \in V$ is defined by $x \in \Omega = \exp(V)$. $\Sigma = \{z \in U : z^* = z^{-1}\}$ is the unit circle in U. We know that $\exp(iV)$ is in the connected unit component of Σ. $L(U) := \{L(z) : z \in U\} \subset \mathcal{L}(U)$ is the set of multiplication operators on U. Since $P(\exp(ix)) = \exp(2iL(x))$ is an isometry of the Banach space U the multiplication operator

(3.1) $L(x)$ is a hermitian operator on U for every $x \in V$.

On the other hand the group $Aut(V)$ of algebra automorphisms of V coincides with the group of surjective ($\mathbb{C}$-linear) isometries of U fixing e, i.e.

(3.2) every derivation of V is a skew-hermitian operator on U.

The strongest finiteness condition on V clearly is that $\dim V < \infty$, i.e. that V is a formally real Jordan algebra of finite dimension. The theorem of Jordan, von Neumann and Wigner gives an explicit classification of all these algebras (compare 1.10). Koecher gave intrinsic descriptions of Ω, D, Δ and Σ for these algebras. On the other hand the bounded symmetric domains $B \subset \mathbb{C}^n$ were completely classified by É. Cartan. Among these Korányi and Wolf characterized the bounded symmetric domains $B \subset \mathbb{C}^n$ of tube type - these are precisely the domains B for which the Šilov boundary $\check{S}$ of B has real dimension n and hence are precisely the domains described by Koecher. From this it was clear that in finite dimensions the symmetric tube domains are in a one-one correspondence to formally real Jordan algebras. But this is also a consequence of our results in 2.8 and 2.9.

An important tool in the theory of formally real Jordan algebras V in finite dimension is the <u>minimal decomposition</u>: Every $a \in V$ has a unique representation

(3.3) $$a = \alpha_1 e_1 + \dots + \alpha_k e_k$$

where $\alpha_1, \dots, \alpha_k \in \mathbb{R}$ and $e_1, \dots, e_k \in V$ satisfy

$$\alpha_i \neq \alpha_j \text{ for } i \neq j \text{ and } e_i e_j = \delta_{ij} e_i, \quad e_i \neq 0 \text{ for all } i,j$$

and

$e = e_1 + \dots + e_k$ ($\{e_1, \dots, e_k\}$ is called a <u>complete orthogonal system of idempotents in</u> V). This is equivalent to saying that the unital subalgebra of V generated by a is isomorphic to the algebra $C(S,\mathbb{R})$ where $S = \{1,\dots,k\}$ - then e_j is the characteristic function of $\{j\} \subset S$. The importance of the minimal decomposition 3.3 stems from the fact that $\{e_1,\dots,e_k\}$ determines a Peirce-decomposition of V which for instance diagonalizes the operator $L(x)$ and hence also $P(x)$. The analogue for an arbitrary JB-algebra V is the fact that for every $a \in V$ the unital closed subalgebra $C(a)$ generated by a is isomorphic to some $C(S,\mathbb{R})$ where S is a compact topological space. But in case S is connected e is the only non-trivial idempotent in $C(a)$ and Peirce decompositions cannot be applied.

A result of Shultz [26] states that for every JB-algebra V the bidual V^{tt} is also a JB-algebra in the Arens-product. Hence every JB-algebra is a norm-closed subalgebra of a JB-algebra which is a dual Banach space. Algebras of this type admit not only a continuous but an L^∞-functional calculus. By definition a JB-algebra V is called a JBW-<u>algebra</u> if V is a dual Banach space (i.e. there exists a Banach space V_t with $V = (V_t)^t$ as dual Banach space - V_t is uniquely determined by V [25] and is called the <u>predual</u> of V). For every a in the JBW-algebra V the w*-closed unital subalgebra $W(a)$ of V generated by a is a commutative von Neumann algebra, i.e. $W(a) \approx C(S,\mathbb{R})$ for S hyperstonian or equivalently $W(a) \approx L^\infty(\mu)$ where μ is a localizable measure (compare [25]). JBW-algebras are the abstract analogues of von Neumann algebras in the Jordan case. Shultz [26] imporoved the Gelfand-Neumark theorem for these algebras:

<u>3.4 Theorem</u>. Every JBW-algebra V is a unique direct sum (as algebra)

$$V = V_{sp} \oplus V_{ex}$$

where V_{sp} is isomorphic to a weakly closed Jordan subalgebra of $\mathcal{H}(H)$ for some complex Hilbert space H and $V_{ex} \approx C(S,\mathcal{H}_3(O))$ with S a compact hyperstonian space.

V_{sp} and V_{ex} are called the <u>special part</u> and the <u>exceptional part</u> of V respectively. Note that every JW-algebra V (i.e. a weakly closed subalgebra of $\mathcal{H}(H)$, H a complex Hilbert space) is a JBW-algebra with $V = V_{sp}$.

A JBW-algebra V is called a <u>factor</u> if the center of V defined by

$$Z(V) := \{x \in V : [L(x),L(y)] = 0 \text{ for all } y \in V\}$$

reduces to $\mathbb{R}e$ or equivalently if there does not exist a direct sum decomposition $V = V_1 \oplus V_2$ of proper ideals. By 3.5 $\mathcal{H}_3(\mathbb{O})$ is the only exceptional JBW-factor, i.e. every other JBW-factor is isomorphic to a JW-algebra.

JBW-algebras are still far from being finite dimensional. But a good substitute for the minimal decomposition (3.3) is the L^∞-functional calculus in JBW-algebras. A simple application thereof is

<u>3.5 Proposition</u>. Suppose V is a JBW-algebra. Then the unit circle $\Sigma = \{z \in U : z^* = z^{-1}\}$ in $U = V \oplus iV$ is connected. Actually

$$\Sigma = \exp(iV)$$

and $GL(\Delta) = P(\Sigma)\ Aut(V)$.

Note that $GL(\Delta)$ is the group of all surjective isometries of the JB*-algebra U. The groups $GL(\Delta)$, $GL(\Omega)$ and $Aut(V)$ are in particular real Banach Lie groups with the same number of connected components. Denote by $\mathfrak{gl}(\Delta)$, $\mathfrak{gl}(\Omega)$ and <u>aut</u>(V) the corresponding Banach Lie algebras. Then for instance <u>aut</u>(V) is the set of all derivations of the algebra V. 2.5 and 3.5 imply together with the formula

$$P(\exp(z)) = \exp(2L(z)) \quad \text{for all} \quad z \in U$$

that $\mathfrak{gl}(\Omega) = \underline{aut}(V) \oplus L(V)$

and $\mathfrak{gl}(\Delta) = \underline{aut}(V) \oplus iL(V)$.

Actually this is true for every JB-algebra V since always $P(\Sigma)Aut(V)$

is an open subgroup of $GL(\Delta)$. For every $x,y \in V$ the commutator $[L(x),L(y)]$ is contained in aut(V) and every finite linear combination of such commutators is called an inner derivation of V. The set int(V) of all inner derivations of V is an ideal of aut(V). In finite dimensions aut(V) = int(V) is known, in the general case this is not always true:

3.6 Counter example. Consider the spin factor $V = V_p = \mathbb{R} \oplus Y$. Then Aut(V) can be identified with the orthogonal group of the Hilbert space Y acting trivially on the center $\mathbb{R}e$. Therefore aut(V) is the set of all skew-symmetric operators on Y. But every commutator $[L(x),L(y)]$ in int(V) has rank ≤ 2 on V, i.e. int(V) is the subalgebra of all finite rank operators in aut(V). The norm closure of int(V) is the ideal of all compact operators in aut(V).

Upmeier [29] showed that 3.6 is typical in some sense:

3.7 Theorem. Suppose V is a JB-algebra. Then aut(V) is the closure of int(V) in $\mathcal{L}(V)$ with respect to the strong operator topology (i.e. the topology of simple convergence on V). If V is a JBW-factor not isomorphic to a spin factor of infinite dimension then aut(V) is the norm closure in $\mathcal{L}(V)$ of int(V).

A JBW-factor V is called of type I (this definition can also be given for arbitrary JBW-algebras in a modified form) if V contains a minimal idempotent. Minimal idempotents $c \in V$ can also be characterized by the condition that the corresponding Peirce-1-space $P(c)V$ has dimension 1. E. Størmer gave a complete classification of JBW-factors of type I:

3.8 Theorem. The JBW-factors of type I are precisely the following algebras with p an arbitrary cardinal number

type	algebra	
I_1	$\mathbb{R}$	
I_2	V_p	$p \geq 3$
I_p	$\mathcal{H}_p(\mathbb{K})$	$\mathbb{K} = \mathbb{R}, \mathbb{C}, \mathbb{H}, \mathbb{O}$ and $p \geq 3$ ($p = 3$ if $\mathbb{K} = \mathbb{O}$)

(For the definition of the spin factor V_p and the algebra $\mathcal{H}_p(\mathbb{K})$ see §1 and §2 respectively).

Let V^t be the dual Banach space of V and denote by

$$C^t := \{\lambda \in V^t : \lambda(C) \geq 0\}$$

the dual cone of $C = V^2$.

$$K := \{\lambda \in C^t : \lambda(e) = 1\}$$

is called the <u>state space of</u> V, the elements of K are called <u>states on</u> V. K is a w*-compact, convex subset and V can be identified (as a Banach space) with the space of all w*-continuous affine functions on K. The bidual of V corresponds to the set of all bounded affine functions on K. Every state $\rho \in K$ defines by

$$(x|y)_\rho := \rho(xy)$$

a positive inner product on V and in particular by

$$|x|_\rho := \rho(x^2)^{1/2}$$

a seminorm on V. ρ is called <u>faithful</u> if $|\ |_\rho$ actually is a norm on V, i.e. if $\rho(x^2) = 0$ implies $x = 0$. A state ρ on a JBW-algebra is called <u>normal</u> if $\lim \rho(x_\alpha) = \rho(x)$ for every increasing net x_α in V with $x = \sup x_\alpha \in V$.

Certain finiteness conditions can be given in terms of the existence of varios types of traces. We discuss a special case here: A normal state ρ on the JBW-algebra V is called a <u>finite trace</u> if it is associative in the sense that

(3.9) $$\rho((xy)z) = \rho(x(yz))$$

for all $x,y,z \in V$. This condition just says that each $L(y)$, $y \in V$, is self-adjoint with respect to the inner product $(\ |\ )_\rho$. With regard to (3.1) such a condition seems to be natural. The condition (3.9) can be restated as

$$\rho\circ\lambda = 0 \quad \text{for all} \quad \lambda \in \underline{\text{int}}(V)$$

which by Upmeier's Theorem 3.7 is equivalent to

$$\rho\circ\lambda = 0 \quad \text{for all} \quad \lambda \in \underline{\text{aut}}(V)$$

and hence to

$$\rho\circ g = \rho \quad \text{for all} \quad g \in \text{Aut}(V)^o$$

where $\text{Aut}(V)^o$ is the norm connected unit component of $\text{Aut}(V)$.

In the following we assume that U is a JBW-algebra with a faithful finite trace ρ (Jordan algebras of this type have been thoroughly studied by Janssen [12]): Then ρ is essentially uniquely determined (every other finite faithfull trace is of the form $\rho\circ P(h) = \rho\circ L(h^2)$ for some $h > 0$ in the center of V) and $U = V\oplus iV$ is a complex pre-Hilbert space with respect to the inner product

$$(z|w) := (z|w)_\rho := \rho(zw^*)$$

where of course ρ is extended $\mathbb{C}$-linearly to U. Denote by $\hat{U}$ the completion of U with respect to the norm

$$\|z\|_2 := \|z\|_\rho := \rho(zz^*)^{1/2}$$

and consider the closures $\hat{V}$ and $\hat{C}$ of V and C in $\hat{U}$. For every $z \in U$ the operators $L(z)$ and $P(z)$ can be continued to bounded operators on $\hat{U}$ satisfying

$$L(z)^* = L(z^*) \quad \text{and} \quad P(z)^* = P(z^*).$$

The cone $\hat{C}$ is selfdual in $\hat{V}$ (i.e.

$$\hat{C} = \{x \in \hat{V} : (x|\hat{C}) \geq 0\}\ ),$$

satisfies a certain geometrical homogeneity condition (which in finite dimensions is equivalent to the condition that $\hat{C}$ has (linearly) homogeneous interior; compare [2]) and has e as trace vector (i.e. a quasi-interior point of $\hat{C}$ which is fixed by every connected set of isometries in $GL(\hat{C})$). Bellisard and Iochum [2] showed that on the other hand every cone of this type in a real Hilbert space is obtained in this way from a JB-algebra with finite faithful trace. This Theorem can be viewed as a generalization to infinite dimensions of the following result due to Koecher: The self-dual cones with homogeneous interior in real Hilbert spaces of finite dimension are precisely (i.e. up to linear equivalence) the cones of squares in formally real Jordan algebras.

<u>3.10 Proposition.</u> The JB*-norm $\| \ \|_\infty$ on U satisfies

$$\| \ \|_2 \le \| \ \|_\infty$$

on U and

$$\Delta = \{z \in U : 1 - P(z)P(z)^* > 0\}$$

$$\Sigma = \exp(iV) = \{z \in U : P(z) \text{ unitary on } \hat{U}\}$$
$$= \{z \in \bar{\Delta} : \|z\|_2 = 1\}$$

In general the norms $\| \ \|_2$ and $\| \ \|_\infty$ are not equivalent and $\hat{C}$ has empty interior. The case $U = \hat{U}$ is almost the finite-dimensional case. The following is not difficult to show:

<u>3.11 Proposition.</u> Let V be a JB-algebra. Then the following conditions are equivalent

(i) There exists a maximal associative subalgebra of finite dimension in V.

(ii) V is locally finite (i.e. every finitely generated subalgebra has finite dimension).

(iii) For every $a \in V$ the operator $L(a) \in \mathcal{L}(V)$ satisfies a polynomial equation over $\mathbb{R}$.

(iv) There exists a natural number r such that every $a \in V$ admits a representation

$$a = \alpha_1 e_1 + \dots + \alpha_r e_r$$

where $\{e_1,\dots,e_r\}$ is a set of orthogonal idempotents (i.e. $e_i \neq 0$ and $e_i e_j = \delta_{ij} e_i$) and $\alpha_1,\dots,\alpha_r \in \mathbb{R}$.

(v) There exists a finite faithful trace ρ on V such that the corresponding Hilbert norm $\|x\|_2 = \rho(x^2)^{1/2}$ on V is equivalent to the JB-norm $\|x\|_\infty$.

(vi) V is reflexive.

We call V a JB-algebra of <u>finite rank</u> if one (and hence all) of the preceding conditions is satisfied. The number $r =: r(V)$ in (iv) is uniquely determined and is called the <u>rank of</u> V. Every JBW-factor of finite rank has minimal idempotents, the classification 3.8 of Størmer therefore implies

<u>3.12 Theorem.</u> Every JB-algebra V of finite rank is a unique direct sum $V = V_1 \oplus \dots \oplus V_k$ of JBW-factors V_j with

$$r(V) = r(V_1) + \dots + r(V_k).$$

The JBW-factors of finite rank are precisely the following algebras

V	$r(V)$	
$\mathbb{R}$	1	
$\mathcal{V}_p$	2	p arbitrary cardinal ≥ 3
$\mathcal{H}_3(\mathbb{O})$	3	
$\mathcal{H}_p(\mathbb{K})$	p	$3 \leq p < \infty$ and $\mathbb{K} = \mathbb{R}, \mathbb{C}$ or $\mathbb{H}$

On a JB-factor V there exists at most one finite faithful trace ρ - in case of $r = r(V) < \infty$ and

$$\alpha = \alpha_1 e_1 + \dots + \alpha_r e_r$$

according to 3.11 (v) we have $\rho(\alpha) = \frac{1}{r}(\alpha_1 + \dots + \alpha_r)$ and hence

$$\|\alpha\|_2 = (\alpha_1^2 + \dots + \alpha_r^2)^{1/2}$$

and

$$\|\alpha\|_\infty = \max(|\alpha_1|,\dots,|\alpha_r|).$$

For $V = \mathcal{H}_p(\mathbb{K})$ with $\mathbb{K} = \mathbb{R}, \mathbb{C}, \mathbb{H}$ or $\mathbb{O}$ we have

$$\rho(a) = \mathrm{Tr}(a) \qquad (= \underline{\text{normalized trace}})$$

and hence $\|a\|_2 = (\frac{1}{p}\sum a_{ij}\overline{a_{ij}})^{1/2}$ for every $a \in \mathcal{H}_p(\mathbb{K})$. For the spin factor $V_p = \mathbb{R} \oplus Y$ we have

$$\rho((t,x)) = t$$

and hence $\|(t,x)\|_2 = (t^2 + \|x\|^2)^{1/2}$, i.e. with respect to $\|\ \|_2$ $\mathbb{R}\oplus Y$ is an orthogonal sum.

JB-algebras of finite rank are except for spin factors just the formally real Jordan algebras. For every formally real Jordan algebra V of finite dimension a finite faithful trace can also be defined by

$$\rho(a) := \mathrm{Tr}(L(a))$$

where $\mathrm{Tr} = (\dim V)^{-1}\cdot$trace is the normalized trace in $\mathcal{L}(V)$. This trace is uniquely determined by the condition that all minimal idempotents in V have the same value (namely $r(V)^{-1}$) under ρ. For $V = \mathcal{H}_p(\mathbb{K})$, the cone C (Ω resp.) is the set of all positive (positive definite resp.) matrices in V. The elements of $U = V \oplus iV$ can be considered as endomorphisms of the complex Hilbertspace $\mathbb{K}^p \oplus i\mathbb{K}^p$ - the JB-norm $\|\ \|_\infty$ on U is precisely the operator norm. Thus in case $\mathbb{K} = \mathbb{R}$ $U = \{z \in \mathbb{C}^{p\times p}: z^t = z\}$ and $\Delta = \{z \in \mathbb{C}^{p\times p} : z^t = z$ and $1 - z^*z > 0\}$, where $\mathbb{C}^{p\times p}$ is set of $p\times p$-matrices with entries in $\mathbb{C}$ and z^t is the transpose of z. The corresponding symmetric tube domain is Siegel's upper half plane

$$D = \{z \in \mathbb{C}^{p\times p} : z^t = z \text{ and } \mathrm{Im}(z) \text{ positive definite}\}.$$

In case $\mathbb{K} = \mathbb{C}$ U can be identified with $\mathbb{C}^{p\times p}$ and

$$\Delta = \{z \in \mathbb{C}^{p\times p} : 1 - z^*z > 0\}.$$

Consider the case $\mathbb{K} = \mathbb{H}$. Then $\mathbb{H} = \mathbb{C} + j\mathbb{C}$ for some $j \in \mathbb{H}$ with $j\lambda = \bar{\lambda}j$ for every $\lambda \in \mathbb{C}$. Every $x \in V = \mathcal{H}_p(\mathbb{H})$ has a unique description as $x = a + jb$ with $a^* = a$ and $b^t = -b$ in $\mathbb{C}^{p\times p}$ and

$$\varphi(x) = \begin{pmatrix} b & -a \\ \bar{a} & \bar{b} \end{pmatrix} \in \mathbb{C}^{q\times q}, \qquad q = 2p,$$

defines a linear injection $\varphi: V \to \mathbb{C}^{q\times q}$ with $\varphi(V) \cap i\varphi(V) = 0$, i.e. φ extends to a $\mathbb{C}$-linear isomorphism of $U = V \oplus iV$ onto

$$W := \{z \in \mathbb{C}^{q\times q} : z^t = -z\}.$$

φ is an isomorphism of JB*-algebras if we define the Jordan product on W by

$$z \circ w := \frac{1}{2}(zJw + wJz) \quad \text{where} \quad J := \varphi(e) = \begin{pmatrix} 0 & -1 \\ 1 & 0 \end{pmatrix}.$$

Therefore $U = V \oplus iV$, $V = \mathcal{H}_p(\mathbb{H})$, can be identified with $W = \{z \in \mathbb{C}^{q\times q} : z^t = -z\}$. In this way Δ corresponds to $\{z \in W : 1 - z^*z > 0\}$.

Now suppose $V_p = \mathbb{R} \oplus Y$ is a spin factor. Then $U_p = V_p \oplus i V_p = \mathbb{C} \oplus X$ where the Hilbert space X is the complexification of Y. Then

$$q(t,x) := t^2 - (x|x^*)$$

defines a non-degenerate quadratic form over $\mathbb{C}$ on U and

$$\Delta = \{z \in U_p : \|z\|_2^2 + \sqrt{\|z\|_2^4 - |q(z)|^2} < 1\}$$

$$\Sigma = \{z \in U_p : |q(z)| = \|z\|_2^2\},$$

i.e.

$$\|z\|_\infty = (\|z\|_2^2 + \sqrt{\|z\|_2^4 - |q(z)|^2})^{1/2}.$$

Another description of Δ can be given in terms of <u>polar coordinates</u>:

<u>3.13 Proposition</u>. Suppose V is a JB-algebra of finite rank. Then every $z \in U = V \oplus iV$ has a representation

$$z = P(\exp(iv))r$$

with $r, v \in V$, $0 \leq v < \pi e$ and $0 \leq r$. This representation is unique if and only if z is invertible. Furthermore

$$\Delta = \{P(\exp(iv))r : 0 \leq r < e\}.$$

By Theorem 2.13 results on JB-algebras can be translated into results on symmetric tube domains. We given two examples

<u>3.14 Proposition</u>. Suppose D_k is a symmetric tube domain in a complex Banach space U_k for $k = 1,2$. Suppose furthermore that U_1 has a pre-dual. Then D_1 and D_2 are biholomorphically equivalent if and only if they are linearly equivalent.

Note that 3.14 does not hold without any condition on U_1 as a counterexample in [3] shows.

<u>3.15 Theorem</u>. Every symmetric tube domain D in a reflexive Banach space U is linearly equivalent to a direct product

$$\text{(i)} \qquad D(V_1) \times D(V_2) \times \ldots \times D(V_k)$$

where $V_1, \ldots, V_k$ are uniquely determined (up to order) algebras from the list in 3.12 and $D(V_j)$ is the upper half plane associated with V_j. On the other hand every domain of the form (i) is linearly equivalent to a symmetric tube domain in a complex Hilbert space.

The algebraic characterization of bounded symmetric domains by hermitian Jordan triple systems easily implies form 3.15.

<u>Corollary</u>. Suppose U is a reflexive complex Banach space. Then there exists an equivalent Hilbert norm on U if and only if there exists a bounded symmetric domain in U.

REFERENCES

[1] Alfsen, E.M., Shultz, F.W., Størmer, E.: A Gelfand-Neumark Theorem for Jordan algebras. Advances in Math. 28, 11-56 (1978).

[2] Bellisard, J., Iochum, B.: Homogeneous self dual cones, versus Jordan algebras. The theory revisited. Ann. Inst. Fourier, 28, 27-67 (1978).

[3] Braun, R., Kaup, W., Upmeier, H.: A holomorphic characterization of Jordan C*-algebras. Math. Z. 161, 277-290 (1978).

[4] Braun, H., Koecher, M.: Jordan-Algebren. Grundlehren der Mathematischen Wissenschaften Bd. 128, Berlin-Heidelberg-New York: Springer 1965.

[5] Cartan, É.: Sur les domaines bornés homogènes de l'espace de n variables complexes. Abh. Math. Sem. Univ. Hamburg 11, 116-162 (1935).

[6] Connes, A.: Caractérisation des espaces vectoriels ordonnés sous-jacents aux algebres de von Neumann. Ann. Inst. Fourier 24, 121-155 (1974).

[7] Harris, L.A.: Operator Siegel domains. Proc. Royal Soc. Edinburgh 79 A, 137-156 (1974).

[8] Helgason, S.: Differential geometry and symmetric spaces. Academic Press 1962.

[9] Helwig, K.H.: Jordan-Algebren und symmetriche Räume I. Math. Z. 115, 315-349 (1970).

[10] Hirzebruch, U.: Halbräume und ihre holomorphen Automorphismen. Math. Ann. 153, 395-417 (1964).

[11] Jacobson, N.: Structure and representations of Jordan algebras. Amer. Math. Soc. Colloq. Publ. 39, AMS 1968.

[12] Janssen, G.: Reelle Jordan-algebren mit endlicher Spur. Manuscripta math. 13, 237-273 (1974).

[13] Jordan, P., von Neumann, J., Wigner, E.: On the algebraic generalization of the quantum mechanical formalism. Ann. of Math. 36, 29-64 (1934).

[14] Kadison, R.V.: Isometries of operator algebras. Ann. of Math. 54, 325-338.

[15] Kaup, W.: Algebraic characterization of symmetric complex Banach manifolds. Math. Ann. 228, 39-64 (1977).

[16] Kaup, W.: Bounded symmetric domains in finite and infinite dimensions - a review. Proc. Int. Conf. Cortona 1976-77, Scuola Norm. Sup. Pisa, 180-191 (1978).

[17] Kaup, W., Upmeier, H.: Jordan algebras and symmetric Siegel domains in Banach spaces. Math. Z. 157, 179-200 (1977).

[18] Koecher, M.: Positivitätsbereiche im R^n. Am. J. Math. 79, 575-596 (1953).

[19] Koecher, M.: Jordan algebras and their applications. Lecture notes. Univ. Minnesota 1962.

[20] Koecher, M.: An elementary approach to bounded symmetric domains. Lecture notes. Rice University 1969.

[21] Korányi, A., Wolf, J.: Realization of hermitian symmetric space as generalized halfplanes. Ann. of Math. 81, 256-288 (1976).

[22] Loos, O.: Bounded symmetric domains and Jordan pairs. Irvine Univ. of California 1977.

[23] McCrimmon, K.: Jordan algebras and their applications. Bull. AMS 84, 612-627 (1978).

[24] Pyatetskii-Shapiro, I.I.: Automorphic functions and the geometry of classical domains. Gordon and Breach 1969.

[25] Sakai, S.: C*-algebras and W*-algebras. Ergebnisse der Mathematik und ihrer Grenzgebiete 60. Springer 1972.

[26] Shultz, F.W.: On normed Jordan algebras which are Banach dual spaces. Journal of Functional Analysis, 31, 360-376 (1979).

[27] Størmer, E.: Jordan algebras of type I. Acta Math. 115, 165-184 (1965).

[28] Topping, D.: Jordan algebras of self-adjoint operators. Mem. AMS 53, 1965.

[29] Upmeier, H.: Derivations of Jordan C*-algebras. To appear in Mathematica scandinavica.

[30] Vigué, J.P.: Le groupe des automorphismes analytiques d'un domaine borné d'un espace de Banach complexe. Ann. Sc. Ec. Norm. Sup. 9, 203-282 (1976).

[31] Wright, J.D.M.: Jordan C*-algebras. Mich. Math. J. 24, 291-302 (1977).

HOW TO RECOGNIZE SUPPORTS FROM THE GROWTH OF FUNCTIONAL TRANSFORMS IN REAL AND COMPLEX ANALYSIS

Christer O. Kiselman
Department of Mathematics
Thunbergsvägen 3
S-752 38 Uppsala Sweden

The purpose of this note is to exhibit an important difference between real and complex analysis when it comes to reconstructing the support of a functional from its Fourier transform.

In the real case the support of a distribution in $\mathbb{R}^n$ is determined by the growth of a non-linear Fourier transform defined on an $(n+1)$-dimensional space, i.e. only one extra dimension is needed. In contrast to this, it becomes necessary in the complex case to study the non-linear Fourier-Laplace transform on an infinite-dimensional space in order to have the corresponding information about the support. Thus, we shall see that holomorphic and plurisubharmonic functions defined on infinite-dimensional spaces appear in what one would like to call a "natural" context, assuming, of course, an interest in finite-dimensional complex analysis.

I prepared this note as a partial answer to the question: "Why infinite-dimensional holomorphy?", posed by Leopoldo Nachbin. It was meant to be included as one of several motivations in his paper [4]. However, it reached him too late and is now presented here, at his kind invitation. Of course, as an answer to his question beyond the universal "Why not?", it should be read together with [4].

This paper could have been written in 1965 or 1966 when I was preparing [2]. The fact that many people have asked me over the years about the real analogues of the results in [2] has convinced me that it is, after all, worthwhile to make explicit this simple comparison of real vs. complex analysis.

1. RECOGNIZING THE SUPPORT OF A DISTRIBUTION

It is well known that the convex hull of the support of a function (or distribution) with compact support is determined by the growth of its Fourier transform. In fact, if $f \in C(\mathbb{R}^n, \mathbb{C})$ is a continuous function with support contained in a compact set K, then the Fourier transform of f,

$$\hat{f}(\zeta) = \int_{\mathbb{R}^n} f(x) e^{-i\langle x, \zeta\rangle} dx, \qquad \zeta \in \mathbb{C}^n,$$

can be estimated as follows:

$$|\hat{f}(\zeta)| \leq \int |f| dx \sup_{x \in K} \exp\,(\operatorname{Im}\langle x, \zeta\rangle) = C \exp H_K(\operatorname{Im} \zeta),$$

where

$$H_K(\xi) = \sup_{x \in K} \langle x, \xi \rangle, \qquad \xi \in \mathbb{R}^n,$$

is the support function of K. The function $u = \log|\hat{f}|$ is plurisubharmonic in $\mathbb{C}^n$ and satisfies

$$u(\zeta) \leq \log C + H_K(\operatorname{Im} \zeta).$$

The indicator of u (or of f) is

$$v(\zeta) = \limsup_{t \to +\infty} u(t\zeta)/t \leq H_K(\operatorname{Im} \zeta). \tag{1}$$

If f is a distribution with support in K the estimate (1) is still valid although its proof is somewhat more involved. Conversely, assume that $f \in \mathcal{E}'(\mathbb{R}^n)$ and that (1) holds for its indicator. Then $\operatorname{supp} f \subset \operatorname{cvx} K$, the convex hull of K. This is, modulo some technicalities on the growth of plurisubharmonic functions, the content of the Paley-Wiener theorem.

It is not possible to reconstruct the support of f, just its convex hull, from the indicator v. However, we can recover supp f from the growth of a kind of non-linear Fourier transform of f. Let

(2) $$\tilde{f}(\zeta) = \int f(x)\exp(-ix_1\zeta_1 - \dots - ix_n\zeta_n - i|x|^2\zeta_{n+1})dx, \quad \zeta \in \mathbb{C}^{n+1}.$$

Then we can reconstruct supp f from the indicator of $\tilde{f}$. This is seen as follows. Let $p: \mathbb{R}^n \to \mathbb{R}^{n+1}$ be the mapping

$$p(x) = (x_1, x_2, \dots, x_n, x_1^2 + \dots + x_n^2).$$

We can pull a function φ back from $\mathbb{R}^{n+1}$ to $\mathbb{R}^n$ by

$$p^*: \mathcal{D}(\mathbb{R}^{n+1}) \to \mathcal{D}(\mathbb{R}^n), \quad p^*: \mathcal{E}(\mathbb{R}^{n+1}) \to \mathcal{E}(\mathbb{R}^n), \quad p^*(\varphi) = \varphi \circ p;$$

and we can push a distribution f forward from $\mathbb{R}^n$ to $\mathbb{R}^{n+1}$ by

$$p_*: \mathcal{D}'(\mathbb{R}^n) \to \mathcal{D}'(\mathbb{R}^{n+1}), \quad p_*: \mathcal{E}'(\mathbb{R}^n) \to \mathcal{E}'(\mathbb{R}^{n+1}), \quad p_*(f) = f \circ p^*.$$

Then if $f \in \mathcal{E}'(\mathbb{R}^n)$ we have $p_*(f) \in \mathcal{E}'(\mathbb{R}^{n+1})$ and $\operatorname{supp} p_*(f) = p(\operatorname{supp} f)$ is contained in the paraboloid

$$p(\mathbb{R}^n) = \{x \in \mathbb{R}^{n+1};\ x_{n+1} = x_1^2 + \dots + x_n^2\}.$$

Clearly the Fourier transform of $p_*(f)$ is $\tilde{f}$. By the Paley-Wiener theorem we can reconstruct $\operatorname{cvx}(\operatorname{supp} p_*(f))$ from the indicator of $\tilde{f}$. However, we know that $\operatorname{supp} p_*(f)$ is contained in $p(\mathbb{R}^n)$. Now

$$p^{-1}(\operatorname{cvx}(p(X))) = X$$

for any set X in $\mathbb{R}^n$. (This follows from the strict convexity of the epigraph

$$\{x \in \mathbb{R}^{n+1};\ x_{n+1} \geq x_1^2 + x_2^2 + \dots + x_n^2\}.)$$

Hence, if we know $\operatorname{cvx}(p(\operatorname{supp} f))$ we also know supp f.

Now (2) is just a special case of a non-linear Fourier transformation:

(3) $$\tilde{f}(\varphi) = \int f(x)e^{-i\varphi(x)}dx, \quad \varphi \in C(\mathbb{R}^n, \mathbb{C}) = E,$$

and $\tilde{f}: E \to \mathbb{C}$ is an entire function on the infinite-dimensional vector space E (cf. the discussion of $\tilde{T}$ below). However, from what we have just seen, it is sufficient to consider the restriction of $\tilde{f}$ to the (n+1)-dimensional space spanned by the linear functions $x \mapsto x_j$ and the single non-linear function $x \mapsto |x|^2$ if we just want to get information on the support of f from the indicator. Thus, in this respect, real analysis gives us no motivation for infinite-dimensional holomorphy!

2. SUPPORTS OF ANALYTIC FUNCTIONALS

We now turn to a complex analogue of distributions: the analytic functionals. Let $\mathcal{O}(\mathbb{C}^n)$ denote the space of all entire functions on $\mathbb{C}^n$ with the topology of uniform convergence on compact sets in $\mathbb{C}^n$ and let $\mathcal{O}'(\mathbb{C}^n)$ be the dual space; an element T of $\mathcal{O}'(\mathbb{C}^n)$ is called an analytic functional. We say that T is carried by a compact set K in $\mathbb{C}^n$ if for every neighborhood ω of K there is a constant C such that

$$|T(\varphi)| \le C \sup_{\omega} |\varphi|, \qquad \varphi \in \mathcal{O}(\mathbb{C}^n).$$

If $\mathfrak{F}$ is a family of subsets of $\mathbb{C}^n$ we call $K \in \mathfrak{F}$ an $\mathfrak{F}$-support if K carries T and K is minimal with respect to inclusion among the carriers in $\mathfrak{F}$. In this way we get e.g. the convex supports and the polynomially convex supports. A functional may have infinitely many convex supports.

Let $P(D_1,\ldots,D_n)$ be a differential operator with constant coefficients in $\mathbb{R}^n$ which is hyperbolic with respect to the hyperplane $H = \{x;\ x_n=0\}$. For every distribution $u \in \mathcal{E}'(\mathbb{R}^n)$ there are distributions $u_0,\ldots,u_{m-1} \in \mathcal{E}'(\mathbb{R}^{n-1})$ such that

$$u(\varphi) = \sum_{0}^{m-1} u_j(D_n^j\varphi|_{x_n=0}) \tag{4}$$

for all $\varphi \in \mathcal{E}(\mathbb{R}^n)$ solving $P(D)\varphi = 0$ (m is the order of P). If P is not hyperbolic with respect to H this is no longer true; (4) just expresses the fact that the Cauchy problem is well posed. But there are analytic functionals $u_j \in \mathcal{O}'(\mathbb{C}^{n-1})$ such that (4) holds for all entire solutions $\varphi \in \mathcal{O}(\mathbb{C}^n)$ of $P(D)\varphi = 0$, assuming only that H is non-characteristic for P. Thus analytic functionals appear even if we start with a real Cauchy problem $P(D)\varphi = 0$, $D_n^j\varphi = g_j$ in H, $j = 0,\dots,m-1$, and have $u \in \mathcal{E}'(\mathbb{R}^n)$. For a discussion of (4), see [1].

Now a non-linear Fourier transform of an analytic functional can be defined in complete analogy with (3): we put

$$\tilde{T}(\varphi) = T(e^{\varphi}), \qquad \varphi \in \mathcal{O}(\mathbb{C}^n).$$

It is easy to estimate $\tilde{T}$ in terms of a carrier K of T:

$$|\tilde{T}(\varphi)| \le C \sup_{z\in\omega} |e^{\varphi(z)}| = C \exp H_{\omega}(\operatorname{Re}\varphi)$$

where

$$H_{\omega}(\psi) = \sup_{z\in\omega} \psi(z)$$

is the support function of ω, a neighborhood of K. The function $\tilde{T}: \mathcal{O}(\mathbb{C}^n) \to \mathbb{C}$ is continuous, in fact we have

$$|\tilde{T}(\psi) - \tilde{T}(\varphi)| \le C \sup_{z\in\omega} |e^{\psi(z)} - e^{\varphi(z)}| =$$

$$= C \sup_{z\in\omega} e^{\operatorname{Re}\varphi(z)} |e^{\psi(z)-\varphi(z)} - 1| \le$$

$$\le C\, H_{\omega}(|\psi-\varphi|) \exp(H_{\omega}(\operatorname{Re}\varphi) + H_{\omega}(|\psi-\varphi|))$$

where the last step follows from the well-known inequality

$$|e^z-1| \le |z|e^{|z|}, \qquad z \in \mathbb{C}.$$

Now $\varphi \mapsto H_{\omega}(|\varphi|)$ is by definition a continuous seminorm on $\mathcal{O}(\mathbb{C}^n)$; hence $\tilde{T}$ is continuous. Next we shall see that $\tilde{T}$ is Gâteaux-analytic:

$$\tilde{T}(\varphi+t\psi) = T(e^{\varphi+t\psi}) = T(e^{\varphi} \sum_0^{\infty} (t\psi)^k/k!) = \sum_0^{\infty} T_k(\varphi,\psi)t^k \tag{5}$$

where

$$T_k(\varphi,\psi) = T(e^{\varphi}\psi^k)/k!,$$

and

$$|T_k(\varphi,\psi)| \le C[\exp H_{\omega}(\operatorname{Re}\varphi)]\, H_{\omega}(|\psi|)^k/k!$$

so that the series $\Sigma\, T_k(\varphi,\psi)t^k$ converges for all $t \in \mathbb{C}$; this estimate also justifies the calculation in (5).

Thus $\tilde{T}$ is both continuous and Gâteaux-analytic on $\mathcal{O}(C^n)$; in other words it is holomorphic, and $u = \log|\tilde{T}|$ is plurisubharmonic. The indicator of u is

$$v(\varphi) = \limsup_{t\to+\infty} u(t\varphi)/t \le \limsup_{t\to+\infty} (t^{-1}\log C + H_\omega(\mathrm{Re}\,\varphi)) = H_\omega(\mathrm{Re}\,\varphi),$$

provided ω is a neighborhood of a carrier of T. By a general result (called théorème de convergence or the lim-sup-star theorem; see [3] for a discussion) the upper regularization v^* of v is also plurisubharmonic in $\mathcal{O}(C^n)$, where

$$v^*(\varphi) = \limsup_{\psi\to\varphi} v(\psi), \qquad \varphi \in \mathcal{O}(C^n).$$

As in the real case, knowledge of $v(\varphi)$ for all φ in a subspace Φ of $\mathcal{O}(C^n)$ containing the linear functions $z \mapsto z_j$ is sufficient to reconstruct the set of all Φ-convex supports, where the Φ-hull is defined by

$$\Phi\text{-hull }(K) = \bigcap_{\varphi\in\Phi} \{z \in C^n;\ \mathrm{Re}\,\varphi(z) \le \sup_K \mathrm{Re}\,\varphi\},$$

and K is called Φ-convex if Φ-hull (K) = K; see [2]. In the real case we saw that the hull of any closed set X in R^n with respect to a certain (n+1)-dimensional space of functions is equal to X. In the complex case, no finite-dimensional space will suffice to describe e.g. the polynomially convex compact subsets of C^n, $n \ge 1$. To illustrate this, take $A = \{1,\theta,\theta^2,\ldots,\theta^{m-1}\} \subset C$ where m is a prime and θ a root of $\theta^m = 1$, $\theta \ne 1$, and let P_k denote the linear subspace of $\mathcal{O}(C)$ consisting of the polynomials of degree at most k. Then the P_{m-1}-hull of A contains the origin, since

$$p(0) = \frac{1}{m}\sum_0^{m-1} p(\theta^k)$$

for all $p \in P_{m-1}$; on the other hand A is P_m-convex. This shows that the polynomially convex supports of an analytic functional T cannot be obtained from knowledge of the growth of $\tilde{T}$ on a finite-dimensional space of polynomials: It becomes necessary to study $\tilde{T}$

on all of $\mathcal{O}(\mathbb{C}^n)$, i.e. to study entire functions and plurisubharmonic functions on infinite-dimensional spaces.

REFERENCES

[1] Kiselman, C.O., Existence and approximation theorems for complex analogues of boundary problems. Ark. Mat. 6 (1967), 193-207.

[2] Kiselman, C.O., On entire functions of exponential type and indicators of analytic functionals. Acta Math. 117 (1967), 1-35.

[3] Kiselman, C.O., Plurisubharmonic functions and plurisubharmonic topologies. Advances in Holomorphy (Ed. J.A. Barroso), 431-449. North-Holland Publishing Company, 1979.

[4] Nachbin, L., Warum unendlichdimensionale Holomorphie? Jahrbuch Überblicke Mathematik 1979, 9-20, Bibliographisches Institut Mannheim-Zürich.

LINEAR DIFFERENTIAL OPERATORS ON VECTOR SPACES

Paul Krée

Institut de Mathématiques Pures et Appliquées
Université Pierre et Marie Curie
(Paris VI)
France

Today phenomenas concerning the algebra $\mathrm{Diff}(\Omega)$ of all linear differential operators on an open subset Ω of $X = \mathbb{R}^d$ (or on a finite dimensional manifold) are generally well understood. The basic concept is the symbol $P(x,\xi)$ of a linear operator $\hat{P}$ of $C^\infty(\Omega)$: this is a function on $\Omega \times X'$ such that

$$(0.1) \qquad \hat{P}_x(\exp i\langle x,\xi\rangle) = P(x,\xi)\exp i\langle x,\xi\rangle .$$

In particular for an arbitrary multi-index $\alpha = (\alpha_1, \alpha_2, \ldots, \alpha_d)$, and $a_\alpha \in C^\infty(\Omega)$, we set $(i^{-1}\,\partial/\partial x)^\alpha = (i^{-1}\,\partial/\partial x^1)^{\alpha_1} \ldots (i^{-1}\,\partial/\partial x_d)^{\alpha_d}$ and the symbol of $\varphi \to (a_\alpha\, i^{-1}\,\partial/\partial x)^\alpha$ is $a_\alpha\, \xi^\alpha$ with $\xi^\alpha = \xi_1^{\alpha_1} \ldots \xi_d^{\alpha_d}$. Any element $\hat{P}$ of the subalgebra $\mathrm{Diff}(\Omega)$ of $\mathrm{End}(C^\infty(\Omega))$, generated by the product operators and directional derivation $\partial/\partial x_i$, can be written:

$$(0.2) \qquad \hat{P}(x, i^{-1}\,\partial/\partial x) = \sum_{|\alpha| \le m} a_\alpha(x)(i^{-1}\,\partial/\partial x)^\alpha$$

hence the symbol of $\hat{P}$ is the function

$$(0.3) \qquad P(x,\xi) = \sum_{|\alpha| \le m} a_\alpha(x)\, \xi^\alpha .$$

For example, the Laplacian $\Delta = -\sum_{j=1}^{d} (i^{-1}\,\partial/\partial x_j)^2$ has the symbol $-|\xi|^2 = -\sum \xi_j^2$. The linear map

$$(0.4) \qquad \begin{array}{ccc} C^\infty(\Omega) & \to & \mathcal{D}'(\Omega) \\ \varphi & \longmapsto & \varphi\, dx \end{array}$$

is an injection. The image of this injection is denoted $C^\infty(\Omega)dx$.

This injection identifies $\mathrm{Diff}(\Omega)$ to a subalgebra of $\mathrm{End}(\mathcal{D}'(\Omega))$, and this is very convenient. We review bellow some other important and usefull technical tools:

- the Fourier transform of tempered distributions
- the Plancherel isometry induced by the F.T.
- the Sobolev spaces H^s
- the kernel theorem
- pseudo differential operators, Fourier integral operators ...

For $\varphi \in C_o^\infty(\Omega)$ and $\hat{P} \in$ Pseudo $\mathrm{Diff}(\Omega)$, we recall that $\hat{P}(\varphi dx) = (P\varphi)(x)dx$, where the function $(P\varphi)(x)$ can be written the followay, in terms of the symbol of $\hat{P}$

$$(0.5) \qquad (P\varphi)(x) = (2\pi)^{-d} \int P(x,\xi) e^{i\langle \xi, x\rangle} \hat{\varphi}(\xi) d\xi .$$

Concerning the applications of this theory in physical theories developped at least fifty years ago (heat, waves, potential, ordinary quantum mechanics) the situation is wonderfull i.e. these applications are obtained for $d = 1,2,3$ or 4, the theory is also valid for $d = 5,6...$

But modern physics (solide state physics, elementary particles, quantum electro dynamics, quantum fields ...) introduce differential operators of a completely different nature. For example

- in the theory of superconductivity, arise the Fröhlich hamiltonian

$$H = p^2/2m + \Sigma_k\, w\, a_k^* a_k + \Sigma_k V_k a_k e^{ikr} + \text{Im Conj}$$

where the a_k and a_k^* are the annihilators and creators of a boson field.

- the Thirring model of a scalar boson field use the free field $\Phi(x)$, the free Hamiltonian H_o, the perturbed hamiltonian $H_o + H_g$ written in the following way

$$(0.6) \qquad \Phi(x) = (2\pi)^{-s/2} \int_{k\in H} (a^*(k)e^{ikx} + a(k)e^{-ikx})\, d\mu(k)$$

$$(0.7) \qquad \hat{H}_o = 2^{-1} \int_{\mathbb{R}^s} a^*(k)\, a(k)\, dk$$

$$(0.8) \qquad \hat{H}_g = \int_{\mathbb{R}^s} g(\vec{x}) : \Phi(0,\vec{x})^n : d\vec{x} \;\ldots\ldots$$

The usual theory of linear differential operators cannot be applied to these operators. The scope of this paper is to given an idea of some mathematical structures developped for the treatment of these operators, and to introduce to constructive quantum field theory.

In the present work, the index $\epsilon = \pm$ is free - The expression "ϵ-symmetric" means symmetric if $\epsilon = -$, antisymmetric if $\epsilon = +$. In mathematical models, $\epsilon = -$ for bosons, $\epsilon = +$ for fermions. In fact bosons and fermions interact in the nature. Hence the physics need a common formulation for the study of symmetric algebras and of Grassmann algebras.

1. SYMMETRIC ALGEBRAS AND GRASSMANN ALGEBRAS

Letters X, Y, Z ... denote vector spaces on a given field $\mathbb{K}$ of characteristic zero. For any $k > 0$, $T^\epsilon_k(X)$ denotes the subspace of $\otimes_k X$ consisting of ϵ-symmetric tensors. Setting $T^\epsilon_o(X) = X$, $T^\epsilon(X) = \otimes^\infty_{k=0} T^\epsilon_k(X)$ is an algebra for the ϵ-symmetrized tensor product $(t;\ t') \mapsto tt' = \mathrm{Sym}_\epsilon(t \otimes t')$. By definition the space $\mathcal{F}^\epsilon_k(X,Z)$ of ϵ-symmetric forms on X, homogeneous of degree k, with values in Z, in the vector space of all ϵ-symmetric and k-linear maps $X^k \to Z$. We set

$$(1.1) \qquad \mathcal{F}^\epsilon(X,Z) = \prod_{k=0}^{\infty} \mathcal{F}^\epsilon_k(X,Z).$$

This is the space of all ϵ-symmetric forms $f = \sum_{k=0}^{\infty} f_k$ on X with "values" in Z. We write $\mathcal{F}^\epsilon(X) = \mathcal{F}^\epsilon(X,\mathbb{K})$, $\mathcal{F}^\epsilon_k(X) = \mathcal{F}^\epsilon_k(X,\mathbb{K})$. We consider bellow, forms defined on different vector spaces; hence the form $f = \Sigma f_k \in \mathcal{F}^\epsilon(X,Z)$ is usually denoted $f(x) = \sum_{k=0}^{\infty} f_k(x)$, even if the value of f (or f_k) in a point $x \in X$, this value is not defined in general. Using this convention, the inverse image of a form $f(y) \in \mathcal{F}^{\mathfrak{C}}(Y,Z)$ by a linear map $\ell: X \to Y$, can be denoted by $f(\ell x)$. For any bilinear map $\gamma: Z_1 \times Z_2 \to Z$, $f_k \in \mathcal{F}^{\mathfrak{C}}_k(X,Z_1)$, $g_\ell \in \mathcal{F}^{\mathfrak{C}}_\ell(X,Z_2)$, the product $\gamma[f_k(x)\cdot g_\ell(x)]$ of f_k and g_ℓ is the element $h \in \mathcal{F}^{\mathfrak{C}}_{k+\ell}(X,Z)$ such that

(1.2) $h(x_1,\ldots,x_{k+\ell}) = \mathrm{Sym}_{\mathfrak{C}}[\gamma(f_k(x_1,\ldots,x_k),g(x_{k+1},\ldots,x_{k+\ell}))]$.

By linearity, this gives a bilinear map $\mathcal{F}^{\mathfrak{C}}(X,Z_1) \times \mathcal{F}^{\mathfrak{C}}(X,Z_2) \to \mathcal{F}^{\mathfrak{C}}(X,Z)$. For scalar forms f and $g \in \mathcal{F}^{\mathfrak{C}}(X)$, γ is the product in $\mathbb{K}$, this product is simply denoted $f(x)g(x)$. The twist $t \to t^{\vee}$ in $T^{\mathfrak{C}}(X)$ (resp. $f \to f^{\vee}$ in $\mathcal{F}^{\mathfrak{C}}(X,Z)$) is an involution such that $(x_1,x_2,\ldots,x_k)^{\vee} = x_k x_{k-1}\ldots x_1$ (resp. $f^{\vee}(x_1;\ldots;x_k) = f(x_k,x_{k-1},\ldots,x_1)$. For $f = \Sigma f_k \in \mathcal{F}^{-}(X,Z)$, $\hat{f}$ denotes the formal series $\Sigma\, \hat{f}_k(x)$ such that for any k, $\hat{f}_k$ is the homogeneous polynomial $x \to f_k(x,x,\ldots,x)$; and $\hat{\mathcal{F}}^{-}(X,Z)$ denotes the vector space of all formal series on X, with "values" in Z. By polarisation, the map $f \to \hat{f}$ is a linear isomorphism $\mathcal{F}^{-}(X,Z) \to \hat{\mathcal{F}}^{-}(\hat{X},Z)$, this map is an isomorphism of algebras if $Z = \mathbb{K}$.

(1.3) <u>Duality</u>

The algebraic dual of $T^{\mathfrak{C}}_k(X)$ is $\mathcal{F}^{\mathfrak{C}}_k(X)$; the natural duality $\langle\ ,\ \rangle$ $\langle\ ,\ \rangle_{k,nat}$ between these two spaces is such that

$$\langle f_k(x),\ x_1x_2\ldots x_k\rangle_{k,nat} = f_k(x_1,\ldots,x_k).$$

The algebraic dual of $T^{\mathfrak{C}}(X)$ is $\mathcal{F}^{\mathfrak{C}}(X)$. The natural duality $\langle\ ,\ \rangle_{nat}$ (resp. the twisted duality $\langle\ ,\ \rangle$) between these two spaces is such that $\forall f = \Sigma f_k \in \mathcal{F}^{\mathfrak{C}}(X)$, $\forall t = \Sigma t_k \in T^{\mathfrak{C}}(X)$

(1.4) $$\langle f,t\rangle_{nat} = \sum_{k=0}^{\infty} k!\ \langle f_k,t_k\rangle_{k,nat}$$

(1.5) $$(\text{resp } \langle f,t\rangle = \langle f^{\vee},t\rangle_{nat} = \langle f,t\rangle_{nat} = \langle f,t^{\vee}\rangle_{nat}).$$

The twisted duality is systematically used bellow. In the same way, if X and X' are two vector spaces in duality, the twisted duality

$\langle\ ,\ \rangle$ is defined between $T^{\mathfrak{e}}(X)$ and $T^{\mathfrak{e}}(X')$. Hence we have a canonical imbedding $T^{\mathfrak{e}}(X') \subset \mathfrak{F}^{\mathfrak{e}}(X)$. The image $\mathfrak{F}^{\mathfrak{e}}_{cyl}(X)$ of this injection is called the space of cylindrical forms on X. If $P = X\text{-}X'$ and $P' = Y\text{-}Y'$ are two paars of vector spaces in duality, a morphism $P \to P'$ is a linear map $\ell: X \to Y$ admitting a transposed map $\ell': Y' \to X'$. The category of paars of vector spaces in duality is denoted (v.s.d.). Then, $T^{\mathfrak{e}}$ (resp. $\mathfrak{F}^{\mathfrak{e}}$) is a covariant (resp. contravariant) functor of (v.s.d.); the linear map $T^{\mathfrak{e}}(X) \to T^{\mathfrak{e}}(Y)$ defined by ℓ is denoted $T^{\mathfrak{e}}\ell = L$. The transpose of L is $g(y) \to$ $\to g(\ell x)$: $\mathfrak{F}^{\mathfrak{e}}(Y) \to \mathfrak{F}^{\mathfrak{e}}(X)$; this last map extends $L' = T^{\mathfrak{e}}\ell'$.
If E^* denotes the algebraic dual of a vector space E, the weak_d topology on E^* is the weakest topology on E^* such that $\forall$ x, the linear form $\xi \to \xi(x)$ is continuous, if $\mathbb{K}$ is equipped with the discrete topology.

(1.6) Basis. The sets $J(\mathfrak{e},d)$ of multi-indices

Let X and X' be two vector spaces in duality, and let I be a finite or countable set, d is the cardinal of I and I is totally ordered. We suppose that there is algebraic basis $\{\xi_i, i \in I\}$ and $\{x_i ; i \in I\}$ in X and X' resp. such that $\langle x_i, \xi_j\rangle = \delta_{i,j}$.
The set $J(+,d)$ of multi-indices is the set of all finite subsets j of I. If $j \neq \Phi$, j is characterized by the ordered list of his elements $j = \{j_1, j_2, \dots, j_k\}$ and we set: $\xi^j = \xi_{j_1} \wedge \xi_{j_2} \dots \wedge \xi_{j_k}$; $|j| = k$; $j! = 1$. For j and $j' \in J(+,d)$, the scalars (j) and (j,j') are defined by $(\xi^j)^{v} = \xi^{j^{v}} = (j)\xi^j$ and $\xi^j \xi^{j'} = (j,j')\xi^{j\cup j'}$.
The set $J(-,d)$ of multi-indices, is the set of families $j =$ $= (j_n)_{n \in I}$ of integers $j_n \geq 0$, vanishing for n sufficiently large. With the conventions $(\xi_j)^0 = 1$; $0! = 1$, we set $j! = j_1! j_2! \dots$; $|j| = j_1 + j_2 + \dots$

$$(1.7) \qquad \xi^j = (\xi_1^{\odot j_1}) \odot (\xi_2^{\odot j_2}) \odot \dots . = (\xi_1)^{j_1} (\xi_2)^{j_2} \dots .$$

For arbitrary p, $J^p(\mathfrak{e},d)$ is the set of indices $j \in J(\mathfrak{e},d)$ with lenght $|j| = p$. For $\varepsilon = \pm$, the tensors (ξ^{j}) and $(x^{j'})$, j and $j' \in J(\varepsilon, d)$ are basis of $T^{\varepsilon}(X)$ and $T^{\varepsilon}(X')$ resp. and

(1.8) $$\langle (x^{j})^{v}, \xi^{j'} \rangle = \delta_{j,j'} = \begin{cases} 1 & \text{if } j=j' \\ 0 & \text{if not} \end{cases}$$

Moreover $(x^{j})_{j}$ is a topological basis of $\mathfrak{F}^{\varepsilon}(X)$ equipped with the weak_d topology.

(1.9) <u>The notation</u> $X_{\alpha} \Subset X$ means that X_{α} is a finite dimensional subspace of X. For $X_{\alpha} \Subset X_{\beta} \Subset X$, the transpose of the canonical injections $i_{\alpha\beta} : X_{\alpha} \to X_{\beta}$ and $i_{\beta} : X_{\beta} \to X$ are the canonical surjections $s_{\beta\alpha} : X'_{\beta} \to X'_{\alpha}$ and $s_{\beta} : X' \to X'_{\beta} = X'/X_{\beta}^{\perp}$. We have $T^{\varepsilon}(X) = \varinjlim T^{\varepsilon}(X_{\alpha})$. Hence any cylindrical form φ on X' can be written $\varphi = \varphi_{\alpha}(s_{\alpha}x)$ for some $X_{\alpha} \Subset X$ and some form φ_{α} of finite degree on X'_{α}.

(1.10) <u>Forms on a product</u>. Let Z be the product of two vector spaces X and Y. The canonical projection $Z \to X$ (resp. injection $X \to Z$) is denoted p_1 (resp. i_1). The tensor $t \in T^{\varepsilon}(X)$ (resp. a form $f \in \mathfrak{F}^{\varepsilon}(X)$) is sometimes identified with $(T^{\varepsilon}i_1)(t)$ (resp. with $f(\pi_1 z)$). The bilinear map $(t;u) \mapsto ut$, defines by the universal property of tensor products, a linear isomorphism

(1.11) $$T^{\varepsilon}(X) \otimes T^{\varepsilon}(Y) \xrightarrow{\beta} T^{\varepsilon}(Z)$$

and $\mathfrak{F}_{i,j}(X;Y)$ denotes the space of $(i+j)$-linear forms on $X^{i} \times X^{j}$, separately ε-symmetric with respect to their i first arguments, and the j last arguments. For arbitrary $h = (h_{k,\ell}) \in \prod_{k \text{ and}=0}^{\infty} \mathfrak{F}_{k,\ell}(X;Y)$ and $t = \sum_{i,j} (x_1^{i} \ldots x_1^{i}) \otimes (y_1^{j} \ldots y_j^{j}) \in \oplus_{i,j} T_i^{\varepsilon}(X) \oplus T_j^{\varepsilon}(Y)$ we define

(1.12) $$\langle h, t \rangle = \sum_{i,j} i!j!h_{i,j}(x_i^{i}, \ldots x_1^{i};\ y_j^{j}, \ldots y_1^{j}).$$

This is a duality between $T^{\varepsilon}(X) \otimes T^{\varepsilon}(Y)$ and the algebraic dual $\prod_{k,\ell} \mathfrak{F}_{k,\ell}(X;Y)$. For $f_{i,j} \in \mathfrak{F}_{i,j}^{\varepsilon}(X;Y)$, the following $(i+j)$-linear form on Z: $(z_1; \ldots; z_{i+j}) \mapsto f_{i,j}(\pi_1 z_1, \ldots \pi_i z_i, \pi_2 z_{i+1}, \ldots \pi_2 z_{i+j})$ is denoted $\underline{f}_{i,j}$.

(1.3) Proposition

a) The transpose of β is a linear isomorphism

(1.14) $$\mathcal{F}^{\varepsilon}(X\times Y) \xrightarrow{\beta'} \prod_{i,j} \mathcal{F}^{\varepsilon}_{i,j}(X;Y)$$

s.t. $\forall\, n$, $\forall\, R_n \in \mathcal{F}^{\varepsilon}_n(Z)$, we have $\beta'(R_n) = (R_{i,j})$ with $R_{i,j} = 0$ for $i+j \neq n$ and for $i+j = n$:

(1.15) $$R_{i,j}(x_1,\dots,x_i;y_{i+1}\dots y_n) = \binom{n}{i} R_n(i_1x_1;\dots i_1x_i;i_2y_{i+1}\dots i_2y_n)$$

b) $\forall\, f \in \mathcal{F}^{\varepsilon}(X)$ $\forall\, g \in \mathcal{F}^{\varepsilon}(Y)$ $\beta'(fg) = f\otimes g$

c) $\forall\, (i,j) \in \mathbb{N}^2$, $R_{i,j} \in \mathcal{F}^{\varepsilon}_{i,j}(X;Y)$, $\beta'(\mathrm{Sym}_{\varepsilon}\, \underline{R}_{i,j}) = R_{i,j}$

d) Finally for arbitrary $t \in T^{\varepsilon}(X)$, $u \in T^{\varepsilon}(Y)$

(1.16) $$\langle gf, tu\rangle = \langle g,u\rangle\langle f,t\rangle.$$

Proof

$$R_{i,j}(x_1,\dots,x_i;\ y_{i+1},\dots,y_{i+j})$$
$$= i!^{-1}\ j!^{-1}\ \langle \beta' R_n,\ (x_i\dots x_1)\otimes(y_{1+j}\dots y_{i+1})\rangle$$
$$= i!^{-1}\ j!^{-1}\ \langle R_n,\ \mathrm{Sym}_{\varepsilon}\ I_2(y_{i+j}\dots y_{i+1})\otimes I_1(x_{i+j}\dots x_1)\rangle$$
$$= i!^{-1}\ j!^{-1}\ \langle R_n,\ i_2(y_{i+1})\dots i_2(y_{i+1})i_1(x_i)\dots i_1(x_1)\rangle.$$

This expression vanishes if $i+j \neq n$ and gives (1.15) for $i+j = n$. Then b) follows from a) by a combinatorial computation. If X and Y are finite dimensional, c) follows from b) using coordinates; and the general case can be reduced to this particular case. Finally for $f_k \in \mathcal{F}^{\varepsilon}_k(X)$, $g_\ell \in \mathcal{F}^{\varepsilon}_\ell(Y)$, $t_i \in T^{\varepsilon}_i(X)$, $u_j \in T^{\varepsilon}_j(Y)$:

$$\langle f_kg_\ell, u_jt_i\rangle = \langle f_kg_\ell, \beta(t_iu_j)\rangle = \langle \beta'(f_kg_\ell), t_iu_j\rangle$$
$$= \langle f_k\otimes g_\ell,\ t_i\otimes u_j\rangle = \langle f_k,t_i\rangle\langle g_\ell,u_j\rangle$$

and (1.16) follows by linearity.

(1.17) For example, the ε-canonical two form $x\cdot\xi$ on $E = X\times X'$ is defined by the following bilinear form on $E\times E$

(1.18) $$(e = (x;\xi);\ e_1 = (x_1;\xi_1)) \mapsto (\langle x,\xi_1\rangle - \varepsilon\langle x_1,\xi\rangle)/2.$$

The exponential of this form in the algebra $\mathcal{F}^{\varepsilon}(X \times X')$ is defined by $\exp(x,\xi) = \sum_{k=0}^{\infty} k!^{-1}(x\cdot\xi)^k$; and with the hypothesis of (1.6)

(1.19)
$$\exp x\cdot\xi = \sum_{\alpha\in J(\varepsilon,d)} \alpha!^{-1}\, x^{\alpha}(\xi^{\alpha})^{v}.$$

The sum converges for the weak$_d$ topology. In particular, for $\varepsilon = -$, the formal series corresponding to $\exp x\cdot\xi$ is the familiar exponential "function" $\exp\langle x,\xi\rangle$ defined on $X \times X'$.

(1.20) <u>Directional derivatives.</u>

For any fixed $t \in T^{\varepsilon}(X)$, the left directional derivation $f \to \partial_t f$ in $\mathcal{F}^{\varepsilon}(X)$ is defined by transposition of the operator of right product by t in $T^{\varepsilon}(X)$, i.e. $\forall\, t' \in T^{\varepsilon}(X)$

(1.21)
$$\langle \partial_t f, t'\rangle = \langle f, t't\rangle.$$

If the same way the left derivation $f \to f\,\partial_t$ is s.t. $\langle f\,\partial_t, t'\rangle = \langle f, tt'\rangle$. Then

(1.22)
$$\partial_{tt'} f = \partial_t(\partial_{t'} f) \qquad (\partial_t f)^v = f^v\,\partial_{t^v}.$$

If ℓ is a linear map $X \to Y$, we have for arbitrary $t \in T^{\varepsilon}(X)$ and $g \in \mathcal{F}^{\varepsilon}(Y)$

(1.23)
$$\partial_t(g(\ell x)) = (\partial_{L(t)} g)(\ell x).$$

In fact we have a commuting diagram

(1.24)
$$\begin{array}{ccc} T^{\varepsilon}(Y) & \xrightarrow{\ .L(t)\ } & T^{\varepsilon}(Y) \\ L \uparrow & & \uparrow L \\ T^{\varepsilon}(X) & \xrightarrow{\ .t\ } & T^{\varepsilon}(X) \end{array}$$

Hence (1.23) follows from the commutativity of the transposed diagram. If dual basis (ξ_i) and (x_i) exist in tho vector spaces X and X' in duality, we set for arbitrary $a \in J(\varepsilon,d)$, $f \in \mathcal{F}^{\varepsilon}(X)$

(1.25)
$$(\partial/\partial x)^a f = \partial^a f = \partial_{\xi^a} f.$$

In the particular case finite, $\varepsilon = -$, $\mathbb{K} = \mathbb{R}$, this general convention agrees with the convention of multi-indices in distribution theory.

For $\varepsilon = +$, $b \in J(\varepsilon, d)$

$$\partial^b x^a = \begin{cases} 0 & \text{if } b \not\subset a \\ (b,b')(b)x^{b'} & \text{if } b \subset a \end{cases} ;$$

(1.26)

$$x^a \partial_b = \begin{cases} 0 & \text{if } b \not\subset a \\ (b)(b',b)x^{b'} & \text{if } b \subset a \end{cases}$$

Let $t = y_1 y_2 \ldots y_\ell \in T^\varepsilon_\ell(X)$, $f_k \in \mathfrak{F}^\varepsilon_k(X)$. Then the left derivative $\partial_t f_k$ vanishes for $k < \ell$. For $k \geq \ell$, $\partial_t f_k$ is an homogeneous form of degree $k-\ell$ such that

(1.27) $(\partial_{y_1 \ldots y_\ell} f_k)(x_1, \ldots, x_{k-\ell}) = k!(k-\ell)!^{-1} f_k(y_\ell, \ldots, y_1, x_1, \ldots, x_{k-\ell})$.

This expression permits the definition for any $t \in T^\varepsilon(X)$ of the derivative $\partial_t f$ of any vectorial form $f \in \mathfrak{F}(X,G)$. In some applications, left convolution operators are used:

(1.28) <u>Proposition</u>

Let X be a vector space, $\varepsilon = \pm$. For arbitrary $g \in \mathfrak{F}^\varepsilon(X)$, the linear map $A: f \to fg$ in $\mathfrak{F}^\varepsilon(X)$ admits a transposed map $A': t \to \partial_g t$ in $T^\varepsilon(X)$. This linear map is called a left convolution operator in $T^\varepsilon(X)$.

By construction, for any vector space X' in duality with X, any $g \in T^\varepsilon(X') \cong \mathfrak{F}^\varepsilon_{cy\ell}(X)$ the map A' agrees the directional left derivation; hence A' can be denoted $t \to \partial_g t$.

(1.29) As a corollary, the space of all left convolution operators in $T^\varepsilon(X)$ is a subalgebra of End $(T^\varepsilon(X))$, isomorphic with $\mathfrak{F}^\varepsilon(X)$, for $f,g \in \mathfrak{F}^\varepsilon(X)$, $t \in T^\varepsilon(X)$:

$$\partial_f(\partial_g t) = \partial_{fg} t.$$

The left convolution operator ∂_f is invertible in this subalgebra iff $f_o \neq 0$.

(1.30) <u>Definition of the global left derivatives</u>

Let X and G be two vector spaces, $\epsilon = \pm$; j denotes an integer > 0. For any vectorial form $f \in \mathfrak{F}^{\epsilon}(X,G)$, the global left derivative $D^j f$ of order j, is the linear map

$$T^{\epsilon}_j(X) \xrightarrow{D^j f} \mathfrak{F}^{\epsilon}(X,G)$$
$$t \longmapsto \partial_{t^v} f$$

If G is the algebraic dual of some vector space E, $D^j f$ can be viewed as a vectorial form

$$D^j f \in (T^{\epsilon}(X) \otimes T^{\epsilon}_j(X) \otimes E)^* \cong \mathfrak{F}^{\epsilon}(X, \mathfrak{F}^{\epsilon}_j(X,E^*)).$$

In the same way, fD^j is defined by the linear map $t \to f\,\partial_{t^v}$. The twisted duality $\langle\ ,\ \rangle_j$ is used between $T^{\epsilon}_j(X)$ and $\mathfrak{F}^{\epsilon}_j(X)$.

(1.31) In view of (1.23), for any linear map $\ell: Y \to X$, the left derivative $D^j(f(\ell x))$ is the linear map

$$T^{\epsilon}(Y) \ni t \mapsto \partial_{t^v}(f(\ell x)) = (\partial_{L(t^v)} f)(\ell x) = (\partial_{L(t)^v} f)(\ell x).$$

(1.32) In particular, for arbitrary $X_\alpha \subseteq X$, the restriction of $D^j f$ to $T^{\epsilon}_j(X_\alpha)$ is the left derivative of the restriction of f.

If coordinates are used (1.6), the definition of $D^j f$ gives directly

(1.33) $$D^j f = j! \sum_{|\alpha|=j} \frac{(\partial^\alpha f)\otimes x^\alpha}{\alpha!}\ ; \qquad fD^j = j! \sum_{|\alpha|=j} \frac{(f\partial^\alpha)\otimes x^\alpha}{\alpha!}\ .$$

For $G = K$, $t \otimes u \in T^{\epsilon}(X) \otimes T^{\epsilon}_j(X)$:

$$\langle D^j f,\ t\otimes u\rangle = \langle \partial_{u^v} f, t\rangle = \langle f,\ tu^v\rangle.$$

(1.34) Hence the operator of left derivation of order j:

$$\mathfrak{F}^{\epsilon}(X) \to \mathfrak{F}^{\epsilon}(X, \mathfrak{F}^{\epsilon}_j(X))$$

appears as the transpose of the linear map

$$T^{\epsilon}(X) \otimes T^{\epsilon}_j(X) \to T^{\epsilon}(X)$$
$$t \otimes u \longmapsto tu^v.$$

This last map can be viewed as a contracted right product with

of $u = u(\eta) \in T^{\mathfrak{C}}(Y)$ with the right argument y of $\tilde{Q}(x,y)$ is defined by

(2.2) $$\langle \tilde{Q}(x,y), u(\eta) \rangle = (\hat{Q}u)(x).$$

For any $t \in T^{\mathfrak{C}}(X)$, the relation $\langle \hat{Q}u, t\rangle = \tilde{Q}(u \otimes t) = \langle \tilde{Q}, tu\rangle$ can be written

(2.3) $$\langle \langle \tilde{Q}(x,y), u(\eta)\rangle, t(\xi)\rangle = \langle \tilde{Q}(x,y), u(\eta)t(\xi)\rangle .$$

Combining this with (1.16): $\forall\, A \in \mathfrak{Z}^{\mathfrak{C}}(X)$, $\forall\, B \in \mathfrak{Z}^{\mathfrak{C}}(Y)$

(2.4) $$\langle A(x)B(y), u(\eta)\rangle = A(x)\langle B(y), u(\eta)\rangle .$$

The left hand side of (2.2) is called the right contraction of the forms $\tilde{Q}$ and u one their dual arguments y and η, $\tilde{Q}$ is called the left kernel of $\hat{Q}$. Sometimes, different conventions are useful:

(2.5) <u>left contractions and right kernels.</u>

We set

(2.6) $$\langle t(\xi), \tilde{Q}(x,y)\rangle = \hat{Q}'(t)$$

where $\hat{Q}'$ is the transpose of $\hat{Q}$. Hence

(2.7) $$\langle u(\eta), \langle t(\xi), \tilde{Q}(x,y)\rangle\rangle = \langle u(\eta)t(\xi), \tilde{Q}(x,y)\rangle .$$

The left contraction of t and $\tilde{Q}$ on their dual arguments ξ and x; $\tilde{Q}$ is the right kernel of $\hat{Q}'$.

(2.8) <u>Regular operators</u>

Let X-X', Y-Y', Z-Z' be three pairs of vector spaces in duality. Let $\hat{L}$ and $\hat{Q}$ be two linear maps

(2.9) $$T^{\mathfrak{C}}(Y) \xrightarrow{\hat{L}} \mathfrak{Z}^{\mathfrak{C}}(X); \qquad T^{\mathfrak{C}}(X') \xrightarrow{\hat{Q}} \mathfrak{Z}^{\mathfrak{C}}(Z).$$

The linear operator $\hat{L}$ is called regular if $\operatorname{Im} \hat{L} \subset T^{\mathfrak{C}}(X')$. Then $\hat{Q}\hat{L}$ is defined; moreover $\hat{L}'$ can be extended in a linear map $\mathfrak{Z}^{\mathfrak{C}}(X') \to \mathfrak{Z}^{\mathfrak{C}}(Y)$. Conversely, if $\hat{L}'$ is regular, $\hat{L}$ is extended by $\hat{L}''$. If $\hat{R} = \hat{Q}\hat{L}$ is defined, the kernel of $\hat{R}$ is denoted

(2.10) $$\langle \tilde{Q}(z,\xi), \tilde{L}(x,y)\rangle = \tilde{R}(z,y).$$

We say that $\tilde{R}$ is the contraction of $\tilde{Q}$ and $\tilde{R}$ on their dual arguments ξ and x. This new definition of contraction extends (2.1) (2.5).

(2.11) Proposition. Let X and X' be two vector spaces in duality. Then the kernel of the canonical imbedding $T^{\mathbb{C}}(X') \to \mathfrak{F}^{\mathbb{C}}(X)$ is $\exp x\cdot\xi$. This map and the transposed map are regular. Hence $\forall f \in \mathfrak{F}^{\mathbb{C}}(X)$

(2.12) $$\langle \exp x\cdot\xi', f(x')\rangle = f(x).$$

Principle of the proof.

If X has finite dimension d, coordinates and (2.3) can be used.

In the general case we must prove:

$$\forall u \in T^{\mathbb{C}}(X') \quad \forall t \in T^{\mathbb{C}}(X) \quad \langle e^{x\cdot\eta}, u(y)t(\xi)\rangle = \langle u,t\rangle.$$

By linearity, we may suppose $t = x_1\ldots x_\ell$ and $u = \xi_1\ldots\xi_k$, with some $x_i \in X$, $\xi_j \in X'$. Let $X_\alpha \Subset X$ be such that $t \in T^{\mathbb{C}}(X_\alpha)$. If σ_α denotes the canonical surjection $X' \to X'_\alpha$:

$$\langle \exp x\cdot\eta, u(y)t(\xi)\rangle = \langle \exp x\cdot\eta, \sigma_\alpha(\xi_1\ldots\xi_k)\; x_1\ldots x_\ell\rangle$$

and we are reduced to the finite dimensional case.

(2.13) We note that the order of terms in the left hand side of (2.12) is important and:

(2.14) $$\langle e^{\xi'\cdot x}, f(x')\rangle = f(-\varepsilon x).$$

(2.15) Proposition. a) Let $\hat{Q}$ be a linear operator $T^{\mathbb{C}}(Y) \to \mathfrak{F}^{\mathbb{C}}(X)$, $f \in \mathfrak{F}^{\mathbb{C}}(X)$, $g \in \mathfrak{F}^{\mathbb{C}}(Y)$, $t \in T^{\mathbb{C}}(X)$, $t' \in T^{\mathbb{C}}(Y)$. Then the kernel of the linear map

(2.16) $u \mapsto \hat{Q}(\partial_g u)$ is $\tilde{Q}(x,y)g(y)$

(2.17) $u \mapsto \hat{Q}(t'u)$ is $\tilde{Q}(x,y)\partial_{t'}$

(2.18) $u \mapsto f(\hat{Q}u)$ is $f(x)\tilde{Q}(x,y)$

(2.19) $u \mapsto \partial_t(\hat{Q}u)$ is $\partial_t \tilde{Q}(x,y)$

b) For any linear map $m: Z \to Y$, the kernel of $\hat{Q}\circ(T^{\varepsilon}m)$ is $\tilde{Q}(x,my)$.

c) For any linear map $\ell: V \to X$, the kernel of the map u $(\hat{Q}u)(\ell v)$ is $\tilde{Q}(\ell v,y)$.

We compute for example the kernel of $\hat{L}: u \to \hat{Q}(\partial_g u)$. In view of (2.3) we have for arbitrary $t \in T^{\varepsilon}(X)$ $\langle \tilde{L},ut\rangle = \langle \hat{Q}(\partial_g u),t\rangle = \langle \tilde{Q},(\partial_g u)t\rangle =$ $= \langle \tilde{Q},\partial_g(ut)\rangle = \langle \tilde{Q}g,ut\rangle$. This proves (2.16). The parts b) and c) are proven using the matricial decomposition of $\hat{Q}$.

(2.20) Corollaries

a) For arbitrary $X_{\alpha} \Subset X$, $X_{\beta} \Subset Y$, the restriction to $X_{\alpha} \times X_{\beta}$ of the kernel of $\hat{Q}: T^{\varepsilon}(Y) \to \mathfrak{F}^{\varepsilon}(X)$, is the kernel of the product of following maps

$$T^{\varepsilon}(Y_{\beta}) \subset T^{\varepsilon}(Y) \xrightarrow{\hat{Q}} \mathfrak{F}^{\varepsilon}(X) \xrightarrow{\text{restr}^{n}} \mathfrak{F}^{\varepsilon}(X_{\alpha}).$$

b) Let X and X' be two vector spaces in duality. For any linear map $\ell: Y \to X'$, the corresponding map $L = T^{\varepsilon}(\ell): T^{\varepsilon}(Y) \to T^{\varepsilon}(X')$ has the kernel $L(x,y) = \exp(x\cdot my)$.

c) Let X and X' be vector spaces in duality. For any linear operator $m: Z \to X$, the kernel of the linear operator $A: f(x) \to f(mx)$: $\mathfrak{F}^{\varepsilon}(X) \to \mathfrak{F}^{\varepsilon}(Z)$ is $\exp((mz)\cdot\xi)$.

d) Let X and X' be vector spaces in duality, $f \in \mathfrak{F}^{\varepsilon}(X)$, $g \in \mathfrak{F}^{\varepsilon}(X')$. Hence

(2.21) $$(\text{Kernel of } u \mapsto f\,\partial_g u) = f(x)e^{x\cdot\xi}\,g(\xi).$$

(2.22) Extension of a linear map $\hat{Q}: T^{\varepsilon}(X') \to \mathfrak{F}^{\varepsilon}(X)$

a) Trivially, for any vector space E, the tensor product of $\hat{Q}$ with the identity map of E is a linear operator $T^{\varepsilon}(X') \otimes E \to$ $\to \mathfrak{F}^{\varepsilon}(X) \otimes E$. For example $E = T^{\varepsilon}(X)$ we obtain the following extension of $\hat{Q}$: $T^{\varepsilon}(X') \otimes T^{\varepsilon}(X) = T^{\varepsilon}(X'\times X) \to \mathfrak{F}^{\varepsilon}(X) \otimes T^{\varepsilon}(X')$. This extension is denoted $\hat{Q}_x$. If X is finite dimensional, $\varepsilon = +$, this permits the definition of $\hat{Q}_x(\exp x\cdot\xi)$. In general, $\exp x\cdot\xi$ can be viewed as a vectorial form:

$$g_o = \exp x\cdot\xi = \sum_{k=0}^{\infty} k!^{-1} \vec{x}^k \in \mathfrak{F}^{\epsilon}(X,\mathfrak{F}^{\epsilon}(X'))$$

and $\hat{Q}_x(\exp(x\cdot\xi))$ is defined using a more sophisticated argument:

b) Let g be a form $\in \mathfrak{F}^{\epsilon}(X,E^*)$, with values in some algebraic dual E^*, such that the transpose of the following map, is regular

$$\begin{array}{ccc} T^{\epsilon}(X'\times X) & \xrightarrow{A(g)} & \mathfrak{F}^{\epsilon}(X,E^*) \\ \tilde{Q}(x,y) & \longrightarrow & \langle \tilde{Q}(x,y), g(\eta)\rangle \end{array}$$

Hence $A(g)' : T^{\epsilon}(X)\otimes E \to T^{\epsilon}(X\times X')$; $A(g)$ is extended by $A(g)''$: $\mathfrak{F}^{\epsilon}(X\times X') \to \mathfrak{F}^{\epsilon}(X,E^*)$, we set $\hat{Q}(g) = A(g)''.\tilde{Q}$.

In our particular case $E = T^{\epsilon}(X')$, $g = g_o$, $A(g)$ is the identity operator of $T^{\epsilon}(X'\times X)$. Hence $A(g_o)''$ is the identity operator of $\mathfrak{F}^{\epsilon}(X\times X')$. Hence

(2.23) $$\hat{Q}_x(\exp x\cdot\xi) = \tilde{Q}(x,\xi).$$

By analogy with (0.1), we set:

(2.24) <u>Definition</u>. For $\epsilon = \pm$, the symbol of any linear operator $\hat{Q}$: $T^{\epsilon}(X') \to \mathfrak{F}^{\epsilon}(X)$ is the following ϵ-symmetric form on $X\times X'$:

(2.25) $$Q(x,\xi) = \tilde{Q}(x,\xi)\exp(-x\cdot\xi).$$

In view of (2.23):

(2.26) $$\hat{Q}_x(\exp x\cdot\xi) = Q(x,\xi)\exp x\cdot\xi.$$

In view of (1.28) and (2.1) the symbol map $\hat{Q} \to Q$ is an homeomorphism.

The following theorem gives for any linear operator $\hat{Q}$ a form "similar" to (0.2).

(2.27) <u>Theorem</u>. Let $\hat{Q}$ be a linear operator: $T^{\epsilon}(X') \to \mathfrak{F}^{\epsilon}(X)$ with symbol

$$Q(x,\xi) = \sum_{i,j} \mathrm{Sym}_{\epsilon}\, Q_{i,j}$$

with $Q_{i,j} \in \mathcal{J}^{\epsilon}_{i,j}(X;X')$. Let $Q^{\sim}_{i,j}$ be the bilinear form on $T^{\epsilon}_i(X) \times T^{\epsilon}_j(X')$ s.t.

$$Q^{\sim}_{i,j}(x_1 x_2 \dots x_i; \xi_1 \xi_2 \dots \xi_j) = Q_{i,j}(x_1; \dots x_i; \xi_1 \dots \xi_j).$$

Then for arbitrary $u \in T^{\epsilon}(X')$

(2.28) $$\hat{Q}u = \sum_{i,j} Q^{\sim}_{i,j} [\vec{x}^i \cdot (D^j u)(x)].$$

In fact

$$Q^{\sim}_{i,j} [\vec{x}^i \cdot D^j_x e^{x\cdot\xi}] = Q^{\sim}_{i,j} [\vec{x}^i \cdot \vec{\xi}^j] \exp x\cdot\xi$$
$$= (\mathrm{Sym}_{\epsilon}\, \underline{Q}_{i,j})(\exp x\cdot\xi).$$

Hence the result follows from (2.27) and (1.13).

(2.29) <u>Examples</u>.

(a) If the symbol of $\hat{Q}$ depends only from the argument ξ, then $Q_{i,j} = 0$ for $i \neq 0$. Hence a comparison of (1.36) and (2.28) shows that $\hat{Q}$ is then the left convolution operator $u \to \partial_Q u$.

(b) In a similar way, if $Q = Q(x)$, then $j \neq 0$ implies $Q_{i,j} = 0$. Hence $\hat{Q}$ is the operator $u \to Q(x)u(x)$ of left product by the form $Q(x)$.

(c) For arbitrary $f \in \mathcal{J}^{\epsilon}(X)$, $g \in \mathcal{J}^{\epsilon}(X')$ we have

(2.30) $$(\text{Symbol of } u \mapsto f\partial_g u) = f(x)g(\xi).$$

(2.31) <u>Definition</u>. If A and B are two endomorphisms of a vector space E, the following linear operator of E is called the ϵ-commutator of A and B:

$$[A,B] = AB + \epsilon\, BA.$$

(2.32) For example $E = T^{\epsilon}(X')$; x; $x' \in X$; $\xi, \xi' \in X'$; denoting ξ (resp. ∂_x) the operator of left product by ξ (resp. of left derivation $u \to \partial_x u$) we have the ϵ-commutation relations

$$[\partial_x\cdot, \partial_{x'}\cdot] = [\xi\cdot, \xi'\cdot] = 0; \quad [\xi\cdot, \partial_x\cdot] = \epsilon\langle\xi, x\rangle.$$

(2.33) <u>Theorem</u>. Let X and X' be two vector spaces in duality,

$\epsilon = \pm$, $u \in T^{\epsilon}(X')$. Then the following linear maps $\mathfrak{F}^{\epsilon}(X \times X') \to \mathfrak{F}^{\epsilon}(X)$:

$$\tilde{Q} \xrightarrow{\hat{A}} \hat{Q}(u) \quad \text{and} \quad Q \xrightarrow{\hat{B}} \hat{Q}(u)$$

are regular. The <u>right</u> kernels of these maps are resp.:

$$A(x', \xi'', x) = u(x')\exp x\cdot\xi'', \quad B(x', \xi''; x) = u(x'+x)\exp x\cdot\xi''.$$

For an arbitrary linear operator $\hat{Q}\colon T^{\epsilon}(X') \to \mathfrak{F}^{\epsilon}(X)$:

(2.34) $$(\hat{Q}u)(x) = \langle \tilde{Q}(x', \xi''), u(x'')e^{\xi'\cdot x}\rangle$$

(2.35) $$(\hat{Q}u)(x) = \langle Q(x', \xi''), u(x''+x)e^{\xi'\cdot x}\rangle.$$

In particular, if $\hat{Q}$ is a left convolution operator

(2.36) $$(\hat{Q}u)(x) = \langle Q(\xi''), u(x''+x)\rangle.$$

We gives for example the principle of the proof of (2.34). For a given $\tilde{Q} \in T^{\epsilon}(X'\times X)$, let $X'_{\alpha} \Subset X$ be such that $u \in T^{\epsilon}(X'_{\alpha})$ and $\tilde{Q} \in T^{\epsilon}(X'_{\alpha}) \otimes T^{\epsilon}(X)$. Then $\hat{Q}(u) \in T^{\epsilon}(X'_{\alpha})$; hence the formula (2.34) can be verified in this case using coordinates in X'_{α}. Hence $\hat{A}$ induces a linear operator $\hat{C}\colon T^{\epsilon}(X'\times X) \to \mathfrak{F}^{\epsilon}(X)$ with right kernel $A(x', \xi''; x)$. The transposed operator $\hat{C}'\colon T^{\epsilon}(X) \to \mathfrak{F}^{\epsilon}(X'\times X)$ maps $t \in T^{\epsilon}(X)$ on:

$$\langle u(x)e^{\xi'\cdot x''}, t(\xi'') = u(x)\ t(\xi').$$

Hence $\hat{C}'$ is regular. Hence $\hat{C}$ is extended by $\hat{C}''$. This proves (2.34).

3. DERIVATIONS OF SYMMETRIC COFORMS

We give a new presentation of some results of [3], and new results.

(3.1) <u>Definition</u>. Let X' be a vector space, $\epsilon = \pm$. For <u>any</u> vector space X in duality with X', a cylindrical ϵ-symmetric coform M on X is a linear form on $\mathfrak{F}^{\epsilon}_{cyl}(X) \simeq T^{\epsilon}(X')$.
The action of M on $u \in T^{\epsilon}(X')$ is written

(3.2) $\langle M,u\rangle = M(x)u(x)$ or $\langle u,M\rangle = u(x)M(x)$.

The vector space $\mathcal{F}^{\varepsilon\,\prime}_{cyl}(X)$ of all cylindrical coforms M on X is isomorphic with $\mathcal{F}^{\varepsilon}(X')$. This isomorphism $M \to \hat{M}$ is called the Laplace transform:

(3.3) $$\langle M,u\rangle = \langle u,M\rangle = \langle u,\hat{M}\rangle.$$

A normal space of ε-symmetric forms on X is a l.c.h.s. $F(X)$ of ε-symmetric forms on X (in this case we suppose $K = R$ or C) such that the injection $T^{\varepsilon}(X') \to F(X)$ is continuous with dense range. Hence by transposition $F(X)' \subset \mathcal{F}^{\varepsilon\,\prime}_{cyl}(X)$. Hence $\mathcal{F}^{\varepsilon\,\prime}_{cyl}(X)$ appears as a very big container of spaces $F(X)'$ of coforms.

(3.4) The image by the transpose of a linear map

Let X' and Y' be vector spaces, and let $\ell: Y' \to X'$ be a linear map. The transpose of $L = T^{\varepsilon}(\ell): T^{\varepsilon}(Y') \to T^{\varepsilon}(X')$ is a linear map $M \to \ell'(M): \mathcal{F}^{\varepsilon\,\prime}_{cyl}(X) \to \mathcal{F}^{\varepsilon\,\prime}_{cyl}(Y)$. This map is defined for arbitrary X and Y in duality with X' and Y'. Moreover

(3.5) $$\ell'(M)(\eta) = \hat{M}(\ell\eta).$$

(3.6) Projective representation of a cylindrical coform.

Suppose $X'_{\alpha} \Subset X'$. By application of (3.4) to the canonical injection i_{α} of X'_{α} in X', the image by $s_{\alpha}: X \to X_{\alpha}$ of a cylindrical coform M on X, is a coform $M_{\alpha} = s_{\alpha}(M)$ on $X_{\alpha} = X/X'^{\perp}_{\alpha}$. Moreover, the Laplace transform of M_{α} is the restriction of $\hat{M}$ to X'_{α}.

(3.7) Left integral - Right integrals

Let Y and Y' be two vector spaces in duality; X denotes a third vector space. Let $f \in \mathcal{F}^{\varepsilon}(X\times Y)$ and $M \in \mathcal{F}^{\varepsilon\,\prime}_{cyl}(Y)$ be such that the right contraction $\langle f(x,y), \hat{M}(\eta)\rangle$ is defined. We set

(3.8) $$\int f(x,y)\, M(y) = \langle f(x,y), \hat{M}(\eta)\rangle.$$

For example, this gives for arbitrary vector spaces X and X' in duality, for any cylindrical coform M on X:

(3.10) $$\int M(x)e^{x\cdot\xi} = \int e^{\xi\cdot x} M(x) = \hat{M}(\xi).$$

Using (2.34) and (2.36) we obtain the following proposition.

(3.11) <u>Proposition</u>. Let X and X' be vector spaces in duality, $\epsilon = \pm$. Let $\hat{Q}$ be a linear operator $T^{\epsilon}(X') \to \mathfrak{Z}^{\epsilon}(X)$. If $\delta\tilde{Q}$ denotes the coform on $X' \times X$ with Laplace transform $\tilde{Q}$, for arbitrary $u \in T^{\epsilon}(X')$

(3.12) $$(\hat{Q}u)(x) = \int \delta\tilde{Q}(\xi', x'') \, u(x'')e^{\xi'\cdot x} .$$

In particular, if $\hat{Q}$ is a left convolution operator

(3.13) $$(\hat{Q}u)(x) = \int \delta Q(x') \, u(x'+x)$$

where δQ denotes the coform on X s.t. $\widehat{\delta Q} = Q$.

The formula (3.12) is a dimension free version of (0.5). Now we consider only the case $\epsilon = -$; hence the index ϵ will be sometimes suppressed. We consider an algebraic complex triplet

(3.14) $$A = (G \subset Z \subset {}^*G)$$

with a conjugation c: G is a dense subspace of the complex separable Hilbert space Z, G is invariant by a conjugation c of Z, ${}^*G$ is the algebraic semi dual of G. The semilinear map $T^-(Z) \to T^-(Z)$ defined by c is denoted C. The antispace $\bar{Z}$ of Z is defined by the additive group of Z and the following product by scalars: $(\lambda;z)$ λz. Hence the identity map of Z defines a semi linear bijection $Z \to \bar{Z}$ denoted z z^*. With this last convention the element of $\bar{Z}$ denoted before λz is written now $(\lambda z)^* = \lambda z^*$. The vector space $X = \{z \in Z, \ z = c(z)\}$ is a real hilbertian space. We suppose that $E = \{z \in G; \ z = c(z)\}$ is dense in X. Hence the algebraic triplet A may be considered as a complexification of the real algebraic triplet $E \subset X \subset E^*$. We need a semi linear variant of the definition of coforms. The scalar product of Z induces a semi duality on $G \times G$, hence a semi duality on $T^-(G) \times T^-(G)$. This defines an imbedding $T^-(G) \simeq \mathcal{F}_{cyl}(\bar{G}) \subset \mathcal{F}(\bar{G})$. The space $'\mathcal{F}_{cyl}(\bar{G})$ of cylindrical semi coforms M on $\bar{G}$, in duality with G, is the semi dual of $T^-(G)$. We write in the following way the action of M on $u = u(q^*) \in \mathcal{F}_{cyl}(\bar{G})$:

$$(u,M) = \int \overline{u(q^*)}\, M(q^*). \tag{3.15}$$

The Laplace transform maps $'\mathcal{F}_{cyl}(\bar{G})$ on $\mathcal{F}(\bar{G})$

$$\hat{M}(z^*) = \int e^{\overline{(q,z)}}\, M(q^*). \tag{3.16}$$

For example, let P be the real coform on X with Laplace transform $\exp \|\xi\|^2/2$ on X: P is the canonical gaussian cylindrical measure on X. By semi linear extension, P defines a complex semi coform on $\bar{G}$ with Laplace transform $\exp(z,c(z))/2$ on $\bar{G}$. Any $t \in T^-(G)$ defines the semi coform tP on $\bar{Z}$ s.t.

$$\forall\, u \in T^-(G) \qquad (u,tP) = (uC(t),P). \tag{3.17}$$

On the space $\mathcal{F}_{cyl}(\bar{G})P$ of these coforms, the following scalar product is defined

$$(tP,\ t'P) = (tC(t'),P). \tag{3.18}$$

The completion $L^2(X)P$ of this prehilbertian space is independent from G, but depends only from Z or X. Hence we have the following algebraic triplet of coforms

$$CF(A) = (\mathcal{F}_{cyl}(\bar{G})P \subset L^2(X)P \subset {}'\mathcal{F}_{cyl}(\bar{G})). \tag{3.19}$$

In the same way $\mathcal{F}_{cyl}(E)P$ denotes the space of real coforms tP on E, $t \in T^-(E)$. Let $L^2_r(X)P$ be the completion of $\mathcal{F}_{cyl}(E)P$ for the scalar product $\langle tP,t'P\rangle = \langle tt',P\rangle$. The triplet $CF(A)$ is a complexification of the following real triplet

$$CF_r(A) = (\mathcal{F}_{cyl}(E)P \subset L^2_r(X)P \subset \mathcal{F}'_{cyl}(E)). \tag{3.20}$$

(3.21) **Remark.** Using the Gram Schmidt procedure, for arbitrary finite subset $\{t,u\ldots\} \subset T^-(G)$, there exists an orthonormal basis (q_j) of X, $(q_j) \subset E$, and an integer n such that $\{t,u,\ldots\} \subset \subset T^-(Z_n)$, where Z_n denotes the complex subspace of Z generated by $q_1,q_2,\ldots,q_n$. Let P_n be the canonical gaussian measure on the real space X_n generated by $q_1\ldots q_n$. Let d be the cardinal of the basis (q_j). For arbitrary finite $n \leq d$, we have a triplet

$$(3.22) \qquad \mathfrak{F}_{cyl}(\boldsymbol{\Sigma}_n)P_n \subset L^2(X_n)P_n \subset {}'\mathfrak{F}_{cyl}(\bar{Z}_n).$$

We have $\mathfrak{F}_{cyl}(\bar{Z}_n)P_n \subset \mathfrak{F}_{cyl}(\bar{Z}_{n+1})P_{n+1}$, $L^2(X_n)P_n$ is isometrically imbedded in $L^2(X_{n+1})P_{n+1}$. The Hilbert space $L^2(X)P$ is the completion of $\varinjlim L^2(X_n)P_n$. Hence any $fP \in L^2(X)P$ is represented by a coherent family $(f_nP_n)_n \in \Pi\, L^2(X_n)P_n$ such that $\|fP\| = \sup_n \|f_nP_n\| < \infty$. For $\alpha \in J(-,d)$ the Hermite polynomial on X

$$(3.23) \qquad H_\alpha(q) = (-1)^{|\alpha|}(\exp \|q\|^2/2)\, \partial^\alpha(\exp -\|q\|^2/2)$$

defines one element of $T^-(E) \subset T^-(G)$. The set $\{\alpha!^{-1/2} H_\alpha P,\ \alpha \in \in J(-,d)\}$ is an orthonormal basis of the real Hilbert space $L^2_r(X)P$, ad of the complex Hilbert space $L^2(X)P$.

(3.24) <u>Triplets of vectorial semi coforms</u>

For arbitrary $j = 0,1,\ldots,$ the closure of $T^-(Z)$ in ${}_H\hat{\otimes}_j Z$ is denoted Z_j. The adjoint of the canonical injection $T_j^-(G) \subset Z_j$ identifies $f_j \in Z_j$ with the following j-homogeneous symmetric form on $\bar{G}$: $(z_1^*,\ldots,z_j^*) \mapsto (z_1z_2\ldots z_j, f_j)$. Hence we have the complex triplet

$$(3.25) \qquad A_j = (T_j^-(G) \subset Z_j \subset \mathfrak{F}_j(\bar{G})).$$

By "tensor product" of the complex triplets $CF(A)$ and A_j, we obtain the following algebraic triplet

$$(3.26) \qquad CF(A,j) = (T^-(G)\otimes T_j^-(G) \subset L^2(X)P\ {}_H\hat{\otimes}\ Z_j \subset {}^*(T^-(G) \otimes T_j^-(G))).$$

In general, if E and F are two complex spaces, $\mathfrak{L}_s(E,{}^*F) = = \mathfrak{L}(\bar{E},{}^*F)$ denotes the space of all semi linear maps $E \to {}^*F$. In particular, the space $'\mathfrak{F}_{cyl}(\bar{G},j)$ of all semi coforms on $\bar{G}$ with values in $\mathfrak{F}_j(\bar{G})$ is by definition

$$\mathcal{L}_s(T^-(G), \mathfrak{F}_j(\bar{G})) \simeq {}^*(T^-(G) \otimes T_j(G)) \simeq \mathcal{L}_s(T_j^-(G), '\mathfrak{F}_{cyl}(\bar{G})). \tag{3.27}$$

For $t \otimes t'_j \in T^-(G) \otimes T_j^-(G)$ the vectorial semi coform $(t\otimes t'_j)P$ is s.t.

$$\forall\, u \in T^-(G) \qquad (u, (t\otimes t'_j)P) = (uC(t),P)t'_j. \tag{3.28}$$

The space $\mathfrak{F}_{cyl}(\bar{G},j)P$ of these semi coforms has a natural scalar product. The completion $L^2(X,j)P$ of this space is a space of vectorial semi coforms. Finally, the triplet $CF(A,j)$ appears as a triplet of spaces of vectorial semi coforms

$$CF(A,j) \simeq (\mathfrak{F}_{cyl}(\bar{G},j)P \subset L^2(X,j)P \subset '\mathfrak{F}_{cyl}(\bar{G},j)).$$

(3.29) <u>Directional derivative. Directional absolute derivative.</u>

a) Let $t \in T_j^-(G)$. For a semi coform $uP \in \mathfrak{F}_{cyl}(\bar{G})P$ on $\bar{G}$, the directional derivative $\partial_{C(t)}(uP)$ is defined by derivation of the density u with respect to $C(t)$:

$$\partial_{C(t)}(uP) = (\partial_{C(t)}u)P. \tag{3.30}$$

b) For arbitrary $M \in '\mathfrak{F}_{cyl}(\bar{G})$, the directional absolute derivative $\tilde{\partial}_{C(t)}M \in '\mathfrak{F}_{cyl}(\bar{G})$ is defined by adjunction of the directional derivation defined in point a):

$$\forall\, t' \in T^-(\bar{G}) \qquad (t'P, \tilde{\partial}_{C(t)}M) = (-1)^j(\partial_{C(t)}(t'P),M). \tag{3.31}$$

With the notations (3.21), we set for $\alpha \in J^j(-,d)$

$$\partial^{\alpha}(uP) = \partial_{q^*\alpha}(uP); \qquad \tilde{\partial}^{\alpha}(M) = \tilde{\partial}_{q^*\alpha}(M). \tag{3.32}$$

c) The linear map $M \to \tilde{\partial}_{C(t)}M$ of $'\mathfrak{F}_{cyl}(\bar{G})P$. In fact for arbitrary u and $t' \in T^-(G)$, $\alpha \in J(-,d)$ we choose $n \le d$ such that $\alpha \in J(-,n)$; $\{u;t'\} \subset T^-(Z_n)$. Hence by integration by parts

$$(t'P, \tilde{\partial}^{\alpha}(uP)) = (-1)^j \int \overline{(\partial^{\alpha} t')(q)(Cu)(q)}\ P(q)$$

$$= (-1)^j\ (2\pi)^{-n/2} \int_{X_n} \overline{(\partial^{\alpha} t')(q)}\ u(q)\ e^{-\|q\|^2/2}\ dq$$

$$= (2\pi)^{-n/2} \int \overline{t'(q)}\ \partial^{\alpha}(u\ e^{-\|q\|^2/2})\ dq.$$

Hence by Leibniz formula and (3.23):

(3.33) $$\tilde{\partial}^{\alpha}(uP) = \sum_{\beta \le \alpha} \binom{\alpha}{\beta} (-1)^{|\beta|}\ (\partial^{\alpha-\beta}\ u) H_{\beta} P.$$

d) In view of b) and c), the adjoint L^* of L is a linear operator of $'\mathfrak{F}_{cyl}(\bar{G})$ extending the directional derivation $uP \to \partial_{C(t)}(uP)$. We set $L^*(M) = \partial_{C(t)} M$.

(3.34) <u>Derivation of order</u> j; <u>absolute derivation of order</u> j.

For any semi coform M on $\bar{G}$, for any integer $j \geq 0$, the absolute derivative $\tilde{D}^j M$ (resp. the derivative of order j $D^j M$) is defined by the semi linear map $T_j^-(G) \ni t \mapsto \tilde{\partial}_{C(t)} M$ (resp. $\mapsto \partial_{C(t)} M$). Hence:

$$D^j\ M \quad \text{and} \quad \tilde{D}^j\ M \in {}'\mathfrak{F}_{cyl}(\bar{G}, j).$$

For example, if G is algebraically generated by the orthonormal basis $(x_j)_j$ of Z

(3.35) $$D^j\ M = j! \sum_{|\alpha|=j} \alpha!^{-1}\ (\partial^{\alpha}\ M) \otimes x^{\alpha}$$

(3.36) $$\tilde{D}^j\ M = j!\ \Sigma\ \alpha!^{-1}\ (\tilde{\partial}^{\alpha}\ M) \otimes x^{\alpha}.$$

These sums converge in $'\mathfrak{F}_{cyl}(\bar{G}, j)$ for the weak topology. In view of (3.33), $\tilde{\partial}^{\alpha}\ P = (-1)^{|\alpha|}\ H_{\alpha} P$. Hence (3.36) gives

(3.37) $$\tilde{D}^j\ P = j!\ (-1)^j \sum_{|\alpha|=j} \alpha!^{-1}\ H_{\alpha}(q) \otimes x^{\alpha}.$$

Hence for any j, the absolute derivative of order j of the cylindrical gaussian measure P defines a linear continuous map $L^2(X)P \to Z_j$.

4. THE NORMALIZED LAPLACE TRANSFORM

We consider the complex triplet A. The holomorphic picture of the symmetric Fock space of Z is

(4.1) $$FH(\bar{Z}) = \{f = \Sigma f_\ell,\ f_\ell \in Z_\ell;\ \|f\|^2 = \Sigma\ \ell!\ \|f_\ell\|^2 < \infty\}.$$

Using the weight $\ell!$, the "sum" of the algebraic triplets A_ℓ, $\ell = 0,1,\ldots$ is the following complex triplet of symmetric forms:

(4.2) $$F(A) = (T^-(G) \subset FH(\bar{Z}) \subset \mathfrak{F}(\bar{G})).$$

The spaces $X' = G$ and $X = \bar{G}$ are is duality; this imbedding $T^-(G) \subset \mathfrak{F}^-(\bar{G})$ is similar to the imbedding $T^{\mathbb{C}}(X') \subset \mathfrak{F}^{\mathbb{C}}(X)$ used in the paragraph 1. Hence the symbolic calculus can be applied to the space $Op = \mathcal{L}(T^-(G), \mathfrak{F}(\bar{G}))$ endowed with his weak topology. Hence we have two homeomorphisms:

(4.3) $$Op \to \mathfrak{F}^-(\bar{G}\times G) \to \mathfrak{F}^-(\bar{G},G)$$

$$\hat{Q} \mapsto \tilde{Q}(\bar{z},z') \mapsto Q(\bar{z},z') = \tilde{Q}\exp -(z,z').$$

The normalized Laplace transform θ gives an isomorphism of the triplet $F(A)$ of spaces of forms, on the triplet $CF(A)$ of spaces of coforms. Hence θ extends the symbolic calculus to the triplet $CF(A)$.

(4.4) <u>Theorem</u>. For any complex triplet A with conjugation, the map

(4.5) $$M \xrightarrow{\theta} (\theta M)(z^*) = \hat{M}(z^*)\exp -(z,cz)/2$$

realizes one isomorphism of the triplet $CF(A)$ on the triplet $F(A)$. Moreover if $\hat{W}$ denotes the convolution operator in $T^-(G)$ with symbol $\exp (cz,z)/2 = W(z)$ on G:

(4.6) $$\forall\, u \in T^-(G) \qquad \theta(uP) = \hat{W}(u).$$

<u>Proof</u>. The map θ is the product of two linear maps: a) the Laplace transform $M \to \hat{M}$ is a bijection $'\mathfrak{F}_{cyl}(\bar{G}) \to \mathfrak{F}(\bar{G})$; b) the product by $\exp(-(z,cz)/2)$ is a bijection of $\mathfrak{F}(\bar{G})$. Hence θ defines a

bijection $'\mathfrak{F}_{cyl}(\bar{G}) \to \mathfrak{F}(\bar{G})$. The operator $\hat{W}$ is the complexification of the convolution operator $\hat{V}$ in $T^-(E)$ with symbol $V(q) = \exp(\|q\|^\varepsilon/2)$. In view of (3.21) for any $u \in T^-(G)$, we choose an orthonormal basis (q_j) of X, $(q_j) \subset E$, such that $u \in T^-(Z_n)$ for some finite $n \leq d$. An integration by parts gives $\theta(H_\alpha P) = \bar{z}^\alpha$ for arbitrary $\alpha \in J(-,d)$. Hence θ induces a bijective isometry $L^2(X)P \to FH(\bar{Z})$. By linearity, θ maps $T^-(G)$ on $T^-(G)$. Finally we have to show $\theta(H_\alpha P) = \bar{z}^\alpha = \hat{V}(H_\alpha)$ for arbitrary $\alpha \in J(-,d)$. The Taylor's formula gives

(4.7) $$\Sigma_\alpha \alpha!^{-1} H_\alpha(q)y^\alpha = \exp(-\|y\|^2/2 - \langle y,q\rangle).$$

But

$$\langle e^{-\|\xi'\|^2/2}, \Sigma_\alpha \alpha!^{-1}(x+x')^\alpha y^\alpha\rangle$$

$$= \langle e^{-\|\xi'\|^2/2}, e^{\langle x',y\rangle} e^{\langle x,y\rangle}\rangle = \exp(-\|y\|^2/2 - \langle x,y\rangle).$$

(4.8) Hence $H_\alpha(x) = \langle e^{-\|\xi'\|^2/2}, (x+x')^\alpha\rangle$. In view of (2.36), $H_\alpha = \hat{V}^{-1}(x^\alpha)$.

(4.9) <u>Corollary</u>. By tensor product of the algebraic triplets $F(A)$ and A_j, we obtain the following triplet of vectorial forms

(4.10) $$F(A,j) = (T^-(G,j) \subset FH(\bar{Z},j) \subset \mathfrak{F}(\bar{G},j) \text{ with } T^-(G,j) = T^-(G) \otimes T_j^-(G),\quad \mathfrak{F}(\bar{G},j) = \mathfrak{F}(\bar{G},\mathfrak{F}_j(\bar{G})),\quad FH(\bar{Z},j) = FH(\bar{Z})_H\hat{\otimes}Z_j.$$

Hence the tensor product θ_j of θ with the identity of A_j gives an isomorphism $CF(A,j) \to F(A,j)$ of algebraic triplets.

(4.11) <u>The operators</u> D^j <u>and</u> D^{*j}.

a) A semi linear map $*: \mathfrak{F}(\bar{G},\mathfrak{F}_j(\bar{G})) \to \mathfrak{F}(G,\mathfrak{F}_j(G))$ is defined such that for any ℓ, $f_\ell \in \mathfrak{F}_\ell(\bar{G},F_j(\bar{G}))$:

(4.12) $$f_\ell^*(z_1,\ldots,z_\ell;\, y_1,\ldots,y_j) = \overline{f_\ell(z_1^*,\ldots,z_\ell^*;\, y_1^*,\ldots,y_j^*)}.$$

This map induces a semi linear isomorphism $T^-(G) \otimes T_j^-(G) \to T^-(\bar{G}) \otimes T_j^-(\bar{G})$ such that $\forall \ell$, $\forall\, t = (z_1\ldots z_\ell) \otimes y_1\ldots y_\ell$, $t^* = (z_1^*\ldots z_\ell^*) \otimes (y_1^*\ldots y_j^*)$. The interest of these star maps is the following relation connecting the semi duality of $F(A,j)$ and the dualities used in §1 and 2; for $t \in T^-(G) \otimes T_j^-(G)$; $f \in \mathfrak{F}(\bar{G},j)$

(4.13) $$(t,f) = \langle t^*,f\rangle = \overline{\langle t,f^*\rangle}.$$

b) For example in view of (1.34), the operators D^j: $T^-(G) \to T^-(G) \otimes T_j(G)$ and

(4.14) $$T^-(G) \otimes T_j^-(G) \xrightarrow{D^{*j}} T^-(G)$$
$$(z_1 \ldots z_\ell) \otimes (y_1 \ldots y_j) \longmapsto z_1 \ldots z_\ell \; y_1 \ldots y_j$$

are connected in the following way. For $u \in T^-(G)$, $f \in T^-(G) \otimes T_j^-(G)$

(4.15) $$(u, D^{*j}f) = (D^j u, f).$$

Hence, there exists linear extensions:

$$\mathfrak{F}(\bar{G}) \xrightarrow{D^j} \mathfrak{F}(\bar{G},j); \quad \mathfrak{F}(\bar{G},j) \xrightarrow{D^{*j}} \mathfrak{F}(\bar{G}).$$

c) For arbitrary semi coform $M \in {}'\mathfrak{F}_{cyl}(\bar{G})$

(4.16) $$\theta_j(D^j M) = D^j(\theta M).$$

In fact, this is true for $M \in \mathfrak{F}_{cyl}(\bar{G})$: use the properties of Hermite polynomials. Hence (4.16) holds, by weakly continuous extensions of the two maps D^j.

d) In the L^2 picture, the map D^{*j} is defined as the adjoint of the map D^j.

e) Let $a_{k\ell} \in \mathfrak{L}(\mathfrak{F}_\ell^-(G), \mathfrak{F}_k(\bar{G}))$. We denote $D^{*k} a_{k\ell} D^\ell$ the element $\in$ Op, obtained in the holomorphic picture by composition of three operators

(4.17) $$T^-(G) \xrightarrow{D^\ell} T^-(G,\ell) \xrightarrow{Id\otimes a_{k\ell}} \mathfrak{F}(\bar{G},k) \xrightarrow{D^{*k}} \mathfrak{F}(\bar{G}).$$

The Theorem (2.27) gives the structure of all linear operators $\in$ Op:

(4.18) Theorem. Any $\hat{Q} \in$ Op can be written in a unique way

(4.19) $$\hat{Q} = Q(D^*,D) = \sum_{k,\ell} D^{*k} a_{k\ell} D^\ell$$

for some $a_{k,\ell}$. The formal series defining the symbol of $\hat{Q}$ is:

(4.20) $$Q(z^*,z') = \sum_{k,\ell} (\otimes_k z, a_{k\ell}(\otimes_\ell z')).$$

If the sum is finite, $\hat{Q}$ is called a differential operator. In view of (0.2), (0.3), this theorem shows a formal analogy with the usual theory of differential operators. A priori, it seems that a natural class of differential operators $Q(D^*,D)$ are the operators corresponding to linear maps of Hilbert Schmidt type $a_{k,\ell}: Z_\ell \to Z_k$. In fact physics use differential operatos corresponding to unbounded operators $a_{k\ell}: Z_\ell \to Z_k$. The general theory of these differential operators seems very difficult.

(4.21) <u>Coherent states</u> $\exp z$

For any $z \in Z$, $\exp z$ denotes the element $\Sigma\, k!^{-1} \otimes_k z$ of $FH(\bar{Z})$. In view of (2.26)

(4.22) $$\theta^{-1}(\exp z) = P \exp((q\cdot z) - (c(z),z)/2).$$

For arbitrary $\hat{Q} \in Op$:

(4.23) $$\tilde{Q}(z,z') = \langle e^{\bar{z}u}, \tilde{Q}(u^*,z')\rangle = (e^z, \hat{Q}(e^{z'})).$$

The corresponding operator in L^2 picture is $\hat{Q}_q = \theta^{-1}\hat{Q}\theta$. Hence

$$\tilde{Q}(z^*,z') = (e^z, \theta\hat{Q}_q\theta^{-1}(e^{z'})) = (\theta^{-1} e^z, \hat{Q}_q \theta^{-1} e^{z'})$$
$$= (P e^{z-(cz,z)/2}, \hat{Q}_q e^{z'-(cz',z')/2})$$
$$= [\exp(-(z,cz)/2 - (cz',z')/2)](P e^z, \hat{Q}_q(P e^{z'})).$$

(4.24) <u>Proposition</u>. The symbol of any $\hat{Q} \in Op$ is expressed in the following way, in terms of $\hat{Q}_q = \theta\hat{Q}\theta^{-1}$:

(4.25) $$Q(z^*,z') = (Pe^z, \hat{Q}_q(Pe^{z'}))\exp-(cz'+z, c(cz'+z))/2.$$

For example, let $\hat{Q}_q: uP \to Mu$ be the operator of product with a given semi coform $M \in {'\mathfrak{F}}_{cyl}(\bar{G})$. Then

$$(Pe^{z}, \hat{Q}_q(Pe^{z'})) = (Pe^{z}, Me^{z'})$$

$$= (\exp(z+cz'), M) = \hat{M}((cz'+z)^*).$$

Hence $Q(z^*,z') = (\theta M)((cz'+z)^*)$. We have the following proposition

(4.26) Proposition. Let M be a given semi coform. Then the symbol of the operator of product by M in the L^2 picture is $(\theta M)((cz'+z)^*)$.

We recall that the Wick product: $\hat{Q}_1 \ldots \hat{Q}_n$: of n operators $\hat{Q}_i \in Op$, is the operator with symbol $Q_1 Q_2 \ldots Q_n$

(4.27) Application of (4.26)

For any integer n, any unit vector $e \in X$ $H_n(e)$ denotes the cylindrical function $q \to H_n(\langle e,q \rangle)$ on X. We have $\theta(H_n(e)P) = (z,e)^n$. Hence the operator $\hat{Q}_n$ of product with $H_n(e)$ has symbol $(z+cz', e)^n$. Hence

(4.28) $$\hat{Q}_n =: (\hat{Q}_1)^n :$$

5. APPLICATION TO THE SCALAR BOSON FIELD

Let M be a Minkowski space of dimension $s+1$, $s = 1, 2$ or 3. Choosing an origin, the affine space M is identified with his tangent space. This vector space is endowed with a quadratic form x^2 of type $+-\ldots$; hence M is identified with his dual. Using coordinates in M with respect to a basis (e_j) s.t. $e_j e_k = \epsilon\, \delta_{jk}$ ($\epsilon = 1$ for $j = 1$, $\epsilon = -1$ for $j > 1$), we use the standart notations $x = (x_o, \vec{x}) = (t, \vec{x})$, $k = (k_o, \vec{k})$, $kx = k_o x_o - \vec{k}\vec{x}$. The Fourier transform of a function $f(x)$ defined on M is denoted simply

$$f(k) = (2\pi)^{-(s+1)/2} \int f(x) e^{ikx}\, dx.$$

The constant $m > 0$ is fixed. A function $f(x)$ satisfies the Klein Gordon equation $(\Box - m^2)f = 0$ iff $f(k)(m^2-k^2) = 0$. Setting $\omega(\vec{k}) = (m^2+\vec{k}^2)^{1/2}$, the positive hyperboloid of mass m

$$H\uparrow = \{k = (k_o,\vec{k}) \in M, \quad k_o = \omega(\vec{k})\}$$

has a natural chart ch: $\mathbb{R}^s \to H\uparrow$. This hyperboloid is endowed with his invariant measure $\mu = ch(\underline{\mu})$, with $\underline{\mu} = (2\omega(\vec{k}))^{-1}\, d\vec{k}$, $\mathcal{D}(H\uparrow)\mu$ is the space of complex measures m on $H\uparrow$ s.t. $ch^{-1}(m) = \varphi\mu$, with $\varphi \in \mathcal{D}(\mathbb{R}^s)$. The results of paragraph 4 are applied to the complex triplet $A = (G \subset Z \subset {}^*G)$ with $G = \mathcal{D}(H\uparrow)\mu$; $Z = L^2(H\uparrow)\mu$; and the conjugation $z \mapsto cz$ such that $\forall\, k \in \mathbb{R}^s$ $(cz)(ch\,\vec{k}) = z(\overrightarrow{-ch\,\vec{k}})$. For any $k \in H$, $a(k)$ (resp. $a^*(k)$) denotes the operator with symbol $(z^*;z') \to z'(k)$ (resp. $\overline{z(k)}$). In view of §2, $a(k)$ (resp. $a^*(k)$) is in the holomorphic picture a derivation operator (resp. the product with the linear form $z^* \to \overline{z(k)}$ on $\bar{Z}$). The product $a^*(k)\, a(k)$ is defined in Op: see (2.21). The maps $k \to a(k)$ (resp. $a^*(k)$, resp. $a^*(k)\, a(k)$) are continuous functions on $H\uparrow$, with values in the complete space Op. Hence the formulas (0.7) and (0.8) define $\Phi(x)$ and H_o as Dunford integrals. The symbol of $\Phi(x)$ is the following linear form on $\bar{G}\times G$

$$(5.1) \qquad \Phi(x;z^*,z') = (2\pi)^{-s/2} \int_H (\overline{z(k)}e^{ikx} + z'(k)e^{-ikx})\, d\mu(k).$$

The map $x \to \text{Symb}\,\Phi(x)$ is an entire function $M^c \to \mathfrak{F}(\bar{G}\times G)$. In view of paragraph 2, the free field $x \to \Phi(x)$ defines an entire function $M^c \to$ Op. The symbol of H_o is

$$(5.2) \qquad H(z^*,z') = 2^{-1} \int \overline{z(k)}z'(k)dk = (z,\omega z')$$

where ω denotes the operator of product with $\omega\circ ch^{-1}$, in G. In view of (4.18), H is the differential operator $D^*\omega\, D$. We denote $\exp it\,\omega$ the operator of product by $\exp(it(\omega\circ ch^{-1}))$ in G, and his natural extension in ${}^*G$. A direct application of (2.20 b and c) gives the following equality in Op:

$$(5.3) \qquad (\exp it\, H_o)\, \Phi(0,\vec{x})\, (\exp -it\, H_o) = \Phi(t,\vec{x})$$

with $\exp it\, H_o = F(\exp it\,\omega)$. The space $M_E = \{(it,\vec{x} \in M^c,\ t \in \mathbb{R};$

$\vec{x} \in \mathbb{R}^s\}$ is euclidean. The restriction of Φ to this space satisfies

$$(5.4) \qquad (\exp -t\, H_o)\, \Phi(0,\vec{x})(\exp t\, H_o) = \Phi(it,\vec{x}).$$

With the notation of [2], we have for $t \geq 0$, z and $z' \in Z$

$$(5.5) \qquad (j_t z,\, j_o z') = (z,\, (\exp -t\, \omega)z').$$

Hence a direct application of (2.20) gives $(\exp - t\, H_o) = J_t^* J_o$. This is the starting point of Nelson's euclidean formalism for scalar bosons. Finally we define the unformal operator (0.8).

The Fourier transform of a temperate distribution on R^s is defined by transposition, using the convention

$$(5.6) \qquad \forall \varphi \in \mathcal{S}(\mathbb{R}^s) \qquad \varphi(\vec{p}) = \int \varphi(\vec{x})\, e^{-i\vec{p}\vec{x}}\, d\vec{x}\ .$$

Then $H_g \in \mathrm{Op}$ is defined by his symbol:

$$(5.7) \qquad H_g(z^*,z') = \int_{\mathbb{R}^s} \Phi(0,\vec{x};z^*,z')^n\, g(\vec{x})\, d\vec{x}\ .$$

Let P be the following polynomial, homogeneous of degree n on $\bar{G}$:

$$(5.8) \quad P(z^*) = (2\pi)\int_{\mathbb{R}^{sn}} (\mathfrak{F}g)(-\sum_1^n p_\ell)(cz)(\vec{p}_1)\ldots(cz)(\vec{p}_n)d\underline{\mu}(p_1)\ldots d\underline{\mu}(p_n)$$

with the convention $(cz)(\vec{p}_\ell) = (cz)\circ(ch)(\vec{p}_\ell)$.

(5.9) <u>Proposition</u>. Let M be the semi coform on $\bar{G}$ with normalized Laplace transform $\theta(M) = P$. Then $\hat{H}_g \in \mathrm{Op}$ is represented in the L^2 picture by the operator of product with M.

<u>Principle of the proof</u>: substitue in (5.7) the expression of $\Phi(0,\vec{x};z^*,z')$ given by (5.1); apply (4.26).

(5.10) <u>Corollary</u>. A sufficient condtion in order that $\hat{H}_g$ is represented in the L^2 picture by the operator of product with some element $\in L^2(X)P$, is that $P \in Z_n$, i.e.

$$(5.11) \qquad \int |\mathfrak{F}_g(\sum_1^n p_\ell)|^2\, d\underline{\mu}(p_1)\ \ldots\ d\underline{\mu}(p_n) < \infty.$$

This is one result of [1]. In the present formalism, this condition is necessary.

(5.12) Remarks.

a) See [4] for topological and nuclear counterparts of the results of this paper.

b) §4 and §5 can be extended for $\varepsilon = +$, i.e. for fermions (to appear).

REFERENCES

[1] Glimm, J. Boson fields with non linear self interaction in two dimensions. Comm. Math. Phys. 8 (1968) 12-25.

[2] Guerra, F., Rosen, L., Simon B., Boundary conditions for the $P(\Phi)_2$ Euclidean quantum field theory, Ann. of Math. 101 (1975) 111-259.

[3] Krée, P. Solutions faibles d'equations aux dérivées fonctionnelles, Seminaire P. Lelong I (Lecture Notes nº 410 p. 142-181) et II (Lecture Notes nº 474).

[4] Krée, P., Seminaire (3e année 1976-1977) sur les e.d.p. en dimension infinie. Secrétariat mathematique de l'Institut H. Poincaré - Paris.

SOLUTIONS FAIBLES ET SOLUTIONS FORTES DU PROBLÈME $\bar{\partial}u = f$ OÙ f EST UNE FONCTION À CROISSANCE POLYNOMIALE SUR UN ESPACE DE HILBERT

Bernard Lascar

Centre de Mathématiques - Ecole Polytechnique

91128 Palaiseau Cedex - France

Le problème de l'étude du système $\bar{\partial}u = f$ sur un espace de Hilbert B déjà fait l'objet d'un certain nombre de travaux, citons [4], [5] et [9]. On retiendra que le phénomène observé est que d'une part on peut trouver des solutions au sens faible (distribution) sur tout l'espace [9] et [5], et d'autre part ou peut trouver une solution au sens fort (dérivable au sens de Fréchet) définie seulement sur un sous espace (l'image de l'espace hilbertien un opérateur de Hilbert Schmidt) Z.

Précisons tout d'abord la notion de solution faible. Une fonction $u \in L^2_\nu(B)$ est une solution faible si on a:

$$\int f\cdot\phi \, d\nu(x) = \int u \, \delta\phi \, d\nu(x)$$

pour un espace de fonction ϕ cylindrique, $\delta\phi$ désignant le transposé formel de $\bar{\partial}$ relativement à la mesure gaussienne.

Par un contre exemple, [2] montre qu'on ne peut espérer pouvoir toujours trouver une solution qui soit continue sur B (ni même localement bornée). Il se pose donc le problème d'étudier comment on peut "prolonger" les solutions fortes définies sur l'espace Z, espace reproduisant de la mesure gaussienne ν sur B, en des solutions faibles aussi régulières que possible.

On fait remarquer que prolonger une fonction définie sur l'espace Z en une fonction de $L^2_\nu(B)$ n'a pas de sens en général car Z est de mesure nulle. On connaît pourtant de nombreux exemples de fonctions f définies sur Z qui ne sont pas continues pour la topologie induite par B et pour lesquelles les fonctions $f \circ s_n(x)$ convergent en probabilité vers un "prolongement" $\tilde{f}$ de f (les s_n sont une suite croissante de projecteurs orthogonaux de rang fini).

C'est cette notion de prolongement que nous utiliserons. Nous partons de la solution exprimée à l'aide d'un noyau noté L_o calculé dans [4], qui donne une solution définie et régulière sur l'espace Z. En construisant d'autres noyaux notés L_u qui définissent moralement la même solution que L_o mais qui fournissent des solutions dans $L^2_{\nu_u}(B)$, on pourra obtenir les propriétés de prolongement.

Nous utiliserons alors essentiellement les théorèmes de convergence des martingales vectorielles pour obtenir la meilleure régularité locale pour la solution. Le résultat essentiel est exprimé dans le théorème à la fin de ce travail. Donnons quelques notations.

On utilise un triplet d'espaces de Hilbert complexes $B' \hookrightarrow Z \hookrightarrow B$ que l'on suppose obtenu par complexification d'un triplet réel. Les injections sont de Hilbert-Schmidt. Les normes sont notées $\| \|'$, $| |$, $\| \|$. On sait qu'on peut construire une base hilbertienne $(e_n)_{n\in\mathbb{N}}$ de Z formée de vecterus deux à deux orthogonaux pour le produit scalaire de B. On désignera ainsi par Z_n le sous espace engendré par $\{e_1,\ldots,e_n\}$ et par s_n la projection orthogonale (qui se prolonge $B \mapsto Z_n$). On désignera par ν_t la mesure gaussienne de transformée de Fourier de la fonction $(\exp - \frac{|k|^2}{t})$, k désigne la variable duale de la variable $z \in Z$. Le produit tensoriel hilbertien est noté $\hat{\otimes}$.

On note:

$$e_\alpha = \frac{\alpha!}{j!} \sum_{(i_1,\ldots,i_j)\in I_\alpha} e_{i_1} \otimes \ldots \otimes e_{i_j} = \mathrm{sym}(e_1^{\otimes\alpha_1} \otimes \ldots \otimes e_n^{\otimes\alpha_n} \ldots)$$

pour un multi indice α de longueur j. Si $x \in B$, $x^{\alpha} = (\underset{j}{\otimes} x, e_{\alpha})$.

On note:

$$P_{\alpha\beta}(x) = \sum_{\rho \leq \inf(\alpha,\beta)} \frac{\alpha!}{(\alpha-\rho)!} \frac{\beta!}{(\beta-\rho)!} \frac{(-1)^{|\rho|}}{\rho!} x^{\alpha-\rho} \bar{x}^{\beta-\rho}$$

un polynôme d'Hermite relatif à la mesure gaussienne complexe de variante 1/2. On a montré dans que les $H_{\alpha\beta}(x) = P_{\alpha\beta}(x)/(\alpha!\beta!)^{1/2}$ forment une base hilbertienne de $L^2_{\nu}(B)$. Si:

$$t \in \hat{\underset{k}{\odot}} \hat{Z} \hat{\underset{l}{\odot}} Z \qquad t = \sum_{|\alpha|=k, |\beta|=l} t_{\alpha\beta} \frac{t!l!}{\alpha!\beta!} e_{\alpha} \otimes e_{\beta}$$

on note:

$$P_{kl}(x)\cdot t = \sum_{\alpha,\beta} \frac{k!l!}{\alpha!\beta!} t_{\alpha\beta} P_{\alpha\beta}(x).$$

On désignera par $\mathcal{C}^m_1(B)$ l'espace des fonctions continues $B \mapsto B'$ bornées par $(1+\|x\|)^m$, $\mathcal{C}^m_o(B)$ les fonctions continues $B \mapsto C$ bornées par $(1+\|x\|)^m$.

Soient $t > 0$ et $u \in L^2_{\nu_t}(B)$, la "t dérivée de densité" de la distribution cylindrique sur B $u\nu_t$ est la forme anti-linéaire sur $O_{c\ cyl}(B) \otimes (B' \otimes B')$ définie par:

$$\phi = \phi' \otimes \phi'' \in O_{c\ cyl}(B) \otimes (B' \oplus B') \mapsto (Du,\phi) = -\int u\, \overline{\delta_t \phi}\, d\nu_t$$

où:

$$\delta_t \phi = 2 \sum_{i=1}^{q} \left(\frac{\partial}{\partial \bar{z}_i} - z_{i/t}\right)\phi'_i + \left(\frac{\partial}{\partial z_i} - \bar{z}_{i/t}\right)\phi''_i$$

qui est définie lorqu'on a posé:

$$\phi' = \sum_{i=1}^{q} \phi'_i e_i, \qquad \phi'' = \sum_{i=1}^{q} \phi''_i e_i$$

les e_i représentant un système orthonormé dans Z formé de vecteurs réels.

Le $\bar{\partial}$ au sens faible de la fonction $u \in L^2_{\nu_t}(B)$ est ainsi la forme linéaire sur $O_{c\ cyl}(B) \otimes B'$ obtenue par la restriction de la "t dérivée de densité" aux éléments $\phi = 0 \oplus \phi''$ où $\phi'' \in O_{c\ cyl}(B) \otimes B'$.

Si u est de la plus la classe d'une fonction dérivable au sens de Fréchet et dont la dérivée est dans $L^2_{\nu_t}(B, B' \oplus B')$, "$\bar{\partial}(u\nu_t)$" = $= (\bar{\partial}u)\nu_t$, où cette fois $\bar{\partial}u$ est la partie anti-linéaire de la dérivée de Fréchet de u.

Dans [4] C.J. Henrich a introduit un opérateur noté L_o de $\mathcal{C}_1^{m-1-\delta}(B)$ dans $\mathcal{C}_o^m(Z)$ (si $m \in \mathbb{N}$ et $\delta > 0$) tel que: $G \in \mathcal{C}_1^{m-1-\delta}(B)$, G de classe C^1, dG à croissance au plus exponentielle, entraînent $L_o(G) \in C^1(Z)$ et $\bar{\partial}(L_o(G)) = G|_Z$.

Ces conditions ne suffisent évidemment pas à définir uniquement l'opérateur L_o; aussi on mentionne que la solution trouvée par l'opérateur est "intuitivement" orthogonale aux polynômes holomorphes de degré plus petit ou égal à m pour le produit scalaire défini par la mesure:

$$q_m = \int_0^{+\infty} e^{-1/t}\, t^{-2-m}\, \nu_t \frac{dt}{m!}.$$

Nous allons montrer que l'on peut en fait "prolonger" la solution $L_o(G)$ en une solution faible, ce que est le résultat de notre théorème:

Théorème. Soient $m \in \mathbb{N}$ et $\delta > 0$. Soit $G \in \mathcal{C}_1^{m-1-\delta}(B)$, une fonction C^1 dont la dérivée dG croît moins vite qu'une exponentielle, $\bar{\partial}G = 0$. Il existe une fonction $LG(x)$ définie sur B ayant les propriétés suivantes:

a) $LG(x)$ est ν_t mesurable $t > 0$. Classe $_t(LG) = L_tG \in$ $\in \bigcap_{1 \leq p < \infty} L^p_{\nu_t}(B)$; L_tG est une solution au sens des distributions de l'équaiton $\bar{\partial}u = G$.

b) Il existe une partie $\Omega \subset B$, telle que $\nu_t(\Omega^c) = 0$ $t > 0$ et telle que:

- $0 \in \Omega$; $x \in \Omega \Rightarrow x + Z \subset \Omega$
- $x \in \Omega$; $h \in Z \mapsto (LG)(x+h)$ est de classe C^1 et vérifie $\bar{\partial}(LG)(x+h) = G(x+h)$
- $LG|_Z = L_oG$ (solution construite par C.J. Henrich [4]);

$\lim_{n \to +\infty} (L_oG)\, Z_n \circ s_n(x) = (LG)(x)$ si $x \in \Omega$; la convergence ayant lieu également dans $L^p_{\nu_t}$ $\forall\, t > 0$, $\forall\, p < +\infty$.

La preuve du théorème se fait par étapes; la première, qui est l'objet du paragraphe 1, est de construire une famille de noyaux $L_u(w,k)$ définis en fait comme des applications $L_u(k)$, $k \in B' \to$ $\to L_u(\cdot,k) \in \mathcal{L}(B', L^2_{\nu_u}(B))$. Posons $F = L^2_{\nu_u}(B)$. On constuira alors un opérateur $L_u: \mathcal{C}^m_1(B) \mapsto F$ ayant la propriété de continuité suivante:

Soient $G_n \in \mathcal{C}^m_1(B)$; $G_n(z) \mapsto G(z)$ dans B' $\forall z \in B$ et $\sup_{n \in N} \|G_n(z)\| < C(1+\|z\|)^m$ entraînent $L_u(G) = \lim_{n \to +\infty} L_u(G_n)$ (dans F). Il est clair que la fonction $L_u(k)(f) = L_u(e^{-i(w\bar{k}+\bar{w}k)}f) \in F$ pour $f \in B'$, $k \in B'$ détermine alors complètement l'opérateur L_u. Ceci fait que l'on déterminera d'abord les noyaux $L_u(k)$.

Dans une seconde partie, on étudie la régularité $L^p_{\nu_u}(B)$ des opérateurs L_u; dans une troisième on étudie la convergence presque sûre des $L_u(G_n)$ vers $L_u(G)$ si G_n est une suite convenablement choisie de fonctions de $\mathcal{C}^{m-1-\delta}_1(B)$ qui convergent vers G. Ceci conduira au théorème annoncé.

1. CONSTRUCTION DU NOYAU L_u $(0 < u \leq 1)$

$L_u(w,k)$ est recherché comme une fonction de $k \in Z \to$ $\to \mathcal{L}(Z, L^2_{\nu_u}(B))$ en fait $B' \mapsto \mathcal{L}(B', L^2_{\nu_u})$ suffit, telle que:

$$-L_u(w,k)\cdot ik = e^{-i(w\bar{k}+\bar{w}k)} - q_m(w,k)$$

où $q_m(w,k)$ est défini par:

$$q_m(w,k) = \int_0^{+\infty} e^{-1/t}\ t^{-2-m}\ e^{-t|k|^2}\ (1-it(w\bar{k}))^m \frac{dt}{m!}\,.$$

On rappelle, en effet, que l'on peut définir grâce à un prolongement continu les fonctions $k \to e^{-k(w\bar{k}+\bar{w}k)} \in L^\infty_{\nu_u}$ et $k \to w\bar{k} \in \bigcap_{+\infty > p \geq 1} L^p_{\nu_u}$ pour $k \in Z$. On posera donc:

$$A_u(w,k) = \sum_{0 \leq p+q \leq m} (-1)^{p+q}\, u^{p+q/2}\, P_{p\cdot q}(w, \bar{w}/u^{+1/2})\cdot\frac{(i\bar{k})^p}{p!q!}(ik)^q$$
$$\frac{e^{-u|k|^2}}{(m-(p+q))!} \int_0^{+\infty} e^{-1/t}\ t^{-2-m+p+q}\ e^{-t|k|^2}\ dt,$$

que l'on écrira aussi: $A_u(w,k,k)$,

$$L''_{1u}(w,k)\cdot h = \sum_{\substack{0\le p+q\le m\\ q\ge 1}} (-1)^{p+q}\, u^{p+q/2}\, P_{p\cdot q}(w,\bar{w}/u^{1/2})\left(\frac{(i\bar{k})^p}{p!q!},(ik)^{q-1}\otimes h\right)$$

$$\frac{e^{-u|k|^2}}{(m-p-q)!}\int_0^{+\infty} e^{-1/t}\, t^{-2-m+p+2}\, e^{-t|k|^2}\,dt,$$

$$L''_{2u}(w,k)\cdot h = (i\bar{k}\cdot h)\,\frac{1-e^{-u|k|^2}}{|k|^2}\, q_m(w,k) =$$

$$u\sum_{j=0}^{m} (-1)^j \frac{(wi\bar{k})^j}{j!(m-j)!}(i\bar{k},h)\left(\int_0^1 e^{-su|k|^2}ds\right)\int_0^{+\infty} e^{-1/t}\, t^{-2-m+j}e^{-t|k|^2}dt$$

De sorte que $-L''_{1u}(w,k)(ik)+A_u(w,k) = e^{-u|k|^2}\, q_m(w,k)$. On posera:

$$L_u(w,k) = \frac{i\bar{k}}{|k|^2}\left(e^{-i(w\bar{k}+\bar{w}k)} - A_u(w,k)\right) - L''_{2u}(w,k),$$

et donc:

$$-L_u(w,k)\cdot ik = e^{-i(w\bar{k}+\bar{w}k)} - q_m(w,k)$$

$$L''_u(w,k) = L''_{1u}(w,k) + L''_{2u}(w,k).$$

Par ailleurs:

$$e^{-i(w\bar{k}+\bar{w}k)} =$$

$$= \sum_{0\le p+q\le m} u^{p+q/2}\,\frac{(-1)^{p+q}}{p!q!}\, P_{p,q}(w,\bar{w}/u^{1/2})\cdot(i\bar{k})^p(ik)^q\, e^{-u|k|^2} + R_{m,u}(w,k),$$

$$R_{m,u}(w,k) =$$

$$= \int_0^1 \frac{(1-\tau)^m}{m!}\sum_{p+q=m+1} u^{p+q/2}\,\frac{(-1)^{p+q}}{p!q!}\, P_{p,q}(w+i\tau k,\ \bar{w}+i\tau\bar{k}/u^{1/2})\ \cdot$$

$$\cdot\,[(i\bar{k})^p,(ik)^q]\ e^{-i\tau(w\bar{k}+\bar{w}k)}\ e^{-u(1-\tau^2)|k|^2}\,dt,$$

$P_{p,q}\left(w+\frac{i\tau k}{u^{1/2}},\ \bar{w}+\frac{i\tau\bar{k}}{u^{1/2}}\right)\frac{(i\bar{k})^p(ik)^q}{p!q!}$ est défini par:

$$\sum_{\substack{0\le p'\le p\\ 0\le q'\le q}} P_{p'q'}\,\frac{(w,\bar{w}/u^{1/2})}{p'!q'!}\,(i\bar{k})^{p'}(ik)^{q'}\,\frac{(-\tau u^{-1/2}|k|^2)^{p+q-(p'+q')}}{(p-p')!(q-q')!},$$

ce qui permet d'écrire:

$$\frac{ik}{|k|^2}\left(e^{-i(w\bar{k}+\bar{w}k)} - A_u(w,k)\right) = \int_0^{+\infty} L_u^o(t,w,k)\,dt.$$

$$L_u^o(t,w,k) = i\bar{k}\ e^{-t|k|^2}\ (e^{-i(w\bar{k}+\bar{w}k)} -$$

$$- \sum_{0\leq p+q\leq m} (-1)^{p+q}\ u^{p+q/2}\ P_{p,q}(w,\bar{w}/u^{1/2})\ \frac{(i\bar{k})^p(ik)^q}{p!q!}$$

$$e^{-u|k|^2} \int_0^t e^{-1/s}\ s^{-2-m+p+2}\ \frac{ds}{(m-p-q)!})$$

car:

$$\int_0^{+\infty} e^{-t|k|^2}\ dt \int_0^t e^{1/s}\ s^{-2+m+p}\ ds = \int_0^{+\infty} e^{-1/s}\ s^{-2-m+p}\ e^{-s|k|^2}\ \frac{ds}{|k|^2},$$

ce qui permet d'exprimer:

$$L_u^o(t,w,k) = i\bar{k}\ e^{-t|k|^2}\ (R_{mu}(w,k) +$$

$$+ \sum_{0\leq p+q\leq m} (-1)^{p+q}\ u^{p+q/2}\ P_{p,q}(w,\bar{w}/u^{1/2})\ \frac{(i\bar{k})^p(ik)^q}{p!q!}$$

$$e^{-u|k|^2} \int_t^{+\infty} e^{-1/s}\ s^{-2-m+p+q}\ \frac{ds}{(m-p-q)!})\ .$$

Pour t petit on a recours à la première expression de L_u^o, pour t grand à la seconde. Si G est une fonction borélienne à croissance de type m $B \mapsto B'$, on posera:

$$(L''_{1u}G)(w) = \sum_{\substack{0\leq p+q\leq m \\ q\geq 1}} (-1)\ \frac{u^{p+q/2}}{p!q!}\ \frac{1}{(m-p-q)!}\ P_{p,q}(w,\bar{w}/u^{1/2}).$$

$$\int_0^{+\infty} e^{-1/t}\ t^{-2-m+\frac{p+q}{2}+\frac{1}{2}}\ dt\ \ P_{q-1,p}(z,\bar{z}/t^{1/2})G(z+Z)d\nu_t(z)d\nu_u(Z).$$

On a vu [5] que $P_{p,q}(w,\bar{w}/u^{1/2})$ applique $\hat{\odot}_p Z \otimes \hat{\odot}_q Z\ \ L^2_{\nu_u}(B)$.
Il suffit donc de vérifier que la forme:

$$(h,k) \mapsto \int_0^{+\infty} e^{-1/t}\ t^{-2-m+\frac{p+q}{2}+\frac{1}{2}}\ dt\ P_{q-1,p}(z,\bar{z}/t^{1/2})(\bar{k}^{q-1},h^p)(G(z+Z)\cdot\bar{k})$$
$$d\nu_t(z)\ d\nu_u(Z)$$

est continue pour les topologies de Hilbert-Schmidt, ce qui est une conséquence du fait que:

$$\sum_{\substack{|\alpha|=p,i \\ |\beta'|=q-1}} \left|\int_0^{+\infty} e^{-1/t}\ t^{-2-m+\frac{p+q}{2}+\frac{1}{2}}\ dt\ P_{\beta'\alpha}(z,\bar{z}/t^{1/2})G_i(z+Z)d\nu_t(z)d\nu_u(Z)\right|^2 \leq$$

$$C\int_0^{+\infty} e^{-1/t}\ t^{-2-2m+p+q+\frac{1}{2}}\ dt \iint |G(z+Z)|^2\ d\nu_t(z)\ d\nu_u(Z).$$

On a donc prouvé:

Lemme 1. Si G est dans $\mathcal{C}^1_m(B)$, $(L''_{1u}G)(w) \in L^2_{\nu_u}(B)$ et l'on a:

$$\|L''_{1u}G\|^2_u \le C \int_0^{+\infty} e^{-1/t}\, t^{-2-m+\frac{1}{2}} (1+t^{-1})^m\, dt \iint |G(z+Z)|^2\, d\nu_t(z)\, d\nu_u(Z),$$

où C ne dépend que de m pour $0 < u \le 1$.

De même, on définit:

$$(L''_{2u}G)(w) = u \sum_{j=0}^{m} \frac{(-1)}{j!(m-j)!} w^j \int_0^{+\infty} e^{-1/t}\, t^{-2-m+\frac{j}{2}-\frac{1}{2}} dt$$

$$\int_0^1 ds\ \bar{z}^j/t^{j/2}\, (G(z+Z), \bar{z}/t^{1/2})\, d\nu_t(z)\, d\nu_{su}(Z).$$

Lemme 2. Si G est dans $\mathcal{C}^1_m(B)$, $L''_{2u}G(w) \in L^2_{\nu_u}(B)$ et l'on a:

$$\|L''_{2u}G\|^2_u \le Cu^2 \int_0^{+\infty} e^{-1/t}\, t^{-2-m-\frac{1}{2}} (1+t^{-1})^m\, dt$$

$$\int_0^1 ds\ |(G(z+Z), \bar{z}/t^{1+2})|^2\, d\nu_t(z)\, d\nu_{su}(Z).$$

Si $t \le 2$ on pose pour $G \in \mathcal{C}^1_m(B)$

$$L^o_u(G)(t,w) = t^{1/2} \quad (G(z+w), \bar{z}/t^{1/2})\, d\nu_t(z) -$$

$$\sum_{0 \le p+q \le m} (-1) \frac{u^{p+q/2}}{p!q!} P_{p,q}(w, \bar{w}/u^{1/2}) \cdot \int_0^t e^{-1/s}\, s^{-2-m+p+q} \frac{ds}{(m-p-q)!}$$

$$\iint (P_{q,p}(z, \bar{z}/t^{1/2})(G(z+Z), \bar{z}/t^{1/2}) - qG(z+Z) \otimes P_{q-1,p}())\, t^{-p+q+1/2}\, d\nu_t(z)\, d\nu_u(Z).$$

Pour $t \ge 1$, on posera au contraire:

$$L^o_u(G)(t,w) = \sum_{0 \le p+q \le m} (-1)\, u^{p+q/2}\, P_{p,q/p!q!}(w, \bar{w}/u^{1/2}) \int_t^{\infty} e^{-1/s}\, s^{-2-m+p+q} \cdot$$

$$\frac{ds}{(m-p-q)} \iint (\bar{z}/t^{1/2}, G(z+Z))\, P_{q,p}(z, z/t^{1/2})\, t^{-p+q+\frac{1}{2}}\, d\nu_t(z)\, d\nu_u(Z) +$$

$$\sum_{\substack{0 \le p+q \le m \\ q \ge 1}} (-1)^{p+q}\, u^{p+q/2} \sum_{\substack{|\alpha|=p \\ |\beta|=q}} t^{-p+q+\frac{1}{2}}\, P_{\alpha\beta}(w, \bar{w}/u^{1/2}) \Big(\int_t^{\infty} e^{-1/s}$$

$$s^{-2-m+p+q} \frac{ds}{(m-p-q)!}\Big) \sum_{(i) \le \beta} \beta_i\, G_i(z+Z)\, P_{\beta-(i)\alpha}(z, \bar{z}/t^{1/2})$$

$$d\nu_t(z)\, d\nu_u(Z) + R_{mu}\, G(w).$$

Lemme 3. Si $t \leq 2$

$$\|L_u^o(G)(t,w)\|_u^2 \leq Ct^{-1}\Big(\iint|(G(z+w),\bar z/t^{1/2})|^2\, d\nu_u(w)d\nu_t(z) + \iint|(G(z+Z),\bar z/t^{1/2})|^2\, d\nu_t(z)d\nu_u(Z)\Big).$$

On prouve également que pour $t \geq 1$:

$$\|L_u^o(G)(t,w)-(R_{mu}\ G)(w)\|_u^2 \leq Ct^{-(m+q)}\iint|(G(z,Z),\bar z/t^{1/2})|^2\, d\nu_t(z)d\nu_u(Z).$$

Seul, en fait, le terme $R_{m,u}G$ pose un problème.

On a vu que:

$$R_{m,u}(w,k) = \int_0^1 \frac{(1-\tau)^m}{m!} R^u_{m+1}(\tau,w,k)e^{-i\tau(w\bar k+\bar w k)}{}_{i\bar k}\ e^{-t|k|^2}\ e^{-(1-\tau^2)u|k|^2}d\tau$$

où tous calculs faits:

$$R^u_m(\tau,w,k) = \sum_{0\leq j+1\leq m}(-1)^{j+1}\frac{(\tau|k|^2)^{m-(j+1)}}{(m-(j+1))!}P_{j\ell}(w,\bar w/u^{1/2})\cdot(i\bar k)^j(ik)^\ell\ 2^{m-(j+1)}\ u^{(j+1)/2},$$

$$e^{-i\tau(w\bar k+\bar w k)}\ R^u_{m+1}\ (\tau,w,k)\ e^{-(1-\tau^2)u|k|^2} = \Big(\frac{\partial}{\partial\tau}\Big)^{m+1}\big(e^{-i\tau(w\bar k+\bar w k)-(1-\tau^2)u|k|^2}\big).$$

$e^{-i\tau(w\bar k+\bar w k)-(1-\tau^2)u|k|^2}$ est l'opérateur f $\quad f(\tau w+Z)d\nu_{(1-\tau^2)u}(Z) =$

$= H_\tau(f)(w)$. On observe que:

$$\Big(\frac{\partial}{\partial\tau}\Big)H_\tau(f)(w) = -\tau^{-1}\ (u\tfrac{1}{2}\Delta_w - wD_w - \bar w D_{\bar w})\cdot H_\tau(f)(w)$$

or:

$$\tau^m\Big(\frac{\partial}{\partial\tau}\Big)^m = \sum_{\ell=0}^m (-1)^\ell\ \Big(\frac{m!}{(m-1)!}\Big)^2\frac{1}{\ell!}\ \Big(\tau\frac{\partial}{\partial\tau}\Big)^{(m-1)},$$

donc:

$$\Big(\frac{\partial}{\partial\tau}\Big)^{m+1}\ \big(e^{-i\tau(w\bar k+\bar w k)-(1-\tau^2)u|k|^2}\big)$$

est le symbole de l'opérateur: $f\mapsto Q_{m+1}(f)$

$$f\mapsto\tau^{-(m+1)}\sum_{\ell=0}^{m+1}(-1)^{m+1-1}\ (u\tfrac{1}{2}\Delta_w - wD_w - \bar w D_{\bar w})^{(m+1-\ell)}\cdot(H_\tau f)\Big(\frac{(m+1)!}{(m+1-\ell)!}\Big)^2\frac{(-1)}{\ell!}.$$

Si $g = f*\nu_t$, $\tilde\tau = (1-\tau^2)u$

$$(H_\tau g)(w) = (f*\nu_{\tilde\tau})*\nu_t(\tau w) = [(f*\nu_{\tilde\tau})(\tau\,.)*\nu_{t/\tau^2}](w).$$

On utilise l'identité:

$$\frac{n!}{(n-m)!} = \sum_{\ell=0}^m (-1)^\ell\ \Big(\frac{m!}{(n-\ell)!}\Big)^2\frac{1}{\ell!}\ n^{m-\ell}.$$

$$\| \ \|_u^2 = \sum_{|\alpha|+|\beta|\geq m} \frac{(|\alpha|+|\beta|)!}{(|\alpha|+|\beta|-m)!} \left| \int g*\nu_t, \bar{H}_{\alpha,\beta}(y/u^{1/2}) d\nu_u(y) \right|^2,$$

or:

$$\int g*\nu_t, \bar{H}_{\alpha,\beta}(y/u^{1/2}) d\nu_u(y) =$$

$$= \iint (ut'^{-1})^{\frac{|\alpha|+|\beta|}{2}} H_{\beta,\alpha}(x/t'^{1/2}) g(x+y) d\nu_u(x) d\nu_{t'}(y).$$

Ici:

$$t' = t/\tau^2, \quad u^{\frac{|\alpha|+|\beta|}{2}} t'^{\frac{|\alpha|+|\beta|}{2}} \tau^{-(m+1)} =$$

$$= t^{-\frac{|\alpha|+|\beta|}{2}} \tau^{|\alpha|+|\beta|-(m+1)} u^{\frac{|\alpha|+|\beta|}{2}},$$

par ailleurs: $t^{-\frac{|\alpha|+|\beta|}{2}} \frac{(|\alpha|+|\beta|!}{(|\alpha|+|\beta|-m+1)!} \leq Ct^{-\frac{(m+1)}{2}}$ pour $|\alpha|+|\beta|\geq m+1$

et $t \geq 2$ et $u \leq 1$.

On a donc obtenu:

$$\|\tau^{-(m+1)}(\tfrac{u}{2}\Delta_w - wD_w - \bar{w}D_w)^{m+1}(g*\nu_{t'})\|_u^2 < Ct^{-(m+1)} \ |g(x+y)|^2 \ d\nu_u(x) d\nu_{t'}(y).$$

L'opérateur de symbole $\tilde{R}^m_{u,\tau} = (\frac{\partial}{\partial t})^{m+1}(e^{-i\tau(w\bar{k}+\bar{w}k)-(1-\tau^2)u|k|^2-t|k|^2})$ applique donc $G^m_o(B)$ dans $L^2_{\nu_u}$, $t \geq 2$, avec une inégalité:

$$\|\tilde{R}^m_{u,\tau} \ f\|_u^2 \leq Ct^{-(m+1)} \iiint |f(x+\tau y+z)|^2 \ d\nu_u(x) d\nu_{\tilde{\tau}}(y) d\nu_t(z).$$

On calcule donc: en notant $R^m_{u,\tau} = i\bar{k} \ \tilde{R}^m_{u,\tau}$

$$\|R^m_{u,\tau} \ G(t,w)\|_u^2 =$$

$$\Big\|\sum_{\ell=0}^{m+1} (\tfrac{u}{2}\Delta_w - wD_w - \bar{w}D_{\bar{w}})^{(m+1-\ell)} \cdot [(G*\nu_{\tilde{\tau}})(\tau .)*\bar{y}/t'^{1/2}\nu_{t'}](w) \frac{(-1)^{m+1}}{\ell!} \frac{(m+1)!}{(m+1-\ell)!})^2 \tau^{-(m+1)} t^{-1/2}\Big\|_u^2$$

avec $t' = t/\tau^2$:

$$= t^{-1} \sum_{|\alpha|+|\beta|\geq m+1} \frac{(|\alpha|+|\beta|!}{(|\alpha|+|\beta|-m-1)!} \cdot$$

$$\cdot \left|\iint (H(x+y)\cdot\bar{y}/t'^{1/2}) \bar{H}_{\alpha,\beta}(x/u^{1/2}) d\nu_{t'}(y) d\nu_u(x)\right|^2$$

avec $H(x) = \int G(\tau x+z) d\nu_{(1-\tau^2)u}(z)$.

Or:

$$\iint (H(x+y)\cdot\bar{y}/t'^{1/2})\bar{\mathcal{H}}_{\alpha,\beta}(x/u^{1/2})d\nu_{t'}(y)d\nu_u(x) =$$

$$\iint (H(x+y)\cdot\bar{y}/t'^{1/2})(ut'^{-1})^{\frac{|\alpha|+|\beta|}{2}}\cdot\mathcal{H}_{\beta\alpha}(y/t'^{1/2})d\nu_{t'}\,d\nu_u +$$

$$\sum_{(i)\leq\beta} \beta_i^{1/2}\iint (ut'^{-1})^{\frac{|\alpha|+|\beta|}{2}} H_i(x+y)\mathcal{H}_{\beta-(i),\alpha}(y/t'^{1/2})d\nu_{t'}\,d\nu_u$$

$$\tau^{-(m+1)}\, u^{(|\alpha|+|\beta|)/2}\, t^{-\frac{|\alpha|+|\beta|}{2}}\, \tau^{|\alpha|+|\beta|}\,(|\alpha|+|\beta|)^{m+3/2} \leq Ct^{-(m+1)/2}$$

pour $t \geq 2$, $\tau \leq 1$, $u \leq 1$. Et donc:

$$\|R^m_{u\tau}(t,w)\cdot G\|^2_u \leq Ct^{-(m+2)}\int (|(G(y+\tau x+z),y/t^{1/2})|^2 + |G(y+\tau x+z)|^2)\, d\nu_t(y)d\nu_u(x)d\nu_{(1-\tau^2)u}(z)\,.$$

On a donc prouvé:

Lemme 4. Si G est une fonction de $\mathcal{C}^m_1(B)$, $t \geq 2$, $R_{m,u}(G) \in L^2_{\nu_u}(B)$ et on a:

$$\|R_{m,u}\,G\|^2_u \leq Ct^{-(m+2)}\int_0^1 d\tau \int (|(G(y+\tau x+z),y/t^{1/2})|^2 + |G(y+\tau x+z)|^2)\, d\nu_t(y)d\nu_u(x)d\nu_{(1-\tau^2)u}(z)$$

$$\leq Ct^{-(m+2)}\iint |(G(y+z),\bar{y}/t^{1/2})|^2 + |G(y+z)|^2)d\nu_t(y)d\nu_u(z)\,.$$

On a donc obtenu le résultat suivant:

Proposition 1. Le noyau L_u défini plus haut applique l'espace $\mathcal{C}^{m-1-\delta}_1(B)$ $(\delta > 0)$ dans $L^2_{\nu_u}(B)$ et on a une inégalité:

$$\|L_u G\|^2_u \leq C\int_0^{+\infty} \phi(t)dt \int (|(G(z+Z),\bar{z}/t^{1/2})|^2 + |G(z+Z)|^2)d\nu_t(z)d\mu_u(Z)$$

avec: .

$$\mu_u = \nu_u + u_0\int_0^1 \nu_{su}\, ds;\quad \phi \in L^1_{dt}$$

$$\int_0^{+\infty} \phi(t)t^{m-1-\delta}\, dt < +\infty$$

C est indépendant de $u \leq 1$.

2. LA SOLUTION $\mathcal{L}G$

On désigne par Z_n un sous espace réel de dimension finie de

Z et $s_n Z \mapsto Z_n$ la projection hermitienne.

On constate que:

$$s_n(L_u(w,k))\Big|_{Z_n \times Z_n} = L_u^n(w,k)$$

où $L_u^n(w,k)$ est le noyau construit sur l'espace Z_n. On constate également que:

$$s_n L_u(w,k) = (s_n L_u)(s_n(w),k)$$

pour $k \in Z_n$. Donc si $G_n(z) = s_n G\big|_{Z_n}$ qui appartient à $G_1^m(Z_n)$ si $G \in G_1^m(B)$:

$$(L_u^n G_n)(s_n(w)) = L_u(G_n \cdot s_n)(w)$$

pour $w \in B$. Ainsi $L_u(G_n \cdot s_n)$ est une fonction cylindrique basée sur Z_n. On a:

<u>Lemme 2.1</u>. Si $G \in G_1^{m-1-\delta}(B)$:

$$\sup_{0<u\leq 1} \|L_u(G-G_n \cdot s_n)\|_u^2 \mapsto 0 \quad \text{quand} \quad n \mapsto +\infty .$$

<u>Démonstration</u>. Vu la Proposition 1, il suffit de voir que:

$$\sup_{0<u\leq 1} \int_0^{+\infty} \phi(t)dt \quad \|s_n G s_n(z)-G(z)\|'^2 \, [(1+\|z\|^2/t)\nu_t]*\mu_u \mapsto 0.$$

On observe d'abord que sur toute la partie K compacte de B, $\sup_{z\in K} \|s_n G s_n(z)-G(z)\|' \to 0$. Posant:

$$K_\rho = \{z \in B \mid \sum_{j=1}^{\infty} |z_j|^2/c_j^2 = \|z\|^{\sim 2} \leq \rho^2\}$$

on note que:

$$\int_{K_\rho^c} \|s_n G s_n - G(z)\|'^2 \, [(1+\|z\|^2/t)\nu_t]*\mu_u \leq$$

$$\leq C/\rho^{\epsilon} \quad (1+\|z\|^{\sim})^{m-1-\delta+\epsilon}(1+\|Z\|^{\sim})^{m-1-\delta+\epsilon}(1+\|z\|^{\sim}/t)^2 \, d\nu_t(z) \, d\mu_u(Z) \leq$$

$$\leq C'/\rho^{\epsilon}(1+t)^{m-1-\delta+\epsilon}.$$

Fixant $\epsilon < \delta$ et $\int_0^{+\infty} (1+t)^{m-1-\delta+\epsilon} \phi(t)dt < +\infty$, on obtient facilement la conclusion. On peut donc trouver une sous suite n_j telle que:

$$0<u\leq 1 \quad \sum_{j=0}^{\infty} \|L_u(G_{n_j} \cdot s_{n_j} - G_{n_{j+1}} \cdot s_{n_{j+1}})\|_u < +\infty .$$

Si de plus, G est de classe C^1 avec $\bar{\partial} G = 0$, il est clair que

$\bar\partial G_n = 0$. Si L_o^n désigne le noyau construit dans [4] relativement à Z_n, on a:

$$\bar\partial(L_o^n G_n) = G_n \quad \text{et} \quad L_o^n G_n \in \mathcal{G}_m^o .$$

La relation $e^{-i(w\bar k+\bar w k)} = q_m(w,k) - L_u^n(w,k)\cdot ik$ pour $(w,k) \in Z_n \times Z_n$ montre que:

$$L_o^n G_n = P^n(L_o^n G_n) + L_u^n(G_n)$$

où $P^n(f)(w) = \int_{Z_n} (1+w\bar z)^m f(z)\, dq_m(z)$ est le projecteur sur les polynômes de degré m. Comme $P^n(L_o^n G_n) = 0$, on obtient:

$$L_o^n(G_n)(w) = L_u^n(G_n)(w) \qquad w \in Z_n, \quad 0 < u \le 1.$$

On va donc définir une fonction $\mathcal{L}G(x)$ par:

- $\lim_{j\to\infty} L_u(G_{n_j}\cdot s_{n_j})(x)$ quand la suite converge
- 0 quand ce n'est pas le cas.

On définit donc ainsi une fonction sur B qui est ν_u mesurable $0 < u \le 1$, qui est dans $\mathcal{L}^2_{\nu_u}(B)$ $0 < u \le 1$ et dont la classe modulo les fonctions ν_u négligeables est $L_u(G)$.

Il se pose en particulier la question de faire le lien entre cette fonctions $\mathcal{L}G(x)$ et la solution $(L_oG)(x)$ définie sur Z obtenue par [4].

On introduit l'opérateur $\Phi_u(f)(x) = \int f(x+y)d\nu_u(y)$ défini $\text{Polcyl}(Z) \mapsto \text{Polcyl}(Z)$ cet opérateur se prolonge de $L^2_{\nu_u}(B) \mapsto C^\infty(Z)$. De plus cet opérateur se prolonge également en un opérateur continu $L^2_{\nu_t}(B) \mapsto L^2_{\nu_{t-u}}(B)$ pour $u < t$ noté Φ_u^t. Il suffit en effet d'observer que $\nu_u * \nu_{t-u} = \nu_t$ et donc que $f \in \mathcal{L}^2_{\nu_t}$ entraîne que $(x,y) \to f(x+y)$ est donc dans $\mathcal{L}^2_{\nu_u \otimes \nu_{t-u}}$.

On remarque que:

$$\Phi_u(P_{\alpha\beta}(w,\bar w/u^{1/2})\, u^{|\alpha|+|\beta|/2}) = w^\alpha \bar w^\beta$$

et que:

$$\Phi_u(e^{-i(w\bar k+\bar w k)}) = e^{-u|k|^2}\, e^{-i(w\bar k+\bar w k)}$$

donc:

$$\Phi_u(L_u(w,k)) = e^{-u|k|^2}\frac{i\bar{k}}{|k|^2}(e^{-i(w\bar{k}+\bar{w}k)}-A(w,k))-L''_o(w,k)e^{-u|k|^2}-L''_{2u}(w,k)$$

$$= e^{-u|k|^2}\ L_o(w,k) - i\bar{k}\ \frac{1-e^{-u|k|^2}}{|k|^2}\,q_m(w,k).$$

On en déduit que:

$$\Phi_u(L_uG)(w) = L_o(\Phi_uG)(w) + (L''_{2u}G)(w)$$

pour $w \in Z$ et pour G continue cylindrique $(\sigma_n G S_n = G)$ dans $\mathcal{G}_1^{m-1-\delta}(B)$. On en déduit aisément que l'égalité est verifiée pour $G \in \mathcal{G}_1^{m-1-\delta}(B)$, $w \in Z$, en notant les continuités respectives de ces différents opérateurs. Il est clair que:

$$\lim_{u\to 0}\ (L''_{2u}G)(w) = 0 \quad \forall\ w \in Z.$$

Pour estimer le terme $L_o(\Phi_uG)$, on utilise l'estimation: si $G \in \mathcal{G}_1^{m-1-\delta}(B)$, $L_oG \in \mathcal{G}_o^m$ et on a une estimation de la forme:

$$\|(L_oG)(w)\|^2 < C(1+|w|)^M\Big(\int_0^{+\infty}\phi(t)dt\ \ \big(|(G(z+w),\bar{z}/t^{1/2})|^2 +$$

$$+\ |G(z)|^2 + |(G(z),\bar{z}/t^{1/2})|^2\big)d\nu_t(z)$$

$$+\int_0^{+\infty}\psi(t)dt\int_0^1 |(G(\tau w+z),\bar{z}/t^{1/2})|^2\ d\tau\ d\nu_t(z)\Big).$$

Il est donc clair que:

$$\lim_{u\to 0} L_o(\Phi_uG)(w) = (L_oG)(w)$$

pour $w \in Z$, car les $\Phi_uG \to G$ en tout point de B out en restant uniformément majorés en norme dans B' par $(1+\|z\|)^{m-\delta}$.

Vu ce qui précède:

$$\Phi_u(L_uG)(w) = \int(\mathcal{L}G)(w+z)d\nu_u(z) \qquad 0 < u \leq 1, \quad w \in Z$$

et donc:

$$\lim_{u\to 0}\int \mathcal{L}G(w+z)d\nu_u(z) = (L_oG)(w)$$

pour $w \in Z$.

On a donc montré comment la fonction $\mathcal{L}G(z)$ détermine la solution $L_oG(z)$, $z \in Z$ si $\bar{\partial}G = 0$. Remplaçant éventuellement la fonction G par s_qGs_q, on suppose que G est cylindrique:

$$\mathcal{E}_u(L_uG/s_n)(w_n) = \int (L_uG(w_n+w')d\nu'_u(w')$$

est l'espérance conditionnelle par rapport à s_n.

$$\mathcal{E}_u(s_nG/s_n)(w_n) = \int (s_nG)(w_n+w')d\nu'_u(w').$$

Il est clair que:

$$\bar{\partial}\, \mathcal{E}_u(s_nG/s_n) = 0 \quad \text{et} \quad \bar{\partial}(\mathcal{E}_u(L_uG/s_n)) = \mathcal{E}_u(s_nG/s_n)$$

donc:

$$\mathcal{E}_u(L_uG/s_n) = L_u^n(\mathcal{E}_u(s_nG/s_n)) + P^n(\mathcal{E}_u(L_uG/s_n)).$$

On va maintenant évaluer $P^n(\mathcal{E}_u(L_uG/s_n))$

Notons d'abord que:

$$P^n(\mathcal{E}_u(L_u(w,k)(ik)/s_n)) = q_m(w_n,k) - e^{-u|k'|^2}\, q_m(w_n,k_n)$$

pour $w_n \in Z_n$ et $k \in Z$.

On va trouver un noyau $K^n_{tu}(W_n,k)$ associé à un opérateur continu $A_1^{m-\delta}(B) \to L^2_{\nu_t}(Z_n)$ $t > u$, qui vérifie également

$$K^n_{tu}(w_n,k)ik = q_m(w_n,k) - e^{-u|k'|^2}\, q_m(w_n,k_n).$$

On introduit donc le noyau $K^n_{t,u}(w_n,k)$ pour $t > u$:

$$K^n_{t,u}(w_n,k) = \frac{i\bar{k}}{|k|^2}\,(q_m(w_n,k) - (1-i(w_n\bar{k}))^m\, e^{-t|k|^2}$$

$$+ \frac{i\bar{k}_n}{|k_n|^2}\,(q_m(w_n,k_n) - (1-i\; w_n\; \bar{k})^m\, e^{-t|k_n|^2})e^{-u|k'|^2}$$

$$+ (1-iw_n\bar{k})^m\, e^{-t|k_n|^2-u|k'|^2}\,(1-e^{-(t-u)|k'|^2})\,\frac{i\bar{k}'}{|k'|^2}.$$

Le noyau $K^n_{t,u}(w_n,k)$ vérifie:

$$K^n_{t,u}(w_n,k)\cdot ik = q_m(w_n,k) - e^{-u|k'|^2}\, q_m(w_n,k_n).$$

On remarque par aillcrus que:

$$\frac{i\bar{k}}{|k|^2}\,(q_m(w_n,k) - (1-i(w_n\bar{k}))^m\, e^{-t|k|^2}) = P_t^{n,m}\circ L_t(w,k)$$

où $P_t^{n,m}$ est le projecteur orthogonal dans l'espace $L^2_{\nu_t}$ sur le sous espace engendré par les polynômes holomorphes basés sur Z_n de degré au plus m.

De même:

$$\frac{i\bar{k}_n}{|k_n|^2}\left(q_m(w_n,k_n) - (1-i(w_n\bar{k})^m e^{-t|k_n|^2}\right)e^{-u|k'|^2} = P_t^m L_t^n(w_n,k_n)e^{-|k'|^2}$$

où P_t^m est le projecteur orthogonal dans $L^2_{\nu_t}(Z_n)$ sur le sous espace des polynômes holomorphes de degré au plus m.

L'opérateur:

$$R^n_{t,u}(w_n,k) = \frac{i\bar{k}'}{|k'|^2}\left(1-e^{-(t-u)|k'|^2}\right)(1-i(w_n\cdot\bar{k}))^m e^{-t|k_n|^2-u|k'|^2}$$

s'exprime par:

$$(R^n_{tu}G)(w_n) = \sum_{0\leq|\alpha|\leq m} w_n^{\alpha}\, t^{-|\alpha|/2} \frac{m!}{(m-|\alpha|)!\,|\alpha|!}\int_{Z_n} \bar{z}^{\alpha}/t^{|\alpha|/2}\, d\nu_t(z)$$

$$\int_0^1\int_{Z^n}\int_{Z^n}(G(z+Z'+Z),Z'/s(t-u)^{1/2})(t-u)^{1/2}s^{-1/2}\, d\nu'_{s(t-u)}(Z')d\nu'_u(Z)ds.$$

On note donc μ'_{tu} la mesure sur Z^n définie par $Z^n = (1-s_n)(B)$

$$\mu'_{tu} = \int_0^1 s^{-1/2}\,(\|z'\|/s(t-u)^{1/2}\,\nu'_{s(t-u)})*\nu'_u ds$$

pour voir que:

$$\|R_{tu}G\|_t^2 \leq C(t-u)\,\|\mu'_{tu}\|\iint\|(1-\sigma_n)G(z+Z)\|'^2\, d\nu_t(z)\, d\mu'_{tu}(Z).$$

Comme $P_t^m\circ L_t^n(w_n,k_n)e^{-u|k'|^2}$ est associé à l'opérateur $P_t^m(L_t^n(\mathcal{E}_u(\sigma_n G/s_n))$, on a l'inégalité:

$$\|K^n_{tu}G\|_t^2 \leq C(\|L_tG\|_t^2 + \|L_t^n(\mathcal{E}_u(\sigma_n G/s_n))\|_t^2 + (t-u)\|\mu'_{tu}\|$$
$$\iint\|(1-\sigma_n)G(z+Z)\|'^2\, d\nu_t(z)\, d\mu'_{tu}(Z)).$$

Si $G\in\partial C_o^\infty(Z_q)$ on a bien:

$$\mathcal{E}_u(L_uG/s_n) = L_u^n(\mathcal{E}_u(\sigma_n G/s_n)) + K^n_{tu}G(w_n).$$

Il reste à voir que cette égalité est en fait valable pour $G\in\mathcal{G}_1^{m-1-\delta}(Z_q)$ avec $\bar{\partial}G = 0$.

Dans ces conditions, on trouve, $F_1\in\mathcal{G}_o^m(Z_q)$ $\bar{\partial}F_1 = G$ et les

$$G_i = \bar{\partial}(\chi_i F_1 * \rho_{1/i}) = (\chi_i G)*\rho_{1/i} + (\bar{\partial}\chi_i)F_1*\rho_{1/i}$$

appartiennent à $\bar\partial C_o^\infty(Z_q)$; convergent vers G en chaque point tout en restant uniformément majorés par $(1+\|z\|)^{m-1}$. Les $L_u G_i$ convergent vers $L_u G$ dans $L^2_{\nu_u}$ tandis que $K^n_{tu} G_i \to K^n_{t,u} G$ en chaque point et que $L^n_u(\mathcal{E}_u(\sigma_n G_i/s_n)) \to L^n_u(\mathcal{E}_u(\sigma_n G/s_n))$ dans $L^2_{\nu_u}$. Donc l'identité est encore valable pour G.

On en déduit alors que:

$$\mathcal{E}_u(L_u G/s_n) = L^n_t(\mathcal{E}_u(\sigma_n G/s_n)) + K^n_{tu} G(w_n) \quad \text{puisque} \quad \bar\partial\mathcal{E}_u(\sigma_n G/s_n) = 0.$$

D'où

$$\|\mathcal{E}(L_u G/s_n)\|^2_t \le$$

$$\le C\Big(\int_0^\infty \varphi(\sigma)d\sigma \int (\|\mathcal{E}_u(\sigma_n G/s_n)\|'^2 + \|G\|'^2)(1+\|z\|^2/\sigma\nu_\sigma * \mu t +$$

$$+ (t-u)\|\mu'_{tu}\| \iint \|(1-\sigma_n)G(z+2)\|'^2 \, d\mu'_{tu}(2)\, d\nu_t(z)\Big).$$

D'autre part

$$\|L''_{2u} G\|^2_t \le C\, u^2 \int_0^{+\infty} \varphi(\sigma)d\sigma \int \|G(z)\|'^2 (1+\|z\|^2/\sigma)\nu_\sigma * \mu_u$$

enfin

$$\Phi_u(L_u G)|_{Z_n} = \Phi_u(\mathcal{E}_u(L_u G/s_n))$$

et donc

$$\|\Phi_u(L_u G)|_{Z_n}\|^2_t \le C\|\mathcal{E}_u(L_u G/s_n)\|^2_{t+u} .$$

Donc $L_o(\Phi_u G)|_{Z_n} \in L^2_{\nu_t}(Z_n)$ et on a l'inégalité:

$$\|L_o(\Phi_u G)|_{Z_n}\|^2_t \le C\Big(\int_0^{+\infty} \varphi(\sigma)d\sigma \int (\|\mathcal{E}_u(\sigma_n G/s_n)\|'^2)(1+\|z\|^2/\sigma)\nu_\sigma * (\mu_{t+u}+\mu_u) +$$

$$+ \|\mu'_{t+u,u}\| \iint \|(1-\sigma_n)G(z+Z)\|'^2 \, d\mu'_{t+u,u}(Z)\, d\nu_{t+u}(z)\Big).$$

On a établi cette inégalité pour G cylindrique dans $G_1^{m-1-\delta}(B)$ avec $\bar\partial G = 0$. On passe aisément à la limite et on obtient une inégalité valable pour $G \in G_1^{m-1-\delta}(B)$ $\bar\partial G = 0$.

Remplaçant maintenant G par $G - \sigma_n G|_{Z_n} \circ s_n = G - G_n$ on va établir que

$$\lim_{n\to\infty} \sup_{0<u\le 1} \|L_o(\Phi_u(G-G_n)|_{Z_n})\|^2_t = 0.$$

On remarque d'abord que $\|\mu'_{t+u,u}\| \leq C\|j\circ(1-s_n)\|_2$ et $\lim_{n\to\infty} \|j\circ(1-s_n)\|_2 = 0$. On a vu plus haut que

$$\lim_{n\to\infty} \sup_{0<u\leq 1} \int_0^{+\infty} \varphi(\sigma)d\sigma \int \|G-G_n\|'^2(1+\|z\|^2/\sigma)\nu_\sigma * \mu_u = 0.$$

Comme

$$\sup_{z\in K} \|\mathcal{E}_u(\sigma_n(G-G_n)/s_n)\circ s_n(z)\|'^2 \leq \sup_{z\in\tilde{K}} \int_{Z^n} \|G(z+u^{1/2}Z')-G(z)\|'^2 \, d\nu(Z')$$

pour $K \subset\subset B$.

Il suffit donc de voir que ce terme tend vers zéro uniformément avec $u \leq 1$. Soient:

$$B_\delta = \{Z \in B \mid \|Z\| \leq \delta \text{ et } B'^n_\delta = (1-s_n)(B_\delta).$$

Si $\varepsilon > 0$ est donné, on détermine δ de sorte que:

$$\sup_{z\in K} \|G(z) - G(z+Z)\|' \leq \varepsilon$$

dès que $\|Z\| \leq \delta$ et il suffit donc de vor que pour tout $\delta > 0$ on a:

$$\lim_{n\to\infty} \sup_{u\leq 1} \int_{(B'^n_\delta)^c} (1+\|u^{1/2}Z'\|)^m \, d\nu'(Z') = 0$$

ce qui est une conséquence évidente du fait que:

$$\int_{(B'^n_\delta)^c} d\nu' \leq \frac{1}{\delta} \|j\circ(1-s_n)\|_2^2 .$$

Les $L_o(\Phi_u(G-G_n))|_{Z_n}$ formant avec $0 < u \leq 1$ un ensemble borné de fonctions de $L^2_{\nu_t}(Z_n)$ qui converge simplement $(u \to 0)$ vers $L_o(G-G_n)|_{Z_n}$ on en déduit que:

$$L_o(G-G_n)|_{Z_n} \in L^2_{\nu_t}(Z_n)$$

et que l'on a:

$$\|L_o(G-G_n)|_{Z_n}\|_t^2 \leq \sup_{0<u\leq 1} \|L_o(\Phi_u(G-G_n))|_{Z_n}\|_t^2$$

et donc:

$$\lim_{n\to\infty} \|(L_oG)|_{Z_n}\cdot s_n - (LG_n)\cdot s_n\|_t^2 = 0$$

donc les $(L_oG)|_{Z_n}\cdot s_n$ convergent quand $n \to \infty$ vers L_tG dans $L^2_{\nu_t}$ $\forall$ t.

3. RÉGULARITÉ L^p DES SOLUTIONS

On va montrer que $L_u G \in L^p_{\nu_u}$ $1 \leq p < \infty$ si $G \in G_1^m(B)$. On prouve tout d'abord qu'un polynôme:

$$\tilde{p}(z,\bar{z}) = \sum_{\substack{|\alpha|=k \\ |\beta|=\ell}} p_{\alpha\beta} H_{\alpha\beta}(z,\bar{z}) \in L^p_\nu \quad 1 \leq p < \infty .$$

On écrit:

$$\tilde{p} = p \cdot H_{k\ell} \quad \text{avec} \quad p \in \hat{\odot}_k Z \hat{\otimes} \hat{\odot}_\ell Z .$$

<u>Lemme 5</u>. Soient $\tilde{p} = p \cdot H_{k\ell}$ et $\tilde{q} = q \cdot H_{mn}$, le produit $\tilde{p}(z,\bar{z})\tilde{q}(z,\bar{z}) = \tilde{r}(z,\bar{z})$ s'exprime par:

$$\tilde{r}(z,\bar{z}) = \sum_{\substack{r \leq \inf(k,n) \\ r' \leq \inf(\ell,n)}} \frac{1}{r!}\frac{1}{r'!} (D_z^r D_{\bar{z}}^{r'} p \cdot D_z^{r'} D_{\bar{z}}^r q) \cdot H_{k+m-r-r' \;\; \ell+n-r-r'} .$$

On a l'estimation:

$$\| D_z^r D_{\bar{z}}^{r'} p \cdot D_z^{r'} D_{\bar{z}}^r q \|_{\underset{k+m-r-r'}{\hat{\odot} Z} \;\; \underset{\ell+n-r-r'}{\hat{\odot} Z}} \leq$$

$$\frac{k!\,m!\,\ell!\,n!}{(k-r)!\,(m-r')!\,(\ell-r')!\,(n-r)!} \|p\|_{k,\ell} \|q\|_{m,n} .$$

On en déduit:

$$\|\tilde{r}\|_2^2 \leq 3^{k+\ell+m+n} \|\tilde{p}\|_2^2 \|\tilde{q}\|_2^2 .$$

<u>Démonstration du lemme</u>

Pour z fixé dans Z, $D_z^r D_{\bar{z}}^{r'} p(z,\bar{z}) \in \hat{\odot}_r Z \hat{\otimes} \hat{\odot}_{r'} Z$ car la forme bilinéaire sur $\hat{\odot}_r Z \otimes \hat{\odot}_{r'} Z$:

$$D_z^r D_{\bar{z}}^{r'} p(z,\bar{z}) = \sum_{\substack{|\lambda|=r \\ |\mu|=r'}} \frac{k!}{(k-r)!} \frac{\ell!}{(\ell-r')!} \frac{r!\,r'!}{\lambda!\,\mu!} (p, \mathrm{sym}(z^{k-r} \otimes e_\lambda) \otimes \mathrm{sym}(\bar{z}^{\ell-r'} \otimes e_\mu)) \; e_\lambda \otimes e_\mu$$

$$(p, \mathrm{sym}(z^{k-r} \otimes e_\lambda) \otimes \mathrm{sym}(\bar{z}^{\ell-r'} \otimes e_\mu)) =$$

$$(p, \sum_{\substack{|\alpha|=k-r \\ |\beta|=\ell-r'}} \frac{(k-r)!}{\alpha!} \frac{(\ell-r')!}{\beta!} z^\alpha \bar{z}^\beta \; e_{\alpha+\lambda} \otimes e_{\beta+\mu})$$

$$D_z^r D_{\bar z}^{r'} p(z,\bar z) = \sum_{|\alpha|=k-r \ |\beta|=l-r'} (e_\alpha \otimes e_\beta , z^{k-r} \otimes \bar z^{\ell-r'})$$

$$\sum_{\substack{|\lambda|=r\\|\mu|=r'}} \frac{r!r'!}{\lambda!\mu!} \frac{k!\ell!}{\alpha!\beta!} (p, e_{\alpha+\lambda} \otimes e_{\beta+\mu}) \cdot e_\lambda \otimes e_\mu .$$

On calcule:

$$\sum_{\alpha,\beta,\lambda,\mu} \frac{1}{\alpha!\beta!} \frac{r!r'!(k!\ell!)^2}{(k-r)!(\ell-r')!\lambda!\mu!} |(p,e_{\alpha+\lambda}\ e_{\beta+\mu})|^2 =$$

$$\sum_{\substack{|\gamma|=k\\|\delta|=\ell}} |(p,e_\gamma \otimes e_\delta)|^2 \sum_{\substack{\alpha+\lambda=\gamma\\ \beta+\mu=\delta}} \frac{1}{\alpha!\beta!\lambda!\mu!} (\ )!$$

qui représente la norme de $D_z^r D_{\bar z}^{r'} p$ comme opérateur de $(\hat{\odot}_{k-r} Z \hat{\otimes} \hat{\odot}_{\ell-r'} Z)$ $\otimes (\hat{\odot}_r Z \otimes \hat{\odot}_{r'} Z)$ soit:

$$\sum_{|\gamma|=k,\ |\delta|=\ell} P(p,e_\gamma \otimes e_\delta)|^2 (\gamma!\delta!)^{-1} \frac{(k!\ell!)^2}{(k-r)!^2} \frac{k!\ell!}{(\ell-r')!^2} = \left(\frac{k!}{(k-r)!} \frac{\ell!}{(\ell-r')!}\right)^2 \|p\|^2_{k\ell}$$

donc la norme de $D_z^r D_{\bar z}^{r'} p \cdot D_z^{r'} D_{\bar z}^r q$ dans $(\hat{\odot}_{k-r} Z \hat{\otimes} \hat{\odot}_{\ell-r} Z) \hat{\otimes} (\hat{\odot}_{m-r'} Z \hat{\otimes} \hat{\odot}_{n-r} Z)$

est au plus:

$$\frac{k!\ell!m!n!}{(k-r)!(\ell-r')!(m-r')!(n-r)!} \|p\|_{k\ell} \|q\|_{mn}$$

comme les opérateurs de symétrisation de $\hat{\odot}_{k-r} Z \hat{\otimes} \hat{\odot}_{m-r'} Z \to \hat{\odot}_{k+m-r-r'} Z$ sont de norme 1. On en déduit que le polynôme de Hulbert Schmidt:

$$Z^{k+m-r-r'} \times Z^{\ell+n-r-r'} \ni (z,\bar z) \to \langle D_z^r D_{\bar z}^{r'} p(z,\bar z) \cdot D_z^{r'} D_{\bar z}^{r} q(\ )\rangle$$

a une norme majorée par:

$$\frac{k!\ell!m!n!}{(k-r)!(\ell-r')!} \frac{1}{(m-r')!(n-r)!} \|p\|_{k\ell} \|q\|_{mn}$$

dans $\hat{\odot}_{k+m-r-r'} Z \hat{\otimes} \hat{\odot}_{\ell+n-r-r'} Z$.

Le fait que r est effectivement donné par la formule indiquée resulte de ce que:

$$H_{\alpha\beta}\ H_{\gamma\delta} = \sum_{\substack{p \le \inf(\alpha,\delta)\\ q \le \inf(\beta,\gamma)}} \frac{\alpha!\gamma!\beta!\delta!}{(\alpha-p)!(\beta-q)!(\gamma-q)!(\delta-p)!} \frac{1}{p!q!} H_{\alpha+\gamma-p-q\ \beta+\delta-p-q}$$

donc:

$$\|\tilde r\|_2^2 = \sum_{r,r'} \left(\frac{k!m!\ell!n!}{(k-r)!(n-r')!(\ell-r')!(n-r)!r!r'!}\right)^2$$

$$\frac{(k+m-r-r')!}{k!\ell!}\ \frac{(\ell+n-r-r')!}{m!n!}\ \|\tilde p\|_2^2\ \|\tilde q\|_2^2$$

$$\|\tilde r\|_2^2 = \sum_{\substack{r\leq \inf(k,n)\\ r'\leq \inf(\ell,n)}} \binom{k+m-r-r'}{k-r}\binom{\ell+n-r-r'}{\ell-r'}\binom{k}{k-r}\binom{m}{m-r'}\binom{\ell}{\ell-r'}\binom{n}{n-r}\|\tilde p\|_2^2\ \|\tilde q\|_2^2$$

$$\|\tilde r\|_2^2 \leq \sum_{r,r'} \frac{(k+m)!}{(k-r)!(m-r)!(r+r')!}\ \frac{(\ell+n)!}{(\ell-r')!(n-r)!(r+r')!}\ \|\tilde p\|_2^2\ \|\tilde q\|_2^2 \leq$$

$$3^{k+m+\ell+n}\ \|\tilde p\|_2^2\ \|\tilde q\|_2^2 .$$

<u>Lemme 6</u>. Soit $1 \leq p < \infty$. Il existe $C(p) > 0$ telle que:

$$\|\tilde f\|_p < C(p)^{\ell+k}\ \|\tilde f\|_2 \quad \text{si} \quad \tilde f = f_{k\ell}\cdot H_{k\ell} .$$

<u>Démonstration</u>: Si $p = 2j$, $j \in N$, on exprime $\tilde f^j = \tilde f^{j-1}\ \tilde f$:

$$\tilde f^{j-1} = \sum_{r,r'} \tilde f_{j-1rr'}$$

où $\tilde f_{j-1rr'} = f_{j-1rr'}\ H_{(j-1)k-r-r',(j-1)\ell-r-r'}$ pour $r+r' \leq (j-1)k$ et $r+r' \leq (j-1)\ell$

$$\tilde f^j = \sum_{r,r'} \tilde f_{j-1rr'}\ \tilde f$$

donc:

$$\|\tilde f^j\|_2 \leq \sum_{r,r'} \|\tilde f_{j-1rr'}\|_2\ 3^{\frac{j(k+\ell)}{2} - (r+r')}$$

et:

$$\|\tilde f^j\|_2 = \|\tilde f\|_{2j}^j \leq C\ 3^{\frac{j(k+\ell)}{2}}\ \|\tilde f\|_2\ \|\tilde f^{j-1}\|_2$$

donc:

$$\|\tilde f^j\|_2 \leq C^j\ \|\tilde f\|_2^j\ 3^{\frac{j(j+1)}{4}(k+\ell)}$$

soit:

$$\|\tilde f\|_{2j} \leq C^{k+\ell}(j)\ \|\tilde f\|_2 .$$

On adopte désormais la notation $\|\ \|_{p,u}$ pour la norme de $L^p_{\nu_u}(B)$.

On prouve:

<u>Proposition 2</u>. Le noyau L_u applique $\mathcal{C}_1^{m-1-\delta}(B)$ dans $L^p_{\nu_u}$ $1\leq p<\infty$ et on a une inégalité:

$$\|L_u G\|_{p,u} \leq C(p)\ |(\int_0^{+\infty} \phi(t)dt$$

$$\int (|(G(z+Z),\bar{z}/t^{1/2})|^2 + |G(z+Z)|^2)\, d\nu_t(z)\, d\mu_u(Z))^{1/2} +$$

$$+(\int_0^{+\infty} \psi(t)dt\ |(G(z+Z),\bar{z}/t^{1/2})|^p\, d\nu_t(z)\, d\nu_u(Z))^{1/p}| \quad (*)$$

avec: $\mu_u = \nu_u + u \int_0^1 \nu_{su} ds$, $\int_0^{+\infty} \phi(t)(1+t)^{m-\delta} dt < +\infty$, $\psi(t) \in L^1_c(\mathbb{R})$ (à support compact).

__Démonstration__. Vu le Lemme 6 et la preuve de la Proposition 1, les estimations de normes $L^p_{\nu_u}$ se déroulent de la même façon que les normes $L^2_{\nu_u}$ pour les termes $L''_u(G)(w)$. Pour le terme $L^o_u(G)(t,w)$ on distinguera cette fois les contributions correspondant à $t \leq C'(p)$ ($C'(p)$ sera déterminé plus loin). Pour $t \leq C'(p)$ seule l'estimation de $t^{-1/2}$ $(G(z+w),\bar{z}/t^{1/2})d\nu_t(z)$ change et donne le second terme de (*). En fait, il faut seulement revenir sur l'estimation de $R_{mu}(G)(w)$:

$$(R^m_{u\tau}G)(t,w) = \sum_{\ell+k\geq m+1} t^{-1/2} \frac{(\ell+k)!}{(\ell+k-(m+1))!}$$

$$\sum_{\substack{|\alpha|=\ell \\ |\beta|=k}} (\iint (H(x+y)\cdot\bar{y}/t'^{1/2})\bar{\mathcal{H}}_{\alpha\beta}(x/u^{1/2})d\nu_{t'}(x))\mathcal{H}_{\alpha\beta}(w/u^{1-2})$$

On a vu que:

$$(H(x+y),\bar{y}/t'^{1/2})\bar{\mathcal{H}}_{\alpha\beta}(x/u^{1/2})d\nu_{t'}(y)d\nu_u(x) =$$

$$(ut'^{-1})^{\frac{\ell+k}{2}} (\iint (H(x+y),\bar{y}/t'^{1/2})\mathcal{H}_{\beta\alpha}(yt'^{-1/2})d\nu_{t'}(y)d\nu_u(x) +$$

$$\sum_{(i)\leq\beta} \beta_i^{1/2} \iint H_i(x+y)\ \mathcal{H}_{\beta-(i),\alpha}(yt'^{-1/2})d\nu_{t'}(y)\ d\nu_u(x)).$$

Utilisant le Lemme 6, on obtient:

$$\|R^m_{u\tau}G\|_{pu} \leq \sum_{\ell+k\geq m+1} t^{-1/2} \frac{(\ell+k)!}{(\ell+k-(m+1))!} C(p)^{\ell+k}$$

$$(\sum_{\substack{|\alpha|=\ell \\ |\beta|=k}} |\iint (H(x+y),\bar{y}/t'^{1/2})\bar{\mathcal{H}}_{\alpha\beta}(x/u^{1/2})d\nu_u(x)|^2)^{1/2}.$$

La seconde somme est majorée par:

$$(ut'^{-1})^{k+\ell/2}\ \Big(\sum_{\substack{|\alpha|=\ell\\|\beta|=k}}\ \Big(\Big|\iint (H(x+y)\bar{y}/t'^{1/2})\mathcal{H}_{\beta\alpha}(yt'^{-1/2})d\nu_{t'}d\nu_u\Big|^2 +$$

$$+\ k\sum_{(i)\le\beta}\Big|\iint H_i(x+y)\mathcal{H}_{\beta-i,\alpha}(yt'^{-1/2})d\nu_{t'}d\nu_u(x)\Big|^2\Big)\Big)^{1/2}$$

$$\|R^m_{u\tau}G\|_{pu} \le C\tau^{-1/2}\lambda(t,\tau)\Big(\iint |(H(x+y)\bar{y}/t'^{1/2})|^2 d\nu_{t'}d\nu_u(x) +$$

$$+\iint |H(x+y)|^2\, d\nu_{t'}d\nu_u\Big)^{1/2}$$

avec:

$$\lambda^2(t,\tau) = \sum_{\ell+k\ge m+1} C(p)^{2(\ell+k)}\Big[\frac{(\ell+k)!}{((\ell+k-(m+1))!}\Big]^2\ (ut'^{-1})^{k+\ell}(1+k)$$

$t' = t\tau^{-2}$, $\lambda^2(t,\tau) \le C(t^{-1}\tau^2)^{m+1}$ si $t \ge C'(p)$ $0 < u \le 1$ $0 \le \tau < 1$.

Donc:

$$\|R_{mu}G\|^2_{pu} \le Ct^{-(m+2)}\iint (|(G(y+z),\bar{y}/t^{1/2})|^2 + |G(y+z)|^2)d\nu_t(y)d\nu_u(z)$$

pour $t \ge C'(p)$.

4. CONVERGENCE PRESQUE SÛRE ET RÉGULARITÉ DE LA SOLUTION

Dans les paragraphes précédents on a vu que les $f_n(X) = (L_oG)|_{Z_n}\cdot s_n$ convergent dans $L^p_{\nu_t}$ vers une solution L_tG au sens des distributions de $\bar{\partial}f = G$. On va prouver maintenant la convergence presque sûre.

<u>Proposition 4.1</u>. Soient $G \in G_1^{m-1-\delta}(B)$ de classe C^1, dG à croissance exponentielle et $\bar{\partial}G = 0$ $\forall\, t \in R^+$ $(L_oG)|_{Z_n}\cdot s_n(x)$ converge ν_t presque sûrement.

<u>Démonstration</u>. Soit $H_{n,u}(w_n,k)$ le noyau de l'opérateur $\mathcal{E}_u(L_uG/s_n)$: $G_1^{m-1-\delta}(B) \to L^2_{\nu_u}(Z_n)$. On vérifie que:

$$H_{nu}(w_n,k)\cdot ik = e^{-i(w_n\bar{k}+\bar{w}_nk)}\ e^{-u|k'|^2} - q_m(w_n,k).$$

Mais le noyau de l'opérateur $G \to (L_oG)|_{Z_n}$ $G_1^{m-1-\delta}(B) \to L^2_{\nu_u}(Z_n)$ (la continuité résulte du fait que l'orsq'une suite G_i G en tout point et en restant uniformément majorée par $(1+\|z\|)^{m-1-\delta}$, les $(L_oG_i)(z) \to (L_oG)(z)$ $\forall\, z \in Z$ et restent uniformément majorés par $(1+|z|)^m$)

est donné par: $H_{no}(w_n,k)$ où

$$H_{no}(w_n,k)\cdot ik = e^{-i(w_n\bar{k}+\bar{w}_n k)} - q_m(w_n,k)$$

donc:

$$-H_{no}(w_n,k)\cdot ik + H_{nu}(w_n,k)\cdot ik = N_u^n(w_n,k)\cdot ik$$

avec:

$$N_u^n(w_n,k) = \frac{i\bar{k}'}{|k'|^2}(1-e^{-u|k'|^2})\, e^{-i(w_n\bar{k}+\bar{w}_n k)} .$$

On constate que l'opérateur de symbole N_u^n s'écrit:

$$(N_u^n G)(w_n) = \int_0^1 \int_{Z^n} (G(w_n+Z),\bar{Z}/(su)^{1/2})u^{1/2}\, s^{-1/2}\, d\nu'_{su}(Z)ds.$$

Introduisant la mesure $\mu'_u = \int_0^1 s^{-1/2}\,\|Z\|/(su)^{1/2}.\nu'_{su}ds$ on constate que:

$$|N_u^n G(w_n)|^2 \le u\|\mu'_u\| \int_{Z^n} \|(1-\sigma_n)G(w_n+Z)\|'^2\, d\mu'_u(Z).$$

Puis on note que $\int_{Z^n} (1+\|Z\|)^{2m}\, d\mu'_u(Z) \le C\|j\circ(1-s_n)\|_2$ et tend donc vers zéro quand $n \to \infty$:

$$|(N_u^n G)(w_n)| \le u\, \varepsilon_n(1+\|w_n\|)^m \quad \text{avec} \quad \lim_{n\to\infty} \varepsilon_n = 0.$$

Ceci permet donc de conclure que pour $G \in G_1^{m-1-\delta}(B)$ de classe C^1, $\bar{\partial}G = 0$ entraîne:

$$L_o G|_{Z_n}\cdot s_n = \mathcal{E}_u(L_u G/s_n)\cdot s_n + (N_u^n G)\cdot s_n .$$

Or, ν_u presque sûrement $\phi_n^u = \mathcal{E}_u(L_u G/s_n)\cdot s_n$ converge à cause du théorème de convergence de martingales (on rappelle que $L_u G \in L^2_{\nu_u}$); on a donc la proposition. On va maintenant étudier la convergence des fonctions $h \in Z \to \phi_n^u(x+h)$, ce qui permettra de prouver des résultats de Z continuité pour la fonction $\lim_{n\to\infty} (L_o G)|_{Z_n}\cdot s_n(x)$ définie sur B là ou la limite existe.

<u>Proposition 4.2</u>. Soient $G \in G_1^{m-1-\delta}(B)$ de classe C^1, dG à croissance exponentielle et $\bar{\partial}G = 0$. Pour tout $t > 0$, la propriété suivante notée P est vérifiée ν_t presque sûrement:

- les fonctions $h \in Z \to (L_o G)|_{Z_n}\cdot s_n(x+h)$ convergent uniformément sur les compacts de Z; de plus

• les dérivées $d(L_o G|_{Z_n} \cdot s_n)(x+h)$ convergent uniformément dans $Z' \oplus Z'$ sur les compacts de Z.

Démonstration. On va prouver que ν_u presque sûrement $\phi_n^u(x+h) \to \phi_\infty^u(x+h)$ uniformément sur les compacts de Z et qu'il en est de même des dérivées.

$\phi_n^u(x+h)$ s'exprime comme une somme de termes correspondant à L''_{1u}, L''_{2u}, $\int_0^\infty L_{ou}(t.)dt$. Pour ce qui est d'un terme comme:

$$L''_{1u}(G)(x) = \sum_{\alpha,\beta \quad |\alpha|+|\beta|\le m} a_{\alpha\beta} \, H_{\alpha\beta}(xu^{-1/2})$$

où $\sum_{\alpha,\beta} |a_{\alpha\beta}|^2 < +\infty$

$$\mathcal{E}(L''_{1u}G \; s_n)\cdot s_n(x+h) = \sum_{p,q \;\; \alpha,\beta\in I_n} H_{\alpha\beta}(xu^{-1/2}) \sum_{\substack{|\gamma|=p \\ |\delta|=q}} \; \sum_{(\gamma,\delta)\in I_n} a_{\alpha+\gamma \; \beta+\delta} \, h^\gamma \bar{h}^\delta$$

que l'on écrit plutôt:

$$\mathcal{E}_u(L''_{1u}G \; s_n)\cdot s_n(x+h) = \sum_{p,q} (Y'_n(x), \sigma_n(h)^p \otimes \sigma_h(\bar{h})^q)$$

avec:

$$Y'_n(x) = \sum_{\alpha,\beta\in I_n} H_{\alpha\beta}(xu^{-1/2}) \Big(\sum_{\gamma,\delta} a_{\alpha+\gamma \; \beta+\delta} \, e_\gamma \otimes e_\delta \Big)$$

I_n désigne l'ensemble des multi-indices α tels que $\alpha_i = 0$ si $i > n$.

Si $Z_{pq} = \hat{\bigodot}_p Z \hat{\otimes} \hat{\bigodot}_q Z$ les $Y'_n(x)$ forment une martingale à valeurs dans Z_{pq} et:

$$\sup_n \int |Y'_n(x)|^2_{pq} \, d\nu_u(x) \le \sum_{\alpha,\beta,\gamma,\delta} |a_{\alpha+\gamma,\beta+\delta}|^2 < +\infty.$$

Un théorème de convergence de martingales vectorielles permet de conclure que ν_u presque partout $Y'_n(x) \to Y'_\infty(x)$ dans Z_{pq}. On note:

$$\sigma_n^{p,q} = \hat{\otimes}_p \, \sigma_n \, \hat{\otimes} \, \hat{\otimes}_q \, \sigma_n .$$

Comme:

$$|\mathcal{E}_u(L''_{1u}G/s_n)\cdot s_n(x+h) - \Sigma\ (Y'_\infty(x),\ h^p \otimes \bar{h}^q)| \le$$

$$\sum_{p,q} |h|^p\ |h|^q\ (|Y'_n(x)-Y'_\infty(x)|_{p,q} + |\sigma_n^{p,q}\ Y'_\infty(x)-Y'_\infty(x)|.$$

On en déduit que ν_u presque partout la convergence est uniforme sur les parties bornées de Z. Pour ce qui est des dérivées, on déduit de même la convergence. Un terme comme L''_{2u} donne lieu à une expression de la même forme.

On étudie maintenant un terme comme $\int_0^c L_{o,u}(t.)dt$ (c est une constante assez grande). La contribution a deux formes différentes, la première est de la forme:

$$\sum_{\substack{\alpha,\beta\\ |\alpha|+|\beta|\le m}} \left(\int_0^c a_{\alpha\beta}(t)dt\right) H_{\alpha\beta}(xu^{-1/2})$$

et on a déjà vu que:

$$\int_0^c \left(\sum_{\alpha,\beta} |a_{\alpha\beta}(t)|^2\right)^{1/2} dt < +\infty$$

et comme:

$$\left(\sum_{\alpha,\beta} \left|\int_0^c a_{\alpha\beta}(t)dt\right|^2\right)^{1/2} \le \int_0^c \left(\sum_{\alpha,\beta} |a_{\alpha\beta}(t)|^2\right)^{1/2} dt$$

on est ramené au cas précédent. L'autre terme est:

$$\alpha(x) = \int_0^c t^{-1/2}(\ (G(x+y),\bar{y}/t^{1/2})d\nu_t(y))dt$$

$$\mathcal{E}_u(\alpha/s_n)\cdot s_n(x+h) = \alpha_n(x+h) =$$

$$= \int_0^c t^{-1/2}dt \int_{Z^n} (G(s_n(x+h)+y+Z),\bar{y}/t^{1/2})d\nu_t(y)d\nu'_u(Z)$$

$$\alpha(x+h) = \int_0^c t^{-1/2} \int (G(x+y+h),\bar{y}/t^{1/2})d\nu_t(y)dt.$$

On peut estimer $\alpha(x) - \alpha_n(x+h)$ en notant que $\forall\ x \in B$, $\forall\ R > 0$, $\forall\ \epsilon > 0$, $\forall\ \rho > 0$, $\exists\ \delta > 0$, $\exists\ n_o$ tel que:

$$\sup_{z\in K_\rho,\|Z\|<\delta,|h|\le R} \|G(x+h+z)-G(s_n(x)+s_n(h)+z+Z\|' \le \epsilon$$

si $n \ge n_o$. Ce qui résulte de la continuité de G et de la compacité de $\{h \in Z,\ |h| \leq R\}$.

$$|\alpha(x+h)-\alpha_n(x+h)| \leq C \int_0^c t^{-1/2}(\epsilon + \frac{1}{\rho} + \frac{1}{\delta}\|j\circ(1-s_n)\|_2)(1+\|x\|)^{m-1}\,dt$$

et donc:

$$\forall\ x \in B,\ \forall\ R > 0,\ \lim_{n\to\infty}\ \sup_{|h|\leq R}\ |\alpha(x+h)-\alpha_n(x+h)| = 0.$$

Montrons maintenant la convergence de la dérivée première:

$$(d\alpha_n)(x+h)\cdot k = \int_0^c t^{-1/2}dt \int((dG)(s_n(x+h)+y+z)\cdot s_n(k),\bar{y}/t^{1/2})d\nu_t(y)d\nu'_u(Z)$$

$$(d\alpha)(x+h)\cdot k = \int_0^c t^{-1/2}dt \int((dG)(x+h+y)\cdot k,\bar{y}/t^{1/2})d\nu_t(y)$$

$$|(d\alpha)(x+h)-(d\alpha_n)(x+h)| \leq \int_0^c t^{-1/2}\,dt$$

$$\int(\|dG(x+h+y)-(dG)(s_n(x+h)+y+Z)\|_{\mathcal{L}(B,B')}\ \|y\|/t^{1/2}\ d\nu_t(y)d\nu'_u(Z)$$

$$+ \int_0^c t^{-1/2}\,dt \int \|dG(x+h+y)\|\ \|y\|/t^{1/2}\ d\nu_t(y)(\sup_{|k|\leq 1}\|(1-s_n)k\|).$$

Pour le premier termc, on raisonnc comme plus haut; pour le second on note que:

$$\lim_{n\to\infty}\ \sup_{|k|\leq 1}\ \|(1-s_n)k\| = 0$$

donc $\forall\ x \in B,\ \forall\ R > 0\ \lim_{n\to\infty}\ \sup_{|h|\leq R}\ |(d\alpha_n)(x+h)-(d\alpha)(x+h)| = 0.$

On considère maintenant la contribution $\int_c^\infty L_{o,u}(t.)dt$, bien entendu seul le terme:

$$Z(x) = \int_c^{+\infty}\int_0^1 (1-\tau)^m\quad R^m_{u,\tau}(t,G)(x)\,dt\,d\tau$$

doit être analysé précisément. On pose:

$$Z(x,t,\tau) = R^m_{u,\tau}(t,G)(x)$$

$A(x+h,t,\tau)$ a pour noyau:

$$(\frac{\partial}{\partial t})^{m+1}(\bar{e}^{i\tau(x\bar{k}+\bar{x}k)-(1-\tau^2)u|k|^2}\ \bar{e}^{i\tau(h\bar{k}+\bar{h}k)})\bar{e}^{t|k|^2}\ i\bar{k} =$$

$$\sum_{j=0}^{m+1} \frac{(m+1)!}{(m+1-j)!\,j!}\ \bar{e}^{i\tau(h\bar{k}+\bar{h}k)}\ R^{m-j}_{u,\tau}(x,k)\ \tau^j(h\bar{k}+\bar{h}k)^j$$

ce qui s'écrit:

$$\sum_{0\le p+q\le m+1} c_{p,q}(h\bar{k})^p(\bar{h}k)^q\, R_{u,\tau}^{m-(p+q)}(x,k) \quad \tau^{p+q}\, e^{-i\tau(h\bar{k}+\bar{h}k)}.$$

On exprime donc:

$$Z(x+h,t,\tau) = \sum_{0\le p+q\le m+1} Z_{pq}(x,t,\tau,h)$$

$$Z_{p,q}(x,t,\tau,h) = \sum_{|\gamma|=p,\,|\delta|=q} p!q!\,\frac{h^{\gamma}\bar{h}^{\delta}}{\gamma!\delta!} \sum_{\substack{\alpha,\beta \\ |\alpha|+|\beta|\ge m+1-(p+q)}} \rho_{p,q}(t,\tau,|\alpha|,|\beta|)$$

$$\mathcal{H}_{\alpha\beta}(xu^{-1/2})\,\frac{(\gamma+\beta)!(\delta+\alpha)!}{\beta!\alpha!}^{1/2}\,(\lambda_{\alpha,\beta,\gamma,\delta}(x'+h)d\nu_u(x'))$$

avec:

$$\rho_{p,q}(t,\tau,|\alpha|,|\beta|) = (ut'^{-1})^{\frac{|\alpha|+|\beta|}{2}}\; t^{-\frac{p+q+1}{2}}\; \tau^{-(m+1)-p-q}$$

$$\frac{(|\alpha|+|\beta|)!}{(|\alpha|+|\beta|-(m+1-p-q))!}\;\tau^{p+q}$$

$$\lambda_{\alpha\beta\gamma\delta}(x'+h) = (\mathcal{H}_{\gamma+\beta,\delta+\alpha}(yt'^{-1/2})(H(x'+y+h),\bar{y}/t'^{1/2}) -$$

$$\sum_{(i)\le\gamma+\beta}(\gamma_i+\beta_i)^{1/2}\,H_i(x'+h+y)\,\mathcal{H}_{\gamma+\beta-(i),\delta+\alpha}(yt'^{-1/2}))d\nu_{t'}(y)$$

où, comme plus haut:

$$t' = t\tau^{-2} \qquad H(x) = G(\tau x+z)d\nu_{(1-\tau^2)u}(z).$$

On va poser:

$$Y_n(t,\tau,x) = \sum_{\alpha,\beta\in I_n,\,|\alpha|+|\beta|\ge m+1-(p+q)} \rho_{p,q}(t,\tau,|\alpha|,|\beta|)\mathcal{H}_{\alpha\beta}(xu^{-1/2})$$

$$\left(\frac{(\gamma+\beta)!(\delta+\alpha)!}{\beta!\alpha!}\right)^{1/2} \sum_{\gamma,\delta}\;\sum_{\substack{|\gamma|=p\\|\delta|=q}} \lambda_{\alpha\beta\gamma\delta}(x')\,\frac{p!q!}{\gamma!\delta!}\,e_\gamma\otimes e_\delta$$

$$Y_n(t,\tau,x)\in C^{o}_{t,\tau}(L^2_{\nu_u}(B)\,\hat{\otimes}\,Z_{p,q})).$$

Si $Z_n(t,\tau,a,h) = \langle\, Y_n(t,\tau,x)e^{-(x'\bar{h}_n+\bar{x}'h_n)u^{-1}-|h_n|^2u^{-1}} d\nu_u(x'),\ \sigma_n(h)^p \otimes \sigma_n(\bar{h})^q\rangle$

on constate que:

$$Z_n(t,\tau,\alpha,h) = \mathcal{E}_u(Z_{p,q}(t,\tau,x)/s_n)\cdot s_n(x+h)$$

$$Z_\infty(t,\tau,x,h) = \langle\ Y_\infty(t,\tau,x)e^{-(x'\bar{h}+\bar{x}'h)u^{-1}-|h|^2\ u^{-1}}\ d\nu_u(x'),\ h^p \otimes \bar{h}^q\rangle$$
$$= Z_{p,q}(t,\tau,x+h).$$

Pour t et τ fixés, les $Y_n(t,\tau,x)$ forment une martingale vectorielle à valeurs dans l'espace de Hilbert: $H = L^2_{\nu_u}(B,Z_{p,q})$. Il suffit de vérifier que

$$\sup_n \int |Y_n|^2_H\ d\nu_u(x) < +\infty$$

c'est-à-dire:

$$\sum_{\alpha,\beta,\gamma,\delta} \rho_{p,q}()^2\ \frac{(\gamma+\beta)!\,(\delta+\alpha)!}{\gamma!\beta!\ \alpha!\delta!}\Big(\int d\nu_u(x)\big(\Big|\ \mathcal{H}_{\gamma+\beta\ \ \delta+\alpha}(H(x'+y),\bar{y}/t'^{1/2})\ d\nu_{t'}\Big|^2 +$$
$$(|\gamma|+|\beta|)\ \sum_{(i)\le()}\ \Big|\int H_i()\ \mathcal{H}_{\beta+\gamma-(i),\delta+\alpha}\ d\nu_t\Big|^2\big).$$

Observant que $\sum_{\ell,k} \rho_{p,q}(t,\tau,\ell,k)^2(1+k+\ell)^N \le Ct^{-(m+2)}$ si $t \ge c$, puis que:

$$\sum_{\substack{|Z|=k+p\\|\eta|=\ell+k}} \Big(\sum_{\substack{\gamma+\beta=Z\\ \alpha+\delta=\eta}} \frac{(\gamma+\beta)!}{\alpha!\beta!}\ \frac{(\delta+\alpha)!}{\alpha!\delta!}\ \Big|\int \mathcal{H}_{Z,\eta}(H(x'+y),\bar{y}/t'^{1/2}))d\nu_{t'}\Big|^2 +$$

$$\sum_{i=1}^{\infty}\ \sum_{Z,\eta}\ \Big|\int H_i(x'+y)\mathcal{H}_{Z,\eta}\ d\nu_{t'}\Big|^2\ \Big(\sum_{\substack{\gamma+\beta=Z+(i)\\ \alpha+\delta=\eta}} \frac{(\gamma+\beta)!}{\gamma!\beta!}\ \frac{(\alpha+\delta)!}{\alpha!\delta!}$$

est majorée par:

$$\frac{(k+p)!}{k!p!}\ \frac{(\ell+q)!}{\ell!q!}\ \Big(\int |H(x'+y),\bar{y}/t'^{1/2})|^2\ d\nu_{t'} + \int |H(x'+y)|^2\ d\nu_{t'}\Big)$$

on a obtenu que:

$$\sup_n \int |Y_n|^2 d\nu_u \le Ct^{-(m+2)}\Big(\iiint \big(|(G(y+z),\bar{y}/t^{1/2})|^2+|G(y|z)|^2\big)\ d\nu_t(y)d\nu_u(z)\Big).$$

En tous cas $\forall\ t,\tau\ \nu_u$ presque sûrement $Y_n(t,\tau,x) \to Y_\infty(t,\tau,x)$ dans $L^2_{\nu_u}(B,Z_{pq})$. Soit:

$$N=\{(t,\tau,x)\ |\ 0\le\tau\le 1\ t>c \text{ et } Y_n(t,\tau,x) \text{ ne converge pas vers } Y_\infty(t,\tau,x)\}$$

N est $dt \otimes d\tau \otimes \nu_u$ mesurable. Or $(t,\tau)N_{t,\tau}$ est ν_u négligeable donc N est $dt \otimes d\tau \otimes \nu_u$ négligeable.

Donc, si $x \notin N'$, $Y_n(t,\tau,x) \to Y_\infty(t,\tau,x)$ $dt \otimes d\tau$ presque partout. Il résulte également des théorèmes de convergence de sous martingales que:

$$(\sup_n |Y_n(t,\tau,x)|_H) \in L^2_{\nu_u}(B)$$

et donc:

$$\int_c^\infty \int_0^1 (\sup_n |Y_n(t,\tau,x)|)^2 d\nu_u(x)dtd\tau < \int_0^{+\infty}\int_0^1 (\sup_n |Y_n|^2 d\nu_u)dtd\tau < +\infty$$

$\sup_n Y_n(t,\tau,x)_H$ est donc ν_u presque partout dt dτ intégrable sur $[c,\infty[\times[0,1]$; le théorème de convergence dominée assure donc que ν_u presque partout:

$$f_n = \int_c^\infty \int_0^1 Y_n(t,\tau,x)dtd\tau \to \int_c^\infty \int_0^1 Y_\infty(t,\tau,x)dtd\tau = f_\infty$$

dans $L^2_{\nu_u}(B,Z_{pq})$. Mais si $Z_n(x+h) = T_{n,h}\cdot \int_c^\infty \int_0^1 Y_n(t,\tau,x)dtd\tau =$

$= \mathcal{E}_u(Z/s_n)\cdot s_n(x+h)$ où T_{nh} est l'opérateur:

$$f \in L^2_{\nu_u}(B,Z_{pq}) \to \int (\sigma^n_{pq}(f)(x'),h^p\otimes\bar{h}^q)e^{-(x'\bar{h}_n+\bar{x}'h_n)u^{-1}-|h_n|^2u^{-1}} d\nu_u(x').$$

Lorsque h décrit un borné de Z, les $T_{n,h}$ forment un ensemble borné de formes linéaires sur $L^2_{\nu_u}(B,Z_{pq})$. On écrit:

$$T_{n,h}(f_n) - T_{\infty,h}(f_\infty) = T_{n,h}(f_n-f_\infty) + T_{n,h}(f_\infty) - T_{\infty,h}(f_\infty)$$

$$|T_{n,h}(f_n) - T_{\infty,n}(f_\infty)| \le C\|f_n-f_\infty\| + |T_{n,h}(f_\infty) - T_{\infty,h}(f_\infty)|.$$

On a vu que le premier terme converge vers zéro ν_u presque sûrement. On majore le deuxième par:

$$|h|^{p+q}(\int\|(1-\sigma^{p,q}_n)f_\infty\|(x'+h)d\nu_u(x')+ \int\|e_{h_n}(x')-e_h(x')| d\nu_u(x')).$$

Mais on peut majorer le second terme par:

$$\int\|f_\infty\|^2 d\nu_u(x') \int |e_{h_n}(x') - e_h(x')|^2 d\nu_u(x')$$

avec:

$$e_h(x') = \exp(\frac{-(x',\bar{h})-(\bar{x}',h)}{u}) \exp(-\frac{|h|^2}{u}).$$

La conclusion résulte de ce que les $h_n = s_n(h)$ convergent vers h uniformément sur tout compact et du fait que $h \in Z \to e_h \in L^2_{\nu_u}$ est continue. On a donc la convergence uniforme sur les compacts de Z de $Z_n(x+h)$.

Pour la dérivée 1er on exprime $(dZ)(x+h)\cdot\alpha$ à l'aide d'une formule analogue aux précédentes, mais $h^p \otimes \bar{h}^q$ sera remplacé par $\mathrm{sym}(h^p \otimes \alpha) \otimes \bar{h}^q$ ou par $h^p \otimes \mathrm{sym}(\bar{h}^q \otimes \bar{\alpha})$.

On obtient donc que ν_u presque sûrement $dZ_n(x+h)$ $dZ(x+h)$ dans $Z' + Z'$ uniformément sur les compacts de Z; les $dZ_n(x+h)$ sont, en outre, uniformément majorés quand h décrit un borné de Z.

On obtient la proposition en notant que $N_u^n(G)(w_n) \to 0$ uniformément sur les bornés de B et qu'il en est de même de la dérivée première.

<u>Corollaire</u>. On désigne par Ω la partie de B où P est vérifiée et on définit une fonction $LG(x)$ sur Ω par:

$$\lim_{n\to\infty} (L_o G)\big|_{Z_n} \cdot s_n(x) = LG(x).$$

Or $\forall\, x \in \Omega$, $x+Z \subset \Omega$; la fonction $LG(x)$ est donc $Z\cdot C^1$ sur Ω.

<u>Théorème</u>. Soit $m \in \mathbb{N}$, $\delta > 0$. Soit $G \in \mathcal{Q}^{m-1-\delta}(B)$ de classe C^1, on suppose que dG croît mons vite qu'une exponentielle et que $\bar{\partial}G=0$. Il existe une fonction $LG(x)$ définie sur B ayant les propriétés suivantes:

α) $LG(x)$ est ν_t mesurable $t > 0$. $\mathrm{Classe}_t(LG) = L_tG \in \bigcap_{1\le p<+\infty} L^p_{\nu_t}$; L_tG est une solution au sens des distributions de $\bar{\partial}u = G$.

β) Il existe une partie Ω B $\nu_t^*(\Omega^c) = 0$ $\forall\, t > 0$ telle que:

- $0 \in \Omega$; $x \in \Omega \to x+Z \subset \Omega$
- $\forall\, x \in \Omega$; $h \in Z \to LG(x+h)$ est de classe C^1 et vérifie $\bar{\partial}(LG)(x+h) = G(x+h)$
- $LG\big|_Z = L_o(G)$ (solution trouvée dans [4]); $\lim_{n\to\infty} (L_oG)\big|_{Z_n} \cdot s_n(x) = $ $= LG(x)$ si $x \in \Omega$, la convergence a également lieu dans $L^p_{\nu_t}$ $\forall\, t$, $\forall\, p$.

BIBLIOGRAPHIE

[1] Chatterji, S.D., Martingale convergence and the Radon Nicodym theorem. Math. Scand. 22, p. 21-41 (1968).

[2] Coeuré, G., Pathologie de la d" cohomolojie en dim. infinie. Compte Rendus Acad. Sciences Paris, à paraître.

[3] Gross, L., Abstract Wiener Spaces. Fifth Berkeley Symposium in Probability, 1965, p. 30-42.

[4] Henrich, C.J., The $\bar{\partial}$ equation with polynomial growth. on Hilbert space. Duke Math. Journal, p. 279-306, 1973.

[5] Krée, P., Séminaire sur les e.d.p. en dim. infinite multigraphié, 1976-1976.

[6] Lascar, B., C.N.S. d'ellipticité en dim. infinie. Com. in P.D.E. (2) 1 - 1977, p. 31-67.

[7] Nachbin, L., Topology on spaces of holomorphic mappings. Springer Verlag, Berlin, 1969.

[8] Neveu, J., Calcul des probabilités, Masson, 1970.

[9] Raboin, P., Etude de l'équation $\bar{\partial} f = g$ sur un espace de Hilbert. C.R. Acad. Sci. t 282, 1976 et décembre 1977.

[10] Schwartz, L., Radon measures on arbitrary topological spaces and cylindrical measures. Oxford, University press, 1973.

SILVA-HOLOMORPHY TYPES

Mário C. Matos and Leopoldo Nachbin

Instituto de Matemática, Estatística e
Ciência da Computação - UNICAMP
Campinas, São Paulo, Brasil

Instituto de Matemática - UFRJ
Rio de Janeiro, RJ, Brasil

0. INTRODUCTION

This paper deals with the concept of Silva-holomorphy type which is the natural generalization to locally convex spaces of the nition of holomorphy type from one Banach space to another. This notion was introduced by L. Nachbin in [1] and [2]. He was motivated by previous works on nuclearly entire functions: [3] and [4]. Similarly the motivation for the concept of Silva-holomorphy type lies in papers dealing with nuclearly Silva-entire functions: [5], [6] and [7]. In [5] and [6] the space $H_{N,b}(E)$ has in general Silva-holomorphic functions which are not continuous. Under this point of view the introduction of the Silva-holomorphy type looks natural.

The α-holomorphy type for Banach space was introduced and studied by S. Dineen in [8]. For the so-called α-Silva-holomorphy type see the work of M. Bianchini [9] in these Proceedings.

The Silva-holomorphy type for locally convex spaces is introduced in §4. Paragraph 5 deals with the differentiation of Silva-holomorphy types proving results which are important for the development of the remaining paragraphs. In §6 a natural topology is introduced in spaces of Silva-holomorphic mappings of a given type. The bounded and relatively compact subsets for this topology are studied in §7 and §8.

We give some basic results with proofs in the first three paragraphs. The reason for the inclusion of these results in this paper is that they are not found in this form in any book or paper and their inclusion makes this paper self-contained.

1. MULTILINEAR MAPPINGS AND POLYNOMIALS

E and F will be complex locally convex spaces and U will be a non-void open subset of E. The set of all bounded closed absolutely convex subsets of E will be denoted by $\mathcal{B}_E$. If $B \in \mathcal{B}_E$ E_B will denote the vector subspace of E generated by B and normed by the Minkowsky functional associated to B which is denoted by $\|\cdot\|_B$.

__1.1 Definition__. A mapping f from U into F is S-continuous in U if for each $B \in \mathcal{B}_E$ the restriction of f to $U \cap E_B$ is continuous in $U \cap E_B$ for the induced topology of E_B. $C_S(U;F)$ will denote the vector space of all S-continuous mappings from U into F.

If E is metrizable and $(x_n)_{n=0}^{\infty}$ is a sequence in U converging to a x of U it is known that there is $B \in \mathcal{B}_E$ such that $x_n, x \in U \cap E_B$ $(n \in \mathbb{N})$ and $\lim_{n\to\infty} x_n = x$ for the topology induced by the norm of E_B. Hence for metrizable spaces continuity and S-continuity are equivalent concepts.

__1.2 Definition__. Let $m \in \mathbb{N}^* = \{1,2,...\}$. $\mathcal{L}_a(^mE;F)$ denotes the vector space of all m-linear mappings from $E^m = E \times ... \times E$ (m times) into F. The vector subspace of $\mathcal{L}_a(^mF;F)$ formed by the symmetric mappings is indicated by $\mathcal{L}_{as}(^mE;F)$. If $A \in \mathcal{L}_a(^mE;F)$ its symmetrization is the element A_s of $\mathcal{L}_{as}(^mE;F)$ defined by

$$A_s(x_1,\dots,x_m) = \sum_{\sigma\in\mathcal{T}_m} A(x_{\sigma(1)},\dots,x_{\sigma(m)})$$

for $x_1,\dots,x_m \in E$. $\mathcal{T}_m$ denotes the symmetric group of order m. $\mathcal{L}_b({}^mE;F)$ and $\mathcal{L}_{bs}({}^mE;F)$ indicate the respective vector subspaces of $\mathcal{L}_a({}^mE;F)$ and $\mathcal{L}_{as}({}^mE;F)$ formed by the mappings which are bounded on the bounded subsets of E^m. $\mathcal{L}({}^mE;F)$ and $\mathcal{L}_s({}^mE;F)$ are respectively the vector subspaces of $\mathcal{L}_a({}^mE;F)$ and $\mathcal{L}_{as}({}^mE;F)$ formed by the continuous mappings in E^m. If m is zero all the above vector spaces are defined to be equal to F. If A belongs to $\mathcal{L}_a({}^oE;F)$ we define $A_s = A$. When $F = C$ the above spaces are denoted by ommiting the letter F. Hence, for instance, we write $\mathcal{L}_a({}^mE)$ for $\mathcal{L}_a({}^mE;C)$.

<u>1.3 Proposition</u>. For $m \in \mathbb{N}^*$ $\mathcal{L}_b({}^mE;F)$ coincides with the vector space of all m-linear S-continuous mappings from E^m into F.

<u>Proof</u>. Each bounded subset of E^m is contained in some $B^m = B\times\dots\times B$ (m times) with $B \in \mathcal{B}_E$. On the other hand we have the topological isomorphism between $(E_B)^m$ and $(E^m)_{B^m}$. Thus the proof of our result is completed if it is proved that the m-linear mappings from the normed space $(E_B)^m$ into F are continuous if and only if they are bounded in the cartesian product of the unit balls B^m. But this is proved with no big efforts. Q.E.D.

<u>1.4 Proposition</u>. If $m \in \mathbb{N}$, the mapping $A \in \mathcal{L}_a({}^mE;F) \mapsto A_s \in \mathcal{L}_{as}({}^mE;F)$ is a linear projection onto $\mathcal{L}_{as}({}^mE;F)$ which takes $\mathcal{L}_b({}^mE;F)$ and $\mathcal{L}({}^mE;F)$ onto $\mathcal{L}_{bs}({}^mE;F)$ and $\mathcal{L}_s({}^mE;F)$ respectively.

<u>1.5 Definition</u>. Let $m \in \mathbb{N}$, $x \in E$ and $A \in \mathcal{L}_a({}^mE;F)$. Ax^m is defined in the following way. If $m = 0$ $Ax^o = A$. If $m \geq 1$ $Ax^m = A(x,\dots,x)$ (x repeated m times). Let $m \in \mathbb{N}$, $k \in \mathbb{N}^*$, $n_1,\dots,n_k \in \mathbb{N}$, $n = n_1+\dots+n_k \leq m$, $x_1,\dots,x_k \in E$ and $A \in \mathcal{L}_a({}^mE;F)$. $Ax_1^{n_1}\dots x_k^{n_k} \in \mathcal{L}_a({}^{m-n}E;F)$ is defined in the following way. If $m=0$, $Ax_1^{n_1}\dots x_k^{n_k} = A$. If $m = n > 0$, $Ax_1^{n_1}\dots x_k^{n_k} = A(x_1,\dots,x_1,\dots,x_k,\dots,x_k)$ (x_i repeated n_i times $i=1,\dots,k$). If $m \geq n$ $Ax_1^{n_1}\dots x_k^{n_k}(y_1,\dots,y_{m-n}) =$

$= Ax_1^{n_1} \ldots x_k^{n_k} \; y_1 \ldots y_{m-n}$ for all $y_1, \ldots, y_{m-n} \in E$.

<u>1.6 Newton's Formula</u>. If $m, n \in \mathbb{N}$, $n \leq m$, $k \in \mathbb{N}^*$, $A \in \mathcal{L}_{as}(^mE;F)$, $x_1, \ldots, x_k \in E$, then

$$A(x_1 + \ldots + x_k)^n = \Sigma\, n!(n_1! \ldots n_k!)^{-1} \, Ax_1^{n_1} \ldots x_k^{n_k}$$

where the sum is taken for all $n_1, \ldots, n_k \in \mathbb{N}$ such that $n_1 + \ldots + n_k = n$.

<u>1.7 Definition</u>. A mapping P from E into F is called an m-homogeneous polynomial if there is $A \in \mathcal{L}_a(^mE;F)$ such that $P(x) = Ax^m$ for all $x \in E$. In order to indicate such a relation between P and A we write $P = \hat{A}$. $\mathcal{P}_a(^mE;F)$ denotes the vector space of all m-homogeneous polynomials from E into F. $\mathcal{P}_b(^mE;F)$ and $\mathcal{P}(^mE;F)$ denote respectively the vector subspaces of $\mathcal{P}_a(^mE;F)$ formed by the mappings bounded on the bounded subsets of E and by the continuous mappings.

<u>1.8 Proposition</u>. $\mathcal{P}_b(^mE;F)$ coincides with the vector space of all m-homogeneous S-continuous polynomials from E into F.

<u>Proof</u>. Like in Proposition 1.3 it is enough to prove the result for the normed case which it is not difficult. Q.E.D.

<u>1.9 Polarization formula</u>. If $m \in \mathbb{N}^*$, $A \in \mathcal{L}_{as}(^mE;F)$ and $x_1, \ldots, x_m \in E$, then

$$A(x_1, \ldots, x_m) = \Sigma\, (m!2^m)^{-1} \, \varepsilon_1 \ldots \varepsilon_m \, \hat{A}(\varepsilon_1 x_1 + \ldots + \varepsilon_m x_m)$$

where the sum is taken for all ε_i which are equal to either 1 or -1, $i = 1, \ldots, m$.

<u>1.10 Definition</u>. If $m \in \mathbb{N}^*$, $B_1, \ldots, B_m \in \mathcal{B}_E$, $T \in \mathcal{L}_b(^mE;F)$, $\beta \in SC(F)$ = set of all continuous seminorms in F, we set

$$\|T\|_{B_1, \ldots, B_m, \beta} = \sup\{\beta(T(x_1, \ldots, x_m));\; x_i \in B_i, \;\; i = 1, \ldots, m\}.$$

It is easy to see that for every $x_i \in E_{B_i}$, $i = 1, \ldots, m$, we have

$$\beta(T(x_1, \ldots, x_m)) \leq \|T\|_{B_1, \ldots, B_m, \beta} \, \|x_1\|_{B_1} \ldots \|x_m\|_{B_m} .$$

If $B_1 = \ldots = B_m = B$ we denote $\|T\|_{B_1, \ldots, B_m, \beta}$ by $\|T\|_{B, \beta}$.

If $m = 0$ we set $\|T\|_{B,\beta} = \beta(T)$. The locally convex topology generated by all seminorms $\|\cdot\|_{B,\beta}$, with $B \in \mathcal{B}_E$ and $\beta \in SC(F)$, is called the strong topology. When we write $\mathcal{L}_b({}^mE;F)$ it is understood that we consider in it the strong topology.

<u>1.11 Proposition</u>. If F is complete $\mathcal{L}_b({}^mE;F)$ is complete. In general $\mathcal{L}_{bs}({}^mE;F)$ is a closed subspace of $\mathcal{L}_b({}^mE;F)$.

<u>Proof</u>. Let $(T_\alpha)_{\alpha\in I}$ be a Cauchy net in $\mathcal{L}_b({}^mE;F)$. It follows that the net $(T_\alpha(x_1,\dots,x_m))_{\alpha\in I}$ is a Cauchy net in F for all $x_1,\dots,x_m \in E$. Since F is complete, this net converges to an element $T(x_1,\dots,x_m)$ of F. It is not difficult to prove that T is m-linear and that $(T_\alpha)_{\alpha\in I}$ converges to T uniformly over the bounded subsets of E^m. It follows that T is bounded over the bounded subsets of E^m. The proof of the second part of the proposition is very simple. Q.E.D.

<u>1.12 Definition</u> - If $m \in \mathbb{N}$, $P \in \mathcal{P}_b({}^mE;F)$, $B \in \mathcal{B}_E$ and $\beta \in SC(F)$, we define

$$\|P\|_{B,\beta} = \sup\{\beta(P(x));\ x \in B\}.$$

It is obvious that $\beta(P(x)) \leq \|P\|_{B,\beta}\,\|x\|_B^m$ for every $x \in E_B$. The locally convex topology generated by all seminorms of the type $\|\cdot\|_{B,\beta}$, with $B \in \mathcal{B}_E$ and $\beta \in SC(F)$ is called the strong topology. When we write $\mathcal{P}_b({}^mE;F)$ we consider this space with the strong topology.

<u>1.13 Proposition</u>. Let $m \in \mathbb{N}^*$. The mapping $A \in \mathcal{L}_a({}^mE;F) \mapsto \hat{A} \in \mathcal{P}_a({}^mE;F)$ is linear and onto $\mathcal{P}_a({}^mE;F)$. It induces vector spaces isomorphisms between (i) $\mathcal{L}_{as}({}^mE;F)$ and $\mathcal{P}_a({}^mE;F)$, (ii) $\mathcal{L}_{bs}({}^mE;F)$ and $\mathcal{P}_b({}^mE;F)$, (iii) $\mathcal{L}_s({}^mE;F)$ and $\mathcal{P}({}^mE;F)$. On the other hand for each $\beta \in SC(F)$ and $B \in \mathcal{B}_E$ we have

$$(1) \qquad \|\hat{A}\|_{B,\beta} \leq \|A\|_{B,\beta} \leq m^m(m!)^{-1}\,\|\hat{A}\|_{B,\beta}\,.$$

<u>Proof</u>. The only part of this proposition whose proof is not a routine task is the inequality on the right-hand side of (1). But the use of

the Polarization Formula makes this inequality quite easy. Q.E.D.

<u>1.14 Corollary</u>. If F is complete $\mathcal{P}_b({}^mE;F)$ is complete.

<u>1.15 Remark</u>. As L. Nachbin has remarked in [1], the best universal constant in (1) is $m^m(m!)^{-1}$.

<u>1.16 Definition</u>. A mapping P from E into F is a polynomial if there are m in $\mathbb{N}$ and $P_k \in \mathcal{P}_a({}^kE;F)$, $k = 0,\ldots,m$ such that $P = P_0 +\ldots+ P_m$. We denote by $\mathcal{P}_a(E;F)$ the vector space of all polynomials from E into F. $\mathcal{P}_b(E;F)$ and $\mathcal{P}(E;F)$ indicate respectively the vector subspaces of the S-continuous and continuous mappings from E into F.

<u>1.17 Proposition</u>. $\mathcal{P}_a(E;F)$, $\mathcal{P}_b(E;F)$ and $\mathcal{P}(E;F)$ are respectively the direct sums of the families $(\mathcal{P}_a({}^mE;F))_{m\in\mathbb{N}}$, $(\mathcal{P}_b({}^mE;F))_{m\in\mathbb{N}}$ and $(\mathcal{P}({}^mE;F))_{m\in\mathbb{N}}$.

<u>Proof</u>. By definition $\mathcal{P}_a(E;F) = \sum_{m\in\mathbb{N}} \mathcal{P}_a({}^mE;F)$. Now we show that if $P_k \in \mathcal{P}_a({}^kE;F)$ for $k = 0,\ldots,m$ and $P_0 +\ldots+ P_m = 0$ then $P_0 =\ldots= P_m = 0$. If $m = 0$ this is clear. We prove by induction the case $m \geq 1$. For each x in E and $\lambda \in \mathbb{C}$ we have $P_0(\lambda x) + +\ldots+ P_m(\lambda x) = 0$. Thus it is not difficult to show that $(\lambda^m-1)P_0 + +\ldots+ (\lambda^m-\lambda^{m-1})P_{m-1} = 0$. If $\lambda^m-1 \neq 0,\ldots,\lambda^m-\lambda^{m-1} \neq 0$ it follows from the induction hypothesis that $P_0 = 0,\ldots,P_{m-1} = 0$. Since $P_0 +\ldots+ P_{m-1} + P_m = 0$ it follows that $P_m = 0$. In order to complete the proof of this proposition it is enough to show that $\mathcal{P}_b(E;F)$ is contained in $\sum_{m\in\mathbb{N}} \mathcal{P}_b({}^mE;F)$ and $\mathcal{P}(E;F) \subset \sum_{m\in\mathbb{N}} \mathcal{P}({}^mE;F)$. Let P be an element of $\mathcal{P}_b(E;F)$ (respectively, $\mathcal{P}(E;F)$) with $P = P_0+\ldots+P_m$, P_k in $\mathcal{P}_a({}^kE;F)$, $k = 0,\ldots,m$. We must show that each $P_k \in \mathcal{P}_b({}^kE;F)$ (respectively, $\in \mathcal{P}({}^kE;F)$). For $m = 0$ this is obvious. For $m \geq 1$ the proof follows by induction. As before $\lambda^m P(x) - P(\lambda x) = = (\lambda^m-1)P_0(x) +\ldots+ (\lambda^m-\lambda^{m-1})P_{m-1}(x)$ for every $x \in E$ and $\lambda \in \mathbb{C}$. If $\lambda^m-1 \neq 0,\ldots,\lambda^m-\lambda^{m-1} \neq 0$ the induction hypothesis imply that $P_0,\ldots,P_{m-1}$ are S-continuous (respectively, continuous). Thus

$P_m = P - P_o - \dots - P_{m-1}$ is S-continuous (respectively, continuous).

2. COMPACT AND NUCLEAR POLYNOMIALS AND MULTILINEAR MAPPINGS

We use the following notations: $E^* = \mathcal{L}_b(^1E;C)$ and $E' = \mathcal{L}(^1E;C)$.

2.1 Definition. Let $m \in \mathbb{N}^*$. For each $i = 1,\dots,m$, φ_i indicates a linear mapping from E into C. For each $b \in F$ we have an m-linear mapping $(x_1,\dots,x_m) \in E^m \mapsto \varphi_1(x_1)\dots\varphi_m(x_m)b \in F$ denoted by $\varphi_1 \times\dots\times \varphi_m b$. If $\varphi_1 = \dots = \varphi_m = \varphi$, this mapping is indicated by $\varphi^m b$. The vector subspace of $\mathcal{L}_a(^mE;F)$ generated by all the mappings of the type $\varphi_1 \times\dots\times \varphi_m b$ is denoted by $\mathcal{L}_{af}(^mE;F)$ and $\mathcal{L}_f(^mE;F)$ denote respectively the subspaces of $\mathcal{L}_{af}(^mE;F)$ formed by the S-continuous and continuous mappings. We use the following notations $\mathcal{L}_{afs}(^mE;F) = \mathcal{L}_{af}(^mE;F) \cap \mathcal{L}_{as}(^mE;F)$, $\mathcal{L}_{bfs}(^mE;F) = \mathcal{L}_{bf}(^mE;F) \cap \mathcal{L}_{bs}(^mE;F)$ and $\mathcal{L}_{fs}(^mE;F) = \mathcal{L}_f(^mE;F) \cap \mathcal{L}_s(^mE;F)$. For $m = 0$ we set all the above spaces as being equal to the vector space F. For each $m \in \mathbb{N}$, if $F = C$, we also may use the notations of the above spaces without the letter F.

2.2 Remark. It is not difficult to show that, for $m \in \mathbb{N}^*$, $\mathcal{L}_{bf}(^mE;F)$ is generated by all mappings of the type $\varphi_1 \times\dots\times \varphi_m b$ with $b \in F$ and $\varphi_1,\dots,\varphi_m \in E^*$. Similarly, for $m \in \mathbb{N}^*$, $\mathcal{L}_f(^mE;F)$ is generated by all the mappings of the type $\varphi_1 \times\dots\times \varphi_m b$ with $b \in F$ and $\varphi_1,\dots,\varphi_m \in E'$. It is easy to see that if $A \in \mathcal{L}_{af}(^mE;F)$ (respectively, $\mathcal{L}_{bf}(^mE;F)$, $\mathcal{L}_f(^mE;F)$), then $A_s \in \mathcal{L}_{afs}(^mE;F)$ (respectively, $\mathcal{L}_{bfs}(^mE;F)$, $\mathcal{L}_{fs}(^mE;F)$).

2.3 Definition. If $m \in \mathbb{N}$, $P \in \mathcal{P}_a(^mE;F)$ is of finite type if $P = \hat{A}$ with $A \in \mathcal{L}_{af}(^mE;F)$. The vector subspace of $\mathcal{P}_a(^mE;F)$ formed by all polynomials of finite type is denoted by $\mathcal{P}_{af}(^mE;F)$. The vector subspaces of $\mathcal{P}_{af}(^mE;F)$ formed by the S-continuous and continuous polynomials are denoted by $\mathcal{P}_{bf}(^mE;F)$ and $\mathcal{P}_f(^mE;F)$ respectively.

<u>2.4 Definition</u>. For m in $\mathbb{N}$, $\mathcal{L}_{bc}(^mE;F)$ denotes the closure of $\mathcal{L}_{bf}(^mE;F)$ in $\mathcal{L}_b(^mE;F)$. The elements of $\mathcal{L}_{bc}(^mE;F)$ are called S-continuous m-linear mappings of compact type. $\mathcal{L}_c(^mE;F)$ stands for the vector space $\mathcal{L}_{bc}(^mE;F) \cap \mathcal{L}(^mE;F)$ and its elements are the continuous m-linear mappings of compact type. Let $\mathbb{L}_c(^mE;F)$ denote the closure of $\mathcal{L}_f(^mE;F)$ in $\mathcal{L}_b(^mE;F)$. The usual conventions are made for the definitions of the spaces $\mathcal{L}_{cs}(^mE;F)$, $\mathcal{L}_{bcs}(^mE;F)$ and $\mathbb{L}_{cs}(^mE;F)$. The usual notations are used in the case $F = \mathbb{C}$. All the above vector spaces are considered with the topology induced by the strong topology. If any other topology is considered in these spaces this will be pointed out explicitly.

<u>2.5 Proposition</u>. The spaces $\mathcal{L}_{bc}(^mE;F)$, $\mathcal{L}_{bcs}(^mE;F)$, $\mathbb{L}_c(^mE;F)$ and $\mathbb{L}_{cs}(^mE;F)$ are complete for F complete and $m \in \mathbb{N}$.

<u>Proof</u>. It is enough to use 2.4 and 1.11.

<u>2.6 Definition</u>. For m in $\mathbb{N}$ we set $\mathcal{P}_{bc}(^mE;F)$ as the closure of $\mathcal{P}_f(^mE;F)$ in $\mathcal{P}_b(^mE;F)$. The elements of $\mathcal{P}_{bc}(^mE;F)$ are called S-continuous m-homogeneous polynomials of compact type. The vector subspace of $\mathcal{P}_{bc}(^mE;F)$ formed by the continuous polynomials is denoted by $\mathcal{P}_c(^mE;F)$. $\mathbb{P}_c(^mE;F)$ indicates the closure of $\mathcal{P}_f(^mE;F)$ in $\mathcal{P}_b(^mE;F)$. The usual conventions are made for the notations when $F = \mathbb{C}$. All the above spaces are considered with the topology induced by the strong topology. If any other topology is considered in these spaces this will be pointed out explicitly.

<u>2.7 Remark</u>. It is clear that $\mathbb{P}_c(^mE;F) \cap \mathcal{P}(^mE;F) \subset \mathcal{P}_c(^mE;F)$. The reverse inclusion is also true if E is bornological. Similar remarks may be made for the corresponding spaces of multilinear mappings.

<u>2.8 Proposition</u>. The mapping $A \in \mathcal{L}_{bs}(^mE;F) \mapsto \hat{A} \in \mathcal{P}_b(^mE;F)$ induces topological and algebraic isomorphisms between: (i) $\mathcal{P}_{bc}(^mE;F)$ and $\mathcal{L}_{bcs}(^mE;F)$; (ii) $\mathcal{P}_c(^mE;F)$ and $\mathcal{L}_{cs}(^mE;F)$; (iii) $\mathbb{P}_c(^mE;F)$ and $\mathbb{L}_{cs}(^mE;F)$.

Proof. It is enough to use 1.13.

2.9 Corollary. $\mathcal{P}_{bc}(^mE;F)$ and $\mathbb{P}_c(^mE;F)$ are complete for F complete and m in $\mathbb{N}$.

2.10 Definition. The direct sum of the families $(\mathcal{P}_{bc}(^mE;F))_{m\in\mathbb{N}}$, $(\mathcal{P}_c(^mE;F))_{m\in\mathbb{N}}$ and $(\mathbb{P}_c(^mE;F))_{m\in\mathbb{N}}$ are denoted by $\mathcal{P}_{bc}(E;F)$, $\mathcal{P}_c(E;F)$ and $\mathbb{P}_c(E;F)$ respectively.

2.11 Proposition. For $m \in \mathbb{N}$, the (m+1)-linear mapping

$$\alpha_m: (\varphi_1,\ldots,\varphi_m,b) \in (E^*)^m \times F \mapsto \varphi_1 \times\ldots\times \varphi_m b \in \mathcal{L}_b(^mE;F)$$

is continuous when the cartesian product topology is considered in its domain.

Proof. The (m+1)-linearity of α_m is easily proved. The continuity of α_m follows from the following equality:

$$\|\varphi_1 \times\ldots\times \varphi_m b\|_{B_1,\ldots,B_m,\beta} = \beta(b)\, \|\varphi_1\|_{B_1}\cdots\|\varphi_m\|_{B_m}$$

for all $\varphi_1,\ldots,\varphi_m \in E^*$; $b \in F$; $B_1,\ldots,B_m \in \mathcal{B}_E$; $\beta \in SC(F)$. Here we set $\|\varphi_i\|_{B_i} = \sup\{|\varphi_i(t)|;\ t \in B_i\}$ for $i = 1,\ldots,m$.

2.12 Remark. There is a unique continuous linear mapping χ_m from the projective tensor product $E^* \otimes_\pi \ldots \otimes_\pi E^* \otimes_\pi F$ into $\mathcal{L}_b(^mE;F)$ such that $\alpha_m = \chi_m \circ \psi_m$ where ψ_m is the natural (m+1)-linear mapping from $E^* \times\ldots\times E^* \times F$ into $E^* \otimes_\pi \ldots \otimes_\pi E^* \otimes_\pi F$. The mapping χ_m is 1-1 and its image is $\mathcal{L}_{bf}(^mE;F)$. The image of $E' \otimes_\pi \ldots \otimes_\pi E' \otimes_\pi F$ by χ_m is $\mathcal{L}_f(^mE;F)$.

2.13 Definition. Let $m \in \mathbb{N}^*$. Given $B_1,\ldots,B_m \in \mathcal{B}_E$ and $\beta \in SC(F)$ we have the nuclear seminorm in $\mathcal{L}_{bf}(^mE;F)$ defined by

$$\|T\|_{N,B_1,\ldots,B_m,\beta} = \inf\{ \sum_{j=0}^{n} \|\varphi_{1j}\|_{B_1} \cdots \|\varphi_{mj}\|_{B_m} \beta(b_j);$$

$$T = \sum_{j=0}^{n} \varphi_{1j} \times\ldots\times \varphi_{mj} b_j \quad \text{with } \varphi_{ij} \in E^*,\ b_j \in F;\ i=1,\ldots m;\ j=0,\ldots,n\} .$$

If $B_1 = \ldots = B_m = B$ this seminorm is denoted by $\|\cdot\|_{N,B,\beta}$. The nuclear topology in $\mathcal{L}_{bf}(^mE;F)$ is the locally convex topology generated by all nuclear seminorms. If $m = 0$ the topology of F is the nuclear topology and we denote $\|A\|_{N,B} = \beta(A)$ for every $A \in F$, $B \in \mathcal{B}_E$, and $\beta \in SC(F)$.

<u>2.14 Proposition.</u> Let $m \in \mathbb{N}^*$. For $T \in \mathcal{L}_{bf}(^mE;F)$, $B_1,\ldots,B_m \in \mathcal{B}_E$ and $\beta \in SC(F)$ we have

$$\|T\|_{B_1,\ldots,B_m,\beta} \leq \|T\|_{N,B_1,\ldots,B_m,\beta} .$$

<u>Proof.</u> If $T = \sum_{j=1}^{n} \varphi_{1j} \times\ldots\times \varphi_{mj} b_j$ it is clear that

$$\|T\|_{B_1,\ldots,B_m,\beta} \leq \sum_{j=1}^{n} \|\varphi_{1j}\|_{B_1} \cdots \|\varphi_{mj}\|_{B_m} \beta(b_j).$$

This inequality and 2.13 imply our result.

<u>2.15 Remark.</u> Let $m \in \mathbb{N}^*$. The nuclear topology in $\mathcal{L}_{bf}(^mE;F)$ makes this space homeomorphic to $E^* \otimes_\pi \ldots \otimes_\pi E^* \otimes_\pi F$ in such way that for $q_i = \|\cdot\|_{B_i} \in SC(E^*)$, $i = 1,\ldots,m$ and $\beta \in SC(F)$

$$\|T\|_{N,B_1,\ldots,B_m,\beta} = \|\chi_m(A)\|_{N,B_1,\ldots,B_m,\beta} = q_1 \otimes\ldots\otimes q_m \otimes \beta(A)$$

and

$$\|\varphi_1 \times\ldots\times \varphi_m b\|_{N,B_1,\ldots,B_m,\beta} = q_1(\varphi_1)\ldots q_m(\varphi_m)\beta(b) =$$

$$= \|\varphi_1 \times\ldots\times \varphi_m b\|_{B_1,\ldots,B_m,\beta} .$$

<u>2.16 Definition.</u> For every $B \in \mathcal{B}_E$, $\beta \in SC(F)$, $m \in N^*$, a nuclear seminorm is defined in $\mathcal{P}_{bf}(^mE;F)$ in the following way:

$$\|P\|_{N,B} = \inf\{ \sum_{j=1}^{n} \|\varphi_j\|_B^m \beta(b_j);\ P = \sum_{j=1}^{n} \varphi_j^m b_j,\ \varphi_j \in E^*,\ b_j \in F,\ j=1,\ldots,n\}.$$

The nuclear topology in $\mathcal{P}_{bf}(^mE;F)$ is defined by all nuclear seminorms. For $m = 0$ we set $\|\cdot\|_{N,B,\beta} = \beta$ for every $B \in \mathcal{B}_E$ and $\beta \in SC(F)$.

<u>2.17 Proposition.</u> If $m \in \mathbb{N}^*$, $P \in \mathcal{P}_{bf}(^mE;F)$, $B \in \mathcal{B}_E$, $\beta \in SC(F)$, then

$$\|P\|_{N,B,\beta} \geq \|P\|_{B,\beta} .$$

<u>2.18 Proposition.</u> If $m \in \mathbb{N}^*$, $T \in \mathcal{L}_{bfs}(^mE;F)$, $B \in \mathcal{B}_E$, $\beta \in SC(F)$, then

$$\|T\|_{N,B,\beta} \leq \|\hat{T}\|_{N,B,\beta} \leq m^m (m!)^{-1} \|T\|_{N,B,\beta} .$$

Proof. The inequality on the left hand side is quite obvious from the definitions. It is not difficult to see we do not lose generality by considering $T = \sum_{j=1}^{n} \varphi_{1j} \times \ldots \times \varphi_{mj} b_j$, where $\beta(b_j) \neq 0$ and $\|\varphi_{ij}\|_B \neq 0$ for $j = 1,\ldots,n$ and $i = 1,\ldots,m$. We have

$$T = \sum_{j=1}^{n} \|\varphi_{1j}\|_B \ldots \|\varphi_{mj}\|_B \, \beta(b_j) \, \psi_{1j} \times \ldots \times \psi_{mj} c_j$$

where $b_j = \beta(b_j)c_j$ and $\varphi_{ij} = \|\varphi_{ij}\|_B \, \psi_{ij}$ for $i=1,\ldots,m$ $j=1,\ldots,n$. By the Polarization Formula we have

$$\|\hat{T}\|_{N,B,\beta} \leq (m!2^m)^{-1} \sum_{j=1}^{n} \|\varphi_{1j}\|_B \ldots \|\varphi_{mj}\|_B \, \beta(b_j) \cdot$$

$$\cdot \sum_{\varepsilon_i=\pm 1} |\varepsilon_1 \ldots \varepsilon_m| \, \|\sum_{i=1}^{m} \varepsilon_i \psi_{ij}\|_B^m \, \beta(c_j) \leq$$

$$\leq m^m (m!)^{-1} \sum_{j=1}^{n} \|\varphi_{1j}\|_B \ldots \|\varphi_{mj}\|_B \, \beta(b_j).$$

Thus the result follows.

From now up to the end of this paragraph F will be complete.

2.19 Remark. Let $m \in \mathbb{N}^*$. The mapping χ_m may be extended to a continuous mapping $\hat{\chi}_m$ from the completion $E^* \hat{\otimes}_\pi \ldots \hat{\otimes}_\pi E^* \hat{\otimes}_\pi F$ of the space $E^* \otimes_\pi \ldots \otimes_\pi E^* \otimes_\pi F$ into $\mathcal{L}_b(^mE;F)$. The image of $\hat{\chi}_m$ is denoted by $\mathcal{L}_{bN}(^mE;F)$ and its elements are called S-continuous m-linear mappings of nuclear type. The vector subspace of $\mathcal{L}_{bN}(^mE;F)$ formed by the continuous mappings is indicated by $\mathcal{L}_N(^mE;F)$. We denote by $\mathbb{L}_N(^mE;F)$ the vector space obtained by taking the image through $\hat{\chi}_m$ of the closure of $E' \otimes_\pi \ldots \otimes_\pi E' \otimes_\pi F$. The natural definitions of $\mathcal{L}_{bNs}(^mE;F)$, $\mathcal{L}_{Ns}(^mE;F)$ and $\mathbb{L}_{Ns}(^mE;F)$ are considered. The usual convention is adopted for the notations when $F = C$. For $m = 0$ all these spaces are considered to be equal to F.

2.20 Definition. The nuclear topology in $\mathcal{L}_{bN}(^mE;F)$ is defined in the following way. Let $m \in \mathbb{N}^*$ and let $B_1,\ldots,B_m \in \mathcal{B}_E$. If $\beta \in SC(F)$ and $q_i = \|\cdot\|_{B_i} \in SC(E^*)$ for $i = 1,\ldots,m$ we denote by

$q_1 \hat{\otimes} \ldots \hat{\otimes}\, q_m \hat{\otimes}\, \beta$ the continuous extension of $q_1 \otimes \ldots \otimes q_m \otimes \beta$ to $E^* \hat{\otimes}_\pi \ldots \hat{\otimes}_\pi E^* \hat{\otimes}_\pi F$. Given $T \in \mathcal{L}_{bN}(^mE;F)$ we denote

$$\|T\|_{N,B_1,\ldots,B_m,\beta} = \inf \{q_1 \hat{\otimes} \ldots \hat{\otimes}\, \beta(A);\ T = \chi_m(A)\}.$$

The nuclear topology is the topology generated by all seminorms of the above type. These seminorms are called nuclear seminorms. For $m = 0$ we set the nuclear topology to be equal to the topology of F and we write $\|A\|_{N,B,\beta} = \beta(A)$ for $A \in F$, $B \in \mathcal{B}_E$ and $\beta \in SC(F)$.

<u>2.21 Remark</u>. For $m \in \mathbb{N}^*$ the nuclear topology in $\mathcal{L}_{bN}(^mE;F)$ makes this space isomorphic to $E^* \hat{\otimes}_\pi \ldots \hat{\otimes}_\pi E^* \hat{\otimes}_\pi F/\ker \hat{\chi}_m$. We also have

$$\|T\|_{N,B_1,\ldots,B_m,\beta} = q_1 \tilde{\otimes} \ldots \tilde{\otimes}\, q_m \tilde{\otimes}\, \beta(\tilde{A})$$

where, if $T = \hat{\chi}_m(A)$, $\tilde{A}$ denotes the equivalence class defined by A and $q_1 \tilde{\otimes} \ldots \tilde{\otimes}\, q_m \tilde{\otimes}\, \beta$ denotes the natural seminorm defined in the quotient space by the seminorm $q_1 \hat{\otimes} \ldots \hat{\otimes}\, q_m \hat{\otimes}\, \beta$.

<u>2.22 Proposition</u>. For $m \in \mathbb{N}^*$, $T \in \mathcal{L}_{bN}(^mE;F)$, $B_i \in \mathcal{B}_E$ $i=1,\ldots,m$, $\beta \in SC(F)$

$$\|T\|_{B_1,\ldots,B_m,\beta} \le \|T\|_{N,B_1,\ldots,B_m,\beta}\,.$$

<u>2.23 Remark</u>. By the above definitions it is clear that $\mathcal{L}_{bN}(^mE;F)$ is complete and that $\mathcal{L}_{bf}(^mE;F)$ is dense in $\mathcal{L}_{bN}(^mE;F)$. If E^* has the approximation property the $\hat{\chi}_m$ is 1-1. See Grothendieck [2].

<u>2.24 Definition</u>. Let $m \in \mathbb{N}$. We denote by $\mathcal{P}_{bN}(^mE;F)$ the image of $\mathcal{L}_{bN}(^mE;F)$ through the mapping $A \mapsto \hat{A}$ from $\mathcal{L}_b(^mE;F)$ into $\mathcal{P}_b(^mE;F)$. The elements of $\mathcal{P}_{bN}(^mE;F)$ are called S-continuous m-homogeneous polynomials of nuclear type. The images of $\mathcal{L}_N(^mE;F)$ and $\mathbb{L}_N(^mE;F)$ through the same mapping are denoted respectively $\mathcal{P}_N(^mE;F)$ and $\mathbb{P}_N(^mE;F)$.

<u>2.25 Definition</u> - The nuclear topology in $\mathcal{P}_{bN}(^mE;F)$ is defined in the following way. If $m = 0$ this topology is the topology of F. If $m \neq 0$ the nuclear topology is the topology making $\mathcal{L}_{bNs}(^mE;F)$ homeomorphic to $\mathcal{P}_{bN}(^mE;F)$. Since $\mathcal{L}_{bfs}(^mE;F)$ is dense in $\mathcal{L}_{bNs}(^mE;F)$, $\mathcal{P}_{bN}(^mE;F)$ is one

completion of $\mathcal{P}_{bf}(^mE;F)$ endowed with the nuclear topology. If B is in $\mathcal{B}_E$ and $\beta \in SC(F)$ then $\|\cdot\|_{N,B,\beta}$ may be extended in a continuous way to $\mathcal{P}_{bN}(^mE;F)$. We denote this continuous extention by the same symbol. The collection of all these seminorms define the nuclear topology in $\mathcal{P}_{bN}(^mE;F)$.

<u>2.26 Proposition</u>. For $m \in \mathbb{N}$, $P \in \mathcal{P}_{bN}(^mE;F)$, $B \in \mathcal{B}_E$, $\beta \in SC(F)$

$$\|P\|_{B,\beta} \leq \|P\|_{N,B,\beta} .$$

<u>2.27 Proposition</u>. If $m \in \mathbb{N}^*$, $T \in \mathcal{L}_{bNs}(^mE;F)$, $B \in \mathcal{B}_E$ and $\beta \in SC(F)$ then

$$\|T\|_{N,B,\beta} \leq \|\hat{T}\|_{N,B,\beta} \leq m^m(m!)^{-1} \|T\|_{N,B,\beta} .$$

<u>Proof</u>. Use 2.18 and a limit argument.

<u>2.28 Remark</u>. $\mathbb{P}_N(^mE;F)$ and $\mathbb{L}_N(^mE;F)$ for $m \in \mathbb{N}$ and the induced topologies by the nuclear topology are complete locally convex spaces. They are completions of $\mathcal{P}_f(^mE;F)$ and $\mathcal{L}_f(^mE;F)$ respectively when these spaces have their nuclear topologies.

<u>2.29 Remark</u>. All the spaces considered in 2.19 and in 2.24 are always considered with the induced topologies by the nuclear topology. If any other topology is considered in one of those spaces then this is said explicitly. We have the following continuous inclusions.

$$\begin{array}{ccccc}
\mathcal{P}_{bN}(^mE;F) & \hookrightarrow & \mathcal{P}_{bc}(^mE;F) & \hookrightarrow & \mathcal{P}_b(^mE;F) \\
\cup & & \cup & & \parallel \\
\mathbb{P}_N(^mE;F) & \hookrightarrow & \mathbb{P}_c(^m;EF) & \hookrightarrow & \mathcal{P}_b(^mE;F) \\
\cup & & \cup & & \cup \\
\mathcal{P}_N(^mE;F) & \hookrightarrow & \mathcal{P}_c(^mE;F) & \hookrightarrow & \mathcal{P}(^mE;F)
\end{array}$$

$$\begin{array}{ccccc}
\mathcal{L}_{bN}(^mE;F) & \hookrightarrow & \mathcal{L}_{bc}(^mE;F) & \hookrightarrow & \mathcal{L}_b(^mE;F) \\
\cup & & \cup & & \parallel \\
\mathbb{L}_N(^mE;F) & \hookrightarrow & \mathbb{L}_c(^mE;F) & \hookrightarrow & \mathcal{L}_b(^mE;F) \\
\cup & & \cup & & \cup \\
\mathcal{L}_N(^mE;F) & \hookrightarrow & \mathcal{L}_c(^mE;F) & \hookrightarrow & \mathcal{L}(^mE;F).
\end{array}$$

<u>2.30 Definition</u>. The direct sums of the families $(\mathcal{P}_{bN}(^mE;F))_{m\in\mathbb{N}}$

$(P_N({}^m E;F))_{m \in \mathbb{N}}$ and $(\mathcal{P}_N({}^m E;F))_{m \in \mathbb{N}}$ are denoted by $\mathcal{P}_{bN}(E;F)$ $\mathbb{P}_N(E;F)$ and $\mathcal{P}_N(E;F)$ respectively.

3. HOLOMORPHIC AND S-HOLOMORPHIC MAPPINGS

3.1 Definition. A formal power series from E into F around $\xi \in E$ is a series in the variable $x \in E$ of the form

$$\sum_{m \in \mathbb{N}} A_m (x-\xi)^m$$

where $A_m \in \mathcal{L}_{as}({}^m E;F)$ for every $m \in \mathbb{N}$. The set of all these series form a vector space $F_{a\xi}[[E]]$. The vector subspace formed by the series which have $A_m \in \mathcal{L}_{bs}({}^m E;F)$ is denoted by $F_{b\xi}[[E]]$ and the vector subspace of the series which have $A_m \in \mathcal{L}_s({}^m E;F)$ is indicated by $F_\xi[[E]]$.

3.2 Proposition. Let $\beta \in SC(F)$. If, in the notation of 3.1, for each B in $\mathcal{B}_E$ there is $\rho_B > 0$ such that

$$\lim_{M \to \infty} \beta[\sum_{m=0}^{M} \hat{A}_m (x-\xi)] = 0$$

for each $x \in \xi + \rho_B B$, then $\beta \circ \hat{A}_m = 0$ for every $m \in \mathbb{N}$.

Proof. If $x = \xi$ it is clear that $\hat{A}_m(x-\xi) = 0$ for $m \in \mathbb{N}$. From the hypothesis it follows that $\beta \circ \hat{A}_o(0) = 0$. Since $\hat{A}_o(x) = \hat{A}_o(0)$ for each x in E it follows that $\beta \circ \hat{A}_o = 0$. We prove by induction that $\beta \circ \hat{A}_m = 0$ for $m \geq 1$. We suppose that it has been proved that $\beta \circ \hat{A}_k = 0$ for $k = 1, \ldots, m-1$. We have

$$\beta(\sum_{k=0}^{M} \hat{A}_k (x-\xi)) = \beta(\sum_{k=m}^{M} \hat{A}_k (x-\xi)) \qquad \forall\, x \in E$$

since we have

$$|\beta(\sum_{k=0}^{M} \hat{A}_k (x-\xi)) - \beta(\sum_{k=m}^{M} \hat{A}_k (x-\xi))| \leq \sum_{k=0}^{m-1} \beta \circ \hat{A}_k (x-\xi) = 0$$

for every x in E. Thus we may write

$$\lim_{M \to \infty} \beta(\sum_{k=m}^{M} \hat{A}_k (x-\xi)) = 0$$

for every x in $\xi + \rho_B B$. Therefore

$$\lim_{M\to\infty} |\lambda|^m \beta(\sum_{k=m}^{M} \lambda^{k-m} \hat{A}_k(x-\xi)) = 0$$

for every $\lambda \in C$, $|\lambda| \leq 1$ and $x \in \xi + \rho_B B$. It follows that

$$\lim_{M\to\infty} \beta(\sum_{k=m}^{M} \lambda^{k-m} \hat{A}_k(x-\xi)) = 0$$

for each $\lambda \in C$, $\lambda \neq 0$, $|\lambda| \leq 1$, $x \in \xi + \rho_B B$. From the hypothesis it is easy to prove that

$$\lim_{\ell\to\infty} \beta\circ\hat{A}_\ell(x-\xi) = 0$$

for every $x \in \xi + \rho_B B$. If we fix $x \in \xi + \rho_B B$ we have

$$(\exists\, L > 0) \qquad \beta\circ\hat{A}_\ell(x-\xi) \leq L \qquad (\forall\, \ell \geq m).$$

Now it follows that

$$\beta\circ\hat{A}_m(x-\xi) = \beta(\sum_{k=m}^{M} \lambda^{k-m} \hat{A}_k(x-\xi) - \sum_{k=m+1}^{M} \lambda^{k-m} \hat{A}_k(x-\xi)) \leq$$

$$\leq \beta(\sum_{k=m}^{M} \lambda^{k-m} \hat{A}_k(x-\xi)) + \sum_{k=m+1}^{M} |\lambda|^{k-m} \beta\circ\hat{A}_k(x-\xi)$$

for every $M > m$, $\lambda \in C$, $\lambda \neq 0$, $|\lambda| \leq 1$. Thus

$$\beta\circ\hat{A}_m(x-\xi) \leq \lim_{M\to\infty} \sum_{k=m+1}^{M} |\lambda|^{k-m} \beta\circ\hat{A}_k(x-\xi) \leq$$

$$\leq \lim_{M\to\infty} \sum_{k=m+1}^{M} |\lambda|^{k-m} L = L\, |\lambda|(1-|\lambda|)^{-1}$$

for $\lambda \in C$, $\lambda \neq 0$, $|\lambda| \leq 1$. Now if we make $|\lambda| \to 0$ we have $\beta\circ\hat{A}_m(x-\xi) = 0$. Consequently $\beta\circ\hat{A}_m$ vanishes in $\rho_B B$, hence in E_B. Since B is arbitrary it follows that $\beta\circ\hat{A}_m = 0$.

<u>3.2 Definition</u>. $f: U \to F$ is holomorphic in U if for every $\xi \in U$ there is $A_m \in \mathcal{L}_s(^mE;F)$ $(m \in \mathbb{N})$ such that for each $\beta \in SC(F)$ there exists an open subset V of U containing $\{\xi\}$ with

$$\lim_{M\to\infty} \beta(f(x) - \sum_{m=0}^{M} A_m(x-\xi)^m) = 0$$

uniformly for $x \in V$. $\mathcal{H}(U;F)$ denotes the vector space of all holomorphic mappings from U into F. We note that in this definition V depends on β. If it is possible to get V independent of β we have

$$f(x) = \sum_{m=0}^{\infty} A_m(x-\xi)^m$$

uniformly for x in V. This is possible if F is seminormed. If we suppose F separated it is easy to show that the A_m $(m \in \mathbb{N})$ are unique once f and ξ are given. We write $d^m f(\xi) = m!\, A_m$ and $\hat{d}^m f(\xi) = m!\, \hat{A}_m$ for every $m \in \mathbb{N}$. They are called the differentials of order m of f at ξ. It follows that we have the differentials of order m of f in U:

$$d^m f\colon x \in U \mapsto d^m f(x) \in \mathcal{L}_s({}^m E;F)$$

$$d^m f\colon x \in U \mapsto \hat{d}^m f(x) \in \mathcal{P}({}^m E;F)$$

and the Taylor series of f at ξ:

$$f(x) \cong \sum_{m\in\mathbb{N}} (m!)^{-1}\, d^m f(\xi)(x-\xi)^m = \sum_{m\in\mathbb{N}} (m!)^{-1}\, \hat{d}^m f(\xi)(x-\xi).$$

The Taylor polynomial of f or order m at ξ is

$$\tau_{m,f,\xi}(x) = \sum_{k=0}^{m} (k!)^{-1}\, d^k f(\xi)(x-\xi)^k =$$

$$= \sum_{k=0}^{m} (k!)^{-1}\, \hat{d}^k f(\xi)(x-\xi).$$

<u>3.4 Remark</u>. If F is not separated we may consider its associated separated space $F_s = F/\overline{\{0\}}$ and then $f \in \mathcal{H}(U;F)$ if and only if $\pi\circ f$ is in $\mathcal{H}(U;F_s)$, where π is the quotient mapping from F onto F_s. Thus in this sense we can always take the holomorphic functions as being valued in a separated space.

<u>3.5 Definition</u>. $F\colon U \to F$ is S-holomorphic in U if for every $\xi \in U$ there is $A_m \in \mathcal{L}_{bs}({}^m E;F)$ $(m \in \mathbb{N})$ such that for every $\beta \in SC(F)$ and for every $B \in \mathcal{B}_E$ there exists $\rho_B > 0$ with

$$\lim_{m\to\infty} \beta\Big(f(x) - \sum_{k=0}^{m} A_k (x-\xi)^k\Big) = 0$$

uniformly for $x \in \xi + \rho_B B$. $\mathcal{H}_S(U;F)$ denotes the vector space of all S-holomorphic mappings from U into F. We suppose F separated. Then it is easy to show that for each f and ξ given the A_m $(m\in\mathbb{N})$ are unique. We write $\delta^m f(\xi) = m!\, A_m$ and $\hat{\delta}^m f(\xi) = m!\, \hat{A}_m$ for $m\in\mathbb{N}$. We have the S-differentials:

$$\delta^m f: x \in U \mapsto \delta^m f(x) \in \mathcal{L}_{bs}(^m E;F)$$

$$\hat{\delta}^m f: x \in U \mapsto \hat{\delta}^m f(x) \in \mathcal{P}_b(^m E;F).$$

The Taylor series of f at ξ is the series:

$$f(x) \cong \sum_{m=0}^{\infty} (m!)^{-1}\, \delta^m f(\xi)(x-\xi)^m =$$

$$\cong \sum_{m=0}^{\infty} (m!)^{-1}\, \hat{\delta}^m f(\xi)(x-\xi).$$

The Taylor polynomial of f at ξ with degree m is given by:

$$\tau_{m,f,\xi}(x) = \sum_{k=0}^{\infty} (k!)^{-1}\, \hat{\delta}^k f(\xi)(x-\xi).$$

We can make here an analogous remark to 3.4: $f \in \mathcal{H}_S(U;F)$ if and only if $\pi \circ f \in \mathcal{H}_S(U;F_s)$, where π is the quotient mapping from F onto F_s.

<u>3.6 Proposition.</u> Let f be a mapping from U into F. Then $f \in \mathcal{H}_S(U;F)$ if and only if $f|U \cap E_B \in \mathcal{H}(U \cap E_B;F)$ for every $B \in \mathcal{B}_E$.

<u>Proof.</u> We may suppose F separated. If $f: U \to F$ is such that $f|U \cap E_B = f_B \in \mathcal{H}(U \cap E_B;F)$ for each $B \in \mathcal{B}_E$ we may define in a coherent way

$$\hat{\delta}^m f(\xi)(x) = \hat{d}^m f_B(\xi)(x) \qquad (m \in \mathbb{N})$$

for every $B \in \mathcal{B}_E$ such that $\xi, x \in E_B$. It is not difficult now to see that $\hat{\delta}^m f(\xi) \in \mathcal{P}_b(^m E;F)$ and that $f \in \mathcal{H}_S(U;F)$ by using the holomorphy of the mappings f_B for $B \in \mathcal{B}_E$.

<u>3.7 Definition.</u> $H_S(U;F) = \mathcal{H}_S(U;\hat{F}) \cap F^U$, $H(U;F) = \mathcal{H}(U;\hat{F}) \cap F^U$ where $\hat{F}$ is a completion of F.

<u>3.8 Proposition.</u> $\mathcal{P}_b(E;F) \subset \mathcal{H}_S(E;F)$ and $\mathcal{P}(E;F) \subset \mathcal{H}(E;F)$.

<u>Proof.</u> It is enough to show the second inclusion since the other follows from this one. We have to show that $\mathcal{P}(^k E;F) \subset \mathcal{H}(E;F)$ for every $k \in \mathbb{N}$. Let ξ be in E and P in $\mathcal{P}(^k E;F)$. If $A \in \mathcal{L}_s(^k E;F)$ is given by $\hat{A} = P$ we use Newton's Formula to get

$$P(x) = A\, x^k = A(\xi + (x-\xi))^k = \sum_{j=0}^{k} \binom{k}{j} A\, \xi^j (x-\xi)^{k-j}$$

where $A\xi^j \in \mathcal{L}_s({}^{k-j}E;F)$ for $j = 0,\dots,k$. Thus $P \in \mathcal{H}(E;F)$.

<u>3.9 Proposition</u>. $\mathcal{H}(U;F) \subset C(U;F)$ and $\mathcal{H}_S(U;F) \subset C_S(U;F)$. Here $C(U;F)$ denotes the vector space of all continuous mappings from U into F.

<u>Proof</u>. By the definition of S-continuity and by 3.6 it is enough to show the first inclusion. Let $f \in \mathcal{H}(U;F)$ and $\xi \in U$. We choose $A_m \in \mathcal{L}_s({}^mE;F)$ $(m \in \mathbb{N})$ as in Definition 3.3. For each $\beta \in SC(F)$ we choose V as in 3.3. Since for x in V

$$\beta(f(x)-f(\xi)) \le \beta(f(x) - \sum_{m=0}^{M} A_m(x-\xi)^m) + \beta(\sum_{m=1}^{M} A_m(x-\xi)^m)$$

we get $\beta(f(x)-f(\xi)) < \epsilon$ for $x \in V_o = V \cap V_1$ a neighborhood of ξ which depends on ϵ. In order to show this, first we choose M such that the first term in the right hand side is smaller than $\epsilon/2$ for every x in V and then we choose V_1 open, $\xi \in V_1$, such that for $x \in V_1$ and $m = 1,\dots,M$ $\beta(A_m(x-\xi)^m) < \epsilon/2M$. Thus f is continuous in ξ.

<u>3.10 Remark</u>. We can adapt the above proof in order to show that $H(U;F)$ is contained in $C(U;F)$ and $H_S(U;F) \subset C_S(U;F)$.

<u>3.11 Definition</u>. Let F be separated and let Γ be an oriented retificable simple Jordan curve in $\mathbb{C}$. Then

$$\int_\Gamma f(z)dz \in \hat{F}$$

exists for every continuous mapping from Γ into F. The same is true if Γ is closed. We also have

$$(*) \qquad \beta[\int_\Gamma f(z)dz] \le |\Gamma| \sup\{\beta(f(z));\ z \in \Gamma\}$$

where $\beta \in SC(F)$ and $|\Gamma|$ is the lenght of Γ. In a more general way:

$$(**) \qquad \beta[\int_\Gamma f(z)dz] \le \int_\Gamma \beta(f(z))|dz|.$$

<u>3.12 Cauchy's integral formula</u>. If F is separated, $f \in \mathcal{H}_S(U;F)$,

$\xi \in U$, $\rho > 1$ with $(1-\lambda)\xi + \lambda x \in U$ for every $\lambda \in C$, $|\lambda| \leq \rho$, then

$$f(x) = (2\pi i)^{-1} \int_{|\lambda|=\rho} f((1-\lambda)\xi + \lambda x))(\lambda-1)^{-1}\, d\lambda.$$

Proof. (1) Let V be a non-void open subset of C. If g is a holomorphic mapping from V into F, τ is an element of $B_\rho(\zeta) \subset \subset \overline{B_\delta(\zeta)} \subset V$ with $\delta > 0$ then we may write:

$$g(\tau) = (2\pi i)^{-1} \int_{|\lambda-\zeta|=\delta} g(\lambda)(\lambda-\tau)^{-1}\, d\lambda.$$

(2) Now we consider:

$$V = \{\lambda \in C ;\ (1-\lambda)\xi + \lambda x \in U\}$$

$$g(\lambda) = f((1-\lambda)\xi + \lambda x) \quad \text{for } \lambda \in V.$$

Then $\overline{B_\rho(0)} \subset V$ and $1 \in B_\rho(0)$. Thus, by (1),

$$g(1) = (2\pi i)^{-1} \int_{|\lambda|=\rho} g(\lambda)(\lambda-1)^{-1}\, d\lambda$$

and this gives the result.

We should remark here the following: Although Proposition 3.12 is given with the hypothesis that f is an S-holomorphic mapping from U into F, the result is still true for weaker conditions on f. It is enough to consider f finitely holomorphic, that is: holomorphic when restricted to the finite dimensional subspaces of E intercepted by U. The same remark hold for the next proposition. A simple exam of the proof shows that this is true.

3.13 Cauchy's integral formulas for differentials. Let F be a separated space. If f is in $\mathcal{H}_S(U;F)$, $\xi \in U$, $x \in E$, $\rho > 0$ with $\xi + \lambda x \in U$ for every $\lambda \in C$, $|\lambda| \leq \rho$, then

$$(m!)^{-1}\, \hat{\delta}^m f(\xi)(x) = (2\pi i)^{-1} \int_{|\lambda|=\rho} f(\xi+\lambda x)\lambda^{-m-1}\, d\lambda.$$

Proof. Let V be a non-void open subset of C. If $g \in \mathcal{H}(V;F)$, $0 < r \leq R$, $\zeta \in C$ and if we suppose that the closed ring with center

ζ and radius r and R is contained in V, then

$$\int_{|\lambda-\zeta|=r} g(\lambda)d\lambda = \int_{|\lambda-\zeta|=R} g(\lambda)d\lambda.$$

Now we consider $V = \{\lambda \in \mathbb{C} ; \lambda \neq 0, \xi + \lambda x \in U\}$, $\overline{B_\rho(0)} - \{0\} \subset V$, $0 < \epsilon \leq \rho$ and $g(\lambda) = \lambda^{-m-1} f(\xi+\lambda x)$. Thus

$$\int_{|\lambda|=\rho} f(\xi+\lambda x)\lambda^{-m-1}\, d\lambda = \int_{|\lambda|=\epsilon} f(\xi+\lambda x)\lambda^{-m-1}\, d\lambda.$$

Let $P_n = (n!)^{-1}\, \hat{\delta}^n f(\xi)$ and $s_M(t) = \sum_{n=0}^{M} P_n(t-\xi)$ for every $t \in U$. For $M \geq m$ it follows that

$$\int_{|\lambda|=\epsilon} s_M(\xi+\lambda x)\lambda^{-m-1}\, d\lambda = \sum_{n=0}^{M} P_n(x) \int_{|\lambda|=\epsilon} \lambda^{-m-1+n}\, d\lambda = 2\pi i\, P_m(x).$$

Thus

$$\int_{|\lambda|=\epsilon} f(\xi+\lambda x)\lambda^{-m-1}\, d\lambda - 2\pi i\, P_m(x) =$$

$$= \int_{|\lambda|=\epsilon} (f(\xi+\lambda x) - s_M(\xi+\lambda x))\lambda^{-m-1}\, d\lambda.$$

It follows that for every $\beta \in SC(F)$ we have:

$$\beta\Big(\int_{|\lambda|=\epsilon} f(\xi+\lambda x)\lambda^{-m-1}\, d\lambda - 2\pi i\, P_m(x)\Big) \leq$$

$$\leq 2\pi\epsilon^{-m} \sup\{\beta(f(\xi+\lambda x) - s_M(\xi+\lambda x));\ |\lambda| = \epsilon\}.$$

For each $B \in \mathcal{B}_E$ there is $\rho_B > 0$ such that $\xi + \rho_B B \subset U$ and

$$\lim_{M\to\infty} \beta\Big(f(x) - \sum_{n=0}^{M} P_n(x-\xi)\Big) = 0$$

uniformly for $x \in \xi + \rho_B B$. Now we choose $B \in \mathcal{B}_E$ such that $x \in E_B$ and $\epsilon > 0$ such that $\xi + \lambda x \in \xi + \rho_B B$ for every $\lambda \in \mathbb{C}$, $|\lambda| = \epsilon$. Thus

$$\lim_{M\to\infty} \sup\{\beta(f(\xi+\lambda x) - s_M(\xi+\lambda x));\ |\lambda| = \epsilon\} = 0.$$

Thus we can write that for every $\beta \in SC(F)$

$$\beta\Big(\int_{|\lambda|=\epsilon} f(\xi+\lambda x)\lambda^{-m-1}\, d\lambda - 2\pi i\, P_m(x)\Big) = 0.$$

This implies our result.

3.14 Definition. If $\alpha \in SC(E)$, $\beta \in SC(F)$, $A \in \mathcal{L}_a(^mE;F)$, $m \in \mathbb{N}^*$, we define

$$\|A\|_{\alpha,\beta} = \sup\{\beta(A(x_1,\ldots,x_m));\ \alpha(x_j) \leq 1,\quad j=1,\ldots,m\}.$$

For $m = 0$ we set $\|A\|_{\alpha,\beta} = \beta(A)$. It is clear that $A \in \mathcal{L}(^mE;F)$ if and only if for every $\beta \in SC(F)$ there is $\alpha \in SC(E)$ such that $\|A\|_{\alpha,\beta} < +\infty$. If $P \in \mathcal{P}_a(^mE;F)$, $m \in \mathbb{N}$, we define

$$\|P\|_{\alpha,\beta} = \sup\{\beta(P(x));\ \alpha(x) \leq 1\}.$$

It is clear that $P \in \mathcal{P}(^mE;F)$ if and only if for every $\beta \in SC(F)$ there is $\alpha \in SC(E)$ such that $\|P\|_{\alpha,\beta} < +\infty$. If $P = \hat{A}$ and A is in $\mathcal{L}_s(^mE;F)$ we have

$$\|P\|_{\alpha,\beta} \leq \|A\|_{\alpha,\beta} \leq m^m(m!)^{-1}\|P\|_{\alpha,\beta}.$$

It is enough to use 1.13 with $E = E_\alpha$ (i.e.: E seminormed by α) and $B = \overline{B_{\alpha,1}(0)}$.

3.15 Cauchy's inequalities. Let f be an element of $\mathcal{H}(U;F)$ with F separated. If $\xi \in U$, $\rho > 0$, $\alpha \in SC(E)$, $\beta \in SC(F)$, $\overline{B_{\alpha,\rho}(\xi)} \subset U$, then

$$\|(m!)^{-1}\, d^m f(\xi)\|_{\alpha,\beta} \leq \rho^{-m} \sup\{\beta(f(t));\ \alpha(t-\xi) \leq \rho\}.$$

Proof. It is a direct consequence of the Cauchy's Integral Formula for differentials.

3.16 Corollary. If F is separated, $\xi \in U \cap E_B$ for some $B \in \mathcal{B}_E$ and $\xi + \rho_B B \subset U$ for some $\rho_B > 0$, then

$$\|(m!)^{-1}\, \hat{\delta}^m f(\xi)\|_{B,\beta} \leq \rho_B^{-m} \sup\{\beta(f(t));\ t \in \xi + \rho_B B\}$$

for every f in $\mathcal{H}_S(U;F)$.

Proof. It is enough to use the fact that $f|U \cap E_B \in \mathcal{H}(U \cap E_B;F)$ and 3.15 with $\alpha = \|\ \|_B$.

3.17 Remark. If $\Gamma_1,\ldots,\Gamma_k$ are oriented, rectifiable, simple, (either closed or not closed) Jordan curves and f is continuous from $\Gamma_1 \times \ldots \times \Gamma_k$ into F then

$$\int_{\Gamma_1 \times \ldots \times \Gamma_k} f(z_1,\ldots,z_k)\,dz_1 \ldots dz_k =$$

$$= \int_{\Gamma_1} \{\ldots\{\int_{\Gamma_k} f(z_1,\ldots,z_k)\,dz_k\}\ldots\}\,dz_1$$

is an element of $\hat{F}$.

<u>3.18 Cauchy's multiple integral formula.</u> If F is separated, $f \in \mathcal{H}_S(U;F)$, $\xi \in U$, $x_i \in E$, $i = 1,\ldots,k$, $m_i \in \mathbb{N}$, $i = 1,\ldots,k$, $m_1 + \ldots + m_k = |m|$. Then for $m = (m_1,\ldots,m_k)$

$$(m!)^{-1}\,\delta^{|m|} f(\xi)\, x_1^{m_1} \ldots x_k^{m_k} =$$

$$(2\pi i)^{-k} \int_{\substack{|\lambda_i| = \rho_i \\ i=1,\ldots,k}} \frac{f(\xi + \lambda_1 x_1 + \ldots + \lambda_k x_k)}{\lambda_1^{m_1+1} \ldots \lambda_k^{m_k+1}}\, d\lambda_1 \ldots d\lambda_k$$

where $\rho_i > 0$, $i = 1,\ldots,k$ are such that $\xi + \lambda_1 x_1 + \ldots + \lambda_k x_k \in U$ for every $|\lambda_i| \leq \rho_i$, $i = 1,\ldots,k$. $(m! = m_1! \ldots m_k!)$.

<u>Proof.</u> Use 3.17 and the same type of reasoning used in 3.13, that is: first prove the result for the finite-dimensional case and then transpose it to the infinite dimensional case.

<u>3.19 Definition.</u> We consider F separated, K a compact space and the spaces $C(K;\mathbb{C})$ and $C(K;F)$ endowed with the locally convex topology of the uniform convergence on K. If μ is a Radon measure (i.e. a continuous linear functional on $C(K;\mathbb{C})$) then there is a unique linear mapping $\vec{\mu}$ from $C(K;F)$ into $\hat{F}$ denoted by $\vec{\mu}(f) = \int_K f\, d\vec{\mu} \in \hat{F}$ such that

$$\int_K (af)\,d\vec{\mu} = a \int_K f\, d\mu$$

for every f of $C(K;\mathbb{C})$ and a in F. We have:

$$\hat{\beta}\left(\int_K f\, d\vec{\mu}\right) \leq \int_K (\beta \circ f) d\mu \leq \|\mu\| \sup\{\beta(f(x));\ x \in K\}$$

for each f in $C(K;F)$ and β in $SC(F)$, $\hat{\beta}$ denoting the continuous extension of β to $\hat{F}$, $\|\mu\|$ being the norm of μ related to the norm of $C(K;\mathbb{C})$. $\vec{\mu}$ is called a vector Radon measure in K.

3.20 Definition. F is called integrally complete if for every compact K, every $f \in C(K;F)$, every Radon measure μ in K, $\int_K f d\vec{\mu} \in F$.

If F is quasi-complete or, more generally, if the closed convex hull of every compact of F is compact, then F is integrally complete.

If K is a metrizable compact space and F is sequentially complete then $\int_K f\, d\vec{\mu} \in F$ for every $f \in C(K;F)$ and every Radon measure in K: μ.

3.21 Proposition. If F is sequentially complete or integrally completo, $\mathcal{H}(U;F)$ is closed in $C(U;F)$ for the compact-open topology τ_0.

3.22 Lemma. If F is separated, $f \in \mathcal{H}_S(U;F)$, $\xi \in U$, $x \in U$, $\rho > 1$ with $(1-\lambda)\xi + \lambda x \in U$ for every $\lambda \in \mathbb{C}$, $|\lambda| \leq \rho$, then

$$f(x) - \tau_{m,f,\xi}(x) = (2\pi i)^{-1} \int_{|\lambda|=\rho} f((1-\lambda)\xi+\lambda x)\lambda^{-m-1}(\lambda-1)^{-1}\, d\lambda .$$

Proof. For $\lambda \in \mathbb{C}$, $0 \neq \lambda \neq 1$, we have:

$$(\lambda-1)^{-1} = \sum_{k=0}^{m} \lambda^{-k-1} + \lambda^{-m-1}(\lambda-1)^{-1}.$$

Now, if we multiply term by $f((1-\lambda)\xi+\lambda x)$ and if we integrate over $|\lambda| = \rho$ we get after division by $2\pi i$:

$$f(x) = \sum_{k=0}^{m} (k!)^{-1} \hat{\delta}^m f(\xi)(x-\xi) + (2\pi i)^{-1} \int_{|\lambda|=\rho} \frac{f((1-\lambda)\xi+\lambda x)\lambda^{-m-1}}{\lambda-1}\, d\lambda .$$

Proof of 3.21. Let f be an element of the closure of $\mathcal{H}(U;F)$ in $C(U;F)$ for τ_0 and let ξ be an element of U. We define $A_0 = f(\xi)$.

For $m \in \mathbb{N}^*$, $x_i \in E$, we consider $\rho_i > 0$, $i = 1,\dots,m$ such that $\xi + \lambda_1 x_1 + \dots + \lambda_m x_m \in U$ for $|\lambda_i| \le \rho_i$, $i = 1,\dots,m$. We define

$$A_m(x_1,\dots,x_m) = (m!)^{-1} \int_{|\lambda_i|=\rho_i} f(\xi+\lambda_1 x_1+\dots+\lambda_m x_m)(\lambda_1\cdots\lambda_m)^{-2}\, d\lambda_1\cdots d\lambda_m$$

which is an element of F by the hypothesis of our proposition. The value of $A_m(x_1,\dots,x_m)$ does not depend on the choice of the ρ_i's since this holds true when f is in $\mathcal{H}(U;F)$ and, by passage to the limit, for f in the closure of $\mathcal{H}(U;F)$ in $C(U;F)$. It is not difficult to see that A_m is continuous in E^m. Also A_m is symmetric and m-linear, since this holds in $\mathcal{H}(U;F)$ and, by passage to the limit, in its closure in $C(U;F)$. Again this type of reasoning grants that for $x \in U$, $\rho > 1$, with $(1-\lambda)\xi + \lambda x \in U$ for $\lambda \in \mathbb{C}$, $|\lambda| \le \rho$, we have

$$f(x) = \sum_{m=0}^{M} A_m(x-\xi)^m = \frac{1}{2\pi i} \int_{|\lambda|=\rho} \frac{f((1-\lambda)\xi+\lambda x)}{\lambda^{M+1}\,(\lambda-1)}\, d\lambda .$$

If $\beta \in SC(F)$ let $\alpha \in SC(E)$ such that $\overline{B_{\alpha,1}(\xi)} \subset U$ and $C = \sup \beta(f(t))$, for t varying in the α-unit ball, finite. For $\rho > 1$ fixed we consider $V = \{t \in E;\ \alpha(t-\xi) \le 1/\rho\} \subset B_{\alpha,1}(\xi) \subset U$. Thus for every x in V $(1-\lambda)\xi + \lambda x \in U$ if $|\lambda| \le \rho$. Thus

$$\beta(f(x) - \sum_{m=0}^{M} A_m(x-\xi)) \le C\rho^{-M}(\rho-1)^{-1}$$

which tends to zero as M goes to $+\infty$.

3.23 Definition. A subset $K \subset E$ is a strict compact set if there is B in $\mathcal{B}_E$ such that $K \subset E_B$ and it is compact for the norm of E_B.

3.24 Definition. τ_{oe} indicates the locally convex topology in $C_S(U;F)$ generated by the seminorms of the type:

$$p_{K,\beta}(f) = \sup\{\beta(f(t));\ t \in K\}$$

for $K \subset U$ a strict compact set and $\beta \in SC(F)$.

3.25 Corollary. If F is either sequentially complete or integrally complete then $\mathcal{H}_S(U;F)$ is closed in $C_S(U;F)$ for τ_{oe}.

__Proof.__ It is enough to note that $\mathcal{H}(U \cap E_B;F)$ is closed in $C(U \cap E_B;F)$ for the compact-open topology.

__3.26 Corollary.__ With no restrictive hypothesis we have:

1) $\overline{\mathcal{H}(U;F)}^{\tau_o} \subset H(U;F)$ and $\overline{\mathcal{H}_S(U;F)}^{\tau_{oe}} \subset H_S(U;F)$

2) $H(U;F)$ is closed in $C(U;F)$ for τ_o and $H_S(U;F)$ is closed in $C_S(U;F)$ for τ_{oe}.

__Proof.__ It is enough to follow the pattern of the proofs of 3.21 and 3.25.

__3.27 Proposition.__ Let K be a compact space and let F be either integrally complete or sequentially complete. If $f: U\times K \to F$ is continuous and such that: $x \in U \to f(x,y) \in F$ is holomorphic for every y in K, then $g: U \to C(K;F)$ given by $g(x)(y) = f(x,y)$ for every x in U and y in K is holomorphic. $C(K;F)$ is considered with the topology of the uniform convergence over K.

__3.28 Lemma.__ If K is compact and F is either sequentially complete or integrally complete, $C(K;F)$ is either sequentially complete or integrally complete.

__Proof.__ If F is sequentially complete the proof is simple and follows the usual pattern. Now we suppose that F is integrally complete. We must show that for J compact and $g \in C(J;C(K;F))$, $\int_J g\, d\vec{\mu} \in C(K;F)$. We have a bijection between $C(J;C(K;F))$ and $C(J\times K;F)$ given by

$$g \in C(J;C(K;F)) \longleftrightarrow f \in C(J\times K;F)$$

where $f(x,y) = g(x)(y)$ for every x in J and y in K. We know that $\int_J f(x,y)d\vec{\mu}(x) \in F$ for every y in K since F is integrally complete. We also know that $\varphi: y \in K \to \varphi(y) = \int_J f(x,y)d\vec{\mu}(x) \in F$ is continuous. Thus $\varphi \in C(K;F)$. But it is clear that $\varphi = \int_J g\, d\vec{\mu} \in C(K;F)$.

Proof of 3.27. Let J be a compact subset of U. Then $f|J\times K$ is uniformly continuous. It follows that for every $\beta \in SC(F)$ and $\varepsilon > 0$ there is a finite cover $K = V_1 \cup \ldots \cup V_n$ by open sets such that for $x \in J$ and $y_1, y_2 \in V_r$ $\beta(f(x,y_1) - f(x,y_2)) \leq \varepsilon$. We may suppose that $V_r \neq \emptyset$. Let $\eta_r \in V_r$. We consider $1 = \alpha_1 + \ldots + \alpha_n$ a continuous partition of the unity subordinated to this cover. Each $\alpha_r \in C(K;R^+)$ and α_r vanishes outside V_r. We define $h(x) = \sum_{r=1}^{n} f(x,\eta_r)\alpha_r$ for every x in U. $h \in \mathcal{H}(U;C(K;F))$ since the mapping $x \in U \to f(x,\eta_r) \in F$ is holomorphic for each r. If x is J and y is in K we have:

$$\beta(g(x)(y) - h(x)(y)) = \beta(f(x,y) - \sum_{r=1}^{n} f(x,\eta_r)\alpha_r(y)) \leq$$

$$\leq \beta(\sum_{r=1}^{n} (f(x,y) - f(x,\eta_r))\alpha_r(y)) \leq$$

$$\leq \sum_{r=1}^{n} \beta(f(x,y) - f(x,\eta_r))\alpha_r(y) \leq \varepsilon .$$

The above result implies that g is in the τ_o-closure of $\mathcal{H}(U;C(K;F))$. Since $C(K;F)$ is either integrally complete or sequentially complete by 3.28, we have the τ_o-closure of $\mathcal{H}(U;C(K;F))$.

3.29 Proposition. We suppose that either K is compact and F is integrally complete or K is metrizable compact and F is sequentially complete. Let f be a continuous mapping from $U\times K$ into F and such that for each y in K the mapping $x \in U \to f(x,y) \in F$ is holomorphic. For every Radon measure μ in K the formula

$$\varphi(x) = \int_K f(x,y)\, d\vec{\mu}(y),$$

for $x \in U$ defines a holomorphic mapping from U into F.

Proof. In the notation of the Proof of 3.27 $g \in \mathcal{H}(U;C(K;F))$ and $\varphi = \vec{\mu}\circ g\colon U \to F$. Since $\vec{\mu}$ is linear and continuous it is easy to prove that $\vec{\mu}\circ g$ is holomorphic.

3.30 Definition. A subset X of E is ξ-equilibrated if $\xi \in X$ and for each $x \in X$ and $\lambda \in C$, $|\lambda| \leq 1$, we have $(1-\lambda)\xi + \lambda x \in X$.

There is the largest ξ-equilibrated subset of a set A containing $\{\xi\}$. If A is an open set this largest ξ-equilibrated set is open. Given $K \subset E$ compact there is the smallest ξ-equilibrated subset of E containing K. This subset is compact.

3.31 Proposition. Let F be separated. If $f \in \mathcal{H}(U;F)$, $\xi \in U$, $K \subset U$ is ξ-equilibrated and compact then, for every $\beta \in SC(F)$ there is an open subset V of U containing K such that:

$$\lim_{M\to\infty} \beta(f(x) - \tau_{M,f,\xi}(x)) = 0$$

uniformly for x in V.

Proof. Let ρ be a real number greater than 1 and let V be an open subset of U containing K such that $(1-\lambda)\xi + \lambda x \in U$ for $x \in V$ and $\lambda \in \mathbb{C}$, $|\lambda| \le \rho$ and such that $C = \sup\{\beta(f((1-\lambda)\xi+\lambda x));\ x \in V, |\lambda| \le \rho\} < +\infty$. We have:

$$f(x) - \tau_{m,f,\xi}(x) = \frac{1}{2\pi i}\int_{|\lambda|=\rho} \frac{f((1-\lambda)\xi+\lambda x)}{\lambda^{m+1}(\lambda-1)}\, d\lambda .$$

Thus

$$\beta(f(x) - \tau_{m,f,\xi}(x)) \le C\, \rho^{-m}(\rho-1)^{-1} \to 0 \quad \text{as} \quad m \to +\infty$$

uniformly for x in V.

3.32 Corollary. If F is separated, $f \in \mathcal{H}(U;F)$ and $\xi \in U$, then

$$f(x) = \sum_{m=0}^{\infty} (m!)^{-1}\, d^m f(\xi)(x-\xi)^m$$

uniformly over every compact subset of U contained in the largest ξ-equilibrated open subset of U.

3.33 Proposition. If F is separated, $f \in \mathcal{H}_S(U;F)$, $\xi \in U$, $K \subset U$ is strict compact and ξ-equilibrated, then for every $\beta \in SC(F)$ and for every $B \in \mathcal{B}_E$ there is $\rho_B > 0$ such that $K + \rho_B B \subset U$ and

$$\lim_{M\to\infty} \beta(f(x) - \tau_{M,f,\xi}(x)) = 0$$

uniformly for $x \in K + \rho_B B$.

Proof. It is enough to use 3.31 for $f|U \cap E_B$.

<u>3.34 Corollary</u>. If F is separated, $f \in \mathcal{H}_S(U;F)$, $\xi \in U$, then

$$f(x) = \sum_{m=0}^{\infty} (m!)^{-1} \, \delta^m f(\xi)(x-\xi)^m$$

uniformly over every strict compact subset of U contained in the largest ξ-equilibrated open subset of U.

<u>3.35 Definition</u>. Let α be in $SC(E)$ and β in $SC(F)$. Then $\mathcal{L}_s({}^m E_\alpha ; F_\beta)$ and $\mathcal{P}({}^m E_\alpha ; F_\beta)$ are seminormed by

$$A \in \mathcal{L}_s({}^m E_\alpha ; F_\beta) \rightarrow \|A\|_{\alpha,\beta} \in \mathbb{R}^+$$

$$P \in \mathcal{P}({}^m E_\alpha ; F_\beta) \rightarrow \|P\|_{\alpha,\beta} \in \mathbb{R}^+.$$

Moreover $\mathcal{L}_s({}^m E_\alpha ; F_\beta)$ and $\mathcal{P}({}^m E_\alpha ; F_\beta)$ are isomorphic and homeomorphic. We have

$$\bigcup_\alpha \mathcal{L}_s({}^m E_\alpha ; F_\beta) = \mathcal{L}_s({}^m E ; F_\beta)$$

$$\bigcap_\beta \mathcal{L}_s({}^m E ; F_\beta) = \mathcal{L}_s({}^m E ; F)$$

and the analogous identities for the spaces of polynomials. For β fixed we consider in $\mathcal{L}_s({}^m E ; F_\beta)$ (resp. $\mathcal{P}({}^m E ; F_\beta)$) the locally convex topology which is the inductive limit of the topologies of $\mathcal{L}_s({}^m E_\alpha ; F_\beta)$ (resp. $\mathcal{P}({}^m E_\alpha ; F_\beta)$). In $\mathcal{L}_s({}^m E ; F)$ (resp. $\mathcal{P}({}^m E ; F)$) we consider the locally convex topology which is the projective limit of the topologies of $\mathcal{L}_s({}^m E ; F_\beta)$ (resp. $\mathcal{P}({}^m E ; F_\beta)$). These topologies are called the limit topologies and they are denoted by τ_ℓ. It is easy to show that τ_ℓ contains the strong topology and this one contains the compact-open topology. Moreover $\mathcal{L}_s({}^m E ; F)$ and $\mathcal{P}({}^m E ; F)$ are homeomorphic if we consider the limit topologies on them.

<u>3.36 Proposition</u>. If F is separated, $f \in \mathcal{H}(U;F)$ and $m \in \mathbb{N}$, then $d^m f$ is in $\mathcal{H}(U;(\mathcal{L}_s({}^m E ; F), \tau_\ell))$ and $\hat{d}^m f$ $\mathcal{H}(U;(\mathcal{P}({}^m E ; F), \tau_\ell))$. Moreover, if

$$f(x) \cong \sum_{k=0}^{\infty} P_k(x-\xi)$$

is the Taylor series of f at ξ, then the Taylor series of $d^m f$ and $\hat{d}^m f$ at ξ are

$$d^m f(x) \cong \sum_{k=0}^{\infty} d^m P_{k+m}(x-\xi)$$

$$\hat{d}^m f(x) \cong \sum_{k=0}^{\infty} \hat{d}^m P_{k+m}(x-\xi).$$

3.37 Lemma. If $P \in \mathcal{P}_a({}^kE;F)$ then $d^m P \in \mathcal{P}({}^{k-m}E;(\mathcal{L}_s({}^mE;F),\tau_\ell))$ and $\hat{d}^m P \in \mathcal{P}({}^{k-m}E;(\mathcal{P}({}^mE;F),\tau_\ell))$ where $m = 0,\dots,k$.

Proof. We have $d^m P \in \mathcal{P}_a({}^{k-m}E;(\mathcal{L}_s({}^mE;F),\tau_\ell))$ since $d^m P(\xi) = m!\binom{k}{m} A\xi^{k-m}$ where $\hat{A} = P$. For every $\beta \in SC(F)$ there is $\alpha \in SC(E)$ such that $\|A\|_{\alpha,\beta} \le 1$. Thus $\beta(A(x_1,\dots,x_k)) \le \alpha(x_1)\dots\alpha(x_k)$ for every $x_1,\dots,x_k \in E$. Therefore $\beta(A\xi^{k-m},x_1,\dots,x_m) \le (\alpha(\xi))^{k-m}\alpha(x_1),\dots,\alpha(x_m)$ and $A\xi^{k-m} \in \mathcal{L}_s({}^mE_\alpha,F_\beta)$ with $\|A\xi^{k-m}\|_{\alpha,\beta} \le (\alpha(\xi))^{k-m}$. Thus the polynomial

$$\xi \in E_\alpha \longmapsto A\xi^{k-m} \in \mathcal{L}_s({}^mE_\alpha,F_\beta)$$

is continuous. Therefore $\xi \in E \mapsto A\xi^{k-m} \in \mathcal{L}_s({}^mE;F_\beta)$ is continuous for the inductive limit topology in the range space. This is true for all β in $SC(F)$. Thus our lemma follows.

Proof of 3.36. We consider

$$P_k = (k!)^{-1}\hat{d}^k f(\xi) \in \mathcal{P}({}^kE;F) \quad \text{and} \quad r_M(x) = f(x) - \sum_{k=0}^{M} P_k(x-\xi)$$

for every x in U, $M \in \mathbb{N}$. For each $\beta \in SC(F)$ and $\gamma \in SC((\mathcal{P}({}^mE;F_\beta),\tau_\ell)$ there is $\alpha \in SC(E)$ such that $\overline{B_{\alpha,1}(\xi)} \subset U$ and the supremum of $\beta\circ f$ over this closed ball is a number $C < +\infty$. On the other hand there is $D \ge 0$ such that $\gamma(P) \le D\|P\|_{\alpha,\beta}$ for every P in $\mathcal{P}({}^mE_a;F_\beta)$. We fix $\theta > 0$ and $\rho > 0$ such that $\theta + \rho < 1$ and $(\theta+\rho)^{-1} = \sigma > 1$. If $x \in B_{\alpha,\theta}(\xi)$, $y \in \overline{B_{\alpha,1}(0)}$, $|\lambda| \le \rho$ we have $\alpha(x+\lambda y-\xi) \le \alpha(x-\xi) + |\lambda|\alpha(y) < \theta + \rho < 1$. Thus $x + \lambda x \in B_{\alpha,1}(\xi)$ and

$$(m!)^{-1}\hat{d}^m r_M(x)(y) = (2\pi i)^{-1}\int_{|\lambda|=\rho} r_M(x+\lambda x)\lambda^{-m-1}\,d\lambda,$$

with

$$r_M(x+\lambda y) = (2\pi i)^{-1}\int_{|\mu|=\sigma} \frac{f((1-\mu)\xi + \mu(x+\lambda y))}{\mu^{M+1}(\mu-1)}\,d\mu$$

since if $|\mu| \le \sigma$, $(1-\mu)\xi + \mu(x+\lambda y) \in B_{\alpha,1}(\xi) \subset U$. Thus

$$\beta((m!)^{-1}\hat{d}^m r_M(x)(y)) \le C\rho^{-m}\sigma^{-M}(\sigma-1)^{-1}$$

and

$$\|\hat{d}^m r_M(x)\|_{\alpha,\beta} \le C\rho^{-m}\sigma^{-M}(\sigma-1)^{-1}\, m!$$

for every $x \in B_{\alpha,\theta}(\xi)$. Thus $\hat{d}^m r_M(x) \in \mathcal{P}(^mE_\alpha;F_\beta)$ for $x \in B_{\alpha,\theta}(\xi)$. It follows that $\gamma(\hat{d}^m r_M(x)) \le m!\, CD\rho^{-m}\sigma^{-M}(\sigma-1)^{-1}$ for each $x \in B_{\alpha,\theta}(\xi)$. But

$$\hat{d}^m r_M(x) = \hat{d}^m f(x) - \sum_{k=m}^{M} \hat{d}^m P_k(x-\xi) = \hat{d}^m f(x) - \sum_{k=0}^{M-m} \hat{d}^m P_{k+m}(x-\xi)$$

if $m \le M$. By 3.37 $\hat{d}^m P_{k+m} \in \mathcal{P}(^kE;(\mathcal{P}(^mE;F),\tau_\ell))$. We also have

$$\lim_{M\to+\infty} \gamma(\hat{d}^m f(x) - \sum_{k=0}^{M} \hat{d}^m P_{k+m}(x-\xi) = 0$$

uniformly for $x \in B_{\alpha,\theta}(\xi)$.

<u>3.38 Proposition</u>. Let F be separated, $f \in \mathcal{H}_S(U;F)$, $m \in \mathbb{N}$. Thus $\delta^m f \in \mathcal{H}_S(U;\mathcal{L}_{bs}(^mE;F))$ and $\hat{\delta}^m f \in \mathcal{H}_S(U;\mathcal{P}_b(^mE;F))$. If $f(x) \cong \sum_{k=0}^{\infty} P_k(x-\xi)$ is the Taylor series of f at ξ, then the Taylor series of $\delta^m f$ and $\hat{\delta}^m f$ at ξ are

$$\delta^m f(x) \cong \sum_{k=0}^{\infty} \delta^m P_{k+m}(x-\xi)$$

and

$$\hat{\delta}^m f(x) \cong \sum_{k=0}^{\infty} \hat{\delta}^m P_{k+m}(x-\xi).$$

<u>3.39 Lemma</u>. If $P \in \mathcal{P}_b(^kE;F)$ then $\delta^m P \in \mathcal{P}_b(^{k-m}E;\mathcal{L}_{bs}(^mE;F))$ and $\hat{\delta}^m P \in \mathcal{P}_b(^{k-m}E;\mathcal{P}_b(^mE;F))$ for $m = 0,\dots,k$.

<u>Proof</u>. We have $\delta^m P \in \mathcal{P}_a(^{k-m}E;\mathcal{L}_{bs}(^mE;F))$ since $\delta^m P(\xi) = m!\binom{k}{m}A\xi^{k-m}$ for $\hat{A} = P$. In fact: we have

$$\beta(A(x_1,\dots,x_k)) \le \|A\|_{B,\beta}\, \|x_1\|_B \cdots \|x_k\|_B$$

for $x_1,\dots,x_k \in E_B$. Thus

$$\beta(A\xi^{k-m}x_1\dots x_m) \le \|A\|_{B,\beta}\, \|\xi\|_B^{k-m}\, \|x_1\|_B \cdots \|x_m\|_B$$

for all $\xi,x_1,\dots,x_m \in E_B$. Hence $A\xi^{k-m} \in \mathcal{L}_{bs}(^mE;F)$ and

$$\|A\xi^{k-m}\|_{B,\beta} \le \|A\|_{B,\beta}\, \|\xi\|_B^{k-m}$$

if $\xi \in E_B$. Thus the lemma follows.

Proof of 3.38. We take $P_k = (k!)^{-1} \hat{\delta}^k f(\xi) \in \mathcal{P}_b({}^kE;F)$ and $r_M(x) = f(x) - \sum_{k=0}^{M} P_k(x-\xi)$ for $x \in U$ and $M \in \mathbb{N}$. If $\beta \in SC(F)$, $B \in \mathcal{B}_E$ with $\xi \in E_B$ there is $\rho_B > 0$ such that $\xi + \rho_B B \subset U$ and $\beta(f(t)) \leq C < +\infty$ for every $t \in \xi + \rho_B B$. Now we fix $\theta > 0$, $\rho > 0$, $\theta + \rho < 1$, $\sigma = (\theta+\rho)^{-1} > 1$. If we take $x \in \xi + \theta\rho_B B$, $y \in \rho_B B$, $|\lambda| \leq \rho$ we have $x + \lambda y$ belonging to $\xi + (\theta+\rho)\rho_B B \subset \xi + \rho_B B$.

Thus we may write:

$$(m!)^{-1} \hat{\delta}^m r_M(x)(y) = (2\pi i)^{-1} \int_{|\lambda|=\rho} \frac{r_M(x+\lambda y)}{\lambda^{m+1}} d\lambda .$$

We also have

$$r_M(x+\lambda y) = (2\pi i)^{-1} \int_{|\mu|=\sigma} \frac{f((1-\mu)\xi + \mu(x+\lambda y))}{\mu^{M+1} (\mu-1)} d\mu$$

since $\xi + (1-\mu)\xi + \mu(x+\lambda y) \in \xi + \rho_B B \subset U$ if $|\mu| \leq \sigma$, thus

$$\beta((m!)^{-1} \hat{\delta}^m r_M(x)(y)) \leq \rho^{-m} \sigma^{-M}(\sigma-1)^{-1} C$$

and

$$\rho_B^m \| (m!)^{-1} \hat{\delta}^m r_M(x)\|_{B,\beta} \leq C\rho^{-m} \sigma^{-M}(\sigma-1)^{-1}$$

for every $x \in \xi + \theta\rho_B B$. Hence

$$\| \hat{\delta}^m r_M(x)\|_{B,\beta} \to 0 \quad \text{as} \quad M \to \infty$$

uniformly for x in $\xi + \theta\rho_B B$. Since

$$\hat{\delta}^m r_M(x) = \hat{\delta}^m f(x) - \sum_{k=0}^{M-m} \hat{\delta}^m P_{k+m}(x-\xi) \quad \text{for} \quad M \geq m$$

and $\hat{\delta}^m P_{k+m} \in \mathcal{P}_b({}^kE; \mathcal{P}_b({}^mE;F))$ by 3.39, the result of Proposition follows.

3.40 Corollary. If $f \in \mathcal{H}_S(U;F)$ then $\tau_{k,\delta^m f,\xi} = \delta^m(\tau_{k+m,f,\xi})$ and $\tau_{k,\hat{\delta}^m f,\xi} = \hat{\delta}^m(\tau_{k+m,f,\xi})$.

3.41 Proposition. (Uniqueness of the holomorphic extension). Let F be separated $f \in \mathcal{H}(U;F)$ with U connected. Then:

(a) If f vanishes in a non-void open subset of U then it vanishes in U.

(b) If $\exists \xi \in U$ and $\hat{d}^m f(\xi) = 0$ for every $m \in \mathbb{N}$, then $f = 0$ in U.

Proof. We know the following facts:

(1) If $V \subset U$ is open and $f = 0$ in V then $d^m f = 0$ in V for every $m \in \mathbb{N}$. (Use the Cauchy integral formulas).

(2) If $\xi \in U$ and $d^m f(\xi) = 0$ for every $m \in \mathbb{N}$, then $f = 0$ in the largest ξ-equilibrated (open) subset of U. (Use the Taylor series of f at ξ).

(a) Let V be the interior of $\{x \in U;\ f(x) = 0\}$. By hypothesis $V \neq \emptyset$. By (1) $d^m f(x) = 0$ for every x in V and every $m \in \mathbb{N}$. If $\xi \in U \cap \bar{V}$ then it follows that $d^m f(\xi) = 0$ for every $m \in \mathbb{N}$. (Use the continuity of $d^m f$). Now (2) implies that $\xi \in V$. Hence V is closed in U and $U = V$.

(b) The hypothesis and (2) imply that f vanishes in a non-void open subset of U. By (a) f vanishes in U.

3.42 Proposition (Uniqueness of the S-holomorphic extension). Let F be separated, $f \in \mathcal{H}_S(U;F)$ with U connected. Then:

(a) If there exist $\xi \in U$ such that for each $B \in \mathcal{B}_E$, with $\xi \in E_B$, there is $\rho_B > 0$ with f vanishing in $\xi + \rho_B B \subset U$, then f vanishes in U.

(b) If there is $\xi \in U$ with $\delta^m f(\xi) = 0$ for every $m \in \mathbb{N}$, then $f = 0$ in U.

Proof. (a) Let $\xi \in E_B \cap U$ such that f vanishes in $\xi + \rho_B B \cap U$. Let x be a point of U. Since U is open and connected U is connected by polygonal lines. Thus there exist a polygonal line P such that ξ and x are joined by it. Let B_o be the closed convex balanced hull of $B \cup P$. $B_o \in \mathcal{B}_E$. By the hypothesis of (a) f vanishes in $\xi + \rho_{B_o} B_o$. Since ξ and x are in the same connected component of $U \cap E_{B_o}$ the preceding result 3.41 applied to $f|U \cap E_{B_o}$

gives $f(x) = 0$.

(b) Use the same type of reasoning used in part (a).

4. SILVA HOLOMORPHY TYPES

From now on F is complete and separated.

4.1 Definition. A Silva holomorphy type (S-holomorphy type) from E into F is a sequence of complete locally convex spaces $\mathcal{P}_{b\theta}(^m E;F)$, for $m \in \mathbb{N}$, the topology of $\mathcal{P}_{b\theta}(^m E;F)$ being generated by the family of seminorms $P \mapsto \|P\|_{\theta,B,\beta}$ indexed by $B \in \mathcal{B}_E$ and $\beta \in SC(F)$. These spaces must have the following properties:

(1) $\mathcal{P}_{b\theta}(^m E;F)$ is a vector subspace of $\mathcal{P}_b(^m E;F)$ for each $m \in \mathbb{N}$.

(2) $\mathcal{P}_{b\theta}(^o E;F)$ coincides with F as a locally convex space. In this case $\|\cdot\|_{\theta,B,\beta} = \beta$ for each $B \in \mathcal{B}_E$.

(3) There is $\sigma \geq 1$ such that for every $B \in \mathcal{B}_E$, $\beta \in SC(F)$, $\ell \in \mathbb{N}$, $m \in \mathbb{N}$, $\ell \leq m$, $x \in E_B$ and $P \in \mathcal{P}_{b\theta}(^m E;F)$ we have

$$\hat{\delta}^\ell P(x) \in \mathcal{P}_{b\theta}(^\ell E;F) \quad \text{and} \quad \|(\ell!)^{-1} \hat{\delta}^\ell P(x)\|_{\theta,B,\beta} \leq$$

$$\leq \sigma^m \|P\|_{\theta,B,\beta} \|x\|_B^{m-\ell}.$$

(4) If $B,D \in \mathcal{B}_E$ and $B \subset D$ then

$$\|P\|_{\theta,B,\beta} \leq \|P\|_{\theta,D,\beta}$$

for every $\beta \in SC(F)$, $P \in \mathcal{P}_{b\theta}(^m E;F)$, $m \in \mathbb{N}$.

4.2 Examples

(1) $\mathcal{P}_{b\theta}(^m E;F) = \mathcal{P}_b(^m E;F)$ $(m \in \mathbb{N})$ (current S-holomorphy type)

(2) $\mathcal{P}_{b\theta}(^m E;F) = \mathcal{P}_{bc}(^m E;F)$ $(m \in \mathbb{N})$ (compact S-holomorphy type)

(3) $\mathcal{P}_{b\theta}(^m E;F) = \mathcal{P}_{bN}(^m E;F)$ $(m \in \mathbb{N})$ (nuclear S-holomorphy type)

(4) $\mathcal{P}_{b\theta}(^m E;F) = \mathbb{P}_c(^m E;F)$ $(m \in \mathbb{N})$ (compact c-S-holomorphy type)

(5) $\mathcal{P}_{b\theta}(^m E;F) = \mathbb{P}_N(^m E;F)$ $(m \in \mathbb{N})$ (nuclear c-S-holomorphy type).

<u>4.3 Proposition</u>. The inclusion mapping from $\mathcal{P}_{b\theta}(^mE;F)$ into $\mathcal{P}_b(^mE;F)$ is continuous and

$$\|P\|_{B,\beta} \leq \sigma^m \|P\|_{\theta,B,\beta}$$

for every $B \in \mathcal{B}_E$, $\beta \in SC(F)$ and $P \in \mathcal{P}_{b\theta}(^mE;F)$.

<u>Proof</u>. If we put $\ell = 0$ in (3) of 4.1 and if we use (1) and (2) of 4.1 we get $\beta(P(x)) \leq \sigma^m \|P\|_{\theta,B,\beta} \|x\|_B^m$ for every $x \in E_B$. Thus $\|P\|_{B,\beta} \leq \sigma^m \|P\|_{\theta,B,\beta}$.

<u>4.4 Definition</u>. A mapping $f \in \mathcal{H}_S(U;F)$ is of S-holomorphy type θ at $\xi \in U$ if the following conditions are satisfied:

(1) $\hat{\delta}^m f(\xi) \in \mathcal{P}_{b\theta}(^mE;F)$ for every $m \in \mathbb{N}$.

(2) For every $B \in \mathcal{B}_E$ and every $\beta \in SC(F)$ there are $C \geq 0$, $c \geq 0$ (depending on B and β) such that

$$\|(m!)^{-1} \hat{\delta}^m f(\xi)\|_{\theta,B,\beta} \leq C c^m \quad \text{for every} \quad m \in \mathbb{N}.$$

f is of S-holomorphy type θ in U if f is of S-holomorphy type θ at every point of U. $\mathcal{H}_{S\theta}(U;F)$ denotes the vector space of all S-holomorphic mappings from U into F which are of S-holomorphy type θ in U.

<u>4.5 Proposition</u>. Let f be an element of $\mathcal{H}_{S\theta}(U;F)$. For every strict compact subset K of U and for every $B \in \mathcal{B}_E$ such that K is contained and compact in E_B and for every $\beta \in SC(F)$ there are $C \geq 0$, $c \geq 0$, $\rho > 0$ satisfying $K + \rho B \subset U$ and

$$\|(m!)^{-1} \hat{\delta}^m f(x)\|_{\theta,B,\beta} \leq C c^m$$

for each x in $K + \rho B$ and $m \in \mathbb{N}$.

<u>4.6 Lemma</u>. Let $f \in \mathcal{H}_S(U;F)$ of S-holomorphy type θ at $\xi \in U$. If $B \in \mathcal{B}_E$ is such that $\xi \in E_B$ and $\beta \in SC(F)$ then there is $\rho > 0$ such that $\xi + \rho B \subset U$ and

(1) $\hat{\delta}^m f(x) \in \mathcal{P}_{b\theta}(^mE;F)$ for every x in $\xi + \rho B$ and m in $\mathbb{N}$.

(2) There are $C \geq 0$, $c \geq 0$ such that

$$\|(m!)^{-1}\hat{\delta}^m f(x)\|_{\theta,B,\beta} \le C\,c^m$$

for every x in $\xi + \rho B$ and m in $\mathbb{N}$.

(3) For each $x \in \xi + \rho B$ and ℓ in $\mathbb{N}$

$$\hat{\delta}^\ell f(x) = \sum_{m=\ell}^{\infty} \hat{\delta}^\ell P_m(x-\xi)$$

where $P_m = (m!)^{-1}\hat{\delta}^m f(\xi)$ for every $m \in \mathbb{N}$ and the convergence is in the sense of $\mathcal{P}_{b\theta}({}^\ell E;F)$.

<u>Proof</u>. We consider the Taylor series of f at ξ

$$f(x) \cong \sum_{m=0}^{\infty} P_m(x-\xi).$$

If B and β are as in the statement of the lemma there is $\rho > 0$ such that $\xi + 2\rho B \subset U$ and $\beta \circ f$ is bounded in $\xi + 2\rho B$. Moreover we may take $\rho > 0$ such that if $x \in \xi + \rho B$ and $\ell \in \mathbb{N}$ we have

$$(*) \qquad f(x) = \sum_{m=\ell}^{\infty} \hat{\delta}^\ell P_m(x-\xi)$$

the convergence being taken in the sense of $\mathcal{P}_b({}^\ell E;F)$. By (1) of 4.4 we have $P_m \in \mathcal{P}_{b\theta}({}^m E;F)$. By (3) of 4.1 we have $\hat{\delta}^\ell P_m(x-\xi) \in \mathcal{P}_{b\theta}({}^\ell E;F)$ and $\sum_{m=\ell}^{\infty} \|\hat{\delta}^\ell P_m(x-\xi)\|_{\theta,B,\beta} \le \ell!\sum_{m=\ell}^{\infty} \sigma^m \|P_m\|_{\theta,B,\beta}\|x-\xi\|_B^{m-\ell}$ for every x in $\xi + \rho B$. By (2) of 4.4 there are $C \ge 0$, $c \ge 0$ such that for every $m \in \mathbb{N}$ we have $\|P_m\|_{\theta,B,\beta} \le C\,c^m$. Now we take $\rho > 0$ in such a way that $\sigma c\rho < 1$ in addition to the previous conditions. We have

$$\sum_{m=\ell}^{\infty} \|\hat{\delta}^\ell P_m(x-\xi)\|_{\theta,B,\beta} \le \ell!\; C(\sigma c)^\ell\,(1-\sigma c\rho)^{-1}$$

for every x in $\xi + \rho B$ and every $\ell \in \mathbb{N}$. By the above results and by (4) of 4.4 we can see that the series (*) converges in the sense of $\mathcal{P}_{b\theta}({}^\ell E;F)$. Since $\mathcal{P}_{b\theta}({}^\ell E;F)$ is complete we have $\hat{\delta}^\ell f(x) \in \mathcal{P}_{b\theta}({}^\ell E;F)$ and (*) holds with the convergence in the sense of $\mathcal{P}_{b\theta}({}^\ell E;F)$. Moreover we have

$$\|(\ell!)^{-1}\hat{\delta}^\ell f(x)\|_{\theta,B,\beta} \le C(\sigma c)^\ell\,(1-\sigma c\rho)^{-1}$$

for every $x \in \xi + \rho B$ and for every $\ell \in \mathbb{N}$.

Proof of 4.5. Apply condition (2) of 4.6.

5. DIFFERENTIATION OF SILVA HOLOMORPHY TYPES

5.1 Definition. Let θ be an S-holomorphy type from E into F and let ℓ be in $\mathbb{N}$. By (3) of 4.1 the vector isomorphism

$$(*)\qquad P \in \mathcal{P}_b({}^{\ell+m}E;F) \longmapsto (\ell!)^{-1}\, \hat{\delta}^{\ell} P \in \mathcal{P}_b({}^mE;\, \mathcal{P}_b({}^{\ell}E;F))$$

(for each m in $\mathbb{N}$) induces a vector isomorphism from $\mathcal{P}_{b\theta}({}^{\ell+m}E;F)$ onto a vector subspace of $\mathcal{P}_b({}^mE;\mathcal{P}_{b\theta}({}^{\ell}E;F))$. This vector subspace will be denoted by $(\ell!)^{-1}\, \hat{\delta}^{\ell}\, \mathcal{P}_{b\theta}({}^{\ell+m}E;F) = \mathcal{P}_{b\tau}({}^mE;\, \mathcal{P}_{b\theta}({}^{\ell}E;F))$. If $m \geq 1$ we consider in this space the locally convex topology generated by the seminorms

$$\|(\ell!)^{-1}\, \hat{\delta}^{\ell} P\|_{\tau,D,\|\cdot\|_{\theta,B,\beta}} = \|P\|_{\theta,\widehat{BUD},\beta}$$

for $P \in \mathcal{P}_{b\theta}({}^{\ell+m}E;F)$, B and D in $\mathcal{B}_E$, β in $SC(F)$. If $m = 0$ we set

$$\|(\ell!)^{-1}\, \hat{\delta}^{\ell} P\|_{\tau,D,\|\cdot\|_{\theta,B,\beta}} = \|P\|_{\theta,B,\beta}$$

for every $P \in \mathcal{P}_{b\theta}({}^{\ell}E;F)$ and every $B, D \in \mathcal{B}_E$ and every $\beta \in SC(F)$.

5.2 Remark. In the conditions of 5.1 the restriction of (*) to the subspace $\mathcal{P}_{b\theta}({}^{\ell+m}E;F)$ is a topological isomorphism from this space onto $\mathcal{P}_{b\tau}({}^mE;\, \mathcal{P}_{b\theta}({}^{\ell}E;F))$. Hence this last space is complete.

5.3 Proposition. The sequence $(\mathcal{P}_{b\tau}({}^mE;\, \mathcal{P}_{b\theta}({}^{\ell}E;F)))_{m\in\mathbb{N}}$ is an S-holomorphy type from E into $\mathcal{P}_{b\theta}({}^{\ell}E;F)$ and we use the notation $(\ell!)^{-1}\, \hat{\delta}^{\ell}\theta = \tau$.

Proof. The conditions (1), (2) and (4) of 4.1 are easily verified. We show that condition (3) also is satisfied. For $m = 0$ the proof is simple. Let $m \geq 1$. We take $k \in \mathbb{N}$, $k \leq m$, $D \in \mathcal{B}_E$, $\|\cdot\|_{\theta,B,\beta} \in SC(\mathcal{P}_{b\theta}({}^{\ell}E;F))$, $x \in E_D$, $P \in \mathcal{P}_{b\tau}({}^mE;\, \mathcal{P}_{b\theta}({}^{\ell}E;F))$. We have $P = (\ell!)^{-1}\, \hat{\delta}^{\ell} Q$ with Q in $\mathcal{P}_{b\theta}({}^{\ell+m}E;F)$ and $(k!)^{-1}\, \hat{\delta}^k P(x) = (k!)^{-1}\, \hat{\delta}^k((\ell!)^{-1}\, \hat{\delta}^{\ell} Q)(x) = (\ell!)^{-1}\, (((\ell+k)!)^{-1}\, \hat{\delta}^{\ell+k}\, Q(x))$. By condition (3) of the definition of S-holomorphy type, since $\hat{\delta}^{\ell+k} Q(x) \in \mathcal{P}_{b\theta}({}^{\ell+k}E;F)$, we have $\hat{\delta}^k P(x)$ in $(\ell!)^{-1}\hat{\delta}^\ell \mathcal{P}_{b\theta}({}^{\ell+k}E;F)$ and

$$\|(k!)^{-1}\hat{\delta}^k P(x)\|_{\tau,D,\|\cdot\|_{\theta,D,\beta}} =$$

$$= \|((\ell+k)!)^{-1}\delta^{\ell+k}Q(x)\|_{\theta,\widehat{BUD},\beta} \leq \sigma^{\ell+m}\|Q\|_{\theta,\widehat{BUD},\beta}\|x\|_{\widehat{B+D}}^{m-k} \leq$$

$$\leq (\sigma^{\ell+1})^m \|(\ell!)^{-1}\hat{\delta}^\ell Q\|_{\tau,D,\|\cdot\|_{\theta,B,\beta}}\|x\|_D^{m-k}.$$

5.4 Proposition. If $\tau = (\ell!)^{-1}\hat{\delta}^\ell\theta$ and $f \in \mathcal{H}_{S\theta}(U;F)$ then $\hat{\delta}^\ell f$ is in $\mathcal{H}_{S\tau}(U;\mathcal{P}_{b\theta}({}^\ell E;F))$.

Proof. We first show that $\hat{\delta}^\ell f \in \mathcal{H}_S(U;\mathcal{P}_{b\theta}({}^\ell E;F))$. Let $\xi \in U$, $\|\cdot\|_{\theta,D,\beta} \in SC(\mathcal{P}_{b\theta}({}^\ell E;F))$ and $B \in \mathcal{B}_E$ with $\xi \in E_B$. Let $C \geq 0$, $c \geq 0$, be the constants relative to f and ξ of Definition 4.4. Let $\rho > 0$ be the positive number corresponding to ξ, B, β given in 4.6. We have

$$\delta^\ell f(x) = \sum_{m=0}^{\infty} \hat{\delta}^\ell P_{m+\ell}(x-\xi) \tag{1}$$

for $x \in \xi + \rho B$, the convergence being in the sense of $\mathcal{P}_{b\theta}({}^\ell E;F)$. If we make $\rho c\sigma < 1$ we may observe that we have:

$$\|(\ell!)^{-1}\hat{\delta}^\ell P_{\ell+m}(x-\xi)\|_{\theta,D,\beta} \leq \sigma^{\ell+m}\|P_{\ell+m}\|_{\theta,D,\beta}\|x-\xi\|_D^m \leq$$

$$\leq C(\sigma c)^\ell(\sigma c)^m\|x-\xi\|_B^m$$

where with no loss of generality it was supposed that $B \subset D$ and $\xi \in E_D$. Thus

$$\|(\ell!)^{-1}\hat{\delta}^\ell P_{\ell+m}(x-\xi)\|_{\theta,D,\beta} \leq C(\sigma c)^\ell(\sigma c)^m\rho^m \tag{2}$$

for every $x \in \xi + \rho B$ and $m \in \mathbb{N}$. By (1) and (2) it follows that

$$\lim_{M\to\infty}\|\hat{\delta}^\ell f(x) - \sum_{m=0}^{M}(\ell!)^{-1}\hat{\delta}^\ell P_{\ell+m}(x-\xi)\|_{\theta,D,\beta} \leq$$

$$\leq \lim_{M\to\infty}\sum_{m=M}^{\infty}\ell!\|(\ell!)^{-1}\hat{\delta}^\ell P_{\ell+m}(x-\xi)\|_{\theta,D,\beta} \leq$$

$$\leq \lim_{M\to\infty}\sum_{m>M}\ell!\,C(\sigma c)^\ell(\sigma c\rho)^m = 0$$

uniformly for x in $\xi + \rho B$. Thus it was proved that $\hat{\delta}^{\ell} f \in \mathcal{H}_S(U; \mathcal{P}_{b\theta}(^{\ell}E;F))$. Let $\xi \in U$ be arbitrary. For each m in $\mathbb{N}$ we have $(m!)^{-1}\, \hat{\delta}^m((\ell!)^{-1}\, \hat{\delta}^{\ell} f)(\xi) = (\ell!)^{-1}\, \hat{\delta}^{\ell} P_{m+\ell}$ where $P_k = (k!)^{-1}\, \hat{\delta}^k f(\xi)$ for k in $\mathbb{N}$. Thus $\hat{\delta}^m(\hat{\delta}^{\ell} f)(\xi) = m!\, \hat{\delta}^{\ell}\, P_{m+\ell} \in \mathcal{P}_{b\tau}(^mE; \mathcal{P}_{b\theta}(^{\ell}E;F))$. Now we consider C_o and c_o as the constants corresponding to ξ, $\widehat{B \cup D}$ and β for f as in 4.4. We have

$$\|(\ell!)^{-1}\, \hat{\delta}^{\ell} P_{\ell+m}\|_{\theta, D, \|\cdot\|_{\theta,B,\beta}} \leq C_o\, c_o^{\ell}\, c_o^m$$

for every m in $\mathbb{N}$. Thus $\hat{\delta}^{\ell} f \in \mathcal{H}_{S\tau}(U; \mathcal{P}_{b\theta}(^{\ell}E;F))$.

<u>5.5 Corollary</u>. If $f \in \mathcal{H}_{S\theta}(U;F)$, $\xi \in U$, $B \in \mathcal{B}_E$ with $\xi \in \mathcal{B}_E$ with $\xi \in E_B$, $\beta \in SC(F)$, $\rho > 0$ with $\xi + \rho B \subset U$, then

$$\|(\ell!)^{-1}\, \hat{\delta}^{\ell} f(x)\|_{\theta,B,\beta} \leq \sum_{m=\ell}^{\infty} \sigma^m \rho^{m-\ell}\, \|(m!)^{-1}\, \hat{\delta}^m f(\xi)\|_{\theta,B,\beta}$$

for every x in $\xi + \rho B$ and every $\ell \in \mathbb{N}$.

<u>Proof</u>. If $\tau = (\ell!)^{-1}\, \hat{\delta}^{\ell}\theta$ we have that $(\ell!)^{-1}\, \hat{\delta}^{\ell} f \in \mathcal{H}_{S\tau}(U;\mathcal{P}_{b\theta}(^{\ell}E;F))$ (by 5.4). If $P_m = (m!)^{-1}\, \hat{\delta}^m f(\xi)$ we have

$$(\ell!)^{-1}\, \hat{\delta}^{\ell} f(x) = \sum_{m=\ell}^{\infty} (\ell!)^{-1}\, \hat{\delta}^{\ell}\, P_m(x-\xi)$$

for x in $\xi + \rho B$. The convergence is taken in the sense of $\mathcal{P}_{b\theta}(^{\ell}E;F)$. Thus

$$\|(\ell!)^{-1}\, \delta^{\ell} f(x)\|_{\theta,B,\beta} \leq \sum_{m=\ell}^{\infty} \|(\ell!)^{-1}\, \hat{\delta}^{\ell} P_m(x-\xi)\|_{\theta,B,\beta} \leq$$

$$\leq \sum_{m=\ell}^{\infty} \sigma^m \|P_m\|_{\theta,B,\beta}\, \|x-\xi\|_B^{m-\ell} \leq \sum_{m=\ell}^{\infty} (\sigma\rho)^m \|(m!)^{-1}\, \hat{\delta}^m f(\xi)\|_{\theta,B,\beta}$$

for every x in $\xi + \rho B$.

<u>5.6 Corollary</u>. If $f \in \mathcal{H}_{S\theta}(U;F)$, $X \subset E_B$ and $X + \rho B \subset U$ for some $\rho > 0$, then for $\epsilon > 0$ and $\beta \in SC(F)$

$$\sum_{\ell=0}^{\infty} \epsilon^{\ell} \sup\{\|(\ell!)^{-1}\, \hat{\delta}^{\ell} f(x)\|_{\theta,B,\beta};\ x \in X + \rho B\} \leq$$

$$\sum_{m=0}^{\infty} [\sigma(\rho+\epsilon)]^m \sup\{\|(m!)^{-1}\, \hat{\delta}^m f(x)\|_{\theta,B,\beta};\ x \in X\}.$$

<u>Proof</u>. Apply 5.6 for every $x \in X$.

6. TOPOLOGIES ON SPACES OF SILVA HOLOMORPHIC MAPPINGS

6.1 Lemma. Let β be a seminorm in F, K a strict compact subset of U, B an element of $\mathcal{B}_E$ with K contained and compact in E_B, p a seminorm in $\mathcal{H}_{S\theta}(U;F)$. Then the following statements are equivalent:

(1) For every $\epsilon > 0$ there is $c(\epsilon) \geq 0$ such that

$$p(f) \leq c(\epsilon) \sum_{m=0}^{\infty} \epsilon^m \sup_{x\in K} \|(m!)^{-1} \hat{\delta}^m f(x)\|_{\theta,B,\beta}$$

for every f in $\mathcal{H}_{S\theta}(U;F)$.

(2) For every $\epsilon > 0$ there is $\rho > 0$ with $K + \rho B \subset U$ and there is $c(\epsilon,\rho) \geq 0$ such that

$$p(f) \leq c(\epsilon,\rho) \sum_{m=0}^{\infty} \epsilon^m \sup_{x\in K+\rho B} \|(m!)^{-1} \hat{\delta}^m f(x)\|_{\theta,B,\beta}$$

for every f in $\mathcal{H}_{S\theta}(U;F)$.

Proof. It is clear that (2) implies (1). In order to prove that (1) implies (2) it is enough to apply 5.6 with $X = K$ and $\rho < \epsilon$.

6.2 Definition. Let K be a strict compact subset of U, $B \in \mathcal{B}_E$ with K contained and compact in E_B, p a seminorm in $\mathcal{H}_{S\theta}(U;F)$. We say that p is K-B ported if there is $\beta \in SC(F)$ such that (1) [and (2)] is satisfied. $\tau_{\omega\theta}$ denotes the locally convex topology in $\mathcal{H}_{S\theta}(U;F)$ generated by all K-B ported seminorms with K and B varying. When θ is the current Silva holomorphy type we use the notation $\tau_{\omega S}$ for this topology.

6.3 Proposition. Let $(f_\lambda)_{\lambda\in\Lambda}$ be a net of elements of $\mathcal{H}_{S\theta}(U;F)$ and f an element of $\mathcal{H}_{S\theta}(U;F)$. If for every $\beta \in SC(F)$, K a strict compact subset of U, $B \in \mathcal{B}_E$ with K contained and compact in E_B, there is $\epsilon > 0$ such that

$$(*) \qquad \lim_{\lambda\in\Lambda} \sum_{m=0}^{\infty} \epsilon^m \sup_{x\in K} \|(m!)^{-1} \hat{\delta}^m (f_\lambda - f)\|_{\theta,B,\beta} = 0$$

then $\lim_{\lambda\in\Lambda} f_\lambda = f$ in the sense of $\tau_{\omega\theta}$.

6.4 Remark. (*) is equivalent to

$$\lim_{\lambda\in\Lambda} \sum_{m=0}^{\infty} \epsilon^m \sup_{x\in K+\rho B} \|(m!)^{-1} \hat{\delta}^m(f_\lambda - f)\|_{\theta,B,\beta} = 0$$

for some $\rho > 0$ depending on ϵ, K, B and β. Use 6.1 to prove this.

6.5 Proposition. Let $f \in \mathcal{H}_{S\theta}(U;F)$. If $\xi \in U$ and U is ξ-equilibrated, the Taylor series of f at ξ converges to f in the sense of $\tau_{\omega\theta}$.

6.6 Lemma. Under the conditions of 6.5, given K a strict compact subset of U, $B \in \mathcal{B}_E$ with K contained and compact in E_B, $\beta \in SC(F)$, thre is $\rho > 0$ such that $K + \rho B \subset U$ and there are constants $\gamma(0 < \gamma < 1)$, $C' \geq 0$, $c' \geq 0$ satisfying

$$\sup_{x\in K+\rho B} \|(m!)^{-1} \hat{\delta}^m(f - \tau_{\ell,f,\xi})(x)\|_{\theta,B,\beta} \leq C'c'^m \cdot \gamma^\ell$$

for every $\ell, m \in \mathbb{N}$.

Proof. By 4.5 there is $\rho_o > 0$ such that for every $m \in \mathbb{N}$

$$\|(m!)^{-1} \hat{\delta}^m f(x)\|_{\theta,B,\beta} \leq C\, c^m$$

for each $x \in \hat{K}_\xi + \rho_o B \subset U$ and for some positive constants C and c. $\hat{K}_\xi$ denotes the smallest ξ-equilibrated subset of U containing K. Now we take $\rho_1 > 1$ and $\rho > 0$ such that $(1-\lambda)\xi + \lambda x \in \hat{K}_\xi + \rho_o B$ if $|\lambda| \leq \rho_1$ and $x \in K + \rho B$. By 5.4 $\hat{\delta}^m f \in \mathcal{H}_S(U;P_{b\theta}(^mE;F))$. By 3.22

$$\hat{\delta}^m f(x) - \tau_{\ell,\hat{\delta}^m f,\xi}(x) = (2\pi i)^{-1} \int_{|\lambda|=\rho_1} \frac{\hat{\delta}^m f((1-\lambda)\xi + \lambda x)}{\lambda^{\ell+1}(\lambda-1)}\, d\lambda .$$

Thus

$$\|\hat{\delta}^m f(x) - \tau_{\ell,\hat{\delta}^m f,\xi}(x)\|_{\theta,B,\beta} \leq C\, c^m\, m!\, \rho_1^{-\ell}\, (\rho_1 - 1)^{-1}$$

for every $x \in K + \rho B$, $\ell \in N$, $m \in N$. Since $\tau_{\ell,\hat{\delta}^m f,\xi} = \hat{\delta}^m \tau_{\ell+m,f,\xi}$ we get:

$$\|(m!)^{-1} \hat{\delta}^m(f - \tau_{\ell+m,f,\xi})(x)\|_{\theta,B,\beta} \leq C\, c^m\, \rho_1^{-\ell}\, (\rho_1 - 1)^{-1}$$

for every $x \in K + \rho B$, $\ell \in \mathbb{N}$, $m \in \mathbb{N}$. We also have

$$(1) \qquad \|(m!)^{-1} \hat{\delta}^m (f-\tau_{\ell,f,\xi})(x)\|_{\theta,B,\beta} \le C(\rho_1 c)^m \rho_1^{-\ell} (\rho_1-1)^{-1}$$

for every $x \in K + \rho B$, $m, \ell \in \mathbb{N}$, $m \le \ell$. (1) holds true for $\ell < m$ since in this case $\hat{\delta}^m \tau_{\ell,f,\xi} = 0$ and since $K + \rho B \subset \hat{K}_\xi + \rho_o B$ we have

$$\|(m!)^{-1} \hat{\delta}^m f(x)\|_{\theta,B,\beta} \le C c^m \le C(\rho_1 c)^m \rho_1^{-\ell} (\rho_1-1)^{-1}$$

for $x \in K + \rho B$ since $\rho_1 - 1 < \rho_1^{m-\ell}$. The lemma is proved with $C' = C(\rho_1-1)^{-1}$, $c' = c\rho_1$ and $\gamma = \rho_1^{-1}$.

<u>Proof of 6.5.</u> We shall use 6.6. Corresponding to K strict compact subset of U, $B \in \mathcal{B}_E$ with K contained and compact in E_B, $\beta \in SC(F)$, there are $0 < \gamma < 1$ and $\rho > 0$ with $K + \rho B \subset U$ such that, if $\epsilon > 0$ satisfies $\epsilon c' < 1$, then

$$\sum_{m=0}^{\infty} \epsilon^m \sup_{x \in K+\rho B} \|(m!)^{-1} \hat{\delta}^m (f-\tau_{\ell,f,\xi})(x)\|_{\theta,B,\beta} \le C' \gamma^\ell (1-\epsilon c')^{-1}$$

for every $\ell \in \mathbb{N}$. By 6.3 and 6.4 the result follows.

<u>6.7 Proposition.</u> The inclusion mapping from $\mathcal{H}_{S\theta}(U;F)$ into $\mathcal{H}_S(U;F)$ is continuous if we consider the topologies $\tau_{\omega\theta}$ and $\tau_{\omega S}$ in $\mathcal{H}_{S\theta}(U;F)$ and $\mathcal{H}_S(U;F)$ respectively.

<u>Proof.</u> It is enough to use the definition of the topologies and 3.4.

<u>6.8 Proposition.</u> Let $\tau = (\ell!)^{-1} \hat{\delta}^\ell \theta$, $\ell \in \mathbb{N}$. The linear mapping

$$(\ell!)^{-1} \hat{\delta}^\ell : \mathcal{H}_{S\theta}(U;F) \to \mathcal{H}_{S\tau}(U; \mathcal{P}_{b\theta}(^\ell E;F))$$

$$f \longrightarrow (\ell!)^{-1} \hat{\delta}^\ell f$$

is continuous for the topologies $\tau_{\omega\theta}$ and $\tau_{\omega\tau}$.

<u>Proof.</u> Let p be K-B ported and $\tau_{\omega\tau}$-continuous. Thus there is $\|\cdot\|_{\theta,D,\beta} \in SC(\mathcal{P}_{b\theta}(^\ell E;F))$ such that for every $\epsilon > 0$ there is $c(\epsilon) \ge 0$ satisfying

$$p(g) \le c(\epsilon) \sum_{m=0}^{\infty} \epsilon^m \sup_{x \in K} \|(m!)^{-1} \hat{\delta}^m g(x)\|_{\tau,B,\|\cdot\|_{\theta,D,\beta}}$$

for every $g \in \mathcal{H}_{S\tau}(U; \mathcal{P}_{b\theta}({}^{\ell}E;F))$. If $f \in \mathcal{H}_{S\theta}(U;F)$ we have:

$$\|(m!)^{-1}\hat{\delta}^m((\ell!)^{-1}\hat{\delta}^{\ell}f(x))\|_{\tau,B,\|\cdot\|_{\theta,D,\beta}} =$$

$$= \|((m+\ell)!)^{-1}\hat{\delta}^{\ell+m} f(x)\|_{\theta,\widehat{DUB},\beta}$$

for every $m \in \mathbb{N}$ and for every x in K. Thus

$$p((\ell!)^{-1}\hat{\delta}^{\ell}f) \le c(\varepsilon) \sum_{m=0}^{\infty} \varepsilon^m \sup_{x\in K} \|((\ell+m)!)^{-1}\hat{\delta}^{\ell+m}f(x)\|_{\theta,\widehat{DUB},\beta} \le$$

$$\le c(\varepsilon)\varepsilon^{-\ell} \sum_{m=0}^{\infty} \varepsilon^m \sup_{x\in K} \|(m!)^{-1}\hat{\delta}^m f(x)\|_{\theta,\widehat{BUD},\beta}.$$

Hence $q(f) = p((\ell!)^{-1}\hat{\delta}^{\ell}f)$ defines a $K - \widehat{BUD}$ ported seminorm in $\mathcal{H}_{S\theta}(U;F)$ and the result follows.

7. BOUNDED SUBSETS

7.1 Proposition. The following conditions are equivalent for a subset $\mathfrak{X}$ of $\mathcal{H}_{S\theta}(U;F)$:

(1) $\mathfrak{X}$ is $\tau_{\omega\theta}$-bounded.

(2) For every $\xi \in U$, every B in $\mathcal{B}_E$ with $\xi \in E_B$ and every $\beta \in SC(F)$, there are $C \ge 0$ and $c \ge 0$ such that:

$$\|(m!)^{-1}\hat{\delta}^m f(\xi)\|_{\theta,B,\beta} \le C c^m$$

for each f in $\mathfrak{X}$ and m in $\mathbb{N}$.

(3) For every strict compact subset K of U, every B in $\mathcal{B}_E$ with K contained and compact in E_B, for every $\beta \in SC(F)$, there are $C \ge 0$ and $c \ge 0$ such that

$$\|(m!)^{-1}\hat{\delta}^m f(x)\|_{\theta,B,\beta} \le C c^m$$

for each f in $\mathfrak{X}$, x in K and m in $\mathbb{N}$.

(4) For every strict compact subset K of U, every B in $\mathcal{B}_E$ with K contained and compact in E_B, every $\beta \in SC(F)$, there are $\rho > 0$, $C \ge 0$ and $c \ge 0$ such that

$$\|(m!)^{-1}\hat{\delta}^m f(x)\|_{\theta,B,\beta} \le C c^m$$

for every f in $\mathfrak{X}$, m in $\mathbb{N}$ and x in $K + \rho B \subset U$.

Proof. (4) implies (3) and (3) implies (2) trivially. We show that (2) implies (4). If $\xi \in U$, $B \in \mathcal{B}_E$ with $\xi \in E_B$ and $\beta \in SC(F)$, we use 5.5 in order to get

$$\|(\ell!)^{-1} \hat{\delta}^{\ell} f(x)\|_{\theta,B,\beta} \leq C(1-\sigma\rho c)^{-1} (\sigma c)^{\ell}$$

for every x in $\xi + \rho B$, $\ell \in \mathbb{N}$, $f \in \mathfrak{X}$ if we choose $\rho > 0$ satisfying $\sigma c \rho < 1$ and $\xi + \rho B \subset U$. The compacticity of K in E_B and the above fact applied to every $\xi \in K$ imply (4).

(3) implies (1) since we may choose $\epsilon > 0$ such that $c\epsilon < 1$. This fact together with (3) imply the boundedness of $\mathfrak{X}$ for every K-B ported seminorm.

Now we show that (1) implies (3). For every strict compact subset K of U, every B in $\mathcal{B}_E$ with K contained and compact in E_B, every $\beta \in SC(F)$, every $(\alpha_m)_{m \in N}$ with $\alpha_m \geq 0$ and $\lim \alpha_m^{1/m} = 0$, the seminorm on $\mathcal{H}_{S\theta}(U;F)$ given by

$$(*) \qquad p(f) = \sum_{m=0}^{\infty} \alpha_m \sup_{x \in K} \|(m!)^{-1} \hat{\delta}^m f(x)\|_{\theta,B,\beta}$$

is K-B ported, hence bounded in $\mathfrak{X}$. We recall that if $s_{m,\lambda} \geq 0$ for $m \in \mathbb{N}$ and $\lambda \in \Lambda$, then

$$\sup_{\lambda \in \Lambda} \sum_{m=0}^{\infty} \alpha_m s_{m,\lambda} < +\infty$$

for every sequence $(\alpha_m)_{m \in N}$ with $\alpha_m \geq 0$ and $\lim \alpha_m^{1/m} = 0$ if and only if there are $C \geq 0$, $c \geq 0$ such that $s_{m,\lambda} \leq C c^m$ for every $\lambda \in \Lambda$ and $m \in N$. Thus the boundedness of every p of the form (*) over $\mathfrak{X}$ implies (3).

7.2 Definition. Given K a strict compact subset of U, $B \in \mathcal{B}_E$ with K contained and compact in E_B, $\beta \in SC(F)$ we may define the following seminorm in $\mathcal{H}_{S\theta}(U;F)$:

$$p(f) = \sup_{x \in K} \|\hat{\delta}^m f(x)\|_{\theta,B,\beta} .$$

$\tau_{\infty\theta}$ denotes the locally convex topology generated by all seminorms of this kind with K, B and β varying. If θ is the current S-holomorphy type we denote this topology by $\tau_{\infty S}$. Clearly $\tau_{\infty\theta} \subset \subset \tau_{\omega\theta}$.

<u>7.3 Proposition</u>. Let $\mathfrak{X}$ be a $\tau_{\omega\theta}$-bounded subset of $\mathcal{H}_{S\theta}(U;F)$. Then $\tau_{\infty\theta}$ and $\tau_{\omega\theta}$ induce in $\mathfrak{X}$ the same topologies.

<u>Proof</u>. First we suppose that $0 \in \mathfrak{X}$ and we prove that a subset of $\mathfrak{X}$ is a neighborhood of 0 for the topology induced by $\tau_{\omega\theta}$ if and only if it is a neighborhood of 0 for the topology induced by $\tau_{\infty\theta}$. One part of this statement is trivial since $\tau_{\infty\theta} \subset \tau_{\omega\theta}$. Now let p be a seminorm $\tau_{\omega\theta}$-continuous in $\mathcal{H}_{S\theta}(U;F)$ which is K-B ported. Let $\beta \in SC(F)$ be the corresponding seminorm in F by Definition 6.2. By 6.1 to each $\epsilon > 0$ we associate $c(\epsilon) \geq 0$ with the properties stated in part (1) of 6.1. Let $C \geq 0$ and $c \geq 0$ be the constants corresponding to $\mathfrak{X}$ in 7.1 part (3). If we choose $\epsilon > 0$ with $\epsilon c < 1$ and $\mu \in N$ such that

$$C\ c(\epsilon) \sum_{m\geq\mu} (\epsilon c)^m \leq 1/2$$

the seminorm defined in $\mathcal{H}_{S\theta}(U;F)$ by

$$q(f) = c(\epsilon) \sum_{m=0}^{\mu} \epsilon^m \sup_{t\in K} \|(m!)^{-1} \hat{\delta}^m f(t)\|_{\theta,B,\beta}$$

is $\tau_{\infty\theta}$-continuous and we can see easily that

$$f \in \mathfrak{X} \text{ and } q(f) \leq 1/2 \text{ imply } p(f) \leq 1.$$

This proves the second part of the statement written above. If $\mathfrak{X}$ is arbitrary $\mathfrak{X}-\mathfrak{X} = \{f-g;\ f \text{ and } g \text{ in } \mathfrak{X}\}$ is $\tau_{\omega\theta}$-bounded and contains 0. Since the neighborhoods of 0 induced by $\tau_{\omega\theta}$ and $\tau_{\infty\theta}$ in $\mathfrak{X}-\mathfrak{X}$ are the same, given ξ in $\mathfrak{X}$, a neighborhood $\mathcal{V}$ of ξ for the topology induced by $\tau_{\omega\theta}$ in $\mathfrak{X}$ is such that $\mathcal{U} = \mathcal{V}-\mathcal{V}$ is a neighborhood of 0 for $\tau_{\infty\theta}$ induced in $\mathfrak{X}-\mathfrak{X}$. Thus $\mathcal{V}$ is a neighborhood of ξ for the topology induced by $\tau_{\infty\theta}$ in $\mathfrak{X}$.

<u>7.4 Proposition</u>. If $\mathfrak{X}$ is a $\tau_{\omega\theta}$-bounded subset of $\mathcal{H}_{S\theta}(U;F)$, then $\mathfrak{X}$ is equi-S-continuous at each point of U.

Proof. First we suppose that ξ, B, β, $C \geq 0$, $c \geq 0$ are as in 7.1 in (2). We have for f in $\mathcal{H}_{S\theta}(U;F)$ and x in $\xi + \rho B \subset U$:

$$\beta(f(x)-f(\xi)) \leq \sum_{m=1}^{\infty} \beta((m!)^{-1}\hat{\delta}^m f(\xi)(x-\xi)) \leq$$

$$\leq \sum_{m=1}^{\infty} \sigma^m \|(m!)^{-1}\hat{\delta}^m f(\xi)\|_{\theta,B,\beta} \|x-\xi\|_B^m \leq$$

$$\leq \sum_{m=1}^{\infty} \sigma^m C\, c^m \|x-\xi\|_B^m = C\sigma c\|x-\xi\|_B\, (1-\sigma c\|x-\xi\|_B)^{-1}$$

if $x \in \xi + \rho B$, $\sigma c\rho < 1$ and $f \in \mathfrak{X}$. Now the result follows easily.

7.5 Proposition. A seminorm p in $\mathcal{H}_S(U;F)$ is K-B ported if and only if there is $\beta \in SC(F)$ such that for each $\rho > 0$ there is $c(\rho) \geq 0$ satisfying

$$p(f) \leq c(\rho) \sup_{x\in K+\rho B} \beta(f(x))$$

for every f in $\mathcal{H}_S(U;F)$.

Proof. As in 6.1 part (1) for every $\epsilon > 0$ thre is $c(\epsilon) \geq 0$ such that

$$p(f) \leq c(\epsilon) \sum_{m=0}^{\infty} \epsilon^m \sup_{x\in K} \|(m!)^{-1}\hat{\delta}^m f(x)\|_{B,\beta}$$

for every f in $\mathcal{H}_S(U;F)$. If $\rho > 0$ is such that $K + \rho B \subset U$ the Cauchy's inequalities give

$$\sup_{x\in K} \|(m!)^{-1}\hat{\delta}^m f(x)\|_{B,\beta} \leq \rho^{-m} \sup_{x\in K+\rho B} \beta(f(x)).$$

If we take $\epsilon = \rho/2$ we get

$$p(f) \leq c(\rho/2) \sum_{m=0}^{\infty} (\rho/2)^m \rho^{-m} \sup_{x\in K+\rho B} \beta(f(x)) \leq$$

$$\leq c(\rho/2)2 \sup_{x\in K+\rho B} \beta(f(x))$$

for every f in $\mathcal{H}_S(U;F)$. The converse is obvious by (2) of 6.1.

7.6 Proposition. If $\mathfrak{X} \subset \mathcal{H}_S(U;F)$, the following conditions are equivalent:

(1) $\mathfrak{X}$ is $\tau_{\omega S}$-bounded.

(2) $\mathfrak{X}$ is τ_{oe}-bounded.

(3) For every strict compact subset K of U, B in $\mathcal{B}_E$ with K contained and compact in E_B, $\beta \in SC(F)$, there are $C \geq 0$ and $\rho > 0$ such that

$$\sup \{\beta(f(x));\ x \in K + \rho B \text{ and } f \in \mathfrak{X}\} \leq C.$$

<u>Proof</u>. (3) implies (1) because 7.5. (1) implies (2) since $\tau_{oe} \subset \tau_{\omega S}$. We show that (2) implies (3). Let $B \in \mathcal{B}_E$ and $\beta \in SC(F)$. By (2) for each compact subset K of U which is contained and compact in E_B $\beta \circ \mathfrak{X} = \{\beta \circ f.\ f \in \mathfrak{X}\}$ is bounded over K. Since $U \cap E_B$ is metrizable $\beta \circ \mathfrak{X}$ is locally bounded for the normed topology of $U \cap E_B$. This implies (3).

8. RELATIVELY COMPACT SUBSETS

<u>8.1 Definition</u>. $\mathfrak{X} \subset \mathcal{H}_{S\theta}(U;F)$ is $\tau_{\omega\theta}$-relatively compact if and only if $\mathfrak{X}$ is $\tau_{\infty\theta}$-relatively compact and $\tau_{\omega\theta}$-bounded.

<u>Proof</u>. Since $\tau_{\infty\theta} \subset \tau_{\omega\theta}$, if $\mathfrak{X}$ is $\tau_{\omega\theta}$-relatively compact then it is $\tau_{\omega\theta}$-bounded and $\tau_{\infty\theta}$-relatively compact. Now we suppose that $\mathfrak{X}$ is $\tau_{\omega\theta}$-bounded and $\tau_{\infty\theta}$-relatively compact. Let $\bar{\mathfrak{X}}$ be the $\tau_{\omega\theta}$-closure of $\mathfrak{X}$ and $\tilde{\mathfrak{X}}$ the $\tau_{\infty\theta}$-closure of $\mathfrak{X}$. If we show that $\bar{\mathfrak{X}} = \tilde{\mathfrak{X}}$ the result follows from 7.3. It is clear that $\bar{\mathfrak{X}} \subset \tilde{\mathfrak{X}}$. Let f be in $\tilde{\mathfrak{X}}$. There is a net $(f_\alpha)_{\alpha \in I}$ in $\mathfrak{X}$ converging to f for $\tau_{\infty\theta}$. Thus $\mathfrak{Y} = \mathfrak{X} \cup \{f\}$ is $\tau_{\omega\theta}$-bounded. By 7.3 this net converges to f for $\tau_{\omega\theta}/\mathfrak{Y}$, hence for $\tau_{\omega\theta}$. It follows that f is in $\bar{\mathfrak{X}}$.

<u>8.2 Proposition</u>. Let $\mathfrak{X}$ be in $\mathcal{H}_{S\theta}(U;F)$. Then $\mathfrak{X}$ is $\tau_{\omega\theta}$-relatively compact if and only if $\mathfrak{X}$ is $\tau_{\omega\theta}$-bounded and for each m in $\mathbb{N}$ and u in U the set $\hat{\delta}^m \mathfrak{X}(u) = \{\hat{\delta}^m f(u);\ f \in \mathfrak{X}\}$ is realtively compact in $\mathcal{P}_{b\theta}({}^mE;F)$.

<u>Proof</u>. For each m in $\mathbb{N}$ and $u \in U$ the mapping

$$f \in \mathcal{H}_{S\theta}(U;F) \longmapsto \hat{\delta}^m f(u) \in \mathcal{P}_{b\theta}({}^mE;F)$$

is linear and continuous for the topology $\tau_{\omega\theta}$ (respectively $\tau_{\infty\theta}$) in $\mathcal{H}_{S\theta}(U;F)$. This follows from the fact that $p(f) = \|\hat{\delta}^m f(u)\|_{\theta,B,\beta}$ defines a $\tau_{\infty\theta}$-continuous seminorm in $\mathcal{H}_{S\theta}(U;F)$ for B in $\mathcal{B}_E$ and $\beta \in SC(F)$. Thus if $\mathfrak{X}$ is $\tau_{\omega\theta}$-relatively compact it is $\tau_{\omega\theta}$-bounded and $\hat{\delta}^m\mathfrak{X}(u)$ is relatively compact in $\mathcal{P}_{b\theta}(^mE;F)$ for each m in $\mathbb{N}$ and $u \in U$. In order to show the converse we define

$$\psi: (\mathcal{H}_{S\theta}(U;F),\tau_{\infty\theta}) \longmapsto (C_S(U; \prod_{m=0}^{\infty} \mathcal{P}_{b\theta}(^mE;F)), \tau_{oe})$$

$$f \longmapsto \psi(f) = ((m!)^{-1}\hat{\delta}^m f)_{m=0}^{\infty}$$

ψ is well defined since $\hat{\delta}^m f \in \mathcal{H}_{S\tau}(U; \mathcal{P}_{b\theta}(^mE;F)) \subset \mathcal{H}_S(U; \mathcal{P}_{b\theta}(^mE;F)) \subset$ $\subset C_S(U; \mathcal{P}_{b\theta}(^mE;F))$ for every m in $\mathbb{N}$. It is clear that ψ is linear and injective. The mappings ψ and ψ^{-1} are continuous since for each strict compact subset K of U, $B \in \mathcal{B}_E$ such that K is contained and compact in E_B, and $\beta \in SC(F)$ we have:

$$p(\psi(f)) = \sup\{\|(j!)^{-1}\hat{\delta}^j f(t)\|_{\theta,B,\beta};\ j=0,\dots,k,\ t \in K\} = q(f)$$

for every f in $\mathcal{H}_{S\theta}(U;F)$ and k in $\mathbb{N}$. Thus p and q define continuous seminorms in $(\psi(\mathcal{H}_{S\theta}(U;F)),\tau_{oe})$ and $(\mathcal{H}_{S\theta}(U;F),\tau_{oe})$ respectively, and then ψ^{-1} and ψ are continuous. Since for each $\xi \in U$ and $m \in \mathbb{N}$ $\hat{\delta}^m\mathfrak{X}(\xi)$ is relatively compact in $\mathcal{P}_{b\theta}(^mE;F)$ it follows that $\psi(\mathfrak{X})(\xi) = \{\psi(f)(\xi);\ f \in \mathfrak{X}\}$ is relatively compact in $\prod_{m=0}^{\infty} \mathcal{P}_{b\theta}(^mE;F)$. By Ascoli's Theorem $\psi(\mathfrak{X})$ is τ_{oe}-relatively compact in $C_S(U; \prod_{m=0}^{\infty} \mathcal{P}_{b\theta}(^mE;F))$ if it is equi-S-continuous in U. This is equivalent to ask for equi-S-continuity in U of $\hat{\delta}^m\mathfrak{X}$ for each m in $\mathbb{N}$. Since $\hat{\delta}^m\mathfrak{X}$ is $\tau_{\omega\tau}$-bounded it follows from 7.4 that $\hat{\delta}^m\mathfrak{X}$ is equi-S-continuous in U. The theorem will be proved if we show that the τ_{oe}-closure of $\psi(\mathfrak{X})$ is contained in the image of ψ. Let $g = (g_m)_{m\in\mathbb{N}}$ be the limit of a net $(\psi(f_\alpha))_{\alpha\in I}$ of elements of $\psi(\mathfrak{X})$ for the topology τ_{oe}. Thus $g_n = \lim_\alpha (n!)^{-n}\hat{\delta}^n f_\alpha$ for τ_{oe} in $C_S(U,\mathcal{P}_{b\theta}(^nE;F)$ for each $n \in \mathbb{N}$. Thus $g_o = \lim_\alpha f_\alpha$ in $C_S(U;F)$ for τ_{oe}. Since $\mathcal{H}_S(U;F)$ is τ_{oe} closed in $C_S(U;F)$ we have $g_o \in \mathcal{H}_S(U;F)$. Using the Cauchy's inequalities it follows that

$(n!)^{-1}\,\hat{\delta}^n g_o(x)(y) = g_n(x)(y)$ for each x in U, y in E and n in $\mathbb{N}$. Thus $(n!)^{-1}\,\hat{\delta}^n g_o(x) \in \mathcal{P}_{b\theta}({}^nE;F)$ for n in $\mathbb{N}$ and x in U. On the other hand by 7.1, part (2), we have

$$\|(n!)^{-1}\,\hat{\delta}^n f_\alpha(x)\|_{\theta,B,\beta} \le C\, c^n$$

for each n in $\mathbb{N}$ and $\alpha \in I$. (C and c depending on B, x and β). If we take the limit for $\alpha \in I$ we get

$$\|(n!)^{-1}\,\hat{\delta}^n g_o(x)\|_{\theta,B,\beta} \le C\, c^n.$$

Thus $g_o \in \mathcal{H}_{S\theta}(U;F)$ and $\psi(g_o) = g$.

<u>8.3 Proposition</u>. Let U be connected and let $\mathfrak{X}$ be a subset of $\mathcal{H}_{S\theta}(U;F)$. $\mathfrak{X}$ is $\tau_{\omega\theta}$-relatively compact if and only if $\mathfrak{X}$ is $\tau_{\omega\theta}$-bounded and there is $\xi \in U$ such that $\hat{\delta}^m\,\mathfrak{X}(\xi)$ is relatively compact in $\mathcal{P}_{b\theta}({}^mE;F)$ for every m in $\mathbb{N}$.

<u>8.4 Lemma</u>. Let $\mathfrak{X}$ be a subset of $\mathcal{H}_{S\theta}(U;F)$ such that:

(1) there is $\xi \in U$ with $\hat{\delta}^m\,\mathfrak{X}(\xi)$ relatively compact for every $m \in \mathbb{N}$,

(2) for each $B \in \mathcal{B}_E$ with $\xi \in E_B$ and for every $\beta \in SC(F)$ there are $\rho > 0$ and $C \ge 0$ such that $\xi + \rho B \subset U$ and

$$\|(m!)^{-1}\,\hat{\delta}^m f(\xi)\|_{\theta,B,\beta} \le C(\sigma\rho)^{-m}$$

for every f in $\mathfrak{X}$ and m in $\mathbb{N}$.

If $0 < r < 1$ then $\hat{\delta}^m\,\mathfrak{X}(x)$ is relatively compact for every x in $\xi + r\rho B$ and m in $\mathbb{N}$.

<u>Proof</u>. We know that for every $f \in \mathcal{H}_{S\theta}(U;F)$

$$(*) \qquad \hat{\delta}^\ell f(x) = \sum_{m=\ell}^{\infty} \hat{\delta}^\ell\, P_{m,f}(x-\xi)$$

where $P_{m,f} = (m!)^{-1}\,\hat{\delta}^m f(\xi)$, $x \in \xi + \rho B$, $\ell \in \mathbb{N}$ and the convergence of (*) is in the sense of $\mathcal{P}_{b\theta}({}^\ell E;F)$. By the condition (2) of the lemma and by (3) of 4.1 we have

$$\|\hat{\delta}^\ell P_{m,f}(x-\xi)\|_{\theta,B,\beta} \le \sigma^m\, C(\rho\sigma)^{-m}\,\|x-\xi\|_B^{m-\ell}$$

for each f in $\mathfrak{X}$, $\ell \in \mathbb{N}$, $m \in \mathbb{N}$ with $\ell \le m$, $x \in \xi + \rho B$. If we use these estimates in (*) we get

$$\|\hat{\delta}^{\ell} f(x)\|_{\theta,B,\beta} \leq \sum_{m=\ell}^{\infty} C\, r^{m-\ell} < +\infty$$

for f in $\mathfrak{X}$, x in $\xi + r\rho B$, $0 < r < 1$. Since $\mathcal{P}_{b\theta}({}^{\ell}E;F)$ is complete it is enough to show that $\hat{\delta}^{\ell}\, \mathfrak{X}(x)$ is precompact for x in $\xi + r\rho B$. Let $\mathcal{V}$ be the set of all $P \in \mathcal{P}_{b\theta}({}^{\ell}E;F)$ such that $\|P\|_{\theta,B,\beta} \leq \epsilon$. Let $M \in \mathbb{N}$ be such that

$$(**) \qquad C \sum_{m>M} r^{m-\ell} < \epsilon/2.$$

Since

$$\{\hat{\delta}^{\ell} P_{m,f}(x-\xi);\ f \in \mathfrak{X}\} = \{\hat{\delta}^{\ell}((m!)^{-1}\, \delta^{m} f(\xi))(x-\xi);\ f \in \mathfrak{X}\},$$

$(m!)^{-1}\, \hat{\delta}^{m}\, \mathfrak{X}(\xi)$ is relatively compact and the mapping

$$P \in \mathcal{P}_{b\theta}({}^{m}E;F) \longmapsto \hat{\delta}^{\ell} P(x-\xi) \in \mathcal{P}_{b\theta}({}^{\ell}E;F)$$

is continuous it follows that for $m = \ell,\dots,m = M$ the set of the $\hat{\delta}^{\ell} P_{m,f}(x-\xi)$ such that $f \in \mathfrak{X}$ is relatively compact, hence precompact in $\mathcal{P}_{b\theta}({}^{\ell}E;F)$. Thus there are $f_1,\dots,f_k \in \mathfrak{X}$ such that for each f in $\mathfrak{X}$ there is $j \in \{1,\dots,k\}$ satisfying

$$\|\hat{\delta}^{\ell} P_{m,f}(x-\xi) - \hat{\delta}^{\ell} P_{m,f_j}(x-\xi)\|_{\theta,B,\beta} \leq 2^{-1}(M-\ell+1)^{-1}.$$

(*), (**) and this fact imply that for each f in $\mathfrak{X}$ there is $j \in \{1,\dots,k\}$ such that

$$\|\hat{\delta}^{\ell} f(x) - \hat{\delta}^{\ell} f_j(x)\|_{\theta,B,\beta} \leq \epsilon.$$

Thus $f \in f_j + \mathcal{V}$.

Proof of 8.3. We consider $\mathfrak{X}$ $\tau_{\omega\theta}$-bounded and $\xi \in U$ such that $\hat{\delta}^{m}\, \mathfrak{X}(\xi)$ is relatively compact for every m in $\mathbb{N}$. Let $B \in \mathcal{B}_E$ with $\xi \in E_B$. We denote by X_B the set of all u in $U \cap E_B$ such that $\hat{\delta}^{m}\, \mathfrak{X}(u)$ is relatively compact for each m in $\mathbb{N}$. By 8.4 $X_B \neq \emptyset$ is open in $U \cap E_B$ for the norm topology. Let η be a point of the closure of X_B in $U \cap E_B$ (closure for the normed topology). Since $\mathfrak{X}$ is $\tau_{\omega\theta}$-bounded, for each $\beta \in SC(F)$ there are $\rho > 0$ and $C \geq 0$

such that $\eta + \rho B \subset U \cap E_B$ and

$$\| (m!)^{-1} \hat{\delta}^m f(x) \|_{\theta,B,\beta} \leq C(\sigma\rho)^{-m}$$

for every $f \in \mathfrak{X}$, $m \in \mathbb{N}$ and $x \in \xi + \rho B$. Let s be such that $0 < s < 1$ and $r = s\rho/2$. Let $y \in X_B \cap (\eta + rB)$. We have

$$\| (m!)^{-1} \hat{\delta}^m f(y) \|_{\theta,B,\beta} \leq C(\sigma\rho)^{-m} \leq C(\sigma r)^{-m}$$

for each f in $\mathfrak{X}$ and m in $\mathbb{N}$. By 8.4 $\hat{\delta}^m \mathfrak{X}(u)$ is relatively compact for each $m \in \mathbb{N}$ and $u \in y + rB$. Since $\eta \in y + rB$ we have $\eta \in X_B$. Thus $X_B \neq \emptyset$ is open and closed in $U \cap E_B$ for the normed topology for every $B \in \mathfrak{B}_E$ such that $\xi \in E_B$. Let $y \in U$ be arbitrary and let Δ be the polygonal line joining y to ξ with $\Delta \subset U$. Let $B_o \in \mathfrak{B}_E$ be such that $\Delta \subset E_{B_o}$. Since y is in the same connected component of $U \cap E_{B_o}$ as $\xi \in X_{B_o}$ and since X_{B_o} is closed and open in $U \cap E_{B_o}$ for the normed topology, it follows that $y \in X_{B_o}$. Thus $\hat{\delta}^m \mathfrak{X}(y)$ is relatively compact for every $m \in \mathbb{N}$. By 8.2 the set $\mathfrak{X}$ is $\tau_{\omega\theta}$-relatively compact.

REFERENCES

[1] Nachbin, L. On spaces of holomorphic functions of a given type. Proceedings of the Conference on Functional Analysis, University of California at Irvine, 1966, Thompson Book Company, USA (1967).

[2] Nachbin, L. Topology on spaces of holomorphic mappings. Ergebnisse der Mathematik und ihrer Grenzgebiete, Band 47, Springer-Verlag, New York (1969).

[3] Gupta, C.P., Malgrange theorem for nuclearly entire functions of bounded type on a Banach space. Thesis. University of Rochester. Notas de Matemática nº 37, IMPA, Rio de Janeiro, Brasil (1966).

[4] Nachbin, L. and Gupta, C.P., On Malgrange theorem for nuclearly entire functions. Preprint (1966).

[5] Matos, M.C., Holomorphic mappings and domains of holomorphy. Thesis University of Rochester. Monografias do Centro Brasileiro de Pesquisas Físicas, nº 27. Rio de Janeiro, Brasil, (1970).

[6] Matos, M.C., Sur le théorème d'approximation et d'existence de Malgrange-Gupta, C.R. Acad. Sci. Paris, Sér. A-B, 271 (1970).

[7] Matos, M.C. and Nachbin, L., Entire functions on locally convex spaces and convolution operators (to appear).

[8] Dineen, S., Holomorphy types on a Banach space. Studia Mathematica. T. XXXIX (1971).

[9] Bianchini, M., Silva-holomorphy types Borel transforms and partial differential operators. Thesis. UNICAMP. These Proceedings.

ENVELOPES FOR TYPES OF HOLOMORPHY

Luiza A. Moraes

Departamento de Matemática Pura
Universidade Federal do Rio de Janeiro
Ilha do Fundão
Rio de Janeiro, RJ, Brasil

School of Theoretical Physics
Dublin Institute for Advanced Studies
Dublin 4, Ireland

ABSTRACT

In this article we define and construct the θ-envelope of holomorphy of X for every type θ where X is a Riemann domain over a complex Banach space E.

We first show that if X is an open connected subset of a Banach space E, then there exists a θ-envelope of holomorphy, unique up to isomorphism. Next we extend the result to any connected Riemann domain over a Banach space E.

The method is about the same as that used by P. Noverraz in [6] to construct the envelope of holomorphy of a Riemann surface over a locally convex space.

Definition 1. A couple (X,φ) is a Riemann domain over a locally convex space E if X is a Hausdorff topological space and $\varphi: X \to E$ is a local homeomorphism.

Remark 1. Given $X' \subset X$, we will write $X' \sim \varphi(X')$ to indicate that X' is homeomorphic to $\varphi(X')$ under $\varphi|_{X'}$.

Definition 2. Let (X,φ) be a Riemann domain over a Banach space E. A function $f: X \to \mathbb{C}$ is holomorphic of type θ at $x \in X$ if there exists a neighbourhood V of x such that $V \sim \varphi(V)$ and a function $\psi \in \mathcal{H}_\theta(\varphi(V))$ such that $f = \psi \circ \varphi$ on V. We say that f is holomorphic of type θ on X if f is holomorphic of type θ at every $x \in X$.

Remark 2. We can say equivalently that $f\colon X \to \mathbb{C}$ is holomorphic of type θ at $x \in X$ if $f\circ[\varphi/V]^{-1}$ is holomorphic of type θ at $\varphi(x)$ for each open subset V of X where φ is a homeomorphism.

Notation. $\mathcal{H}_\theta(X)$ = set of all holomorphic functions of type θ on X.

Definition 3. Let (X,φ) and (X',φ') be two Riemann domains over the same Banach space E. A morphism from (X,φ) over (X',φ') is an analytic function u which is a local isomorphism, $u\colon X \to X'$, such that $\varphi = \varphi'\circ u$

$$\begin{array}{ccc} X & \xrightarrow{\;u\;} & X' \\ {\scriptstyle\varphi} & \searrow \quad \swarrow & {\scriptstyle\varphi'} \\ & E & \end{array}$$

Remark 3. In Definition 3, we say that u is analytic if $\varphi'\circ u$ is analytic.

Definition 4. Let (X,φ) and (X',φ') be Riemann domains over a Banach space E. A morphism u from (X,φ) over (X',φ') is a θ-holomorphic extension of (X,φ) if to every $f \in \mathcal{H}_\theta(X)$ we can associate one and only one $f' \in \mathcal{H}_\theta(X')$ such that $f = f'\circ u$.

$$\begin{array}{ccc} X & \xrightarrow{\;u\;} & X' \\ {\scriptstyle f} & \searrow \quad \swarrow & {\scriptstyle f'} \\ & \mathbb{C} & \end{array}$$

Definition 5. A θ-envelope of holomorphy of a Riemann domain (X,φ) is a Riemann domain $(\tilde{X},\tilde{\varphi})$ and an extension $u\colon X \to \tilde{X}$ such that if (X',φ') is a Riemann domain and $u'\colon X \to X'$ is a θ-holomorphic extension of X, then there exists a θ-holomorphic extension of X', $\tilde{u}\colon X' \to \tilde{X}$, such that $u = \tilde{u}\circ u'$.

$$\begin{array}{ccc} X & \xrightarrow{\;u\;} & \tilde{X} \\ {\scriptstyle u'} & \searrow \quad \nearrow & {\scriptstyle \tilde{u}} \\ & X' & \end{array}$$

Let $I = \mathcal{H}_\theta(X)$, where X is an open subset of the Banach space E. We call $\mathcal{B}$ the collection of all open subsets of E and to each $a \in E$ we associate the set $\mathcal{U}_a = \{U \in \mathcal{B} : a \in U\}$.

We fix $a \in E$.

Let $g = (g_i)_{i\in I}$ where, $\forall\, i \in I$, $g_i \in \mathcal{H}_\theta(U)$ for some $U \in \mathcal{U}_a$ (the same U for all $i \in I$).

Let $\mathcal{H}^*_{\theta a}$ be the set of all such $g = (g_i)_{i\in I}$. We define in $\mathcal{H}^*_{\theta a}$ the following equivalence relation $\sim$: $(f_i)_{i\in I} \sim (g_i)_{i\in I}$ if there exists a neighbourhood V of a, independent of $i \in I$, such that $\forall\, i\in I$ the mappings f_i and g_i belong to $\mathcal{H}_\theta(V)$ and $f_i|V = g_i|V$. We call $(g_i)_a$ the equivalence class of $(g_i)_{i\in I}$ given by the equivalence relation defined above. If $g = (g_i)_{i\in I}$ is such that $g_i: U \to \mathbb{C}$ $\forall\, i \in I$ and $U \in \mathcal{U}_a$ and $g_i \in \mathcal{H}_\theta(U)$, we will say that $g \in \mathcal{H}_\theta(U;\mathbb{C}^I)$.

With this notation we will have:

$$\mathcal{H}^*_{\theta a} = \bigcup_{U\in\mathcal{U}_a} \mathcal{H}_\theta(U;\mathbb{C}^I).$$

Let $\mathcal{O}_{Ea} = \mathcal{H}^*_{\theta a}/\sim$, i.e., let $\mathcal{O}_{Ea}$ be the quocient set corresponding to the equivalence relation defined above.

Let $\mathcal{O}_E = \bigcup_{a\in E} \mathcal{O}_{Ea}$.

We define $\pi: \begin{array}{l}\mathcal{O} \longrightarrow E\\ (f_i)_a \longmapsto a\end{array}$ i.e. $\pi(f) = a$ where a is such that $f \in \mathcal{O}_{Ea}$. We observe that $\mathcal{O}_{Ea} \cap \mathcal{O}_{Eb} = \emptyset$ if $a \neq b$. So, π is well defined.

We want to define a suitable topology on $\mathcal{O}_E$. If $U \in \mathcal{B}$ and $(f_i)_{i\in I} \in \mathcal{H}(U;\mathbb{C}^I)$, then $\{(f_i)_a : a \in U\} \subset \mathcal{O}_E$. The collection of all these sets when U runs over $\mathcal{B}$ is a base of neighbourhoods for a topology τ on $\mathcal{O}_E$.

<u>Lemma 1</u>. $\pi: \mathcal{O}_E \to E$ is continuous.

<u>Proof</u>. We consider $(f_i)_a \in \mathcal{O}_{Ea}$. Then $\pi[(f_i)_a] = a$. We take $V \in \mathcal{U}_a$ and let $(f_i)_{i\in I} \in \mathcal{H}_\theta(U;\mathbb{C}^I)$ be a family representing $(f_i)_a$ (it is clear that $U \in \mathcal{U}_a$).

Then $V \cap U \in \mathcal{U}_a$ and $A = \{(f_i)_b : b \in V \cap U\}$ is an element of the basis of τ such that $(f_i)_a \in A$ and $\pi(A) \subset V$.

Lemma 2. The topology τ defined on $\mathfrak{G}_E$ is Hausdorff.

Proof. We take $(f_i)_a, (g_i)_b \in \mathfrak{G}_E$, $(f_i)_a \neq (g_i)_b$. If $\pi[(f_i)_a] \neq \pi[(g_i)_b]$, we can find disjoint open subsets Ω_a, Ω_b of E containing $\pi[(f_i)_a]$, $\pi[(g_i)_b]$ respectively.

Then $\pi^{-1}(\Omega_a)$ and $\pi^{-1}(\Omega_b)$ are disjoint open subsets of $\mathfrak{G}_E$ containing $(f_i)_a$ and $(g_i)_b$ respectively.

If $\pi[(f_i)_a] = \pi[(g_i)_b]$, then $(g_i)_b, (f_i)_a \in \mathfrak{G}_{Ea}$. (From our definition of π we have $a = b$ in this case).

Let $(f_i)_{i\in I} \in \mathcal{H}_\theta(U;C^I)$ and $(g_i)_{i\in I} \in \mathcal{H}_\theta(U';C^I)$ be the families representing $(f_i)_a$ and $(g_i)_a$ respectively, and let V be a connected neighbourhood of a such that $V \subset U \cap U'$.

We claim that $\{(f_i)_c : c \in V\} \cap \{(g_i)_c : c \in V\} = \emptyset$. Indeed, if there exists $x \in \{(f_i)_c : c \in V\} \cap \{(g_i)_c : c \in V\}$, then f_i and g_i are equal for all $i \in I$ in a neighbourhood of $\pi(x)$ and so $f_i \equiv g_i \ \forall\, i$ on V (since V is connected).
In particular: $(f_i)_a = (g_i)_b$ and this is a contradiction.

Lemma 3. Let $X' = \{(a,(h_i)); a \in E\ (h_i) \in \mathfrak{G}_{Ea}\}$. We define $N(V; (g_i)_{i\in I}) = \{(a,(g_i)_a); a \in V\}$ for all $V \in \mathfrak{B}$ and $(g_i)_{i\in I} \in \mathcal{H}_\theta(V;C^I)$ respectively. The sets $N(V; (g_i)_{i\in I})$ determine a topology on X' which is stronger than the product topology (and consequently it is a Hausdorff topology).

Proof. Let $\mathfrak{U}$ be a base of absolutely convex neighbourhoods in E and $\mathfrak{V}$ be a base of absolutely convex neighbourhoods in $\mathfrak{G}_E$. We take

$$\begin{cases} p_1(a,(h_i)) = a \in E \\ p_2(a,(h_i)) = (h_i) \in \mathfrak{G}_E \end{cases}$$

The set of all non-void intersections $p_1^{-1}(U) \cap p_2^{-1}(V) \cap X'$ when U runs over $\mathfrak{U}$ and V runs over $\mathfrak{V}$, is a base of neighbourhoods in X' (for the product topology) $p_1^{-1}(U) = \{(a,(h_i)) : a \in U, (h_i) \in \mathfrak{G}_E\}$ $p_2^{-1}(V) = \{(a,(h_i)) : a \in E, (h_i) \in V\}$.

Now, $V = \{(h_i)_b : b \in W\}$ (see the construction of the topology τ on $\mathcal{O}_E$).

If $p_1^{-1}(U) \cap p_2^{-1}(V) \cap X' \neq \emptyset$ then $U \cap W \neq \emptyset$ and $N(U \cap W; (g_i)_{i\in I}) \subset p_2^{-1}(V) \cap X'$.

Lemma 4. Let $X' = \{(a,(h_i)); a \in E, (h_i) \in \mathcal{O}_{Ea}\}$. We define $\varphi' : X' \to E$ such that $\varphi'((a,(h_i))) = a$ for all $(a,(h_i)) \in X'$. Then (X',φ') is a Riemann domain over E.

Proof. 1) X' is Hausdorff (Lemma 3).

2) φ' is continuous: let $(a,(h_i)) \in X'$, and let $(h_i)_{i\in I} \in \mathcal{H}_\theta(V;C^I)$ where $V \subset \mathfrak{U}_a$ is a representative of $(h_i) \in \mathcal{O}_{Ea}$. Given any $U \in \mathfrak{U}_a$ we have $N(U \cap V; (h_i)_{i\in I}) = \{(b,(h_i)_b); b \in U \cap V\} \in \tau$. Since $a \in U \cap V$, we have $(a,(h_i)_a) \in N(U \cap V, (h_i)_{i\in I})$ and, on the other hand, $\varphi'[N(U \cap V; (h_i)_{i\in I})] = U \cap V \subset U$. It follows that φ' is continuous.

3) Given $(a,(g_i)_a) \in X'$, let $(g_i)_{i\in I} \in \mathcal{H}_\theta(U;C^I)$, where $U \in \mathfrak{U}_a$, be a fixed representative of $(g_i)_a$. We claim that $\varphi'|_{N(U;(g_i)_{i\in I})}$ is injective from $N(U,(g_i)_{i\in I})$ onto U.

Let $\psi : U \longrightarrow N(U;(g_i)_{i\in I})$
$\qquad a \longmapsto (a,(g_i)_a)$

We can easily see that ψ is a function such that

$$\psi \circ [\varphi'|_{N(U;(g_i)_{i\in I})}] = \text{identity on } N(U;(g_i)_{i\in I})$$

$$[\varphi'|_{N(U;(g_i)_{i\in I})}] \circ \psi = \text{identity on } U.$$

We conclude that $\varphi'|_{N(U;(g_i)_{i\in I})}$ is an injectives function from $N(U;(g_i)_{i\in I})$ on U and its inverse is just the function ψ.

It is easy to see that ψ is continuous. So, φ' is a local homeomorphism and consequently (X',φ') is a Riemann domain over E.

Lemma 5. Let X be a connected open subset of E and $I = \mathcal{H}_\theta(X)$. Let (X',φ') be as in the statement of Lemma 4. Then there exists a subset $\tilde{X}$ of X' and a function $\tilde{\varphi} : \tilde{X} \to E$, $\tilde{\varphi} = \tilde{\varphi}|\tilde{X}$, such that

$(\tilde{X},\tilde{\varphi})$ is a connected Riemann surface over E; and there exists a morphism between (X,id_X) and $(\tilde{X},\tilde{\varphi})$ that is a θ-holomorphic extension of (X,id_X) (where id_X = identity on X).

<u>Proof</u>. Let $(h_i)_{i\in I} \in \mathcal{H}(X,C^I)$ where $h_i = i \;\forall\; i \in I$. If $x \in X$, we denote by $[(i)_x]_{i\in I}$ the equivalence class of this family at x.

We define $u': X \to X'$ such that

$$u'(x) = (x, [(i)_x]_{i\in I}) \in X' \quad \forall\; x \in X.$$

It is easy to see that u' is injective. We claim that u' is continuous. Let $a \in X$. Then $u'(a) = (a, [(i)_a]_{i\in I})$. Let $\Gamma(U;(g_i)_{i\in I})$ be an element of the basis of τ such that at contains $u'(a)$ (hence $U \in \mathfrak{U}_a$). It follows that $(a, [(i)_a]_{i\in I}) = (b,(g_i)_b)$ for some $b\in U$. But then it is clear that $a = b$ and $(g_i)_a = (g_i)_b = [(i)_a]_{i\in I}$.

But $(g_i)_a = [(i)_a]_{i\in I}$ and X connected imply $(g_i)_x = [(i)_x]_{i\in I} \;\forall\; x \in X$.

So $\Gamma(U,(g_i)_{i\in I}) = \Gamma(U,(i)_{i\in I})$. Then there exists an open set $U \in \mathfrak{U}_a$ such that $u'(U) = \{(a,[(i)_a]_{i\in I}) : a \in U\} = \Gamma(U,(i)_{i\in I})$ and u' is continuous.

Finally $id_X(x) = x = (\varphi' \circ u')(x) \;\forall\; x \in X$. So, u' is a morphism between (X,id_X) and (X',φ').

As X is connected, the same occurs with $u'(X)$ in X'. We will denote by $\tilde{X}$ the connected component of X' that contains $u'(X)$ and let

$$u: X \to \tilde{X}$$
$$x \mapsto u'(x)$$

The restriction $\tilde{\varphi}$ of φ' to $\tilde{X}$, the mapping $(a,(h_i)) \mapsto a$, makes $(\tilde{X},\tilde{\varphi})$ a connected Riemann domain over E and u is therefore a morphism from (X,id_X) into $(\tilde{X},\tilde{\varphi})$.

Now we claim that u is a θ-holomorphic extension of (X,id_X). Let us fix $f \in \mathcal{H}_\theta(X)$. We want to show that there exists one and only one $\tilde{f} \in \mathcal{H}_\theta(\tilde{X})$ such that $f = \tilde{f}\circ u$. We define $\tilde{f}$ in the following way: if $(a,(g_i)_a) \in \tilde{X}$, let $(g_i)_{i\in I} \in \mathcal{H}_\theta(U;C^I)$, where

$U \in \mathfrak{U}_a$, be a representative of $(g_i)_a$. By definition we have $\tilde{f}[(a,(g_i)_a)] = g_f(a)$. It is clear that $\tilde{f}[u(x)] = \tilde{f}[(x,[(i)_x]_{i\in I})] = f(x) \quad \forall\, x \in X$, that is, $f = \tilde{f}\circ u$.

Now, $\tilde{f} \in \mathcal{H}_\theta(\tilde{X})$ if and only if for each $\alpha \in \tilde{X}$ there exists a neighbourhood V of α such that $V \sim \tilde{\varphi}(V)$ and there exits a mapping $f_1 \in \mathcal{H}_\theta(\tilde{\varphi}(V))$ such that $\tilde{f} = f_1\circ\tilde{\varphi}$ in $\tilde{\varphi}(V)$. Let $\alpha \in \tilde{X}$. Hence $\alpha = (a,(g_i)_a)$ and $\tilde{\varphi}(\alpha) = a$. We know already that if $(g_i)_{i\in I} \in \mathcal{H}_\theta(U;\mathbb{C}^I)$, where $U \in \mathfrak{U}_a$, is a representative of $(g_i)_a$, then

$$N(U;(g_i)_{i\in I}) \sim \tilde{\varphi}[N(U;(g_i)_{i\in I})]$$

$$\tilde{\varphi}[N(U;(g_i)_{i\in I})] = U$$

$$\alpha \in N(U;(g_i)_{i\in I}).$$

Moreover, if $\beta \in N(U;(g_i)_{i\in I})$ we have $\beta = (b,(g_i)_b)$ for some $b \in U$. It follows from this fact and from the definition of $\tilde{f}$ that $\tilde{f}(\beta) = g_f(b) = (g_f\circ\tilde{\varphi})(\beta)$ where $g_f \in \mathcal{H}_\theta(U) = \mathcal{H}_\theta(\tilde{\varphi}[N(U;(g_i)_{i\in I})])$.

We infer from this that $\tilde{f} \in \mathcal{H}_\theta(\tilde{X})$.

Finally we claim that if $\tilde{f}_1 \in \mathcal{H}_\theta(\tilde{X})$ and is such that $f = \tilde{f}_1\circ u$, then $\tilde{f}_1 \equiv f$ on $\tilde{X}$. For all $x \in X$ we have $\tilde{f}(u(x)) = f(x) = \tilde{f}_1(u(x))$, that is, $\tilde{f} = \tilde{f}_1$ on $u(X)$.

Now, $u(X) = \Gamma(X,(i)_{i\in I})$ is an open set contained in $\tilde{X}$ which is connected.

It follows that $\tilde{f} \equiv \tilde{f}_1$ on $\tilde{X}$.

<u>Lemma 6</u>. Under the conditions of Lemma 5, the $(\tilde{X},\tilde{\varphi})$ constructed there factors through every θ-holomorphic extension of X, i.e., $(\tilde{X},\tilde{\varphi})$ is a θ-envelope of holomorphy of X.

<u>Proof</u>. Let $(\bar{X},\bar{\varphi})$ be a Riemann domain over E and $v\colon (X,id_X) \to (\bar{X},\bar{\varphi})$ be a θ-holomorphic extension of X. We want to show that there exists $\bar{u}\colon \bar{X} \to \tilde{X}$, a θ-holomorphic extension of $\bar{X}$, such that $u = \bar{u}\circ v$.

Let $x \in \bar{X}$ and U an open neighbourhood of x such that $U \sim \bar{\varphi}(U)$. The restriction of $\bar{\varphi}$ to U will be denoted by $\bar{\varphi}_U$.

If $i \in \mathcal{H}_\theta(X)$, since $v\colon (X, id_X) \to (\bar{X}, \bar{\varphi})$ is a θ-holomorphic extension of X, there exists $\bar{i} \in \mathcal{H}_\theta(\bar{X})$ such that $i = \bar{i} \circ v$.

Now $i \in \mathcal{H}_\theta(\bar{X})$ implies $\bar{i} \circ \bar{\varphi}_U^{-1} \in \mathcal{H}_\theta(\bar{\varphi}(U))$ (see Remark 2). But $\bar{\varphi}(U)$ is an open subset of E and we have $\bar{\varphi}(x) \in \bar{\varphi}(U)$ if $x \in U$. Then we can say that $(\bar{i} \circ \bar{\varphi}_U^{-1})_{i \in I} \in \mathcal{H}_\theta(\bar{\varphi}(U); \mathbb{C}^I)$ and $[(\bar{i} \circ \bar{\varphi}_U^{-1})_{\bar{\varphi}(x)}]_{i \in I} \in \mathcal{O}_{E, \bar{\varphi}(x)}$.

So we can define $\bar{u}' : \bar{X} \to X'$

$$x \mapsto (\bar{\varphi}(x), [(\bar{i} \circ \bar{\varphi}_U^{-1})_{\bar{\varphi}(x)}]_{i \in I})$$

1) We claim $u' = \bar{u}' \circ v$. For all $x \in X$ we have $\bar{u}'(v(x)) =$ $= (\bar{\varphi}(v(x)), [(\bar{i} \circ \bar{\varphi}_U^{-1})_{\bar{\varphi}(v(x))}]_{i \in I}) = (x, [(i)_x]_{i \in I}) = u'(x)$ since $\bar{i} \circ \bar{\varphi}_U^{-1} \circ \bar{\varphi} \circ v = \bar{i} \circ v = i$.

2) We claim that $\bar{u}'$ is continuous. Let $x \in \bar{X}$. We want to prove that $\bar{u}'$ is continuous at x. It suffices to show that given any element of the basis of τ which contains $u'(x)$, there exists an open set containing $x \in \bar{X}$ such that $\bar{u}'(x)$ belongs to this element of the basis of τ. Let $x \in \bar{X}$ and U be an open neighbourhood of x such that $U \sim \bar{\varphi}(U)$. We take $U_o \subset E$ such that $U_o = \bar{\varphi}(U)$ is an open neighbourhood of $\bar{\varphi}(x)$ and let $N(U_1; (g_i)_{i \in I})$ be an element of the basis of τ such that $\bar{u}'(x) \in$ $\in N(U_1; (g_i)_{i \in I})$. It follows that $\bar{\varphi}(x) \in U_1$ and $(g_i)_{\bar{\varphi}(x)} =$ $= (\bar{i} \circ \bar{\varphi}_U^{-1})_{\bar{\varphi}(x)}$. If W is a connected component of $U_1 \cap U_o \subset E$ such that $\bar{\varphi}(x) \in U_1 \cap U_o$, then $V_1 = \bar{\varphi}_U^{-1}(W)$ is a connected open set (since $\bar{\varphi}_U$ is continuous) and $x \in V_1$ since $\bar{\varphi}(x) \in W$. We have $g_i/W = \bar{i} \circ \bar{\varphi}_U^{-1}/W \ \forall\, i \in I$ since $(g_i)_{\bar{\varphi}(x)} = (\bar{i} \circ \bar{\varphi}_U^{-1})_{\bar{\varphi}(x)}$ and we can infer

$$N(W; (\bar{i} \circ \bar{\varphi}_U^{-1}/W)_{i \in I}) \subset N(U_1; (g_i)_{i \in I}).$$

Now $(\bar{\varphi}(x), [(\bar{i} \circ \bar{\varphi}_U^{-1})_{\bar{\varphi}(x)}]_{i \in I}) \in N(W; (\bar{i} \circ \bar{\varphi}_U^{-1}/W)_{i \in I})$ as $\bar{\varphi}(x) \in W$. So, for all $\alpha \in V_1$ we have $\bar{u}'(a) = (\bar{\varphi}(\alpha), [(\bar{i} \circ \bar{\varphi}_U^{-1})_{\bar{\varphi}(\alpha)}]_{i \in I}) \in$ $\in N(W; (\bar{i} \circ \bar{\varphi}_U^{-1}/W)_{i \in I}) \subset N(U_1; (g_i)_{i \in I})$ since $\bar{\varphi}(\alpha) \in W$; and we conclude that there exists an open neighbourhood V_1 of x such that $\bar{u}'(V_1) \subset N(U_1; (g_i)_{i \in I})$, and so $\bar{u}'$ is continuous.

3) We claim that $\bar{u}'$ is a local isomorphism. Let $x \in \bar{X}$ (and $\bar{\varphi}(x) \in X'$) and W be the open neighbourhood of x constructed in (2). We will prove that $\bar{u}' : W \to N(W; (\bar{i}\circ\bar{\varphi}_U^{-1}/W)_{i\in I})$ is an isomorphism.

$$\text{Let } t' : N(W; (\bar{i}\circ\bar{\varphi}_U^{-1}/W)_{i\in I}) \longrightarrow W$$
$$(w,[(\bar{i}\circ\bar{\varphi}_U^{-1})_w]_{i\in I}) \longmapsto w$$

It is easy to see that

$$\begin{cases} \bar{u}'\circ t' = \text{identity of } N(W; (\bar{i}\circ\bar{\varphi}_U^{-1}/W)_{i\in I}) \\ t'\circ\bar{u}' = \text{identity of } W \end{cases}$$

We conclude that $\bar{u}'$ is injctive and onto, and that its inverse is t'. That is: u' is a local isomorphism.

4) We claim that $\bar{\varphi} = \varphi'\circ\bar{u}'$

$$\begin{array}{ccc} \bar{X} & \xrightarrow{\bar{u}'} & X' \\ & {\scriptstyle\bar{\varphi}}\searrow \quad \swarrow{\scriptstyle\varphi'} & \\ & E & \end{array}$$

For all $x \in \bar{X}$ we have

$$(\varphi'\circ\bar{u}')(x) = \varphi'(\bar{u}'(x)) = \varphi'[(\bar{\varphi}(x), [\bar{i}\circ\bar{\varphi}_U^{-1}]_{\bar{\varphi}(x)})_{i\in I}] = \bar{\varphi}(x)$$

and so $\varphi'\circ\bar{u}' = \bar{\varphi}$.

We infer from (2), (3) and (4) that $\bar{u}'$ is a morphism of Riemann domains. As $\bar{X}$ is connected, it follows that $\bar{u}'(X)$ is connected. Now $v^{-1}(x) \in X$ if $x \in \bar{X}$ and wo have $\tilde{X} \ni u'(v^{-1}(x)) = u'\circ v^{-1}(x) = \bar{u}'\circ v\circ v^{-1}(x) = \bar{u}'(x)$.

It follows that $\bar{u}'(\bar{X}) \subset \tilde{X}$.

Wo define $\bar{u}: (\bar{X},\bar{\varphi}) \to (\tilde{X},\tilde{\varphi})$
$$x \longmapsto \bar{u}'(x)$$

It is clear that $\bar{u}$ is a θ-holomorphic extension of $\bar{X}$ such that $u = \bar{u}\circ v$.

This completes the proof that $(\tilde{X},\tilde{\varphi})$ is a θ-envelope of holomorphy of X.

<u>Theorem 1</u>. Let E be a Banach space an X an open connected subset of E. Then there exists a θ-envelope of holomorphy of X, which is unique up to isomorphism.

<u>Proof</u>. We have already showed the existence in the lemmas. We will prove that it is unique up to isomorphism.

Let $(\tilde{X},\tilde{\varphi})$ and $(\tilde{X}',\tilde{\varphi}')$ be two θ-envelopes of holomorphy of X. From the definition we know that:

1) there exists a θ-envelope of holomorphy of $\tilde{X}'$, $\sigma: \tilde{X}' \to \tilde{X}$, such that $u = \sigma \circ u'$, and

2) there exists a θ-envelope of holomorphy of $\tilde{X}$, $\sigma': \tilde{X} \to \tilde{X}'$, such that $u' = \sigma' \circ u$

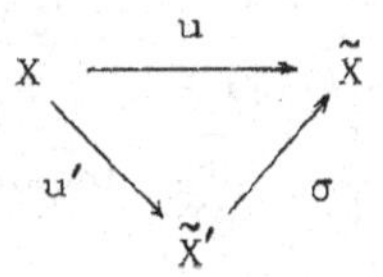
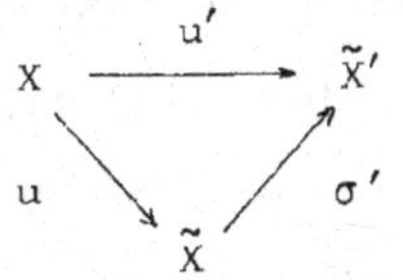

Let $y \in u(X) \subset \tilde{X}$. Then there exists $x \in X$ such that $y = u(x)$ and so $(\sigma \circ \sigma')(y) = (\sigma \circ \sigma')(u(x)) = (\sigma \circ \sigma' \circ u)(x) = (\sigma \circ u')(x) = u(x) = y = id_{u(X)}(y)$.

Let $y' \in u'(X) \subset \tilde{X}'$. Then there exists $x \in X$ such that $y' = u'(x)$ and so $(\sigma \circ \sigma)(y') = (\sigma' \circ \sigma)(u'(x)) = (\sigma' \circ \sigma \circ u')(x) = (\sigma' \circ u)(x) = u'(x) = y' = id_{u'(X)}(y')$. So, there exist two morphisms $\sigma: X \to \tilde{X}$ and $\sigma': \tilde{X} \to \tilde{X}'$ such that $\sigma \circ \sigma' = id_{u(X)}$ and $\sigma' \circ \sigma = id_{u'(X)}$. We extend $id_{u(X)}$ and $id_{u'(X)}$ to $\tilde{X}$ and $\tilde{X}'$, respectively, and we conclude that $(\tilde{X},\tilde{\varphi})$ and $(\tilde{X}',\tilde{\varphi}')$ are isomorphics.

<u>Remark</u>. Our construction for an open set $X \subset E$ is still valid if (X,φ) is a connected Riemann domain over E. So, we have the following

<u>Theorem 2</u>. Every connected Riemann domain over a Banach space E has a θ-envelope of holomorphy, unique up to isomorphism.

<u>Proof</u>. Since the proof is similar to the proof of Theorem 1, we will give only a sketch.

Let (X,φ) be the connected Riemann domain over E and $I = \mathcal{H}_\theta(E)$. We use the same notation as in Theorem 1 to define $X' = \{(a,(h_i));\ a \in E,\ (h_i) \in \mathcal{O}_{Ea}\}$. We define $N(V;(g_i)_{i\in I}) = \{(a,(g_i)_a);\ a \in V\}$ for all $V \in \mathcal{B}$ and for all $(g_i)_{i\in I} \in \mathcal{H}_\theta(V;\mathbb{C}^I)$. The sets $N(V;(g_i)_{i\in I})$ determine a stronger topology on X' than the product topology, and consequently it is a Hausdorff topology (Lemma 3). If we define $\varphi'((a,(h_i))) = a$, (X',φ') is a Riemann domain over E (Lemma 4).

We define

$$u' : X \longrightarrow X'$$
$$x \longmapsto (\varphi(x), [(i\circ\varphi_U^{-1})_{\varphi(x)}]_{i\in I})$$

If $x \in X$, there exists a neighbourhood U of x such that $\varphi_U = \varphi|_U$ is a homeomorphism between U and $\varphi(U)$; for all $f \in \mathcal{H}_\theta(X)$, the equivalence class of $f\circ\varphi_U^{-1}$ in $\varphi(x)$ doesn't depend on U but only on x. Analogously we verify that $[(i\circ\varphi_U^{-1})_{\varphi(x)}]_{i\in I}$ doesn't depend on U but only on x.

If we consider the definition of u' and the topology of X', we see that u' is continuous and that it is a morphism between Riemann domains.

$u'(X)$ is connected in X' since X is connected.

$\tilde{X}$ will denote the connect component of X' that contains $u'(X)$ and we define

$$u: X \to \tilde{X}$$
$$x \mapsto u'(x)$$

The restriction $\tilde{\varphi}$ of φ' to $\tilde{X}$ is the mapping $(a,(h_i))$ a and $(\tilde{X},\tilde{\varphi})$ is a Riemann domain over E; u is a morphism from (X,φ) in $(\tilde{X},\tilde{\varphi})$.

It remains to show that u is a θ-holomorphic extension of (X,φ).

Let $f \in \mathcal{H}_\theta(X)$. We want to show that there exists a unique $\tilde{f} \in \mathcal{H}_\theta(\tilde{X})$ such that $f = \tilde{f}\circ u$. We define $\tilde{f}$ in the following way: If $(a,(g_i)_a) \in \tilde{X}$, let $(g_i)_{i\in I} \in \mathcal{H}_\theta(U;\mathbb{C}^I)$ (where $U \in \mathcal{U}_a$) be a representative of $(g_i)_a$. We define $\tilde{f}((a,(g_i)_a)) = g_f(a)$. It is clear that $\tilde{f}[u(x)] = \tilde{f}[(\varphi(x), [(i\circ\varphi_U^{-1})_{\varphi(x)}]_{i\in I})] = (f\circ\varphi_U^{-1})(\varphi(x)) =$ $= f(x) \quad \forall\, x \in X$. So $f = \tilde{f}\circ u$.

We can show the uniqueness of $\tilde{f}$ and that $\tilde{f} \in \mathcal{H}_\theta(\tilde{X})$ (in the same way as in Lemma 5).

Finally we show (as in Lemma 6) that $(\tilde{X},\tilde{\varphi})$ factors through any θ-holomorphic extension of (X,φ).

This completes our proof of the existence of a θ-envelope of holomorphy of (X,θ). The proof of the uniqueness is the same as in Theorem 1.

REFERENCES

[1] Alexander, H., Analytic Functions on Banach Spaces, Thesis, University of California at Berkeley (1968).

[2] Hirschowitz, A., Prolongement Analytique en Dimension Infinie, C.R. Acad. Sc. Paris, t. 270 (1970) Série A, pp. 1736-1737.

[3] Matos, M.C., Holomorphic Mappings and Domains of Holomorphy, Monografias do Centro Brasileiro de Pesquisas Físicas, nº27, Rio de Janeiro (1970).

[4] Nachbin, L., Holomorphic Functions, Domains of Holomorphy and Local Properties, North-Holland, 1970.

[5] Narasimhan, R., Several Complex Variables, Chicago Lectures in Mathematics - The University of Chicago Press (1971).

[6] Noverraz, P., Pseudo-Convexité, Convexité Polynomiale et Domaines d'Holomorphie en Dimension Infinie, Notas de Matemática, Vol. 48 (1973), North-Holland.

DOMAINS OF HOLOMORPHY IN (DFC)-SPACES

Jorge Mujica

Instituto de Matemática
Universidade Estadual de Campinas
Caixa Postal 1170
13100 Campinas, SP, Brasil

INTRODUCTION

This paper, devoted to the study of domains of holomorphy in infinite dimensional spaces, is divided into two essentially different parts. In the first part, comprising Sections 1 through 6, we recall the basic properties of domains of holomorphy in arbitrary locally convex spaces, and their relationship with various convexity properties. The first part of the paper is mainly addressed to the students, in an attempt to motivate the study of the Levi problem and of other problems on holomorphic continuation or holomorphic approximation which are nowadays subject of active research.

The second part of the paper is devoted to the study of domains of holomorphy in those spaces of the form $E = F'_c$ where F is a separable Fréchet space. This class includes all (DFM)-spaces and in particular all Silva spaces. The spaces in this class have nice properties: if E is any such a space, then any open set in E is uniformly open and is a hemicompact k-space. These properties are very useful, for they help us to prove the Cartan-Thullen theorem: every holomorphically convex open set in E is a domain of existence. And if we assume, furthermore, that E has the approximation property, then we can also solve the Levi problem: every pseudoconvex open set in E is a domain of existence.

CONTENTS

1. HOLOMORPHIC FUNCTIONS

The letters $\mathbb{N}$, $\mathbb{R}$ and $\mathbb{C}$ will denote the set of all non-negative integers, real numbers and complex numbers, respectively.

The letter E will always represent a complex locally convex space and the letter Ω will always represent a nonvoid open subset of E. We will denote by $cs(E)$ the set of all continuous seminorms on E.

For each $m \in \mathbb{N}$ we will denote by $\mathcal{P}(^mE)$ the space of all continuous m-homogeneous polynomials on E, and we will denote by $\mathcal{P}(E)$ the space of all continuous polynomials on E, i.e. the space $\oplus \mathcal{P}(^mE)$. We recall that $\mathcal{P}(^oE)$ consists of the constant functions $P: E \to \mathbb{C}$ and that for $m \geq 1$, $\mathcal{P}(^mE)$ consists of the functions $P: E \to \mathbb{C}$ of the form $P(x) = A(x,\ldots,x)$, where $A: E^m \to \mathbb{C}$ is a continuous m-linear form on E^m.

Finally we will represent by $\mathcal{H}(\Omega)$ the space of all holomorphic functions on Ω. We recall that a function $f: \Omega \to \mathbb{C}$ is holomorphic if for each point $\xi \in \Omega$ there is a sequence (P_m), with $P_m \in \mathcal{P}(^mE)$, such that the series $\Sigma P_m(x-\xi)$ converges to $f(x)$ uniformly on a neighbourhood of ξ. We will often use the fact that a function $f: \Omega \to \mathbb{C}$ is holomorphic if and only it is continuous and its restriction to each complex line is holomorphic in the classical sense. Let us remark also that $\mathcal{P}(E) \subset \mathcal{H}(\Omega)$.

We refer to Nachbin [22] or Noverraz [24] for the basic properties of polynomials and holomorphic functions on infinite dimensional spaces.

2. DOMAINS OF HOLOMORPHY

2.1 Definition (a) Ω is a domain of holomorphy if there are no connected open sets Ω_1 and Ω_2 in E, with $\Omega_1 \not\subset \Omega$ and $\emptyset \neq \Omega_2 \subset \Omega \cap \Omega_1$, and such that for each $f \in \mathcal{H}(\Omega)$ there exists $f_1 \in \mathcal{H}(\Omega_1)$ with $f_1 = f$ on Ω_2.

(b) Ω is the domain of existence of a function $f \in \mathcal{H}(\Omega)$ if there are no connected open sets Ω_1 and Ω_2 in E, and a function $f_1 \in \mathcal{H}(\Omega_1)$, such that $\Omega_1 \not\subset \Omega$, $\emptyset \neq \Omega_2 \subset \Omega \cap \Omega_1$ and $f_1 = f$ on Ω_2.

We will say that Ω is a domain of existence if it is the domain of existence of some function $f \in \mathcal{H}(\Omega)$. Clearly every domain of existence is a domain of holomorphy. In this section we give sufficient conditions for Ω to be a domain of holomorphy or a domain of existence. We begin with the lemma below which assures existence of certain boundary sequences.

<u>2.2 Lemma</u>. Let Ω_1 and Ω_2 be connected open sets in E with $\Omega_1 \not\subset \Omega$ and $\emptyset \neq \Omega_2 \subset \Omega \cap \Omega_1$. Let $\tilde{\Omega}_2$ denote the connected component of $\Omega \cap \Omega_1$ which contains Ω_2. Then there exists a point $x_o \in \partial\Omega \cap \Omega_1 \cap \partial\tilde{\Omega}_2$ which is limit of a sequence (x_n) in $\tilde{\Omega}_2$.

<u>Proof</u>. Let $\gamma: [0,1] \to \Omega_1$ be a path in Ω_1 joining a point in $\tilde{\Omega}_2$ to a point in $\Omega_1 \setminus \Omega$. Set

$$t_o = \inf\{t: \gamma(t) \notin \tilde{\Omega}_2\}, \qquad t_1 = \inf\{t: \gamma(t) \notin \Omega\}.$$

Clearly $\gamma(t_o) \in \Omega_1 \cap \partial\tilde{\Omega}_2$, $\gamma(t_1) \in \Omega_1 \cap \partial\Omega$ and $0 < t_o \leq t_1$. We claim that $t_o = t_1$. If $t_o < t_1$ then $\gamma([0,t_o]) \subset \Omega$ by the definition of t_1. Thus the restriction of γ to $[0,t_o]$ would yield a path in $\Omega \cap \Omega_1$ joining a point in $\tilde{\Omega}_2$ to a point outside $\tilde{\Omega}_2$. But this is impossible for $\tilde{\Omega}_2$ is a connected component of $\Omega \cap \Omega_1$. Thus $t_o = t_1$. If $x_n = \gamma(t_o - 1/n)$ then $x_n \in \tilde{\Omega}_2$ and (x_n) converges to $\gamma(t_o) \in \Omega_1 \cap \partial\Omega \cap \partial\tilde{\Omega}_2$.

Q.E.D.

<u>2.3 Lemma</u>. Suppose that for each sequence (x_n) in Ω which converges to a point $x_o \in \partial\Omega$, there exists a function $f \in \mathcal{H}(\Omega)$ with $\sup|f(x_n)| = +\infty$. Then Ω is a domain of holomorphy.

<u>Proof</u>. Suppose that there are connected open sets Ω_1 and Ω_2 in E, with $\Omega_1 \not\subset \Omega$ and $\emptyset \neq \Omega_2 \subset \Omega \cap \Omega_1$, and such that to each $f \in \mathcal{H}(\Omega)$ corresponds an $f_1 \in \mathcal{H}(\Omega_1)$ with $f_1 = f$ on Ω_2. Without loss of generality we may assume that Ω_2 is a connected component of $\Omega \cap \Omega_1$.

Then by Lemma 2.2 we can find a sequence (x_n) in Ω_2 which converges to a point $x_o \in \partial\Omega \cap \Omega_1 \cap \partial\Omega_2$. Then on one hand the sequence $f_1(x_n)$ converges to $f_1(x_o)$ and on the other hand the sequence $f_1(x_n) = f(x_n)$ is unbounded. This is impossible.

Q.E.D.

2.4 Lemma. Let $f \in \mathcal{H}(\Omega)$ and suppose that for each convex open set V in E with $V \cap \partial\Omega \neq \emptyset$, f is unbounded on each connected component of $\Omega \cap V$. Then Ω is the domain of existence of f.

Proof. Suppose that there exist connected open sets Ω_1 and Ω_2 in E, and a function $f_1 \in \mathcal{H}(\Omega_1)$, such that $\Omega_1 \not\subset \Omega$, $\emptyset \neq \Omega_2 \subset \Omega \cap \Omega_1$ and $f_1 = f$ on Ω_2. Without loss of generality we may assume that Ω_2 is a connected component of $\Omega \cap \Omega_1$. By Lemma 2.2 we can find a point $\xi \in \partial\Omega \cap \Omega_1 \cap \partial\Omega_2$. Let V be any convex open neighbourhood of zero in E such that $\xi + V \subset \Omega_1$. We will show that f_1 is unbounded on $\xi + V$, thus contradicting the local boundedness of f_1. Choose a point $\eta \in \Omega_2 \cap (\xi+V)$ and let ω denote the connected component of $\Omega \cap (\xi+V)$ which contains η. Then we see at once that $\omega \subset \Omega_2$. Since by hypothesis f is unbounded on ω, and since $f_1 = f$ on Ω_2, we conclude that f_1 is also unbounded on ω, and hence on $\xi + V$.

Q.E.D.

3. HOLOMORPHIC CONVEXITY

3.1 Definition. Given a set $A \subset \Omega$ and a family $\mathfrak{F} \subset \mathcal{H}(\Omega)$ we define the $\mathfrak{F}$-hull of A by

$$\hat{A}_{\mathfrak{F}} = \{x \in \Omega : |f(x)| \le \sup_A |f| \text{ for all } f \in \mathfrak{F}\}.$$

If $\mathcal{Q}(E) = \mathbb{C} \oplus E'$ denotes the space of all continuous affine forms on E, then we have the following result, whose proof can be found in Noverraz [24, p.40, Lemme 2.1.2].

3.2 Lemma. For each compact subset K of E, the $\mathcal{Q}(E)$-hull of K coincides with the closed, convex hull $\bar{\Gamma}(K)$ of K.

3.3 Corollary. If E is quasi-complete then for every compact subset

K of E, $\hat{K}_{\mathfrak{G}(E)}$ is also compact.

With the aid of Lemma 3.2 we can generalize to infinite dimensions a result in Grauert and Fritzsche [8, p.40, Theorem 3.2].

<u>3.4 Proposition</u>. The following conditions are equivalent:

(a) Ω is convex.

(b) For each compact set $K \subset \Omega$, the set $\hat{K}_{\mathfrak{G}(E)} \cap \Omega$ is bounded away from Ω, i.e. there exists a 0-neighorhood V in E such that $\hat{K}_{\mathfrak{G}(E)} \cap \Omega + V \subset \Omega$.

(c) For each compact set $K \subset \Omega$, the set $\hat{K}_{\mathfrak{G}(E)}$ is contained in Ω and is bounded away from $\complement\Omega$.

<u>Proof</u>. (a) $\Rightarrow$ (c) Let K be a compact subset of Ω and let V be a convex, balanced 0-neighborhood in E such that $K + V \subset \Omega$. Then certainly $\Gamma(K) + V \subset \Omega$ too and hence

$$\bar{\Gamma}(K) + 1/2\, V \subset \Gamma(K) + V \subset \Omega .$$

Thus by Lemma 3.2, $\hat{K}_{\mathfrak{G}(E)} + 1/2\, V \subset \Omega$.

It is obvious that (c) $\Rightarrow$ (b), so that we only have to show that (b) $\Rightarrow$ (a). Let $K = \{x,y\} \subset \Omega$. We must show that the line segment

$$\Gamma(K) = \{(1-t)x + ty\colon \; 0 \leq t \leq 1\}$$

is contained in Ω. Set

$$L_1 = \hat{K}_{\mathfrak{G}(E)} \cap \Omega , \qquad L_2 = \hat{K}_{\mathfrak{G}(E)} \setminus \Omega .$$

Since the set $\hat{K}_{\mathfrak{G}(E)} = \bar{\Gamma}(K) = \Gamma(K)$ is clearly compact, L_2 is also compact. But then L_1 is also compact, for being closed in Ω and by hypothesis bounded away from $\complement\Omega$, it is actually closed in E. Thus the convex set $\Gamma(K)$ is the union of the disjoint compact subsets L_1 and L_2. But this is impossible unless L_2 is empty. Thus $\Gamma(K) = \hat{K}_{\mathfrak{G}(E)} \subset \Omega$.

Q.E.D.

Proposition 3.4 motivates the following definition.

<u>3.5 Definition</u>. Let $\omega \subset \Omega$ be two open sets in E and let $\mathfrak{F} \subset \mathcal{H}(\Omega)$.

(a) ω is said to be $\mathfrak{F}$-convex if for each compact set $K \subset \omega$, the set $\hat{K}_{\mathfrak{F}} \cap \omega$ is bounded away from $\complement\omega$.

(b) ω is said to be strongly $\mathfrak{F}$-convex if for each compact set $K \subset \omega$, the set $\hat{K}_{\mathfrak{F}}$ is contained in ω and is bounded away from $\complement\omega$.

In particular, an open set ω in E is said to be:

(c) holomorphically convex if it is $\mathcal{H}(\omega)$-convex;

(d) polynomially convex if it is $\mathcal{P}(E)$-convex;

(e) strongly polynomially convex if it is strongly $\mathcal{P}(E)$-convex.

3.6 Example. Let ω be a convex open set in E. Then ω is strongly $\mathcal{A}(E)$-convex and hence is strongly $\mathfrak{F}$-convex for any open set Ω with $\omega \subset \Omega \subset E$ and any family $\mathfrak{F}$ with $\mathcal{A}(E) \subset \mathfrak{F} \subset \mathcal{H}(\Omega)$.

3.7 Example. Let $\mathfrak{F} \subset \mathcal{H}(\Omega)$ and assume that Ω is $\mathfrak{F}$-convex. Then any open subset ω of the form

$$\omega = \{x \in \Omega: |f(x)| < c\}$$

where $f \in \mathfrak{F}$ and $c \in \mathbb{R}$, is strongly $\mathfrak{F}$-convex.

3.8 Remark. Let $\mathfrak{F} \subset \mathcal{H}(\Omega)$. Then the intersection of finitely many (strongly) $\mathfrak{F}$-convex open subsets of Ω is a (strongly) $\mathfrak{F}$-convex open subset of Ω.

The connection between holomorphic convexity and domains of holomorphy is given by the following theorem, established by Hirschowitz [11, Théorème 1.6] in the case of Banach spaces and by Noverraz [24, p.62, Théorème 3.5] in the general case.

3.9 Theorem. Every domain of holomorphy is holomorphically convex.

4. PLURISUBHARMONIC FUNCTIONS

Let Δ denote the open unit disc in $\mathbb{C}$. We recall the following definitions.

4.1 Definition. Let Ω be an open set in $\mathbb{C}$. A function $u:\Omega \to [-\infty,+\infty)$ is said to be subharmonic if:

(a) u is upper semicontinuous;

(b) for each $a \in \Omega$ and each $r > 0$ such that $a + r\bar{\Delta} \subset \Omega$, then

$$u(a) \leq \frac{1}{2\pi}\int_0^{2\pi} u(a + re^{it})dt.$$

<u>4.2 Definition</u>. Let Ω be an open set in E. A function $u: \Omega \to [-\infty,+\infty)$ is said to be <u>plurisubharmonic</u> if:

(a) u is upper semicontinuous;

(b) the restriction of u to each complex line is subharmonic.

We denote by $\mathcal{P}s(\Omega)$ the set of all plurisubharmonic functions on Ω and we denote by $\mathcal{P}sc(\Omega)$ the subset of all $u \in \mathcal{P}s(\Omega)$ which are continuous.

We refer to Noverraz [24] for the basic properties of plurisubharmonic functions on infinite dimensional spaces.

5. PSEUDOCONVEXITY

If $\alpha \in cs(E)$, $x \in E$ and $r > 0$ then we set

$$B_\alpha(x;r) = \{y \in E: \alpha(x-y) < r\}.$$

<u>5.1 Definition</u>. (a) For each $\alpha \in cs(E)$ we define a function

$$x \in \Omega \to d_\alpha(x; \complement\Omega) \in [0,+\infty]$$

by

$$d_\alpha(x; \complement\Omega) = \sup\{0\} \cup \{r > 0: B_\alpha(x;r) \subset \Omega\}$$
$$= \inf\{\alpha(x-y): y \notin \Omega\}.$$

(b) We also define a function

$$(x,a) \in \Omega \times E \to \delta(x,a; \complement\Omega) \in (0,+\infty]$$

by

$$\delta(x,a; \complement\Omega) = \sup\{r > 0: x + r\Delta a \subset \Omega\}$$
$$= \inf\{|\lambda|: \lambda \in C, \quad x + \lambda a \notin \Omega\}.$$

<u>5.2 Remark</u>. Given $A \subset \Omega$ and $\alpha \in cs(E)$ we write

$$d_\alpha(A; \complement\Omega) = \inf\{d_\alpha(x; \complement\Omega): x \in A\}$$
$$= \inf\{\alpha(x-y): x \in A, \; y \notin \Omega\}.$$

Then A is bounded away from $\complement\Omega$ if and only if there exists $\alpha \in cs(E)$ such that $d_\alpha(A; \complement\Omega) > 0$.

The following lemma can be found in Noverraz [24, p.41, Lemme 2.1.4 and p.44].

<u>5.3 Lemma</u>. (a) For $x,y \in \Omega$:

$$|d_\alpha(x; \complement\Omega) - d_\alpha(y; \complement\Omega)| \leq \alpha(x-y).$$

In particular $d_\alpha(x; \complement\Omega)$ is continuous on Ω.

(b) $\delta(x,a; \Omega)$ is lower semicontinuous on $\Omega \times E$ for any topology on E that makes Ω open.

(c) For any $x \in \Omega$

$$d_\alpha(x; \complement\Omega) = \inf\{\delta(x,a; \complement\Omega): \alpha(a) \leq 1\}.$$

<u>5.4 Definition</u>. Ω is said to be <u>pseudoconvex</u> if the function $-\log \delta(x,a; \complement\Omega)$ is plurisubharmonic on $\Omega \times E$.

<u>5.5 Definition</u>. Given a set $A \subset \Omega$ and a family $\mathcal{U} \subset \mathcal{P}s(\Omega)$ we define the $\mathcal{U}$-<u>hull</u> of A by

$$\hat{A}_{\mathcal{U}} = \{x \in \Omega: u(x) \leq \sup_A u \text{ for all } u \in \mathcal{U}\}.$$

<u>5.6 Definition</u>. Let $\omega \subset \Omega$ be two open sets in E and let $\mathcal{U} \subset \mathcal{P}s(\Omega)$.

(a) ω is said to be $\mathcal{U}$-<u>convex</u> if for each compact set $K \subset \omega$, the set $\hat{K}_{\mathcal{U}} \cap \omega$ is bounded away from $\complement\omega$.

(b) ω is said to be <u>strongly</u> $\mathcal{U}$-<u>convex</u> if for each compact set $K \subset \omega$, the set $\hat{K}_{\mathcal{U}}$ is contained in ω and bounded away from $\complement\omega$.

Let $|\mathcal{G}|(E) = \{|f|: f \in \mathcal{G}(E)\}$. Since $f \in \mathcal{H}(\Omega)$ implies $|f| \in \mathcal{P}sc(\Omega)$ we see that $|\mathcal{G}|(E) \subset \mathcal{P}sc(\Omega)$.

<u>5.7 Example</u>. Let ω be a convex open set in E. Then ω is strongly $|\mathcal{G}|(E)$-convex and hence is strongly $\mathcal{U}$-convex for any open set Ω with $\omega \subset \Omega \subset E$ and any family $\mathcal{U}$ with $|\mathcal{G}|(E) \subset \mathcal{U} \subset \mathcal{P}s(\Omega)$.

<u>5.8 Example</u>. Let $\mathcal{U} \subset \mathcal{P}s(\Omega)$ and assume that Ω is $\mathcal{U}$-convex. Then any open subset ω of the form

$$\omega = \{x \in \Omega: u(x) < c\}$$

where $u \in \mathcal{U}$ and $c \in \mathbb{R}$, is strongly $\mathcal{U}$-convex.

<u>5.9 Remark</u>. Let $\mathcal{U} \subset \mathcal{P}s(\Omega)$. Then the intersection of finitely many (strongly) $\mathcal{U}$-convex open subsets of Ω is a (strongly) $\mathcal{U}$-convex open subset of Ω.

The connection between pseudoconvexity and $\mathcal{P}s(\Omega)$-convexity is given by the following theorem, whose proof can be found in Noverraz [24, p.50, Corollaire 2.3.7].

<u>5.10 Theorem</u>. The following conditions are equivalent:

(a) Ω is pseudoconvex.

(b) Ω is finitely pseudoconvex, i.e. $\Omega \cap M$ is pseudoconvex in M for each finite dimensional subspace M of E.

(c) Ω is $\mathcal{P}s(\Omega)$-convex.

<u>5.11 Remark</u>. If Ω is $\mathcal{P}sc(\Omega)$-convex then it is obviously $\mathcal{P}s(\Omega)$-convex, and hence pseudoconvex, but it is not known whether the converse holds in general. See Proposition 9.1 for a partial converse.

<u>5.12 Remark</u>. Since $f \in \mathcal{H}(\Omega)$ implies $|f| \in \mathcal{P}sc(\Omega)$, we see that every holomorphically convex open set Ω is $\mathcal{P}sc(\Omega)$-convex and hence pseudoconvex.

6. THE CARTAN-THULLEN THEOREM AND THE LEVI PROBLEM

The following theorem is just a summary of previous results.

<u>6.1 Theorem</u>. Consider the following conditions:

(a) Ω is a domain of existence;

(b) Ω is a domain of holomorphy;

(c) Ω is holomorphically convex;

(d) Ω is pseudoconvex.

Then (a) $\Rightarrow$ (b) $\Rightarrow$ (c) $\Rightarrow$ (d).

In the case where $E = \mathbb{C}^n$, the classical Cartan-Thullen Theorem [2], published in 1932, asserts that (c) $\Rightarrow$ (a). The much harder implication (d) $\Rightarrow$ (a) was established much later, in 1953-1954, by Oka [28], Bremermann [1] and Norguet [23], thus solving a problem posed by Levi [19] in 1911.

Many efforts have been devoted to the study of these implications in infinite dimensional spaces. The usual technique for studying the Levi problem in infinite dimensional spaces has been to assume a suitable approximation property in the space and then to use the known finite dimensional results. In this way, in 1972 Gruman [9] succeded in solving the Levi problem in separable Hilbert spaces, and immediately afterwards Gruman and Kiselman [10] generalized the result to all Banach spaces having a Schauder basis. Later on these results have been generalized to various infinite dimensional spaces having a Schauder basis or having the Banach approximation property; see Noverraz [24], [27], Pomes [29], Dineen [5], Schottenloher [30], [31] and Dineen-Noverraz-Schottenloher [7]. The Levi problem has been recently solved by Colombeau and the author [4] in all nuclear Silva spaces, and these spaces do not necessarily have a Schauder basis or the Banach approximation property (though they always have the approximation property). We should also point out that Josefson [16] has found a nonseparable Banach space with the approximation property where even the implication (c) $\Rightarrow$ (b) is false.

Thus in all of the aforementioned cases, the Cartan-Thullen Theorem has been obtained as a consequence of the corresponding solution of the Levi problem. But in 1972 Noverraz [25] established directly a version of the Cartan-Thullen Theorem ((c) $\Rightarrow$ (b)) in all Silva spaces, with no assumption of a Schauder basis or any other kind of approximation property. Silva spaces have special features that make them look very much like finite dimensional spaces, e.g. every open set in a Silva space is uniformly open and is a hemicompact k-space. Motivated by these properties of Silva spaces, we have undertaken the study of domains of holomorphy in those spaces of the form $E = F'_c$ where F is a separable Fréchet space. This class includes all (DFM)-spaces, in particular all Silva spaces, and all the spaces in this class share the aforementioned properties of Silva spaces. Those properties prove to be very useful for the study of domains of holomorphy, for not only they help us to prove the Cartan-Thullen Theorem in all the spaces in this class, but they also serve us to solve the Levi problem in all those spaces in this class which have the approximation property.

7. (DFC)-SPACES

7.1 Definition. A (DFC)-space is any space of the form $E = F'_c$ where F is a Fréchet space, i.e. the dual of a Fréchet space, endowed with the topology of compact convergence.

(DFC)-spaces ((DCF)-Räume in German) have been extensively studied by Hollstein [12], [13] in connection with the theory of topological tensor products. We refer to [12], [13] for a number of linear properties of these spaces. Since we are chiefly interested in the study of holomorphic functions on (DFC)-spaces, we will mainly establish nonlinear properties of these spaces.

We recall that a topological space X is hemicompact if it has a sequence of compact subsets (K_m) such that each compact subset of X is contained in some K_m. We also recall that a topological space X is Lindelöf if each open cover of X admits a countable subcover. Clearly every hemicompact space is Lindelöf.

7.2 Proposition. Let E be a (DFC)-space. Then:

(a) E is a semi-Montel space, in particular quasi-complete.

(b) E is hemicompact, in particular Lindelöf.

Proof. Let $E = F'_c$ where F is a Fréchet space. Let (W_m) be a countable 0-neighborhood base in F. Then $K_m = W_m^o$ is a compact subset of E and each bounded subset of E is an equicontinuous subset of F' and hence is contained in some K_m.

Q.E.D.

7.3 Remark. If $E = F'_c$ where F is a Fréchet space, then we see at once that $E'_b = E'_c = F$.

Proposition 7.2 tells us that every (DFC)-space is hemicompact. But we do not know whether every open set in a (DFC)-space is hemicompact. Theorem 7.4 below provides a partial answer to this question.

7.4 Theorem. Let $E = F'_c$ where F is a separable Fréchet space. Then:

(a) Each compact subset of E is metrizable and separable.

(b) Each open subset of E is hemicompact, in particular Lindelöf.

Proof. Let D be a countable dense subset of F and let $\sigma(F',D)$ denote the topology of uniform convergence on the finite subsets of D. This is a metrizable topology, weaker than the topology of E. Let ρ be a metric on F' generating the topology $\sigma(F',D)$. Then E and the metric ρ induce the same topology on each compact subset of E. In particular each compact subset of E is metrizable, and hence separable. Let (K_m) be an increasing sequence of compact subsets of E such that each compact subset of E is contained in some K_m. Given an open subset Ω of E we define

$$L_m = \{x \in K_m \cap \Omega: \rho(x,K_m \setminus \Omega) \geq 1/m\}$$

where

$$\rho(x,K_m \setminus \Omega) = \inf\{\rho(x,y);\ y \in K_m \setminus \Omega\}.$$

Then one can readily see that each L_m is a compact subset of Ω and that each compact subset of Ω is contained in some L_m. Thus Ω is hemicompact.

Q.E.D.

7.5 Corollary. Let $E = F'_c$ where F is a separable Fréchet space. Then each open subset of E is sequentially separable, i.e. each open set Ω has a countable subset D such that each point of Ω

is the limit of a sequence of points of D.

We could have proved Theorem 7.4 and Corollary 7.5 using the theory of Suslin spaces (see Schwartz [34, Chapter II]), but we have preferred a direct proof.

We recall that a k-space is a topological space X with the property that a set $U \subset X$ is open if $U \cap K$ is open in K for each compact subset K of X, or with the equivalent property that a mapping $f: X \to Y$ into a topological space Y is continuous if the restriction of f to K is continuous for each compact subset K of X. It follows at once from the Banach-Dieudonné Theorem (see Horváth [15, p.245, Theorem 1]) that each (DFC)-space is a k-space. Since a subset of a k-space need not be a k-space under the induced topology (see Kelley [17, p.240]), it does not follow immediately that each open subset of a (DFC)-space is a k-space. But the method of proof of the Banach-Dieudonné Theorem yields also a proof of this.

7.6 Theorem. Let E be a (DFC)-space. Then:

(a) Each closed subset of E is a k-space.

(b) Each open subset of E is a k-space.

Proof. (a) Let $E = F'_c$ where F is a Fréchet space. Let X be a closed subset of E. Let U be a subset of X with the property that $U \cap K$ is open in K for each compact subset K of X. We want to show that U is open in X. Thus for each point $\xi \in U$ we want to find a compact subset L of F such that

$$(\xi + L^o) \cap X \subset U.$$

Let (W_m) be a decreasing countable 0-neighborhood base in F with $W_o = F$. Set

$$K_m = (\xi + W_m^o) \cap X.$$

We claim that there exists a sequence of finite sets (A_m) with $A_m \subset W_m$ for every $m \in \mathbb{N}$, and such that, if we write

$$B_m = A_o \cup \ldots \cup A_m$$

then

$$(\xi + B_m^o) \cap K_{m+1} \subset U$$

for every $m \in \mathbb{N}$.

Indeed each K_m is a compact subset of X and hence E induces on each K_m the weak topology $\sigma(F',F)$. Since $U \cap K_1$ is a neighborhood of ξ in K_1 we can find a finite set $A_o \subset F = W_o$ such that

$$(\xi + A_o^o) \cap K_1 \subset U.$$

Suppose that we have already found finite sets $A_o \subset W_o, \dots, A_m \subset W_m$ such that

$$(\xi + B_m^o) \cap K_{m+1} \subset U$$

but suppose that

$$(\xi + (B_m \cup P)^o) \cap K_{m+2} \not\subset U$$

for each finite set $P \subset W_{m+1}$. Set $K = K_{m+2} \setminus U$. Then K is compact, being a closed subset of K_{m+2} and for each finite set $P \subset W_{m+1}$ the set

$$(\xi + (B_m \cup P)^o) \cap K = (\xi + B_m^o \cap P^o) \cap K$$

is a nonvoid closed subset of K. One readily verifies that these sets have the finite intersection property, and therefore

$$\cap \{(\xi + B_m^o \cap P^o) \cap K: P \subset W_{m+1}, \ P \text{ finite}\} \neq \emptyset$$

hence

$$(\xi + B_m^o \cap W_{m+1}^o) \cap K \neq \emptyset$$

and this yields

$$(\xi + B_m^o) \cap K_{m+1} \not\subset U$$

a contradiction. Thus the existence of the sequence (A_m) is proven.

Set $B = \bigcup A_m = \bigcup B_m$. Since $A_m \subset W_m$ for every m and since each A_m is finite, we see that B is the range of a null sequence in F. Set $L = B \cup \{0\}$. Then L is a compact subset of F and

$$(\xi + L^o) \cap K_{m+1} \subset (\xi + B_m^o) \cap K_{m+1} \subset U$$

for every m. And since $X = \bigcup K_m$ we conclude that

$$(\xi + L^{o}) \cap X \subset U$$

as we wanted.

(b) Let Ω be an open subset of E and let $f: \Omega \to Y$ be a mapping into a topological space Y whose restriction to each compact subset of Ω is continuous. It follows from (a) that the restriction of f to each set $X \subset \Omega$, X closed in E, is continuous. But since each point of Ω admits a neighborhood base formed by closed sets, we conclude that f is locally continuous, hence continuous.

Q.E.D.

The proof of Theorem 7.6 yields the following well-known lemma. (See Horváth [15, p.247]).

<u>7.7 Lemma</u>. Each compact subset of a Fréchet space is contained in the closed, convex, balanced hull of a suitable null sequence.

<u>Proof</u>. Let F be a Fréchet space, let L be a compact subset of F and let $U = L^{o}$. Then U is a 0-neighborhood in $E = F'_c$ and hence $U \cap K$ is a 0-neighborhood in K for each compact subset K of E which contains 0. Then the proof of Theorem 7.6 (a) with $X = E$ and $\xi = 0$ yields a null sequence (b_m) in F such that if B is the range of (b_m) then $B^{o} \subset U$. Hence $L \subset B^{oo}$ = closed, convex, balanced hull of B.

Q.E.D.

It is well-known (see Horváth [15, p.116, Proposition 3]) that given a sequence (B_m) of bounded subsets of a Fréchet space, one can find a sequence (ϵ_m) of strictly positive numbers such that the set $B = \bigcup \epsilon_m B_m$ is also a bounded set. With the aid of Lemma 7.7 we can prove a similar result for compact sets.

<u>7.8 Lemma</u>. Given a sequence (L_m) of compact subsets of a Fréchet space F, there exists a sequence (ϵ_m) of strictly positive numbers such that the set $L = \bigcup \epsilon_m L_m$ is a relatively compact subset of F.

<u>Proof</u>. By Lemma 7.7 for each $m \in \mathbb{N}$ we can find a null sequence

$(x_{mn})_{n\in\mathbb{N}}$ in F such that L_m is contained in the closed, convex, balanced null of the range X_m of $(x_{mn})_{n\in\mathbb{N}}$. Let $(W_m)_{m\in\mathbb{N}}$ be a decreasing, countable 0-neighborhood base in F. Then for each $m\in\mathbb{N}$ we can find $\varepsilon_m > 0$ such that

$$\varepsilon_m x_{mn} \in W_m \quad \text{for all} \quad n \in \mathbb{N}.$$

Then for each $k \in \mathbb{N}$:

(1) $$\varepsilon_m x_{mn} \in W_k \quad \text{for all} \quad m \geq k \quad \text{and} \quad n \in \mathbb{N}.$$

But since $\varepsilon_m x_{mn} \to 0$ when $n \to +\infty$ we can inductively find a sequence of nonnegative integers $n_1 < n_2 < \ldots < n_k < \ldots$ such that

(2) $$\varepsilon_m x_{mn} \in W_k \quad \text{for all} \quad m < k \quad \text{and} \quad n \geq n_k.$$

Then we define a sequence of finite sets $Y_1 \subset Y_2 \subset \ldots \subset Y_k \subset \ldots$ by

$$Y_k = \{\varepsilon_m x_{mn} : \ m < k, \ n < n_k\}.$$

If $Y = \bigcup Y_k$ then it follows from (1) and (2) that $Y \setminus Y_k \subset W_k$ for each $k \geq 1$ and it is then clear that Y is the range of a null sequence. And since $n_k \to +\infty$ when $k \to +\infty$ we see that $Y = \bigcup \varepsilon_m X_m$. It is then clear that

$$L = \bigcup \varepsilon_m L_m \subset \text{closed, convex, balanced null of } \bigcup \varepsilon_m X_m = Y$$

and hence L is relatively compact in F.

Q.E.D.

<u>7.9 Corollary</u>. Given a sequence (V_m) of 0-neighborhoods in a (DFC)-space, there exists a sequence (λ_m) of strictly positive numbers such that $V = \bigcap \lambda_m V_m$ is also a 0-neighborhood.

<u>7.10 Definition</u>. An open set Ω in a locally convex space E is said to be <u>uniformly open</u> if there exists $\alpha \in cs(E)$ such that Ω is open in the seminormed space (E,α).

<u>7.11 Theorem</u>. Let $E = F'_c$ where F is a separable Fréchet space. Then every open set in E is uniformly open.

<u>Proof</u>. Let Ω be an open set in E. By Theorem 7.4 Ω is Lindelöf and hence we can find a sequence $(x_m) \subset \Omega$ and a sequence (V_m) of convex, balanced 0-neighborhoods in E such that

$$\Omega = \bigcup (x_m+V_m) \quad \text{and} \quad x_m + 2V_m \subset \Omega.$$

By Corollary 7.9 we can find a sequence (λ_m) of strictly positive numbers such that $V = \bigcap \lambda_m V_m$ is also a 0-neighborhood. Thus

$$\Omega = \bigcup (x_m+V_m) \quad \text{and} \quad (x_m+V_m) + \lambda_m^{-1} V \subset \Omega$$

and we conclude that Ω is open in (E,α) where α denotes the Minkowski functional of V.

Q.E.D.

Since every Fréchet-Montel space is separable (see Köthe [18, p.370]), the class of all spaces of the form $E = F'_c$ where F is a separable Fréchet space, includes all (DFM)-spaces, and in particular, all Silva spaces.

The main results of this section, namely Theorems 7.4, 7.6 and 7.11, will be essential in subsequent sections. Theorems 7.4 and 7.6 had been established for Silva spaces and (DFM)-spaces by Noverraz [25] and Dineen [6], respectively. Theorem 7.11 had been established for (DFM)-spaces by Colombeau and the author [4]. We do not know whether Lemma 7.8 has appeared before, but in any case the proof we have given is inspired on the proof of a similar result of Colombeau [3, Lemma 1].

8. THE CARTAN-THULLEN THEOREM IN (DFC)-SPACES

An immediate consequence of Theorems 7.4 and 7.6 is the following:

<u>8.1 Proposition</u>. Let $E = F'_c$ where F is a separable Fréchet space. If Ω is any open set in E, then $\mathcal{H}(\Omega)$ is a Fréchet space for the topology of compact convergence.

Now we can easily prove a weak version of the Cartan-Thullen theorem:

8.2 Proposition. Let $E = F'_c$ where F is a separable Fréchet space. Then each holomorphically convex open set in E is a domain of holomorphy.

Proof. Let Ω be a holomorphically convex open set in E, and let (x_m) be a sequence in Ω which converges to a point $x_o \in \partial\Omega$. We will prove that there exists a function $f \in \mathcal{H}(\Omega)$ such that $\sup|f(x_m)| = +\infty$. Otherwise we may define a seminorm p on $\mathcal{H}(\Omega)$ by $p(f) = \sup|f(x_m)|$. As a supremum of continuous functions, p is lower semicontinuous, and hence is continuous, for $\mathcal{H}(\Omega)$ is barrelled. Hence there exists a compact set $K \subset \Omega$ and a constant $c > 0$ such that

$$p(f) \leq c \cdot \sup_K |f|$$

for all $f \in \mathcal{H}(\Omega)$. Replacing f by its mth power, taking mth root and letting $m \to +\infty$ we see that $c = 1$. Then we see that the sequence (x_m) is contained in $\hat{K}_{\mathcal{H}(\Omega)}$. But this is impossible, for $x_m \to x_o \in \partial\Omega$ and $\hat{K}_{\mathcal{H}(\Omega)}$ is bounded away from Ω.

Q.E.D.

But we can actually prove the much stronger result:

8.3 Theorem. Let $E = F'_c$ where F is a separable Fréchet space. Then each holomorphically convex open set in E is a domain of existence.

To prove the theorem we will need the following lemma:

8.4 Lemma. Let $E = F'_c$ where F is a separable Fréchet space. Then there exists a translation-invariant metric ρ on E of the form

$$\rho(x,y) = \sum_{j\in\mathbb{N}} 2^{-j} \frac{\alpha_j(x-y)}{1 + \alpha_j(x-y)}$$

where each $\alpha_j \in cs(E)$. If E_ρ denotes the vector space E endowed with the metric ρ, then:

(a) The identity mapping $E \to E_\rho$ is continuous.

(b) E and E_ρ induce the same topology on each compact subset of E. In particular each compact subset of E is metrizable and separable.

Let Ω be an open subset of E and let K be a convex, compact subset of E such that both $K \cap \Omega$ and $K \setminus \Omega$ are nonvoid. Let $\xi \in K \cap \Omega$. Then:

(c) If we define

$$\rho(\xi) = \inf \{\rho(\xi, y): y \in K \setminus \Omega\}$$

then $\rho(\xi) > 0$.

(d) If we define

$$B(\xi) = \{x \in K: \rho(\xi, x) < \rho(\xi)\}$$

then $B(\xi) \subset K \cap \Omega$ and $\overline{B(\xi)} \not\subset \Omega$. In particular $d_{\alpha}(B(\xi): \complement\Omega) = 0$ for every $\alpha \in cs(E)$.

Proof. Existence of the metric ρ and assertions (a) and (b) were already established in the proof of Theorem 7.4. It is also clear that $\rho(\xi) > 0$ for the set $K \setminus \Omega$ is closed in K for the metric ρ. It is also obvious that $B(\xi) \subset K \cap \Omega$. To prove that $\overline{B(\xi)} \not\subset \Omega$ we choose a sequence $(y_k) \subset K \setminus \Omega$ such that

$$\rho(\xi, y_k) < \rho(\xi) + \frac{1}{k}.$$

Since K is compact the sequence (y_k) admits a subsequence which converges to a point $\eta \in K$. Then $\eta \in K \setminus \Omega$ and $\rho(\xi, \eta) \leq \rho(\xi)$. But since $\eta \notin \Omega$ it is clear that actually $\rho(\xi, \eta) = \rho(\xi)$. We claim that $\eta \in \overline{B(\xi)}$. To show this it suffices to prove that

$$\rho(\xi, x_t) < \rho(\xi, \eta) \quad \text{for every } t, \quad 0 < t < 1$$

where

$$x_t = \xi + t(\eta - \xi).$$

It is clear that each $x_t \in K$ for K is convex. And since the function f defined by

$$f(s) = \frac{s}{1+s}$$

is strictly increasing for $s > 0$ we see that

$$\rho(x_t,\xi) = \rho(t(\eta-\xi),0) < \rho(\eta-\xi,0) = \rho(\eta,\xi)$$

for every t, $0 < t < 1$.

Q.E.D.

Proof of Theorem 8.3. Let Ω be a holomorphically convex open set in E, with $\emptyset \neq \Omega \neq E$. By Proposition 7.2 we can find an increasing sequence (K_m) of convex, compact subsets of E such that each compact subset of E is contained in some K_m. Without loss of generality we may assume that $K_m \cap \Omega \neq \emptyset$ and $K_m \setminus \Omega \neq \emptyset$ for every m. By Theorem 7.4 we can find an increasing sequence (L_m) of compact subsets of Ω such that each compact subset of Ω is contained in some L_m. Since Ω is holomorphically convex and E is quasi-complete we may assume that $(\hat{L}_m)_{\mathcal{H}(\Omega)} = L_m$ for every m. Let ρ be the translation-invariant metric on E described in Lemma 8.4. For each $m \in \mathbb{N}$, $\xi \in K_m$ and $r > 0$, set

$$B_m(\xi;r) = \{x \in K_m : \rho(\xi,x) < r\}.$$

For each $m \in \mathbb{N}$ and $\xi \in K_m \cap \Omega$ set

$$\rho_m(\xi) = \inf\{\rho(\xi,y) : y \in K_m \setminus \Omega\}$$

and

$$B_m(\xi) = B_m(\xi;\rho_m(\xi)) = \{x \in K_m : \rho(\xi,x) < \rho_m(\xi)\}.$$

Then by Lemma 8.4, $d_\alpha(B_m(\xi);\ \Omega) = 0$ for every $\alpha \in cs(E)$ and therefore $B_m(\xi) \not\subset L$ for every compact subset L of Ω.

For each $m \in \mathbb{N}$ let D_m be a countable dense subset of $K_m \cap \Omega$. Let D denote the set of all pairs (m,ξ) with $m \in N$ and $\xi \in D_m$. Let (m_j,ξ_j) be a sequence of elements of D such that each member of D appears in the sequence infinitely many times.

Since $B_{m_j}(\xi_j) \not\subset L$ for every $j \in \mathbb{N}$ and every compact subset L of Ω, after replacing (L_m) by a subsequence which we still call (L_m), we can find a sequence $(\eta_j) \subset \Omega$ such that

$$\eta_j \in B_{m_j}(\xi_j), \quad \eta_j \notin L_j, \quad \eta_j \in L_{j+1}.$$

Then, since $(\hat{L}_j)_{\mathcal{H}(\Omega)} = L_j$ for every j, we can inductively find a sequence $(f_j) \subset \mathcal{H}(\Omega)$ such that

$$\sup_{L_j} |f_j| \le 2^{-j}$$

and

$$|f_j(\eta_j)| \ge j + 1 + |\sum_{k<j} f_k(\eta_j)|.$$

Thus the series Σf_j converges uniformly on the compact subsets of Ω to a function f. And since by Theorem 7.6, Ω is a k-space, we conclude that $f \in \mathcal{H}(\Omega)$. It is also clear that $|f(\eta_j)| \ge j$ for every $j \in \mathbb{N}$.

We claim that Ω is the domain of existence of f. To prove this we will show that for each convex open set V in E with $V \cap \partial\Omega \neq \emptyset$, f is unbounded on each connected component of $\Omega \cap V$. Let ω be a connected component of $\Omega \cap V$. Choose points $a \in \omega$ and $b \in V \setminus \Omega$. Choose $m \in N$ such that $a, b \in K_m$. Set $\Omega_m = K_m \cap \Omega$ and $V_m = K_m \cap V$ and let ω_m denote the connected component of $\Omega_m \cap V_m$ which contains the point a. Then clearly $\omega_m \subset \omega$. The proof of Lemma 2.2 shows the existence of a point $\zeta \in \partial\Omega_m \cap V_m \cap \partial\omega_m$. Choose $r > 0$ such that $B_m(\zeta;2r) \subset V_m$ and next choose a point $\xi \in D_m \cap \omega_m \cap B_m(\zeta;R)$. Then

$$B_m(\xi) \subset B_m(\xi;r) \subset B_m(\zeta;2r) \subset V_m.$$

Thus $B_m(\xi) \subset \Omega_m \cap V_m$ and therefore

$$B_m(\xi) \subset \omega_m \subset \omega.$$

From the definition of the sequence (m_j, ξ_j) we see that there exists a strictly increasing sequence $(j_k) \subset \mathbb{N}$ such that

$$(m, \xi) = (m_{j_k}, \xi_{j_k}) \quad \text{for every} \quad k \in \mathbb{N}.$$

Then we see that $\eta_{j_k} \in B_m(\xi)$ for every $k \in \mathbb{N}$ and we conclude that f is unbounded on $B_m(\xi)$ and hence on ω.

Q.E.D.

As we mentioned before, Noverraz [25] proved Proposition 8.2 for Silva spaces, and later on Schottenloher [30] established Theorem 8.3 for Silva spaces. Actually, Theorem 8.3 had been announced by

Noverraz [27] and Dineen [5], for Silva spaces and (DFM)-spaces, respectively, but their proofs were incomplete. Both Noverraz's proof and Dineen's proof contained essentially the same gap, and since this gap has also appeared in several other papers, we will describe it here in some detail. Given a holomorphically convex open set Ω in a Silva space or a (DFM)-space, Noverraz and Dineen found a sequence (x_m) in Ω such that each point in $\partial\Omega$ is the limit of a subsequence of (x_m), and constructed a function $f \in \mathcal{H}(\Omega)$ with $\lim|f(x_m)| = +\infty$. And then Noverraz and Dineen concluded that Ω was the domain of existence of f. But the conclusion is not correct, for one can even find an open set $\Omega \subset \mathbb{C}$ and a function $f \in \mathcal{H}(\Omega)$ satisfying the aforementioned condition, but such that Ω is not the domain of existence of f.

9. PSEUDOCONVEX UNIFORMLY OPEN SETS

<u>9.1 Proposition</u>. Let Ω be a pseudoconvex open set which is uniformly open. Then Ω is $\mathcal{P}sc(\Omega)$-convex.

Proposition 9.1 is an immediate consequence of the following more precise lemma.

<u>9.2 Lemma</u>. Let Ω be a pseudoconvex open set which is uniformly open. Then there exists a function $u \in \mathcal{P}sc(\Omega)$ such that, if we set

$$\Omega_c = \{x \in \Omega : u(x) \leq c\}$$

for $c \in \mathbb{R}$, then:

(a) Each compact subset of Ω is contained in some Ω_c.

(b) $(\Omega_c)_{\mathcal{P}sc(\Omega)} = \Omega_c$.

(c) Ω_c is bounded away from $\complement\Omega$.

<u>Proof</u>. Choose $\alpha \in cs(E)$ such that Ω is open in (E,α). Then $0 < d_\alpha(x; \complement\Omega) \leq +\infty$ for every $x \in \Omega$ and hence the function

$$u: x \in \Omega \to -\log d_\alpha(x; \complement\Omega) \in [-\infty, +\infty)$$

belongs to $\mathcal{P}sc(\Omega)$, for, according to Lemma 5.3, u is continuous and it is equal to the supremum of the plurisubharmonic functions

$$x \in \Omega \to -\log \delta(x,a; \ \Omega) \in [-\infty,+\infty)$$

with $a \in E$, $\alpha(a) \leq 1$. Assertions (a) and (b) are obvious. And (c) is also clear for

$$d_\alpha(\Omega_c; \ \Omega) = \inf_{x \in \Omega_c} d_\alpha(x; \ \Omega) \geq e^{-c}.$$

Q.E.D.

The main result in this section is the following:

9.3 Theorem. Let E be a quasi-complete locally convex space. Let Ω be a pseudoconvex open set in E which is uniformly open. Let K be a compact subset of Ω such that $\hat{K}_{\mathcal{P}sc(\Omega)} = K$. Then each open neighborhood of K in Ω contains another open neighborhood of K in Ω which is strongly $\mathcal{P}sc(\Omega)$-convex.

Proof. Let $u \in \mathcal{P}sc(\Omega)$ be the function given by Lemma 9.3. Then

$$K \subset \{x \in \Omega: u(x) < c\}$$

for some $c \in R$. Set

$$L = \bar{\Gamma}(K) \cap \{x \in \Omega: u(x) \leq c\}.$$

Since the set $\{x \in \Omega: u(x) \leq c\}$ is bounded away from $\complement\Omega$, L is a compact subset of Ω. Let ω be an open neighborhood of K in Ω. Since $\hat{K}_{\mathcal{P}sc(\Omega)} = K$, for each $a \in L\setminus\omega$ we can find a function $v_a \in \mathcal{P}sc(\Omega)$ such that

$$v_a(a) > 0 \quad \text{and} \quad v_a(x) < 0 \quad \text{for all} \quad x \in K.$$

Then by compactness of $L\setminus\omega$ we can find functions $v_o,\dots,v_m \in \mathcal{P}sc(\Omega)$ such that

$$v_j(x) < 0 \quad \text{for all} \quad x \in K \quad \text{and} \quad j = 0,\dots,m$$

and

$$L\setminus\omega \subset \bigcup_{j \leq m} \{x \in \Omega: v_j(x) > 0\}$$

so that

$$L \cap \{x \in \Omega: v_j(x) \leq 0 \quad \text{for} \quad j \leq m\} \subset \omega.$$

Let $w = \sup\{u-c, v_o,\dots,v_m\} \in \mathcal{P}sc(\Omega)$. Then

$$w(x) < 0 \quad \text{for all} \quad x \in K$$

and

(1) $$\bar{\Gamma}(K) \cap \{x \in \Omega : w(x) \leq 0\} \subset \omega .$$

We claim that there exists a convex, open 0-neighborhood V in E such that

(2) $$(\bar{\Gamma}(K) + V) \cap \{x \in \Omega : w(x) \leq 0\} \subset \omega .$$

Otherwise, if (V_α) is a base of convex, open neighborhood of zero in E, then for each α we can find a point x_α with

$$x_\alpha \in (\bar{\Gamma}(K) + V_\alpha) \cap \{x \in \Omega : w(x) \leq 0\} \cap \complement\omega .$$

For each α we choose $y_\alpha \in \bar{\Gamma}(K)$ with $x_\alpha - y_\alpha \in V_\alpha$. Since $\bar{\Gamma}(\Omega)$ is compact, the net (y_α) admits a subset (y_β) which converges to a point $y \in \bar{\Gamma}(K)$. But since $x_\alpha - y_\alpha \to 0$ we see that the corresponding subnet (x_β) of (x_α) also converges to y. And then

$$y \in \bar{\Gamma}(K) \cap \{x \in \Omega : w(x) \leq 0\} \cap \complement\omega$$

but this set is empty, by (1). Thus (2) is proven. We then define

$$\sigma = (\bar{\Gamma}(K) + V) \cap \{x \in \Omega : w(x) < 0\} .$$

Then $K \subset \sigma \subset \omega$ and it follows from Examples 5.7 and 5.8 and Remark 5.9 that σ is a strongly $\mathcal{P}sc(\Omega)$-convex open set.

Q.E.D.

The proof of Theorem 9.3 was inspired on the proof of a similar result of Ligocka on polynomially convex open sets: see Ligocka [20, Proposition 2.1]. See also the author's paper [21] where Ligocka's result is applied to the proof of a polynomial Oka-Weil Theorem.

10. HOLOMORPHIC APPROXIMATION IN (DFC)-SPACES WITH THE APPROXIMATION PROPERTY.

We recall that a locally convex space E has the <u>approximation property</u> if for each compact set K in E and each 0-neighborhood V in E there exists a continuous linear operator of finite rank

$T: E \to E$ such that $T(x) - x \in V$ for every $x \in K$.

By considering the transpose, it is a routine matter the proof of the following well-known lemma, which is implicitly contained in Schwartz [33, Exposé nº 14, Théorème 2].

<u>10.1 Lemma</u>. Let $E = F'_c$ where F is a Fréchet space. Then E has the approximation property if and only if F has the approximation property.

The main result of this section is the following Oka-Weil Theorem.

<u>10.2 Theorem</u>. Let $E = F'_c$ where F is a separable Fréchet space, and assume that E (or equivalently F) has the approximation property. Let Ω be a pseudoconvex open set in E and let K be a compact subset of Ω such that $\hat{K}_{\mathcal{P}sc(\Omega)} = K$. Then each function which is homorphic on an open neighborhood of K, can be approximated uniformly on K by functions belonging to $\mathcal{H}(\Omega)$.

<u>Proof</u>. Ω is uniformly open by Theorem 7.11 and hence $\mathcal{P}sc(\Omega)$-convex by Proposition 9.1. Then using Theorem 7.4 we can find an increasing sequence (K_m) of compact subsets of Ω such that each compact subset of Ω is contained in some K_m and such that $K = K_o$ and $(\hat{K}_m)_{\mathcal{P}sc(\Omega)} = K_m$ for every m.

Let f be a function which is holomorphic on an open neighborhood of K. By Theorem 9.3 we may assume that $f \in \mathcal{H}(\omega)$ where ω is open, $K \subset \omega \subset \Omega$ and ω is strongly $\mathcal{P}sc(\Omega)$-convex. Let $\epsilon > 0$ be given. Choose a 0-neighborhood V_1 in E such that

$$K + V_1 \subset \omega, \qquad K_1 + V_1 \subset \Omega$$

and

$$|f(y)-f(x)| \le \frac{\epsilon}{7} \quad \text{for} \quad x \in K, \quad y-x \in V_1.$$

Let $T_1: E \to E$ be a continuous linear operator of finite rank such that

$$T_1(x) - x \in V_1 \quad \text{for all} \quad x \in K_1.$$

Then

$$T_1(K) \subset K + V_1 \subset \omega, \qquad T_1(K_1) \subset K_1 + V_1 \subset \Omega$$

and

$$|f \circ T_1(x) - f(x)| < \frac{\varepsilon}{4} \quad \text{for all} \quad x \in K.$$

Thus $T_1(K) \subset \omega \cap T_1(E) \subset \Omega \cap T_1(E)$ and hence

$$\widehat{T_1(K)}_{\mathcal{P}sc(\Omega \cap T_1(E))} \subset \widehat{T_1(K)}_{\mathcal{P}sc(\Omega)} \cap T_1(E) \subset \omega \cap T_1(E)$$

since ω is strongly $\mathcal{P}sc(\Omega)$-convex. Thus f is holomorphic on a neighborhood of the compact set $L = \widehat{T_1(K)}_{\mathcal{P}sc(\Omega \cap T_1(E))}$ in $\Omega \cap T_1(E)$. Then by the classical Oka-Weil Theorem (see Hormander [14, p.91, Theorem 4.3.2]), there exists a function $h_1 \in \mathcal{H}(\Omega \cap T_1(E))$ such that

$$\sup_{T_1(K)} |h_1 - f| \leq \sup_{L} |h_1 - f| \leq \frac{\varepsilon}{4}.$$

Since $T_1(K_1) \subset \Omega$ the function $g_1 = h_1 \circ T_1$ is holomorphic on a neighborhood of K_1 and

$$\sup_{K} |g_1 - f| \leq \sup_{K} |h_1 \circ T_1 - f \circ T_1| + \sup_{K} |f \circ T_1 - f| \leq \frac{\varepsilon}{2}.$$

Since $(\hat{K}_1)_{\mathcal{P}sc(\Omega)} = K_1$ we may apply the same argument to g_1 and find a function g_2, holomorphic on a neighborhood of K_2, and satisfying

$$\sup_{K_1} |g_2 - g_1| \leq \frac{\varepsilon}{4}.$$

By induction we can find a sequence (g_m) where g_m is holomorphic on a neighborhood of K_m, and $g_o = f$, and satisfying

$$\sup_{K_{m-1}} |g_m - g_{m-1}| \leq \varepsilon \cdot 2^{-m}.$$

If $m > n \geq N$ then

$$\sup_{K_N} |g_m - g_n| \leq \sum_{n+1}^{m} \sup_{K_{j-1}} |g_j - g_{j-1}| \leq \sum_{n+1}^{\infty} \varepsilon \cdot 2^{-j} = \varepsilon \cdot 2^{-n}.$$

Thus the sequence (g_m) converges uniformly on the compact subsets of Ω to a function g. And since by Theorem 7.6 Ω is a k-space, we see that $g \in \mathcal{H}(\Omega)$. We also see that

$$\sup_{K_N} |g-g_n| \le \epsilon \cdot 2^{-n}$$

for all $n \ge N$ and in particular

$$\sup_K |g-f| < \epsilon .$$

Q.E.D.

Since $\hat{K}_{\mathcal{P}sc(\Omega)} \subset \hat{K}_{\mathcal{H}(\Omega)}$ for every $K \subset \Omega$, se wee that Theorem 10.2 is also valid if we replace $\hat{K}_{\mathcal{P}sc(\Omega)}$ by $\hat{K}_{\mathcal{H}(\Omega)}$. And then we can prove the following result.

<u>10.3 Corollary</u>. Let $E = F'_c$ where F is a separable Fréchet space, and assume that E (or equivalently F) has the approximation property. Let Ω be a holomorphically convex open set in E. Then for an open set $\omega \subset \Omega$ the following conditions are equivalent:

(a) ω is $\mathcal{H}(\Omega)$-convex.

(b) ω is strongly $\mathcal{H}(\Omega)$-convex.

(c) ω is holomorphically convex and $\mathcal{H}(\Omega)$ is dense in $\mathcal{H}(\omega)$ for the topology of compact convergence.

<u>Proof</u>. (a) $\Rightarrow$ (c). Obviously ω is holomorphically convex, so we only have to show that $\mathcal{H}(\Omega)$ is dense in $\mathcal{H}(\omega)$. Let $f \in \mathcal{H}(\omega)$. Let K be a compact subset of ω and let $\epsilon > 0$ be given. Set

$$L = \hat{K}_{\mathcal{H}(\Omega)}, \qquad L_1 = L \cap \omega, \qquad L_2 = L \backslash \omega .$$

Since Ω is holomorphically convex, L is a compact subset of Ω and hence L_2 is a compact subset of $\Omega \backslash \omega$. And since ω is $\mathcal{H}(\Omega)$-convex, L_1 is a compact subset of ω. We define a function $\tilde{f}$, holomorphic on a neighborhood of L, by

$$\tilde{f} = f \text{ on a neighborhood of } L_1$$
$$\tilde{f} = 1 \text{ on a neighborhood of } L_2 .$$

Since $\hat{L}_{\mathcal{H}(\Omega)} = L$ an application of Theorem 10.2 yields a function $g \in \mathcal{H}(\Omega)$ such that

$$(*) \qquad \sup_L |g-\tilde{f}| < \epsilon$$

and hence

$$\sup_K |g-f| < \epsilon$$

and (c) is proven.

(c) ⇒ (a). Since $\mathcal{H}(\Omega)$ is dense in $\mathcal{H}(\omega)$ for the topology of compact convergence, we immediately get that

$$\hat{K}_{\mathcal{H}(\Omega)} \cap \omega = \hat{K}_{\mathcal{H}(\omega)}$$

for each compact subset K of ω. And since ω is holomorphically convex, we conclude that $\hat{K}_{\mathcal{H}(\Omega)} \cap \omega$ is compact. This shows (a).

(a) ⇒ (b). Let K be a compact subset of ω and apply the proof of (a) ⇒ (c) to $f = 0$ and $\epsilon = 1/2$. Then (*) yields

$$\sup_K |g| < 1/2 < |g(a)|$$

for every $a \in L_2$. But this is impossible, unless L_2 is empty. Hence $L \subset \omega$ and (b) is proven. Since obviously (b) ⇒ (a) the proof of the corollary is now complete.

Q.E.D.

We refer to Noverraz [26] and Schottenloher [31] for similar results on holomorphic approximation. Noverraz deals with the case of Banach spaces with the Banach approximation property and Schottenloher deals with certain locally convex spaces admitting a finite dimensional Schauder decomposition.

11. THE LEVI PROBEM IN (DFC)-SPACES WITH THE APPROXIMATION PROPERTY

<u>11.1 Theorem</u>. Let $E = F'_c$ where F is a separable Fréchet space, and assume that E (or equivalently F) has the approximation property. Let Ω be a pseudoconvex open set in E. Then

$$\hat{K}_{\mathcal{P}s(\Omega)} = \hat{K}_{\mathcal{P}sc(\Omega)} = \hat{K}_{\mathcal{H}(\Omega)}$$

for each compact subset K of Ω. In particular Ω is holomorphically convex, and hence a domain of existence.

For the proof of the theorem we will need the following lemma, which is essentially due to Schottenloher: see the proof of [32,

Proposition 2].

<u>11.2 Lemma.</u> Let E be a locally convex space with the approximation property. Let Ω be a pseudoconvex open set in E. Let K be a compact subset of Ω and let $a \in \Omega \setminus \hat{K}_{\mathcal{P}s(\Omega)}$. Then for each compact set L with $K \cup \{a\} \subset L \subset \Omega$, there exists a function f, holomorphic on a neighborhood of L, and satisfying

$$|f(a)| > \sup_K |f|.$$

<u>Proof</u>. Since $a \in \Omega \setminus \hat{K}_{\mathcal{P}s(\Omega)}$ there exists a function $u \in \mathcal{P}s(\Omega)$ such that

$$u(a) > 0 \quad \text{and} \quad u(x) < 0 \quad \text{for all} \quad x \in K.$$

Since u is upper semicontinuous and since K and L are compact, we can find a convex, balanced 0-neighborhood V such that

$$L + V \subset \Omega$$

and

$$u(x) < 0 \quad \text{for all} \quad x \in K + V.$$

Let $T: E \to E$ be a continuous linear operator of finite rank such that

$$T(x) - x \in 1/2\, V \quad \text{for all} \quad x \in L.$$

Define a continuous affine operator $S: E \to E$ by

$$S(x) = T(x) + a - T(a) \qquad (x \in E).$$

Then

$$S(a) = a, \qquad S(K) \subset K + V, \qquad S(L) \subset L + V \subset \Omega$$

and we see that

$$u(S(a)) = u(a) > 0, \qquad u(S(x)) < 0 \quad \text{for all} \quad x \in K.$$

Let M be the vector subspace of E generated by $S(E)$. Then $u \in \mathcal{P}s(\Omega \cap M)$ and

$$S(a) \notin \widehat{S(K)}_{\mathcal{P}s(\Omega \cap M)}.$$

An application of Hormander [14, p.91, Theorem 4.3.4] tells us that

$$\widehat{S(K)}_{\rho s(\Omega \cap M)} = \widehat{S(K)}_{\mathcal{H}(\Omega \cap M)}$$

and we conclude that

$$S(a) \notin \widehat{S(K)}_{\mathcal{H}(\Omega \cap M)}.$$

Hence there exists a function $g \in \mathcal{H}(\Omega \cap M)$ such that

$$|g(S(a))| > \sup_{S(K)} |g|.$$

Let $f = g \circ S$. Since $S(L) \subset \Omega$ we see that f is holomorphic on a neighborhood of L and

$$|f(a)| > \sup_{K} |f|.$$

Q.E.D.

Proof of Theorem 11.1. It suffices to show that $\hat{K}_{\mathcal{H}(\Omega)} \subset \hat{K}_{\rho s(\Omega)}$. Let $a \in \Omega$, $a \notin \hat{K}_{\rho s(\Omega)}$. Let $L = \widehat{(K \cup \{a\})}_{\rho sc(\Omega)}$. Since Ω is $\rho sc(\Omega)$-convex, L is a compact subset of Ω. By Lemma 11.2 there exists a function f, holomorphic on a neighborhood of L, and satisfying

$$(*) \qquad |f(a)| > \sup_{K} |f|.$$

Since $\hat{L}_{\rho sc(\Omega)} = L$ we may apply the Oka-Weil Theorem 10.2 and approximate f, uniformly on L, by functions belonging to $\mathcal{H}(\Omega)$. Thus using $(*)$ we can find a function $g \in \mathcal{H}(\Omega)$ such that

$$|g(a)| > \sup_{K} |g|.$$

Hence $a \in \hat{K}_{\mathcal{H}(\Omega)}$.

Q.E.D.

Since every nuclear space has the approximation property, Theorem 11.1 implies in particular the main result of a recent article of Colombeau and the author: every pseudoconvex open set in a nuclear Silva space is a domain of existence; see [4, Theorem 2.1 and Example 3.2]. But of course there are spaces to which [4, Theorem 2.1] applies but Theorem 11.1 does not apply, e.g. any nonclosed vector subspace of a nuclear Silva space: see [4, Example 3.4].

ACKNOWLEDGEMENTS

I would like to thank Mário C. Matos for several helpful discussions on the subject of this paper.

REFERENCES

[1] Bremermann, H., Über die Äquivalenz der pseudokonvexen Gebiete und der Holomorphiegebiete im Raum von n Komplexen Veränderlichen. Math. Ann. 128 (1954), 63-91.

[2] Cartan, H. - Thullen, P., Regularitäts - und Konvergenzbereiche Math. Ann. 106 (1932), 617-647.

[3] Colombeau, J.F., Infinite dimensional C^∞ mappings with a given sequence of derivatives at a given point. J. Math. Anal. Appl. 71 (1979), 95-104.

[4] Colombeau, J.F. - Mujica, J., The Levi problem in nuclear Silva spaces. Ark. Mat. (to appear)

[5] Dineen, S., Surjective limits of locally convex spaces and their applications to infinite dimensional holomorphy. Bull. Soc. Math. France 103 (1975), 441-509.

[6] Dineen, S., Holomorphic functions on strong duals of Fréchet-Montel spaces. In: Infinite Dimensional Holomorphy and Applications, edited by M.C. Matos, pp. 147-166. North Holland, Amsterdam, 1977.

[7] Dineen, S. - Noverraz, Ph. - Schottenloher, M., Le problème de Levi dans certains espaces vectoriels topologiques localement convexes. Bull. Soc. Math. France 104 (1976), 87-97.

[8] Grauert, H. - Fritzsche, K., Several Complex Variables, Springer, New York, 1976.

[9] Gruman, L., The Levi problem in certain infinite dimensional vector spaces. Illinois J. Math. 18 (1974), 20-26.

[10] Gruman, L. - Kiselman, C.O., Le problème de Levi dans les espaces de Banach à base. C.R. Acad. Sc. Paris 274 (1972), Série A, 1296-1298.

[11] Hirschowitz, A., Prolongement analytique en dimension infinie. Ann. Inst. Fourier (Grenoble) 22 (1972), 255-292.

[12] Hollstein, R., (DCF)-Räume und lokalkonvexe Tensorprodukte. Arch. Math. (Basel) 29 (1977), 524-531.

[13] Hollstein, R., Tensorprodukte von stetigen linearen Abbildungen in (F) - und (DCF)-Räumen. J. reine angew. Math. 301 (1978), 191-204.

[14] Hörmander, L., An Introduction to Complex Analysis in Several Variables. North-Holland, Amsterdam, 1973.

[15] Horváth, J., Topological Vector Spaces and Distributions, Vol.I. Addison-Wesley, Reading, Massachusetts, 1966.

[16] Josefson, B., A counterexample in the Levi problem. In: Proceedings on Infinite Dimensional Holomorphy, edited by T.L. Hayden and T.J. Suffridge, pp. 168-177. Lecture Notes in Mathematics 364. Srpinger, Berlin, 1974.

[17] Kelley, J.L., General Topology. Van Nostrand, Princenton, New Jersey, 1955.

[18] Köthe, G., Topological Vector Spaces I. Springer, Berlin, 1969.

[19] Levi, E.E., Sulle ipersuperfici dello spazio a 4 dimensioni che possono essere frontiera del campo di esistenza di una funzioni analitica di due variabili complesse. Ann. Math. Pura Appl. (3) 18 (1911), 69-79.

[20] Ligocka, E., A local factorization of analytic functions and its applications. Studia Math. 47 (1973), 239-252.

[21] Mujica, J., The Oka-Weil theorem in locally convex spaces with the approximation property. Séminaire Paul Krée 1977-1978, exposé n° 3. Institut Henri Poincaré, Paris, 1979.

[22] Nachbin, L. Topology on Spaces of Holomorphic Mappings, Springer, Berlin, 1969.

[23] Norguet, F., Sur les domaines d'holomorphie des fonctions uniformes de plusieurs variables complexes. Bull. Soc. Math. France 82 (1954), 137-159.

[24] Noverraz, Ph., Pseudo-Convexité, Convexité Polynomiale et Domaines d'Holomorphie en Dimension Infinie. North-Holland, Amsterdam, 1973.

[25] Noverraz, Ph., Sur le théorème de Cartan-Thullen-Oka en dimension infinie. In: Séminaire Pierre Lelong Année 1971-1972, pp. 59-68. Lecture Notes in Mathematics 332. Springer, Berlin, 1973.

[26] Noverraz, Ph., Approximation of holomorphic or plurisubharmonic functions in certain Banach spaces. In: Proceedings on Infinite Dimensional Holomorphy, edited by T.L. Hayden and T.J. Suffridge, pp. 178-185. Lecture Notes in Mathematics 364. Springer, Berlin, 1974.

[27] Noverraz, Ph., Pseudo-convexité et base de Schauder dans les elc. In: Séminaire Pierre Lelong Année 1973-1974, pp. 63-82. Lecture Notes in Mathematics 474. Springer, Berlin, 1975.

[28] Oka, K., Sur les fonctions analytiques de plusieurs variables complexes. IX. Domaines finis sans point critique intérieur. Japan J. Math. 23 (1953), 97-155.

[29] Pomes, R., Solution du problème de Levi dans les espaces de Silva à base. C.R. Acad. Sc. Paris 278 (1974), Série A, 707-710.

[30] Schottenloher, M., Das Leviproblem in unendlichdimensionalen Räumen mit Schauderzerlegung. Habilitationsschrift, Universität München, 1974.

[31] Schottenloher, M., The Levi problem for domains spread over locally convex spaces with a finite dimensional Schauder decomposition. Ann. Inst. Fourier (Grenoble) 26 (1976), 207-237.

[32] Schottenloher, M., Polynomial approximation on compact sets. In: Infinite Dimensional Holomorphy and Applications, edited by M.C. Matos, pp. 379-391. North-Holland, Amsterdam, 1977.

[33] Schwartz, L., Produits tensoriels topologiques d'espaces vectoriels topologiques. Espaces vectoriels topologiques nucléaires. Applications. Séminaire Schwartz 1953-1954. Faculté des Sciences de Paris, 1954.

[34] Schwartz, L., Radon Measures on Arbitrary Topological Spaces and Cylindrical Measures. Tata Institute of Fundamental Research, Bombay, and Oxford University Press, London, 1973.

NOTE ADDED IN PROOF. Richard Aron and Philip Boland have kindly pointed out to me that the presentation in Section 7 could be shortened considerably. On one hand, Theorem 7.6 is a direct consequence of the Banach-Dieudonné Theorem, since each open (resp. closed) subset of a Hausdorff k-space is also a k-space for the induced topology. On the other hand, Lemma 7.8 can be easily proved directly, without any recourse to Lemma 7.7.

HOMOMORPHISMS OF ALGEBRAS OF GERMS OF HOLOMORPHIC FUNCTIONS

Olympia Nicodemi

Rochester Institute of Technology
Rochester, New York 14623 USA

1. INTRODUCTION

Let E and F be complex Banach spaces. Let K be a compact subset of E. We denote the algebra of germs of complex valued holomorphic functions defined on a neighborhood of K by $\mathcal{O}(K)$. When K contains only the origin of E, we denote $\mathcal{O}(K)$ by $\mathcal{O}(E)$. When K contains only the point x, we denote $\mathcal{O}(K)$ by $\mathcal{O}(x)$.

Let L be a compact subset of F. We say that an (algebra) homomorphism $A: \mathcal{O}(K) \to \mathcal{O}(L)$ is <u>induced</u> if there is an open set V containing L and a holomorphic map $g: V \to E$ such that $g(L) \subset K$ and, for each holomorphic function f defined on a neighborhood of K, the germ of $f \circ g$ (g perhaps restricted to a smaller neighborhood of L) is equal to the image of f under A. In this paper we determine conditions which insure that A is induced. When E and F are of finite dimension, K and L point sets, and A local, it is a straightforward exercise to show that A is necessarily induced. But we show that this is not so in general by establishing a procedure by which we can, for instance, extend any continuous linear map $A: E' \to F'$ between the duals of E and F to a homomorphism $\tilde{A}: \mathcal{O}(E) \to \mathcal{O}(F)$. The extension of any map A which is not the transpose of a continuous linear map from F to E cannot be induced.

Let U be an open subset of E and denote the space of holomorphic functions on U by $H(U)$. Let $H_c(U)$ denote $H(U)$ endowed with the compact open topology. Let $\mathcal{O}_c(K)$ denote the space $\mathcal{O}(K)$

endowed with the locally convex inductive limit topology with respect to the inclusions $H_c(U) \hookrightarrow \mathcal{O}(K)$, $K \subset U$. Any homomorphism $A: \mathcal{O}_c(K) \to \mathcal{O}_c(L)$ which is induced is continuous. Conversely, let A be continuous. We show that if E has the approximation property and K and L are point sets, then A is induced. If E has a Schauder basis, we show more generally that A is induced if K is polynomially convex.

The results of this paper appeared as part of the author's docotoral thesis undertaken at the University of Rochester under the direction of Professor David Prill.

2. THE SPACE $\mathcal{O}_\infty(K)$

Let $H_\infty(U)$ denote the subset of $H(U)$ of functions bounded on U. We endow $H_\infty(U)$ with the topology of uniform convergence on U whereby $H_\infty(U)$ is a Banach space. We let $\mathcal{O}_\infty(K)$ denote the space $\mathcal{O}(K)$ endowed with the locally convex inductive limit topology with respect to the inclusions $H_\infty(U) \hookrightarrow \mathcal{O}(K)$, $K \subset U$. (In [3], Mujica shows that $\mathcal{O}_\infty(K)$ is identical to the space $\mathcal{O}_\omega(K) = \varinjlim H_\omega(U)$, $K \subset U$, where $H_\omega(U)$ is $H(U)$ endowed with the ported topology.) We remark that the topological spaces $\mathcal{O}_c(K)$ and $\mathcal{O}_\infty(K)$ are identical if and only if E is of finite dimension.

Let $x \in X$ and $r > 0$. We let $B(x,r;E)$ denote the open ball of radius r about x. We let $K(r)$ denote the set $\bigcup \{B(x,r;E) \mid x \in K\}$. Let N denote the set of positive integers. Then for $\alpha = c$ or ∞ we have $\mathcal{O}_\alpha(K) = \bigcup H_\alpha(K(\frac{1}{i}))$, $i \in N$, $i > 0$.

<u>Proposition 1</u>. The spaces $\mathcal{O}_c(K)$ and $\mathcal{O}_\infty(K)$ are Hausdorff.

<u>Proof</u>. Let $x \in K$, $y \in E$ and $i \in N$. For all complex numbers t sufficiently small, the function $f(x+ty)$ is well-defined and analytic. Define a semi-norm p on $\mathcal{O}(K)$ by $p(f) = \left|\frac{d^i}{dt^i} f(x+ty)\Big|_{t=0}\right|$. Then p is continuous on $\mathcal{O}_\infty(K)$ and $\mathcal{O}_c(K)$. The set of all such semi-norms defines a Hausdorff topology on $\mathcal{O}(K)$ which proves the proposition.

3. A BASIC CONSTRUCTION

Let U be a neighborhood of the origin of F and let $A: E' \to H_\infty(U)$ be a continuous linear map of norm M such that $A(\ell)(0) = 0$ for each $\ell \in E'$. In this section we construct a continuous homomorphism $\tilde{A}: \mathcal{O}_\infty(E) \to \mathcal{O}_\infty(F)$ which extends A. We proceed inductively and construct maps $A_n: P(^nE) \to H_\infty(U)$, $n \in N$, where $P(^nE)$ is the space of continuous n-homogeneous polynomials from E to C. Let $f \in \mathcal{O}_\infty(E)$ and let f_i denote $\frac{d^i f(0)}{i!}$ for each $i \in N$. We show that $\sum_{i=0}^{\infty} A_i(f_i)$ converges in $\mathcal{O}_\infty(F)$. We define $\tilde{A}(f)$ to be $\sum_{i=0}^{\infty} A_i(f_i)$ and show that $\tilde{A}$ is continuous and multiplicative.

We begin by establishing our notation (in so far as possible we follow Nachbin in [5]) and by constructing an auxiliary map ψ_p. Let ^nE denote the n-fold Cartesian product of E with itself, $P(^nE;F)$ the space of n-homogeneous continuous polynomials from E to F, and $L(^nE;F)$ the space of n-linear continuous symmetric maps from ^nE to F. (We write simply $P(^nE)$ and $L(^nE)$ when $F = C$.) We let $\rho: P(^nE;F) \to L(^nE;F)$ denote the isomorphism obtained via the polarization formula and let r denote its inverse. Let $n > 1$ and let $i: L(^nE) \to L(^{n-1}E;E')$ denote the standard inclusion. For each $p \in P(^nE)$, let $\bar{p} = i \circ \rho(p)$. We have $\|\bar{p}\| \leq \|\rho(p)\|$.

Let $n > 1$ and $p \in P(^nE)$. We define $\psi_p: U \to L(^{n-1}E)$ by $\psi_p(y)(x) = A(p(x))(y)$ for each $x \in {}^{n-1}E$ and $y \in U$. We show that ψ_p is holomorphic by exhibiting its Taylor series about each point in U. Let $z \in U$ and choose $r > 0$ such that the closure of $B(z,r;F)$ is contained in U. Define $\psi_p(i,z) \in P(^1F;L(^{n-1}E))$ by $\psi_p(i,z)(y)(x) = \frac{d^iA}{i!}(\bar{p}(x))(z)(y)$ for each $x \in {}^{n-1}E$ and $y \in F$. We apply the Cauchy inequalities to $A(\bar{p}(x))$ to obtain that $\|\psi_p(i,z)\| \leq M \cdot \|\rho(p)\| \cdot (\frac{1}{r})^i$. Thus for any s, $0 < s < r$, the series $\sum_{i=0}^{\infty} \psi_p(i,z)(y-z)$ converges uniformly to ψ_p on $B(z,s;F)$ and we conclude that ψ_p is holomorphic.

We now construct the maps A_n. Recalling that $P(^0E) = C =$

$= P({}^0F)$, we let A_0 be the identity map on $\mathbb{C}$. We set $A_1 = A$ and proceed inductively. Let $n > 1$ and suppose, for each integer i, $1 \le i < n$, we have constructed A_i such that $\|A_i\| \le M^i \cdot \|\rho(p)\|$. Let $p \in P({}^nE)$ and $y \in U$. Let $\tilde{y}$ denote the map which sends each element f of $H_\infty(U)$ to $f(y)$. Define $G_p: U \times U \to \mathbb{C}$ by $G_p(x,y) = (A_{n-1} \circ r \circ \tilde{y} \circ A \circ \bar{p})(x)$ for each pair $(x,y) \in U \times U$. Fixing $y \in U$, the map which sends each $x \in U$ to $G_p(x,y)$ is holomorphic by the induction hypothesis. Fixing $x \in U$, the map which sends each $y \in U$ to $G_p(x,y)$ is just the composition of ψ_p with continuous linear maps and hence holomorphic. Since it is holomorphic in each variable separately, G_p is holomorphic. We define $A_n(p)(y)$ to be $G_p(y,y)$ for each $y \in U$. Since $\|\rho \circ r \circ \tilde{y} \circ A \circ p\| = \|\tilde{y} \circ A \circ \bar{p}\|$, it follows from the induction hypothesis that $\|A_n(p)\| \le M^n \cdot \|\rho(p)\|$.

We need the following three lemmas to show that $\tilde{A}$ is well-defined, continuous and multiplicative. We recall that for each $n \in N$ and $p \in P({}^nE)$, we have $\|\rho(p)\| \le \frac{n^n}{n!} \|p\| \le c^n \cdot \|p\|$.

<u>Lemma 2</u>. Let r and s be real numbers such that $0 < r < s$ and suppose that the closure of $B(0,s;F)$ is contained in U. Let $n \ge 0$ and $p \in P({}^nE)$. Then $\sup\limits_{\|x\|<r} |A_n(p)(x)| \le \left(\frac{M \cdot r}{s-r}\right)^n \cdot \|\rho(p)\| \le \left(\frac{M \cdot r \cdot e}{s-r}\right)^n \cdot \|p\|$.

<u>Proof</u> (by induction on n). The lemma is trivial for $n=0$. Let $n=1$ and let $\ell \in E'$. Since $A(\ell)(0) = 0$ we can find $f_i \in P({}^iF)$ for each $i \in N$ such that $A(\ell)(y) = \sum\limits_{i=1}^{\infty} f_i(y)$ for all $y \in B(0,s;F)$. By the Cauchy inequalities, we have $\|f_i\| \cdot r^i \le \sup\limits_{y \in U} |A(\ell)(y)| \cdot \left(\frac{r}{s}\right)^i \le M \cdot \|\ell\| \cdot \left(\frac{r}{s}\right)^i$. Thus $\sup\limits_{y \in B(0,r;F)} |A(\ell)(y)| \le \sum\limits_{i=1}^{\infty} \|f_i\| \cdot r^i \le M \cdot \|\ell\| \cdot \frac{r}{s-r}$ and the lemma is proved for $n=1$. Suppose $n > 1$ and the lemma holds for $n-1$. Let $p \in P({}^nE)$, $y \in B(0,r;F)$. From the first part of the proof we conclude that $\|\tilde{y} \circ A \circ \bar{p}\| \le \frac{r \cdot M}{s-r} \cdot \|\rho(p)\|$ where the norm of $\tilde{y} \circ A \circ \bar{p}$ is taken in $L({}^{n-1}E)$. From the induction hypothesis we have

$$|A_{n-1}(r \circ \tilde{y} \circ A \circ \bar{p})(y)| \le \left(\frac{M \cdot r}{s-r}\right)^{n-1} \cdot \|\rho \circ r \circ \tilde{y} \circ A \circ \bar{p}\| = \left(\frac{M \cdot r}{s-r}\right)^{n-1} \cdot \|\tilde{y} \circ A \circ \bar{p}\|.$$

Substituting, the lemma follows.

In what follows we shall denote the norm of $H_\infty(B(0,q;E))$ by $\|\cdot\|_q$.

<u>Lemma 3</u>. Let $f = \sum_{i=0}^{\infty} f_i$ be an element of $H_\infty(B(0,w;E))$ and let $0 < t < w$. Let $s > 0$ be such that the closure of $B(0,s;F)$ is contained in U. Let $r > 0$ be such that $0 < r < s$ and $\frac{M\cdot r\cdot e}{s-r} < t$. Then $\sum_{i=0}^{\infty} A_i(f_i)$ converges in $H_\infty(B(0,r;F))$ and $\|\sum_{i=0}^{\infty} A_i(f_i)\|_r \leq$ $\leq \|f\|_w \cdot \frac{w}{w-t}$.

<u>Proof</u>. By the Cauchy inequalities, $\sum_{i=0}^{\infty} \|f_i\|\cdot t^i \leq \|f\|_w \cdot \frac{w}{w-t}$. By Lemma 2, we have $\|A_i(f_i)\|_r \leq \left[\frac{M\cdot r\cdot e}{s-r}\right]^i \cdot \|f_i\| \leq \|f_i\|\cdot t^i$. Thus $\sum_{i=0}^{\infty} A_i(f_i)$ converges absolutely in $H_\infty(B(0,r;F))$ and $\|\sum_{i=0}^{\infty} A_i(f_i)\|_r$ $\leq \|f\|_w \cdot \frac{w}{w-t}$.

<u>Lemma 4</u>. Let $p \in P(^nE)$ and $q \in P(^mE)$. Then $A_{n+m}(p\cdot q) =$ $A_n(p)\cdot A_m(q)$.

<u>Proof</u> (by induction on the sum $k=m+n$). The lemma is trivial when $k = 0$ or 1. Let $k > 1$ and assume the lemma holds for $k-1$. Let $p \in P(^nE)$ and $q \in P(^mE)$ and assume $k = n+m$. Let $y \in U$. From the definition of A_m we have $A_n(p)\cdot A_m(q)(y) = A_n(p)(y)\cdot$ $A_{m-1}(r\circ\tilde{y}\circ A\circ\bar{q}))(y)$, the induction hypothesis, we obtain that $A_n(p)\cdot A_m(q)(y) =$

(1) $A_{n+m-1}(p(r\circ\tilde{y}\circ A\circ\bar{p})\cdot\bar{q})(y)$.

Similarly, from the definition of A_n, we obtain that $A_n(p)\cdot A_m(q)(y) =$

(2) $A_{n+m-1}((r\circ\tilde{y}\circ A\circ\bar{p})\cdot q)(y)$.

Let A denote the expression in line (1) and B that in line (2). Then $A_n(p)\cdot A_m(q)(y) = \frac{m}{n+m} A + \frac{n}{n+m} B$.

Let $(x_1,\ldots,x_{n+m})$ be an $n+m$ tuple such that $x_1 = x_2 =$ $=\ldots= x_{n+m-1} = x$ and let $x_{n+m} = z$. We note that $\rho(p\cdot q)(x,\ldots,x,z)$ $= \frac{n}{n+m}(\rho(p)(x,\ldots,z))\cdot q(x) + \frac{m}{n+m} p(x)\cdot(\rho(q)(x,\ldots,x,z))$. Composing with A we obtain $A\circ\overline{p\cdot q}\,(x,\dots,x)(y) = \frac{m}{n+m}\cdot p(x)\cdot(A\circ\bar{q}(x,\dots,x))(y)+$ $\frac{n}{n+m}\cdot q(x)\cdot A(\circ\bar{p}(x,\dots,x))(y)$. Thus $r\circ\tilde{y}\circ A\circ\overline{p\cdot q}(x) = \frac{m}{n+m}\cdot p(x)\cdot$ $(r\circ\tilde{y}\circ A\circ\bar{q}(x)) + \frac{n}{n+m}\cdot q(x)\cdot(r\circ\tilde{y}\circ A\circ\bar{p}(x))$. Finally, we have $A_{n+m}(p\cdot q)(y) = A_{n+m-1}(r\circ\tilde{y}\circ A\circ\overline{p\cdot q})(y) = \frac{n}{n+m}A_{n+m-1}(p\cdot(r\circ\tilde{y}\circ A\circ\bar{q}))(y) +$ $\frac{n}{n+m}A_{n+m-1}(q\cdot(r\circ\tilde{y}\circ A\circ p))(y) = \frac{m}{n+m}A + \frac{n}{n+m}B$. The lemma is thus proved.

<u>Proposition 5</u>. The map $\tilde{A}\colon \mathfrak{G}_\infty(E)\to\mathfrak{G}_\infty(F)$ defined by $\tilde{A}(f) = \sum_{i=0}^{\infty} A_i(f_i)$ for each $f = \sum_{i=0}^{\infty} f_i$ in $\mathfrak{G}_\infty(E)$ is a continuous homomorphism.

<u>Proof</u>. That $\tilde{A}$ is a homomorphism follows from Lemma 4. Let $w > 0$ and choose r as in Lemma 3. Then $\tilde{A}(H_\infty(B(0,w;E))\subset H_\infty(B(0,r;F))$ and the restriction of $\tilde{A}$ to $H_\infty(B(0,w;E))$ is continuous. Since $\mathfrak{G}_\infty(E)$ carries the inductive limit topology with respect to the inclusions $H_\infty(B(0,q;E))\subset\mathfrak{G}_\infty(E)$, $q > 0$, we conclude that $\tilde{A}$ is continuous.

We call $\tilde{A}$ the homomorphism constructed from A. We remark that if A is induced by composition with a holomorphic map, then $\tilde{A}$ is also.

4. A COUNTEREXAMPLE

Let E be a Banach space and let $A\colon E'\to E'''$ be the standard inclusion of E' into its double dual. It follows from the Hahn-Banach Theorem and the Banach Homomorphism Theorem that A is the transpose of a continuous linear map $f\colon E''\to E$ if and only if E is reflexive. Thus $\tilde{A}$ is induced if and only if E is reflexive.

5. INDUCED HOMOMORPHISMS

In this section we establish conditions on K and E which guarantee that a continuous homomorphism $A\colon \mathfrak{G}_c(K)\to\mathfrak{G}_c(L)$ is induced. We shall nee the following theorem whose proof is found in [1], page 268. It shall be referred to as Grothendieck's theorem.

<u>Theorem</u>. Let L be a Hausdorff, locally convex space and supposo

$L = \bigcup_{n \in N} H_n$ where each H_n is a Fréchet space and each inclusion map $i_n: H_n \to L$ is continuous. Let G be a Fréchet space and suppose that $f: G \to L$ is continuous. Then there is an $m \in N$ and a continuous map $\bar{f}: G \to H_m$ such that $i_m \circ \bar{f} = f$.

We note that $\mathfrak{G}_c(K) = \bigcup_{n=1}^{\infty} H_\infty(K(\frac{1}{n}))$ and that each inclusion map $i_n: H_\infty(K(\frac{1}{n})) \to \mathfrak{G}_c(K)$ is continuous. Let $K \subset E$ and $L \subset F$ be compact sets and let $A: \mathfrak{G}_c(K) \to \mathfrak{G}_c(L)$ be a continuous homomorphism. Fix $x \in L$. The map which sends each $\ell \in E'$ to $A(\ell)(x)$ is well-defined and continuous when E' carries the compact open topology. By the Mackey-Arens Theorem ([7], page 131), there is a unique $y(x) \in E$ such that $A(\ell)(x) = \ell(y(x))$ for each $\ell \in E'$. We define $\hat{A}: L \to E$ by $\hat{A}(x) = y(x)$ for each $x \in L$.

We say that a polynomial $p \in \mathfrak{G}(K)$ is of finite type if it is an element of the subalgebra of $\mathfrak{G}(K)$ generated by E'. We denote the subspace of polynomials of finite type by $P_f(E)$.

<u>Theorem 6</u>. Let $K \subset E$ and $L \subset F$ be compact sets and let $A: \mathfrak{G}_c(K) \to \mathfrak{G}_c(L)$ be a continuous algebra homomorphism. Suppose that (a) $P_f(E)$ is dense in $\mathfrak{G}_c(K)$ and (b) $\hat{A}(L) \subset K$. Then A is induced.

<u>Proof</u>. Let $i: E' \to \mathfrak{G}_c(K)$ denote the standard inclusion. The map $A \circ i$ is then continuous when E' carries the norm topology. We can thus apply Grothendieck's theorem to obtain an $m \in N$ and a continuous map $\bar{A}: E' \to H_\infty(L(\frac{1}{m}))$ such that $i_m \circ \bar{A} = A \circ i$. Denote $L(\frac{1}{m})$ by U and let $x \in U$. Define $\tilde{x}: H_\infty(U) \to C$ by $\tilde{x}(f) = f(x)$ for each $f \in H_\infty(U)$. We will show that $\tilde{x} \circ \bar{A}$ is weakly continuous for each $x \in U$. Since E is complete we need only show that $x \circ \bar{A}$ is continuous on equicontinuous subsets of E'. (See [7], page 149.) We recall that the equicontinuous subsets of E' are precisely the strongly bounded subsets of E'. Let $(\ell_\alpha)_{\alpha \in A}$ be a bounded net in E' which converges weakly to 0. The net $(\bar{A}(\ell_\alpha))_{\alpha \in A}$ converges in

$\mathcal{O}_c(L)$ and is bounded in $H_\infty(U)$ since $\bar{A}$ is continuous and $(\ell_\alpha)_{\alpha \in A}$ strongly bounded. Let t be such that $0 < t < \frac{1}{m}$. We apply a theorem proved by Mujica in [3], page 144, to conclude that the net $(\bar{A}(\ell_\alpha))_{\alpha \in A}$ converges to 0 in $H_\infty(L(t))$. Thus for each $x \in L(t)$, $(\tilde{x} \circ \bar{A}(\ell_\alpha))_{\alpha \in A}$ converges to 0. Since each $x \in U$ is contained in $L(t)$ for some t, $0 < t < \frac{1}{m}$, $\tilde{x}$ is weakly continuous for each $x \in U$.

Let $x \in U$. Since each weakly continuous linear functional on E' is given by evaluation at a point of E, we can find an element $y(x) \in E$ such that $\tilde{x} \circ \bar{A}(\ell) = \ell(y(x))$ for each $\ell \in E'$. Let $\phi: U \to E$ be defined by $\phi(x) = y(x)$ for each $x \in U$. The composition $\ell \circ \phi$ is holomorphic for each $\ell \in E'$ since $\ell \circ \phi = \bar{A}(\ell)$. Thus ϕ is holomorphic. Now ϕ agrees with $\hat{A}$ on L and so by hypothesis (b), $\phi(L) \subset K$. Let $\gamma: \mathcal{O}_c(K) \to \mathcal{O}_c(L)$ be defined by $\gamma(f) = f \circ \phi$ for each $f \in \mathcal{O}_c(K)$. Then γ is continuous and agrees with A on E' and hence on $P_f(E)$. By hypothesis (a), $P_f(E)$ is dense in $\mathcal{O}_c(K)$ and we have that γ and A agree everywhere. We conclude that A is induced.

For the rest of this section we investigate conditions which guarantee that conditions (a) and (b) of Theorem 6 hold.

Let $P(E)$ denote the set of polynomials on E and let $K \subset E$ be compact. The polynomial hull of K is the set $\hat{K} = \{x \in E \mid |p(x)| \le \sup_{y \in K} |p(y)|$ for all $p \in P(E)\}$. The set K is said to be polynomially convex if $K = \hat{K}$. We note that if E has the approximation property, then $\hat{K} = \{x \in E \mid |p(x)| \le \sup_{y \in K} |p(y)|$ for all $p \in P_f(E)\}$.

The following proposition is proved by Mujica in [4] for E a Fréchet space.

<u>Proposition 10</u>. Suppose that E has the approximation property and that $K \subset E$ is compact and polynomially convex. Then any homomorphism $\phi: \mathcal{O}(K) \to \mathbb{C}$ is given by evaluation at a point of K.

Corollary 8. Suppose that E has the approximation property and that $K \subset E$ is compact and polynomially convex. Let $L \subset F$ be compact and $A: \mathcal{O}_c(K) \to \mathcal{O}_c(L)$ be a continuous homomorphism. Then $\hat{A}(L) \subset K$.

Proposition 9. Let E have the approximation property and let $A: \mathcal{O}_c(E) \to \mathcal{O}_c(F)$ be a continuous homomorphism. Then A is induced.

Proof. We need only remark that in this case $P_f(E)$ is dense in $\mathcal{O}(E)$ and that a point set is polynomially convex.

We now show that if E has a Schauder basis and $K \subset E$ is compac t and polynomially convex, then $P_f(E)$ is dense in $\mathcal{O}_c(K)$. Let $\underline{N}$ denote the natural numbers. Let $\{e_i\}_{i \in \underline{N}}$ be a Schauder basis for E and let E_j denote the span of $\{e_1, \ldots, e_j\}$. Let $\pi_j: E \to E_j$ be the natural projection. We may assume that E is normed so that $\|\pi_i\| = 1$ for each $i \in \underline{N}$. Let $\hat{K}^j = \{x \in E \mid |p \circ \pi_j(x)| \leq \sup_{y \in K} |p \circ \pi_j(y)|$ for all $p \in P(E)\}$. Let $K_j = \{x \in E_j \mid p(x) \leq \sup_{y \in \pi_j(K)} |p(y)|\}$. The following two propositions are found in [6], the first on page 73, the second on page 76.

Proposition 10. $\hat{K}^j = K_j \oplus \ker \pi_j$.

Proposition 11. For each $i \in \underline{N}$, $\pi_i(K) = \bigcap_{j \in N} \pi_i(\hat{K}^j)$.

Lemma 12. Let E have a Schauder basis and $K \subset E$ be compact. The functions of the form $f \circ \pi_i$, $i \in \underline{N}$, $f \in \mathcal{O}_c(K)$, are dense in $\mathcal{O}_c(K)$.

The lemma follows from a straightforward compactness argument once we notice that we can find a neighborhood U of K and an $m > 0$ such that for all $i > m$, $f \circ \pi_i$ is defined on U.

Theorem 13. Let E have a Schauder basis and let $K \subset E$ be compact and polynomially convex. The polynomials of finite type are dense in $\mathcal{O}_c(K)$.

Proof. Let $f \in \mathcal{O}(K)$ and suppose that $i \in N$ is such that $f \circ \pi_i \in \mathcal{O}(K)$. It follows from the preceding lemma that the theorem will be proved if we can find a neighborhood W of K such that $f\circ\pi_i$ can be uniformly approximated on compact subsets of W by polynomials of finite type.

Suppose that $f\circ\pi_i$ is defined on a neighborhood U of K. By Proposition 11, we have $\pi_i(K) = \bigcap_{j\in\underline{N}} \pi_i(\hat{K}^j)$ which is the intersection of a nested sequence of sets. For $j \geq i$, $\pi_i(\hat{K}^j)$ is compact and we can thus find $k > i$ such that $\pi_i(\hat{K}^k) \subset \pi_i(U)$. For $j \geq i$, we have $\pi_i(\hat{K}^j) = \pi_i(K_j)$ and so we have $K_k \subset \pi_i(U) \oplus Ce_{i+1} \cdots \oplus Ce_k$. Considered as a subset of the space E_k spanned by $\{e_1,\ldots,e_k\}$, K_k is polynomially convex. Thus given any open set $G \subset E_k$ which contains K_k, we can find a polynomially convex open set V such that $K_k \subset V \subset G$. In particular, we can find V such that $V \subset \pi_i(U) \oplus Ce_{i+1} \cdots \oplus Ce_k$. We note that $K \subset V \oplus \ker(\pi_k)$. By a theorem of Oka, (see [6], page 76), we have that the restriction of $f\circ\pi_i$ to V can be uniformly approximated on compact subsets of V by polynomials defined on E_k. We show that $f\circ\pi_i$ can be approximated on compact subsets of $V \oplus \ker \pi_k$ by polynomials of finite type.

Let $L \subset V \oplus \ker \pi_k$ be compact. Then $\pi_k(L) \subset V$ and is compact. Let $r > 0$. We can find a polynomial of the form $p\circ\pi_k$ such that $|f\circ\pi_i(x) - p\circ\pi_k(x)| < r$ for each $x \in \pi_k(L)$. Since whenever $k \geq i$, $\pi_i = \pi_i\circ\pi_k$, we have that for any $x \in L$, $|f\circ\pi_i(x) - p\circ\pi_k(x)| = |f\circ\pi_i(\pi_k(x)) - p\circ\pi_k(\pi_k(x))| < r$. Since $p\circ\pi_k$ is of finite type, the theorem is proved.

6. GERMS OF THE NUCLEAR TYPE

The theory of holomorphy types allows one to consider natural subspaces of the space of all holomorphic functions on a given open set. Results often obtain for these subspaces which do not do so generally. With this in mind we consider homomorphisms of rings of germs of the nuclear type. For notation, definitions and general results regarding holomorphy types, we refer the reader to [5], and for results specific to the nuclear type, to [2].

Let θ be a holomorphy type from E to $\mathbb{C}$, and let $\mathcal{O}_\theta(E)$ denote the subspace of $\mathcal{O}(E)$ of germs of the type θ at the origin of E. Let $\sigma \geq 1$ be as per the definition of θ (see [5], page 34), and let $r > 0$. Let $H_\theta(r;E) = \{f \mid f \in \mathcal{O}_\theta(E)$ and $\sum_{i=0}^{\infty} \|\frac{d^i f(0)}{i!}\|_\theta r^i \sigma^i < \infty\}$. It is routine to check that $H_\theta(r;E)$ is a Banach space under the norm $\|f\|_{r,\theta} = \sum_{i=0}^{\infty} \|\frac{d^i f(0)}{i!}\|_\theta r^i \sigma^i$ for each $f \in H_\theta(r;E)$. We have $\mathcal{O}_\theta(E) = \bigcup_{r>0} H_\theta(r;E)$. We endow $\mathcal{O}_\theta(E)$ with the locally convex inductive limit topology with respect to the inclusions $H_\theta(r;E) \hookrightarrow \mathcal{O}_\theta(E)$, $r > 0$. Since the inclusions $P_\theta({}^n E) \hookrightarrow P({}^n E)$ are continuous and of norm $\leq \sigma^n$ for each $n \in N$, we have that each $f \in H_\theta(r;E)$ is defined and bounded on $B(0,r;E)$ and that the inclusion $H_\theta(r;E) \hookrightarrow H_\infty(B(0,r;E))$ is continuous. It follows that the inclusion $\mathcal{O}_\theta(E) \hookrightarrow \mathcal{O}_\infty(E)$ is continuous and that $\mathcal{O}_\theta(E)$ is Hausdorff. We note that if $\theta = \omega$, the current holomorphy type, then $\mathcal{O}_\omega(E) = \mathcal{O}_\infty(E)$.

We now sketch results on algebra homomorphisms and the nuclear holomorphy type $\mathfrak{n}$. Let $n \in N$ and $p \in P_\mathfrak{n}({}^n E)$. Suppose that p can be expressed as the product of n linear functionals $t_1, \ldots, t_n$. It follows from the polarization formula that $\|p\|_\mathfrak{n} \leq \frac{n^n}{n!} \prod_{i=1}^{n} \|t_i\|$ (see [2], page 3). More generally, if $p_j \in P_\mathfrak{n}({}^{i_j}E)$, $j = 1, \ldots, q$, and if $t = \sum_{i=1}^{q} i_j$, then $\|\prod_{j=1}^{q} p_j\|_\mathfrak{n} \leq \frac{t^t}{t!} \cdot \prod_{j=1}^{q} \|p_j\|_\mathfrak{n}$. From this and from the fact that $\frac{t^t}{t!} < e^t$, it can be shown that if $f_j \in H_\mathfrak{n}(r;E)$, $j = 1, \ldots, q$, and $f = \prod_{j=1}^{q} f_j$, then $f \in H_\mathfrak{n}(\frac{r}{e};E)$ and $\|f\|_{\mathfrak{n},\frac{r}{e}} \leq \prod_{j=1}^{q} \|f_j\|_{\mathfrak{n},r}$. Thus we obtain:

<u>Proposition 14</u>. $\mathcal{O}_\mathfrak{n}(E)$ is an algebra.

<u>Theorem 15</u>. Let $A: E' \to H_\mathfrak{n}(r;F)$ be a continuous linear map such that $A(\ell)(0) = 0$ for each $\ell \in E'$. Let i denote the inclusion map $H_\mathfrak{n}(r;F) \hookrightarrow H_\infty(r;F)$. Let $\psi: \mathcal{O}_\infty(E) \to \mathcal{O}_\infty(F)$ be the map construct-

ed from $i \circ A$ as in Section 3 and let ψ_n denote its restriction to $\mathcal{O}_n(E)$. Then $\psi_n(\mathcal{O}_n(E)) \subset \mathcal{O}_n(F)$ and $\psi_n: \mathcal{O}_n(E) \to \mathcal{O}_n(F)$ is continuous. Further any continuous homomorphism from $\mathcal{O}_n(E)$ to $\mathcal{O}_n(F)$ which agrees with A on E' agrees with ψ_n on all of $\mathcal{O}_n(E)$.

<u>Proof</u>. Let $n \in N$ and let $M = \|A\|$. A simple density argument will show that if $f \in P_n(^nE)$, then $\psi(f) \in H_n(\frac{r}{e};F)$, $\|\psi(f)\|_{\frac{r}{e},n} \leq M^n \cdot \|f\|_{r,n}$ and that the order of $\psi(f) \geq n$ at 0.

Let $t = \frac{r}{e}$, $f \in H_n(s;E)$, and let $f_i = \frac{d^i f(0)}{i!}$. Choose $w > 0$ so that $\frac{w}{t} \leq \min(1, \frac{s}{M})$. Then $\|\psi(f_i)\|_{n,w} \leq (\frac{w}{t})^i \|\psi(f_i)\|_{n,t} \leq (\frac{s}{M})^i \cdot M^i \cdot \|f_i\|_n$. Thus $\sum_{i=0}^{\infty} \psi(f_i)$ is Cauchy in $H_n(w;F)$. It converges to $\psi(f)$ in $H_\infty(B(0,w;F))$ and hence to $\psi(f)$ in $H_n(w;F)$ also.

We conclude that $\psi_n(\mathcal{O}_n(E)) \subset \mathcal{O}_n(F)$ and that $\psi_n: \mathcal{O}_n(E) \to \mathcal{O}_n(F)$ is continuous. Further, any other algebra homomorphism which agrees with A on E' must agree with ψ_n on the dense set $P_f(E)$ and hence on all of $\mathcal{O}_n(E)$. The theorem is proved.

We note that for any Banach space G, $G' = P_n(^1G)$. Let $A: E' \to F'$ be a continuous linear map which is not the transpose of a map from F to E. Let $\psi_n: \mathcal{O}_n(E) \to \mathcal{O}_n(F)$ be as in Theorem 15. Then ψ_n is not induced. Conversely, we also apply Theorem 15 to prove:

<u>Theorem 16</u>. Let E be reflexive and let $A: \mathcal{O}_n(E) \to \mathcal{O}_n(F)$ be a continuous algebra homomorphism. Then A is induced.

REFERENCES

[1] Grothendieck, A.: Espaces vectoriels topologiques, Second Edition, Sociedade Matemática de São Paulo, 1958.

[2] Gupta, C.P.: Malgrange theorem for nuclearly entire functions of bounded type on a Banach space, Thesis, University of Rochester, 1966.

[3] Mujica, J.: Spaces of Germs of Holomorphic Functions, Advances in Mathematics, to appear.

[4] Mujica, J.: Ideals of holomorphic functions on Fréchet Spaces, Advances in Holomorphy (Editor: J.A. Barroso), North-Holland, to appear

[5] Nachbin, L.: Topology on Spaces of Holomorphic Mappings, Springer Verlag, Berlin, 1969.

[6] Noverraz, Ph.: Pseudo-convexité, Convexité polynomiale et Domaines d'holomorphie en Dimension Infinie, North-Holland, Amsterdam, 1973.

[7] Schaeffer, H.H.: Topological Vector Spaces, Springer-Verlag, Berlin, 1971.

ON THE SPECTRA OF NON-ARCHIMEDEAN FUNCTION ALGEBRAS

João B. Prolla

Instituto de Matemática
Universidade Estadual de Campinas
Campinas, Brasil

0. INTRODUCTION

Throughout this paper X stands for a zero-dimensional Hausdorff space, and F stands for a rank-one non-archimedean valued field.

Let E be a <u>non-archimedean normed algebra over</u> F; that is, E is a linear algebra over F equipped with a non-archimedean norm $v \to \|v\|$ satisfying

$$\|uv\| \leq \|u\| \cdot \|v\|$$

for all $u, v \in E$. When E has an identity e, we shall assume that

$$\|e\| = 1.$$

If X is compact, then the space $C(X;E)$ of all continuous E-valued functions on X, with pointwise operations and sup-norm

$$\|f\| = \sup \{\|f(x)\| ; x \in X\}$$

becomes a non-archimedean normed algebra over F too. Now the problem arises of characterizing the <u>closed right</u> (resp. <u>left</u>) <u>ideals in</u> $C(X;E)$.

If the algebra E has an identity, then <u>all</u> ideals are $C(X;F)$-modules, and we characterized in [4] all closed right (resp. left) ideals in $C(X;E)$.

When the algebra E has no identity, then all <u>regular</u> ideals are $C(X;F)$-modules and an argument similar to that of [4] can be used to characterize all closed regular right (resp. left) ideals. (See Theorem 1.1 below.)

When E has no identity, one seeks conditions on E that imply that all ideals are $C(X;F)$-modules. When studying the problem of characterizing all closed right (resp. left) ideals in $C(X;E)$ in the case of real or complex E, Kaplansky ([1], Theorem 3.1) imposed the following condition:

(*) for every $v \in E$, v lies in the closure of vE.

In this paper we extend (see Theorem 1.4) Kaplansky's theorem to the non-archimedean case. As in Kaplansky's paper, there is no extra difficulty in dealing with vector fibrations and spaces of cross-sections. The argument rests in proving that under condition (*) all closed ideals are $C(X;F)$-modules, and then applying an approximation theorem valid for $C(X;F)$-modules.

In §2 we apply the characterization of all closed two-sided regular ideals to the problem of finding the spectrum of $C(X;E)$. Indeed, the kernel of a continuous algebra homomorphism of E onto F is regular, and we can apply Theorem 1.1 to show that $X \times \Delta(E)$ is homeomorphic to $\Delta(C(X;E))$, where $\Delta(E)$ (resp. $\Delta(C(X;E))$) denotes the set of all non-zero continuous algebra homomorphisms of E (resp. $C(X;E)$) onto F, topologized with the relative weak topology. (See Theorem 2.1.) The only restriction we impose is that the field F be non-trivially valued.

In a subsequent paper [5] we will treat the case of R or C with their usual valuations and weighted algebras of functions. It is to be noticed that we do not use tensor product techniques, so our argument applies to spaces X which are not k-spaces and E need not be complete.

1. CLOSED IDEALS

Let X be compact and assume that, for each $x \in X$, a non-archimedean normed space E_x over F has been given. The product set $\pi\{E_x;\ x \in X\}$ is made into a vector space over F in the usual way, and its elements are called <u>cross-sections over</u> X.

We shall restrict our attention to vector spaces A of cross-sections over X which satisfy the following conditions:

(a) <u>For every</u> $f \in A$, <u>the function</u> $x \mapsto \|f(x)\|$ <u>is upper-semi-continuous</u>;

(b) <u>for every</u> $x \in X$, $A(x) = E_x$, <u>where</u> $A(x) = \{f(x);\ f \in A\}$;

(c) A <u>is a</u> $\mathcal{C}(X;F)$<u>-module, that is, for each</u> $\varphi \in \mathcal{C}(X;F)$ <u>and</u> $f \in A$, <u>the cross-section</u> $x \mapsto \varphi(x)f(x)$ <u>belongs to</u> A.

Since we have assumed that X is compact, it follows from condtion (a) alone that A is a non-archimedean normed space over F, when equipped with the sup-norm

$$f \mapsto \|f\| = \sup\{\|f(x)\|;\ x \in X\}.$$

When each E_x is an algebra over F, the product of two cross-sections f and g is the cross-section $x \mapsto f(x)g(x)$. If each E_x is a non-archimedean normed algebra over F, then every subalgebra A of cross-sections which satisfies condition (a) is a non-archimedean normed algebra over F under the sup-norm.

The following result is an easy consequence of Corollary 9.28 of [3]. Indeed, since X is zero-dimensional, the algebra $\mathcal{C}(X;F)$ is separating over X.

<u>Theorem 1.0.</u> Let A be a vector space of cross-sections over X, satisfying conditions (a) to (c) above, and let $W \subset A$ be a $\mathcal{C}(X;F)$-submodule. Then, for each $f \in A$, f belongs to the closure of W if, and only if, $f(x)$ belongs to the closure of $W(x) = \{g(x);g\in W\}$ in E_x, for each $x \in X$.

Suppose now that, for every $x \in X$, a closed regular right (resp. left) ideal I_x in E_x is given in such a way that we can find a cross-section $g \in A$ such that $g(x)$ is a left (resp. right) identity modulo I_x, for each $x \in X$. Define

$$I = \{f \in A;\ f(x) \in I_x \text{ for all } x \in X\}.$$

Manifestly, I is a closed right (resp. left) ideal in A, and g is a left (resp. right) identity modulo I in A; that is, I is regular. We shall prove that any closed regular right (resp. left) ideal in A has the above form. Namely we have the following.

<u>Theorem 1.1</u>. Let A be a subalgebra of cross-sections satisfying conditions (a) to (c) above. Any closed regular right (resp. left) ideal I in A has the following form: for every $x \in X$ a closed regular right (resp. left) ideal I_x in E_x is given in such a way that we can find $g \in A$ such that $g(x)$ is a left (resp. right) identity modulo I_x in E_x for each $x \in X$; and I consists of all those $f \in A$ such that $f(x) \in I_x$ for each $x \in X$.

<u>Proof</u>. Let I be any closed regular right (resp. left) ideal in A. Let $g \in A$ be a left (resp. right) identity modulo I. Let I_x be the closure of $I(x) = \{f(x);\ f \in I\}$ in E_x; then I_x is a closed right (resp. left) ideal in E_x and $g(x)$ is a left (resp. right) identity modulo I_x in E_x for each $x \in X$. Clearly, I is contained in the set

$$J = \{f \in A;\ f(x) \in I_x \text{ for all } x \in X\}.$$

Conversely, let $f \in J$. By Theorem 1.0, $f \in \bar{I} = I$, if I is a $C(X;F)$-module. This is indeed the case. To see this let $h \in I$ and $a \in C(X;F)$ be given. Then, by condition (c), $ag \in A$, and therefore $(ag)h \in I$. Now g is a right identity modulo I, so the cross-section $g(ah) - ah$ belongs to I. But, since a is scalar-valued, we have $g(ah) = (ag)h$. Therefore $ah \in I$. The case of a left ideal is similar and the proof is complete.

Let us now study the case in which I is not assumed to be regular. Following Kaplansky [1], let us assume that each E_x satisfies condition (*) of §0, i.e.

(*) <u>for every</u> $v \in E_x$, v <u>lies in the closure of</u> vE_x.

Clearly, if E_x has an identity, then condition (*) is satisfied. Let us give an example of an algebra without identity, satisfying (*). Let X be a <u>locally compact</u> Hausdorff space which is <u>not</u> compact. Let $C_\infty(X;F)$ be the space of all continuous functions that vanish at infinity, i.e. for each $f \in C_\infty(X;F)$ given $\epsilon > 0$ there exists a compact subset $K \subset X$ for which $|f(x)| < \epsilon$ whenever $x \notin K$. Then $E = C_\infty(X;F)$ is a non-archimedean normed algebra over F, without identity, when equipped with pointwise operations and sup-norm. If we consider the one point compactification of X, $Y = X \cup \{\infty\}$, the mapping $T: C_\infty(X;F) \to C(Y;F)$ given by

$$Tf(x) = f(x), \quad \text{if} \quad x \in X$$

$$Tf(\infty) = 0$$

is an isometry and an algebra homomorphism. The image of $C_\infty(X;F)$ is the subalgebra of $C(Y;F)$ consisting of all those $g \in C(Y;F)$ such that $g(\infty) = 0$. Using the above isometry and the Stone-Weierstrass theorem for compact spaces (e.g. Theorem 1.3 of Prolla [4]) one has the following.

<u>Theorem</u>. Let $A \subset C_\infty(X;F)$ be a subalgebra and $f \in C_\infty(X;F)$. Then f belongs to the uniform closure of A if, and only if, the following two conditions are satisfied:

(i) given $x \in X$, with $f(x) \neq 0$, there exists $g \in A$ such that $g(x) \neq 0$;

(ii) given $x,y \in X$, with $f(x) \neq f(y)$, there exists $g \in A$ such that $g(x) \neq g(y)$.

Let us prove that any subalgebra $A \subset C_\infty(X;F)$ satisfies condition (*). Let $f \in A$ and $\epsilon > 0$ be given. The set $K = \{x \in X; |f(x)| \geq \epsilon\}$ is compact and open. Hence, its F-characteristic function ξ_K belongs to $C_\infty(X;F)$. Now it is easy to see that ξ_K satisfies (i) and (ii) above. Therefore ξ_K belongs to the closure of A. Let $g \in A$ be chosen so that $\|\xi_K - g\| < \epsilon/(\|f\|+1)$. Then we have

$$\|gf - f\| = \|gf - \xi_K f + \xi_K f - f\| \leq$$

$$\leq \max(\|g - \xi_K\| \cdot \|f\|, \|\xi_K f - f\|) < \epsilon .$$

Hence f belongs to the closure of fA, and A satisfies (*), q.e.d.

<u>Proposition 1.2</u>. Let A be a subalgebra of cross-sections satisfying conditions (a) to (c) above. If each E_x satisfies condition (*), then so does A.

<u>Proof</u>. Let $W = fA$, where $f \in A$ has been given. Since A is a $C(X;F)$-module, so is W. We claim that $f \in \overline{W}$. By Theorem 1.0, it is enough to prove that $f(x) \in \overline{W(x)}$ for each $x \in X$. By condition (b), $A = E_x$. Hence $W(x) = f(x)E_x$. By condition (*), applied to $v = f(x)$, we see that $f(x)$ lies in the closure of $W(x)$ as desired.

<u>Lemma 1.3</u>. Let A be a subalgebra of cross-sections satisfying conditions (a) to (c) above. Let us assume that A satisfies condition (*). Then every closed right (resp. left) ideal in A is a $C(X;F)$-module.

<u>Proof</u>. Let I be a closed right (resp. left) ideal in A. Let $a \in C(X;F)$ and $h \in I$ be given. Then $ah \in A$ and by condition (*), ah lies in the closure of ahA. Since $ahA \subset hA \subset I$ this shows that ah lies in the closure of I. Since I is closed, $ah \in I$. That is I is a $C(X;F)$-module.

<u>Theorem 1.4</u>. Let A be as in Lemma 1.3. Then, any closed right (resp. left) ideal I in A has the following form: for every $x \in X$ a closed right (resp. left) ideal I_x in E_x is given, and I consists of all those $f \in A$ with $f(x) \in I_x$ for each $x \in X$.

<u>Proof</u>. Let I be a closed right (resp. left) ideal in A. Let I_x

be the closure in E_x of $I(x) = \{f(x);\ f \in I\}$. Then I_x is a closed right (resp. left) ideal in E_x. Clearly, I is contained in

$$J = \{f \in A;\ f(x) \in I_x \text{ for all } x \in X\}.$$

Conversely, let $f \in J$. Since I is a $\mathbb{C}(X;F)$-module by Lemma 1.3, then by Theorem 1.0, $f \in \bar{I}$. Since I is closed, $f \in I$.

Corollary 1.5. Under the hypothesis of Theorem 1.4 assume that each E_x is simple (i.e. has no proper closed two-sided ideal). Then a subset $I \subset A$ is a closed two-sided ideal in A if, and only if, there exists a closed subset $N \subset X$ such that

$$I = \{f \in A;\ f(x) = 0 \text{ for all } x \in N\}.$$

Proof. Let $N \subset X$ be a closed subset. Let us define

$$I = \{f \in A;\ f(x) = 0 \text{ for all } x \in N\}.$$

Clearly I is a two-sided ideal in A, which is closed.

Conversely, let $I \subset A$ be a closed two-sided ideal in A and let

$$N = \{x \in X;\ f(x) = 0 \text{ for all } f \in I\}.$$

Clearly N is closed in X, and if we set

$$W = \{f \in A;\ f(x) = 0 \text{ for all } x \in N\},$$

then $I \subset W$. Let now $f \in W$. Assume, by contradiction, that $f \notin I$. By Theorem 1.4 there is some point $x \in X$, such that $f(x) \notin I_x = \overline{I(x)}$. Since E_x is simple, $I_x = \{0\}$. Therefore $f(x) \neq 0$, and $f \in W$ implies that $x \notin N$. However, $I(x) \subset I_x = \{0\}$, so $x \in N$. Hence $f \in I$.

Corollary 1.6. Under the hypothesis of Corollary 1.5, if $M \subset A$ is a proper closed two-sided ideal which is maximal, then $M = \{f \in A;\ f(x) = 0\}$ for some $x \in X$.

Proof. Let $M \subset A$ be a proper closed two-sided ideal which is maximal. Since M is proper, there is some $x \in X$ such that $\overline{M(x)} \neq E_x$.

Since E_x is simple, $\overline{M(x)} = \{0\}$. Let $W \subset A$ be the set $W = \{f \in A;\ f(x) = 0\}$. Clearly, W is a two-sided ideal. Since $M \subset W$, and M is maximal, either $M = W$ or $W = A$. We claim that $W = A$ is imposible. Indeed, if $X = \{x\}$, then $A = E_x$ is simple, and there are no proper closed two-sided ideals in A. Since $M \subset A$ is by hypothesis one such proper closed two-sided ideal, we conclude that $X \neq \{x\}$. Let then $y \in X$ be such that $y \neq x$. Since X is zero-dimensional, there is a clopen neighborhood U of X, with $y \notin U$. Let ξ_U be the F-characteristic function of U. Then $\xi_U \in C(X;F)$. Let $v \in E_x$ be chosen with $v \neq 0$. This is possible, because $\{0\} \subset \overline{M(x)} \neq E_x$. By condition (b), choose $f \in A$, with $f(x) = v$. By condition (c), $g = \xi_U f \in A$. Now $g(x) = v \neq 0$; hence $g \notin W$. Therefore $W \neq A$, as claimed.

<u>Remark</u>. The algebra $E = F$ has an identity and therefore satisfies trivially property (*). On the other hand, F is a field, and thus it has no proper ideals. Therefore we can apply both Corollaries 1.5 and 1.6 to the algebra $A = C(X;F)$. Hence we have

<u>Theorem 1.7</u>. Let X be a compact zero-dimensional Hausdorff space, and let F be a rank-one non-archimedean valued field. A subset $I \subset C(X;F)$ is a closed ideal if, and only if, there exists a closed subset $N \subset X$ such that

$$I = \{f \in C(X;F);\ f(x) = 0 \text{ for all } x \in N\}.$$

Moreover, if X is not a singleton, I is a proper closed maximal ideal if, and only if,

$$I = \{f \in C(X;F);\ f(x) = 0\}$$

for some $x \in X$.

<u>Proof</u>. The only statement that does not follow from Corollaries 1.5 and 1.6 is the last one. Let then $x \in X$ and define

$$I = \{f \in C(X;F);\ f(x) = 0\}.$$

Since $C(X;F)$ is strongly separating,

$$0 \subsetneq I \subsetneq C(X;F)$$

in the case that X is not a singleton. Hence $I = \delta_x^{-1}(0)$ is the kernel of a nontrivial homomorphism of $C(X;F)$ onto its underlying field F and therefore it is a maximal ideal. Since δ_x is continuous, I is closed.

2. THE SPECTRUM OF $C(X;E)$

Throughout this section we assume that the field F is non-trivially valued and that $E \neq \{0\}$.

For any normed algebra E over F, let $\Delta(E)$ be the set of all non-zero continuous algebra homomorphisms of E onto F. The kernel of each $h \in \Delta(E)$ is then a closed two-sided regular maximal ideal. Since the field F is non-trivially valued, the elements of $\Delta(E)$ are bounded (see Lemma 1, section 3.3, [2]) and we define

$$\|h\| = \sup \{\frac{|h(v)|}{\|v\|}; \ v \neq 0\}$$

for all $h \in \Delta(E)$. We claim that $\|h\| \leq 1$.

Let $v \neq 0$. If $v^n = 0$ for some $n \in \mathbb{N}$, then $[h(v)]^n = h(v^n) = 0$. If $v^n \neq 0$ for all $n \in \mathbb{N}$, then $0 < \|v^n\| \leq \|v\|^n$. Hence

$$\frac{|h(v)|^n}{\|v\|^n} \leq \frac{|h(v^n)|}{\|v^n\|} \leq \|h\|.$$

Thus $\frac{|h(v)|}{\|v\|} \leq \|h\|^{1/n}$, for all $n \in \mathbb{N}$. Since h is non-zero, $\|h\| > 0$ and therefore $\|h\|^{1/n} \to 1$, when $n \to \infty$. Hence $\|h\| \leq 1$, and $\Delta(E)$ is a subset of the unit ball of E'. We consider in $\Delta(E)$ the relative weak topology $\sigma(E',E)$.

Let us consider, for each $x \in X$ the evaluation map

$$\delta_x \colon C(X;E) \to E$$

defined by $f(x) = \delta_x(f)$

for all $f \in C(X;E)$. This is an algebra homomorphism of $C(X;E)$ onto E, which is clearly bounded:

$$\|\delta_x(f)\| = \|f(x)\| \leq \|f\|$$

for all $f \in C(X;E)$. This shows that for each $h \in \Delta(E)$, the composition $h\circ\delta_x$ belongs to $\Delta(C(X;E))$. Let us denote by G the map

$$(x,h) \mapsto h\circ\delta_x$$

from $X \times \Delta(E)$ into $\Delta(C(X;E))$. Hence

$$G(x,y)(f) = h(f(x))$$

for all $f \in C(X;E)$. We shall prove that G is a homeomorphism between $X \times \Delta(E)$ and $\Delta(C(X;E))$.

<u>Theorem 2.1</u>. The mapping G is a homeomorphism between $X \times \Delta(E)$ and $\Delta(C(X;E))$.

<u>Proof</u>. The proof will be divided into several steps.

STEP 1: G <u>is onto</u> $\Delta(C(X;E))$.

<u>Proof</u>. Let $H \in \Delta(C(X;E))$ be given. Its kernel $I = H^{-1}(0)$ is a proper closed two-sided maximal ideal in $C(X;E)$, which is regular. By Theorem 1.1, there exists some point $x \in X$ such that $I_x = \overline{I(x)}$ is a proper ideal. Choose $f \in C(X;E)$ such that $H(f) = 1$. Then the function f is an identity modulo I. For each $u \in E$, let $u^* = uf \in C(X;E)$, and define $h: E \to F$ by setting

$$h(u) = H(u^*)$$

for all $u \in E$. Clearly, $h \in E'$, the dual of E. Now if $u,v \in E$, then

$$\begin{aligned} h(uv) &= H((uv)^*) = H(uvf) = \\ &= H(fuvf) = H(fu)h(v) = \\ &= H(fuf)h(v) = h(u)h(v). \end{aligned}$$

Hence h is multiplicative. Let $J \subset E$ be the kernel of h. Let $u \in J$. Then $u^* \in I$. Choose $g \in C(X;E)$ with $g(x) = u$. Then

$gf - g$ belongs to I and therefore $uf(x) - u$ belongs to $I(x)$. Now $uf(x) = u^*(x) \in I(x)$. Hence $u \in I(x)$. This shows that $J \subset I(x)$. Since I_x is proper, $h \neq 0$; that is $h \in \Delta(E)$. Let W be the kernel of $h \circ \delta_x$ in $\mathcal{C}(X;E)$. We claim that $I \subset W$. Indeed, let $g \in \mathcal{C}(X;E)$ be such that $g \notin W$. Hence $g(x) \notin J$. Now, since J is a proper closed maximal ideal and $J \subset I_x$, it follows that $J = I_x$. This shows that $g(x) \notin I_x$. By Theorem 1.1, it follows that $g \notin I$. Since I is maximal, in fact we have $I = W$. This shows that H and $h \circ \delta_x$ have the same kernel. Since both are multiplicative, $H = h \circ \delta_x = G(x,h)$.

STEP 2: G <u>is one-to-one</u>.

<u>Proof</u>. Let $(x,h) \neq (y,k)$. If $x = y$, then $h \neq k$. Choose $u \in E$ with $h(u) \neq k(u)$, and then choose $g \in \mathcal{C}(X;E)$ with $g(x) = u$. It follows that

$$G(x,h)(g) = h(g(x)) = h(u) \neq k(u) =$$
$$= k(g(y)) = G(y,k)(g).$$

If $x \neq y$, choose $u \in E$ with $k(u) = 1$, and then choose $g \in \mathcal{C}(X;E)$ with $g(x) = 0$ and $g(y) = u$. Now we have $G(x,h)(g) = h(g(x)) = 0$ and $G(y,k)(g) = k(g(y)) = k(u) = 1$.

STEP 3: G <u>is continuous</u>.

<u>Proof</u>. Let $(x_o,h_o) \in X \times \Delta(E)$, $\varepsilon > 0$ and $g \in \mathcal{C}(X;E)$ be given. Let

$$N = \{h \in \Delta(E);\ |(h-h_o)(g(x_o))| < \varepsilon\}.$$

Choose a neighborhood U of x_o in X such that $\|g(x)-g(x_o)\| < \varepsilon$ for all $x \in U$. Let $(x,h) \in U \times N$. Then

$$|G(x,h)(g) - G(x_o,h_o)(g)| =$$
$$= |h(g(x)) - h_o(g(x_o))| =$$
$$= |h(g(x)) - h(g(x_o)) + h(g(x_o)) - h_o(g(x_o))| \leq$$
$$\leq \max\{|h(g(x) - g(x_o))|,\ |(h - h_o)(g(x_o))|\} < \varepsilon$$

because $h \in N$ implies $|(h-h_o)(g(x_o))| < \epsilon$, and $x \in U$ implies $|h(g(x) - g(x_o))| \leq \|g(x) - g(x_o)\| < \epsilon$, since $\|h\| \leq 1$.

STEP 4: G^{-1} <u>is continuous</u>.

<u>Proof</u>. Let $H_\alpha \to H$ in $\Delta(C(X;E))$. Since G is onto, there exist nets $\{x_\alpha\}$ in X and $\{h_\alpha\}$ in $\Delta(E)$ such that $H_\alpha = G(x_\alpha, h_\alpha)$; by the same reason there is $(x,h) \in X \times \Delta(E)$ such that $H = G(x,h)$.

Since $H \neq 0$, there is some $f \in C(X;E)$ such that $H(f) = 1$. Choose α_o such that $\alpha \geq \alpha_o$ implies $|H_\alpha(f)| > 0$. Let $g \in C(X;F)$ be given. For all $\alpha \geq \alpha_o$ we have

$$g(x_\alpha) = \frac{g(x_\alpha)h_\alpha(f(x_\alpha))}{h_\alpha(f(x_\alpha))} = \frac{H_\alpha(gf)}{H_\alpha(f)}.$$

Therefore $g(x_\alpha) \to H(gf) = h(g(x)f(x) = g(x)h(f(x)) = g(x)H(f) = g(x)$. Hence $g(x_\alpha) \to g(x)$ for all $g \in C(X;F)$. We claim that $x_\alpha \to x$ in X. Indeed, for any neighborhood N of x in X, let $W \subset N$ be a clopen neighborhood of x. Let $g \in C(X;F)$ be the F-characteristic function of W. Choose β_o such that $\alpha \geq \beta_o$ implies $|g(x_\alpha)-1| < 1$. This is possible because $g(x_\alpha) \to g(x) = 1$. Now $g(x_\alpha) = 0$ is impossible for all $\alpha \geq \beta_o$, and so $x_\alpha \in W \subset N$ for all such $\alpha \geq \beta_o$.

Similarly, given $u \in E$, then

$$h_\alpha(u) = \frac{h_\alpha(u)h_\alpha(f(x_\alpha))}{h_\alpha(f(x_\alpha))} = \frac{H_\alpha(uf)}{H_\alpha(f)}$$

for all $\alpha \geq \alpha_o$. Therefore $h_\alpha(u) \to H(uf) = h(u(f(x)) = h(u)h(f(x)) = h(u)H(f) = h(u)$. Hence $h_\alpha \to h$ in the topology of $\Delta(E)$.

<u>Corollary 2.2</u>. Let $(F,|\cdot|)$ be a non-archimedean non-trivially valued field. Then the map $\delta: X \to \Delta(C(X;F))$ defined by

$$\delta(x)(f) = f(x)$$

for all $f \in C(X;F)$ is a homeomorphism between X and $\Delta(C(X;F))$.

<u>Proof</u>. Take $E = F$ in Theorem 2.1. Since $\Delta(F) = \{h\}$, where $h: F \to F$ is the identity map on F, X is homeomorphic with $X \times \Delta(F)$.

If we define δ as the composition of $x \mapsto (x,h)$ with G, then

$$\delta(x)(f) = G(x,h)(f) = f(x)$$

for all $x \in X$, $f \in C(X;F)$.

Corollary 2.3. Let $(F,|\cdot|)$ be as in Corollary 2.2 and let X and Y be two compact 0-dimensional Hausdorff spaces. If $C(K;F)$ and $C(Y;F)$ are isomorphic as topological algebras, then X and Y are homeomorphic.

Proof. Let T be a topological algebra isomorphism from $C(X;F)$ onto $C(Y;F)$. For each element $h \in \Delta(C(Y;F))$, clearly $h \circ T \in \Delta(C(X;F))$. We claim that the map $\varphi: h \mapsto h \circ T$ is a homeomorphism between the two spectra.

(1) φ is one-to-one, because T is onto.

(2) φ is onto $\Delta(C(X;F))$:

Indeed, let $k \in \Delta(C(X;F))$. By Corollary 2.2, $k = \delta_x$ for some $x \in X$. Let $J = k^{-1}(0)$. J is a maximal closed two-sided ideal in $C(X;F)$. Since T is a topological algebra isomorphism, $T(J)$ is a maximal closed two-sided ideal in $C(Y;F)$. By Corollary 1.6, there exists $y \in Y$ such that

$$T(J) = \{g \in C(Y;F);\ g(y) = 0\}.$$

Hence $T(J) = \delta_y^{-1}(0)$. Let $h = \delta_y$. We claim that $k = h \circ T$. The kernel of $h \circ T$ is $\{f \in C(X;F);\ h(Tf) = 0\} = \{f \in C(X;F);(Tf)(y)=0\}$, which contains J. Hence $(h \circ T)^{-1}(0) = J$, because both J and $(h \circ T)^{-1}(0)$ are maximal closed two-sided ideals. Since k and $h \circ T$ are multiplicative and have the same kernel, they coincide.

(3) φ is continuous.

Let $h_\alpha \to h$ in $\Delta(C(Y;F))$. Then $h_\alpha(f) \to h(f)$, for all f in $C(Y;F)$, and $h_\alpha(Tg) \to h(Tg)$, for all g in $C(X;F)$. Therefore $h_\alpha \circ T \to h \circ T$ in $\Delta(C(X;F))$.

(4) φ^{-1} is continuous.

Since φ^{-1} is the map $k \mapsto k \circ T^{-1}$ from $\Delta(C(X;F))$ onto $\Delta(C(Y;F))$, the proof of (4) is similar to that of (3) and therefore it is omitted.

Hence $\Delta(C(Y;F))$ and $\Delta(C(X;F))$ are homeomorphic. By Corollary 2.2, X and Y are now homeomorphic.

REFERENCES

[1] Kaplansky, I., The structure of certain operator algebras, Trans. Amer. Math. Soc. 70 (1951), 219-255.

[2] Narici, L., Beckenstein, E., and Bachman, G., Functional Analysis and Valuation Theory, Marcel Dekker, Inc. New York 1971.

[3] Prolla, J.B., Approximation of Vector Valued Functions, North Holland Publ. Co. Amsterdam, 1977.

[4] Prolla, J.B., Non-archimedean function spaces. In Linear Spaces and Approximation. Edited by P.L. Butzer and B.Sz.-Nagy, ISNM 40, Birkhäuser Verlag Basel, 1978, 101-117.

[5] Prolla, J.B., Topological algebras of vector-valued continuous functions, to appear in Mathematical Analysis and Applications, Volume 7 of the Advances in Mathematics Supplementary Studies, L. Nachbin, editor. Academic Press, Inc., 1980.

AN EXAMPLE OF THE BARRELLED SPACE ASSOCIATED TO $C(X;E)$

Jean Schmets

Institut de Mathématique
Université de Liège
Avenue des Tilleuls, 15
B-4000 LIEGE/BELGIQUE

SUMMARY

Let X be a Hausdorff completely regular space and E be a Hausdorff locally convex topological vector space. Then $C(X;E)$ [resp. $C_s(X;E)$] denotes the space of the continuous functions on X with values in E when endowed with the compact-open (resp. the simple or pointwise) topology. We essentially show that if E is a Fréchet space, then $C(\mu X;E)$ is the barrelled space associated to $C_s(X;E)$. In particular if X is a μ-space and E a Fréchet space, then $C(X;E)$ is barrelled. We also mention some corresponding results for the evaluable case.

INTRODUCTION AND NOTATIONS

Let X be a Hausdorff completely regular space (in short a c.r. space). Then $\mathcal{C}(X)$ stands for the linear algebra of the continuous functions on X (with values in $\mathbb{R}$ or in $\mathbb{C}$). A subset B of X is <u>bounding</u> if every $f \in \mathcal{C}(X)$ is bounded on B. Of course every relatively compact subset of X is bounding but the converse is not true in general. If every bounding subset of X is relatively compact, X is a μ-<u>space</u>. To X, Buchwalter [2] has associated a μ-space μX as the smallest μ-space containing X as a dense topological space (up to a homeomorphism, it is unique and can be considered as a topological subspace of the Stone-Čech compactification βX of X). Moreover he has shown that every continuous map from X into another c.r. space Y has a unique continuous extension from

μX into μY, and proved that every paracompact space is a μ-space.

In what follows E is a Hausdorff locally convex topological vector space (in short a l.c. space) which system of continuous semi-norms is P.

Then $C(X;E)$ is the linear space of the continuous functions on X with values in E. Of course it can be endowed with many different locally convex topologies to become a l.c. space. We shall essentially consider the following two:

a) $C(X;E)$ is $C(X;E)$ with the compact-open topology.

b) $C_s(X;E)$ is $C(X;E)$ with the simple or pointwise topology.

So a basis of the continuous semi-norms of $C(X;E)$ [resp. $C_s(X;E)$] is given by the set of the semi-norms

$$\|\varphi\|_{p,A} = \sup_{x\in A} p[\varphi(x)], \quad \forall \varphi \in C(X;E),$$

where p runs through P and A through the family of the relatively compact (resp. the finite) subsets of X.

Remark. If E is a μ-space, the linear spaces $C(X;E)$ and $C(\mu X;E)$ are isomorphic and we shall identify them. However let us insist immediately on the fact that the l.c. spaces $C(X;E)$ and $C(\mu X;E)$ [resp. $C_s(X;E)$ and $C_s(\mu X;E)$] are essentially distinct if X is not a μ-space.

Here we deal essentially with the following questions: when is $C(X;E)$ barrelled? What are the barrelled spaces associated (in the sense of Komura [5]) to $C_s(X;E)$ and to $C(X;E)$? and with the corresponding questions relative to the evaluable case.

When E is equal to $\mathbb{R}$ or $\mathbb{C}$, we simply write $C(X)$ and $C_s(X)$ instead of $C(X;E)$ and $C_s(X;E)$. Nachbin [7] and Shirota [14] have proved that $C(X)$ is barrelled if and only if X is a μ-space.

Buchwalter [2] has shown that $C(\mu X)$ is the barrelled space associated to $C(X)$ and Buchwalter-Schmets [3] have got that $C(\mu X)$ is even the barrelled space associated to $C_s(X)$ - therefore $C_s(X)$ is barrelled if and only if every bounding subset of X is finite.

EXAMPLES OF BARRELLED $C(X;E)$ SPACES .

Let us first mention the following result which gives necessary conditions for $C(X;E)$ to be barrelled.

Proposition. If $C(X;E)$ is barrelled, then $C(X)$ and E are barrelled.

It is a part of Proposition V.3 of [11].

The following very interesting example is due to S. Dierolf of the University of Munich (Germany). It shows that by no means the converse of this proposition is true, even when X is compact.

Example (S. Dierolf). a) If E is a linear space of uncountable dimension over $\mathbb{K}$ ($=\mathbb{R}$ or $\mathbb{C}$) provided with the finest locally convex topology, then the space $c_o(E)$ is not evaluable (= quasi-barrelled or infra-barrelled).

b) If moreover $K = \mathbb{N} \cup \{\infty\}$ is the Alexandroff compactification of the discrete space $\mathbb{N}$, then $C(K;E)$ is not evaluable, hence neither barrelled, nor bornological.

Proof. a) Let us recall first that $c_o(E)$ is the linear space

$$\{(x_n)_{n\in\mathbb{N}} \in E^{\mathbb{N}} : \ x_n \to 0 \ \text{ in } \ E\}$$

endowed with the topology of uniform convergence.

By hypothesis, there is an uncountable set I such that E may be identified with the locally convex direct sum $\bigoplus_{i\in I} \mathbb{K}$. For every $j \in I$, let $p_j : \bigoplus_{i\in I} \mathbb{K} \to \mathbb{K}$ denote the j-th canonical projection.

Now let us consider the following set

$$T = \bigcap_{m\in\mathbb{N}} \ \bigcap_{\substack{i_1,\ldots,i_m\in I \\ |\{i_1,\ldots,i_m\}|=m}} \{(x_n)_{n\in\mathbb{N}} \in c_o(E) : |\sum_{\ell=1}^{m} p_{i_\ell}(x_\ell)| \le 1\}.$$

It is clear that T is absolutely convex and closed in $c_o(E)$.

Moreover T is bornivorous in $c_o(E)$. Let A be a bounded subset of $c_o(E)$. Then of course there is a bounded subset B of E such that $A \subset B^{\mathbb{N}} \cap c_o(E)$. Furthermore there exists a finite subset $J \subset I$ such that $p_i(x) = 0$ for every $i \in I \setminus J$ and every $x \in B$. Therefore for every $(x_n)_{n\in\mathbb{N}} \in A$, every $m \in \mathbb{N}$ and every $i_1,\ldots,i_m \in I$ such that $|\{i_1,\ldots,i_m\}| = m$, we get

$$\left|\sum_{\ell=1}^{m} p_{i_\ell}(x_\ell)\right| = \left|\sum_{\substack{1\le \ell\le m \\ i_\ell\in J}} p_{i_\ell}(x_\ell)\right|$$

$$\le |J| \cdot \sup_{i\in J}\sup_{x\in B} |p_i(x)| = M < \infty.$$

Thus $\frac{1}{M+1} A \subset T$.

To conclude, let us show that T is not a neighborhood of 0 in $c_o(E)$. Assume on the contrary the existence of a 0-neighborhood U in E such that $U^{\mathbb{N}} \cap c_o(E) \subset T$. For every $i \in I$, there exists $x_i \in U$ such that $p_i(x_i) > 0$. As I is uncountable, there is then $\varepsilon > 0$ and a sequence $(i_n)_{n\in\mathbb{N}}$ of distinct elements of I such that $p_{i_n}(x_{i_n}) > \varepsilon$ for every $n \in \mathbb{N}$. Now choose $m \in \mathbb{N}$ such that $m > 1/\varepsilon$ for every $z = (z_n)_{n\in\mathbb{N}}$ by $z_\ell = x_{i_\ell}$ $(1 \le \ell \le m)$ and $z_\ell = 0$ $(\ell > m)$. Of course z belongs to $U^{\mathbb{N}} \cap c_o(E)$ but we have

$$\left|\sum_{\ell=1}^{m} p_{i_\ell}(z_\ell)\right| > m\varepsilon > 1,$$

whence $z \notin T$, which is contradictory.

b) This is an easy consequence of a) since the map T from $C(K;E)$ into $c_o(E) \times E$ defined by

$$T\varphi = ((\varphi(n)-\varphi(\infty))_{n\in\mathbb{N}}, \varphi(\infty))$$

is an isomorphism onto, as can be easily verified.

In this paragraph it is now our purpose to list the examples we know so far of barrelled $C(X;E)$ spaces.

Of course if $X = K$ is compact and if E is a Banach or a

Fréchet space, then $C(K;E)$ is a Banach or a Fréchet space and hence is barrelled.

One can also obtain easily that if X is a $k_{\mathbb{R}}$-space which is σ-compact and if E is a Banach or a Fréchet space, then $C(X;E)$ is a Fréchet space.

But leaving these trivial examples aside, it seems rather difficult to get cases when $C(X;E)$ is barrelled.

Mujica [6] has shown that if $X = K$ is compact and if E is a Hausdorff compactly regular inductive limit of an increasing sequence of Banach spaces $(E_n)_{n\in\mathbb{N}}$, then the equality

$$C(K;E) = \operatorname{ind}_{n\in\mathbb{N}} C(K;E_n)$$

holds algebraically and topologically. Moreover his proof does not in fact require that the spaces E_n $(n \in \mathbb{N})$ be Banach spaces. Therefore if $X = K$ is compact and if E is a strict inductive limit of a sequence of Banach or Fréchet spaces, then $C(K;E)$ is barrelled (it is even ultrabornological).

Recently Hollstein [4] obtained the following results:

a) if X is locally compact and if E is a complete bornological (DF) space, then $C(X;E)$ is barrelled and (DF).

b) if X is a $k_{\mathbb{R}}$-space such that every countable union of compact subsets of X is relatively compact in X and if E is an inductive limit of a sequence of quasi-complete spaces E_n "stark beschränkt-retraktiv" (= every bounded subset of E is contained and bounded in one of the E_n's, and for every $n \in \mathbb{N}$ there is $m \geq n$ such that on every bounded subset of E_n the topologies of E_m and E are equivalent. Cf. [1] for this notion), then one has $C(X;E) = \operatorname{ind} C(X;E_n)$.

Finally let us state the following example mentioned already in [12]. It is rather different from the previous ones since X needs no longer be compact, locally compact or a space in which compact subsets play an important role. We shall recall its proof because it will permit us to mention a few results which are helpful in this theory.

<u>Theorem</u>. If $C(X)$ is barrelled (i.e. if X is a μ-space) and if E is a Banach or a Fréchet space, then $C(X;E)$ is barrelled.

The proof is based on the following facts which are of independant interest.

a) A result of Nachbin concerning $C(X)$ (cf. [7]) can be extended to $C(X;E)$ in the following way (cf. Theorem V.4.4 of [9] and Proposition 3.9 of [10]).

Let D be an absolutely convex subset of $C(X;E)$. Then

i) there is a smallest compact subset $K(D)$ of the Stone-Čech compactification βX of X such that $\varphi \in C(X;E)$ belongs to D if its unique continuous extension $\tilde{\varphi}$ from βX into βE is identically 0 on some neighborhood of $K(D)$ in βX.

ii) if there are $p \in P$ and $r > 0$ such that

$$D \supset \{\varphi \in C(X;E) : p[\varphi(x)] \le r, \ \forall x \in X\},$$

then $K(D)$ is also the smallest compact subset K of βX such that $\varphi \in C(X;E)$ belongs to D if $\tilde{\varphi}$ is identically 0 on K. Moreover one has then

$$D \supset \{\varphi \in C(X;E) : [p(\varphi)]^{\sim}(x) \in [0,r[, \ \forall x \in K(D)\},$$

where $[p(\varphi)]^{\sim}$ is the unique continuous extension of $p(\varphi)$ from βX into $\beta \mathbb{R}$.

b) The Singer theorem [13] characterizing the dual of $C(K;E)$ when K is compact and E a Banach space can be easily adapted to the general space $C(X;E)$ (cf. Theorem II.3 of [11]) and yields the following result.

Let τ be a linear functional on $C(X;E)$. Then $p \in P$, $C > 0$ and the compact subset K of X are such that

$$|\tau(\varphi)| \leq C\|\varphi\|_{p,K}, \quad \forall \varphi \in C(X;E),$$

if and only if there is a countably additive regular Borel measure m_τ on X with values in the polar set of $\{e \in E : p(e) \leq 1/C\}$, such that

$$\sup_{\beta \in P(X)} \sum_{b \in \beta} \|m_\tau(b)\|_p \leq C$$

where $P(X)$ is the set of the finite Borel partitions of X, such that $m_\tau(b) = 0$ for every Borel subset b of X disjoint from K and such that $\tau(\varphi) = \int \varphi \, dm_\tau$ for every $\varphi \in C(X;E)$. Moreover the representation of τ by such a measure m_τ is unique.

c) By use of a) and b), it is possible to associate a kind of support to every $\tau \in C(X;E)'$.

In fact, for every $\tau \in C(X;E)'$, the set

$$D_\tau = \{\varphi \in C(X;E) : |\tau(\varphi)| \leq 1\}$$

is absolutely convex and one can the consider the set $\operatorname{supp} \tau = K(D_\tau)$. Theorem IV.1 of [11] states that $\operatorname{supp} \tau$ is a compact subset of X and that $x \in X$ belongs to $\operatorname{supp} \tau$ if and only if every neighborhood V of x in X contains a Borel subset b of X such that $m_\tau(b) \neq 0$.

d) Finally Proposition IV.3 and IV.4 of [11] give the following results about the set

$$\operatorname{supp} \beta = \bigcup_{\tau \in \beta} \operatorname{supp} \tau,$$

where β is a subset of $C(X;E)'$.

If β is a weak*-bounded subset of $C(X;E)'$, then $\operatorname{supp} \beta$ is a bounding subset of X. If β is a strongly bounded subset of $C(X;E)'$, then $\{n \in \mathbb{N} : G_n \cap \operatorname{supp} \beta \neq \emptyset\}$ is finite for every sequence $G_{n\in\mathbb{N}}$ of open subsets of X such that $\{n \in \mathbb{N} : G_n \cap K \neq \emptyset\}$ is finite for every compact subset K of X.

With these results in mind we now come to the

Proof of the theorem. Let T be a barrel of $C(X;E)$.

Since T is absolutely convex and absorbs the fast converging sequences of $C(X;E)$, and since E is a Banach or a Fréchet space, Proposition 5.3 of [10] gives the existence of $p \in P$ and $r > 0$ such that

$$T \supset \{\varphi \in C(X;E) : p[\varphi(x)] \leq r, \quad \forall\, x \in X\}.$$

In this way we are in position to apply the most powerful part of the result of Nachbin and to get therefore a compact subset $K(T)$ of βX such that

$$T \supset \{\varphi \in C(X;E) : [p(\varphi)]^{\sim}(x) \in [0,r[, \quad \forall\, x \in K(T)\}.$$

So all we need to know to conclude is that, in fact, $K(T)$ is a subset of X because then T will contain a semi-ball of $C(X;E)$.

Of course the polar set T° is weak*-bounded in $C(X;E)'$. Therefore $\operatorname{supp} T^{\circ}$ is relatively compact in X.

Now let $\varphi \in C(X;E)$ be equal to 0 on $\overline{\operatorname{supp} T^{\circ}}$. Of course we have then $\tau(\varphi) = 0$ for every $\tau \in T^{\circ}$ and therefore φ belongs to T. This means that $K(T)$ is contained in the compact subset $\overline{\operatorname{supp} T^{\circ}}$ of X.

Hence the conclusion.

AN EXAMPLE OF THE BARRELLED SPACE ASSOCIATED TO $C(X;E)$

Let us first recall the definition given by Komura [5] of the barrelled associated space.

Consider a l.c. space (E,τ), τ being its topology. It is well known that E endowed with the finest l.c. topology is ultra-bornological, hence barrelled. Therefore the inductive limit $(E,\tau)_t$ of the spaces (E,τ') where τ' runs through the family of the l.c. topologies on E which at the same time are finer than τ and make (E,τ') barrelled makes sense. It is a barrelled l.c. space which topology is finer than τ and such that there is no l.c. topology on E which is finer than τ and strictly coarser than the one of $(E,\tau)_t$, for which E is barrelled. The space $(E,\tau)_t$ is called

<u>the barrelled space associated to</u> (E,τ); it can be seen as "the closest barrelled space to (E,τ)".

Moreover Komura has given a characterization of $(E,\tau)_t$ by use of a transfinite recursion, that we shall follow to prove part b) of the following example.

<u>Theorem</u>. a) If $X = K$ is compact and if E is a compactly regular inductive limit of an increasing sequence of Banach or Fréchet spaces, then $C(K;E)$ is the barrelled space associated to $C_s(K;E)$.

b) If E is a Banach or a Fréchet space, then $C(\mu X;E)$ is the barrelled space associated to $C_s(X;E)$ [and therefore to $C(X;E)$].

<u>Proof</u>. a) By Mujica's result we know that $C(K;E)$ is barrelled. Therefore, as $C(K;E)$ has a finer topology than $C_s(K;E)$, the topology of $C_s(K;E)_t$ is coarser then the one of $C(K;E)$. On the opposite, for every $p \in P$ and every $r > 0$, the set

$$\{\varphi \in C(K;E) : \|\varphi\|_{p,K} \leq r\} = \bigcap_{x \in K} \{\varphi \in C(K;E) : p[\varphi(x)] \leq r\}$$

is clearly a barrel of $C_s(K;E)$ and must therefore be a neighborhood of 0 in $C_s(K;E)_t$.

b) Again it is clear that the topology of $C_s(X;E)_t$ is coarser than the one fo $C(\mu X;E)$. Now for every bounding subset B of X, every $p \in P$ and every $r > 0$, the set

$$\{x \in C(X;E) : p[\varphi(x)] \leq r, \quad \forall\, x \in B\}$$

is certainly a barrel of $C_s(X;E)$ and must therefore be a neighborhood of 0 in $C_s(X;E)_t$. So we can proceed as Buchwalter [1] did to get the barrelled space associated to $C(X)$ and claim that one has

$$C_s(X;E)_t = C(X;E)_t = C_b(X;E)_t$$

where $C_b(X;E)$ is the space $C(X;E)$ endowed with the locally convex topology determined by the semi-norms $\|\cdot\|_{p,B}$ where $p \in P$ and B is any bounding subset of X. So we introduce the subspace

$$X_1 = \bigcup \{\bar{B}^{\beta X} : B \text{ is a bounding subset of } X\}$$

of βX and get

$$C_s(X;E)_t \subset C_b(X_1;E) \subset C_s(X;E).$$

We now proceed by transfinite recursion by claiming

$$X_{\alpha+1} = \bigcup \{\bar{B}^{\beta X} : B \text{ is a bounding subset of } X_\alpha\}$$

for every ordinal α and

$$X_\alpha = \bigcup \{X_\beta : \beta < \alpha\}$$

for every limit ordinal α. From what precedes, we always get

$$C_s(X;E)_t \subset C_b(X_\alpha;E) \subset C_s(X;E),$$

but, as Buchwalter noticed it, there is an ordinal α such that $X_\alpha = \upsilon X$. Hence the conclusion.

RESULTS IN THE EVALUABLE CASE

A l.c. space is evaluable if every bornivorous barrel is a neighborhood of 0 (= quasi-barrelled or infra-barrelled).

With minor modifications, what precedes can be adapted to the evaluable case. One gets then the following results.

a) If $C(X;E)$ is evaluable, then $C(X)$ and E are evaluable. [Let us mention that [9] it is possible to introduce a topological subspace X_e of βX, containing X, and a family $\mathcal{P}_e$ of bounding subsets of X_e such that $C_{\mathcal{P}_e}(X_e)$ is the evaluable space associated to $C(X)$].

b) If $C(X)$ is evaluable and if E is a metrizable space, then $C(X;E)$ is evaluable.

c)If E is metrizable, then $C_{\mathcal{P}_e}(X_e;E)$ is the evaluable space associated to $C(X;E)$.

REFERENCES

[1] Bierstedt, K.-D., Meise, R.: Bemerkungen über die Approximationseigenschaft lokalkonvexer Funktionenräume, Math. Ann. 209 (1974), 99-107.

[2] Buchwalter, H.: Parties bornées d'un espace topologique complètement régulier, Sém. Choquet 9 (1969-1970), nº 14-15, 15 pages.

[3] Buchwalter, H., Schmets, J.: Sur quelques propriétés de l'espace $C_s(X)$, J. Math. Pures et Appl. 52 (1973), 335-352.

[4] Hollstein, R.: ε-Tensorprodukte von Homomorphismen, Habilitationsschrift, Paderborn (1978).

[5] Komura, Y.: On linear topological spaces, Kumamoto Math. J. of Sc. 5A (1962), 148-157.

[6] Mujica, J.: Representation of analytic functionals by vector measures, Vector space measures and applications, Proc. Dublin 1977, Lecture Notes in Math. 645 (1978), 147-161.

[7] Nachbin, L.: Topological vector spaces of continuous functions, Proc. Nat. Acad. USA 40 (1954), 471-474.

[8] Noureddine, K., Schmets, J.: Espaces associés à un espace localement convexe et espaces de fonctions continues, Bull. Soc. Roy. Sc. Liège 42 (1973), 116-124.

[9] Schmets, J.: Espaces de fonctions continues, Lecture Notes in Math. 519 (1976), Springer Verlag, Berlin.

[10] Schmets, J.: Bornological and ultrabornological C(X;E) spaces, Manuscripta Math. 21 (1977), 117-133.

[11] Schmets, J.: Spaces of vector-valued continuous functions, Vector space measures and applications, Proc. Dublin 1977, Lecture Notes in Math. 644 (1978), 368-377.

[12] Schmets, J.: Survey on some locally convex properties of the spaces of continuous functions, Bull. Soc. Math. de Belgique, (à paraître).

[13] Singer, I.: Sur les applications linéaires intégrales des espaces de fonctions continues, I, Rev. Roumaine Math. Pures et Appl. 4 (1959), 391-401.

[14] Shirota, T.: On locally convex vector spaces of continuous functions, Proc. Japan Acad. 30 (1954), 294-298.

ON SUPRABARRELLED SPACES

Manuel Valdivia

Facultad de Matemáticas
Burjasot - Valencia
Spain

In this paper we introduce the class of the suprabarrelled spaces. Some properties are studied as well as examples are given.

The linear space we use are defined over the field K of the real or complex numbers. By "space" we mean "separated locally convex topological vector space". If A is a bounded closed absolutely convex subset of a space E, E_A denotes the normed space over the linear hull of A having A as unit ball.

<u>Definition</u>. A space E is suprabarrelled if given any increasing sequence (E_n) of subspaces of E with E as union, then there is a positive integer p such that E_p is dense in E and barrelled.

A space E is a Baire-like space if given any increasing sequence of closed absolutely convex subsets of E, with E as union, then one of them has non-void interior. Every barrelled space whose completion is a Baire space is a Baire-like space [9] and in particular every metrizable barrelled space is Baire-like [1]. In [6] some properties of Baire-like spaces are studied. In [8] the unordered Baire-like spaces are introduced and some properties of them are given: E is an unordered Baire-like space if given any sequence of closed absolutely convex subset of E covering E, then one of them has non-void interior. In [7] and [2] examples of normed Baire-like spaces which are not unordered Baire-like and an unordered Baire-like space which are not Baire are given.

Obviously every suprabarrelled spaces is Baire-like and every unordered Baire-like space is suprabarrelled. We shall show that the class of the suprabarrelled spaces is strictly placed between unordered Baire-like and Baire like spaces.

The following two theorems are of trivial nature:

__Theorem 1.__ Every separated quotient of a suprabarrelled space is suprabarrelled.

__Theorem 2.__ Let F be a dense subspace of a space E. If F is suprabarrelled then E is suprabarrelled.

__Theorem 3.__ If F is a countable codimensional subspace of a suprabarrelled space E, then F is suprabarrelled.

__Proof.__ Let (F_n) be an increasing sequence of subspaces of F covering F. Since E is Baire-like and therefore F, [6], there is an index p such that F_p is dense in F. Let G be an algebraic complement of F in E. Then (F_n+G) is an increasing sequence of subspaces of E and thus there is an index $q \geq p$ such that F_q+G is barrelled. Then F_q is barrelled ([3] and [9]) and obviously dense in F. q.e.d.

A space F is a Γ_r-space, [10], if given any barrelled space E and a linear mapping $f: E \to F$ with closed graph, then f is continuous. We shall use later the following result which was given in [10]: a) If f is a continuous linear mapping from a barrelled space E in a Γ_r-space F, then f can be extended in a continuous way to the completion $\hat{E}$ of E in F.

__Theorem 4.__ Let E be a suprabarrelled space and let F be an inductive limit of an increasing sequence (F_n) of Γ_r-spaces contained in F. If $f: E \to F$ is a linear mapping with closed graph then f is continuous and there is a positive integer p such that $f(E)$ is contained in F_p.

__Proof.__ Since E is a suprabarrelled space there is a positive integer p such that $f^{-1}(F_p)$ is barrelled and dense in E. Then the restriction h of f to $f^{-1}(F_p)$ is continuous. According to result a), h can be extend to a continuous mapping $g\colon E \to F_p$. Let $\{x_i\colon i \in I, \geq\}$ be a net in $f^{-1}(F_p)$ converging in E to a point x. The net $\{g(x_i)\colon i \in I, \geq\}$ converges in F_p to $g(x)$. Since $g(x_i) = f(x_i)$, $i \in I$, and the graph of f is closed then $g(x) = f(x)$ and thus $f^{-1}(F_p) = E$. q.e.d.

Let E be the topological product $\prod\{E_i\colon i \in I\}$, being E_i, $i \in I$, suprabarrelled. Consider E_i, $i \in I$, as a subspace of E and denote by E_o the subspace of E whose elements have all coordinates zero but a countable number of them. Take an increasing sequence (F_n) of subspaces of E_o covering E_o. Let U_n be a barrel in F_n. V_n is the closure of U_n in E_o call G_n the linear hull of V_n.

<u>Proposition 1</u>. There is a positive integer p such that $G_p \supset E_i$, for every $i \in I$.

<u>Proof</u>. Suppose that the property is not true. Given a natural number n_1 take an index $i_1 \in I$ such that $G_{ns} \not\supset E_{i_1}$. Suppose $n_1, n_2, \ldots, n_q$ and the indices $i_1, i_2, \ldots, i_q$ obtained. Since E_{i_q} is a suprabarrelled space and since $\bigcup_{n=1}^{\infty} (F_n \cap E_{i_q}) = E_{i_q}$ there is a natural number $n_{q+1} > n_q$ such that $F_{n_{q+1}} \cap E_{i_q}$ is barrelled and dense in E_{i_q} and therefore $V_{n_{q+1}} \cap E_{i_q}$ is a neighbourhood of the origin in E_{i_q} and thus $G_{n_{q+1}} \supset E_{i_q}$. We obtain now an index $i_{q+1} \in I$ such that $G_{n_{q+1}} \not\supset E_{i_{q+1}}$.

For any positive integer r we take in E_{i_r} an one-dimensional space L_r which is not contained in G_{i_r}. Since $i_1, i_2, \ldots, i_q, \ldots$ are all different the space $L = \prod\{L_r\colon r=1,2,\ldots\}$ is a subspace of E_o. On the other hand the family $\{m\, V_{n_q}\colon m=1,2,\ldots, q=1,2,\ldots\}$ of closed absolutely convex subsets of E_o covers L and since L is a Baire space there is a positive integer s such that $V_{n_s} \cap L$ is a neighbourhood of the origin in L and thus $G_{n_s} \supset L$ which is a contradiction. q.e.d.

Proposition 2. The countable product of suprabarrelled spaces is suprabarrelled.

Proof. The product can be taken as an infinite product because if the product is finit it can be multiplied by a sequence of subspaces equal to $\{0\}$ which is obviously suprabarrelled. Let $E = \prod_{n=1}^{\infty} E_n$, being E_n, $n = 1,2,\ldots$, a suprabarrelled space. Suppose that E is not suprabarrelled. Using the same notation as it was done before suppose that U_m can be taken in F_m such that V_m is not a neighbourhood of the origin in E, $m = 1,2,\ldots$ According to Proposition 1 a positive integer p can be chosen such that $G_p \supset E_n$, $n=1,2,\ldots$ Since E is barrelled, given a positive integer $n_1 > p$ a point $x_1 \in E$ can be found such that $x_1 \notin G_{n_1}$. Proceeding by recurrence, if the positive integers $n_1 < n_2 < \ldots < n_q$ and the points $x_1, x_2, \ldots, x_q$ are taken, let H_q and K_q the following spaces

$$H_q = \prod\{E_j : j=1,2,\ldots,q\}$$

$$K_q = \prod\{E_j : j=q+1,q+2,\ldots\},$$

and take $n_{q+1} > n_q$ so that $G_{n_{q+1}} \supset H_q$ and therefore there is a point $x_{q+1} \in K_q$ which is not in $G_{n_{q+1}}$. The set $A = \{x_1, x_2, \ldots, x_n, \ldots\}$ projects in E_q, $q = 1,2,\ldots$ in a finite set and thus the closed absolutely convex hull B of A is compact and E_B is a Banach space. The family $m\, V_q$: $m=1,2,\ldots,q=p+1,p+2,\ldots$ covers E and therefore covers also E_B. Then there is a positive integer $r \geq p$ such that $V_{n_r} \cap E_B$ is a neighbourhood of the origin in E_B, hence $G_{n_r} \supset A$ which is a contradiction.

Theorem 5. The product of suprabarrelled spaces is suprabarrelled.

Proof. Set $E = \prod\{E_i : i \in I\}$, being $\{E_i : i \in I\}$ a family of suprabarrelled spaces and set E_o for the subspace of E for all those vectors with at most a countable number of non vanishing coordinates. According to Theorem 2 it is enough to show that E_o is suprabarrelled. Let (F_n) be an increasing sequence of subspaces of E_o covering E_o. Using the same notations as in the proof of Proposition 1 and supposing that E_o is not suprabarrelled take V_n which is not a neighbourhood of the origin in E_o, $n=1,2,\ldots$ Then, since E_o is a barrelled space there is a point $x_n \in E_o$, $x_n \notin G_n$. There is a countable subset J of I such that the coordinates of x_n corresponding to the indices not belonging to J are nulle, $n = 1,2,\ldots$ Therefore, $\{x_1,x_2,\ldots,x_n,\ldots\} \in \prod\{E_i : i \in J\}$. On the other hand, according to Proposition 2, there is a positive integer q such that $G_q \supset \prod\{E_i : i \in J\}$ which is a contradiction. q.e.d.

Let E be a metrizable non complete space, let (E_n) be an increasing sequence of subspaces of E covering E and let τ_n be a topology on E_n finer than the topology induced on E_n by the original topology, such that $E_n[\tau_n]$ is a barrelled B_r-complete space. Suppose that E is the inductive limit of the sequence $(E_n(\tau_n))$. It is easy to show that there is a positive integer p such that, for $n \geq p$, E_n is not barrelled, according to Ptak's homomorphism theorem. Since there are metrizable non complete (LF)-spaces, the class of the Baire-like spaces is strictly larger than the class of the suprabarrelled spaces.

In what follows we shall construct a very large class of metrizable suprabarrelled spaces which are not unordered Baire-like spaces. We shall use a construct which is partially in [2] (used to give examples of spaces which are not unordered Baire-like) and partially in [11].

<u>Theorem 6</u>. If E is a separable infinite dimensional Fréchet space there is a dense subspace F of E with the following properties:

1. F is suprabarrelled.
2. F is not an inductive limit of unordered Baire-like spaces.

Proof. According to [4, p.118] there is a biorthogonal system (x_n, u_n) for E such that (x_n) is total in E. We consider on the set N of the positive integers the filter $\mathfrak{F}$ of the parts having finite complement in N and let $\mathfrak{U}$ be an ultrafilter finer than $\mathfrak{F}$. For every $U \in \mathfrak{U}$ let L(U) be the closure in E of the linear hull of $\{x_n : n \notin U\}$ then $L(U_1) \cup L(U_2) \subset L(U_1 \cap U_2)$ and therefore the set $L = \bigcup \{L(U)\ U \in \mathfrak{U}\}$ is a dense subspace of E. We shall see now that L is a suprabarrelled space. Suppose that L is covered by an increasing sequence (H_n) of subspaces of L. If F_n denotes the linear hull of $H_n \cup \{x_1, x_2, \ldots, x_p, \ldots\}$ and H_n is not barrelled, $n=1,2,\ldots$, then F_n is not barrelled (see [3] and [9]) and therefore there is a barrel U_n in F_n which is not neighbourhood of the origin in F_n, $n = 1,2,\ldots$ Call V_n the closure of U_n in L. Our next step is to show that there is a positive integer h such that V_h absorbs every bounded set of the linear hull H of $\{x_1, x_2, \ldots, x_p, \ldots\}$. If this property is true, since H is a bornological space, then $V_h \cap H$ is a neighbourhood of the origin in H and thus V_h is a neighbourhood of the origin in L, which is a contradiction. Therefore, there is a positive integer q so that H_n is barrelled for $n \geq q$. Taking an increasing sequence of subspaces of L all of them coincident with L and according to the discussion above it follows that L is barrelled and therefore Baire-like [1] and thus there is a positive integer $p \geq q$ such that H_q is dense (and obviously barrelled) in L, hence L is suprabarrelled.

In order to show that there is a positive integer h such that V_h absorbs every bounded set of H, we begin by showing that if B is a bounded closed absolutely convex set of L then there is a positive integer k such that V_k absorbs $H \cap B$. We take $y_{11} \notin V_1$, $y_{11} \in H \cap B$. Then

$$y_{11} = a_{11}x_1 + a_{21}x_2 + \ldots + a_{n_1 1} x_{n_1}, \quad a_{11}, a_{21}, \ldots, a_{n_1 1} \in K.$$

Suppose we have already chosen y_{rs}, $r \geq s$, $r,s = 1,2,\ldots,p$, in

$H \cap B$ so that

$$y_{(r+1)s} = a_{n_r s}\, x_{n_r} + a_{(n_r+1)s}\, x_{n_r+1} + \dots + a_{n_{r+1} s}\, x_{n_{r+1}}$$

$$y_{(r+1)s} \notin (r+1)V_s,$$

$a_{n_r s} \in K$, $n_o = 1$, $r = 0, 1,2,\dots,p-1$, $s = 1,2,\dots,p$. Let H_p and K_p be the linear hulls of $\{x_1, x_2, \dots, x_{n_p}\}$ and $\{x_{n_p+1}\ x_{n_p+2}, \dots\}$, respectively. Since H_p is a topological complement of finite dimension of K_p in H, it is obvious that $V_1, V_2, \dots, V_{p+1}$ does not absorb $K_p \cap B$ and therefore there are $p+1$ elements

$$y_{(p+1)1},\ y_{(p+1)2}, \dots, y_{(p+1)(p+1)}$$

in $K_p \cap B$ such that

$$y_{(p+1)s} \notin (p+1)V_s, \quad s = 1,2,\dots,p+1.$$

Then

$$y_{(p+1)s} = a_{n_p s}\, x_{n_p} + a(n_p+1)\, x_{n_p+1} + \dots + a_{n_{p+1} s}\, x_{n_{p+1}},$$

$$a_{n_p s},\ a_{(n_p+1)}, \dots, a_{n_{p+1} s} \in K, \quad s = 1,2,\dots,p+1.$$

We set

$$N_1 = \{1,2,\dots,n_1\}$$

$$N_2 = \{n_1+1,\ n_1+2, \dots, n_2\}$$

$$\dots\ \dots\ \dots\ \dots\ \dots\ ..$$

$$N_{p+1} = \{n_p,\ n_p+1, \dots, n_{p+1}\}$$

$$\dots\ \dots\ \dots\ \dots\ \dots\ \dots.$$

$$M_1 = \bigcup \{N_{2p+1} \quad p=0,1,2,\dots\}$$

$$M_2 = \bigcup \{N_{2p} : p = 1,2,\dots\}.$$

Since $N = M_1 \cup M_2$, $M_1 \cap M_2 = \emptyset$, then either M_1 or M_2 are in $\mathfrak{U}$. Suppose $M_1 \in \mathfrak{U}$. The set

$$D = \{y_{(2r)s} : 2r \geq s, \quad r,s = 1,2,\dots\}$$

is bounded and lies in $L(M_1)$, which is a Fréchet space and thus there is a positive integer m such that $F_m \cap L(M_1)$ is of second category in $L(M_1)$ and therefore $V_m \cap L(M_1)$ is a neighbourhood of the origin in $L(M_1)$, hence V_m absorbs D which is a contradiction since

$$y_{(2r)m} \notin (2r)\, V_m, \quad r = 1,2,\ldots .$$

Suppose finally that there is a bounded set C_n in H such that V_n does not absorb C_n, $n = 1,2,\ldots$. Since E is metrizable there is a sequence of strict positive numbers (λ_n) such that the closed absolutely convex hull C of $\cup\{\lambda_n C_n : n = 1,2,\ldots\}$ in E is a bounded set. As we did before there is a positive integer ℓ such that V_ℓ absorbs $C \cap H$ which is contradiction with V_ℓ does not absorb $\lambda_\ell C_\ell$.

If $x \in L$, there is an element U in $\mathcal{U}$ so that $x \in L[U]$. Since $U \neq \emptyset$ we can take $p_o \in U$. On the other hand, $L[U]$ is the closure in E of the linear hull of $\{x_n : n \notin U\}$ and therefore $\langle x, u_{p_o}\rangle = 0$. Let $T_n = \{x : x \in E, \langle x, u_n\rangle = 0\}$. Since E is a Baire space, we have that $E \neq \cup\{T_n : n=1,2,\ldots\}$ and therefore we can take a vector x_o in E so that $x_o \notin \cup\{T_n : n=1,2,\ldots\}$, hence $\langle x_o, u_n\rangle \neq 0$, $n = 1,2,\ldots$, and thus $x_o \notin L$.

Let F be the subspace of E generated by $L \cup \{x_o\}$. According to Theorem 2, F is suprabarrelled. On the other hand, using the procedure given in ([11], Proposition 5) it follows that F is not an inductive limit of unordered Baire-like spaces. q.e.d.

REFERENCES

[1] Amemiya, I. and Komura, Y.: Über nicht-vollständige Montel-räume, Math. Ann. 177, 273-277 (1968).

[2] Dierolf, P., Dierolf, S. and Drewnowski, L.: Remark and examples concerning unordered Baire-like and ultrabarrelled spaces Coll. Math. Vol. XXXIX, Fasc. 1, 110-117 (1978).

[3] Dieudonne, J.: Sur les propriétés de permanence de certains espaces vectoriels topologiques. Ann. Soc. Polon. Math. 25, 50-55 (1952).

[4] Marti, J.T.: Introduction to the theory of bases, Springer-Verlag Berlin, 1969.

[5] Pták, V.: Completeness and the open mapping theorem. Bull. Soc. Math. France, 86, 41-74, (1958).

[6] Saxon, S.: Nuclear and product spaces, Baire-like spaces, and the strongest locally convex topology, Math. Ann. 197, 87-106, (1972).

[7] Saxon, S.: Some normed barrelled spaces which are not Baire. Math. Ann. 209, 153-160, (1974).

[8] Todd, A. and Saxon, S.: A property of locally convex Baire spaces. Math. Ann. 206, 23-35 (1973).

[9] Valdivia, M.: Absolutely convex sets in barrelled spaces. Ann. Inst. Fourier, Grenoble 21, (2). 3-13, (1971).

[10] Valdivia, M.: Sebre el teorema de la gráfica cerrada. Seminario Matemático de Barcelona, Collectanea Mathematica, Vol. XXII, Fasc. 1, (1977).

[11] Valdivia, M.: On Baire-hyperplane spaces. Proceedings of the Edinburgh Mathematical Society, 22, 247-255 (1979).

ENVELOPES OF SILVA-HOLOMORPHY

Maria Carmelina F. Zaine

Instituto de Matemática
IMECC - UNICAMP
Caixa Postal 1170
13.100 - Campinas - SP
Brasil

1. NOTATIONS AND BASIC DEFINITIONS

Throughout this paper E will represent a complex separated locally convex space (s.l.c.s.), U will denote a non-void open subset of E. The family of all closed convex balanced bounded subsets of E will be denoted by $\mathcal{B}_E$. We will denote by E_B the vector space of E spanned by $B \in \mathcal{B}_E$ and normed by the Minkowski functional p_B of B. We will indicate by $CS(E)$ the family of continuous semi-norms on E.

For each $n \in \mathbb{N}$, we will represent by $\mathcal{P}_b(^nE)$ the vector space of all n-homogeneous polynomials from E to $\mathbb{C}$ which are bounded on the bounded subsets of E and by $\mathcal{P}(^nE)$ the vector subspace of $\mathcal{P}_b(^nE)$ of all continuous n-homogeneous polynomials from E to $\mathbb{C}$.

1.1 Definition. A function P from E to $\mathbb{C}$ is a bounded polynomial if there are $n = 0,1,\ldots,P_k \in \mathcal{P}_b(^kE)$ with $k = 0,1,\ldots,n$, such that $P = P_o + P_1 + \ldots + P_n$. Let $\mathcal{P}_b(E)$ denote the vector space of all bounded polynomials from E to $\mathbb{C}$.

1.2 Definition. A function $f: U \to \mathbb{C}$ is Silva-holomorphic (S-holomorphic) on U if, corresponding to every $\xi \in U$, there are $P_n \in \mathcal{P}_b(^nE)$ $(n=0,1,\ldots)$ such that for all $B \in \mathcal{B}_E$, there is $\rho_B > 0$

with $\xi + \rho_B B \subset U$, such that

$$f(x) = \sum_{n=0}^{\infty} P_n(x-\xi)$$

uniformly on $\xi + \rho_B B$. We will represent by $\mathcal{H}_S(U)$ the vector space of all S-holomorphic functions from U to $\mathbb{C}$.

$\mathcal{H}(U)$ represents the vector space of all holomorphic functions from U to $\mathbb{C}$. Thus $\mathcal{H}_S(U) = \mathcal{H}(U)$ for all non-void open subsets U of a metrizable or Silva space.

For basic results of holomorphic mappings on infinite dimension see [7], [2].

<u>1.3 Proposition</u>. A function $f \in \mathcal{H}_S(U)$ if, and only if,

$$f|_{U \cap E_B} \in \mathcal{H}(U \cap E_B) \quad \text{for all} \quad B \in \mathcal{B}_E.$$

<u>1.4 Corollary</u>. $\mathcal{P}_b(^nE)|_U \subset \mathcal{H}_S(U)$ for all $n \in \mathbb{N}$.

<u>1.5 Definition</u>. A strict compact subset of E is a subset K of E such that there is B in $\mathcal{B}_E$ for which K is contained and compact in E_B.

τ_{oe} will be the locally convex topology on $\mathcal{H}_S(U)$ of the uniform convergence on strict compact subsets of U, i.e., τ_{oe} is the topology generated by all semi-norms p_K defined by $p_K(f) = \sup\{|f(x)|;\ x \in K\}$ for all $f \in \mathcal{H}_S(U)$ and $K \subset U$ strict compact.

For more detailed information on bounded polynomials and Silva -holomorphic functions see [9].

<u>1.6 Definition</u>. (1) A pair (X,φ) is called a domain spread over E if $X \neq \emptyset$ is a connected Hausdorff space and $\varphi: X \to E$ is a local homeomorphism.

(2) Let (X,φ), (Y,φ') be domains spread over E. A map j from X to Y is called a morphism (or a weak inclusion) if j is continuous and satisfies $\varphi = \varphi' \circ j$.

(3) A compact subset K of a domain spread (X,φ) over E is a strict compact set if $\varphi(K)$ is a strict compact subset of E.

(4) If (X,φ) is a domain spread over E, $A \subset X$ and $U \subset E$, then, $A+U = \bigcup_{x \in A} \{x+U\}$ where $x+U = (\varphi|_V)^{-1}(\varphi(x)+U)$ with $V \subset X$ and open, $\varphi|_V$ a homeomorphism, $x \in V$ and $\varphi(x) + U \subset \varphi(V)$.

Instead of (X,φ) we write X if no confusion is likely.

<u>1.7 Definition</u>. Let (X,φ) be a domain spread over E and let f be a function from X to $\mathbb{C}$. We say f is S-holomorphic on X if, for each $x \in X$ there are V, a neighborhood of x in X, and f', an S-holomorphic function from $\varphi(V)$ to $\mathbb{C}$, such that $\varphi|_V$ is a homeomorphism and $f|_V = f' \circ \varphi|_V$.

We will denote by $\mathcal{H}_S(X)$ the vector space of all S-holomorphic functions from X to $\mathbb{C}$.

<u>1.8 Definition</u>. Let (X,φ) be a domain spread over E. For each $B \in \mathcal{B}_E$ we define $X_B = \{x \in X;\ \varphi(x) \in E_B\}$ and let τ_B be the topology on X_B defined in the following way: for each $x \in X_B$, $U \subset X_B$ is a neighborhood of x in X_B if $\varphi|_U$ is 1-1 and $\varphi(U)$ is a neighborhood of $\varphi(x)$ in E_B. If $\mathcal{B}(x)$ is the family of all these neighborhoods, then W is a neighborhood of x if, and only if, there is $V \in \mathcal{B}(x)$ such that $V \subset W$.

<u>1.9 Remark</u>. With the above definitions we have

(1) The mapping $\varphi|_{X_B}$ is a local homeomorphism from X_B to E_B.

(2) Each connected non-void component of X_B is a domain spread over E_B for each $B \in \mathcal{B}_E$.

(3) The topology τ_B is finer than the induced topology.

We have to show that the intersection of each open subset of X with X_B is an open subset of τ_B. Let A be an open subset of X and we consider $A \cap X_B$. From the hypothesis, for each $x \in A \cap X_B$ there is a neighborhood W of x in X such that $\varphi|_W$ is a homeomorphism. We can have $W \subset A$, then $W \cap X_B \subset A \cap X_B$ and the result follows.

(4) i: $X_B \hookrightarrow X$ is continuous.

(5) If $f \in \mathcal{H}_S(X)$ then $f|_{X_B} \in \mathcal{H}(X_B)$ for each $B \in \mathcal{B}_E$.

(6) A function $f \in \mathcal{H}_S(X)$ if, and only if, for all $x \in X$, there is a neighborhood W of x in X such that $\varphi|_W$ is a homeomorphism and $f|_{W\cap X_B} \in \mathcal{H}(W\cap X_B)$ for all $B \in \mathcal{B}_E$.

(7) If K is a strict compact subset of X and $f \in \mathcal{H}_S(X)$, then $\sup\{|f(t)|;\ t \in K\} < +\infty$.

We will consider on $\mathcal{H}_S(X)$ the locally convex topology τ_{oe} of the uniform convergence on strict compact subsets of X.

Let (X,φ) be a domain spread over E and let f be an S-holomorphic function from X to $\mathbb{C}$. If U is an open subset of X such that $\varphi|_U$ is a homeomorphism and $x \in U$, there is a sequence $(P_n)_{n=0}^{\infty}$ of elements $P_n \in \mathcal{P}_b(^nE)$, $n=0,1,\ldots,$ and also a convex balanced open neighborhood V of 0 in E such that

$$f\circ(\varphi|_U)^{-1}(\varphi(x)+y) = \sum_{n=0}^{+\infty} P_n(y),$$

the series converging pointwise for $y \in V$. The sequence $(P_n)_{n=0}^{+\infty}$ is unique and each P_n will be denoted by $\frac{1}{n!}\hat{\delta}^n f(x)$.

By Definition 1.6 we can write

$$f(x+y) = \sum_{n=0}^{+\infty} \frac{1}{n!}\hat{\delta}^n f(x)y.$$

<u>1.10 Proposition</u>. If $f \in \mathcal{H}_S(X)$, X is a domain spread over E and $y \in E$, then the mapping $\hat{\delta}^n f(\ )(y)\colon x \in X \to \hat{\delta}^n f(x)(y) \in \mathbb{C}$ is S-holomorphic on X, $n=0,1,2,\ldots$.

<u>1.11 Proposition</u>. Let (X,φ) be a domain spread over E, $f \in \mathcal{H}_S(X)$, $x \in X$, V an open neighborhood of x such that $\varphi|_V$ is a homeomorphism and there is $f' \in \mathcal{H}_S(\varphi(V))$ such that $f|_V = f'\circ\varphi|_V$. For $\xi \in V$ and $\rho > 1$ such that $(1-\lambda)\varphi(\xi) + \lambda\varphi(x) \in \varphi(V)$ for every $\lambda \in \mathbb{C}$, $|\lambda| \leq \rho$, we have

$$f\circ(\varphi|_V)^{-1}(\varphi(x)) = \frac{1}{2\pi i}\int_{|\lambda|=\rho} \frac{f\circ(\varphi|_V)^{-1}((1-\lambda)\varphi(\xi)+\lambda\varphi(x))}{\lambda-1}\,d\lambda .$$

<u>1.12 Proposition</u>. Let (X,φ) be a domain spread over E, $f \in \mathcal{H}_S(X)$, $\xi \in X$, V an open neighborhood of ξ such that $\varphi|_V$ is a homeomorphism and there is $f' \in \mathcal{H}_S(\varphi(V))$ such that $f|_V = f' \circ \varphi|_V$. For $x \in E$ and $\rho > 0$ such that $\varphi(\xi) + \lambda x \in \varphi(V)$ for every $\lambda \in \mathbb{C}$, $|\lambda| \leq \rho$,

$$\frac{1}{n!}\,\hat{\delta}^n f(\xi)(x) = \frac{1}{2\pi i}\int_{|\lambda|=\rho} \frac{f(\xi+\lambda x)}{\lambda^{n+1}}\,d\lambda$$

for $n = 0,1,\ldots$.

<u>1.13 Proposition</u>. Let (X,φ) be a domain spread over E, $f \in \mathcal{H}_S(X)$, $\xi \in X$, V an open neighborhood of ξ such that $\varphi|_V$ is a homeomorphism and there is $f' \in \mathcal{H}_S(\varphi(V))$ such that $f|_V = f' \circ \varphi|_V$. For $B \in \mathcal{B}_E$ and $\rho_B > 0$ such that $\varphi(\xi) + \rho_B B \subset \varphi(V)$,

$$\left\|\frac{1}{n!}\,\hat{\delta}^n f(\xi)\right\|_B = \sup_{x\in B}\left|\frac{1}{n!}\,\hat{\delta}^n f(\xi)(x)\right|$$

$$< \frac{1}{\rho_B^n}\sup_{t\in\xi+\rho_B B}|f(t)|$$

for $n = 0,1,\ldots$.

<u>1.14 Definition</u>. Let A be a family of S-holomorphic functions defined on the domain X over E.

(1) A morphism $j\colon X \to Y$ of domains over E is said to be a simultaneous S-holomorphic continuation (s. S-h.c.) of A if each $f \in A$ factors S-holomorphically through j.

(2) A s. S-h.c. $j\colon X \to Y$ of A is said to be maximal if it factors through every s. S-h.c. of A as a morphism. If $j\colon X \to \mathcal{E}(X)$ is a maximal s. S-h.c. of $\mathcal{H}_S(X)$, then $\mathcal{E}(X)$ is called an envelope of S-holomorphy of X.

(3) X is said to be an A-domain of S-holomorphy if every s. S-h.c. of A is an isomorphism of domains. The domain X is a domain of S-holomorphy if X is a $\mathcal{H}_S(X)$-domain of S-holomorphy.

2. CONSTRUCTION OF THE ENVELOPE OF S-HOLOMORPHY VIA GERMS OF S-HOLOMORPHIC FUNCTIONS

In this paragraph we use germs of S-holomorphic functions to obtain a construction of the envelope of S-holomorphy of a domain spread over a locally convex Hausdorff space. We show that this envelope has a property of S-holomorphic convexity and that the S-holomorphic functions separate its points.

<u>2.1 Proposition</u>. Let E be a locally convex Hausdorff space and let U be a connected open non-void subset of E. If $f,g \in \mathcal{H}_S(U)$ and if $f = g$ on a non empty open subset of U, then $f \equiv g$.

<u>Proof</u>. Let A be a non empty open subset of U such that $f = g$ on A. Let $a \in A$, $x \in U$ and let γ be a polygonal path in U joining a and x, then $\operatorname{Im}\gamma \subset U$ and $\operatorname{Im}\gamma$ is compact. Let $B \in \mathcal{B}_E$ be such that $\operatorname{Im}\gamma \subset E_B$, then $\operatorname{Im}\gamma \subset U \cap E_B$. The connected component U' of $U \cap E_B$ containing a contains $\operatorname{Im}\gamma$. Since $f = g$ on a neighborhood of a, then they coincide, as holomorphic functions, on a neighborhood of A in U', thus they coincide on U' by the principle of holomorphic continuation and $f(x) = g(x)$ since $x \in U'$.

If $F = \prod_{i\in I} F_i$ is a product of locally convex spaces, if f is a mapping from a locally convex space E to F and if $f = (f_i)_{i\in I}$, where the f_i are the coordinate mappings from E to F_i, then the mapping f is S-holomorphic if, and only if, each f_i is S-holomorphic.

Let E be a locally convex space, let $a \in E$ and let V_1, V_2 be open subsets of E containing a. If

$$f = (f_i)_{i\in I} : V_1 \to \mathbb{C}^I$$

$$g = (g_i)_{i\in I} : V_2 \to \mathbb{C}^I$$

are S-holomorphic mappings, we say that $f \sim g$ if, and only if, there is a neighborhood W of a, $W \subset V_1 \cap V_2$ such that $f_i|_W = g_i|_W$ for all $i \in I$. This is an equivalence relation on

$$H_E(a) = \bigcup_{\substack{U \subset E \text{ open} \\ U \supset \{a\}}} \mathcal{H}_S(U;\mathbb{C}^I)$$

and we denote by $((f_i)_a)_{i\in I}$ the equivalence class determined by $f = (f_i)_{i\in I}$. The equivalence class $((f_i)_a)_{i\in I}$ is called the germ of the family $(f_i)_{i\in I}$ at a.

<u>2.2 Theorem.</u> Every domain spread over a locally convex Hausdorff space admits an envelope of S-holomorphy which is unique up to isomorphism.

<u>Proof.</u> Let (X,φ) be a domain spread over a locally convex Hausdorff space E and also $I = \mathcal{H}_S(X)$. For all open subsets $V \subset E$ such that there is an open subset $U \subset X$ and $\varphi|_U$ is a homomeomorphism between U and V we define

$$\mathcal{O}_E(V) = \{f \in \mathcal{H}_S(V);\ \exists\ g \in \mathcal{H}_S(U) \text{ and } f = g\circ(\varphi|_U)^{-1}\}.$$

Let X' be the set defined by $X' = \{(a,((f_i)_a)_{i\in I});\ a \in E$ and $f = (f_i)_{i\in I}$ is such that there is $V \subset E$ open, $a \in V$ and $f_i \in \mathcal{O}_E(V)\ \forall\ i \in I\}$.

For all open subsets $V \subset E$, for all $(g_i)_{i\in I} \in \prod_{i\in I} \mathcal{O}_E(V)$ we define $N(V,(g_i)_{i\in I}) = \{(a,((g_i)_a)_{i\in I});\ a \in V\}$. This set span on X' a topology finer than the product topology, hence a separated topology. Thus (X',φ') is a manifold spread over E if we define $\varphi'(a,((f_i)_a)_{i\in I}) = a$.

We define a mapping $u' : X \to X'$ by

$$u'(x) = (\varphi(x),\ ((i\circ(\varphi|_U)^{-1})_{\varphi(x)})_{i\in I}),$$

where $x \in X$ and U is a neighborhood of x such that $\varphi|_U$ is a homeomorphism from U onto $\varphi(U)$. It follows from the definition of u' and also from the topology of X' that u' is continuous and is a morphism of manifolds spread. We have $u'(X)$ is connected in X'

since X is connected. We denote by $\tilde{X}$ the connected component of X' containing $u'(X)$, by $\tilde{u}$ the restriction $u': X \to \tilde{X}$ and by $\tilde{\varphi}$ the restriction $\varphi'|_{\tilde{X}}$. We just have $(\tilde{X},\tilde{\varphi})$ is a domain spread over E and $\tilde{u}$ is a morphism from (X,φ) to $(\tilde{X},\tilde{\varphi})$.

The domain $(\tilde{X},\tilde{\varphi})$ is an envelope of S-holomorphy of (X,φ). For each function j S-holomorphic on X we define a function f'_j on X' by $f'_j(a,((g_i)_a)_{i\in I}) = g_j(a)$ for all $(a,((g_i)_a)_{i\in I}) \in X'$. We denote the restriction $f'_j|_{\tilde{X}}$ by $\tilde{f}_j$. The function $\tilde{f}_j$ is the desired extension of j since it is S-holomorphic and $j = \tilde{f}_j \circ \tilde{u}$. It follows from Proposition 2.1 that $\tilde{f}_j$ is unique.

Now, let $v: (X,\varphi) \to (\bar{X},\bar{\varphi})$ be a s. S-h.c. of $\mathcal{H}_S(X)$, then the sets $\mathcal{H}_S(\bar{X})$ and $\mathcal{H}_S(X)$ are isomorphic. Let $\bar{i} \in \mathcal{H}_S(\bar{X})$ be the element corresponding to $i \in \mathcal{H}_S(X)$ by this isomorphism. We define a mapping $\overline{u'}$ from $\bar{X}$ to X' as above, $\overline{u'}(x) =$
$= (\bar{\varphi}(x),(\bar{i}\circ(\bar{\varphi}|_U)^{-1})_{\bar{\varphi}(x)})_{i\in I})$, where $x \in \bar{X}$, $\bar{i} \in \mathcal{H}_S(\bar{X})$ and U is a neighborhood of x such that $\bar{\varphi}|_U$ is a homeomorphism onto $\bar{\varphi}(U)$.

As before, we can verify that $\overline{u'}$ is continuous, is a morphism of manifolds spread and $u' = \overline{u'}\circ v$. Since $u'(\bar{X})$ is connected we have $\overline{u'}(\bar{X}) \subset \tilde{X}$. We denote by $\bar{u}$ the morphism $\overline{u'}: (\bar{X},\bar{\varphi}) \to (\tilde{X},\tilde{\varphi})$, and since $\tilde{u} = \bar{u}\circ v$ it follows that $(\tilde{X},\tilde{\varphi})$ is an envelope of S-holomorphy.

The envelope of S-holomorphy is unique. In fact, let $(\tilde{X},\tilde{\varphi})$ and $(\tilde{X}',\tilde{\varphi}')$ be two envelopes of S-holomorphy, applying the maximality condition for each we get two morphism, $\sigma: \tilde{X} \to \tilde{X}'$ and $\sigma':\tilde{X}' \to \tilde{X}$, such that $\sigma'\circ\sigma$ is the identity on $u(X)$ which is open in $\tilde{X}$ and $\sigma\circ\sigma'$ is the identity on $u'(X)$ which is open in $\tilde{X}'$. Hence, by Proposition 2.1, $\sigma'\circ\sigma$ = identity on $\tilde{X}$ and $\sigma\circ\sigma'$ = identity on $\tilde{X}'$. Thus, $(\tilde{X},\tilde{\varphi})$ and $(\tilde{X},\tilde{\varphi}')$ are isomorphic.

<u>2.3 Proposition</u>. If (X,φ) is a domain of S-holomorphy, the S-holomorphic functions on X separate the points of X.

This proposition follows from a similar result for holomorphic functions.

<u>2.4 Definition.</u> Let (X,φ) be a domain spread over a locally convex space E. For all $x \in X$, $A \subset X$ and $p \in CS(E)$ we define:

$$d_p(x) = \sup\{r > 0;\ \exists\, X'_r \subset X \text{ open such that } x \in X'_r \text{ and } X'_r \text{ is homeomorphic to } B_p(\varphi(x),r)\}$$

$$d_p(A) = \inf_{x\in A} d_p(x).$$

<u>2.5 Theorem.</u> Let (X,φ) be a domain of S-holomorphy spread over a locally convex space E. For every strict compact subset K of X, there is $p \in CS(E)$ such that $d_p(\hat{K}_{\mathcal{H}_S(X)}) = d_p(K)$, where $\hat{K}_{\mathcal{H}_S(X)} = \{x \in X;\ |f(x)| \le \|f\|_K \text{ for all } f \in \mathcal{H}_S(X)\}$.

<u>Proof.</u> Let $p \in CS(E)$ be such that $d_p(K) = \delta > 0$. For each fixed point $y \in E$ such that $p(y) < \delta$, we choose $\delta_1 > 1$ such that if $K_y = \varphi(K) + \{\lambda y;\ |\lambda| \le \delta_1\}$ then $d_p(K + \{\lambda y;\ |\lambda| \le \delta_1\}) > 0$. As K is a strict compact subset of X, there is $B \in \mathcal{B}_E$ such that $\varphi(K)$ is contained and compact in E_B and $\varphi(\hat{K}_{\mathcal{H}_S(X)}) \subset E_B$ since $\varphi(\hat{K}_{\mathcal{H}_S(X)}) \subset \widehat{\varphi(K)}_{\mathcal{H}_S(E)}$. Let $B_1 = [B \cup \{\lambda y;\ |\lambda| \le \delta_1\}]$ be the balanced convex hull of $B \cup \{\lambda y;\ |\lambda| \le \delta_1\}$. We have $E_B \hookrightarrow E_{B_1}$ continuously, hence $\varphi(K)$ is contained and compact in E_{B_1} and $\varphi(\hat{K}_{\mathcal{H}_S(X)}) \subset E_{B_1}$. Since $\{\lambda y;\ |\lambda| \le \delta_1\}$ is contained and compact in E_{B_1} we have K_y is compact in E_{B_1}.

If $f \in \mathcal{H}_S(X)$. Let W be a balanced convex neighborhood of 0 in E such that if $W_{B_1} = W \cap E_{B_1}$ then $\|f\|_{K+\{\lambda y;|\lambda|\le\delta_1\}+\delta_1 W_{B_1}} = M < +\infty$. Thus for every $t \in K$ and for each neighborhood V_t of t such that $\varphi|_{V_t}$ is a homeomorphism and $t + \{\lambda y;\ |\lambda| \le \delta_1\} + \delta_1 W_{B_1} \subset \varphi(V_t)$ we have

$$\sup_{\substack{|\lambda|\le\delta_1 \\ w\in W_{B_1}}} |f\circ(\varphi|_{V_t})^{-1}(t+\lambda y+\delta_1 w)| \le M.$$

We denote $\tilde{f}_t = f\circ(\varphi|_{V_t})^{-1}$.

If $x \in \hat{K}_{\mathcal{H}_S}(X)$ and V_x is neighborhood of x so that $\varphi|_{V_x}$ is a homeomorphism, then for all $\lambda \in \mathbb{C}$, $|\lambda| \leq 1$ and $w \in W_{B_1}$ we have

$$\left|\frac{1}{n!}\, \hat{\delta}^n (f\circ(\varphi|_{V_x})^{-1})(\varphi(x))(\lambda y+w)\right| \leq$$

$$\leq \sup_{t\in K} \left|\frac{1}{n!}\, \hat{\delta}^n (f\circ(\varphi|_{V_t})^{-1})(\varphi(t))(\lambda y+w)\right| =$$

$$= \sup_{t\in K} \left|\frac{1}{n!}\, \hat{\delta}^n \tilde{f}_t(\varphi(t))(\lambda y+w)\right|$$

$$= \sup_{t\in K} \left|\frac{1}{2\pi i} \int_{|\lambda_1|=\delta_1} \frac{\tilde{f}_t(\varphi(t) + \lambda_1(\lambda y+w))}{\lambda_1^{n+1}}\, d\lambda_1\right| ,$$

where V_t is a neighborhood of t and $\varphi|_{V_t}$ is a homeomorphism. But $\varphi(t) + \lambda_1 \lambda y + \lambda_1 w \in K_y + \delta_1 W_{B_1}$, thus

$$\left|\frac{1}{n!}\, \hat{\delta}^n (f\circ(\varphi|_{V_x})^{-1})(\varphi(x))(\lambda y+w)\right| \leq M\, \delta_1^{-n}$$

for all $x \in \hat{K}_{\mathcal{H}_S}(X)$, $\lambda \in \mathbb{C}$ such that $|\lambda| \leq 1$, $w \in W_{B_1}$. This implies

$$\sum_{n=0}^{\infty} \sup_{\substack{|\lambda|\leq 1\\ w\in W_{B_1}}} \left|\frac{1}{n!}\, \hat{\delta}^n (f\circ(\varphi|_{V_x})^{-1})(\varphi(x))(\lambda y+w)\right| < +\infty ,$$

therefore for all $x \in \hat{K}_{\mathcal{H}_S}(X)$

$$\sum_{n\geq 0} \frac{1}{n!}\, \hat{\delta}^n (f\circ(\varphi|_{V_x})^{-1})(\varphi(x))(z)$$

converges uniformly on some neighborhood in E_{B_1} of each $y \in B_p(0,\delta)$. Since this is true for all $B \in \mathcal{B}_E$, we have that the series above defines an S-holomorphic function g_f on $\hat{K}_{\mathcal{H}_S}(X) + B_p(0,\delta)$ such that g_f has the same Taylor expansion as f for all $x \in \hat{K}_{\mathcal{H}_S}(X)$. The ball $B_p(0,\delta)$ does not depend on the function f considered.

Clearly $d_p(K) \geq d_p(K_{\mathcal{H}_S}(X))$. Suppose that strict inequality holds. Then, there is $x_o \in \hat{K}_{\mathcal{H}_S}(X)$ such that $d_p(x_o) = r_o < d_p(K) = \delta$. Let $a = \varphi(x_o)$ and let $P_o = B_p(a,\delta)$. Let X_o be the connected component of $\varphi^{-1}(P_o)$ containing x_o. We assert that $\varphi|_{X_o}$ is injective. In fact, it follows from the first part that for any

$f \in \mathcal{H}_S(X)$ there is $g_f \in \mathcal{H}_S(X_o)$ such that $f|_{X_o} = g_f|_{X_o}$. If $\varphi(x_o) = \varphi(y_o)$ for two different prints x_o, y_o of X_o then $g_f(x_o) = g_f(y_o)$ and $f(x_o) = f(y_o)$ for every $f \in \mathcal{H}_S(X)$. Since, by Theorem 2.4, $\mathcal{H}_S(X)$ separates the points of X, we have a contradiction. Thus $\varphi|_{X_o}$ is injective.

Let $r_o < r < \delta$, let $X_1 = \{x \in X_o;\ p(\varphi(x)-a) < r\}$, and let Y be the disjoint union of X and $P = \{z \in E;\ p(z-a) < r\}$. We define the following equivalence relation on Y:

(1) If $z \in P$ and there is no $x \in X_1$ such that $\varphi(x) = z$, then z is equivalent only to itself.

(2) If $z \in P$ and there is $x \in X_1$ such that $\varphi(x) = z$, then z is equivalent to x.

(3) If $x \in X$ and $x \notin X_1$, then x is equivalent only to itself.

Let Z be the quotient of Y by this equivalence relation. Then Z is Hausdorff, and the map from Y to E which is φ on X and the inclusion of P in E induces a local homeomorphism $\varphi' : Z \to E$. Moreover, the inclusion of X into Y induces a map $\sigma: X \to Z$ such that $\varphi = \varphi' \circ \sigma$. We claim that for any $f \in \mathcal{H}_S(X)$ there is $F_f \in \mathcal{H}_S(Z)$ such that $F_f \circ \sigma = f$. In fact, since there is $g_f \in \mathcal{H}_S(X_o)$ such that $f|_{X_o} = g_f|_{X_o}$, we define a function G_f on Y by $G_f|_X = f$, $G_f|_P = g_f \circ (\varphi|_V)^{-1}$, where $\varphi|_V$ is a homeomorphism and $\varphi(V) \supset P$. This induces $F_f \in \mathcal{H}_S(Z)$ with $F_f \circ \sigma = f$. Hence (Z, φ') is an S-holomorphic extension of (X, φ). Since (X, φ) is, by assumption, a domain of S-holomorphy, it follows that σ is an isomorphism. In particular, since Z contains a p-ball of radius r about $\sigma(x_o)$, X contains an open subset X'_r homeomorphic to this p-ball, with $x_o \in X'_r$, contradicting the assumption that $d_p(x_o) = r_o < r$, and the theorem is proved.

3. NORMAL ENVELOPE OF S-HOLOMORPHY

Let E be a locally convex Hausdorff space where there exists a fundamental system of complete bounded subsets. Let U be a domain in E. In $\mathcal{H}_S(U)$ we consider the topology τ_{oe}. Let $\mathcal{S}$ be the spectrum of $\mathcal{H}_S(U)$, i.e., $\mathcal{S}$ is the set of all continuous nonzero homomorphisms of $\mathcal{H}_S(U)$ into $\mathbb{C}$. We want $\mathcal{S}$ to be a manifold spread over E.

We shall need the complex directional derivative. For $u \in E$ and $f \in \mathcal{H}_S(U)$ we define a function $D_u f$ on U by

$$(D_u f)(x) = \frac{d}{d\lambda} f(x+\lambda u)|_{\lambda=0} .$$

<u>3.1 Theorem</u>. If $f,g \in \mathcal{H}_S(U)$ and $u,v,u_1,\dots,u_n \in E$ with $n \in \mathbb{N}$, then

(a) $f \to D_u f$ is a continuous linear map of $\mathcal{H}_S(U)$ into itself.

(b) $D_u D_v f = D_v D_u f$.

(c) $D^n_{\alpha u} f = \alpha^n D^n_u f$, $\alpha \in \mathbb{C}$.

(d) $D^n_{u+v} f = \sum_{s+t=n} \frac{n!}{s!t!} D^s_u(D^t_v f)$.

(e) $D^n_u(fg) = \sum_{s+t=n} \frac{n!}{s!t!} D^s_u f D^t_u g$.

(f) $(D_{u_1} D_{u_2} \dots D_{u_n} f)(x) = f^{(n)}(x)(u_1,\dots,u_n)$ where $f^{(n)}(x)$ is the n-th Silva derivative of f at x.

<u>Proof</u>. (a) It can easily be shown that $f \mapsto D_u f$ is a linear map. We show that this map is continuous. Let $K \subseteq U$ be a strict compact. We show that there exists a strict compact $L \subseteq U$ and $A > 0$ such that $\|D_u f\|_K \le A\|f\|_L$. If we take A such that

$$L = K + \{\lambda u;\ |\lambda| \le A^{-1},\ \lambda \in \mathbb{C}\}$$

is contained in U, then L is strict compact. If we fix $k \in K$ and we set $g(\lambda) = f(k+\lambda u)$ for $|\lambda| \le A^{-1}$, then by Cauchy's inequality $|g'(0)| \le A\|g\|_{\{\lambda\ ;|\lambda| \le A^{-1}\}}$. Therefore

$$|D_u f(k)| \le A\|f\|_L, \quad \text{and hence} \quad \|D_u f\|_K \le A\|f\|_L .$$

The proofs of (b), (c), (d), (e) and (f) do not present difficulties.

By the definition of the topology τ_{oe}, if $h \in \mathcal{S}$ then there is a strict compact set $K \subseteq U$ and $M > 0$ such that $|h(f)| \le M\|f\|_K$, for all $f \in \mathcal{H}_S(U)$. Applying this to f^n, taking the n-th root, and letting $n \to +\infty$ we get $|h(f)| \le \|f\|_K$ for all $f \in \mathcal{H}_S(U)$. When this holds for $h \in \mathcal{S}$ and the strict compact $K \subset U$, we write $K \succ h$. We want to find a projection $\pi: \mathcal{S} \to E$. Let $h \in \mathcal{S}$ be fixed. We identify the vector space E^* of the linear forms on E which are bounded on the bounded subsets of E with a subspace of $\mathcal{H}_S(U)$ and for each h we find an $a \in E$ such that

(1) $$F(a) = h(F) \qquad \forall F \in E^*.$$

If a exists then it is unique since E^* separates the points of E. Let K be a strict compact subset of U such that $K > h$ and let L be the closed convex hull of K in E. Then L is strict compact. We will find $a \in L$ satisfying (1). Let $\mathfrak{F} = \{F_1,\dots,F_n\}$ be a finite subset of E^*. Define $\mathfrak{a}_{\mathfrak{F}} = \{x \in L;\ F_i(x) = h(F_i),\ 1 \le i \le n\}$. We assert that $\mathfrak{a}_{\mathfrak{F}} \neq \Phi$. For if not, then if $T: E \to C^n$ is defined by $T(x) = (F_1(x),\dots,F_n(x))$, then $(h(F_1),\dots,h(F_n)) = p \notin T(L)$. But since T is linear and S-continuous, and L is a strict compact convex set, so is $T(L)$. By the separation theorem, there is a linear functional G on C^n such that

(2) $$\operatorname{Re} G(p) > \operatorname{Re} G(T(L)).$$

But then setting $f = e^{G(F_1,\dots,F_n)}$, $f \in \mathcal{H}_S(U)$ and (2) implies $|h(f)| > \|f\|_L \ge \|f\|_K$. But this contradicts $K \succ h$. Hence $\mathfrak{a}_{\mathfrak{F}} \neq \Phi$ and $\mathfrak{a}_{\mathfrak{F}}$ is closed in L since L is a strict compact and the F_i's are S-continuous. The family $\{\mathfrak{a}_{\mathfrak{F}};\ \mathfrak{F}$ is a finite subset of $E^*\}$ has the finite intersection property, since

$$\mathfrak{a}_{\mathfrak{F}_1} \cap \ldots \cap \mathfrak{a}_{\mathfrak{F}_q} \supseteq \mathfrak{a}_{\mathfrak{F}_1 \cup \ldots \cup \mathfrak{F}_q} \neq \phi .$$

Thus, by the compactness of L we see that $\cap \mathfrak{a}_{\mathfrak{F}}$ is non-empty, so if $a \in \cap \mathfrak{a}_{\mathfrak{F}}$ then when $\mathfrak{F} = \{F\}$ we have that $F(a) = h(F)$, i.e., a satisfies (1) (hence a is the unique element of $\cap \mathfrak{a}_{\mathfrak{F}}$), and so setting $\pi(h) = a$ defines a map $\pi : \mathcal{S} \to E$.

Topology for $\mathcal{S}$.

Let $h \in \mathcal{S}$ and K be a strict compact in U with $K > h$. We take a convex balanced neighborhood V of 0 in E such that $K + V \subseteq U$. Formally we define a linear functional h_u for $u \in V$ on $\mathcal{H}_S(U)$ by

$$h_u(f) = \sum_{n=0}^{\infty} \frac{1}{n!} h(D_u^n f), \qquad f \in \mathcal{H}_S(U).$$

h_u is a continuous homomorphism of $\mathcal{H}_S(U)$. First we show that the series converges absolutely. Since $u \in V$, there is $\alpha > 1$ such that $\{\lambda u ; |\lambda| \leq \alpha\} \subseteq V$. Since $K > h$

$$(3) \qquad |h(D_u^n f)| \leq \|D_u^n f\|_K .$$

Let $k \in K$ and set $g(\lambda) = f(k + \lambda u)$, $\lambda \in \mathbb{C}$, $|\lambda| \leq \alpha$. By Cauchy

$$(4) \qquad |g^{(n)}(0)| \leq \frac{n!}{\alpha^n} \|g\|_{\{\lambda ; |\lambda| \leq \alpha\}} .$$

Let $K_u = K + \{\lambda u ; |\lambda| \leq \alpha\}$, then K_u is a strict compact set of U and (4) implies

$$|(D_u^n f)(k)| \leq \frac{n!}{\alpha^n} \|f\|_{K_u} ,$$

therefore $\|D_u^n f\|_K \leq n! \, \alpha^{-n} \|f\|_{K_u}$.

This, with (3) gives $|h(D_u^n f)| \leq n! \, \alpha^{-n} \|f\|_{K_u}$, thus

$$\sum_{n=0}^{\infty} \frac{1}{n!} |h(D_u^n f)| \leq \sum_{n=0}^{\infty} \frac{1}{\alpha^n} \|f\|_{K_u} = \frac{1}{1-\alpha^{-1}} \|f\|_{K_u} .$$

Hence the convergence is absolute and $|h_u(f)| \leq (1-\alpha^{-1})^{-1} \|f\|_{K_u}$. Next, we show that h_u is a homomorphism. h_u is clearly a linear map, so we only verify that $h_u(fg) = h_u(f)h_u(g)$. Since the above estimate implies that

$$\sum_{n,m} \frac{1}{n!}\frac{1}{m!} |h(D_u^n f)|\,|h(D_u^m g)| < +\infty,$$

the order of summation can be changed and we obtain

$$\begin{aligned} h_u(fg) &= \sum_{n\geq 0} \frac{1}{n!} h(D_u^n(fg)) = \\ &= \sum_{\substack{n>0\\ s+t=n}} \frac{1}{n!}\frac{n!}{s!t!} h(D_u^s f)h(D_u^t g) \\ &= \sum_{s\geq 0} \frac{1}{s!} h(D_u^s f) \sum_{t\geq 0} \frac{1}{t!} h(D_u^t g) \\ &= h_u(f)\, h_u(g). \end{aligned}$$

h_u is not identically zero since $1 \in \mathcal{H}_S(U)$ and $D_u^n 1 = 0$, for $n \geq 1$, implies that $h_u(1) = h(1) = 1$. Thus h_u is a continuous homomorphism, in fact, $|h_u(f)| \leq (1-\alpha^{-1})^{-1} \|f\|_{K_u}$ implies

(5) $$|h_u(f)| \leq \|f\|_{K_u} .$$

We can now compute $\pi(h_u)$. If $F \in E^*$ then $D_u F \equiv F(u)$ and $D_u^n F \equiv 0$ for $n > 1$ and so

$$\begin{aligned} h_u(F) &= h(F) + h(F(u)) \\ &= F(\pi(h)) + F(u) \\ &= F(\pi(h) + u). \end{aligned}$$

This implies, by the definition of π,

(6) $$\pi(h_u) = \pi(h) + u.$$

If $h \in \mathcal{S}$, $K \succ h$ and V is a balanced convex neighborhood of 0 in E such that $K + V \subseteq U$, then we set $N_{h,V} = \{h_u;\ u \in V\}$. We say that $W \subseteq \mathcal{S}$ is open if $h \in W$ implies that there is some $N_{h,V}$ contained in W. Thus a topology is defined on $\mathcal{S}$.

We now show that $\pi: \mathcal{S} \to E$ is a local homeomorphism. From (6), π on $N_{h,V}$ is given by $\pi(h_u) = \pi(h) + u$ and so if we prove that $N_{h,V}$ is open, it follows that π is a local homeomorphism and that $\pi|_{N_{h,V}}$ is a homeomorphism of $N_{h,V}$ onto $\pi(h) + V$. If $h_u \in N_{h,V}$, then there are $K > h$ with $K + V \subseteq U$ and a strict compact set $L > h_u$. Let W be a balanced convex neighborhood of 0 in E such that $L + W \subseteq U$ and $u + W \subseteq V$. We show that $N_{h_u,W} \subseteq N_{h,V}$, this clearly gives that $N_{h,V}$ is open. Since $w \in W$ implies $u + w \in V$ we have, using the properties of the directional derivative, that

$$\begin{aligned}(h_u)_w(f) &= \sum_{n\geq 0} \frac{1}{n!} h_u(D_w^n f)\\ &= \sum_{n\geq 0} \frac{1}{n!} \Big(\sum_{m\geq 0} \frac{1}{m!} h(D_u^m(D_w^n f))\Big)\\ &= \sum_{s\geq 0} \frac{1}{s!} h\Big(\sum_{n+m=s} \frac{s!}{n!m!} D_u^m(D_w^n f)\Big)\\ &= \sum_{s\geq 0} \frac{1}{s!} h(D_{u+w}^s f) = h_{u+w}(f).\end{aligned}$$

Thus $(h_u)_w = h_{u+w} \in N_{h,V}$ and $N_{h_u,W} \subseteq M_{h,V}$.

Since π is a local homeomorphism, $\mathcal{S}$ is a manifold spread over E. A function F on $\mathcal{S}$ is S-holomorphic when $F\circ(\pi|_{N_{h,V}})^{-1}$ is S-holomorphic on $\pi(h) + V$ for each $N_{h,V}$.

Now for $f \in \mathcal{H}_S(U)$, we define $\hat{f}$ on $\mathcal{S}$ by $\hat{f}(h) = h(f)$. We take $h \in \mathcal{S}$ and $K \subset U$ a strict compact such that $K > h$. Let V be a balanced convex neighborhood of 0 in E such that $K + V \subseteq U$. We will show that $\hat{f}\circ(\pi|_{N_{h,V}})^{-1}$ is S-holomorphic on $\pi(h) + V$. Take $\pi(h) + u_o \in \pi(h) + V$. We know that $K_{u_o} \subseteq U$ is a strict compact. If $B \in \mathcal{B}_E$ and K_{u_o} is contained and compact in $U \cap E_B$, then there are $\rho_B > 0$ and $M_B > 0$ such that $\|f\|_{K_{u_o}+\rho_B B} \leq M_B$. Consider the following neighborhood of $\pi(h) + u_o$ in E_B: $\pi(h) + u_o + \frac{1}{2}\rho_B B$, such that $\{u_o\} + \rho_B B \subset V$. Then

$$\sup_{t\in\frac{1}{2}\rho_B B} |\hat{f}\circ(\pi|_{N_{h,V}})^{-1}(\pi(h)+u_o+t)| = \sup_{t\in\frac{1}{2}\rho_B B} |\hat{f}(h_{u_o+t})| =$$

$$= \sup_{t\in\frac{1}{2}\rho_B B} |h_{u_o+t}(f)| \leq \sup_{t\in\frac{1}{2}\rho_B B} |(h_{u_o})_t(f)| \leq \sup_{t\in\frac{1}{2}\rho_B B.} \|f\|_{(K_{u_o})_t} < M_B$$

since $(K_{u_o})_t \subset K_{u_o} + \rho_B B$, $K_{u_o} > h_{u_o}$ and $(K_{u_o})_t > (h_{u_o})_t$. Next, we show that $\hat{f}\circ(\pi|_{N_{h,V}})^{-1}$ is G-holomorphic on $\pi(h) + V$. Since

$$\hat{f}\circ(\pi|_{N_{h,V}})^{-1}(\pi(h) + u) = \hat{f}(h_u) = h_u(f),$$

it is sufficient to prove that the mapping $u \mapsto h_u(f)$ is G-holomorphic on V. Let $u \in V$, $w \in E$ and set $g(\lambda) = h_{u+\lambda w}(f)$ for $\lambda \in \mathbb{C}$. For $u + \lambda_o w \in V$ we have that g is holomorphic at λ_o. In fact, choose $\epsilon > 0$ such that $u + \lambda_o w + \gamma w \in V$ for $|\gamma| \le \epsilon$. So, if $|\lambda - \lambda_o| < \epsilon$, we have

$$\begin{aligned} g(\lambda) &= h_{u+\lambda_o w+(\lambda-\lambda_o)w}(f) = (h_{u+\lambda_o w})_{(\lambda-\lambda_o)w}(f) = \\ &= \sum_{n\ge 0} \frac{1}{n!} h_{u+\lambda_o w}\left(\frac{D^n_{\frac{(\lambda-\lambda_o)}{\epsilon}\epsilon w}(f)}{}\right) \\ &= \sum_{n\ge 0} \frac{1}{n!} \frac{(\lambda-\lambda_o)^n}{\epsilon^n} h_{u+\lambda_o w}(D^n_{\epsilon w} f). \end{aligned}$$

This series converges uniformly and absolutely for $|\lambda - \lambda_o| \le \epsilon$ since $u + \lambda_o w + \epsilon w \in V$ implies

$$\sum_{n\ge 0} \frac{1}{n!} |h_{u+\lambda_o w}(D^n_{\epsilon w} f)| < +\infty$$

as we saw before. Thus if $f \in \mathcal{H}_S(U)$, then $\hat{f} \in \mathcal{H}_S(\mathcal{S})$.

There is a natural map $i: U \to \mathcal{S}$ defined by $i(x)$ = the homomorphism of evaluation at x. Evidently $\pi \circ i = id_U$. This implies that i is 1-1 and since π is a local homeomorphism defining the S-holomorphic structure of $\mathcal{S}$, i is a homeomorphic local homeomorphism and consequently a biholomorphism of U onto an open subset of $\mathcal{S}$. Since U is connected, there is a connected component $\mathcal{C}$ of $\mathcal{S}$ containing $i(U)$. How the inclusion map $j: U \to \mathcal{C}$ induces $j^*: \mathcal{H}_S(\mathcal{C}) \to \mathcal{H}_S(U)$ where $j^*(F) = F\circ j$ for $F \in \mathcal{H}_S(\mathcal{C})$. j^* is 1-1 since if $F \in \mathcal{H}_S(\mathcal{C})$ and $j^*(F) = 0$, then F vanishes identically on the open set $j(U) \subset \mathcal{C}$ and since $\mathcal{C}$ is connected, $F \equiv 0$. The mapping j^* is onto since for every S-holomorphic function $f \in \mathcal{H}_S(U)$, if $F = \hat{f}|_{\mathcal{E}}$ we have $j^*(F)(x) = F(j(x)) = \hat{f}(j(x)) = j(x)(f) = f(x)$ for all $x \in U$. Evidently j^* is an algebraic isomorphism. Let's now prove that j^* is continuous: let $K \subseteq U$ be a strict compact,

$$p_K(j^*(F)) = \sup_{t\in K} |j^*(F)(t)| = \sup_{t\in K} |(F\circ j)(t)| =$$

$$= \sup_{x\in j(K)} |F(x)| = q_{j(K)}(F),$$

where $q_{j(K)} \in CS(\mathcal{H}_S(\mathcal{E}))$ since $j(K)$ is a strict compact of $\mathcal{E}$, j is continuous and $\pi(j(K)) = K$. We will later prove that j^* is a topological isomorphism.

Finally we note that the S-holomorphic functions on $\mathcal{E}$ separate the points of $\mathcal{E}$: if $h_1, h_2 \in \mathcal{E}$ with $h_1 \neq h_2$, there is $f \in \mathcal{H}_S(U)$ such that $h_1(f) \neq h_2(f)$ and there is $F = \hat{f}|_{\mathcal{E}} \in \mathcal{H}_S(\mathcal{E})$ such that

$$F(h_1) = (\hat{f}|_{\mathcal{E}})(h_1) = h_1(f)$$

$$F(h_2) = (\hat{f}|_{\mathcal{E}})(h_2) = h_2(f),$$

therefore $F(h_1) \neq F(h_2)$ as we wanted.

__3.2 Theorem.__ Let U be a domain in a locally convex Hausdorff space where there exists a fundamental system of complete bounded subsets. Then there is a domain spread $\mathcal{E}$ over E and a biholomorphism $j: U \to U_{\mathcal{E}}$ ($U_{\mathcal{E}}$ = open subset of $\mathcal{E}$) such that:

(a) $\mathcal{H}_S(\mathcal{E})$ separates the points of $\mathcal{E}$.

(b) Every S-holomorphic function on U (identifying U with $U_{\mathcal{E}}$) extends to one S-holomorphic function on $\mathcal{E}$ and the extension map is a topological isomorphism.

Moreover $\mathcal{E}$ is maximal with respect to (a) and (b) in the sense that if M is a domain spread over E and $j': U \to U_M$ (U_M = = open subset of M) is a biholomorphism and (a) and (b) hold with M in place of $\mathcal{E}$, then M can be identified with an open subset of $\mathcal{E}$ by a biholomorphism which preserves the points of U.

Proof. It remains to prove the maximality of $\mathcal{E}$ and the fact that j^* is a topological isomorphism. We prove the former here. Let M be given as in the theorem. Define a map $T: M \to \mathcal{S}$ by $T(x)$ = the homomorphism of evaluation at x ($T(x)(f) = \bar{f}(x)$ where $\bar{f}$ is the extension of f at M and where we identify U both as a subset of M and $\mathcal{S}$). Since (b) holds, that is, $\mathcal{H}_S(M)$ is topologically isomorphic to $\mathcal{H}_S(U)$, evaluation at x is indeed a continuous homomorphism and T is well-defined. Since (a) holds for $\mathcal{H}_S(M)$, T is 1-1.

If we show that T is a local biholomorphism then, as T is 1-1, it follows that T maps M biholomorphically onto an open subset of $\mathcal{E}$ (since M is connected) while preserving points of U. Let $x \in M$ and set $h = T(x) \in \mathcal{S}$. Then there is a strict compact set $K \subseteq U$ such that $K > h$, by (b) for $\mathcal{H}_S(M)$. We choose a convex balanced neighborhood V of 0 in E such that $K + V \subseteq U$ and $x + V \subseteq M$. If $T(x+u) = h_u$ for $u \in V$, then T is a local biholomorphism. In fact, $(\mathcal{S},\pi)$ and (M,π_1) are manifold spread over E, hence if $V \subseteq E$, $x \in M$ with $x + V \subseteq M$, then there is a neighborhood W of x in M such that $\pi_1|_W$ is a homeomorphism of W onto $\pi_1(W)$ and $\pi_1(x) + V \subseteq \pi_1(W)$, with $x+V = (\pi_1|_W)^{-1}(\pi_1(x) + V)$, so that $\pi \circ T \circ \pi_1^{-1}$ and $\pi_1 \circ T^{-1} \circ \pi^{-1}$ are holomorphic. Therefore T and T^{-1} are holomorphic.

If $f \in \mathcal{H}_S(U)$ we denote by $\bar{f}$ its extension to M. $g(\lambda) = \bar{f}(x+\lambda u)$ is holomorphic on a neighborhood of $|\lambda| \leq 1$. As $g(\lambda) = \sum_{n\geq 0} \frac{1}{n!} \lambda^n g^{(n)}(0)$ we get, putting $\lambda = 1$, $\bar{f}(x + u) = \sum_{n\geq 0} \frac{1}{n!} (D_u^n \bar{f})(x)$. But $D_u^n \bar{f} = \overline{(D_u^n f)}$ since the two agree on $U \subseteq M$, so

$$\bar{f}(x+u) = \sum_{n\geq 0} \frac{1}{n!} (\overline{D_u^n f})(x) = \sum_{n\geq 0} \frac{1}{n!} h(D_u^n f) = h_u(f).$$

Hence $T(x+u) = h_u$, which is precisely what we want.

3.3 Definition. Let $(\mathcal{G},U)$ be a pair of domains spread over a locally convex Hausdorff with $U \subseteq \mathcal{G}$. If every complex valued S-holomorphic function on U extends to $\mathcal{G}$, we say $(\mathcal{G},U)$ is an S-extension pair. If the induced algebraic isomorphism $\mathcal{H}_S(U) \to \mathcal{H}_S(\mathcal{G})$ is a topological isomorphism in the topology of uniform convergence on strict compact subsets of the respective spaces, then we say $(\mathcal{G},U)$ is a normal S-extension pair.

3.4 Theorem. Let $(\mathcal{G},U)$ be an S-extension pair. Denote the extension of a function $f \in \mathcal{H}_S(U)$ to $\mathcal{G}$ by $\bar{f}$. If $f \mapsto \bar{f}(x)$ is a continuous linear functional on $(\mathcal{H}_S(U),\tau_{oe})$ for each $x \in \mathcal{G}$, then $(\mathcal{G},U)$ is a normal S-extension pair.

Proof. Let L be a strict compact subset of $\mathcal{G}$. It is sufficient to prove that there exists a strict compact subset K of U such that $\|\bar{f}\|_L \le \|f\|_K$. This implies the topological isomorphism of $(\mathcal{H}_S(\mathcal{G}),\tau_{oe})$ with $(\mathcal{H}_S(U),\tau_{oe})$.

Since $f \mapsto \bar{f}(x)$ is continuous on $(\mathcal{H}_S(U),\tau_{oe})$ for every $x \in L$, it follows that there is a strict compact subset K_x of U such that $|\bar{f}(x)| \le \|f\|_{K_x}$. Choose a closed balanced convex neighborhood V_x of 0 in E such that

(i) $x + V_x \subseteq \mathcal{G}$

(ii) $K_x + 2V_x \subseteq U$.

Then as $x + V_x$ is neighborhood of x in $\mathcal{G}$ and L is strict compact, there exist $x_1+V_1, x_2+V_2,\ldots,x_t+V_t$ covering L (where $V_i = V_{x_i}$ for $1 \le i \le t$). Define for $1 \le i \le t$

$$L_i = K_i + \{\lambda[\pi(L \cap (x_i+V_i)) - \pi(x_i)) \cap V_i]; |\lambda| \le 2\},$$

(where $K_i = K_{x_i}$ for $1 \le i \le t$). Then $L_i \subseteq K_i + 2V_i \subseteq U$ and L_i is a strict compact. V_i is closed and L is compact which implies $\pi(L \cap (x_i+V_i)) - \pi(x_i)$ is compact and since its intersection with V_i is contained in $[\pi(L) - \pi(x_i)] \cap V_i$ which is a strict compact, it follows that L_i is a strict compact. Thus $K = \bigcup_{i=1}^{t} L_i$ is a strict compact subset of U.

If $x \in L$, then $x \in x_i + V_i$ for some i, therefore V_i contains $\pi(x) - \pi(x_i)$ and so $x_i + \lambda(\pi(x)-\pi(x_i)) \in G$ for $|\lambda| \le 1$ by (i). Set $g(\lambda) = \bar{f}(x_i + \lambda(\pi(x)-\pi(x_i)))$ in a neighborhood of $|\lambda| \le 1$. Then $g(\lambda) = \sum_{n\ge 0} \frac{\lambda^n}{n!} g^{(n)}(0)$ implies that $g(1) = \sum_{n\ge 0} \frac{1}{n!} g^{(n)}(0)$ and so

$$\bar{f}(x) = \sum_{n\ge 0} \frac{1}{n!} (D^n_{\pi(x)-\pi(x_i)}\bar{f})(x_i).$$

Hence

$$(1) \qquad |\bar{f}(x)| \le \sum_{n\ge 0} \frac{1}{n!} |(D^n_{\pi(x)-\pi(x_i)}\bar{f})(x_i)| \le \sum_{n\ge 0} \frac{1}{n!} \|D^n_{\pi(x)-\pi(x_i)} f\|_{K_i} .$$

If $k \in K_i$, then $k + \lambda(\pi(x)-\pi(x_i)) \in U$ for $|\lambda| \le 2$ by (ii). So if σ is defined by $\sigma(\lambda) = f(k+\lambda(\pi(x)-\pi(x_i)))$ for $|\lambda| \le 2$, we have by Cauchy

$$|\sigma^n(0)| \le \frac{n!}{2^n} \|\sigma\|_{\{\lambda;|\lambda|\le 2\}} .$$

This implies

$$(2) \qquad |(D^n_{\pi(x)-\pi(x_i)} f)(k)| \le \frac{n!}{2^n} \|f\|_{L_i} ,$$

since $k + \lambda(\pi(x)-\pi(x_i)) \in L_i$ for $|\lambda| \le 2$. It follows from (1) and (2) that

$$|\bar{f}(x)| \le \|f\|_{L_i} \sum_{n\ge 0} \frac{1}{2^n} \le 2\|f\|_K .$$

As usual, applying this to powers of f we get $|\bar{f}(x)| \le \|f\|_K$. This holds for all $x \in L$, for all $f \in \mathcal{H}_S(U)$, hence $\|\bar{f}\|_L \le \|f\|_K$ for all $f \in \mathcal{H}_S(U)$.

<u>3.5 Corollary</u>. j^* is a topological isomorphism.

<u>Proof</u>. We know that $(\mathcal{E},U)$ is an S-extension pair. Since $\mathcal{E} \subseteq \mathcal{S}$, evaluations at points of $\mathcal{E}$ are τ_{oe}-continuous on $\mathcal{H}_S(U)$. Therefore $(\mathcal{E},U)$ is a normal S-extension pair. This implies that j^* is a topological isomorphism.

<u>3.6 Definition.</u> $\mathcal{E}$ is called the normal envelope of S-holomorphy of U.

4. CONSTRUCTION OF THE ENVELOPE OF S-HOLOMORPHY VIA THE SPECTRUM OF REGULAR CLASSES

Let E be a locally convex Hausdorff space over $\mathbb{C}$. For $B \in \mathcal{B}_E$, $z \in E$ and $r > 0$ the B-ball $B_E^B(z,r)$ with center z e radius r is defined by $B_E^B(z,r) = z + rB$.

Let (X,φ) be a domain spread over E. For each $B \in \mathcal{B}_E$ the B-boundary distance $d_X^B(x)$ at a point $x \in X$ is defined by

$$d_X^B(x) = \sup\{r > 0;\ x + rB \subset X\}.$$

We define $d_X^B(V) = \inf\{d_X^B(x);\ x \in V\}$ for $V \subset X$, and $V_r^B = V + rB$ if $r \in (0, d_X^B(V))$.

For $B \in \mathcal{B}_E$ and $P \in \mathcal{P}_b({}^nE;\mathbb{C})$ define $\|P\|_B = \sup\{P(x);x \in B\}$.

If $f: X \to \mathbb{C}$ is a S-holomorphic function, it follows from the definition of S-holomorphic functions that for each $x \in X$ there exists a unique sequence $(P_n)_{n\geq 0}$, $P_n \in \mathcal{P}_b({}^nE)$ such that for each $B \in \mathcal{B}_E$, with $\varphi(x) \in E_B$, there is $r > 0$ satisfying $d_X^B(x) > r$ and

$$\lim_{N\to+\infty} \{|f(y) - \sum_{n=0}^{N} P_n(y-x)|;\ y \in B_X^B(x,r)\} = 0,$$

where P_n is denoted by $\frac{1}{n!}\hat{\delta}^n f(x)$ or $\frac{1}{n!}\hat{\delta}^n(f\circ(\varphi|_V)^{-1})(\varphi(x))$, V is a neighborhood of x so that $\varphi|_V$ is homeomorphism and

$$f(x+h) = (f\circ(\varphi|_V)^{-1}(\varphi(x)+h) =$$

$$= \sum_{n\geq 0} \frac{1}{n!}\hat{\delta}^n(f\circ(\varphi|_V)^{-1})(\varphi(x))(h) = \sum_{n\geq 0}\frac{1}{n!}\hat{\delta}^n f(x)(h).$$

If $f \in \mathcal{H}_S(X)$, for $B \in \mathcal{B}_E$ the B-radius of convergence of f at the point $x \in X$, $\varphi(x) \in E_B$, is defined by

$$0 \leq \rho_f^B(x) = \sup\{r > 0;\ \sum_{n\geq 0} \|P_n\|_B\, r^n < +\infty\} = (\overline{\lim_{n\to+\infty}}\ \|P_n\|_B^{1/n})^{-1}.$$

4.1 Proposition. Given $f \in \mathcal{H}_S(X)$ and $x \in X$, for each $B \in \mathcal{B}_E$ with $\varphi(x) \in E_B$, let $r > 0$ and $(\frac{1}{n!}\hat{\delta}^n f(x))_{n\geq 0}$ be as above. Then for all $s \in (0,r)$ and all $a \in B$

(a) $$\frac{1}{n!}\hat{\delta}^n f(x)(a) = \frac{1}{s^n}\int_0^1 f\circ(\varphi|_{x+rB})^{-1}(\varphi(x)+se^{2\pi it}a)e^{-2\pi int}\,dt$$

(b) $$\left\|\frac{1}{n!}\hat{\delta}^n f(x)\right\|_B \leq \frac{1}{s^n}\|f\|_{B_X^B(x,s)} .$$

Proof. For each y such that $\varphi(y) \in \varphi(x) + sB$ we have that $f(y) = f\circ(\varphi|_{B_X^B(x,r)})^{-1}(\varphi(y))$ and as $f\circ(\varphi|_{B_X^B(x,r)})^{-1}$ is holomorphic on $\varphi(x) + sB$, we obtain the proposition.

4.2 Definition. Let X be a domain over E and $\mathcal{A} \subset \mathcal{H}_S(X)$. A set $D \subset X$ will be called $\mathcal{A}$-bounding if $\|f\|_D = \sup\{|f(x)|;\ x\in D\} < +\infty$ for all $f \in \mathcal{A}$.

4.3 Definition. A set $\mathcal{A} \subset \mathcal{H}_S(X)$ of S-holomorphic functions on the domain X over E will be called a regular class in X if $\mathcal{A} \neq \Phi$ and if

(1) $\lambda f^m \in \mathcal{A}$ for all $f \in \mathcal{A}$, $m \in \mathbb{N}$ and $\lambda \in \mathbb{C}$.

(2) $\frac{1}{n!}\hat{\delta}^n f(\ )(a) \in \mathcal{A}$ for all $f \in \mathcal{A}$, $n \in \mathbb{N}$ and $a \in E\setminus\{0\}$,

where $\frac{1}{n!}\hat{\delta}^n f(\ )(a)$ is the nth derivative of f in the direction of a, defined by

$$\frac{1}{n!}\hat{\delta}^n f(\ )(a):\ X \ni x \mapsto \frac{1}{n!}\hat{\delta}^n f(x)(a) \in \mathbb{C}.$$

(3) For all $x \in X$ and $B \in \mathcal{B}_E$ such that $\varphi(x) \in E_B$

$$\rho_{\mathcal{A}}^B(x) = \inf\{\rho_f^B(x);\ f \in \mathcal{A}\} > 0.$$

4.4 Definition. An open covering ν of the domain X over E will be called admissible if:

(1) For all $U \in \nu$ there is W neighborhood of 0 in E and $V \in \nu$ such that $U + W \subset V$.

(2) $U \cup V \in \nu$ for all $U,V \in \nu$.

It follows from Definition 4.4 that for all $B \in \mathcal{B}_E$ there is

$s > 0$ such that $d_X^B(U) > s$. In fact, for each $x \in U$, $x + 2W' = (\varphi|_{V'})^{-1}(\varphi(x)+2W')$, where $V' \subset X$ is open, $\varphi|_{V'}$ is a homeomorphism, $x \in V'$, $\varphi(x) + 2W' \subset \varphi(V')$, W' is a neighborhood of 0 in E such that $U + 2W' \subset V$. On the other hand, for all $B \in \mathcal{B}_E$ there exists $s > 0$ such that $sB \subset W' \cap E_B$, hence

$$\varphi(x) + 2sB \subset \varphi(x) + 2W' \subset \varphi(V')$$

and $d_X^B(x) \geq 2s$, from which we get $d_X^B(U) > s$.

Let $\eta(X)$ denote the set of all admissible coverings. For $\nu \in \eta(X)$ define

$$\mathcal{A}_\nu = \{f \in \mathcal{H}_S(X);\ \|f\|_{U \cap X_B} = \|f\|_{U_B} < +\infty \quad \forall\, U \in \nu,\ \forall\, B \in \mathcal{B}_E\}.$$

4.5 Lemma. (1) $\mathcal{A}_\nu$ is a complete locally convex Hausdorff algebra in the topology of ν_B-convergence (i.e. the topology of uniform convergence on all sets U_B, where $U \in \nu$, $B \in \mathcal{B}_E$).

(2) $\mathcal{A}_\nu$ is a regular class.

Proof. (1) Evidently $\mathcal{A}_\nu$ is locally convex Hausdorff. $\mathcal{A}_\nu$ is complete since a given Cauchy net on $\mathcal{A}_\nu$ converges pointwise to a map $f: X \to \mathbb{C}$, because of the completeness of $\mathbb{C}$, and the convergence is uniform on every U_B, with $U \in \nu$ and $B \in \mathcal{B}_E$.

Let us now prove that $\mathcal{A}_\nu$ is a regular class.

(a) It is immediate that $\lambda f^m \in \mathcal{A}_\nu$ for all $f \in \mathcal{A}_\nu$, $m \in \mathbb{N}$ and $\lambda \in \mathbb{C}$.

(b) Let U, W and V as in 4.4 with $U + 2W \subset V$. For $a \in E\backslash\{0\}$ and $x \in U_B$, where $B \in \mathcal{B}_E$,

$$\frac{1}{n!}\hat{\delta}^n f(x)(a) = \frac{1}{2\pi i}\int_{|\lambda|=\rho} \frac{f\circ(\varphi|_{V'})^{-1}(\varphi(x)+\lambda a)}{\lambda^{n+1}}\, d\lambda\ ,$$

where $\rho > 0$ is such that $\lambda a \in W$, for all $\lambda \in \mathbb{C}$, $|\lambda| \leq \rho$, $V' \subset X$ is open and $\varphi|_{V'}$ is homeomorphism, $x \in V'$ and $\varphi(x)$ +

$+ 2W \subset \varphi(V')$. Then

$$\left|\frac{1}{n!}\hat{\delta}^n f(x)(a)\right| \le \frac{1}{\rho^n}\|f\|_{V_{B_o}} < +\infty,$$

where $B_o = [B + \{\lambda a;\ |\lambda| \le \rho\}]$ is the convex balanced closed hull of $B + \{\lambda a;\ |\lambda| \le \rho\}$. It is easy to verify that $U_B + \{\lambda a; |\lambda| \le \rho\} \subset \subset V_{B_o}$. Then $\left\|\frac{1}{n!}\hat{\delta}^n f(\ )(a)\right\|_{U_B} < +\infty$ for all $f \in \mathcal{A}_\nu$, $a \in E\backslash\{0\}$ and $n \in \mathbb{N}$, i.e., $\frac{1}{n!}\hat{\delta}^n f(\ )(a) \in \mathcal{A}_\nu$.

(c) For each $x \in X$ there is $U \in \nu$ such that $x \in U$ and by Definition 4.4 there is W neighborhood of 0 in E and also $V \in \nu$ such that $U + 2W \subset V$. For each $B \in \mathcal{B}_E$ there exists $s > 0$ such that $d_X^B(x) > s$. From the fact $x \in U \cap E_B$ and $f \in \mathcal{A}_\nu$ it follows from the Cauchy inequalities that, for $a \in E_B$,

$$\begin{aligned}\left|\frac{1}{n!}\hat{\delta}^n f(x)(a)\right| &\le \left\|\frac{1}{n!}\hat{\delta}^n f(x)\right\|_B \|a\|_B^n \\ &\le \left(\frac{\|a\|_B}{s}\right)^n \|f\|_{x+sB} \\ &\le \left(\frac{\|a\|_B}{s}\right)^n \|f\|_{V\cap E_B}.\end{aligned}$$

Thus for $a \in rB$, $r \in (0,s)$, the series $\sum_{n\ge 0}\left\|\frac{1}{n!}\hat{\delta}^n f(x)\right\|_B r^n \le \le \|f\|_{V\cap E_B}\sum_{n\ge 0}\left(\frac{r}{s}\right)^n$ is convergent, which show that $\rho_f^B(x) \ge s$ for all $f \in \mathcal{A}_\nu$.

Let $\mathcal{A}_\nu$ be the regular class in X defined by an admissible covering ν. For all $B \in \mathcal{B}_E$ we have that

(1) ν gives origin to one admissible covering of X_B:

$$\nu_B = \{U_B = U \cap X_B;\ U \in \nu\}.$$

Obviously ν_B is a covering of X_B. Let us now prove that ν_B is admissible. From the hypothesis, for each $U \in \nu$ there is a W neighborhood of 0 in E and also $V \in \nu$ such that $U + W \subset V$. Taking $U_B, W_B = W \cap E_B$ = neighborhood of 0 in E_B and V_B, it follows that $U_B + W_B \subset V_B$. If $U_B, V_B \in \nu_B$, then

$$U_B \cup V_B = (U \cap X_B) \cup (V \cap X_B) = (U \cup V) \cap X_B \in \nu_B$$

because $U \cup V \in \nu$.

Moreover, the existence of $s > 0$ such that $d_x^B(U) > s$ and $U_s^B \subset V$ implies $d_X^B(U_B) > s$ and $(U_B)_s^B \subset V_B$.

In $\mathcal{A}_{\nu,B} = \{f|_{X_B};\ f \in \mathcal{A}_\nu\}$ we consider the topology of uniform convergence on all sets $U_B \in \nu_B$.

(2) $\mathcal{A}_{\nu,B}$ is a regular class in X_B.

It is easily seen that $\lambda(f|_{X_B})^m \in \mathcal{A}_{\nu,B}$ for all $f \in \mathcal{A}_\nu$, $\lambda \in \mathbb{C}$, $m \in \mathbb{N}$.

The equality

$$\frac{1}{n!}\,\hat{\delta}^n(f|_{X_B})(\ )(a) = \left(\frac{1}{n!}\,\hat{\delta}^n f(\ )(a)\right)\Big|_{X_B}$$

implies that $\frac{1}{n!}\,\hat{\delta}^n(f|_{X_B})(\ )(a) \in \mathcal{A}_{\nu,B}$ for all $f|_{X_B} \in \mathcal{A}_{\nu,B}$, $n \in \mathbb{N}$ and $a \in E\setminus\{0\}$.

Since $\mathcal{A}_\nu$ is a regular class we obtain $\rho^B_{\mathcal{A}_\nu}(x) > 0$ for all $x\in X_B$, and as $\rho^B_{\mathcal{A}_\nu}(x) = \rho^B_{\mathcal{A}_{\nu,B}}(x)$ for all $x\in X_B$, the proof is complete.

(3) Let (Y,φ') be a domain spread over E, $j: X \to Y$ an s. S-h.c of $\mathcal{A}_\nu$, $A' = \{f' \in \mathcal{H}_S(Y);\ f' \circ j \in \mathcal{A}_\nu\}$, $Y_B = \{y\in Y; \varphi'(y)\in E_B\}$ and in Y_B consider a topology analogous to the topology of X_B. Denote by j_B the restriction of j to X_B, $j_B: X_B \to Y_B$. Obviously, $\varphi|_{X_B} = \varphi'|_{Y_B} \circ j_B$.

For $x \in X_B$ let U be an open neighborhood of $j_B(x)$ in Y_B. By the hypothesis, there is $V \in \mathcal{B}(j_B(x))$, i.e., $V \subset Y_B$, $\varphi'|_V$ 1-1 and $\varphi'(V)$ neighborhood of $\varphi(x)$ in E_B, such that $V \subset U$. Since φ is continuous of X_B into E_B, there is a W neighborhood of x in X_B such that $\varphi(W) \subset \varphi'(V)$. We can take W such that $\varphi|_W$ is 1-1. For each $t \in W$

$$j_B(t) = (\varphi'|_V)^{-1}\,((\varphi|_W)(t)) \in V,$$

showing that $j_B(W) \subset V \subset U$ and so j_B is continuous of X_B into Y_B.

Let us now prove that j_B is an s.h.c. of $\mathcal{A}_{\nu,B}$. We have that for each $f \in \mathcal{H}_\nu$ there is $f' \in \mathcal{H}_S(Y)$ such that $f = f' \circ j$, then, for each $x \in X_B$, $f(x) = f'(j(x))$ and as $j(x) \in Y_B$ and $f'|_{Y_B} \in \in \mathcal{H}(Y_B)$ we have that $f|_{X_B}$ factors holomorphicly through j_B: $f|_{X_B} = f'|_{Y_B} \circ j_B$.

4.6 An example of an admissible covering with a barrelled regular class.

Let E_i be a complex Banach space ofr each $i = 1,2,\dots,$ with $E_1 \subset E_2 \subset \dots \subset E_n \subset \dots,$ where the topology induced on E_i by the topology of E_{i+1} coincides with the topology of E_i, and let E be the strict inductive limit of (E_i). If $A_n = \bigcup_{i=0}^{\infty} B_i(0,n)$, where $B_{i+1}(0,n) \cap E_i = B_i(0,n)$, we have that A_n is a convex balanced open subset in E, $\bigcup_{n=1}^{+\infty} A_n = E$ and $\nu = \{A_n\}_{n=1}^{\infty}$ is an admissible covering of E. Then, if

$$\mathcal{H}_\nu = \{f \in \mathcal{H}_S(E);\ \|f\|_{A_n} < +\infty\ \forall\ n\},$$

$\mathcal{H}_\nu$ is a barrelled regular class.

4.7 Proposition. Let ν be an admissible covering of the domain X over E such that $\mathcal{H}_{\nu,B}$ is barrelled for each $B \in \mathcal{B}_E$. Suppose $j: X \to Y$ is a s. S-h.c. of $\mathcal{H}_\nu$ and denote $A' = \{f' \in \mathcal{H}_S(Y);\ f' \circ j \in \mathcal{H}_\nu\}$. Then, if $x \in Y_B$ and $\hat{x}: \mathcal{H}_\nu \ni f \to f'(x) \in \mathbb{C}$, it follows that $\hat{x}: \mathcal{H}_{\nu,B} \ni f \mapsto f'(x) \in \mathbb{C}$ is continuous.

Proof. For each connected component $X'_B \neq \Phi$ of X_B let Y'_B be the corresponding connected component of Y_B.

The set $Z = \{x \in Y'_B;\ \hat{x}$ is continuous$\}$ is open in Y'_B: let $x \in Z$ and $U_B \in \nu_B$ such that $|\hat{x}(g)| \le c\|g\|_{U_B}$ for all $g \in \mathcal{H}_{\nu,B}$ and a suitable $c > 0$. Since $\mathcal{H}_{\nu,B}$ is an algebra and $\hat{x}$ is a homomorphism we assume that $c = 1$. Since ν_B is an admissible covering of X_B, and therefore of X'_B, there are $s > 0$ and $V_B \in \nu_B$ such that $d^B_{X'_B}(x) > s$ and $(U_B)^B_s \subset V_B$. By the Cauchy inequalities we have, for all $r \in (0,s)$ and $a \in rB$,

$$\sum_{n\geq 0} \left|\frac{1}{n!}\hat{d}^n g'(x)a\right| = \sum_{n\geq 0} \left|\left(\frac{1}{n!}\hat{d}^n g(\ )a\right)'(x)\right| =$$

$$= \sum_{n\geq 0} \left|\hat{x}\left(\frac{1}{n!}\hat{d}^n g(\ )a\right)\right| \leq \sum_{n\geq 0} \left\|\frac{1}{n!}\hat{d}^n g(\ )a\right\|_{U_B} \leq$$

$$\leq \|g\|_{V_B} \sum_{n\geq 0} \left(\frac{r}{s}\right)^n < +\infty.$$

Hence $\rho^B_{g'}(x) \geq s$ for all $g \in \mathcal{A}_{\nu,B}$. Let $0 < r < \min\{s, d^B_{Y'_B}(x)\}$. For $x' \in x + rB \subset Y'_B$

$$g'(x') = \sum_{n\geq 0} \frac{1}{n!}\hat{d}^n g'(x)(\varphi'(x') - \varphi'(x)),$$

if $a = \varphi'(x') - \varphi'(x)$, then $\|a\|_B \leq r$,

$$|g'(x')| \leq \sum_{n\geq 0} \left|\frac{1}{n!}\hat{d}^n g'(x)a\right| \leq \|g\|_{V_B} \sum_{n\geq 0} \left(\frac{r}{s}\right)^n.$$

Since $\mathcal{A}_{\nu,B}$ is an algebra we have that $\|g'\|_{x+rB} \leq \|g\|_{V_B}$, thus, for all $y \in x + rB \subset Y'_B$, $|\hat{y}(g)| = |g'(y)| \leq \|g\|_{V_B}$ for all $g \in \mathcal{A}_{\nu,B}$, i.e., $x + rB \subset Z$.

To prove that Z is closed in Y'_B it is enough to show that Z is sequentially closed. Let (x_n) be a sequence in Z converging to $x \in Y'_B$. $(\hat{x}_n)$ is weakly bounded in $(\mathcal{A}_{\nu,B})'$ = dual of $\mathcal{A}_{\nu,B}$, since $\hat{x}_n(g) = g'(x_n) \to g'(x) = \hat{x}(g)$, with $n \to \infty$, for all $g \in \mathcal{A}_{\nu,B}$. From the barrelledness of $\mathcal{A}_{\nu,B}$ it follows now that $(\hat{x}_n)$ is equicontinuous. Thus, we can find $U_B \in \nu_B$ such that $|\hat{x}_n(g)| \leq \|g\|_{U_B}$ for all $n \in N$ and $g \in \mathcal{A}_{\nu,B}$. Hence, $|\hat{x}(g)| \leq \|g\|_{U_B}$ for all $g \in \mathcal{A}_{\nu,B}$ which implies $x \in Z$.

From the connectedness of Y'_B we conclude $Z = Y'_B$ which shows the proposition.

4.8 Remark. To prove that $\hat{x}: \mathcal{A}_\nu \to C$ is continuous for all $x \in Y$ we must show that there exist $U \in \nu$, $B \in \mathcal{B}_E$ and $c > 0$ such that $|\hat{x}(f)| \leq c\|f\|_{U_B}$ for all $f \in \mathcal{A}_\nu$. But for each $x \in Y$ there is $B \in \mathcal{B}_E$ such that $x \in Y_B$, and since $\|f\|_{U_B} = \|f|_{X_B}\|_{U_B}$ for all U_B, $U \in \nu$ and

$$\hat{x}(f|_{X_B}) = (f|_{X_B})'(x) = (f'|_{Y_B})(x) = f'(x) = \hat{x}(f),$$

by Proposition 4.7 there are $U \in \nu$ and $c > 0$ such that

$$|\hat{x}(f)| = |\hat{x}(f|_{X_B})| \le c\|f|_{X_B}\|_{U_B} = c\|f\|_{U_B} .$$

Now, we construct the maximal s. S-h.c. of a regular class $\mathcal{A}_\nu$, so that $\mathcal{A}_{\nu,B}$ is barrelled for each $B \in \mathcal{B}_E$, via the spectrum $\mathcal{S}(\mathcal{A}_\nu)$ of $\mathcal{A}_\nu$. For a complex algebra A the spectrum $\mathcal{S}(A)$ of A is the set of nonzero homomorphisms of A into $\mathbb{C}$. Then we use this result to construct the envelope of S-holomorphy of certain domains X with the aid of the spectrum $\mathcal{S}(\mathcal{H}_S(X))$ of $\mathcal{H}_S(X)$.

Let (X,φ) be a domain over E and let $j: X \to Y$ be a s. S-h.s. of $\mathcal{A}_\nu$, where ν is an admissible covering of X. Then j is also a s. S-h.c. of $\mathcal{A}_\nu \cup \varphi^*E^*$, $\varphi^*E^* = \{\mu\circ\varphi;\ \mu \in E^*\}$, and each point $x \in (Y,\varphi')$ defines an evaluation map $\hat{x}: \mathcal{A}_\nu \cup \varphi^*E^* \to \mathbb{C}$ such that $E^* \ni \mu \to \hat{x}(\mu\circ\varphi) = \mu(\varphi'(x)) \in \mathbb{C}$ is weakly continuous, i.e., it is continuous for the topology $\sigma(E^*,E)$. If $\mathcal{A}_{\nu,B}$ is barrelled for all $B \in \mathcal{B}_E$, then $\hat{x}|_{\mathcal{A}_\nu}$ is continuous by Remark 4.8. Thus, it is reasonable to consider the following set:

$$\mathfrak{m}(X,\nu) = \{h: \mathcal{A}_\nu \cup \varphi^*E^* \to \mathbb{C};\ h|_{\mathcal{A}_\nu} \in \mathcal{S}(\mathcal{A}_\nu) \text{ is continuous and } h\circ\varphi^*: E^* \ni \mu \mapsto h(\mu\circ\varphi) \in \mathbb{C} \text{ is weakly continuous}\}.$$

Evidently, $\hat{X} = \{\hat{x};\ x \in X\} \subset \mathfrak{m}(X,\nu)$. Each $h \in \mathfrak{m}(X,\nu)$ defines a vector $qh \in E$ such that $\mu(qh) = h(\mu\circ\varphi)$ for all $\mu \in E^*$, since $h\circ\varphi^*$ is weakly continuous. $\mathfrak{m}(X,\nu)$ can be endowed with a Hausdorff topology so that $q: \mathfrak{m}(X,\nu) \to E$ is a local homeomorphism. In fact, given $h \in \mathfrak{m}(X,\nu)$, there are $U \in \nu$ and $B \in \mathcal{B}_E$ such that $|h(f)| \le \|f\|_{U_B}$ for all $f \in \mathcal{A}_\nu$. For this U and B let W, V be as in 4.4, with $U + 2W \subset V$. For $a \in W$ we put

$$(1) \qquad h_a(f) = \sum_{n\ge 0} h\left(\frac{1}{n!}\hat{\delta}^n f(\ )(a)\right) \quad \text{for } f \in \mathcal{A}_\nu .$$

We have that $|h(\frac{1}{n!}\hat{\delta}^n f(\ )(a))| \le \|\frac{1}{n!}\hat{\delta}^n f(\ )(a)\|_{U\cap E_B}$. Now, for $x \in U \cap E_B$,

$$\left|\frac{1}{n!}\hat{\delta}^n f(x)(a)\right| = \left|\frac{1}{2\pi i}\int_{|\lambda|=2} \frac{f\circ(\varphi|_{V'})^{-1}(\varphi(x)+\lambda a)}{\lambda^{n+1}}\,d\lambda\right|$$

$$\leq \frac{1}{2^n}\sup_{|\lambda|=2} |f(x+\lambda a)|.$$

If B_o is such that $B_o \supseteq B$ and $a \in E_{B_o}$, for

$$x \in U \cap E_B,\quad x+\lambda a \in (U+2W)\cap E_{B_o} \subset V \cap E_{B_o},$$

thus

$$\left|\frac{1}{n!}\hat{\delta}^n f(x)(a)\right| \leq \frac{1}{2^n}\|f\|_{V\cap E_{B_o}} < +\infty \quad \text{for all } x \in U\cap E_B.$$

Then, for all $f \in \mathcal{A}_\nu$ and $a \in W \cap E_{B_o}$,

$$(2)\qquad |h_a(f)| \leq \|f\|_{V\cap E_{B_o}} \sum_{n\geq 0}\frac{1}{2^n} = 2\|f\|_{V\cap E_{B_o}} < +\infty.$$

Furthermore, h_a is a homomorphism, hence $h_a \in \mathcal{S}(\mathcal{A}_\nu)$, since fo $f,g \in \mathcal{A}_\nu$:

$$\frac{1}{n!}\hat{\delta}^n(fg)(\ )(a) = \sum_{i+j=n}\frac{1}{i!}\hat{\delta}^i f(\ )(a)\,\frac{1}{j!}\hat{\delta}^j g(\ )(a),$$

$$\sum_{i+j\geq 0} \left|h(\tfrac{1}{i!}\hat{\delta}^i f(\ )(a)\right|\,\left|h(\tfrac{1}{j!}\hat{\delta}^j g(\ )(a))\right| \leq$$

$$\leq \sum_{i+j\geq 0}\frac{1}{2^i}\|f\|_{V\cap E_{B_o}}\frac{1}{2^j}\|g\|_{V\cap E_{B_o}} =$$

$$= \|f\|_{V\cap E_{B_o}}\|g\|_{V\cap E_{B_o}}\sum_{i,j=0}^{\infty}\frac{1}{2^{i+j}} < +\infty,$$

$$h_a(fg) = \sum_{n\geq 0} h(\tfrac{1}{n!}\hat{\delta}^n(fg)(\ )(a))$$

$$= \sum_{n\geq 0}\sum_{i+j=n} h(\tfrac{1}{i!}\hat{\delta}^i f(\ )(a))\,h(\tfrac{1}{j!}\hat{\delta}^j g(\ )(a))$$

$$= \Big(\sum_{i\geq 0} h(\tfrac{1}{i!}\hat{\delta}^i f(\ )(a))\Big)\Big(\sum_{j\geq 0} h(\tfrac{1}{j!}\hat{\delta}^j g(\ )(a))\Big)$$

$$= h_a(f)\,h_a(g).$$

For $\mu \in E^*$ we have, if $\mu\circ\varphi \in \mathcal{A}_\nu$,

$$h_a(\mu\circ\varphi) = \sum_{n\geq 0} h(\tfrac{1}{n!}\hat{\delta}^n(\mu\circ\varphi)(\ )(a))$$

$$= h(\mu\circ\varphi) + h(\mu(a)) = \mu(qh) + \mu(a) = \mu(qh+a).$$

Thus, if we define $h_a(\mu\circ\varphi) = \mu(qh) + \mu(a)$ for $\mu \in E^*$, then $h_a \in \mathfrak{m}(X,\nu)$. h_a defines $qh_a \in E$ such that $\mu(qh_a) = h_a(\mu\circ\varphi) = \mu(qh+a)$, hence $qh_a = qh + a$ since E^* separates the point of E.

Now, we show that $(h_a)_b = h_{a+b}$ for $a,b \in E$ so that $\{a,b,a+b\} \subset W$:

$$\frac{1}{n!}\hat{\delta}^n f(\ )(a+b) = \sum_{i+j=n} \frac{1}{i!}\hat{\delta}^i(\frac{1}{j!}\hat{\delta}^j f(\ )(b))(\ )(a)$$

$$(h_a)_b(f) = \sum_{j\geq 0} h_a(\frac{1}{j!}\hat{\delta}^j f(\ )(b))$$

$$= \sum_{j\geq 0}(\sum_{i\geq 0} h(\frac{1}{i!}\hat{\delta}^i(\frac{1}{j!}\hat{\delta}^j f(\ )(b))(\ )(a))) =$$

$$= \sum_{n\geq 0} h(\frac{1}{n!}\hat{\delta}^n f(\ )(a+b) = h_{a+b}(f)$$

for all $f \in \mathcal{A}_\nu$.

Let $\mathfrak{n}$ be the filter of convex balanced open neighborhoods of $0 \in E$. To define a topology on $\mathfrak{m}(X,\nu)$ we define a base of neighborhoods for each $h \in \mathfrak{m}(X,\nu)$:

$\mathfrak{n}(h) = (\{h_a;\ a \in W'\}$, where $U \in \nu$ is determined by the continuity of $h|_{\mathcal{A}_\nu}$, W is neighborhood of $0 \in E$ and $V \in \nu$ as in 4.4, with $U + 2W \subset V$, $W' \in \mathfrak{n}$, $W' \subset W$). For $h^1, h^2 \in \mathfrak{m}(X,\nu)$, we consider $N(h^1)$, $N(h^2)$ where $N(h^i) \in \mathfrak{n}(h^i) = (\{(h^i)_{a_i};\ a_i \in W^i\}$, U^i, W_1^i, V^i are determined as above, and $W^i \in \mathfrak{n}$, $W^i \subset W_1^i$ for $i = 1,2$). If $h \in N(h^1) \cap N(h^2)$, then $h = (h^1)_{a_1} = (h^2)_{a_2}$ with $a_1 \in W^1$, $a_2 \in W^2$. For $U = V^1 \cup V^2 \in \nu$ there is a neighborhood W of $0 \in E$ and also $V \in \nu$ such that $U + 2W \subset V$. Let W' be a neighborhood of $0 \in E$ such that $a_1 + W' \subset W^1$, $a_2 + W' \subset W^2$ and $W' \subset W$. For each $c \in W'$ we have that

$$h_c = ((h^1)_{a_1})_c = (h^1)_{a_1+c} \in N(h^1),$$

$$h_c = ((h^2)_{a_2})_c = (h^2)_{a_2+c} \in N(h^2),$$

hence, $N(h) = \{h_c;\ c \in W'\} \subset N(h^1) \cap N(h^2)$. It is easy to see that the topology defined by this neighborhoods base is a Hausdorff topology on $\mathfrak{M}(X,\nu)$ and that $q: \mathfrak{M}(X,\nu) \to E$ is a local homeomorphism. The extension $\hat{f}$ of a function $f \in \mathcal{A}_\nu$ is given by $\hat{f}: \mathfrak{M}(X,\nu) \ni$ $\ni h \mapsto \hat{f}(h) = h(f) \in \mathbb{C}$. $\hat{f}$ is G-holomorphic since for small $|z|$, $z \in \mathbb{C}$, $\hat{f}(h_{za}) = \sum_{n\geq 0} h(\frac{1}{n!} \hat{d}^n f(\ )(a))z^n$. The map $\hat{f}$ is S-holomorphic: let $N(h) = \{h_a;\ a \in W'\}$ be a neibhborhood of h in $\mathfrak{M}(X,\nu)$, $q|_{N(h)}$ is a homeomorphism, $q(N(h)) = qh + W'$. Let $B_1 \in \mathcal{B}_E$ be such that $qh \in E_{B_1}$, then $(qh + W') \cap E_{B_1}$ is a neighborhood of qh in E_{B_1}, hence there is $\rho > 0$ such that $qh + \rho B_1 \subset (qh+W') \cap E_{B_1}$ and

$$(q|_{N(h)})^{-1}\ (qh + \rho B_1) = \{h_{\rho b};\ b \in B_1\}$$

then

$$|\hat{f}(h_{\rho b})| = |h_{\rho b}(f)| \leq 2\|f\|_{V\cap E_{B_0}} < +\infty$$

by (2), where $B_0 = [B_1 \cup B]$ and B is determined by the continuity of h, showing that $\hat{f}$ is S-holomorphic.

Now let $\mathcal{E}(X,\nu)$ be the componente of $\mathfrak{M}(X,\nu)$ which contains $\hat{X}$ and we define $j_\nu: X \ni x \mapsto j_\nu(x) = \hat{x} \in \mathfrak{M}(X,\nu)$. It is clear that $j_\nu: X \to \mathcal{E}(X,\nu)$ is a s. S-h.c. of $\mathcal{A}_\nu$.

<u>4.9 Proposition</u>. If $\mathcal{A}_\nu$ is such that $\mathcal{A}_{\nu,B}$ is barrelled for all $B \in \mathcal{B}_E$, then $j_\nu: X \to \mathcal{E}(X,\nu)$ is the maximal s. S-h.c. of $\mathcal{A}_\nu$.

<u>Proof</u>. Let $j': X \to X'$ be a s. S-h.c. of $\mathcal{A}_\nu$. Let $f' \in \mathcal{H}_S(X')$ be the S-holomorphic continuation of $\mathcal{A}_\nu$. By Proposition 4.7, each $x \in X'$ defines a continuous homomorphism $\hat{x}: \mathcal{A}_\nu \ni f \mapsto f'(x) \in \mathbb{C}$. Moreover, $\hat{x}(\mu\circ\varphi) = \mu(\varphi'(x))$ for all $\mu \in E^*$. Hence, the mapping $j'': X' \ni x \mapsto \hat{x} \in \mathfrak{M}(X,\nu)$ is well-defined and satisfies $j_\nu = j''\circ j'$ and $\varphi' = q\circ j''$, where φ' is the local homeomorphism from X' into E. Of the continuity of j'' it follows that $j''(X') \subset \mathcal{E}(X,\nu)$. Then, we can conclude that $j_\nu: X \to \mathcal{E}(X,\nu)$ is the maximal s. S-h.c. of $\mathcal{A}_\nu$.

<u>4.10 Definition</u>. Let X be a domain over E. We define

$$\tau(X) = \{\nu \in \eta(X);\ \mathcal{A}_{\nu,B} \text{ is barrelled for all } B \in \mathcal{B}_E\}.$$

X is said to be a Sτ-domain if $\mathcal{H}_S(X) = \bigcup\{\mathcal{A}_\nu;\ \nu \in \tau(X)\}$. For example, every metrizable domain is an $S\tau$-domain.

Define $\mathfrak{m}_\tau(X) = \{h \in \mathcal{S}(\mathcal{H}_S(X));\ h \in \mathcal{E}(X,\nu)$ for all $\nu \in \tau(X)$ and there is a neighborhood W of $0 \in E$ such that $h_a \in \mathcal{E}(X,\nu)$ for all $a \in W$, $\nu \in \tau(X)\}$. For $h \in \mathfrak{m}_\tau(X)$, $\bigcup_{a \in W} h_a = h+W =$ $= (q_\nu|_V)^{-1}(q_\nu h + W)$ where V is such that $V \subset \mathcal{E}(X,\nu)$ is open, $h \in V$, $q_\nu|_V$ is a homeomorphism, $q_\nu h + W \subset q_\nu(V)$, for all $\nu \in \tau(X)$. So, $d^B_{\mathcal{E}(X,\nu)}(h) > 0$ for all $B \in \mathcal{B}_E$, $\nu \in \tau(X)$.

The family of all sets $\{h_a;\ a \in W\} \subset \mathcal{E}(X,\nu)$ for all $\nu \in \tau(X)$ and W as above is a base of a Hausdorff topology on $\mathfrak{m}_\tau(X)$ such that $q\colon \mathfrak{m}_\tau(X) \to E$ is a local homeomorphism ($q = q_\nu$ for all $\nu \in \tau(X)$). The extension of a function $f \in \mathcal{A}_\nu$, $\nu \in \tau(X)$, is given by $\hat{f}\colon \mathfrak{m}_\tau(X) \ni h \mapsto h(f) \in \mathbb{C}$. Let $\mathcal{E}_\tau(X)$ be the component of $\mathfrak{m}_\tau(X)$ which contains $\hat{X}$. Then, $j_X\colon X \ni x \mapsto \hat{x} \in \mathcal{E}_\tau(X)$ is the maximal s. S-h.c. of $\bigcup\{\mathcal{A}_\nu;\ \nu \in \tau(X)\}$.

<u>4.11 Proposition</u>. If X is an $S\tau$-domain, then $j_X\colon X \to \mathcal{E}_\tau(X)$ is the envelope of S-holomorphy of X.

<u>Proof</u>. The maximality of j_X is a consequence of the intersection property (see [13]). Let $j'\colon X \to X'$ be a s. S-h.c. of $\mathcal{H}_S(X)$, then j' is a s. S-h.c. of each $\mathcal{A}_\nu$; $\nu \in \tau(X)$. By Proposition 4.9 there are morphisms of domains $j'_\nu\colon X' \to \mathcal{E}(X,\nu)$ such that $j_\nu = j'_\nu \circ j'$ for all $\nu \in \tau(X)$. j'_ν is defined by $j'_\nu\colon X' \ni x \mapsto \hat{x} \in \mathcal{E}(X,\nu)$. It is easy to see that $j''\colon X' \ni x \mapsto \hat{x} \in \mathfrak{m}_\tau(X)$ is well-defined and it is continuous (see Proposition 4.9), hence $j''(X') \subset \mathcal{E}_\tau(X)$ and $j''\colon X' \to \mathcal{E}_\tau(X)$ is a morphism of domains satisfying $j_X = j'' \circ j'$ which implies the proposition.

REFERENCES

[1] Alexander, H., Analytic functions on Banach spaces, tese, Univ. Berkeley, 1968.

[2] Barroso, J.A., Topologia nos espaços de aplicações holomorfas entre espaços localmente convexos, Anais da Academia Brasileira de Ciências, t. 43 (1971).

[3] Coeuré, G., Fonctions plurisousharmoniques sur les espaces vectoriels topologiques et applications à l'étude des fonctions analytiques, Ann. Inst. Fourier, t. 20, (1970), 361-432.

[4] Hirschowitz, A., Prolongement analytique en dimension infinie, Ann. Inst. Fourier, t. 22, (1972), 255-292.

[5] Hormander, L., An Introduction to Complex Analysis in Several Variables, Van Nostrand, Princeton, 1966.

[6] Matos, M.C., Holomorphic mappings and domains of Holomorphy, thesis, Univ. Rochester, 1970.

[7] Nachbin, L., Topology on spaces of holomorphic mappings, Ergenbnisse der Mateematik, t. 47, Springer-Verlag, Berlin, 1969.

[8] Noverraz, P., Pseudo-convexité, convexité polynomiale et domaines d'Holomorphie en dimension infinie, Notes in Mathematics, 48, North-Holland, Amsterdam, 1973.

[9] Paques, O.W., Produtos tensoriais de funções Silva-holomorfas e a propriedade de aproximação, Thesis, UNICAMP, 1977.

[10] Paques, O.W., Tensor Products of Silva-Holomorphic Functions, Advances in Holomorphy, North Holland (to appear)

[11] Paques, O.W., The approximation property for certain spaces of holomorphic mappings, Approximation Theory and Functional Analysis, Proceedings of the International Symposium on Approximation Theory, to appear, Campinas, 1977.

[12] Schottenloher, M., Analytic continuation and regular classes in locally convex Hausdorff spaces, Portugaliae Mathematica, t. 33, fasc. 4, 1974.

[13] Schottenloher, M., Riemann Domains: Basic Results and Open Problems - Proceedings on Infinite Dimensional Holomorphy, University of Kentucky 1973, Lecture Notes in Mathematics, 364, Springer-Verlag.

DENSE SUBALGEBRAS IN TOPOLOGICAL ALGEBRAS OF DIFFERENTIABLE FUNCTIONS

Guido Zapata*

Instituto de Matemática
Universidade Federal do Rio de Janeiro
Rio de Janeiro, Brasil

1. INTRODUCTION

Description of dense subalgebras in topological algebras of differentiable functions on finite dimensional manifolds started in 1949 with L. Nachbin's paper [8]. The corresponding theorem given there characterizes dense subalgebras in the algebra of C^m functions endowed with the compact open topology of order m. Also, it represented a partial answer to a question posed by I. Segal the same year. That question, concerning the description of closed subalgebras of differentiable functions, in analogy with Stone-Weierstrass theorem, is, up to our knowledge, still unsolved. In fact, it seems that the only known results in that direction are the theorem of H.Whitney on closed ideals, published in 1948 [17], and Nachbin's theorem already mentioned. At this point we should quote that a related result was published in early 1940 in an apparently little known paper by J. Khourguine and N. Tschetinine [6]. These authors gave a characterization of $C^m[0,1]$ among its closed subalgebras, under the influence of Stone's results [15].

Later, other theorems were given by G. Reid in 1966, density conditions remaining the same as in Nachbin's theorem [13]. That

* The author was partially suported by FINEP, Brasil.

author was motivated by attempts of constructing a theory of distributions for, in particular, compact groups, and considered several topological algebras, including for instance $C_c^\infty(X)$ endowed with the inductive limit topology. We remark that Reid was unaware of Nachbin's paper since he obtained also some particular cases of Nachbin's theorem, but using a different approach.

Finally, as a consequence of our own research on the differentiable analogous of the Bernstein-Nachbin problem, we also obtained density theorems for subalgebras in weighted algebras of differentiable functions [19]. We remark that in this setting there is a restriction on the domains of the functions, namely they are open subsets of euclidean space. However, in this restricted context we generalized Nachbin's theorem and almost all related Reid's theorems. In fact, the only exception was concerned with the C^∞ inductive limit topology, in spite of the fact that, when m is finite, the corresponding C^m inductive limit topology is a weighted one. Further, we remark that a theory of weighted spaces of differentiable functions on a manifold is lacking. Having as motivations those facts, we looked for a general setting in order to study the problem of describing dense subalgebras in topological algebras of differentiable functions. A suitable context seems to be that of m-admissible algebras (Definition 1) and, for instance, many important spaces arising in distribution theory are examples of that notion.

In the new context adopted here, we characterize m-admissible algebras among their closed subalgebras (Theorems 1 and 2) and we generalize, in an improved way, all of the above mentioned theorems on dense subalgebras (Theorems 3 and 4). In particular, since the present approach emphasizes the essential contents of those theorems, it consequently allows fairly simple proofs of them. At the same time this approach seems to be useful in the attempts to extend analysis to infinite dimensional spaces, as is the case of Nachbin's theorem. (See [9] and [10] for results and additional references). Finally, the setting adopted here is suitable for the study of other problems in the approximation of differentiable functions.

The final version of this paper was written at the University of Paderborn, West Germany, during our stay under the CNPq-GMD agreement. We would like to express here our acknowledgments to those institutions and also to Professor K. Bierstedt, from Paderborn, for useful conversations on several subjects, including the approximation of functions.

2. THE CLASS $\mathrm{Top}_a^m(X)$.

Here we introduce the main notations and terminology that will be used in the following.

Throughout this paper n denotes an integer ≥ 1. Same for m, also allowed to be ∞. $\mathbb{N}$ is the set of natural numbers and $\mathbb{N}_m^n$ is the set of multi-indexes $\alpha \in \mathbb{N}^n$ such that $|\alpha| = \alpha_1 + \ldots + \alpha_n \leq m$; we put $\mathbb{N}_m = \mathbb{N}_m^1$.

Function will mean $\mathbb{R}$-valued function and algebra of functions an algebra with respect to the pointwise operations. Given G, a nonempty set of functions defined on a set X, $\mathbb{R}[G]$ denotes the algebra generated by G and if $\mathfrak{F}(\mathbb{R}^n)$ is a set of functions on $\mathbb{R}^n$, then $\mathfrak{F}(G^n)$ has the natural meaning as a set of functions on X. We say that: i) G separates points if given $x,y \in X$, $x \neq y$, there exists $g \in G$ such that $g(x) \neq g(y)$; G does not vanish at any point if $G(x) \neq \{0\}$, $\forall\, x \in X$; iii) G is strongly separating if it separates points and does not vanish at any point.

If X is a locally compact Hausdorff space we denote by $C_o(X)$ and $C_c(X)$ the algebras of continuous functions on X which, respectively, vanish at infinity and have compact support. Further, if X is also a C^∞ manifold of finite dimension (without boundary), we denote by $C^m(X)$ the algebra of C^m functions on X and we put

$C_c^m(X) = C^m(X) \cap C_c(X)$. When X is an open subset of $\mathbb{R}^n$ we denote by $C_b^m(X)$ the algebra of all $f \in C^m(X)$ such that f and all its partial derivatives of order $\leq m$ are bounded. Analogous definition for $C_o^m(X)$, where "bounded" is replaced by "vanish at infinity". Also, we denote by $C_{0,1}^\infty(\mathbb{R})$ the algebra of all $f \in C_o^\infty(\mathbb{R})$ such that $|f(x)| \leq \text{constant}\cdot|x|$, $\forall\, x \in \mathbb{R}$. The seminorms $f \in C_b^m(X) \to$ $\to \sup\{|\partial^\alpha f(x)|,\ x \in X\}$, $\alpha \in \mathbb{N}_m^n$, generate the topology τ_u^m of uniform convergence of order m.

If X is a C^∞ manifold of finite dimension and $\mathcal{Q}$ denotes the maximal atlas on X, let $\mathcal{Q}_c$ be the atlas $(\lambda_j, V_j)_{j \in J}$ obtained by restricting the charts of $\mathcal{Q}$ to their relatively compact open subsets. Then it is clear that $(V_j)_{j\in J}$ is a basis for the topology on X. Also, for every $f \in C^m(X)$ and $j \in J$ we have that $f \circ \lambda_j^{-1} \in$ $\in C_b^m(\lambda_j(V_j))$ and for $\ell \in \mathbb{N}_m$ we define

$$p_{j,\ell}(f) = \sum_{|\alpha| \leq \ell} \frac{1}{\alpha!} \sup\{|\partial^\alpha (f\circ\lambda_j^{-1})(y)|,\ y \in \lambda_j(V_j)\}.$$

Then $p_{j,\ell}$ is a seminorm on $C^m(X)$ such that

$$p_{j,\ell}(fg) \leq p_{j,\ell}(f)\cdot p_{j,\ell}(g), \quad \forall\, f,g \in C^m(X).$$

The topology τ_c^m generated by the family of seminorms $p_{j,\ell}$, $j \in J$, $\ell \in \mathbb{N}_m$, is called compact open topology of order m. Further, if K is a nonempty compact subset of X, let $C_c^m(K)$ be the space of all $f \in C_c^m(X)$ such that $\mathrm{Supp}(f) \subset K$, endowed with the topology induced by τ_c^m. Let τ_i^m be the finest linear topology on $C_c^m(X)$ such that all the inclusions $C_c^m(K) \subset C_c^m(X)$ are continuous, called linear inductive limit topology of order m. When X is σ-compact, τ_i^m is a locally convex topology.

We recall that a topological algebra is an algebra endowed with a linear topology such that multiplication is separately continuous. In particular, if A is a topological algebra and $B \subset A$ is a subalgebra, then $\bar{B}$ is a subalgebra too.

Remark 1. $(C^m(X),\tau_c^m)$ and $(C_c^m(X),\tau_i^m)$ are examples of topological algebras. In fact, the topologies under consideration are linear so we need to verify that multiplication is separately continuous. In the case of $C^m(X)$ this is straightforward. In the other case, it is enough to show that for $g \in C_c^m(X)$ and $K \subset X$ compact given, the linear mapping $f \in C_c^m(K) \to fg \in C_c^m(X)$ is continuous. In fact, given a neighbourhood W of 0 in τ_i^m, then $W \cap C_c^m(K)$ is a neighbourhood of 0 in $C_c^m(K)$, so there exist $J_o \subset J$ finite, $\ell \in \mathbb{N}_m$, and $\epsilon > 0$ such that $f \in C_c^m(K)$ and $p_{j,\ell}(f) < \epsilon$, $\forall\ j \in J_o$ imply $f \in W \cap C_c^m(K)$. Taking $\epsilon' > 0$ such that $p_{j,\ell}(g)\cdot\epsilon' < \epsilon$, $\forall\ j \in J_o$, it follows that $fg \in W$ for every $f \in C_c^m(K)$ such that $p_{\ell,j}(f) < \epsilon'$, $j \in J_o$.

Definition 1. Let X be a C^∞ Hausdorff manifold of dimension n. $Top_a^m(X)$ will denote the class of all topological algebras A such that:

i) A is a subalgebra of $C^m(X)$

ii) $C_{0,1}^\infty(A)\cdot C_c^m(A^n) \subset A$

iii) A contains $C_c^m(X)$ as a dense subset

iv) For any $x \in X$ and $v \in T_x(X)$ = tangent space at x, the mappings $f \in A \to f(x) \in \mathbb{R}$ and $f \in A \to df(x)(v) \in \mathbb{R}$ are continuous.

An element $A \in Top_a^m(X)$ will be called m-admissible algebra on X. The topology on A will be denoted by τ_A.

Examples:

1) $C^m(X)$ endowed with the topology τ_c^m.

2) $C_c^m(X)$ endowed with the topology τ_i^m.

3) $C_c^\infty(X)$ endowed with the coarser of the topologies for which the inclusions $C_c^\infty(X) \subset C_c^m(X)$ are continuous for all $m \in \mathbb{N}$, where $C_c^o(X) = C_c(X)$.

4) Let X be an open subset of $\mathbb{R}^n$ and $V = (V_\alpha)$, $\alpha \in \mathbb{N}_m^n$, be a family of sets of weights on X, that is upper-semicontinuous and

positive functions on X. We will assume that V has the following properties: i) $\forall x \in X$ and $\alpha \in \mathbb{N}_1^n$ there exists $v_\alpha \in V_\alpha$ such that $v_\alpha(x) > 0$; ii) $\forall \alpha, \beta \in \mathbb{N}_m^n$ such that $\beta \leq \alpha$ and $v_\alpha \in V_\alpha$ there exist $v_\beta \in V_\beta$ and $v_{\alpha-\beta} \in V_{\alpha-\beta}$ such that $v_\alpha \leq v_\beta \cdot v_{\alpha-\beta}$. Let $C^m V_\infty(X)$ be the algebra of all $f \in C^m(X)$ such that $v_\alpha \partial^\alpha f$ vanishes at infinity for all $\alpha \in \mathbb{N}_m^n$ and $v_\alpha \in V_\alpha$. Every such α and v_α define a seminorm

$$f \to \sup \{v_\alpha(x) \; |\partial^\alpha f(x)|, \; x \in X\}$$

on $C^m V_\infty(X)$. Under the topology generated by those seminorms, $C^m V_\infty(X)$ becomes a topological algebra also called a weighted algebra ([19]). For appropriate choices of V we get as particular cases the topological algebras $C^m(X)$, $\mathcal{D}^m(X)$ $(m < \infty)$, $\mathcal{D}^F(X)$, $\mathcal{B}_o^m(X)$ and $S^m(\mathbb{R}^n)$ (See [5] for the notation). Also if $V_\alpha = C_o^+(\mathbb{R}^n)$, $\alpha \in \mathbb{N}_m^n$, then $C^m V_\infty(\mathbb{R}^n) = C_b^m(\mathbb{R}^n)$ as sets. The corresponding topology is called strict topology of order m, in analogy with the continuous case. Denote by $\dot{C}^m V_\infty(X)$ the closure of $C_c^m(X)$ in $C^m V_\infty(X)$. When $C^m V_\infty(X) = \dot{C}^m V_\infty(X)$, as in the particular cases mentioned above, it follows easily that $C^m V_\infty(X) \in \mathrm{Top}_a^m(X)$.

<u>Remark 2</u>. In general we have that $C^m V_\infty(X) \neq \dot{C}^m V_\infty(X)$ ([1]), but there are known conditions ensuring equality ([2] and [18]). Even when $C^m V_\infty(X) \neq \dot{C}^m V_\infty(X)$, it can be shown that $\dot{C}^m V_\infty(X) \in \mathrm{Top}_a^m(X)$. In fact this is a consequence of the continuity of the mappings $g \in A \to \varphi \circ g \in A$, $\varphi \in C_{0,1}^\infty(\mathbb{R})$ and $g \in A \to gh \in A$, $h \in C_c^m(A^n)$, where $A = C^m V_\infty(X)$.

3. SOME PRELIMINARY LEMMAS

<u>Definition 2</u>. Let X be a locally compact Hausdorff space. A set B of continuous functions on X has partitions of unity on compact subsets of X if for any nonempty compact subset K of X and any finite open covering $V_1, \ldots, V_r$ of K there exist $\theta_1, \ldots, \theta_r \in B$

such that $\theta_1 +\dots+ \theta_r = 1$ on K, $0 \le \theta_i \le 1$ and $\mathrm{Supp}(\theta_i) \subset V_i$, $i = 1,\dots,r$.

Lemma 1. Let X be a locally compact Hausdorff space and $B \subset C_c(X)$ be a subalgebra. Then B has partitions of unity on compact subsets of X if and only if for any $x,y \in X$, $x \neq y$, there exists $g \in B$ such that $g = 1$ on a neighbourhood of x, $g = 0$ on a neighbourhood of y and $0 \le g \le 1$.

Proof. Clearly the condition on B is necessary. The sufficiency will be a consequence of the following result: Given $x \in X$ and V an open neighbourhood of x, there exists $g_x \in B$ such that $g_x = 1$ on a neighbourhood of x, $0 \le g_x \le 1$ and $\mathrm{Supp}(g_x) \subset V$. In fact, from the hypothesis there exists $g \in B$ such that $g = 1$ on a neighbourhood of x and $0 \le g \le 1$. In case $\mathrm{Supp}(g) \subset V$ we take $g_x = g$. Otherwise, let $H = (X\setminus V) \cap \mathrm{Supp}(g)$. For every $y \in H$ there exists $h_y \in B$ such that $h_y = 1$ on a neighbourhood of x, $h_y = 0$ on a neighbourhood V_y of y and $0 \le h_y \le 1$. Since H is compact, there exist $y_1,\dots,y_n \in H$ such that $H \subset V_{y_1} \cup\dots\cup V_{y_n}$. Then it is enough to take $g_x = g\cdot h_{y_1}\cdot\dots\cdot h_{y_n}$.

Now let K be a nonempty compact subset of X and $V_1,\dots,V_r$ be an open covering of K. Using the result already proved it follows that for every $x \in K$ threre exists $i \in \{1,\dots,r\}$, $g_x \in B$ and W_x a compact neighbourhood of x such that $g_x = 1$ on W_x, $0 \le g_x \le 1$ and $\mathrm{Supp}(g_x) \subset V_i$. By compactness there exists $F \subset K$ finite such that $K \subset \bigcup_{x\in F} W_x$. Let $F_i = \{x \in F,\ W_x \subset V_i\}$. We can assume that every F_i is nonempty. For $i \in \{1,\dots,r\}$ fixed, let $F_i = \{x_1,\dots,x_\ell\}$ and define

$$f_1 = g_{x_1}$$

$$f_2 = (1-g_{x_1})g_{x_2}$$

$$\vdots$$

$$f_\ell = (1-g_{x_1})\cdot\dots\cdot(1-g_{x_{\ell-1}})g_{x_\ell}\ .$$

If $\varphi_i = f_1 + \dots + f_\ell$, then $\varphi_i \in B$ and $\mathrm{Supp}(\varphi_i) \subset V_i$. Also $0 \le \varphi_i \le 1$ and $\varphi_i = 1$ on $\bigcup_{x \in F_i} W_x$ since $\varphi_i = 1-(1-g_{x_1})\cdot\ldots\cdot(1-g_{x_\ell})$.
Now we define

$$\theta_1 = \varphi_1$$
$$\theta_2 = (1-\varphi_1)\varphi_2$$
$$\vdots$$
$$\theta_r = (1-\varphi_1)\cdot\ldots\cdot(1-\varphi_{r-1})\varphi_r .$$

Then $\theta_i \in B$, $\theta_1 + \dots + \theta_r = 1$ on K, $0 \le \theta_i \le 1$ and $\mathrm{Supp}(\theta_i) \subset V_i$, $i = 1,\dots,r$, as we wanted.

Lemma 2. Let X be a locally compact Hausdorff space and $B \subset C_o(X)$ be a strongly separating subalgebra. Assume that there exists $\varphi \in C(\mathbb{R})$ and real numbers r, s such that $0 < r < s$, $\varphi(-\infty,r] = 0$, $\varphi[s,+\infty) = 1$, $0 \le \varphi \le 1$ and $\varphi \circ B \subset B$. Then B has partitions of unity on compact subsets of X.

Proof. Let $x,y \in X$, $x \ne y$, be given. Since B is a strongly separating algebra, there exists $g \in G$ such that $g(y) < r$ and $g(x) > s$. Let $g_o = \varphi \circ g$, $B_o = B \cap C_c(X)$. Then $g_o \in B_o$, $g_o = 1$ on a neighbourhood of x, $g_o = 0$ on a neighbourhood of y and $0 \le g_o \le 1$. From Lemma 1 it follows that B_o, and hence B, has partitions of unity on compact subsets of X.

Lemma 3. Let X be a C^∞ Hausdorff manifold of dimension n and let $x \in X$ be given. Assume that $G \subset C^m(X)$ is such that for every $v \in T_x(X)$, $v \ne 0$, there exists $g \in G$ for which $dg(x)(v) \ne 0$. Then there exists $g_x \in G^n$ such that g_x is a local C^m diffeomorphism at x.

Proof. (After [8]). Let $v_1 \in T_x(X)$, $v_1 \ne 0$. Then there exists $g_1 \in G$ such that $dg_1(x)(v_1) \ne 0$. In case $n = 1$ we take $g_x = g_1$. Otherwise we remark the following: Let $g_1,\dots,g_k \in G$, $1 \le k \le n-1$, be such that no $L_i = dg_i(x)$ vanishes identically. If $H_i = L_i^{-1}(0)$, $i = 1,\dots,k$, then $\dim H_1 \cap \dots \cap H_k \ge n-k$. In particular, there exists

$v_2 \in T_x(X)$, $v_2 \neq 0$, such that $dg_1(x)(v_2) = 0$. Using the hypothesis we obtain $g_2 \in G$ such that $dg_2(x)(v_2) \neq 0$. In this way we obtain $v_1,\ldots,v_n \in T_x(X)$ and $g_1,\ldots,g_n \in G$ such that $dg_i(x)(v_i) \neq 0$, $1 \leq i \leq n$ and $dg_i(x)(v_j) = 0$ for $1 \leq i < j \leq n$. Let $g_x = (g_1,\ldots,g_n)$ and consider the linear mapping $L: T_x(X) \to \mathbb{R}^n$ defined by $L(v) = dg_1(x)(v)e_1 + \ldots + dg_n(x)(v)e_n$, where $e_1,\ldots,e_n$ is the canonical basis in $\mathbb{R}^n$. For some constants $\lambda_1,\ldots,\lambda_n$ we have

$$\begin{aligned} L(\lambda_1 v_1) &= e_1 + c_{1,2}\, e_2 + \ldots + c_{1,n}\, e_n \\ L(\lambda_2 v_2) &= \phantom{e_1 + c_{1,2}\,} e_2 + \ldots + c_{2,n}\, e_n \\ &\vdots \\ L(\lambda_n v_n) &= \phantom{e_1 + c_{1,2}\, e_2 + \ldots + c_{2,n}\,} e_n . \end{aligned}$$

From this it follows that the mapping L is onto and therefore it is an isomorphism. By applying inverse function theorem we conclude that g_x is a local C^m diffeomorphism at x.

<u>Lemma 4</u>. Let $G \subset C^m(X)$ be a set satisfying the hypothesis of Lemma 3. Then there exists a neighbourhood V_x of x such that $C^m(X)|_{V_x} \subset C^m_c(G^n)|_{V_x}$.

<u>Proof</u>. From Lemma 3 there exists $g_x \in G^n$ and a neighbourhood W_x of x such that $g = g_x|_{W_x}$ is a bijection onto the open set $g_x(W_x) \subset \mathbb{R}^n$ and g^{-1} is C^m differentiable. Let V_x be a compact neighbourhood of x contained in the interior of W_x. Let $\theta \in C^m_c(\mathbb{R}^n)$ be such that $\theta = 1$ on $g(V_x)$, $\mathrm{Supp}(\theta) \subset g(W_x)$. Given $f \in C^m(X)$, let

$$\varphi(y) = \begin{cases} \theta(y)(f \circ g^{-1})(y), & y \in g(W_x) \\ 0, & y \in \mathbb{R}^n \setminus g(W_x). \end{cases}$$

Then $\varphi \in C^m_c(\mathbb{R}^n)$ and $f|_{V_x} = \varphi \circ g_x|_{V_x}$, as we wanted to prove.

4. MAIN RESULTS

Theorem 1. Let $A \in Top_a^m(X)$ and assume that $B \subset A$ is a closed subalgebra. Then $B = A$ if and only if the following conditions hold:

1) B has partitions of unity on compact subsets of X.

2) For every $x \in X$ and $v \in T_x(X)$, $v \neq 0$, there exists $g \in B$ such that $dg(x)(v) \neq 0$.

3) If $B_c = B \cap C_c^m(X)$ and $n = \dim X$, then $B_c \cdot C_c^m(B_c^n) \subset B_c$.

Proof. Assume that $B = A$. Since $B \supset C_c^m(X)$, from Lemma 1 it follows that B has partitions of unity on compact subsets of X. Also, given $x \in X$ and $v \in T_x(X)$, $v \neq 0$, the set of all $g \in A$ such that $dg(x)(v) = 0$ is a proper closed subalgebra, hence condition 2) holds. Further it is clear that $B_c \cdot C_c^m(B_c^n) \subset C_c^m(X)$. Since the function $\varphi(x) = \dfrac{x}{1+x^2}$, $x \in \mathbb{R}$, belongs to $C_{0,1}^\infty(\mathbb{R})$ it follows that $B \cdot C_c^m(B^n) \subset (1+B^2) \cdot C_{0,1}^\infty(B) \cdot C_c^m(B^n) \subset (1+B^2)B \subset B$, hence $B_c \cdot C_c^m(B_c^n) \subset B_c$.

Conversely, assume that conditions 1), 2) and 3) hold. Since $C_c^m(X)$ is dense in A, it is enough to show that $C_c^m(X) \subset B_c$. Let $f \in C_c^m(X)$ and $K = \mathrm{Supp}(f)$. Condition 2) (Lemma 4) implies that for every $x \in K$ there exists a compact neighbourhood V_x of x such that $f|_{V_x} \in C_c^m(B^n)|_{V_x}$ and by condition 1) we can assume that $f|_{V_x} \in C_c^m(B_c^n)|_{V_x}$. Since K is compact, there exist $x_1,\ldots,x_r \in K$ such that $K \subset \overset{\circ}{V}_{x_1} \cup\ldots\cup \overset{\circ}{V}_{x_r}$. By condition 1) there exist $\theta_1,\ldots,\theta_r \in B$ such that $\theta_1 + \ldots + \theta_r = 1$ on K and $\mathrm{Supp}(\theta_i) \subset V_{x_i}$, in particular $\theta_i \in B_c$, $i = 1,\ldots,r$. Since $f = \theta_1 f + \ldots + \theta_r f$ and $\theta_i f \in B_c \cdot C_c^m(B_c^n)$, $\forall\, i=1,\ldots,r$, from condition 3) it follows that $f \in B_c$.

Theorem 2. Let τ be a topology on $C_c^m(X)$ such that $A = (C_x^m(X),\tau) \in Top_a^m(X)$. Let $B \subset A$ be a closed subalgebra. Then $B = A$ if and only if the following conditions hold:

1) B is strongly separating.

2) For every $x \in X$ and $v \in T_x(X)$, $v \neq 0$, there exists $g \in B$ such that $dg(x)(v) \neq 0$.

3) $C^\infty_{0,1}(B) \cdot C^m_c(B^n) \subset B$, where $n = \dim X$.

<u>Proof</u>. Assume that $B = A$. Since the mapping $f \in A \to f(x) \in \mathbb{R}$ is continuous for every $x \in X$, it follows that $\{f \in A,\ f(x) = 0\}$ and $\{f \in A,\ f(x) = f(y)\}$, $x \neq y$, are proper closed subalgebras, hence condition 1) holds. Condition 2) follows as in the proof of Theorem 1 and condition 3) is clear.

Conversely, assume that conditions 1), 2) and 3) hold. Let $\varphi \in C^m_c(\mathbb{R}^n)$ be such that $\varphi(0,\dots,0) = 1$. Then $C^\infty_{0,1}(B) = C^\infty_{0,1}(B) \cdot \varphi(0,\dots,0) \subset C^\infty_{0,1}(B) \cdot C^m_c(B^n) \subset B$ by condition 3). Since $B \subset C_o(X)$ is a strongly separating subalgebra such that $C^\infty_{0,1}(B) \subset B$, from Lemma 2 it follows that B has partitions of unity on compact subsets of X. Also condition 3) implies that $B \cdot C^m_c(B^n) \subset B$ and from Theorem 1 it follows $B = A$.

<u>Remark 3</u>. Conditions 1) and 2) in the above theorems are independent in the sense that given $A \in \mathrm{Top}^m_a(X)$ there exist closed subalgebras B_1 and B_2 of A such that letting $\{i,j\} = \{1,2\}$, then B_i satisfies condition i) but does not satisfy condition j). For instance, given $x \in X$ and $v \in T_x(X)$, $v \neq 0$, we can take $B_1 = \{g \in A,\ dg(x)(v) = 0\}$ and $B_2 = \{g \in A,\ g(x) = 0\}$.

<u>Definition 3</u>. $\mathrm{Top}^m_{a,c}(X)$ will denote the class of all $A \in \mathrm{Top}^m_a(X)$ such that if $n = \dim X$, then

$$C^\infty_{0,1}(g_o) C^m_c(g_1,\dots,g_n) \subset \overline{\mathbb{R}[g_o,\dots,g_n]}^{\tau_A}, \quad \forall\, g_o,\dots,g_n \in C^m_c(X).$$

An element $A \in \mathrm{Top}^m_{a,c}(X)$ is called m-admissible algebra on X of compact type.

<u>Remark 4</u>. When $A \in \mathrm{Top}^m_{a,c}(X)$, we can omit condition 3) in both Theorems 1 and 2. In fact, this is clear in the case of Theorem 2. As to Theorem 1, let $g_o,\dots,g_n \in C^m_c(X)$ and $\varphi \in C^m_c(\mathbb{R}^n)$. Then

$g_o\varphi(g_1,\dots,g_n) = (1+g_o^2)\dfrac{g_o}{1+g_o^2}\varphi(g_1,\dots,g_n) \in (1+g_o^2)\overline{\mathbb{R}[g_o,\dots,g_n]}^{\tau_A} \subset$ $\subset \overline{\mathbb{R}[g_o,\dots,g_n]}^{\tau_A}$. Further, as a consequence of the next proposition, it follows that all the known examples of m-admissible algebras are of compact type. However, it remains the question whether $\mathrm{Top}^m_{a,c}(X) = \mathrm{Top}^m_a(X)$ or not.

<u>Proposition 1.</u> Let $A \in \mathrm{Top}^m_a(X)$ be such that $\tau_A|C^m_c(X) \leq \tau^m_i$. Then $A \in \mathrm{Top}^m_{a,c}(X)$.

<u>Proof.</u> It is enough to prove the proposition in the case $\tau_A = \tau^m_i$. Let $\varphi_o \in C^\infty_{0,1}(\mathbb{R})$, $\varphi \in C^m_c(\mathbb{R}^n)$, $g_o,\dots,g_n \in C^m_c(X)$ and W a neighbourhood of 0 in τ^m_i be given. Then $K = \mathrm{Supp}(g_o) \cup\dots\cup \mathrm{Supp}(g_n)$ is a compact subset of X and $W_K = W \cap C^m_c(K)$ is a neighbourhood of 0 in $C^m_c(K)$. Hence there exist $J_o \subset J$ finite, $\ell \in \mathbb{N}_m$, and $\varepsilon > 0$ such that $f \in C^m_c(K)$ and $p_{j,\ell}(f) < \varepsilon$, $\forall\, j \in J_o$, imply that $f \in W_K$. Let q be a polynomial on $\mathbb{R}^{n+1}$ without constant term and define

$$\psi(x_1,x_2,\dots,x_{n+1}) = \varphi_o(x_1)\cdot\varphi(x_2,\dots,x_{n+1}) - q(x_1,x_2,\dots,x_{n+1}),$$

$\forall\,(x_1,\dots,x_{n+1}) \in \mathbb{R}^{n+1}$. Then it is clear that $\psi \in C^m(\mathbb{R}^{n+1})$. For $j \in J_o$ fixed, let $h_o = g_o\circ\lambda_j^{-1},\dots,h_n = g_n\circ\lambda_j^{-1}$. Since $h_o,\dots,h_n \in C^m_b(\lambda_j(V_j))$, for every $\alpha \in \mathbb{N}^n_m$ there exists a constant $c_\alpha > 0$ such that on $\lambda_j(V_j)$ we have

$$|\partial^\alpha\{\bar\psi(h_o,\dots,h_n)\}| \leq c_\alpha\, \Sigma|\partial^\gamma\bar\psi(h_o,\dots,h_n)|$$

for all $\bar\psi \in C^m(\mathbb{R}^{n+1})$, where the sum is taken on the set $\mathbb{N}^{n+1}_{|\alpha|}$. From this it follows that there exists a constant $C_{j,\ell} > 0$ such that

$$p_{j,\ell}(\psi(g_o,\dots,g_n)) \leq C_{j,\ell} \sum_{|\alpha|\leq\ell}\frac{1}{\alpha!}\sum_{|\gamma|\leq|\alpha|}\ \sup_{y\in\lambda_j(V_j)} |\partial^\gamma\psi(h_o,\dots,h_n)(y)|.$$

We have that the set $P(\mathbb{R}^{n+1})$ of all polynomials on $\mathbb{R}^{n+1}$ is τ^m_c dense in $C^m(\mathbb{R}^{n+1})$ ([12]), hence it follows that the set of all $q \in P(\mathbb{R}^{n+1})$ without constant term is dense in the set of all $\bar\psi \in C^m(\mathbb{R}^{n+1})$ such that $\bar\psi(0) = 0$. Since $H = \{(g_o(x),\dots,g_n(x)), x\in X\}$

is a compact subset in $\mathbb{R}^{n+1}$, then given $\varepsilon' > 0$ we can choose q such that $|\partial^{\gamma}\psi| \leq \varepsilon'$ on H, for all $\gamma \in \mathbb{N}_{\ell}^{n+1}$. From this it follows that there exists a constant $C'_{j,\ell} > 0$ such that

$$p_{j,\ell}(\psi(g_o,\ldots,g_n)) \leq C'_{j,\ell}\cdot\varepsilon'.$$

If we choose ε' such that $C'_{j,\ell}\cdot\varepsilon' < \varepsilon$, $\forall\, j \in J_o$ it follows that

$$q(g_o,\ldots,g_n) \in (W+\varphi_o(g_o)\cdot\varphi(g_1,\ldots,g_n)) \cap \mathbb{R}[g_o,\ldots,g_n]$$

and the proof is finished.

Definition 4. A subset $G \subset C^m(X)$ satisfies conditions (N) if

1) G is strongly separating

2) For every $x \in X$ and $v \in T_x(X)$, $v \neq 0$, there exists $g \in G$ such that $dg(x)(v) \neq 0$.

Remark 5. Let $A \in \mathrm{Top}_a^m(X)$ and $G \subset A$ be such that $\mathbb{R}[G]$ is dense in A. From the continuity of the mappings $g \in A \to g(x) \in \mathbb{R}$ and $g \in A \to dg(x)(v) \in \mathbb{R}$, $x \in X$, $v \in T_x(X)$, it follows, as in the proof of Theorem 2, that G satisfies conditions (N).

Also, as a straightforward consequence of Remark 4 and Theorem 2 we have

Corollary 1. Let τ be a topology on $C_c^m(X)$ such that $(C_c^m(X),\tau) \in \mathrm{Top}_{a,c}^m(X)$. If $G \subset C_c^m(X)$ is a nonempty subset, $\mathbb{R}[G]$ is τ-dense in $C_c^m(X)$ if and only if G satisfies conditions (N).

Corollary 2. Let $A \in \mathrm{Top}_{a,c}^m(X)$. Then $C_c^{\infty}(X)$ is dense in A.

Corollary 3. Let X and Y be C^{∞} manifolds of finite dimension. If $A \in \mathrm{Top}_{a,c}^m(X\times Y)$, then $C_c^{\infty}(X) \otimes C_c^{\infty}(Y)$ is dense in A.

Lemma 5. Let $U \subset \mathbb{R}^n$ be a nonempty open subset. Then there exists $F \subset C_c^{\infty}(U)$ countable, which satisfies conditions (N).

Proof. Let $\{x_k,\ k \in \mathbb{N}\}$ be the set of points of U with rational coordinates and denote by $B_r(x)$ the open ball with center x and radius r with respect to the euclidean norm $\|\ \|$ in $\mathbb{R}^n$. For any

$k,\ell \in \mathbb{N}$, $\ell \neq 0$, let $\theta_{k,\ell} \in C_c^\infty(U)$ be such that $\theta_{k,\ell}|\bar{B}_{1/2\ell}(x_k) = 1$, $\mathrm{Supp}(\theta_{k,\ell}) \subset B_{1/\ell}(x_k)$, in case $\bar{B}_{1/\ell}(x_k) \subset U$, $\theta_{k,\ell} = 0$ otherwise. If $\varphi_1,\ldots,\varphi_n$ denote the restrictions to U of the projections on $\mathbb{R}^n$, let

$$F = \{\varphi_i \cdot \theta_{k,\ell},\ \ i=1,\ldots,n;\ \ k,\ell \in \mathbb{N},\ \ \ell \neq 0\}.$$

It is clear that F is a countable subset of $C_c^\infty(U)$. Let $x,y \in U$, $x \neq y$ and $v \in \mathbb{R}^n$, $v \neq 0$, be given. There exists $\ell \in \mathbb{N}$, $\ell \neq 0$, such that $1/\ell \leq \|x-y\|$ and $B_{1/\ell}(x) \subset U$. Also, there exists $k \in \mathbb{N}$ such that $x \in B_{1/4\ell}(x_k)$, consequently $\theta_{k,2\ell}(x) = 1$. Since $\|y-x_k\| > \frac{1}{2\ell}$ it follows that $\theta_{k,2\ell}(y) = 0$. If $i \in \{1,\ldots,n\}$ is such that $v_i \neq 0$ then $d(\varphi_i \cdot \theta_{k,2\ell})(x)(v) = v_i \cdot \theta_{k,2\ell}(x) \neq 0$. We have thus proved that F satisfies conditions (N).

Proposition 2. Let $A \in \mathrm{Top}^m_{a,c}(X)$. Then A is separable if and only if X is σ-compact.

Proof. Assume that X is not σ-compact. Then there exists an uncountable family $(K_i)_{i\in I}$ of pairwise disjoint compact subsets of X such that $\overset{o}{K}_i \neq \Phi$, $\forall\, i \in I$. Let $(g_i)_{i\in I}$ be a family on A such that $g_i \in C_c^m(K_i)$ and $g_i \neq 0$, $\forall\, i \in I$. Fixed $j \in I$, let B be the vector space generated by g_i, $i \neq j$. Let $x \in X$ be such that $g_j(x) \neq 0$ and T be the mapping $g \in A \to g(x) \in \mathbb{R}$. Since $T^{-1}(0)$ is a closed subspace of A and $g_i \in T^{-1}(0)$, $\forall\, i \in I$, $i \neq j$, it follows that $\bar{B} \subset T^{-1}(0)$, consequently $g_j \notin \bar{B}$. Then $(g_i)_{i\in I}$ is an uncountable family on A which is topologically linearly independent, that is A is nonseparable.

Conversely, assume that X is σ-compact. Then X is Lindelöf and we can pick a sequence of charts $(\lambda_k, V_k)_{k\in\mathbb{N}}$ such that $(V_k)_{k\in\mathbb{N}}$ is a covering of X. From Lemma 5 it follows that for every $k \in \mathbb{N}$ there exists a countable subset $F_k \subset C_c^m(\lambda_k(V_k))$ which satisfies conditions (N). Then $G = \bigcup_{k\in\mathbb{N}} F_k \circ \lambda_k^{-1}$ is countable and also satisfies conditions (N). From Corollary 1 it follows that $\mathbb{R}[G]$ is τ_A-dense in $C_c^m(X)$, consequently in A. Continuity of multiplication by real numbers implies that $\mathbb{R}[G]$ is contained in the closure of the algebra $\mathbb{Q}[G]$, over the rational numbers, generated by G. Since $\mathbb{Q}[G]$ is countable, it follows that A is separable.

Corollary 4. Let $X \subset \mathbb{R}^n$ be a nonempty open subset. Let A be a topological vector space of functions or distributions on X such that it contains $C_c^m(X)$ as a dense subset. If the inclusion $C_c^m(X) \subset A$ is continuous, when $C_c^m(X)$ is endowed with the inductive limit topology, then A is separable.

Proof. Since X is σ-compact, it is enough to apply Propositions 1 and 2.

Definition 5. Let $A \in \mathrm{Top}_a^m(X)$. A is called a Nachbin m-admissible algebra on X or just a Nachbin m-algebra if dense subalgebras of A are described by conditions (N). Equivalently, given a nonempty subset G of A, then $\mathbb{R}[G]$ is dense in A if and only if G satisfies conditions (N).

Example:

5) Let $A = (C_c^m(X), \tau_i^m)$. Then as a consequence of Proposition 1 and Corollary 1 it follows that A is a Nachbin m-algebra.

Remark 6. The algebras considered in [8] and [13] are also examples of Nachbin m-algebras.

Remark 7. The question whether every $A \in \mathrm{Top}_a^m(X)$ is a Nachbin m-algebra has a negative answer. In fact, let $\varphi \in C_c^\infty(\mathbb{R})$, $\varphi \neq 0$ and $M_k = \sup\{|\varphi^{(k)}(x)|,\ x \in \mathbb{R}\}$, $k \in \mathbb{N}$. Also let $u(x) = \inf\{\frac{M_k}{|x|^k}, k \in \mathbb{N}\}$, $x \in R$ and $U = \{u^{1/k},\ k=1,2,\ldots\}$. Then U is a (directed) set of weights on R such that $U \leq U \cdot U$. If $V_o = V_1 = U$ then $A = C^1V_\infty(\mathbb{R}) \in \mathrm{Top}_a^1(\mathbb{R})$ (see Example 4). Let A_o be the algebra of all polynomials on $\mathbb{R}$. Then A_o is a subalgebra of A which satisfies conditions (N) and as a consequence of Corollary 1 in [20] it follows that A_o is not dense. Hence A is not a Nachbin m-algebra.

<u>Remark 8</u>. If A is a Nachbin m-algebra, then every proper closed subalgebra is contained in a maximal closed subalgebra.

<u>Definition 6</u>. Let A be a topological space of functions defined on another topological space X. A family $(X_j)_{j\in J}$ of nonempty closed subsets of X is called a supporting family for A if given $B \subset A$ closed and $f \in A$ we have $f \in B$ when $f|_{X_j} \in B|_{X_j}$, $\forall\, j \in J$.

Examples:

6) The family which consists only of X is supporting for A. It is called the trivial supporting family for A.

7) If $\mathcal{C}_c = (\lambda_j, V_j)_{j\in J}$, then $(\bar{V}_j)_{j\in J}$ is a supporting family for $C^m(X)$.

8) Let $(X_j)_{j\in J}$ be a directed family of closed subsets of $\mathbb{R}^n$ such that for every $j \in J$ there exists $j' \in J$ for which $X_j \subset \overset{\circ}{X}_{j'}$. For every $\alpha \in \mathbb{N}^n_m$, let V_α denote the set of characteristic functions of the sets X_j, $j \in J$. Then $(X_j)_{j\in J}$ is a supporting family for $C^m V_\infty(\mathbb{R}^n)$.

<u>Theorem 3</u>. Let $A \in \mathrm{Top}^m_a(X)$ and assume that the following holds:

1) There exists a supporting family $(X_j)_{j\in J}$ for A such that $A|X_j \subset C_o(X_j)$, $\forall\, j \in J$.

2) $C^\infty_{0,1}(g_o)C^m_c(g_1,\dots,g_n) \subset \overline{\mathbb{R}[g_o,\dots,g_n]}^{\tau_A}$, $\forall\, g_o,\dots,g_n \in A$.

Then A is a Nachbin m-algebra.

<u>Proof</u>. Let $B \subset A$ be a subalgebra which satisfies conditions (N). Fixed $j \in J$, it is enough to show that $\bar{B}|X_j$ has partitions of unity on compact subsets of X_j. In fact, assuming this, let $f \in C^m_c(X)$ be given. Lemma 4 implies that for every $x \in K = \mathrm{Supp}(f) \cap X_j$, there exists a neighbourhood V_x of x such that $f|_{V_x} \in C^m_c(B^n)|_{V_x}$. By compactness of K there exist $x_1,\dots,x_r \in K$ such that $K \subset W_{x_1} \cup\dots\cup W_{x_r}$ where $W_x = V_x \cap X_j$. Let $\theta_1,\dots,\theta_r \in \bar{B}|X_j$ be a partition of unity on K subordinated to the covering

$W_{x_1},\ldots,W_{x_r}$. If $h = f|X_j$ then $h = \theta_1 h + \ldots + \theta_r h$ and $\theta_i h \in$ $\in (\bar{B}\cdot C_c^m(B^n))|X_j$. Condition 2) implies that $\bar{B}\cdot C_c^m(B^n) \subset \bar{B}$, hence $\theta_i h \in \bar{B}|_{X_j}$, $\forall$ $i=1,\ldots,r$ and $f|_{X_j} \in \bar{B}|_{X_j}$. Consequently $f \in \bar{B}$, since $j \in J$ is arbitrary.

Then let $x,y \in X_j$, $x \neq y$ be given. Since B is a strongly separating algebra, there exists $g \in B$ such that $g(x) = 1$ and $g(y) = 0$. Let $\varphi \in C_{0,1}^{\infty}(\mathbb{R})$ be such that $\varphi = 1$ on a neighbourhood of 1, $\varphi = 0$ on a neighbourhood of 0 and $0 \leq \varphi \leq 1$. If $f = \varphi\circ g|_{X_j}$, then condition 1) implies that $f \in C_c(X_j)$. Also $f = 1$ on a neighborhood of x, $f = 0$ on a neighbourhood of y and $0 \leq f \leq 1$. Since $\varphi\circ g \in \overline{\mathbb{R}[g]}$ from condition 2), it follows that $f \in \bar{B}|_{X_j}$. Hence from Lemma 1, $\bar{B}|_{X_j}$ has partitions of unity on compact subsets of X_j, as we wanted to prove.

__Remark 9__. It is straighforward to prove that $(C^m(X),\tau_c^m),(C_o^m(X),\tau_u^m)$ and $S^m(\mathbb{R}^n)$ satisfy the hypothesis of Theorem 3. In fact 1) is clear. Also with natural modifications in the proof of Proposition 1 it follows that $(C^m(X),\tau_c^m)$ satisfies 2). Since $(C_o^m(X),\tau_u^m)$ and $S^m(\mathbb{R}^n)$ are weighted algebras it is enough to apply Remark 8 and Lemma 1 of [19] to conclude that these algebras also satisfy 2). Hence we have

__Corollary 5__. Let G be a nonempty subset of any of the topological algebras $(C^m(X),\tau_c^m)$, $(C_c^m(X),\tau_i^m)$, $(C_o^m(X),\tau_u^m)$ and $S^m(\mathbb{R}^n)$. Then $\mathbb{R}[G]$ is dense if and only if G satisfies conditions (N).

__Remark 10__. The theorems on dense subalgebras of differentiable functions obtained by Nachbin [8] and Reid [13] are subsumed by Corollary 5. Further, Theorem 3 allows us to obtain other Nachbin m-algebras. For instance the algebra of Example 8. Also let A be the algebra of all $f \in C_o(\mathbb{R}^n) \cap C^m(\mathbb{R}^n)$ such that $\partial^{\alpha} f$ has polynomial growing, $\forall$ $\alpha \in N_m^n$, endowed with the topology generated by the seminorms $g \to \sup\{u^{1/k}(x)\ |\partial^{\alpha} f(x)|,\ x \in \mathbb{R}^n\}$, $\alpha \in \mathbb{N}_m^n$, $k = 1,2,\ldots$ where

$u(x) = e^{-e^{\|x\|}}$, $x \in \mathbb{R}^n$. Condition 2) is a consequence of Remarks 5 and 6, and Lemma 1 of [19].

Remark 11. In order to prove that condition 2 in Theorem 3 holds for the algebras of Corollary 5 and the first example of Remark 10, it is enough to apply the theorem obtained by de la Vallée Poussin [16] that the set $P(\mathbb{R}^n)$ of all polynomials on $\mathbb{R}^m$ is τ_c^n dense in $C^m(\mathbb{R}^n)$. However, to prove in general that condition 2) holds, we need to use solutions of Bernstein approximation problem for differentiable functions ([18] and [21]). Conversely, Theorem 3 gives also some information on Bernstein problem.

Corollary 6. Let $A \in Top_a^m(\mathbb{R}^n)$ be a weighted algebra such that $P(\mathbb{R}^n) \subset A$. Then $P(\mathbb{R}^n)$ is dense in A if and only if $\frac{1}{1+\|x\|^2} \in \overline{P(\mathbb{R}^n)}^{\tau_A}$, where $\| \ \|$ denotes the euclidean norm on $\mathbb{R}^n$.

Proof. Necessity of the above condition is clear. Conversely, assume that $\frac{1}{1+\|x\|^2} \in \overline{P(\mathbb{R}^n)}^{\tau_A}$ and let $G = \{\frac{1}{1+\|x\|^2}, \frac{x_1}{1+\|x\|^2}, \ldots, \frac{x_n}{1+\|x\|^2}\}$. From Remark 8 and Lemma 1 of [19], it follows that condition 2) of Theorem 3 holds for $(C_o^m(\mathbb{R}^n), \tau_A)$. Hence $(C_o^m(\mathbb{R}^n), \tau_A)$ is a Nachbin m-algebra and $\mathbb{R}[G]$ is a dense subalgebra, since G satisfies conditions (N). Consequently, $\mathbb{R}[G]$ is dense in A and this implies that $P(\mathbb{R}^n)$ is dense in A, since $\mathbb{R}[G] \subset \overline{P(\mathbb{R}^n)}^{\tau_A}$.

Definition 7. Let G be a set of continuous functions defined on a locally compact space X and X_o be a nonempty closed subset of X. X_o is called G-regular if for every $x \in X_o$ there exists a finite sequence $g_1, \ldots, g_k$ in G and a neighbourhood W of $(g_1(x), \ldots, g_k(x))$ in $\mathbb{R}^k$ such that $X_o \cap (g_1, \ldots, g_k)^{-1}(W)$ is compact.

Examples:

9) If X_o is compact, then it is trivially G-regular, for any G.

10) If $G \cap C_o(X)$ does not vanish at any point, then X is G-regular.

11) If every $g \in G$ is proper, then X is also G-regular.

<u>Theorem 4</u>. Let $A \in \mathrm{Top}_a^m(X)$ and $G \subset A$ be a nonempty subset. Assume that

1) G satisfies conditions (N).

2) There exists a supporting family $(X_j)_{j\in J}$ for A such that each X_i is G-regular.

3) $C_{0,1}^{\infty}(G)\cdot C_c^m(G^n) \subset \overline{\mathbb{R}[G]}$.

Then $\mathbb{R}[G]$ is dense in A.

<u>Proof</u>. It is enough to show that $B = \overline{\mathbb{R}[G]}|_{X_j}$ has partitions of unity on compact subsets of X_j, for any $j \in J$. In fact assuming this, let $f \in C_c^m(X)$ be given. Lemma 4 implies that for any $x \in X$ there exists a neighbourhood V_x of x such that $f|_{V_x} \in C_c^m(G^n)|_{V_x}$. Since $\mathrm{Supp}(f)$ is compact, there exist $x_1,\dots,x_r \in X$ such that $\mathrm{Supp}(f) \subset V_{x_1} \cup\dots\cup V_{x_r}$. Let $\theta_1,\dots,\theta_r \in B_j$ be a partition of unity on $\mathrm{Supp}(f) \cap X_j$ subordinated to the covering $V_{x_1} \cap X_j,\dots,V_{x_r} \cap X_j$. Also let $\theta \in B_j$ be equal to 1 on $\mathrm{Supp}(f) \cap X_j$. If $h = f|_{X_j}$, then $h = \theta\theta_1 h +\dots+ \theta\theta_r h$ and $\theta\theta_i h \in (\overline{\mathbb{R}[G]}\cdot\overline{\mathbb{R}[G]}\cdot C_c^m(G^n))|_{X_j}$, $i=1,\dots,r$. Let $g_1,g_2 \in \overline{\mathbb{R}[G]}$ and $\varphi \in C_c^m(G^n)$ be given. Since $g_2\varphi \in A$, the mapping $g \in A \to gg_2\varphi \in A$ is continuous, hence $g_1g_2\varphi \in \overline{\mathbb{R}[G]g_2\varphi}$. Also for any $g \in G$ we have that $g\varphi = (1+g^2)\dfrac{g}{1+g^2}\varphi \in (1+g^2)\overline{\mathbb{R}[G]}$, by condition 3). Since $\overline{\mathbb{R}[G]}$ is an algebra, it follows that $g\varphi \in \overline{\mathbb{R}[G]}$, consequently $\overline{\mathbb{R}[G]}g_2\varphi \subset \overline{\mathbb{R}[G]}$ and $g_1g_2\varphi \in \overline{\mathbb{R}[G]}$. In particular $\theta\theta_i h \in \overline{\mathbb{R}[G]}|_{X_j}$, $\forall\ i=1,\dots,r$, hence $f|_{X_j} \in \overline{\mathbb{R}[G]}|_{X_j}$. Since $(X_j)_{j\in J}$ is a supporting family we conclude that $f \in \overline{\mathbb{R}[G]}$.

Then let $j \in J$ and $x,y \in X_j$, $x \neq y$, be given. Since X_j is G-regular, there exists $g_1,\dots,g_k \in G$ and a neighbourhood W of $(g_1(x),\dots,g_k(x))$ in $\mathbb{R}^k$ such that $H = (g_1,\dots,g_k)^{-1}(W) \cap X_j$ is compact. We can assume that $W = W_1 \times\dots\times W_k$, where each W_i is a compact neighbourhood of $g_i(x)$. Let $\psi_i \in C_c^m(\mathbb{R})$ be such that $\psi_i = 1$ on a compact neighbourhood V_i of $g_i(x)$, $\mathrm{Supp}\,\psi_i \subset W_i$ and

$0 \leq \psi_i \leq 1$, $i = 1,\dots,k$. Then $y \in X_j \setminus H$ implies that $\psi_i \circ g_i(y) = 0$ at least for some i. If $V = V_1 \times \dots \times V_k$, then V is a neighbourhood of $(g_1(x),\dots,g_k(x))$, $K = (g_1,\dots,g_k)^{-1}(V) \cap X_j$ is a compact neighbourhood of x in X_j and $\psi_i \circ g_i = 1$ on K, for every $i = 1,\dots,k$. Now we can assume that k is a multiple of n, say $k = \ell n$. Otherwise, let $k = \ell n + r$, $1 \leq r < n$. We define $g_i = g_k$, $W_i = W_k$, $\psi_i = \psi_k$ and $V_i = V_k$ for $i = k+1,\dots,(\ell+1)n$. Summarizing we have the following: There exist $h_1,\dots,h_\ell \in C_c^m(G^n)$ and compact subsets $K \subset H \subset X_j$ such that K is a neighbourhood of x in X_j, $h_1 \cdot \ldots \cdot h_\ell = 1$ on K and 0 on $X_j \setminus H$. Also, $0 \leq h_1 \cdot \ldots \cdot h_\ell \leq$ ≤ 1. Let $f,g \in G$ be such that $f(x) \neq 0$ and $g(x) \neq g(y)$. Also let $\varphi,\psi \in C_c^\infty(\mathbb{R})$ be such that $0 \leq \varphi,\psi \leq 1$, $\varphi = 1$ on a neighbourhood of $f(x)$, $\varphi = 0$ on a neighbourhood of 0, $\psi = 1$ on a neighbourhood of $g(x)$ and $\psi = 0$ on a neighbourhood of $g(y)$. If $h = (\varphi \circ f)^{\ell+1} (\psi \circ g)^n \cdot h_1 \cdot \ldots \cdot h_\ell$, it follows that $h \in \overline{\mathbb{R}[G]}$ by condition 3). Let $\theta = h|_{X_j}$, then $\theta \in C_c(X_j)$, since it vanishes on $X_j \setminus H$. Further $\theta = 1$ on a neighbourhood of x, $\theta = 0$ on a neighbourhood of y and $0 \leq \theta \leq 1$. From Lemma 1 it follows that $\overline{\mathbb{R}[G]}|_{X_j}$ has partitions of unity on compact subsets of X_j as we wanted to prove.

Remark 12. Conditions 1 and 2 in Theorem 4 are not sufficient, in general, for ensuring density of $\mathbb{R}[G]$ in A. This is shown for instance with the example of Remark 7, where $G = \{1,x\}$. Further, Theorem 4 subsumes Theorem 3 in a proper way and essentially all the results on dense subalgebras given in [19].

Remark 13. The proofs of Whitney's theorem on closed ideals of differentiable functions and of the above Theorems 3 and 4 on dense subalgebras depend crucially on carefully chosen partitions of unity. Perhaps this could be an indication of the method to be used in establishing the general analogous results, for differentiable functions, of the Bernstein-Nachbin theory, in particular, the description of closed subalgebras.

5. SOME PROBLEMS

1. Does every $A \in Top_a^m(X)$ have the approximation property?
2. Given $A \in Top_a^m(X)$, describe its maximal closed subalgebras.
3. Does a topology τ exist on $C_c^m(X)$, strictly finer than τ_i^m and such that $(C_c^m(X),\tau)$ is a Nachbin m-algebra?
4. Let X be a nonempty open subset of $\mathbb{R}^n$ and $C_b^m(X)$ be endowed with the m-strict topology that is, generated by the seminorms $f \to \sup\{u(x)\,|\partial^\alpha f(x)|,\ x \in X\}$, $u \in C_o^+(X)$, $\alpha \in \mathbb{N}_m^n$. Describe dense subalgebras in $C_b^m(X)$ (Remark that sometimes $C_c^m(X)$ is not dense in $C_b^m(X)$).
5. Let $A \in Top_a^1(X)$ be a Nachbin algebra and $B \subset A$ be a subalgebra which satisfies conditions (N). Let $W \subset A$ be a vector subspace such that $BW \subset W$. Under such conditions, describe $\bar{W}$.
6. Does it hold a characterization of closed ideals in $A \in Top_a^1(X)$, analogous to Whitney's theorem?
7. Characterize Nachbin m-algebras.
8. Characterize dense subalgebras in a general $A \in Top_a^m(X)$.

REFERENCES

[1] Baumgarten, B., Gewichtete Raüme differenzierbarer Funktionen, Universität Darmstadt, 1976.

[2] Garnir, H., de Wilde, M., Schmets, J., Analyse fonctionnelle, Tome III, Birkhäuser Verlag, 1973.

[3] Gauld, D., Topological properties of manifolds, Amer. Math. Monthly 81 (1974) nº 6, 633-636.

[4] Glaeser, G., Algèbres et sous-algèbres de fonctions différentielles, Anais da Academia Brasileira de Ciências 37 (1965), 395-406.

[5] Horvath, J., Topological vector spaces and distributions, Vol.I, Addison Wesley, 1966.

[6] Khourguine, J., Tschetinine, N., Sur les sous-anneaux fermés do l'anneau des fonctions à n dérivées continues, Doklady Akademi Nauk SSSR 29 (1940), 288-291.

[7] Michael, E., Locally multiplicatively-convex topological algebras, Memoirs Amer. Math. Sco. nº 11 (1952).

[8] Nachbin, L., Sur les algèbres denses de fonctions différentiables sur une variété, C.R. Acad. Sci. Paris, 228 (1949), 1549-1551.

[9] Nachbin, L., Résultats récents et problèmes de nature algebrique en théorie de l'approximation, Proceedings of the International Congress of Mathematicians, Stockholm 1962 (1963), 379-384.

[10] Nachbin, L., Sur la densité de sous-algèbres polynomiales d'applications continûment différentiables, Sém. Pierre Lelong et Henri Skoda (Analyse), 1976/77, Springer Verlag, Lecture Notes in Mathematics 694 (1978).

[11] Nachbin, L., A look at apprximation theory, in Approximation Theory and Functional Analysis (ed. J. Prolla), Notas de Matemática 66, North-Holland (1979), 309-331.

[12] Narasimhan, R., Analysis on real and complex manifolds, North-Holland, 1968.

[13] Reid, G., A theorem of Stone-Weierstrass type, Proceedings of the Cambridge Philosophical Society 62 (1966), 649-666.

[14] Rudin, W., Real and complex analysis, Mc Graw Hill Co., 1966.

[15] Stone, M., Applications of the theory of Boolean rings to general topology, Transactions Amer. Math. Soc. 41 (1937), 375-481.

[16] Vallée Poussin, Ch. de la, Sur l'approximation des fonctions d'une variable rélle et de leurs dérivées par des polynômes et des suites finies de Fourier, Bull. Ac. Sc. Belgique (1908), 193-254.

[17] Whitney, H., On ideals of differentiable functions, American Journal of Mathematics 70 (1948), 635-658.

[18] Zapata, G., Bernstein approximation problem for differentiable functions and quasi-analytic weights, Transactions Amer. Math. Soc. 182 (1973), 503-509.

[19] Zapata, G., Approximation for weighted algebras of differentiable functions, Bollettino della Unione Matematica Italiana 9 (1974), 32-43.

[20] Zapata, G., Weighted approximation, Mergelyan theorem and quasi-analytic weights, Arkiv för Matematik 13 (1975), 252-262.

[21] Zapata, G., Fundamental seminorms, in Approximation Theory and Functional Analysis (Ed. J. Prolla), Notas de Matemática 66, North-Holland (1979), 429-443.